THE STOKES FIELD GUIDE TO THE BIRDS OF NORTH AMERICA

OTHER BIRD BOOKS BY DONALD AND LILLIAN STOKES

Stokes Field Guide to Birds: Eastern Region
Stokes Field Guide to Birds: Western Region
Stokes Field Guide to Warblers
Stokes Guide to Bird Behavior, Volumes 1–3

. . . .

Stokes Beginner's Guide to Bird Feeding
Stokes Beginner's Guide to Birds: Eastern Region
Stokes Beginner's Guide to Birds: Western Region
Stokes Beginner's Guide to Hummingbirds
Stokes Beginner's Guide to Shorebirds

. . . .

Stokes Bird Feeder Book
Stokes Bird Gardening Book
Stokes Birdhouse Book
Stokes Bluebird Book
Stokes Hummingbird Book
Stokes Oriole Book
Stokes Purple Martin Book

OTHER BOOKS BY DONALD AND LILLIAN STOKES

The Natural History of Wild Shrubs and Vines
Stokes Beginner's Guide to Bats (with Kim Williams and Rob Mies)
Stokes Beginner's Guide to Butterflies
Stokes Beginner's Guide to Dragonflies (with Blair Nikula and Jackie Sones)
Stokes Butterfly Book (with Ernest Williams)
Stokes Guide to Animal Tracking and Behavior
Stokes Guide to Enjoying Wildflowers
Stokes Guide to Nature in Winter
Stokes Guide to Observing Insect Lives
Stokes Wildflower Book: East of the Rockies
Stokes Wildflower Book: From the Rockies West

THE STOKES FIELD GUIDE TO THE BIRDS OF NORTH AMERICA

DONALD & LILLIAN STOKES

Photographs by Lillian Stokes and Others

Maps by Paul Lehman

Maps Digitized by Matthew Carey

LITTLE, BROWN AND COMPANY
New York | Boston | London

Little, Brown and Company
Hachette Book Group
237 Park Avenue, New York, NY 10017
www.hachettebookgroup.com

First Edition: October 2010

Library of Congress Cataloging-in-Publication Data
Stokes, Donald W.
The Stokes field guide to the birds of North America / by Donald and Lillian Stokes ; photographs by Lillian Stokes and others ; maps by Paul Lehman ; maps digitized by Matthew Carey. – 1st ed.
p. cm.
Includes index.
ISBN 978-0-316-01050-4
1. Birds – North America – Identification.
I. Stokes, Lillian Q. II. Lehman, Paul E. III. Title.
IV. Title: Field guide to the birds of North America.
QL681.S734 2010
598.097 – dc22 2010019453

10 9 8 7 6 5 4 3 2 1

IM

DESIGNED BY LAURA LINDGREN

Printed in China

Key to Bird Accounts

Caption in left corner tells age, sex, season, morph, and subspecies as appropriate.

Common name at the time this guide was written; these change occasionally.

Scientific name: First is the genus, or group, the bird belongs to; second, the species within that group.

Shape is an extremely helpful identification tool when combined with plumage, behavior, habitat, and voice.

Subspecies are populations within a species with consistent variations in plumage, shape, and/or behavior. A number in parentheses indicates the number of N. American subspecies, and their names and descriptions generally follow. "Monotypic" means there are no subspecies.

Hybrids are a result of breeding between two species. If there are no naturally occurring hybrids, this heading is omitted.

Adult TX/05

Adult, *solitarius* OH/04

Blue-headed Vireo 1

Vireo solitarius L 5½"

Shape: Medium-sized relatively short-tailed vireo; similar shape to Cassin's Vireo, but bill slightly longer, slightly hooked. Primary extension past tertials long (2 x bill length). **Ad:** Bluish-gray to gray head; olive to grayish-olive back and rump (see Subspp.); bright white below with bright yellow sides of breast and flanks. Bright white spectacles and throat contrast strongly with dark gray face; edges of secondaries greenish; 2 bold yellowish-white wingbars. More contrasting plumage overall than similar Cassin's Vireo. Not always possible to distinguish from Cassin's, especially in worn plumage. **Juv:** (May–Aug.) Like ad. but with brownish wash over back. **Flight:** Direct; usually short. Short tail; yellowish flanks. **Hab:** Extensive forests of variable types; adaptable. **Voice:** Song a series of well-spaced, short, whistled phrases like *teeyoo taree tawit teeyoh* (song sweeter than that of Plumbeous and Cassin's Vireos). Calls include a harsh chattered *chechechecheee* and a nasal *neah neah neah.* 🎧**66**.

Subspp: (2) •*solitarius* (s. PA–NJ north to BC–NS) has medium-sized bill, greenish back; •*alticola* (WV–MD south) has longer heavier bill, darker grayish-green back. **Hybrids:** Yellow-throated Vireo.

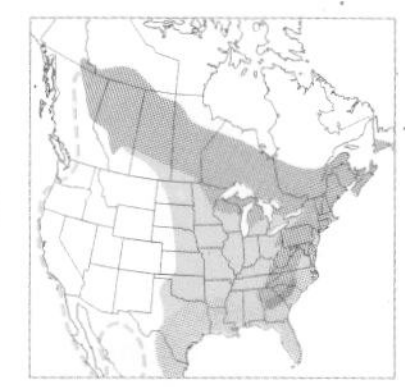

Photographs are typically organized from top left to bottom right and from oldest to youngest, male to female, summer to winter. Flight shots are often last.

Caption in right corner tells location and month of the photograph.

ABA Code (1–6): The American Birding Association rates bird species by ease of seeing them in their normal range. Birds rated 1 are easiest to see.

Length of bird from tip of bill to tip of tail.

Plumage color clues follow Shape, starting with adult and ending with youngest stages.

Flight includes plumage, shape, and behavioral clues to help you identify flying birds.

Habitat lists the bird's preferred environments.

Voice describes calls and songs. (A 🎧 symbol at the end identifies birds on the bonus CD.)

Range maps are provided for species that regularly occur in N. America. For the rest, a sentence at the start of the account outlines their origins and occurrence.

Contents

Preface

Writing and designing this new field guide to the birds of North America has been a huge project and an exciting challenge. We embarked on the work more than six years ago with the knowledge that we could build upon and make valuable additions to the great guides that have come before us. In writing this guide we had several major goals: to produce the most *useful* guide to identifying the birds of North America ever published; to cover more species than almost all other guides; to create the most complete photographic record of these species' plumages and subspecies variations that has ever existed in one guide; to make sure the identification clues were both thorough and up-to-date; to clearly and quantitatively describe each species' shape as a key element of field identification; and to summarize all subspecies ranges and main distinguishing characteristics.

In addition, we have included the most up-to-date maps available, added the American Birding Association (ABA) rarity rating (birding code) for each species, listed known wild hybrids, and reflected the most recent additions and deletions, lumps and splits, and common and scientific name changes of the American Birding Association Checklist as of the writing of this guide.

With the help and support of our publisher, we are thrilled to have completed this new national guide while accomplishing the goals we set for ourselves and bringing to fruition a guide we hope will be both beautiful and useful for all birders and will continue to further the love and conservation of birds in North America.

Bald Eagle

How to Use This Guide

Area and Species Covered

The area of this guide follows that of the American Birding Association Checklist and includes the 49 continental states of the United States, the provinces and territories of Canada, and the French islands of St. Pierre and Miquelon (just south of Newfoundland). It also includes adjacent ocean waters 200 nautical miles from land.

In deciding which birds to include, we have used the ABA Checklist and its "birding codes." The birding codes roughly indicate the difficulty of observing a bird species, with 1 designating those most easily observed and 6 indicating species that are extinct or are unable to self-sustain wild populations (see Birding Code, below, for more). We have included virtually all species on the ABA Checklist (as of the writing of this guide) with ABA birding codes of 1–4. We also included many species with birding codes of 5 (birds having occurred 5 or fewer times in North America or fewer than 3 times in the past 30 years); space limited the number we could include, so we chose those that had been seen more recently, were of particular interest to birders, and/or had increasing or stable populations. Finally, we included 2 species with code 6 (probably extinct or unsustainable populations in the wild)—the California Condor (because condors may be seen in the wild, even though they do not have self-sustaining populations) and the Ivory-billed Woodpecker (because there is still some question as to whether it exists in the wild). Our final total is 854 species, all from the ABA Checklist and all with full accounts (note that other field guides cover many exotic species not on the ABA Checklist and many species with only minor accounts, which are added to their species totals).

The birds in this guide are arranged by phylogenetic order as determined by the American Ornithologists' Union (AOU) Checklist Committee as of the writing of this guide. Phylogenetic order is the order in which various groups of birds are believed to have evolved, from the earliest to the most recent. This order has changed considerably in recent years with the advent of DNA studies, which provide a clearer picture of birds' relationships to each other. Although the groups of birds in this guide follow phylogenetic order, we have occasionally adjusted the order of individual species within groups to show very similar species next to each other for comparison and to meet the needs of the book's layout.

Species Accounts

PHOTOGRAPHS AND CAPTIONS

We are in the midst of a digital revolution. Within the last few years, rapidly advancing technology has led to more powerful, high-speed, high-resolution, digital SLR cameras with image-stabilized telephoto lenses, making it more possible than ever to capture fabulous images of live birds.

Excellent, well-chosen photographs are always more detailed and more accurate than a drawing. This is particularly true when it comes to shape, feather wear, feather detail, and color placement. Because a bird's shape is a composite of so many subtly interacting components, it is rare that a drawing puts them all together accurately enough to get a true representation of a bird's shape. To learn shape well, it is best to both look at the bird in the wild and study good photographs.

Photographs also are truer representations of real feather condition. For most of the year, immature and adult birds have worn feathers, but this is rarely depicted on drawn birds, which are almost always shown in equally fresh plumage, even if this is inaccurate.

If you know when and where they were taken, photographs become even more valuable. Thus we have given the location and month for each photo in abbreviated form in its lower right corner. The month a photo was taken can help indicate the plumage (such as eclipse or immature), the state of the plumage (worn or fresh), and the timing of molt, while the location can help indicate regional variation in plumage or (along with the date) the likely subspecies. For a list of state, province, and country abbreviations used, see page 769; the months are represented by two digits (02, for instance, is February). If either the place or the month is not known, this is indicated by a dash (–).

In the lower left corner of the image are two types of information. First is an indication of the age, sex, season, and morph (if applicable) of the bird if it can be clearly determined. Many images also carry a subspecies name in italics (for example, *carolinensis* for some Green-winged Teals). We have added a subspecies name only if the plumage, shape, time, and location all lead to an accurate identification to subspecies. In cases where a bird seemed to represent a subspecies but the photograph was not taken in the bird's breeding range, we have not added the subspecies name. If there is only one subspecies in the range of this guide, or if a species is monotypic (i.e., has no subspecies), the caption does not include a subspecies designation.

In a few instances, a photo is of a subspecies that does not occur in the range of this guide. We did this when the difference between the North American subspecies and the foreign subspecies was slight and the foreign photo was deemed helpful for the identification of the species. In such cases, the subspecies name on the photo has an asterisk and the foreign subspecies and its range are listed in the Subspp. section of the account.

SPECIES NAME

Each species has a common and a scientific name that is agreed upon by the ABA Checklist Committee, which in turn follows the lead of the American Ornithologists' Union Checklist Committee. From time to time, these names change and may not match ones in older guides. The common and scientific names we have used are the most recent as of the writing of this guide.

The scientific name for each species has two parts. The first is the genus, or group, to which that bird belongs, such as *Buteo* for one group of hawks. The second is the species within that genus, such as *platypterus* for Broad-winged Hawk. When a subspecies exists, it receives a third name; in the case of Broad-winged Hawk, the North American subspecies' name is also *platypterus*—thus, *Buteo platypterus platypterus.* When a shortened version of a subspecies' full name is written, just the initials of the genus and species are given, as in *B. p. platypterus.*

BIRDING CODE

The ABA keeps a checklist of all the documented appearances of bird species in North America north of Mexico. This is updated regularly. The association also has a "birding code," or rating system, that indicates the relative ease of observing each species. This is on a 1–6 scale, with 1 indicating the species easiest to observe and 6 those that are extinct or unable to sustain themselves in the wild. (See inside front flap for more detailed descriptions of these codes.)

Following the common name heading in an account, we have included the ABA birding code for each species. It is helpful for a number of reasons. If you are just starting bird identification or are in an area of the country for the first time and see a bird that has a 1 or 2 rating, you can know that this is a commonly seen bird in its normal range. On the other hand, seeing a 3 bird is a little more unusual. If you see a 4 or 5 bird, this is very unusual, and you may want to check your identification especially carefully, try to get a photo of the bird, write down detailed notes on your sighting, and be sure to report the sighting to bird organizations in the area.

LENGTH

To the right of the scientific name is the length of the bird from the tip of the beak to the tip of the tail. In most cases, this is an average length, for within species and between sexes lengths can vary. In a few cases, when there are large sexual differences in size, we have placed a range of length and have indicated which sex is at either end of the range.

SHAPE

Shape should be the first step in most bird identification, and becoming good at observing and describing shape is one of the best ways to fast-forward your bird identification skills. When birds are backlit, in poor light, or distant, shape is often all you can see. In this guide, shape includes seeing and comparing outlines, volumes, positioning of body parts, and linear measurements of feathers and body parts.

In the history of North American bird identification, shape has taken a backseat to plumage colors; yet shape is often a more consistent clue than plumage. Plumage varies with feather wear, sex, age, and season, but through all of these plumage variations, shape often remains the same. For most of the difficult pairs of species with look-alike plumage, careful analysis of shape is the first and best way to distinguish between them.

Reddish Egret

The description of shape can be approached in roughly two ways: qualitative and quantitative. A qualitative description uses general impressions and describes the "quality" of the shape, often in human terms. For example, qualitative words for the shape of birds' bodies include "stocky," "bulky," "compact," "plump," "elegant," and "athletic"; qualitative words for bills include "heavy," "slim," "chisel-shaped."

A quantitative description (the approach taken in this guide) gives comparative measurements within the bird and attempts to use more precise language to describe shape. Examples of quantitative descriptions of the shape of birds' bodies include "deep-chested," "deep-bellied," "broad-necked," "defined neck"; quantitative descriptions for the shape of birds' bills include "blunt-tipped," "deep-based," "decurved," "conical," "abruptly pointed."

A good friend of ours once said that "birds carry their own rulers with them," meaning you can always compare lengths of various portions of a bird's body with each other. Examples of this include comparing bill length to depth of bill base, or length of primary projection past tertials to length of lore, or wing length to wing width. It is our belief that every qualitative description of bird shape can be analyzed and more accurately conveyed to others through a careful, more quantitative description.

Observing shape carefully can be challenging—birds are always moving, necks can be extended, bodies assume different positions and angles, and feathers may be ruffed or sleeked. But even with all this, it is always possible to see some aspects of size and proportions and begin to piece together an overall understanding of a bird's shape as you continue to watch.

There is also evidence suggesting that humans process shape and color with different parts of their brains. Because of this, we have separated our shape and color descriptions to make each more accessible and memorable, rather than mixing them as most other field guides do.

Developing quantitative ways of recognizing and describing shape is an ongoing and creative process, and it is our hope that it will continue to grow in American birding and become more useful as birders hone their abilities to distinguish and recognize shapes of birds. What is new in this guide is not the looking at shape but the quantitative description of it and the greater use of comparisons within birds. Through this, we hope to heighten the awareness and value of shape as an identification tool and make it accessible to all by using descriptions that are quantifiable, communicable, and objective.

In this guide, shape descriptions always refer to adults; immature and juvenile birds may have slightly different proportions. Also, the use of terms like "long" and "short" or "large" and "small" are relative to that group of birds (a long bill in the heron group is not the same as a long bill in the sparrow group). In this guide, ***bill length*** is from the tip of the bill to the base of the culmen (the lengthwise ridge of the upper mandible); ***head length*** is from the base of the culmen to the back of the nape. The ***lore*** is the distance between the front of the eye and the base of the bill. All of these lengths can be compared to each other when trying to distinguish between similar species.

The primaries are the outermost and longest feathers of the wing. On a perched bird they can be seen on the lowest edge of the folded wing and may extend out over the rump or tail. ***Primary extension*** (also called ***primary projection***)is best seen on perched, sitting, or swimming birds. There are two common measurements of primary extension: how far the primaries project past the tail and how far they project past the tertials. This measurement can then be compared to other parts of the same bird, such as the length of the lore or bill or to the length of the longest tertial. If primaries do not extend past the tail tip, you can observe how far they extend over the tail, such as halfway down or nearly to the tip.

The tertials are important feathers to observe on birds because their length can be compared to other parts of a bird and because they often help in identification and in determining age. They are the innermost feathers of the wing and, on a perched bird, tend to lie over the folded secondaries in a widely overlapping fashion. Estimating the length of the longest tertial is not too hard to do and can be a useful measurement to compare with bill length or length of primary projection past tertials.

When describing the shape of flying hawks and some other large birds, we have used comparisons of tail length to wing width and wing length to wing width. In both of these cases ***wing width*** refers to the greatest width of the extended wing. ***Tail length*** is the distance from the trailing edge of the wing to the tip of the tail. These comparative measurements are not necessarily the same as a bander could take on a bird in hand, but they are useful for quantitative descriptions of birds in the field.

AGE AND SEASONAL PLUMAGE DESCRIPTIONS

In describing birds as they appear at various ages and times of year, we have tried to keep terminology as simple as possible and yet still accurate. There are several systems for describing the plumages and ages of birds. In this guide, we have used the life-year system. It describes birds' ages much as we describe our own—a bird's first year is roughly its first 12 months, second year the next 12 months, and so on. Birds' plumages are described on a seasonal basis, which roughly reflects molt timings: summer (which mostly coincides with breeding) and winter (which usually coincides with nonbreeding). Below are the headings we have used in the identification sections of the guide and brief explanations of their meaning.

- **Ad.** – Adult bird without seasonal plumage change.
- **Ad. Summer** – Adult bird (that has seasonal plumage change) from roughly March to August; similar to "breeding" plumage in some other guides. (Corresponds to the traditional "spring" designation in the case of warblers; see page 601.)
- **Ad. Winter** – Adult bird (that has seasonal plumage change) from roughly September to February; similar to "nonbreeding" plumage in some other guides. (Corresponds to the traditional "fall" designation in the case of warblers; see page 601.)
- **Juv.** – Refers to juvenal plumage, the first full plumage of a bird after it leaves the nest. Some birds keep this plumage only until fall, when they molt into 1st-winter or adult plumage; others may keep it for as long as the first year, as in many hawks. (The abbreviation ***juv.*** is also used in the text to refer to a juvenile, a young bird in juvenal plumage.)
- **1st Winter** – This term is used when the juvenal plumage is partially molted in fall and the resulting new plumage is not yet like the adult. It can continue with 2nd Winter, 3rd Winter, etc., until adult plumage is acquired. In these cases, birds may also have a slightly different summer plumage in each year before becoming adult, and this is called 1st Summer.
- **Juv./1st Winter** – This heading is used when there is a gradual change from juvenal to 1st winter over several months and a clear distinction between the two cannot be made.
- **Imm.** – Short for "immature," this term refers to any distinct plumage between juvenal and adult.
- **1st Yr.** – This plumage is kept for roughly the first 12 months of a bird's life. The term is used often in describing larger birds that take several years to reach adult plumage but generally have a single plumage throughout any given year. This plumage may be followed by 2nd-year, 3rd-year, even 4th-year plumage before adult plumage is reached (as in eagles).
- **Perched/Flying** – Some species that are often seen and identified both perched and flying, such as gulls and hawks, have separate headings for the perched and flying bird. In these cases the normal Flight clues section (see below) may be omitted.

Species that are similar in appearance to the one being described in a particular account are dealt with in the portion of the account where they are most relevant; thus, you may find similar-species information under Shape, Ad., Juv., Flight, etc.

PLUMAGE DATES

After some juvenal and immature headings, we give a range of months in which that plumage can be seen, such as Apr.–Nov. It is important to remember that plumages change gradually, so that at either end of the range you may see a bird molting as it transitions out of an old and into a new plumage. Also, individuals vary in the timing of their transitions between plumages; thus, at either end of a date range you may see one bird still mostly in the previous plumage while another is already fully into the new plumage.

BIRD MOLTS AND AGING

Knowing how to determine the age of birds and understanding their molt schedule is important to identification as well as to other bird studies. Molt schedules vary tremendously among birds, but a few general patterns are helpful to know.

The majority of molting takes place after breeding, in late summer and early fall; another important period of molting for some birds occurs in late winter and early spring. Some larger birds, like gulls and hawks, may be molting at least to some extent in all months.

Thus, most birds undergo a complete molt each fall and a partial molt of just their body feathers in spring. But juveniles, in their first fall, often molt only their body feathers and retain most of their wing and tail feathers until their second fall. This results in a key clue to recognizing birds in their first year: their wing and tail feathers are more worn and faded and often a slightly lighter color than those of the adults.

There are many exceptions to this general pattern, and for a complete account of molt in most North American birds we refer you to Peter Pyle's *Identification Guide to North American Birds, Part I* and *Part II* (Slate Creek Press, Bolinas, CA, 1997 and 2008).

FLIGHT
This section highlights aspects of plumage and shape seen only while the bird is flying. In some birds, like chickadees, clues may help you identify the bird only to group; in others, such as many shorebirds, the clues may indicate the species. Characteristic flight patterns or behaviors are also described when relevant.

VOICE
Sound is a key feature to identifying birds. Our first recommendation is that you get a set of recordings of the birds of North America. Of course, we feel that the *Stokes Field Guide to Bird Songs* CDs are the best recordings available, but there are also many others. The CD accompanying this book includes a sampling from the *Stokes Field Guide to Bird Songs,* with more than 600 sounds (including calls and songs) of 150 common birds. A headphones symbol 🎧 at the end of a Voice section in the text identifies the species as being on the CD and is followed by the track number for the recording. A list of the 150 birds and their tracks is on the last page of the book.

Well-written renditions of birds' songs and calls can sometimes be helpful as a subsidiary clue to confirming an identification. Thus, we have included written descriptions of calls and song where they are appropriate. We generally attempt to describe the quality of the call as well as give a mnemonic rendition of the sound. ***Calls*** are usually innate sounds that a bird uses and are often short; ***song*** is usually a partially learned, more complex vocalization given by birds later in the phylogenetic order (from about shrikes through finches).

HABITAT (Hab.)
This section provides a brief description of the key elements of breeding habitat for each species. In some cases, where particularly important, consistent, and/or strikingly different, winter habitats are also described. When a species uses a wide range of habitats, we try to describe this variety.

SUBSPECIES (Subspp.)
Subspecies designations are given to consistent regional variations in a species' appearance and/or behavior. A subspecies is often defined by these physical characteristics as well as by the geographic region in which it breeds. Subspecies are decided upon through careful examination of birds in the wild and in museum collections, and also through field studies of breeding behavior, ranges, and visual and auditory displays.

Differences between subspecies are often subtle and clinal (changing gradually over a geographic area), and there can be disagreement on the presence or number of subspecies within a species. The American Ornithologists' Union maintains an authoritative list called *Checklist of North American Birds,* and this is periodically revised. Although species classification is fairly regularly updated, the last time subspecies were comprehensively updated by the AOU was in 1957. So much additional research has occurred since that time that we were compelled to reflect that in our guide, and we had to look for a consistent and more current approach to subspecies classification.

One of the most thorough reviews of subspecies of North American birds has been done by Peter Pyle in his *Identification Guide to North American Birds, Part I* and *Part II.* This is one of the great ornithological works of the past one hundred years; it is detailed, dense with information, and thoroughly documented throughout. It is designed primarily for bird banders and people studying birds in the hand, but it is also a thorough guide to the aging of birds and the sequence of their molts and can be extremely useful to any field birder.

In this guide, we have followed Peter Pyle's classification of subspecies and summarized his ranges and descriptions in a way we hope will be useful for the field birder and will further the understanding and exploration of regional variation in birds' appearance and behavior. We encourage all interested birders to consult Pyle for more thorough descriptions of the ranges and differences between subspecies.

Immediately following the Subspp. heading is either the word "monotypic," which means there are no subspecies in the world for this species, or a number in parentheses, which indicates the number of subspecies that occur in the area covered by this guide (there may be more subspecies in other areas and countries). If there is only one subspecies in the area of this guide (but others in the rest of the world), there will be a 1 in parentheses followed by the subspecies name; for example: (1) •*americana*. We give no description of the range or appearance of this subspecies since this is covered by the whole species account and accompanying distribution map. For species with two or more subspecies in the area of the guide, we put the total number in parentheses and include descriptions of the range and main field characteristics for each one.

In describing breeding ranges of subspecies, we have often used the shorthand of a compass direction starting at a certain point, such as "AZ and west." When we do this, it always means west (or whatever the direction might be) within the range of that species. We used this shorthand to save space and limit what would otherwise be long descriptions. In a few cases, we have added the winter range of a subspecies when it is well known (as in some geese) and particularly helpful. In all subspecies descriptions of plumage, we are referring to adults unless we state otherwise.

For certain birds with many subspecies, the subspecies are joined into groups having similar characteristics. In these cases, we describe only the characteristics of the group and then list the subspecies and their ranges.

Certain subspecies have acquired generally accepted common names over the years; we have included these whenever possible, since birders often use them. However, there is no authoritative standard for these names, so you may read or hear variations on them elsewhere.

Identification of subspecies in the field can often only be done reliably by seeing a bird in the heart of its designated subspecies breeding range during breeding. Since there is often a gradual blending of characteristics where subspecies meet and often quite a bit of variation within the range of a subspecies, and since ranges are inexactly known and defined, identification in the field to the subspecies level outside the heart of the breeding range and outside the breeding period is frequently inexact.

We have included photographs of subspecies when they show a marked variation in appearance or are representative of a group of subspecies. Showing all subspecies is beyond the scope of this guide.

As mentioned earlier, in a few instances a subspecies not occurring in North America is shown as an aid to identification. Foreign subspecies are indicated by an asterisk next to their name.

HYBRIDS

We have included hybrid information for species when hybridization is believed to occur under natural conditions. Obviously not all cases in which hybridization occurs are known, for ornithologists cannot follow or test all birds and, in some cases, hybridization between similar-looking species may not be immediately evident. Any present list of hybrids will continue to grow and be modified with time. Our sources for this information come from *The Birds of North America,* edited by A. Poole and F. Gill (a joint project of the AOU, the Cornell Lab of Ornithology, and the Academy of Natural Sciences), Pyle's *Identification Guide to North American Birds,* and the bird hybrids search engine www.bird-hybrids.com, which includes complete references for each hybrid listed.

When there is no Hybrids heading, there are no naturally occurring hybrids known.

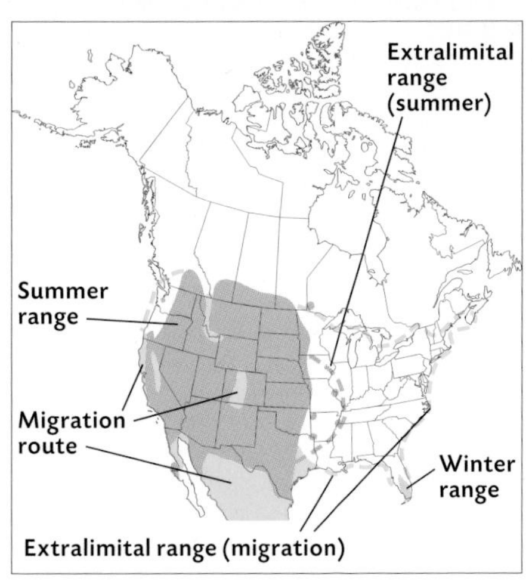

Western Kingbird

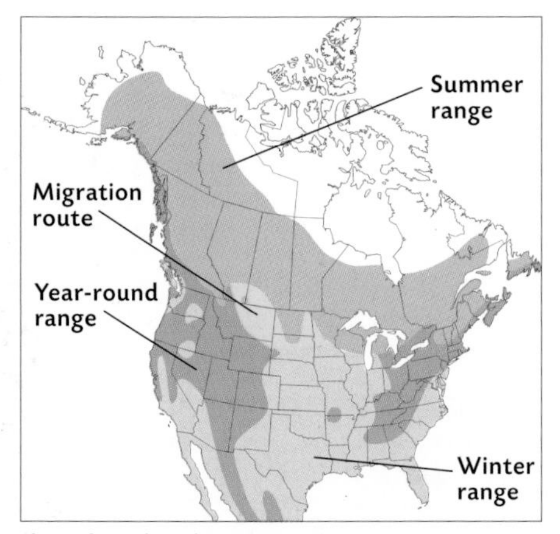

Sharp-shinned Hawk

RANGE MAPS

The range maps are the most up-to-date available and were drawn by Paul Lehman, the widely acknowledged expert on bird distribution. We have included maps for all species with regularly repeating patterns of distribution. When relevant, maps show summer ranges (red), winter ranges (blue), year-round ranges (purple), and migration routes (yellow).

The maps also occasionally include dotted lines to indicate regular extralimital occurrences during the respective season (winter, summer, or on migration). For example, the Western Kingbird's summer range is the western half of the United States, but in late summer these birds regularly wander to the east and along the East Coast. Thus a red dotted line shows where they can occur east of the Mississippi in summer, and a yellow dotted line shows the extent of their wanderings in migration.

The rest of the species in this guide do not regularly occur in any one location and so a map is useless; for these birds a brief description at the beginning of the account outlines their origin, where they tend to show up, and how often they occur. These species wander into our area from nearby continents and countries—European birds tend to come to our East Coast; Caribbean birds tend to come into Florida; Mexican and South American birds tend to come to Texas, the Gulf Coast, and the Southwest; and Asian birds tend to occur in Alaska and along the West Coast. There are also many species of seabirds that wander widely and show up in our surrounding oceans.

Some of these species arrive only once or a few times in North America, others come irregularly every few years, and some show up regularly but in no predictable place. Most birds without maps have ABA birding codes of 3, 4, or 5.

American White Pelican

Photo Key to the Parts of Birds

Learning the names for the parts of a bird and the names of its major feather tracts is extremely important and essential to progressing in bird identification. Only if you know the various parts of a bird can you accurately record for yourself and convey to others what you have seen. The diagrams below and on the next few pages show the most important terms used in this guide and the area on the bird to which each refers.

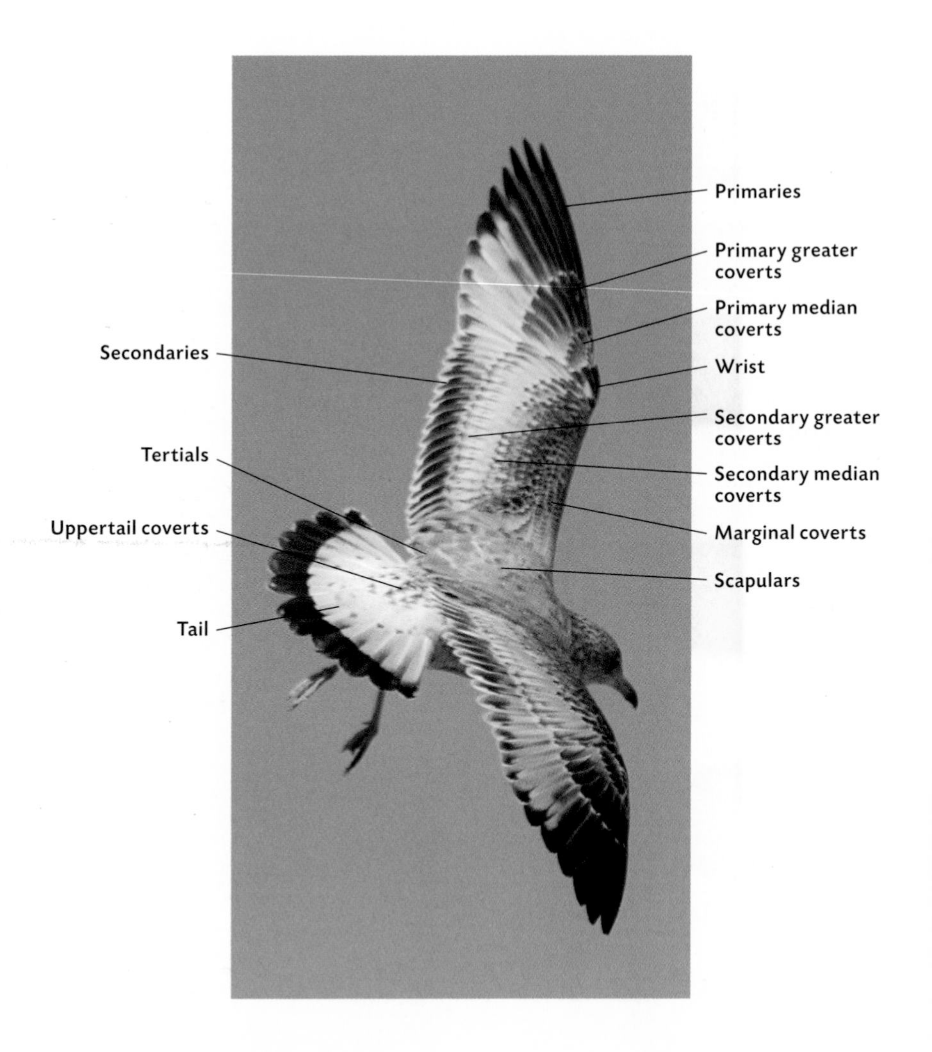

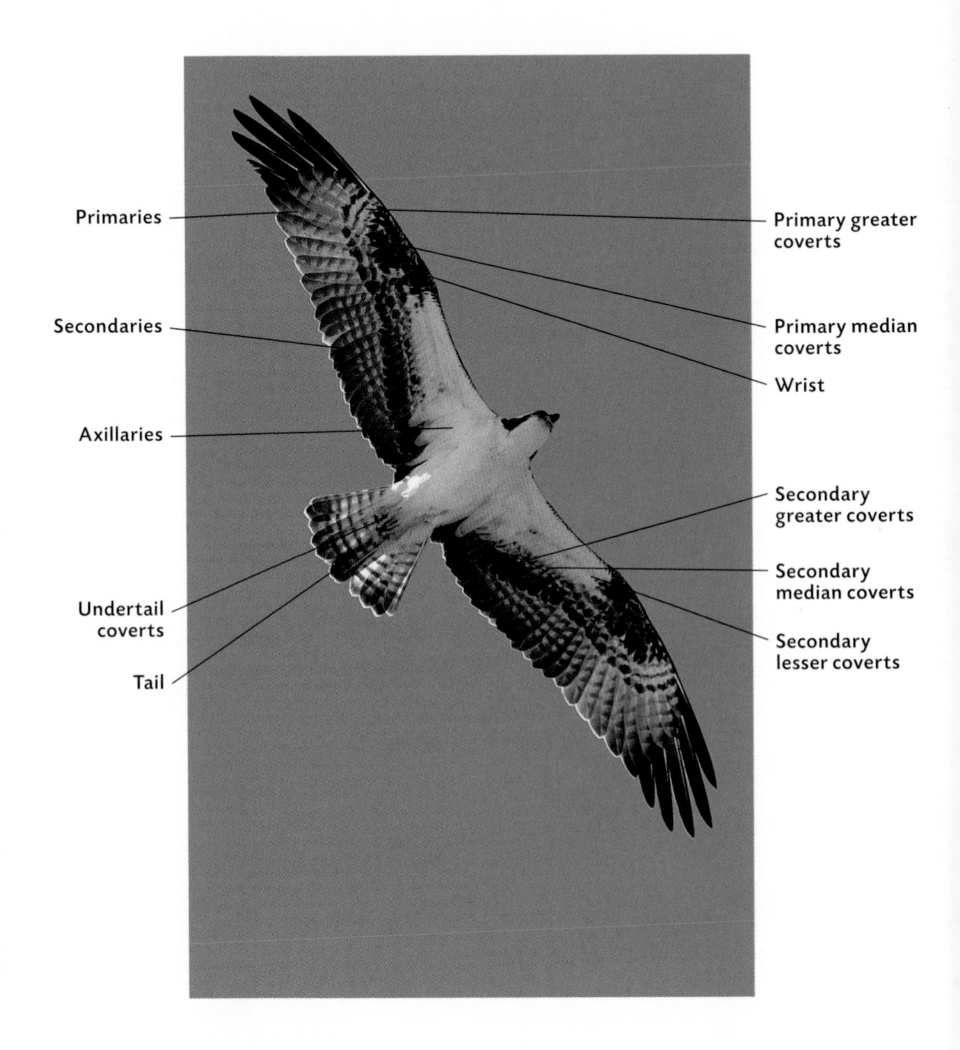
Primaries
Secondaries
Axillaries
Undertail coverts
Tail
Primary greater coverts
Primary median coverts
Wrist
Secondary greater coverts
Secondary median coverts
Secondary lesser coverts

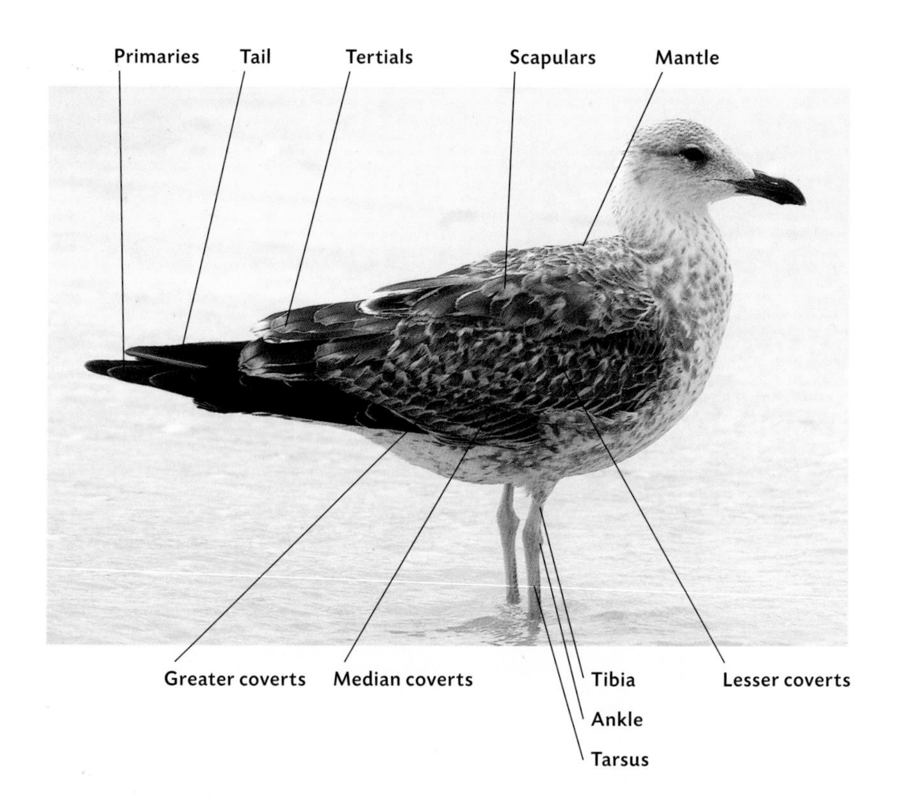
Primaries
Tail
Tertials
Scapulars
Mantle
Greater coverts
Median coverts
Tibia
Lesser coverts
Ankle
Tarsus

Tail
Uppertail coverts
Tertials
Scapulars
Undertail coverts
Primaries
Secondaries
Flank

Primaries
Tertials
Lower scapulars
Upper scapulars
Mantle
Nape
Crown
Forehead
Lore
Tail
Greater coverts
Median coverts
Ankle
Tibia
Tarsus
Belly
Breast
Chin

Orbital ring
Eye
Culmen
Nostril
Upper mandible
Gape
Lower mandible
Gonys

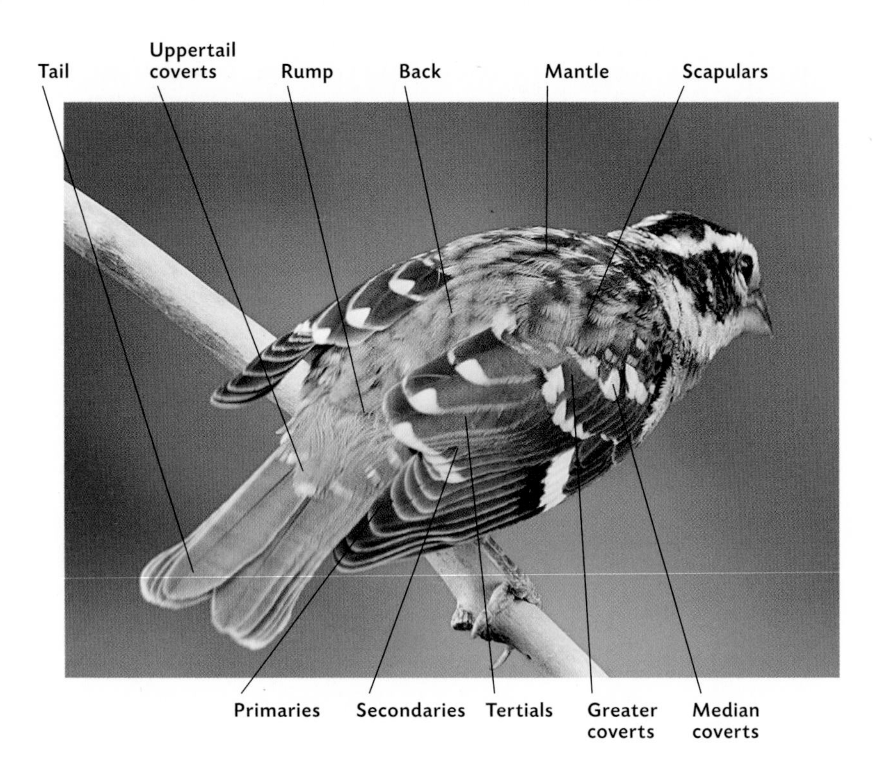
Tail
Uppertail coverts
Rump
Back
Mantle
Scapulars
Primaries
Secondaries
Tertials
Greater coverts
Median coverts

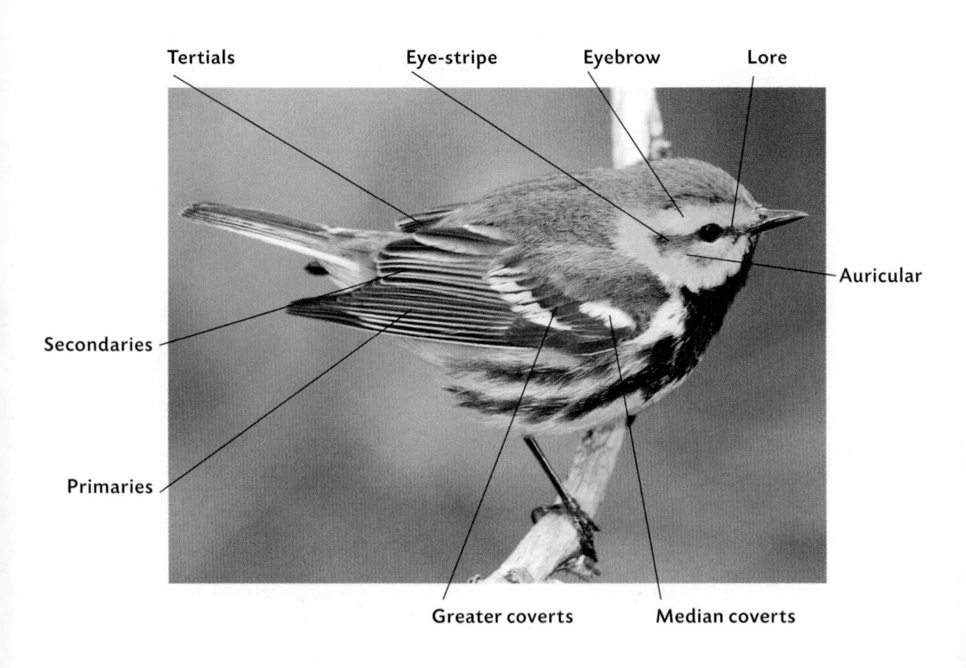
Tertials
Eye-stripe
Eyebrow
Lore
Auricular
Secondaries
Primaries
Greater coverts
Median coverts

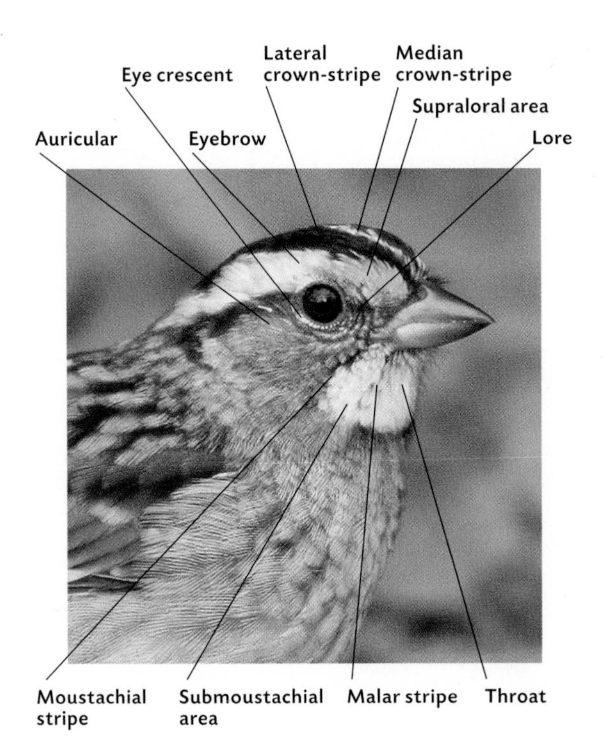
Lateral crown-stripe
Median crown-stripe
Eye crescent
Supraloral area
Auricular
Eyebrow
Lore
Moustachial stripe
Submoustachial area
Malar stripe
Throat

Black Skimmers

The Birds

Whistling-Ducks

Geese

Swans

Ducks

Adult, *fulgens* TX/02

Adult, *fulgens* TX/11

Juv., *fulgens* TX/11

Adult, *fulgens* TX/11

Black-bellied Whistling-Duck 1

Dendrocygna autumnalis L 21"

Shape: Medium-sized, long-legged, very long-necked duck. Bill length ¾ head length; sloping forehead to shallow crown; short tuft on hindcrown creates squarish angle to nape. **Ad:** Bright red bill; gray face with bold white eye-ring; narrow brown stripe through crown and nape. Reddish-brown back and chest; black belly; buff-and-white swath on wing. **Juv:** (Jun.–Nov.) Like ad. but gray belly, bill, and legs. **Flight:** Head and legs often held below body line; white greater coverts and base of primaries create broad white stripe through blackish upperwing; underwing all blackish. **Hab:** Woodland ponds, marshes. **Voice:** Very high-pitched, thin, loud whistle of several notes, *seeetewseeew.*

Subspp: (2) •*fulgens* (NAM); upper back reddish brown, no or indistinct grayish band on breast. •*autumnalis* (possible vag. from SAM); grayish upper back, distinct gray band on breast.

Adult TX/04

Adult FL/02

Adult TX/04

Fulvous Whistling-Duck 1

Dendrocygna bicolor L 20"

Shape: Medium-sized, long-legged, very long-necked duck. Bill length ¾ head length; sloping forehead to flat shallow crown; short tuft on hindcrown may create squarish angle to nape. **Ad:** Pale reddish-brown face, neck, and belly; foreneck blackish with fine white streaks; bill dark gray; narrow blackish stripe through crown and nape. Back blackish with reddish-brown barring; white dashes angle up along flanks; undertail coverts whitish. **Juv:** (Jun.–Sep.) Like ad. but paler with pale buffy barring on back. **Flight:** Long broad wings all black above and below; black tail contrasts with white band of uppertail coverts. **Hab:** Wet agricultural land, ponds, marshes. **Voice:** Repeated, scratchy, high-pitched whistle, *chicheeew.*

Subspp: Monotypic.

Adult JAP/02

Adult JAP/02

Taiga Bean-Goose 3

Anser fabalis L 34"

Rare Eurasian vag. to w. AK, mostly in spring; has occurred in NE, IA, WA. **Shape:** Large goose with a moderately long neck; bill long, thin (especially near tip), and pinched slightly flatter near tip, making tip look slightly bulbous and the top and bottom lines of the bill slightly concave just behind tip; ratio of bill length to depth at base about 2:1. Similar Tundra Bean-Goose slightly smaller and shorter-necked; bill is proportionately deeper-based, blunt-tipped, not concave near tip, and ratio of length to depth at base about 4:3. **Ad:** Bill bicolored, black with narrow orangish (can appear pinkish) band near tip (see Subspp.); legs orange; dark brown head contrasts with paler neck, breast, and belly; back brownish (contrastingly grayish in Pink-footed Goose). Flanks variably barred with black, but belly unbarred. In rare cases, bill has narrow white band around base. Similar Tundra Bean-Goose has different bill shape. Similar Greater White-fronted Goose has all-orange to pink bill with prominent white band around base, barred belly (in ad.), and very different call (see Voice). **Juv:** (Jul.–Dec.) Like ad. but head slightly paler; back and flanks with indistinct barring (distinct in ad.), indistinct feather grooves along neck (distinct in ad.). **Flight:** Wings dark, little contrast between coverts and flight feathers; tail dark with white tip; narrow white band on rump. **Hab:** Open areas, fields, agricultural lands. **Voice:** Deep nasal honk, *unk unk*. Similar Greater White-fronted Goose has high-pitched sibilant call.

Subspp: (1) •*middendorffii* (Eurasian vag. to w. AK); this subsp. has narrowest ring of orange on the bill among the subspp. of Taiga Bean-Goose (European subsp. *fabalis* has much wider band of orange).

Adult, *serrirostris* RUS/–

Adult, *serrirostris* RUS/–

Adult, *rossicus* LIT/05

Tundra Bean-Goose 3

Anser serrirostris L 30"

Rare Eurasian vag. to primarily w. AK, mostly in spring; has occurred in QC. **Shape:** Medium-sized goose with a moderately long neck; bill is proportionately deep-based, rather thick and blunt at the tip, with fairly straight top and bottom lines. Ratio of bill length to depth at base about 4:3. Similar Taiga Bean-Goose is slightly larger; bill is proportionately longer and thinner-based, thinner and more pointed at tip, and ratio of length to depth at base about 2:1. **Ad:** Bill bicolored, black with a narrow orangish (can appear pinkish) band near the tip; legs orange; dark brown head contrasts with paler neck, breast, and belly; back brownish like rest of bird (contrastingly grayish in Pink-footed Goose). In rare cases, bill has narrow white band around base. Similar Taiga Bean-Goose has different bill shape. Similar Greater White-fronted Goose has all-orange to pink bill with prominent white band around base, barred belly (in ad.), and very different call (see Voice). **Juv:** (Jul.–Dec.) Like ad. but head slightly paler; back and flanks with indistinct barring (distinct in ad.), indistinct feather grooves along neck (distinct in ad.). **Flight:** Wings dark, little contrast between coverts and flight feathers; tail dark with white tip; narrow white band on rump. **Hab:** Open areas, fields, agricultural lands. **Voice:** Deep nasal honk, *unk unk*. Similar Greater White-fronted Goose has high-pitched sibilant call.

Subspp: (2) •*serrirostris* (Eurasian vag. to w. AK); •*rossicus* (Eurasian vag.; one record from QC).

Adult SCO/01

Adult ENG/01

Adult SCO/01

Pink-footed Goose 4

Anser brachyrhynchus L 26"

Casual European vag. to Northeast; usually seen with other geese. **Shape:** Small fairly short-necked goose with a rounded head and relatively short stubby bill. Similar in structure to Tundra Bean-Goose. **Ad:** Bill bicolored, black at base and tip with pinkish band between; legs deep pink; brown head blends with paler brown neck and breast; silvery wash over back and wings (seen best in fresh plumage) contrasts subtly with brown tones on rest of body. Similar bean-geese have orange legs, sharper demarcation between dark head and paler neck, and brownish back. Similar Greater White-fronted Goose has orange bill and legs, white band at base of bill, and barred belly. **Juv:** (Jun.–Sep.) More brownish overall; legs may be yellowish pink. **Flight:** Gray wing coverts contrast with black primaries and secondaries (bean-geese have all-dark wings). **Hab:** Marshes, agricultural lands in winter. **Voice:** Nasal honk, *unk unk*.

Subspp: Monotypic.

IDENTIFICATION TIPS

Geese are midsized waterfowl with large bodies, medium to long necks, and moderately long triangular bills; the sexes look alike.

Species ID: Note relative proportions of body size to neck length, and head size and shape to bill size and shape; look at colors on bill, head, and legs.

Swans are our largest waterfowl and have very long necks and long flattened bills; the sexes look alike.

Species ID: Note bill shape and colors; look at pattern of bare skin where it meets the eye and head feathers.

Adult CA/11

Juv. USA/–

Adult, *elgasi* CA/03

Adult, *flavirostris* SCO/–

Adult EUR/11

Greater White-fronted Goose 1

Anser albifrons L 28"

Shape: Medium-sized goose with a relatively large rounded head and long deep-based bill. Structure varies (see Subspp.). **Ad:** Pink to orangish bill (see Subspp.) with variable white band around base; orange legs; grayish-brown head, neck, and back; variable black barring on pale brown breast and belly. Similar domestic Graylag Goose much heavier, deep-bellied, and more evenly medium brown overall with minimal barring on underparts. Bean-geese lack white at base of bill and lack barring on belly. **Juv:** (Aug.–Dec.) Lacks distinctive white on bill base and black barring on underparts; bill pinkish to orange with dark nail (tip); legs yellowish orange; gradually acquires ad. plumage by April. Similar juv. bean-geese and Pink-footed Goose have wide dark base to bill. **Flight:** Gray wing coverts contrast with black primaries and secondaries; barred belly (in ad.). **Hab:** Tundra lakes in summer; marshes, agricultural lands in winter. **Voice:** 2–3-part flight call (higher-pitched and more sibilant than Canada Goose), *cheelee, cheeleelee;* slightly lower-pitched *enk enk*.

Subspp: (4) •*frontalis* (breeds w. AK across n. CAN; winters s.w. BC–s. CA east to LA); small with pale brown head and neck and pink to orangish bill, feeds in fields. •*elgasi* (breeds Cook Inlet, AK; winters cent. CA), "Tule Goose"; large long bill and with darker head and neck than *frontalis*, sometimes dark cap, pink bill, orbital skin sometimes dull yellow, feeds in marshes. •*gambeli* (breeds n. AK, n.w. CAN; winters in AZ to s. LA; vag. to e. NAM); medium to large with medium brown neck, pink to orangish bill. •*flavirostris* (EUR vag. to e. NAM), "Greenland White-fronted Goose"; small to medium with dark brown head and neck and yellowish to orange bill. **Hybrids:** Snow Goose, Canada Goose.

Adult AK/02

Adult AK/02

Juv. WA/08

Adult AK/08

Emperor Goose 2

Chen canagica L 26"

Largely coastal goose of w. AK; casual vag. south to CA. **Shape:** Medium-sized, compact, short-necked goose with a short stubby bill, steep forehead, and high well-rounded crown. **Ad:** Distinctive white head and nape (sometimes stained orangish); black chin and throat; bill pink with black edges and nostrils; legs bright orange. Pale gray body feathers have thin black subterminal, then pale terminal, bands creating strongly scaled effect. **Juv:** (Aug.–Nov.) Dark gray overall except for whitish eye-ring and white tail. Gray body feathers and wing coverts scaled with white as on adult. Gradually acquires ad. plumage by Dec. Similar juv. dark-morph Snow or Ross's Goose lacks white scaling on back and flanks. **Flight:** Wings uniformly gray above and below; tail, nape, and head white in adult. **Hab:** Tundra pools in summer; intertidal zones in winter. **Voice:** Repeated, rapid, high-pitched *cheedee cheedeedee.*

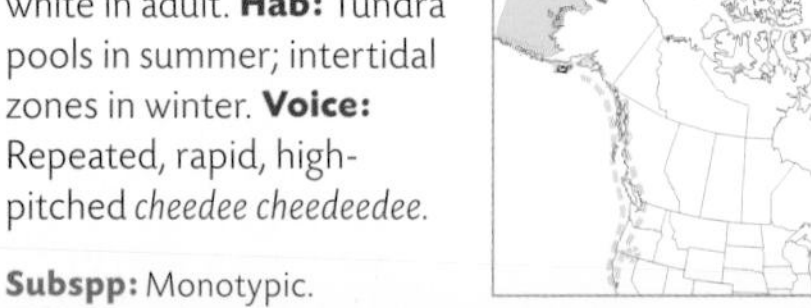

Subspp: Monotypic.

Adult OK/12

Adult OK/12

Adults NET/02

Barnacle Goose 4

Branta leucopsis L 27"

Casual vag. from GRN to n.e. NAM, fall–spring; escapees from captivity may be seen in e. NAM. **Shape:** Medium-sized stout-bodied goose with a relatively short thick neck and a small stubby bill. Culmen slightly concave leading to sloping forehead and rounded, or sometimes slightly squared, crown. **Ad:** Black crown, neck, and breast contrast strongly with white face (sometimes tinged yellowish) and belly; black line through eye; back barred black; upperwing coverts silvery with subterminal dark and then white tips. **Juv:** (Jun.–Nov.) Like adult but with less well defined grayish barring on flanks. **Flight:** Note strong contrast between black breast and whitish underparts; wing coverts silvery barred darker above and all silvery below. **Hab:** Fields in winter. **Voice:** Barklike *caw.*

Subspp: Monotypic. **Hybrids:** Cackling Goose.

Adult, *hrota* NH/11

Adult, *nigricans* AK/06

Juv., *hrota* NH/11

Adult, *bernicla* ENG/02

Adult, *hrota* NJ/03

Adult, captive, "Gray-bellied Brant" WA/06

Brant 1

Branta bernicla L 25"

Shape: Small relatively slim-bodied goose with relatively short neck and small stubby bill. Fairly straight culmen leads to sloping forehead and high crown. **Ad:** Black head, neck, and breast contrast weakly with sooty to blackish back and pale to blackish belly (see Subspp.). Variable whitish crescent on upper neck (see Subspp.); flanks pale with variable barring; rear white; legs and bill black. **Juv:** (Jul.–Mar.) Like ad. but with whitish margins to wing coverts; lacks neck crescent; no barring on flanks. **Flight:** Long thin wings; short body; wings all dark with slightly paler coverts. **Hab:** Coastal waters. **Voice:** Fairly low-pitched croaking *craak craak;* similar to Wood Frogs.

Subspp: (3) •*hrota* (n.cent.–n.e. NU, winters Atlantic Coast, vag. elsewhere), "Atlantic Brant"; black breast contrasts strongly with whitish belly, grayish flanks with wide white barring, neck crescent narrow and broken. •*nigricans* (coastal n. AK–n.cent. NU; winters on Pacific Coast, vag. to mid-Atlantic Coast), "Black Brant"; blackish breast contrasts weakly or blends with blackish-brown belly, brownish flanks with moderately wide whitish barring, neck crescent usually wide and complete in front; darker mantle than Atlantic Brant. •*bernicla* (Eurasia, vag. to e. NAM), "Dark-bellied Brant"; blackish breast contrasts moderately with brownish belly, brownish flanks with narrow whitish barring, narrow broken neck crescent. ▶ Another population does not have subsp. status but is slightly different, having a grayish belly without contrastingly brighter flanks and a broken neck-ring; it is called "Gray-bellied Brant" (High Arctic, winters s.w. BC–n.w. WA; vag. to East Coast).

Adult white morph, *caerulescens* CA/12

Adult dark morph, *caerulescens* IN/08

Juv. white morph, *caerulescens* CA/12

Juv. dark morph, *caerulescens* NM/12

Adult white morph, *caerulescens* NM/12

Adult dark morph, *caerulescens* NM/11

Snow Goose 1

Chen caerulescens L 28–31"

Shape: Medium-sized moderately long-necked goose with a fairly long thick-based bill, sloping forehead, and shallow crown. Line of feathering at base of upper mandible markedly convex. Similar Ross's Goose has short neck, short stubby bill, and straight line of feathering across base of both mandibles. ▶ Occurs in 2 morphs with variations in between. **White Morph – Ad:** All white except for black primaries; bill pink with black edges along gape; legs pink; face often stained with orange wash. **Juv/1st Yr:** (Aug.–Aug.) Like ad. but with extensive gray wash over head, neck, and upperparts; legs dusky. Gradually acquires ad. plumage by 2nd fall. Similar juv./1st yr. Ross's almost all white. **Flight:** Moderately long neck; long pointed wings; bird all white except for black primaries. **Dark Morph** ("Blue Goose") – **Ad:** White head and upper neck; rest of body dark grayish brown overall except for whitish to grayish undertail coverts; bill deep pink with black edges along gape; legs pink; tertials strongly edged with white. Variations on dark morph include variable white on belly. **Juv/1st Yr:** (Aug.–Aug.) Grayish brown overall with slightly paler belly; legs dark. Gradually acquires ad. plumage by 2nd fall. **Flight:** Moderately long neck, long pointed wings; primaries and secondaries dark, wing coverts gray; dark body and white to gray rear. **Hab:** Tundra in summer; marshes, agricultural fields in winter. **Voice:** Well-spaced medium-pitched *aank aank.*

Subspp: (2) Differ mostly in size, which overlaps (and can also be affected by diet). •*caerulescens* (n. AK–n.w. QC; winters CA–NM and e. TX–NE to FL–NJ), "Lesser Snow Goose"; smaller, both white and dark morphs common. •*atlantica* (NU–w. GRN; winters coastal MD–NJ), "Greater Snow Goose"; large, almost all white morphs. **Hybrids:** Greater White-fronted Goose, Ross's Goose, Cackling Goose, Canada Goose.

Adult white morph CA/12

Adult dark morph CA/01

Juv. white morph NY/10

Adult white morph NM/12

Ross's Goose 1

Chen rossii L 23"

Shape: Small, compact, short-necked goose with a small rounded head. Short stubby bill abuts steep forehead and fairly high rounded crown. Line of feathering across base of mandibles straight; gape with little or no dark edging. Similar Snow Goose has longer bill, neck, and wings; longer flatter crown; convex feathering across gape; and conspicuous dark edges to gape. ▶ Occurs in 2 morphs; dark morph very rare. **White Morph – Ad:** All white except for black primaries; bill deep pink with bluish-gray to purple base and no black edges to gape (similar Snow Goose with all-pink bill and black edges to gape); legs pink; rarely or never shows orange staining on head like Snow Goose. **Juv/1st Yr:** (Aug.–Aug.) Like ad. but with slight pale grayish wash over head, neck, and upperparts; legs dusky; bill grayish. Gradually acquires ad. plumage by 2nd fall. Similar juv./1st yr. Snow more extensively gray on head, neck, and back. **Flight:** Short neck, relatively short rounded wings; all white except for black primaries. **Dark Morph – Ad:** Rare. Dark grayish brown over most of body except for white face, throat, belly, undertail coverts; white tertials and wing coverts; bill deep pink with bluish tones to base and no black edges to gape; legs pink. Darker than dark-morph Snow Goose. **Juv/1st Yr:** (Aug.–Aug.) Grayish brown overall except for white belly and undertail coverts and white inner wing; bill dark gray; legs dark. Gradually acquires ad. plumage by 2nd fall. **Flight:** Short neck, relatively short wings. Primaries and secondaries dark, wing coverts gray. **Hab:** Tundra in summer; marshes, agricultural fields in winter. **Voice:** High-pitched single toots, like *teek, teek;* also low grunting *unk;* voice like Snow Goose but higher-pitched.

Subspp: Monotypic. **Hybrids:** Snow Goose.

Adult, *minima* BC/10

Adult, *leucopareia* CA/01

Adult, *hutchinsii* NH/11

Adult, *taverneri* NAM/11

Cackling, *hutchinsii* (l.), and Canada NH/11

Adult, *minima* WA/11

Cackling Goose 1

Branta hutchinsii L 25–27"

Shape: Small, relatively short-legged, short-necked goose with a short stubby bill. Generally, ratio of bill length to depth at base is 3:2 to 1:1; bill length less than 50% length of side of head; primaries almost always project past tail. See similar Canada Goose. **Ad:** Distinctive black head and neck with broadly white cheeks; pale grayish-brown body with white rear and pale to dark breast. **Juv:** (Aug.–Feb.) Similar to adult; cheek patch may be reduced and/or have brownish wash; head and neck may be brownish black; generally lacks white collar and has less distinct patterning on wing coverts. **Flight:** White uppertail coverts form broad U between black tail and rump. **Hab:** Lakes, marshes in summer; lakes, bays, fields, parks in winter. **Voice:** Higher-pitched and squeakier than that of Canada Goose; *eeek* and *eeerik*.

Subspp: (4) From smallest to largest: •*minima* (summer, n.w. AK; winter, w. OR, w. WA, cent. CA; rare e. of Rockies; a popular captive subsp.), "Ridgway's Goose"; 3½ lb. Head rounded, forehead slightly steeper than culmen; breast darkest of subspp., with glossy purple to bronze sheen, flanks and belly paler; wing coverts gray with dark brown subterminal and whitish terminal bands; usually a gular stripe, sometimes a partial collar; ratio of bill length to depth about 1:1. •*leucopareia* (summer, w. Aleutian Is.; winter, cent. CA, w. OR; rarely vagrant), "Aleutian Goose"; 4 lb. Fairly steep forehead rounds off to long flattish crown, which rounds off to nape; medium grayish-brown breast, often paler than belly and flanks; wing coverts brown with well-defined creamy terminal band; usually a gular stripe (often wide) and a broad white collar; ratio of bill length to depth about 4:3. •*hutchinsii* (summer, coastal NU; winter, coastal TX and LA, CO–NM–w. TX; subsp. most likely seen in e. NAM), "Richardson's Goose"; 4½ lb. Head blocky, steep forehead angles to flat crown, which angles to nape; averages whitest breast among subspp., underparts generally of uniform color; wing coverts grayish brown with variable buffy terminal band; cheek patch often with small notch near eye; usually no gular stripe, sometimes a partial collar; ratio of bill length to depth about 3:2. •*taverneri* (summer, n.–n.e. AK; winter, w. OR, w. WA, cent. CA; rare but regular e. of Rockies), "Taverner's Goose"; 5½ lb. Long gently angled forehead blends to flat crown, which rounds off to nape; longer neck than other subspp.; medium gray breast, darker flanks and belly; wing coverts grayish with variable dark subterminal and whitish terminal bands; usually a gular stripe, rarely a collar; ratio of bill length to depth about 3:2; may intergrade with *hutchinsii*. ▶ Subspp. may be difficult to disting. from smallest subsp. of Canada Goose, *B. c. parvipes*. **Hybrids:** Barnacle Goose.

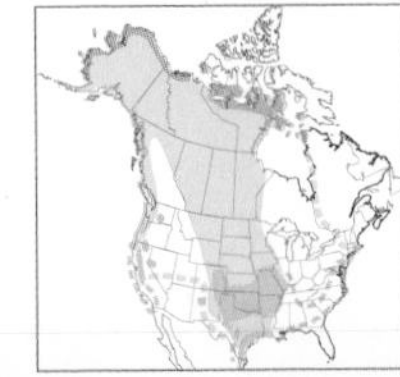

Adult OH/05

Adult, *moffitti* OR/02

Adults, *occidentalis* BC/10

Adult TX/12

Adult and young, *moffitti* CO/06

Adults NJ/10

Canada Goose 1

Branta canadensis L 27–45"

Shape: Medium-sized to large goose with a long neck and large body. Generally, bill length to depth is 2:1 to 3:1; bill length greater than 50% length of side of head; primaries fall short of or just reach tail tip. See similar Cackling Goose. **Ad:** Distinctive black head and neck with broad white cheeks; pale grayish-brown body with white rear; pale to dark breast; sometimes white collar. Rarely, white forehead or no white cheek. **Juv:** (Aug.–Jan.) Like adult but cheek patch may have brownish wash; head and neck may be brownish black; collar and covert patterns often less distinct. **Flight:** Flies in shifting line and V formations; white uppertail coverts form a broad U. **Hab:** Summer, lakes, marshes, fields, parks; winter, lakes, bays, fields, parks. **Voice:** M. low-pitched *ahonk;* f. higher-pitched *hink.* 🎧**1**.

Subspp: (5) From roughly smallest to largest: •*parvipes* (summer, w. CAN–AK; winter, s.e. CO–s.w. NE south to NM–w. TX, also WA–OR), "Lesser Canada Goose"; 7 lb. Short neck and bill (for Canada Geese); forehead shape varies from gradual slope to short and steep; pale brown overall, whitish to grayish breast, darker flanks and belly (WA–OR birds darker overall, often with gular stripe); sometimes a partial white collar, generally no gular stripe; ratio of bill length to depth at base about 2:1. •*occidentalis* (summer, s. AK–BC; winter, WA–OR), "Dusky Canada Goose"; 7½ lb. Long bill, legs, and neck; dark overall, dark breast may be washed reddish brown; occasionally a thin or incomplete white collar. •*interior* (summer, e.–cent. CAN; winter, IA–NY south to s.e. TX–n. FL), "Hudson Bay Canada Goose"; 9 lb. Comparatively long neck and short legs; slightly darker and browner than *canadensis,* especially on back; no distinct white collar. •*canadensis* (summer, n.e. CAN–ME; winter, coastal NF–NC), "Atlantic Canada Goose"; 9½ lb. Comparatively long legs and long neck; pale overall, especially on breast; no distinct white collar. •*moffitti* (summer, BC–s. ON south to n.e. CA–w. TN; winter, s. CA–LA; birds breeding from WI–NH and south are most likely naturalized or introduced populations of *moffitti*), "Large Canada Goose" or "Great Basin Canada Goose"; 10 lb. Large long neck; pale overall; no distinct white collar, often partial or complete gular stripe, occasionally white forehead. ▶ May be hard to distinguish *parvipes* from similar-sized subspp. of Cackling Goose, *taverneri* and *hutchinsii,* because of variation in size and color within subspp. and size variation due to nutrition and sexual dimorphism. Various subspp. have been introduced into many regions of NAM, confusing some subsp. distinctions. **Hybrids:** Greater White-fronted Goose, Snow Goose.

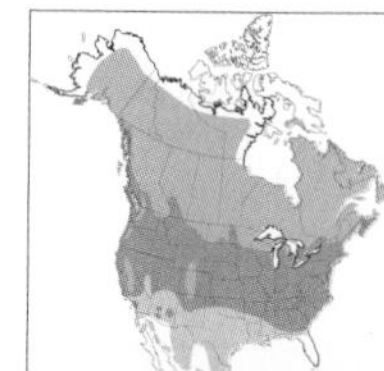

Adult MN/01

1st summer MN/02

Adult (stained) WI/09

Juv. MN/01

Adult MN/01

Adult MN/01

Trumpeter Swan 1

Cygnus buccinator L 60"

Shape: Very large, long-bodied, long-necked swan with a short rounded tail. Bill very long with straight culmen; black lore broadly adjoins eye (slightly narrower than diameter of eye where they meet); feathering across back of gape usually angled (not vertical); feathering on forehead comes to a pointed V (not a rounded U) over top of bill; crown appears shallow and often slightly peaked near rear. Similar Tundra Swan is smaller, has a high-pitched wailing call, and differs in shape of bill, lore, and adjacent feathering. **Ad:** Plumage all white; black bill and lore; legs black; head sometimes stained orangish. Rarely, small greenish-yellow dash in lore before eye. **Juv/1st Spring:** (Aug.–May) Brownish-gray wash over whitish body; bill variably black at base, pink in middle, black at tip (juv. Tundra bill can be similar in winter as it gradually changes to all black). By March, Tundra juv. has completed molt and is mostly white; Trumpeter juv. still molting and retains some grayish feathers into summer. Legs and feet yellowish to black. Small percentage of juv. are all white with yellow feet and all-pink bill. Use shape of bill to distinguish Trumpeter from Tundra juvs. **Flight:** Runs over water at takeoff; wings quiet in flight; rounded tail. **Hab:** Lakes, ponds, large rivers in summer; also along coasts, in wet agricultural fields, pastures in winter. **Voice:** Much like short trumpet toots, *tootoot tootootoot*.

Subspp: Monotypic.

Adult, *columbianus* AK/06

Adult, *bewickii* JAP/–

Adult, *columbianus* PA/03

Juv., *columbianus* OR/01

Adult, *columbianus* AK/06

Adult, *columbianus* YT/04

Tundra Swan 1

Cygnus columbianus L 52"

Shape: Large, long-bodied, long-necked swan with a short rounded tail. Bill long with slightly curved culmen; lores narrowly adjoin eye, leaving eye fairly distinct; feathering across back of gape usually fairly vertical (not angled), and feathering on forehead comes to a rounded U (not a pointed V) over top of bill; crown appears rounded. Similar Trumpeter Swan is larger, has a trumpet-like call, and differs in shape of bill, lore, and adjacent feathering. **Ad:** Plumage all white; black lores and bill; variable yellow dash in lore and onto bill base (can be very small or absent). Head may be slightly stained orangish from minerals. **Juv/1st Winter:** (Aug.–Mar.) Grayish wash over whitish body; bill variable, usually all pinkish, gradually becoming black from base outward over winter. By March, mostly white (Trumpeter juv. will have some grayish feathers into summer); legs and feet gray. Use shape of bill to help distinguish between Trumpeter and Tundra juvs. **Flight:** Runs over water at takeoff; wings quiet in flight; rounded tail. **Hab:** Tundra in summer; lakes, ponds, open marshes in winter. **Voice:** Drawn-out high-pitched wail, *weeel ahweeel;* short low-pitched *wahw.*

Subspp: (2) •*columbianus* (all of NAM range), "Whistling Swan"; small to moderate yellow dash (rarely none) in lore, lower mandible black. •*bewickii* (Asian vag. to Atl. and Pac. coasts of NAM), "Bewick's Swan"; slightly smaller, extensive yellow to orange on lore, yellow on bill base extends to back edge of nostril (similar Whooper Swan has yellow extending past nostril), lower mandible may be yellowish.

Adult NH/02
Adult NH/03

Juv. gray morph NY/08

Adult NY/05

Mute Swan 1

Cygnus olor L 60"

European species introduced into NAM. **Shape:** Very large, long-bodied, long-necked swan with a long pointed tail. Large rounded knob on ad. forehead. Popularly depicted posture with wings held up over back is generally an aggressive display. **Ad:** All white with orangish-red bill; black knob on bill; black lores. Head may be stained orangish to greenish due to water and vegetation. **Juv:** (Jun.–Nov.) At first, juv. has no knob on forehead and lores are feathered; gradually, into winter, the knob grows and lores become naked. Two morphs of juvs. – white and gray. White morph (quite rare) all white except for tan bill and legs. Gray morph grayish brown overall with slate-gray bill and legs; may retain some gray feathers on rump until following fall. **Flight:** Runs over water on takeoff; wings in flight create repeated short humming sound (other swans quieter in flight). Pointed tail helps distinguish it from other swans in flight. **Hab:** Lakes, parks, coastal bays. **Voice:** Common sound is a low-pitched nasal *heeorhh.*

Subspp: Monotypic.
Hybrids: Whooper Swan.

Adult ENG/–

Juv. ENG/11

Adult FIN/09

Whooper Swan 3

Cygnus cygnus L 60"

Winters Aleutian Is. AK; rare winter vag. to AK, WA, OR, n. CA, MT. Records in e. NAM probably pertain to escapees from captivity. **Shape:** Very large, long-bodied, long-necked swan with a short rounded tail. Bill very long; culmen slightly curved; feathering on forehead U-shaped over top of bill; crown rounded; feathering from eye to gape a straight line. **Ad:** All white; bill black with extensive yellow over lore and onto bill base; yellow extends under base of nostrils in a point (similar Tundra Swan subsp. *bewickii* has yellow extending just to back of nostrils and rounded off). **Juv:** (Jul.–May) Pale grayish overall; has bill pattern of ad. in different colors – whitish base, pinkish toward tip. Gradually acquires ad. plumage by late spring. **Flight:** Runs over water at takeoff; wings quiet in flight; rounded tail. **Hab:** Tundra lakes, marshes in summer; open water, marshes, agricultural fields in winter. **Voice:** Low-pitched, repeated in 3's and 4's, *kloo kloo kloo.*

Subspp: Monotypic. **Hybrids:** Mute Swan, Tundra Swan (subsp. *columbianus*).

IDENTIFICATION TIPS

Ducks are our smallest waterfowl, with generally moderate-length bodies, short necks, and fairly long flattened bills; the sexes look different. Ducks can be divided into **dabblers,** which feed by tipping up their rear as they reach underwater with their head and which take off directly from the water surface (Muscovy Duck to Green-winged Teal); and **divers,** which feed by diving underwater and usually take off by first running across the water (Canvasback to Ruddy Duck).

Species ID: Male ducks are generally easy to identify due to their distinctive color patterns; species with similar plumage have fairly marked differences in the shape of the bill and head. Female ducks look much more alike; the shape of head and bill are your best clues to distinguishing most of them; also note the fine details of colors and patterns on head and bill.

Plumage Notes: Female ducks tend to look generally the same through all seasons and all ages. Male ducks start out looking like females and gradually acquire their adult male breeding plumage over the first winter (or first few years in some cases). These immature males have various mixtures of male and female plumage colors, depending on their stage of maturity, and are sometimes difficult to identify by plumage. In these cases, rely most on bill and head shape to solidify your species ID. Adult males have what is termed an eclipse plumage in late summer, when they look much like the female. As they molt into and out of this plumage they are again a mixture of male and female plumages and can look much like the immature male but at a different time of year (late summer, not winter and spring).

Adult m., feral —/04

Adult m., feral FL/03

Muscovy Duck 2

Cairina moschata L 25–31"

Native birds found only in s. TX; domestic birds may be seen in park ponds across NAM. Often perches in trees. **Shape:** Very large, heavy-bodied, short-legged duck with a short thin neck and crested head; when crest raised, crown is triangular. M. much larger than f. and has small knobs at base of bill. **Ad:** All black except for white wing coverts, (less extensive in f.), which may show on resting birds as a sliver of white above flank. Somewhat pied bill; bare skin at base of bill black to pink. M. glossier than f. with hints of iridescent green and purple. **Juv/1st Yr:** (Jul.–Jul.) All black with little or no white at tips of primary coverts. **Flight:** Big-bodied, long tail, very broad wings, slow wingbeats, striking white on upper and lower wing coverts (of ad.). **Hab:** Wild birds prefer wooded streams, swamps, bottomlands; domestic birds frequent parks. **Voice:** Usually silent.

▶ Feral variety widely introduced to parks and ponds, especially in FL. Feral Muscovy is similar in shape but larger; body blotched with white to almost all white; bare skin on face usually red. Wild birds usually seen in flight.

Subspp: Monotypic.

Adult m. CA/01

Juv. m. OH/07

Adult f. CA/01

Adult m. IL/03

Adult m., late eclipse NY/09

Adult f. IL/03

Wood Duck 1

Aix sponsa L 18"

Shape: Medium-sized, thin-necked, short-legged, short-billed duck. Bill length about ½ head length; steep forehead angles to flat crown; long drooping crest off hindcrown creates big-headed look, accentuated by thin neck. Tail relatively long and squared off – a good clue when in flight or at rest. **Ad:** M. (Oct.–Jun.) with unique colorful and intricate head pattern: white neck-ring and chin strap, red eye and base of bill, glossy green crown and crest. Eclipse m. (Jun.–Oct.) mostly brownish but keeps white chin strap and neck-ring, red eye and base of bill. F. mostly brownish with bold white comet-shaped eye patch; dark eye; mostly gray bill encircled by white at base; white throat. **Juv/1st Winter:** (Jul.–Jan.) Brownish with white eye-ring, white eyebrow, dark eye-stripe. Gradually acquires ad. plumage over winter. **Flight:** Agile flier; long tail; upperwing dark brown with white trailing edge to secondaries; central secondaries and their coverts iridescent blue; underwing grayish. **Hab:** Wooded swamps, rivers, lakes. **Voice:** Common flight call of f. a high-pitched rising *oooweeek, oooweeek;* m. a repeated rising *zweeep.* 🎧**2**.

Subspp: Monotypic. **Hybrids:** Mallard, Redhead.

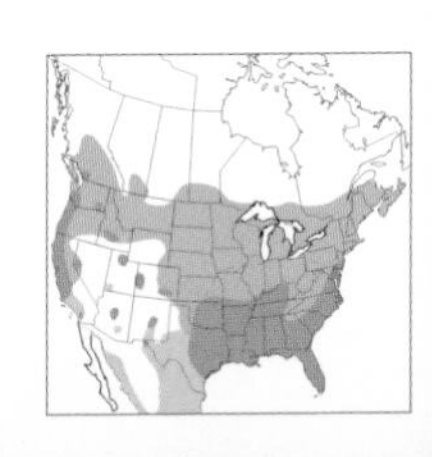

Adult m. CA/01

Adult m. CA/01

Adult f. CA/01

Adult f. IL/06

Adult m., late eclipse ENG/08

Gadwall 1

Anas strepera L 21"

Shape: Medium-sized fairly short-billed duck with a thin neck and rather squarish head. Bill length about 2/3 head length; long steep forehead rounds off to rather high flat crown, which often slopes back to fairly vertical nape. Similar f. Mallard has long sloping forehead and shallow rounded crown. **Ad:** Our only dabbling duck with white speculum. M. (Oct.–May) has slaty to black bill; light brown face; dark brown crown; body grayish except for black rear; cinnamon-edged scapulars. Eclipse m. (Jun.–Sep.) like f., including orangish bill, except for smooth silvery tertials (rather than brown with buff tips as in f.). In Aug.–Sep., male is almost completely molted to breeding plumage but still has traces of "female" bill. F. bill orange on sides, dark gray in center; pale brown head and fairly unmarked face; mottled brown-and-buff body with pale belly; dark tail; white speculum. Similar f. Mallard has whitish tail, distinct dark eyeline and dark crown, dark speculum with white edges, and more sloping forehead and rounded crown. **Juv/1st Winter:** (Jul.–Feb.) Like ad. f.; m. gradually acquires ad. m. plumage by spring. **Flight:** Upperwings dark with white inner secondaries; underwings grayish with white coverts; belly pale. **Hab:** Open lakes, marshes. **Voice:** F. gives low nasal *quahk* (lower-pitched and less strident than Mallard or Black Duck); m. gives repeated low-pitched *arehb*.

Subspp: Monotypic.
Hybrids: Falcated Duck, Eurasian Wigeon, American Wigeon, Mallard, Northern Shoveler, Northern Pintail, Redhead.

Adult m. CA/03

Adult m., eclipse ENG/09

Adult f. WA/10

Adult m. ENG/11

Eurasian Wigeon 2

Anas penelope L 19"

Shape: Nearly identical to American Wigeon (see American Wigeon). **Ad:** M. (Oct.–Jun.) head reddish brown with buffy forehead; bill pale bluish gray with small dark tip; flanks and back pale gray; breast pinkish; hip patch white; rear black. May have some green around eye (not necessarily indicative of hybrid origin). Eclipse m. (Jul.–Sep.) evenly reddish brown on head and body; bill stays the same. F. has pale bluish-gray bill with small dark tip; warm brown to grayish-brown head, breast, and flanks; dark back feathers with reddish-brown to buff edges. Disting. from f. American Wigeon with difficulty. Eurasian f. has gray axillaries seen in flight, is generally warmer brown overall, especially on head (but can be grayish like Am.), has fairly concolor head, breast, and flanks, and lacks black ring around bill base; similar American Wigeon f. has white axillaries seen in flight, is more grayish overall, has gray head contrasting with breast and flanks, and usually has thin black line around bill base. **Juv/1st Winter:** (Jul.–Feb.) Like ad. f.; m. gradually acquires ad. m. plumage by spring but white wing patch reduced. **Flight:** Fast and agile. Above, m. has white secondary coverts, f. gray secondary coverts; below, axillaries and underwing gray, belly white. **Hab:** Lakes, bays, estuaries. **Voice:** M. call a whistled *kaweeer* (slightly higher-pitched and thinner, but more emphatic than American Wigeon call); f. gives low croaking *graahr.*

Subspp: Monotypic. **Hybrids:** Gadwall, Falcated Duck, American Wigeon, Mallard, Northern Pintail, Garganey, Baikal Teal, Green-winged Teal.

Adult m. AZ/11

1st-winter m. IL/04

Adult f. CA/03

Adult m. (r.) and f. BC/12

American Wigeon 1

Anas americana L 20"

Shape: Medium-sized, short-necked, short-billed duck with a usually rounded head and pointed tail. Bill length less than ½ head length. Head shape varies: when forecrown feathers raised, forehead is long and steep, crown strongly downsloping to rear; when forecrown sleeked, head appears smaller and evenly rounded. **Ad:** M. (Oct.–Jun.) has white to creamy forehead and crown; variable-width iridescent green from eye to nape; speckled gray to whitish (occasionally buffy) cheek; bill pale blue-gray with black tip and dark ring around base. Breast and flanks orangish brown; hip patch white; rear black. Eclipse m. (Jul.–Sep.) like ad. f. but may have brighter reddish-brown flanks, also white wing coverts (present all year), which may be seen when bird is at rest. F. has pale blue-gray bill with small dark tip and usually dark ring around bill base; grayish-brown head contrasting with warm brown breast and flanks; dark back feathers with reddish-brown to buff edges. Distinguished from f. Eurasian Wigeon with difficulty (see that sp. for comparison). **Juv/1st Winter:** (Jul.–Apr.) Like ad. f.; m. gradually acquires ad. m. plumage by spring. **Flight:** Fast and agile. Above, m. has white secondary coverts with black tips to greater coverts, f. has grayish-brown coverts with greater secondary coverts tipped with white; below, axillaries and secondary median coverts white, belly white. **Hab:** Lakes, marshes in summer; wet meadows, lakes, parks, protected coastal waters in winter. **Voice:** M. gives 2- or 3-part whistling *ahweeehwoo*; f. gives low croaking *graahr.*

Subspp: Monotypic.
Hybrids: Gadwall, Eurasian Wigeon, Mallard, Northern Pintail, Green-winged Teal.

Adult m., *platyrhynchos* OH/05

Adult f., *platyrhynchos* NH/11

Adult m., eclipse, *platyrhynchos* CA/07

Adult m., *platyrhynchos* NM/11

Adult f., *platyrhynchos* NM/11

Mallard 1

Anas platyrhynchos L 23"

Shape: Large, long-bodied, short-tailed duck with a fairly long bill. Bill length 3/4 head length; sloping forehead leads to rounded crown. On resting bird, primaries extend just past tips of uppertail coverts (only reach base of uppertail coverts on similar Mottled Duck). M. has curlicue tail coverts (Oct.–May). **Ad:** M. (Oct.–May) has bright yellow bill, iridescent green head; also reddish-brown breast, silver body, black rear. Eclipse m. (Jun.–Sep.) like ad. f. but with reddish-brown breast and unmarked dull yellow to olive bill. F. body and wing feathers medium brown with wide buffy margins and internal markings; head and neck paler brown with dark eye-stripe and dark crown; bill orange, mottled black in center; tail feathers strongly edged white. Similar f. Mottled Duck lacks dark mottling on bill, has dark tail and paler buffy head and neck, and blue speculum is bordered in front and back by black lines. Similar f. American Black Duck lacks mottling on bill, has blackish-brown body feathers with thin pale margins, and a dark speculum with no white. Similar f. Gadwall has a steeper forehead and squarer head, lacks a strong facial pattern, and has a white speculum. Domestic Mallards show a wide variety of color patterns and are a result of selective breeding. **Juv/1st Fall:** (Jul.–Nov.) Like ad. f.; m. gradually acquires ad. m. plumage by late fall. **Flight:** Strong and direct; blue speculum bordered in front and back by white; f. with white on tail; grayish underwing coverts do not contrast strongly

Adult m., *diazi* NM/05

Adult f., *diazi* NM/05

Adult m., Domestic Mallard OH/05

with body. **Hab:** Lakes, marshes, rivers, parks, agricultural fields. **Voice:** F. gives low *gwaak* and rapid *gegege;* m. gives slow drawn-out *rhaeb* and whistled *tseep.* 🎧**3**.

Subspp: (2) •*diazi* (s.e. AZ–s. TX), "Mexican Duck"; f. similar to *platyrhynchos* but darker, with less mottled bill, mostly dark tail, dark belly, narrower white bars bordering speculum, dark undertail coverts; m. like f. but with yellow to olive bill, lacks curlicue tail coverts. *Diazi* often interbreeds with *platyrhynchos.* •*platyrhynchos* (rest of range) as described above in main account; f. has whitish undertail coverts. **Hybrids:** Wood Duck, Gadwall, Falcated Duck, American Wigeon, Eurasian Wigeon, American Black Duck, Mottled Duck, Cinnamon Teal, Northern Shoveler, Northern Pintail, Green-winged Teal, Canvasback, Common Pochard, Tufted Duck, Common Eider.

Adult m. FL/02

Adult f. FL/02

Adult f. FL/02

Mottled Duck 1

Anas fulvigula L 23"

Shape: Large long-bodied duck with a long thin neck. Bill length ¾ head length; sloping forehead leads to rounded crown. Primaries usually do not reach tips of uppertail coverts (Mallard and American Black Duck primaries usually extend past uppertail coverts). **Ad:** Body and wing feathers dark brown with wide paler buffy to reddish-brown margins; neck and head contrastingly pale buff and unmarked; tail and belly dark; both sexes have distinctive black mark at base of lower mandible. M. has bright yellow bill. Eclipse m. like breeding male. F. has olive to orange bill; for comparison to similar f. ducks, see Mallard. **Juv:** (Jun.–Sep.) Like ad. f. but with grayish bill. **Flight:** Strong and direct; long pale buff neck; greenish speculum bordered in front and back with black. **Hab:** Fresh- and saltwater marshes near coast. **Voice:** Like Mallard's. F. gives low *gwaak* and rapid *gegege;* m. gives slow drawn-out *rhaeb* and whistled *tseep.*

Subspp: Monotypic.
Hybrids: Mallard.

Adult m. MD/02

Adult m. MN/09

Adult f. NJ/03

Adult m. NF/11

American Black Duck 1

Anas rubripes L 23"

Shape: Large, long-bodied, short-tailed duck with a fairly long bill. Bill length 3/4 head length; sloping forehead leads to rounded crown. On resting bird, primaries extend just past tips of uppertail coverts (only reach base of uppertail coverts on similar Mottled Duck). **Ad:** Both sexes appear mostly blackish brown overall with paler head and neck, dark crown and eye-stripe; feathers of body, wings, and tail dark blackish brown with thin buffy margins. M. has yellow to olive bill. F. has olive bill, usually with gray blotches or shading in center. Eclipse m. similar to breeding male. Similar f. Mottled Duck and f. Mallard have dark brown feathers with wide buffy margins; f. Mallard has whitish tail. **Juv:** (Jul.–Sep.) Like ad. f. **Flight:** Direct with fairly slow wingbeats. Disting. from other regularly occurring ducks by strong contrast between white underwing coverts and blackish body; deep purple speculum with no white borders. **Hab:** Fresh- and saltwater marshes, coastal areas. **Voice:** Like Mallard's but slightly lower-pitched and hoarse. F. gives low *gwaak* and rapid *gegege;* m. gives slow drawn-out *rhaeb* and whistled *tseep.*

Subspp: Monotypic. **Hybrids:** Mallard, Northern Pintail.

Adult m. FL/01

Adult f. FL/03

1st-winter m. FL/01

Adult m. FL/01

Adult f. FL/02

Blue-winged Teal 1

Anas discors L 15"

Shape: Small compact duck with a comparatively large bill and small head. Bill length more than ¾ head length, and bill is fairly deep-based and thick throughout. Sloping forehead blends into short rounded crown. Similar Green-winged Teal bill noticeably shorter and thinner, head proportionately larger; similar Cinnamon Teal bill is wider and more spatulate at tip, and forehead is more sloping. **Ad:** M. (Dec.–Jun.) has dark gray bill, dark eye, and large white crescent across front of face; rest of head and neck bluish to purplish gray with slight iridescence; breast and flanks buffy brown with dark brown arcs on breast and dots on flanks; small white hip patch before black rear. Eclipse m. (Jul.–Nov.) like ad. f. but bill darker gray; may have suggestion of white crescent on face; dark eye may help distinguish from eclipse Cinnamon Teal, which has a red eye. F. a cool grayish brown overall with distinct facial markings including: white eye crescents, dark eyeline, obvious pale spot at base of bill connecting to pale chin. Similar f. Green-winged is smaller, has shorter, thinner bill, steeper forehead, and higher crown, and has beige wedge just under side of tail. Similar f. Cinnamon is warmer brown overall, lacks Blue-winged's combination of distinct facial markings, and bill is more spatulate at tip. **Juv/1st Winter:** (Jul.–Mar.) Like ad. f.; m. gradually acquires ad. m. plumage by spring. **Flight:** Fast; often erratic flight path. Upperwings like Cinnamon Teal and Northern Shoveler – m. secondary coverts pale blue bordered by white, iridescent green speculum; f. secondary coverts mostly pale blue with little white, dull green speculum. Underwing dark with contrasting white median and greater coverts. **Hab:** Small lakes in summer; marshes, coastal coves in winter. **Voice:** M. gives high whistled *wheet wheet*; f. short guttural *kut*.

Subspp: Monotypic.
Hybrids: Cinnamon Teal, Northern Shoveler, Green-winged Teal.

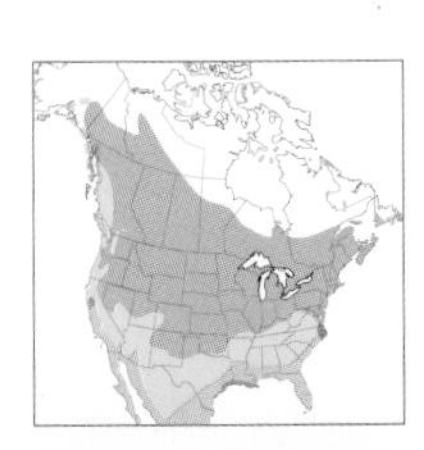

Adult m. CA/02

1st-winter m. BC/02

Adult f. –/02

Adult m. BC/05

Adult m., eclipse NJ/08

Adult f. MB/08

Northern Shoveler 1

Anas clypeata L 19"

Shape: Medium-sized duck with a distinctive long, thick, broad-tipped bill and rather small head. Bill slightly longer than head; sloping forehead leads to short, very shallow, rounded crown. Large bill creates front-heavy look. **Ad:** M. (Nov.–Jun.) with dark iridescent green head, yellowish eye, and black bill; breast white; flanks reddish brown with large white hip patch before black rear. Eclipse m. (Jul.–Oct.) with grayish head and breast; reddish-brown flanks; bill orange along edges (like female); yellowish eye. F. body and wing feathers brown with broad warm buff margins; head warm brown; eye yellowish brown to brown; distinctively shaped bill dark along center, deep orange along edges. **Juv/1st Winter:** (Jul.–Feb.) Like ad. f.; m. has honey-colored eyes, may have thin pale crescent on face, and gradually acquires ad. m. plumage by spring. **Flight:** Strong, direct; large bill creates front-heavy look. Upperwings like those of Blue-winged and Cinnamon Teal (see Blue-winged descrip.), but f. coverts gray rather than pale blue. **Hab:** Shallow lakes, marshes in summer; coastal and inland water areas in winter. **Voice:** M. gives soft, low, guttural *thooktook* and *athook*; f. gives short *kweg geg*.

Subspp: Monotypic. **Hybrids:** Gadwall, Mallard, Blue-winged Teal, Cinnamon Teal, Garganey, Baikal Teal, Green-winged Teal.

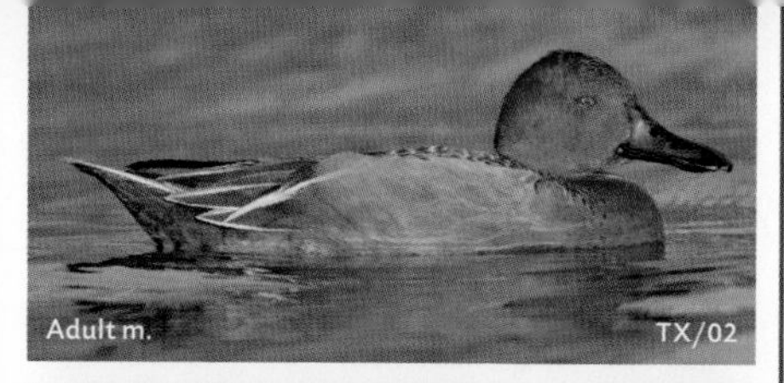
Adult m. TX/02

Adult f. CA/02

Adult m. (l.) and f. CA/11

Cinnamon Teal 1

Anas cyanoptera L 16"

Shape: Fairly small long-bodied duck with a long wide bill that is noticeably spatulate at tip. Bill length ¾–1 x head length; sloping forehead leads to short rounded crown. **Ad:** M. (Feb.–Jun.) deep reddish brown overall with black bill and red eye; blackish rear. Eclipse m. (Jul.–Feb.) like ad. f. but with red eye, warmer brown overall. F. fairly warm brown overall with relatively plain face, occasionally faint pale spot at base of bill (but not connected to pale chin, as in Blue-winged f.), and indistinct buffy eye crescents; indistinct or no eyeline; dark eye; blackish bill. See Blue-winged Teal for distinctions. **Juv/1st Winter:** (Jul.–Feb.) Like ad. f. but paler and less reddish, m. gradually acquires ad. m. plumage by spring. Juv. may have slightly more distinct facial markings and slightly smaller bill, thus more closely resembling ad. f. Blue-winged Teal. **Flight:** Strong, direct, agile, often with sharp turns. Wings like those of Blue-winged Teal and Northern Shoveler (see Blue-winged descrip.). **Hab:** Open shallow lakes, marshes. **Voice:** M. gives short, percussive, clattering sound; f. short guttural *kut*.

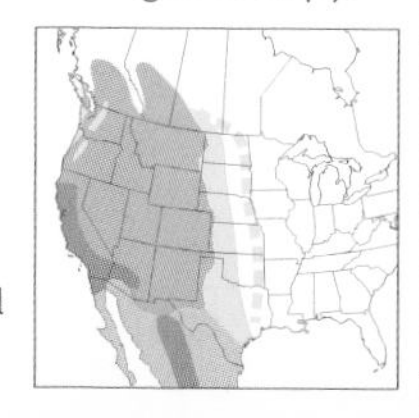

Subspp: (1) •*septentrionalium*. **Hybrids:** Mallard, Blue-winged Teal, Northern Shoveler, Green-winged Teal.

Adult, captive CA/02

Adult PR/04

White-cheeked Pintail 4

Anas bahamensis L 17"

Casual West Indies vag. to cent. and s. FL, usually winter–spring; some may be escapees from captivity. **Shape:** Small, long-bodied, thin-necked, long-billed duck with a long pointed tail. Bill length equals head length. Deep-based bill has strongly curved culmen that leads to sloping forehead and short, shallow, rounded crown. **Ad:** Warm brownish duck with a striking white cheek and chin and bright red base to grayish bill. Tail mostly white. Sexes similar. **Flight:** Fast and agile. Upperwing brown with green speculum bordered in front and back by broad buffy bar. **Hab:** Shallow fresh or salt water. **Voice:** M. gives short whistle; f. a low quack.

Subspp: (1) •*bahamensis*.

Adult m. CA/01

Adult f. CA/01

Adult m., late eclipse JAP/10

Adult m. CA/11

Adult f. ND/06

Northern Pintail 1

Anas acuta L 25"

Shape: Large, elegant, long-tailed, long-necked duck with a relatively small rounded head. Bill length about 3/4 head length; long fairly steep forehead leads to high rounded crown. Tail very long in breeding male (Oct.–Jun.). **Ad:** M. (Oct.–Jun.) with brown head and throat; white on breast and foreneck extends in thin line up side of nape; silvery body; large white hip patch before black rear; upper bill black-centered with pale grayish-blue stripes on either side. Eclipse m. (Jul.–Sep.) with buffy head, whitish neck and breast, and grayish body; bill remains the same. F. body and wing feathers medium brown with buff to reddish-brown to whitish margins; head and neck smooth warm brown; bill like m. or slightly duller, often appearing black. **Juv/1st Fall:** (Jul.–Nov.) Like ad. f.; m. gradually acquires ad. m. plumage by late fall. **Flight:** Agile flight; able to make quick maneuvers; graceful profile with long thin wings, neck, and tail. Upperwings on m. gray with iridescent green speculum bordered in front by buff and behind by white; f. similar but with brown wing and speculum, and speculum bordered by white in front. **Hab:** Open marshes, ponds in summer; coastal bays, lakes, agricultural fields in winter. **Voice:** A liquid *tloloo;* rapid guttural *g'g'g'g;* Mallard-like *rhaeb.*

Subspp: Monotypic. **Hybrids:** Gadwall, Eurasian Wigeon, American Wigeon, American Black Duck, Mallard, Northern Shoveler, Baikal Teal, Green-winged Teal, Redhead, Common Pochard, Common Eider.

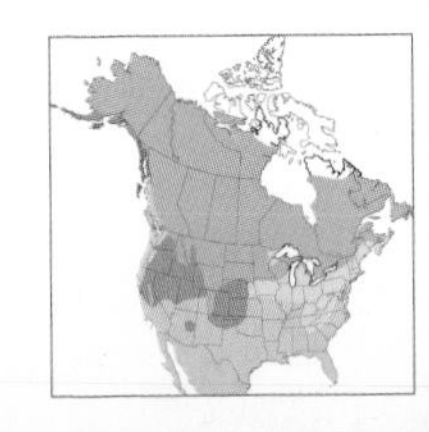

Adult m., *carolinensis* GA/02

Adult m., *carolinensis* CA/10

Adult f., *carolinensis* GA/02

Adult m., *crecca* SWI/–

Adult f., *carolinensis* BC/08

Green-winged Teal 1

Anas crecca L 14"

Shape: Very small compact duck with a short fairly thin bill. Bill length 2/3 head length; long fairly steep forehead leads to high rounded crown. Disting. from other f. teals by shorter thinner bill, steeper forehead, higher crown. **Ad:** M. (Oct.–Jun.) with reddish-brown head; wide iridescent green band surrounding eye and continuing down nape; black bill; body silvery with white shoulder stripe (see Subspp.) and black rear with beige wedge just under side of tail. Eclipse m. (Jul.–Sep.) much like ad. female. F. body and wing feathers dark brown with buffy margins; head and neck paler with dark crown, dark eye-stripe, and lack of distinct pale patch near bill base; distinctive beige wedge just under side of tail; bill gray with orangish tones on basal edges; gray legs. F. sometimes shows second dark stripe through cheek, much like f. Garganey. **Juv/1st Winter:** (Jul.–Dec.) Like ad. f.; m. gradually acquires ad. m. plumage by midwinter. **Flight:** Fast, very agile. Upperwings dark with green speculum bordered in front by white to buffy bar; underwing dark with whitish stripe through secondary coverts (white stripe through secondary and primary underwing coverts in Cinnamon and Blue-winged Teals and Garganey.) **Hab:** Ponds, lakes in summer; ponds, inland and coastal marshes, wet agricultural fields in winter. **Voice:** M. gives a high ringing *peep* much like Spring Peeper (a frog); f. a short quack.

Subspp: (2) Subspecies often intergrade. •*carolinensis* (most of NAM); m. has vertical white shoulder stripe, lacks white scapular line, has little or no white to buffy lines around green eye patch. •*crecca* (EUR vag. to NAM; res. of w. AK islands), "Eurasian Teal" or "Common Teal"; m. lacks vertical shoulder stripe, has thin white horizontal stripe along scapulars, green eye patch with bolder white to buffy outline above and below. ▶ Intergrades may show both white stripes or neither. **Hybrids:** Gadwall, Eurasian Wigeon, American Wigeon, Mallard, Cinnamon Teal, Northern Shoveler, Northern Pintail, Garganey, Baikal Teal.

Adult m. OMA/04

Adult f. ENG/09

Adult f. ENG/09

Garganey 4

Anas querquedula L 15"

Casual Eurasian vag. to NAM (records widely distributed but most often Pac. Coast states in spring). **Shape:** Small, compact, broad-billed duck with a short thick neck. Bill length about ¾ head length; fairly steep forehead leads to rounded crown. **Ad:** M. (Feb.–May) with distinctive wide white eyebrow that extends down along side of nape; rest of face and neck finely streaked purplish brown; brownish body with broad pale gray swath in center. Eclipse m. (Jun.–Feb.) much like ad. female. F. brownish overall with dark crown; broad pale eyebrow, dark eyeline from bill to nape, and faint dark stripe from gape through pale cheek; pale white mark at base of bill and creamy white chin and throat; tertials with distinctive thin, crisp, white margins; bill all dark gray. Similar f. Green-winged Teal (which can have a similar face pattern) has beige wedge along side of tail; f. Cinnamon and Blue-winged have tertials with beige edges (not white). **Juv:** (Aug.–Nov.) Like ad. f. **Flight:** M. upperwing silver with dark subterminal band across secondaries; f. dark gray with white trailing edge to secondaries. Underwing with bold white stripe through primary and secondary coverts like Blue-winged and Cinnamon Teals but with bolder, darker forewing; similar Green-winged Teal has white stripe only through secondary coverts. **Hab:** Shallow ponds, marshes. **Voice:** M. a short clattering call; f. a short high-pitched quack.

Subspp: Monotypic. **Hybrids:** Eurasian Wigeon, Northern Shoveler, Northern Pintail, Green-winged Teal, Tufted Duck.

Adult m. USA/04

Adult f. USA/04

Adult m. JAP/–

Baikal Teal 4

Anas formosa L 17"

Casual E. Asian vag. to AK. **Shape:** Medium-sized duck with a fairly short broad neck and proportionately short bill; male has long pointed scapulars except during eclipse, when they may be absent. **Ad:** M. distinctive with long pink-edged scapulars and bold black-and-yellow face pattern; also has white vertical stripes at shoulder and base of rump. M. eclipse similar to f. but with a reddish-brown wash overall. F. body with dark brown feathers with buff edges; head with dark crown, dark line behind eye, and sometimes dark line through cheek; well-defined white spot at bill base and white chin and throat that sometimes extends up onto the cheek to a variable extent. Similar f. Green-winged Teal is smaller and has distinctive buffy wedge under sides of tail. **Juv:** (Jul.–Feb.) Similar to ad. f. but lacks white chin and white spot at bill base. **Flight:** Underwings dark with paler median coverts. **Hab:** Lakes. **Voice:** M. call a low *wok wok wok;* f. a quack.

Subspp: Monotypic. **Hybrids:** Eurasian Wigeon, Northern Shoveler, Northern Pintail, Green-winged Teal.

Adult m. USA/04

Adult f. JAP/–

Falcated Duck 4

Anas falcata L 19"

Casual E. Asian vag. to AK and occasionally to s. CA; records outside AK may be escapees from captivity. **Shape:** Medium-sized, short-necked, fairly short-bodied duck with a short, thin, straight bill and steep forehead; bill length about ½ depth of head. M. with elongated tertials that drape over sides of tail (not in eclipse); head sometimes with slight corner at hindcrown. **Ad:** In both sexes, bill dark gray. M. head with dark brown forecrown and foreface; green auricular; small white dot on forehead; white chin and throat with black collar; body gray. M. eclipse like f. but with darker crown and nape; tertials shorter. F. warm brown overall with no distinctive marks or contrast; bill all dark gray; body feathers with broad blackish and reddish-brown markings; head fairly uniformly warm brown with fine darker streaking. Similar f. Gadwall has white speculum, paler belly, orangish edges to bill. **Juv/1st Winter:** (Jun.–Nov.) F. like ad. but buffier overall; m. becomes like ad. by winter, except tertials shorter until following summer. **Flight:** Dark secondaries flanked by white lines. **Hab:** Lakes, marshes in summer; wet meadows, lakes, protected coastal waters in winter. **Voice:** M. with short low-pitched whistle; f. a hoarse quack.

Subspp: Monotypic. **Hybrids:** Gadwall, Eurasian Wigeon, Mallard.

Adult HKO/01

Adult JAP/01

Eastern Spot-billed Duck 4

Anas zonorhyncha L 22"

Casual Asian vag. to w. islands of AK. **Shape:** Fairly large, long-bodied, short-tailed duck with a long thin bill. Bill length nearly equals head length; sloping forehead leads to short rounded crown. **Ad:** Distinctive bill in all ages is black with a broad yellow tip. Head and neck pale brown with a dark crown, broad dark eyeline, and a dark bar from the gape extending into the cheek. Breast pale brown with some dark brown spotting; rest of body dark brown with fine paler edging on feathers of flanks and back. Tertials variably edged with white (but not solid white). Sexes and ages similar, but female smaller and sometimes paler. **Flight:** Underwings with dark primaries and secondaries contrasting with pale brown coverts; upperwings show band of white on tertials and blue speculum bordered by thin white lines. **Hab:** Shallow freshwater lakes and marshes. **Voice:** Similar to Mallard's.

Subspp: Monotypic.

Adult m. CA/01

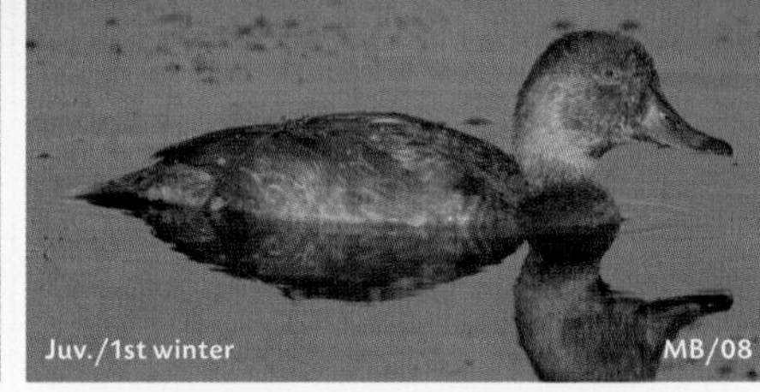
Juv./1st winter MB/08

Adult f. CA/01

Adult m. TX/11

Adult m., eclipse (l.), and f. CA/09

Redhead 1

Aythya americana L 19"

Shape: Medium-sized, long-billed, short-bodied, thick-necked duck. Bill length about 3/4 head length; steep forehead leads to high rounded or somewhat squared-off crown with a high point above or in front of the eye and sloping back (shape more obvious when crown is raised). **Ad:** M. (Nov.–Jun.) with dark reddish-brown head and neck; orange to yellow eye; pale gray bill fades to whitish (can look like white ring) near small black tip; body silvery with black breast and rear. Eclipse m. (Jul.–Oct.) with dull reddish-brown head, reddish eye, dull brown flanks and back. F. pale tawny brown overall with a paler throat and pale area on face next to bill; whitish eye arcs; dark gray bill with a black tip and adjacent paler area (sometimes suggesting a ring); blackish eye. Similar f. Ring-necked Duck and f. Canvasback have differently shaped heads: Canvasback's is large and triangular with a Roman nose; Ring-necked's is peaked at back of crown and crown slopes to front; Redhead's crown is highest at front and slopes to the back. Redhead f. has dark tertials; Canvasback f. has pale tertials. **Juv/1st Winter:** (Jul.–Jan.) Like ad. f.; m. gradually acquires ad. m. plumage by midwinter. **Flight:** Flight fast and strong; runs over water to get airborne. Upperwing with light gray flight feathers, dark gray coverts; underwing with gray flight feathers, whitish coverts. **Hab:** Ponds, lakes, bays. **Voice:** M. a distinctive, mellow, descending *awooorrr*, f. a low grunting *grrnh*.

Subspp: Monotypic. **Hybrids:** Wood Duck, Northern Pintail, Canvasback, Ring-necked Duck, Greater Scaup, Lesser Scaup.

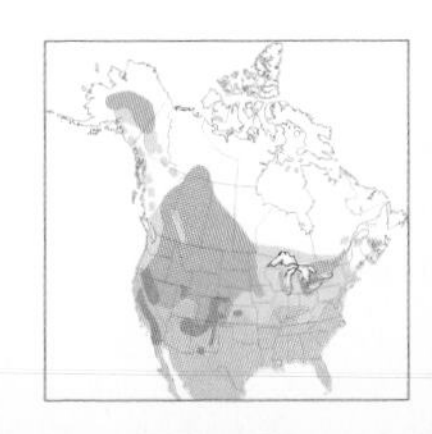

Adult m. AZ/12

Adult f. AZ/12

Adult m. CA/02

Canvasback 1

Aythya valisineria L 21"

Shape: Medium-sized heavily built duck with a strongly arched back and fairly thick neck. Head shape distinctive and triangular. Thin-tipped and deep-based bill leads to strongly sloping forehead and short, high, rounded crown. Bill length over ³/₄ head length. **Ad:** M. (Nov.–Jun.) with dark reddish-brown head and neck; red eye and black bill; body white with breast and rear black. Eclipse m. (Jul.–Oct.) similar but dark breast is mottled with reddish brown. F. with dull brown head, neck, and breast which contrasts with grayish-brown body; black bill and eye. **Juv/1st Winter:** (Jul.–Dec.) Like ad. f.; m. gradually acquires ad. m. plumage by midwinter. **Flight:** Fast and strong; runs over water to get airborne; often flies in irregular V's. M. upperwing has whitish inner wing, gray outer wing; f. upperwing all gray. Underwings of both sexes gray with whitish coverts. **Hab:** Prairie lakes, marshes in summer; lakes, sheltered coastal waters in winter. **Voice:** M. a liquid *wuhoo;* f. a hoarse, low, grating *hrrrrh*.

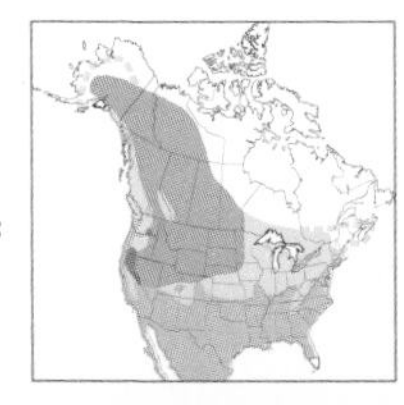

Subspp: Monotypic. **Hybrids:** Wood Duck, Mallard, Redhead, Common Pochard, Tufted Duck, Lesser Scaup.

Adult m. ENG/01

Adult f. ENG/11

Adult m. ENG/11

Adult f. HUN/05

Common Pochard 3

Aythya ferina L 18"

Rare Eurasian visitor primarily to w. AK; has occurred in CA. **Shape:** Small heavily built duck with proportions midway between Canvasback and Redhead. Heavy neck and peaked head like Canvasback's, but thinner-based bill and steep forehead more like Redhead's. **Ad:** M. and f. similar to Redhead (and rare Canvasback x Redhead) except for bill, which is three-toned (gray base, pale middle, blackish tip); back, which is midway in color between Redhead and Canvasback; and eyes, which are deep red in m. and dark brown in f. (in Redhead, eyes are golden in m. and light brown in f.). **Flight:** Short rounded wings. Upperwing and underwing like Redhead's. **Hab:** Lakes, ponds.

Subspp: Monotypic. **Hybrids:** Mallard, Northern Pintail, Canvasback, Ring-necked Duck, Tufted Duck, Greater Scaup, Lesser Scaup, Common Goldeneye.

Adult m. CA/01

Adult m., late eclipse AZ/10

Adult f. CA/01

Adult m. YT/04

Adult f. YT/04

Ring-necked Duck 1

Aythya collaris L 16"

Shape: Small, compact, vertically profiled duck with a relatively short bill. Bill length about 2/3 length of head; strongly curved culmen blends into a steep forehead and high round-peaked crown (higher crown on m.); when crest raised, peak is at back of head and crown slopes forward. Nape fairly flat and vertical. If bird has been diving, crown often lower and more squared at back. **Ad:** M. (Nov.–Jun.) glossy black breast and back; pale gray sides with bright white wedge at shoulder; bill gray with dark tip and encircled by 2 white rings – one at base, the other next to dark tip; golden eye. Common name comes from inconspicuous dark brown ring at base of m. neck. Eclipse m. (Jul.–Oct.) similar, but black areas more dusky and flanks brownish; may have pale patch at base of bill. F. back and rear dark brown; flanks and breast warm brown; head dusky brown and faintly bicolored (darker on crown and nape, paler on cheek and foreneck); white eye-ring with short white eyeline behind; eye dark red; pale patch at base of bill. F. bill similar to m. but lacks narrow white line at base. Similar f. Tufted Duck and scaup have concolor crown and face, lack white eye-ring or white eyeline, and have golden eye; Tufted Duck has little or no white at bill base; f. scaup has extensive white at bill base. **Juv/1st Fall:** (Jul.–Nov.) Like ad. f.; m. gradually acquires ad. m. plumage by late fall. **Flight:** Flight fast and strong; runs over water to get airborne; groups may form compact wedge. Upperwing with light gray flight feathers, darker coverts; underwing with gray flight feathers, whitish coverts. **Hab:** Lakes, woodland ponds in summer; lakes, small ponds, bays in winter. **Voice:** M. a soft short whistle; f. a short gruff *grek*.

Subspp: Monotypic. **Hybrids:** Redhead, Common Pochard, Tufted Duck, Greater Scaup, Lesser Scaup.

Adult m. JAP/02

1st-winter m. ENG/02

Adult f. JAP/02

Adult m. ENG/06

Adult m., eclipse ENG/07

Tufted Duck 3

Aythya fuligula L 17"

Rare Eurasian vag. that shows up in coastal states; accidentally well inland; often with scaup (especially Greater Scaup). **Shape:** Small, compact, round- to slightly square-headed duck with variable tuft at back of crown. Bill length about 2/3 head length and bill is fairly thick; steep forehead leads to fairly high rounded (when crest is relaxed) to squared-off (crest raised) crown. M. tuft long and loose; f. tuft short and neat. **Ad:** M. (Nov.–May) glossy black back and breast with all-white flanks; variable long black tuft off hindcrown; bill pale bluish gray with dark tip; eye golden. Eclipse m. (Jun.–Oct.) similar but black areas more dusky and flanks grayish brown; tuft shorter. F. has variable small tuft; head and neck concolor dark brown and relatively unmarked except for indistinct pale band or patch near bill base (can be absent); back and rear blackish brown (can be some white on rear in fall); flanks brown; bill gray with dark tip and hint of adjacent paler ring; eye golden. **Juv/1st Winter:** (Aug.–Feb.) Like ad. f. but browner and may lack head tuft. Both sexes gradually acquire ad. plumage by early spring. **Flight:** Fast and strong; runs over water to get airborne. Upperwing flight feathers white with dark trailing edge, coverts dark; underwing mostly white. **Hab:** Lakes, rivers in summer; coastal bays, lakes, ponds in winter; often with scaup. **Voice:** M. a soft short whistle; f. a short gruff *grek*.

Subspp: Monotypic. **Hybrids:** Mallard, Common Pochard, Ring-necked Duck, Greater Scaup, Lesser Scaup.

Adult m. NJ/02

Juv./1st-winter m., *marila* ENG/11

Juv./1st-winter f., *marila* ENG/11

Adult f., *marila* ENG/03

Adult m., *marila* NOR/06

Adult m., eclipse, *marila* ENG/08

Adult f., *marila* NOR/06

Greater Scaup 1

Aythya marila L 18"

Shape: Fairly small rather long-bodied duck with a fairly long broad bill that is slightly spatulate at the tip; viewed head-on, the tip is convex. Bill length 3/4 head length; steep forehead rounds out into fairly long crown. Head has a horizontal (sometimes slightly rectangular) appearance. In general, head length is greater than height (the opposite in Lesser Scaup), and this is accentuated by the slightly longer bill. Lesser's taller head and slightly shorter bill create a more vertical appearance. Greater Scaup also appears longer-bodied and lower-riding on water than Lesser. Head shape differences with Lesser negligible when actively feeding; more pronounced when at rest. From head-on, Greater's head is broader-cheeked; Lesser's head is more even width throughout. **Ad:** M. (Oct.–May) head, neck, and breast glossy black; head appears greenish in good light; back gray; flanks white and rear black; bill bluish gray with black nail; eye golden. Eclipse m. (Jun.–Sep.) similar but with brownish wash over back and flanks and some brown on breast. F. with dark brown head and neck, grayish-brown back and flanks, dark rear; oval white patch at base of bill (reduced on worn plumage, Mar.–Sep.); yellow eye; light crescent on cheek usually visible on worn plumage (Mar.–Sep.). See Lesser Scaup for comparison. **Juv/1st Winter:** (Jul.–Feb.) Similar to ad. f. but warmer brown overall, eyes brownish, reduced white loral patch, and buffy flanks. Juv. f. has whitish frosting to flanks and/or upperparts (in fresh plumage); juv. m. does not. Both sexes gradually acquire ad. plumage over winter. **Flight:** Fast and strong; runs over water to get airborne. Dark upperwing coverts contrast with broadly white secondaries and inner primaries, which create a long white stripe along wing; underwing mostly white. **Hab:** Tundra and taiga lakes in summer; ocean, bays, coastal ponds, large inland lakes in winter. **Voice:** M. a soft short whistle; f. a short gruff *grek*.

Subspp: (2) •*nearctica* (all of NAM range) less white on outer 4 primaries, thick black vermiculation on back and scapulars. •*marila* (n. Eurasia, vag. to w. AK and possibly e. NAM) more white on outer 4 primaries, thin black vermiculation on back and scapulars. **Hybrids:** Redhead, Common Pochard, Tufted Duck, Ring-necked Duck, Lesser Scaup, Common Goldeneye.

Adult m. CA/01

Juv./1st-winter m. CA/01

Juv./1st-winter f. BC/03

Adult f. CA/02

Adult m. BC/05

Adult m., eclipse ENG/07

Adult f. NV/11

Lesser Scaup 1

Aythya affinis L 17"

Shape: Fairly small rather compact duck with a medium-length narrow bill. Bill length ⅔ head length; steep forehead rounds off to high, short, somewhat peaked crown with a slight "corner" at the rear. In general, head height is greater than length (the opposite in Greater Scaup), creating a vertical appearance to head. Greater's slightly longer head and slightly longer bill create a more horizontal appearance. Lesser Scaup also appears shorter-bodied and often rides higher on water than Greater. Head differences with Greater negligible when actively feeding; more pronounced when at rest. **Ad:** M. and f. plumages essentially identical to Greater Scaup's except f. Lesser has smaller or no light crescent on cheek (most visible on worn plumage, Mar.–Sep.) and averages slightly darker than Greater f.; m. Lesser Scaup flanks washed with pale gray, head in good light with purplish to greenish gloss (Greater m. has mostly white on flanks and greenish gloss on head). Disting. from Greater, with patience, mostly by shape of head and bill. **Juv/1st Winter:** (Jul.–Dec.) Essentially identical to Greater Scaup except for shape of head and bill. **Flight:** Fast and strong; runs over water to get airborne. Dark upperwing coverts contrast with broadly white secondaries, which creates a white stripe along inner wing (on Greater Scaup white extends onto inner primaries, making stripe longer); underwing mostly pale gray. **Hab:** Prairie and taiga lakes, marshes in summer; generally freshwater lakes and ponds, but also nearshore bodies of water and saltwater bays in winter. **Voice:** M. a soft whistle; f. a short gruff *grek*.

Subspp: Monotypic. **Hybrids:** Canvasback, Redhead, Common Pochard, Ring-necked Duck, Tufted Duck, Greater Scaup.

Adult m. NOR/04

Juv./1st-winter f. AK/02

Adult f. NOR/04

Adult m. NOR/03

Juv./1st-winter m. CA/02

Adult f. NOR/03

Steller's Eider 3

Polysticta stelleri L 17"

Shape: Small eider with a short comparatively thin-based bill. Bill length ⅔ head length; steep forehead leads to shallow fairly flattened crown (often squared off at rear crown because of short tufts). Long, stiff, pointed tail feathers usually held up. **Ad:** M. (Oct.–Jun.) with distinctive white head with wide black ring surrounding dark eye and greenish tufts off hindcrown; black throat and collar; conspicuous dark spot on foreflank; bill gray. Eclipse m. (Jul.–Sep.) much like ad. f. but has white upperwing coverts, which may show as white line across top of flanks. F. (Oct.–May) dark brown overall with faint eye-ring; dark blue on long tertials; usually showing are 2 white wingbars on either side of dark blue speculum (when wing is not covered by flanks). F. (Jun.–Sep.) is more reddish brown with dark mottling on head and breast and darker mottled collar; fairly distinct white circle around eye; gray bill.

Juv/1st Yr: Like ad. f. at first; then m. gradually acquires ad. m. plumage over winter and summer. **Flight:** Strong and agile; long narrow wings. M. has white upperwing secondary coverts and pale rusty belly; f. dark overall with white leading and trailing edge to dark blue speculum. Both sexes with white underwing coverts. **Hab:** Freshwater tundra ponds in summer; shallow coastal waters in winter. **Voice:** M. a cooing *whoooh;* f. a clucking *kukukuk* and grating *grrr.*

Subspp: Monotypic.

Adult m. AK/06

Adult m., eclipse OH/07

Adult f. AK/06

1st-winter m. USA/03

Spectacled Eider 3

Somateria fischeri L 21"

Shape: Medium-sized eider with a large body, long thin neck, and compact triangular head; heavy feathering drapes over top of most of upper mandible (on King and Common Eiders central upper mandible unfeathered). **Ad:** All plumages show distinctive large "spectacles" that are paler than rest of head. M. (Nov.–Jun.) with greenish head, white spectacles, orange bill; back white; upper breast white; lower breast, flanks, and rear blackish gray. Eclipse m. (Jul.–Nov.) much like ad. f. but darker, with a pinkish bill, white line along flanks (sometimes hidden), and gray wash to body feathers. F. has buff head and neck with thin darker streaking; pale large spectacles; reddish-brown wash over foreface; back and flanks barred with light brown, reddish brown, and black. **Juv/Imm:** Like ad. f. at first; then m. gradually develops ad. m. plumage over 2–3 yrs.; starts out mostly dark and gets white first on back, then on breast. Changes take place gradually and vary individually, making identification of imm. m. age classes difficult. F. takes 2 yrs. to mature, looks much like ad. f. throughout. **Flight:** Strong. M. wing coverts white, flight feathers dark gray, back white; f. wings and coverts all dark except for gray axillaries. **Hab:** Coastal wetlands in summer; openings in pack ice, ice floes at sea in winter. **Voice:** M. a cooing *aoooh;* f. a clucking *kukukuk* and grating *geoow.*

Subspp: Monotypic.

Adult m. NOR/03

Juv./1st-winter m. NJ/12

Adult f. NOR/03

Adult m. NOR/03

Adult m., eclipse SCO/09

Adult f. NOR/03

King Eider 2

Somateria spectabilis L 21"

Shape: Medium-sized fairly short-bodied eider with a relatively short thick neck. Obvious back of gape slightly upcurved, creating a smiling expression. Longest scapulars have vertical triangular portions ("sails") that usually project up on lower back (smaller on ad. f.). Ad. m. has distinctive large frontal lobe on forehead; f. has fairly short bill, slightly bulging forehead, and fairly shallow rounded crown. **Ad:** M. (Oct.–Jun.) has distinctive orange frontal lobe, red bill, greenish face, and pale blue crown and nape; breast whitish; rest of body mostly black except for large white hip patch. Eclipse m. (Jul.–Oct.) mostly dark brown but still has orange frontal lobe, though smaller, and bill is orangish; often blotched with white; retains white upperwing patch. F. variably buffy to medium brown overall with reddish-brown tertials; body feathers have U-shaped dark markings (f. Common and Spectacled Eiders are barred on back and flanks); bill dark gray to blackish; may have pale spot on face at base of bill; faint pale line extends from behind eye down side of neck. **Juv/Imm:** Like ad. f. at first; then m. gradually develops ad. m. plumage over 2–3 yrs.; starts out mostly dark and gets white first on back, then on breast. Changes take place gradually and vary individually, making identification of imm. m. age classes difficult. F. takes 2 yrs. to mature, looks much like ad. f. throughout. **Flight:** Strong; may fly in long lines, V's, or clusters. M. has black back and upperwings with white lesser and median secondary coverts; below, white coverts, dark flight feathers. F. wing all dark except for whitish axillaries. **Hab:** Tundra ponds in summer; subarctic coast in winter. **Voice:** M. a cooing *aoooh;* f. a clucking *kukukuk* and grating *geoow.*

Subspp: Monotypic. **Hybrids:** Common Eider.

Adult m. and f., *dresseri* ME/05

Adult f. and m., *sedentaria* MB/06

Adult f., *dresseri* ME/11

Adult m., *v-nigrum* AK/05

Juv./1st-winter m., *dresseri* ME/09

Juv. to 2nd-winter m., *mollissima** NOR/04

Adult f., *mollissima** NOR/03

Common Eider 1

Somateria mollissima L 25"

Shape: Very large, heavyset, long-bodied eider with a long triangular head. Bill very deep-based with straight culmen that meets slightly steeper forehead. F. facial feathering is pointed and extends to nostril (rounded and falls short of nostril in f. King Eider). **Ad:** M. (Nov.–Jun.) has black cap, white face, and greenish nape; flanks and rear black; breast and back white; bill olive to orange (see Subspp.). Eclipse m. (Jul.–Oct.) mostly dark and blackish brown with white flecking on breast, whitish lower back, white upperwing and underwing coverts, and olive to orange bill. F. dull brown or reddish brown overall (see Subspp.) with dark barring on back and flanks; bill gray. **Juv/Imm:** Like ad. f. at first; then m. gradually develops ad. m. plumage over 2–3 yrs.; starts out mostly dark and gets white first on back, then on breast. Changes occur gradually, vary individually. F. takes 2 yrs. to mature, looks much like ad. f. throughout. **Flight:** M. white coverts contrast with black flight feathers. F. (and imm. m.) wing all dark except for grayish axillaries (whitish in f. King). **Hab:** Coastal waters. **Voice:** M. a cooing *aoooh;* f. a clucking *kukukuk* and grating *geoow.*

Subspp: (4) Divided into two groups. **Pacific Group:** •*v-nigrum* (w.cent. AK–NT), "Pacific Eider." Large; m. with thin black V on sides of chin (Oct.–Jun.), long yellow to orange bill with narrow pointed lobes; f. dark brown. **Atlantic Group:** •*borealis* (cent. LB–n. Hudson Bay and north), "Northern Eider." Small; m. with orangish-yellow to greenish-yellow bill with fairly long, tapered, rounded lobes (lacks black V on sides of chin); f. medium brown. •*sedentaria* (Hudson and James bays), "Hudson Bay Eider." Medium-sized; m. with short olive to yellowish bill with long, fairly wide, rounded lobes that generally extend to just below eye; f. grayish brown. •*dresseri* (n. QC–NF–ME), "Atlantic Eider." Medium-sized; m. has short olive bill with long, wide, rounded lobes that generally extend above eye; f. reddish brown. **S. m. mollissima* (n.w. Eurasia) used for good photographs. **Hybrids:** Mallard, Northern Pintail, Steller's Eider, King Eider, White-winged Scoter, Common Merganser.

Adult m. CA/02

1st-winter m. NJ/02

Adult f. QC/07

Adult m. NJ/02

Adult m., early eclipse AK/06

Adult f. NJ/12

Harlequin Duck 1

Histrionicus histrionicus L 17"

Shape: Small, thick-necked, rotund duck with a small rounded head and short deep-based bill. Bill length less than ½ length of head. Steep forehead leads to high well-rounded crown; long pointed tail (longer in m.) often held out of water. **Ad:** M. (Oct.–Jun.) with slate-gray head and breast boldly patterned with white stripes and dots; white shoulder stripe; chestnut flanks; blackish rear. Eclipse m. (Jul.–Oct.) like ad. f. but with dark belly, some white on scapulars and tertials, and sometimes a variable shoulder stripe. F. dark brown with distinctive white earspot and variable white on foreface; upper belly whitish. **Juv/1st Yr:** Like ad. f. at first; m. then gradually acquires ad. m. plumage by summer. **Flight:** Usually low over water with rapid wingbeats; agile with erratic flight path. Wings dark; m. has white stripes across chest and dark belly; f. dark with light belly. **Hab:** Mountain rivers and streams, some along coast, in summer; rocky coasts in winter. **Voice:** A short *ziip;* usually quiet.

Subspp: Monotypic.

Adult m. BC/10

Adult m. NJ/02

Adult f. CA/12

Adult f. NJ/02

1st winter NJ/02

1st winter NJ/02

Surf Scoter 1

Melanitta perspicillata L 20"

Shape: Medium-sized scoter with long deep-based bill. Bill length as long as head. Long sloping forehead leads to rounded crown. No feathering on bill sides (similar White-winged Scoter has feathers extending down side of bill to nostril). **Ad:** M. (Oct.–Jun.) blackish overall; bill orange and white with black side spot; 2 white patches on head, on forehead and nape. Eclipse m. (Jul.–Sep.) blackish overall with slightly more brownish underparts; one white patch (on nape); bill colors muted. F. (Nov.–Jul.) grayish brown overall with darker cap; bill blackish gray; sometimes has white nape patch (variable in size); 2 faint white patches on face, one (mostly vertical) next to bill (flattened on bill side) and one (more horizontal) on auricular. F. (Aug.–Oct.) similar but no white on nape and facial spots very faint. **Juv/1st Winter:** Like ad. f. but with whitish belly, brownish wings, face patches more distinct than on adult, and no white nape patch; m. gradually acquires ad. m. plumage by 2nd summer. **Flight:** Strong. Wings all dark above and below. White-winged Scoter has white patch on upperwing; American Scoter has black coverts and gray primaries and secondaries. **Hab:** Semiwooded arctic lakes, rivers in summer; mostly along coast in winter. **Voice:** A rapid flutelike *wuwuwu*.

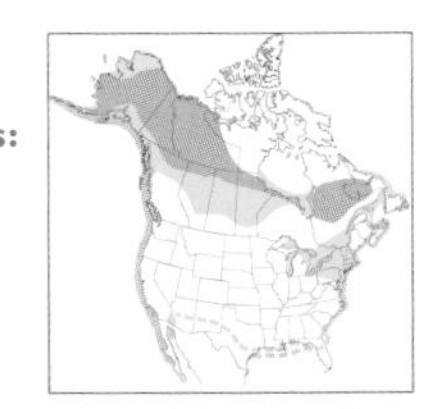

Subspp: Monotypic. **Hybrids:** White-winged Scoter.

Adult m., *deglandi* CA/12

Juv./1st-winter f., *deglandi* NH/10

Juv./1st-winter f., *deglandi* NH/10

Adult m., *deglandi* BC/02

Adult f., *deglandi* BC/11

Adult f., *deglandi* BC/02

1st-winter m., *deglandi* ON/02

White-winged Scoter 1

Melanitta fusca L 21"

Shape: Large long-bodied scoter with a fairly long deep-based bill; ad. m. with hump at base of bill. Fairly steep forehead rounds off to high, sometimes flattened crown. Feathering extends down sides of bill to nostril (stops at base of bill on Surf Scoter). **Ad:** M. black overall with white crescent under eye; white speculum may show on floating bird. Late summer to early fall, white crescent on eye may be faint. F. dark brown; head and crown concolor; 2 faint oval white patches on side of face, one next to bill (oval) and one on auricular; white speculum may show on floating bird. **Juv/1st Yr:** Like ad. f. at first but with pale belly. Then, in winter, m. has black head (no white on face or around eye), blackish-brown body, dark bill. In spring m. develops hump at base of bill, light eye, some white under eye. 1st-winter f. with more distinct white patches on face than ad. f. (most distinct on cheek); after, like ad. f. (dark belly). **Flight:** May run along water to get airborne. White speculum obvious. **Hab:** Lakes, ponds in summer; mostly along coast in winter. **Voice:** A short *gruu*.

Subspp: (2) •*deglandi* (all of NAM range); m. flanks brown contrasting with black back and breast, upper mandible all orange. •*stejnegeri* (Asian vag. to w. AK); m. flanks black and not contrasting with rest of back and breast, upper mandible orange with yellow sides. **Hybrids:** Common Eider, Surf Scoter, Common Goldeneye.

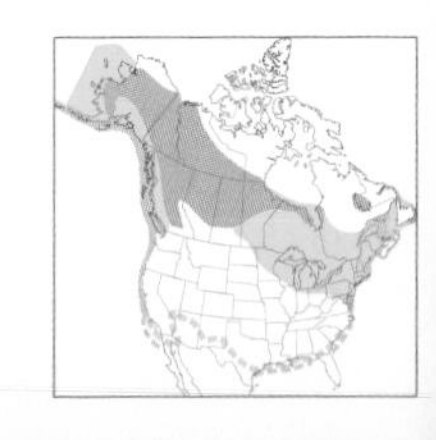

Adult m. NJ/03

Adult f. NJ/02

1st-winter m. NJ/01

Adult m. NJ/02

Adult f. NJ/02

American Scoter 1

Melanitta americana L 19"

Shape: Smallest scoter; compact. M. bill short with swollen knob at base abutting a short steep forehead and smoothly rounded crown. F. ducklike with comparatively thin-based bill, steep forehead, and rounded crown. **Ad:** M. body all black; orange knob at base of black bill. F. brown with two-toned head – dark cap and nape, paler face and foreneck. **Juv/1st Yr:** Like ad. f. at first but with pale belly; m. and f. gradually acquire ad. plumage over 1st yr. and m. attains knob color. **Flight:** Often forms lines in flight. Underwings two-toned with blackish coverts and gray flight feathers; in late winter, primaries pale silver (Surf Scoter underwing all dark). **Hab:** Tundra lakes in summer; mostly coasts in winter. **Voice:** A drawn-out whistled *weeeah* and *aweee;* male's wings create distinct whistling sound during flight.

Subspp: (1) •*americana.*

Adult m., summer MB/06

Juv. ENG/01

Adult f., summer AK/06

1st-winter m. CA/02

Adult m., winter, trans. to summer NOR/04

Adult m., summer NJ/04

Adult f., winter NOR/04

Adult m., winter NOR/04

Long-tailed Duck 1

Clangula hyemalis L 16"

Shape: Small rounded duck with a relatively short thick neck and short bill. Bill is thick-based and is less than ½ length of head. Steep forehead leads to rounded or slightly flattened crown; crown often high in front, strongly sloping to rear. M. tail with long thin streamers; f. tail moderately short. **Ad. Summer:** All ages and seasons have distinctive white rear and undertail coverts. M. bill black with pink band near tip; f. bill dark gray. M. (May–Aug.) has black head and neck (crown and nape whitish Jul.–Aug.), gray face, reddish-brown scapulars. F. has mostly sooty head except for bold white eye patch with pointed extension to rear; blackish back, brownish flanks. **Ad. Winter:** M. (Sep.–Apr.) has white crown and neck, gray face, black patch below auriculars (patch grayish and inconspicuous Sep.–Nov.), silvery scapulars. F. (Sep.–Apr.) has thin dark crown and nape, white face, brownish patch below auriculars, brown back, whitish flanks. **Juv/1st Yr:** Like winter ad. f. at first (Nov.–Apr.) but with grayish back, then gradually acquires ad. plumage by summer. M. acquires pink on bill in late fall and winter, allowing sexing. **Flight:** Swift, agile, with erratic path; body often tilted up in flight; runs along water at takeoff. Short, pointed, all-black wings; white rear. **Hab:** Tundra lakes, coastal inlets in summer; mostly coasts in winter. **Voice:** A gull-like *gah* and *gaguyah.*

Subspp: Monotypic.

Adult m. CA/01

Adult m. BC/01

Adult f. BC/01

1st-winter m. NY/05

Adult f. BC/02

Bufflehead 1

Bucephala albeola L 14"

Shape: Very small, compact, short-bodied, large-headed duck with a tiny bill. Bill length less than 1/3 length of head. Long steep forehead leads to very high rounded crown. **Ad:** M. (Sep.–May) head dark iridescent green or purple (usually looks just dark) with large white wedge from cheek through nape; glossy black back; white breast and underparts. Eclipse m. (Jun.–Aug.) similar but with white wedge on head outlined by dark nape; flanks gray with white line (wing coverts) at top often visible. F. brownish gray on head and back with wide white streak or patch on auriculars; breast and underparts paler, dusky. **Juv/1st Yr:** Like ad. f. at first; by 1st winter, m. still like f. but with wider white auricular patch and white chest and belly; like ad. m. by summer. **Flight:** Very fast wingbeats; short run over water at takeoff. M. upperwing with broad white bar across inner secondaries and adjacent coverts; white head patch. F. with just white inner secondaries and tips of their greater coverts. **Hab:** Wooded lakes in summer; lakes, coastal waters in winter. **Voice:** A breathy *gruuhf gruuhf.*

Subspp: Monotypic.

Adult m., *americana* ID/12

1st-winter m., *americana* BC/02

Adult f., *americana* ME/02

Adult m., *americana* BC/01

Adult f., *americana* ME/02

Adult f., *americana* CO/04

Common Goldeneye 1

Bucephala clangula L 18½"

Shape: Small large-headed duck with a comparatively short bill, sloping forehead, and short rounded crown. Bill length ½ length of head; bill clearly longer than length of lore. Head appears vertically balanced and often somewhat triangular (can be slightly squared off). Similar Barrow's Goldeneye has shorter bill (about length of lore), steeper forehead (almost vertical), long, sloping crown and nape (forecrown generally highest point), giving head and neck a forward-leaning appearance. **Ad:** M. (Dec.–Jun.) has glossy greenish-black head and back, golden eye, white oval dot on face below eye and toward bill; white flanks join broadly white scapulars; no black "spur" on shoulder as on m. Barrow's. Eclipse m. (Jul.–Nov.) like ad. f.; during molts, m. may have only partial or slightly obscured white dot on face, which can look like crescent of Barrow's, but it never extends above eye (as does crescent of m. Barrow's). F. with dark brown head, golden eye, grayish body; bill usually mostly dark with outer third (variable) orangish yellow, sometimes becoming dusky at start of breeding (rarely all or mostly orangish yellow). **Juv/1st Winter:** Like ad. f. but with dark eye; m. gradually acquires ad. m. plumage from late winter to summer. **Flight:** Rapid wingbeats; short run at takeoff; strong high-pitched whistling sound from ad. m. wings. Upperwings of m. with broad white swath through inner wing; f. with partitioned white patches on inner wing. **Hab:** Lakes, woodland ponds in summer; interior and coastal waters in winter. **Voice:** A loud buzzy *babreeent;* f. low grating *brrr brrr.*

Subspp: (2) •*americana* (all of NAM range) bill comparatively larger. •*clangula* (Eurasian vag. to w. AK) bill comparatively smaller. **Hybrids:** Common Pochard, White-winged Scoter, Barrow's Goldeneye, Smew, Hooded Merganser, Common Merganser, Red-breasted Merganser.

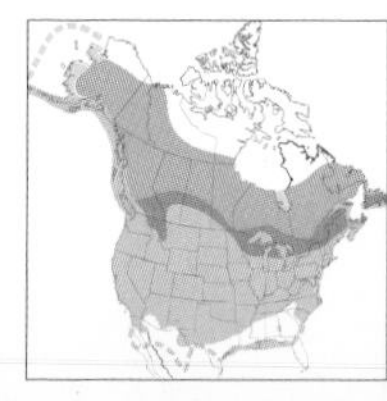

Adult m. AK/03

Juv./1st-winter m. CA/12

Adult f. CA/01

Adult m. YT/05

Adult f., summer BC/07

Adult f. YT/04

Barrow's Goldeneye 1

Bucephala islandica L 18"

Shape: Small large-headed duck with a very short triangular bill, steep forehead, and long sloping crown and nape, with highest point on forecrown. Bill length about ⅓ length of head – about the length of the lore. Head and neck often appear forward-leaning. See Common Goldeneye for comparison. **Ad:** M. (Oct.–Jun.) has glossy purplish-black head and back; golden eye; white crescent on face toward bill extends above eye at highest point; white flanks; scapulars mostly black with line of square white dots; black "spur" on shoulder divides breast from flanks. Eclipse m. (Jul.–Sep.) like ad. f.; during molts, m. may have only partial white crescent on face, which can look like partial dot on Common. F. with dark brown head and golden eye; grayish body; bill usually mostly orangish or yellow during Nov.–May, but may be all dark during Jun.–Oct. **Juv/1st Winter:** Like ad. f. but with dark eye; m. gradually acquires ad. m. plumage by summer but acquires pale crescent on face by early winter. **Flight:** Rapid wingbeats; short run at takeoff; strong whistling sound from ad. m. wings (lower-pitched than with Common Goldeneye). Upperwings with partitioned white patch on inner wing of both sexes. **Hab:** Lakes in summer; interior and coastal waters in winter. **Voice:** Generally silent, except see Flight.

Subspp: Monotypic. **Hybrids:** Common Goldeneye, Hooded Merganser.

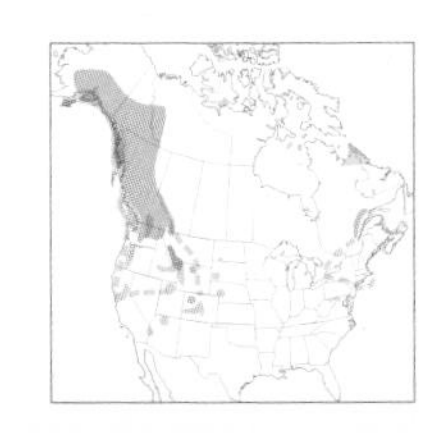

Adult m., *americanus* NH/04

1st-winter m., *americanus* AK/02

Adult f., *americanus* NH/04

Adult m., *americanus* NH/04

Juv., *americanus* MN/10

Adult f., *americanus* USA/–

Common Merganser 1

Mergus merganser L 25"

Shape: Large, long-bodied, fairly thick-necked duck with a thin but thick-based straight bill. Bill length more than ½ length of head. Sloping forehead leads to long gently rounded crown. M. has no obvious crest; f. has longish even-length crest. Feathering across back of gape straight and vertical. Similar Red-breasted Merganser has thinner-based slightly upcurved bill; shaggy uneven crest on both sexes; steeper forehead. **Ad:** M. (Nov.–Jun.) has dark greenish-black head and back; white collar, breast, and flanks; appears crestless; bright red bill; dark eye. Eclipse m. (Jul.–Oct.) like ad. f. but with large white wing patch and darker gray back. F. has reddish-brown head and upper neck with sharply contrasting white chin; clear-cut demarcation between reddish-brown upper neck and whitish lower neck. **Juv/1st Winter:** Like ad. f. at first but with white dash above gape; m. gradually acquires ad. m. plumage by summer, often starting with dark feathers around the eye in winter. **Flight:** Shallow wingbeats; usually runs over water for takeoff but can rise directly; long thin body with wings placed well back. M. upperwing has white secondaries and coverts with single dark dividing bar; f. has white speculum. **Hab:** Wooded lakes, rivers in summer; large lakes, rivers, estuaries, usually freshwater, in winter. **Voice:** M. a snoring *ahwuuuhn;* f. a short grating *grek*.

Subspp: (2) •*americanus* (all of NAM range) has white upperwing patch divided by black bar (tips of greater secondary coverts), bill with thick base and small hook at tip. •*merganser* (Eurasian migrant to w. Aleutians, AK) has solid undivided white upperwing patch, bill thin with shallower base and pronounced hook at tip. **Hybrids:** Common Eider, Common Goldeneye, Red-breasted Merganser.

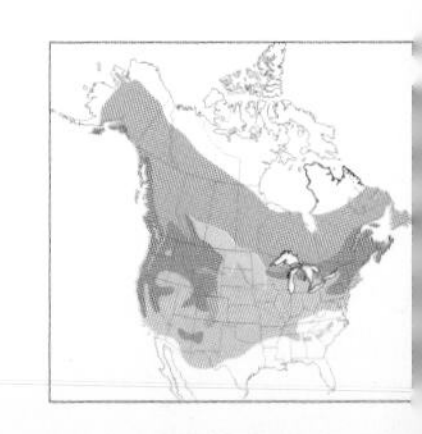

Adult m. CA/04

1st-winter m. FL/02

Adult f. FL/02

Adult m. NJ/01

Adult m., eclipse FIN/06

Juv./1st yr. NJ/01

Juv. FL/12

Red-breasted Merganser 1

Mergus serrator L 23"

Shape: Large, long-bodied, thin-necked, comparatively small-headed duck with an evenly thin slightly upcurved bill. Bill length about 3/4 length of head. Short steep forehead leads to gently rounded crown. Crest bushy, uneven, and untidy on both sexes. Feathering across back of gape sharply angled in a V. Similar Common Merganser has thicker-based straight bill; no crest on m. and an even crest on f.; more sloping forehead. **Ad:** M. (Nov.–Apr.) has dark greenish-black head and back, white collar, mottled rusty breast; jagged untidy crest; bright orangish-red bill and reddish eye. Eclipse m. (May–Oct.) like ad. f. but with large white wing patch and darker gray back. F. has dull reddish-brown head and upper neck blending to grayish lower neck and breast; whitish eye crescents. **Juv/1st Yr:** Like ad. f. at first but with white dash above gape; m. gradually acquires ad. m. plumage by summer, often starting with dark feathers around the eye in winter. **Flight:** Fast flight; runs along water for takeoff. M. upperwing with white secondaries and coverts with 2 dark dividing bars; f. has white speculum. **Hab:** Wooded lakes, rivers in summer; mostly coastal areas, saltwater marshes in winter.

Voice: M. a drawn-out *jidik jadeee;* f. short *uk uk.*

Subspp: Monotypic. **Hybrids:** Common Goldeneye, Hooded Merganser, Common Merganser.

Adult m., crest up FL/03

1st-winter m. NV/03

Adult m., crest down FL/03

Adult m. NY/11

Adult f. FL/03

1st-yr. m. USA/01

Hooded Merganser 1

Lophodytes cucullatus L 18"

Shape: Small-bodied, long-tailed, proportionately large-headed duck with a thin pencil-like bill. Bill length about ½ length of head. Steep forehead leads to long crown with large bushy crest; crown can be long and flattened or, when crest is raised, tall and rounded. Tail often held cocked up at an angle. **Ad:** M. (Oct.–May) has black head with white-centered crest; white crest area conspicuous and fan-shaped when raised, a wide white stripe behind the eye when lowered. Back black; flanks reddish brown; chest white with black vertical stripes on sides; pale yellow eye. Eclipse m. (Jun.–Sep.) like ad. f. but with light eye, all-dark bill. F. has grayish head with paler cheek; brownish crest; flanks and breast gray; back dark; eye dark brown; bill dark with orange variably on base and lower mandible. **Juv/1st Yr:** Like ad. f. at first; m. gradually acquires ad. m. plumage by summer, often starting with dark feathers around the eye in winter and dark feathers on shoulder in spring. **Flight:** Flight fast and agile with shallow stiff wingbeats; appears small, thin-necked, large-headed, long-tailed. Upperwing of both sexes with whitish secondaries and greater covert tips; brighter in m. **Hab:** Wooded ponds, sluggish rivers, lakes in summer; ponds, shallow fresh- and saltwater bays, salt marshes in winter. **Voice:** M. a short *grek;* f. a low, grating, drawn-out *aahwaaah.*

Subspp: Monotypic. **Hybrids:** Wood Duck, Gadwall, Bufflehead, Common Goldeneye, Barrow's Goldeneye, Red-breasted Merganser.

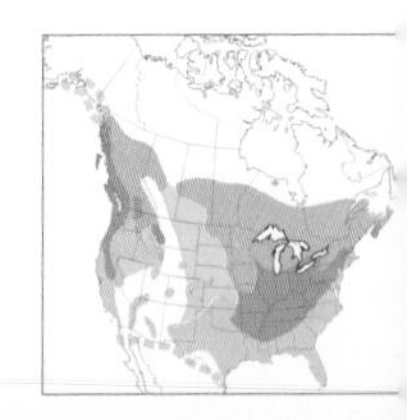

Adult m., captive CA/01

Adult f., captive CA/01

Imm. m. ENG/11

Smew 3

Mergellus albellus L 16"

Rare Eurasian migrant to w. AK, spring and fall; casual vag. elsewhere fall–spring. **Shape:** Small, short-bodied, relatively long-tailed duck with a very short fine-pointed bill, sloping forehead, and flattened crown that is rounded when crest is raised. **Ad:** M. mostly white with distinctive oval black eye patch that extends to bill base; also black lateral hindcrown stripes. Eclipse m. similar to ad. female. F. with reddish-brown head except for white lower cheek, chin, and throat; upperparts mostly gray; belly pale. **Juv/1st Yr:** Like ad. f. at first; m. gradually acquires ad. m. plumage by summer. **Flight:** Fast, agile. White wing patches on upperwing secondary coverts. **Hab:** Wooded lakes, rivers in summer; lakes, coastal bays in winter. **Voice:** Generally silent.

Subspp: Monotypic. **Hybrids:** Common Goldeneye.

Adult m. TX/05

Adult f. TX/12

Adult m. and f. TX/02

Masked Duck 3

Nomonyx dominicus L 13½"

Rare Mexican visitor to mostly s. TX but occasionally elsewhere in e. NAM. **Shape:** Small, short-bodied, long-tailed, large-headed duck with a wide deep-based bill. Bill length about ⅔ length of head. Sloping forehead leads to very shallow somewhat flattened crown. Eye proportionately large; tail often held up and slightly fanned. **Ad:** M. (May–Oct.) has black head, reddish-brown nape, pale blue bill; body reddish brown mottled black on back. Eclipse m. (Nov.–Apr.) like ad. female. F. with a dark crown and 2 dark bars across buffy face; grayish bill. Similar f. Ruddy Duck lacks the pale eyebrow and the white patch on the upper secondary coverts of the f. Masked Duck. **Juv/1st Yr:** Like ad. f. at first; m. gradually acquires ad. m. plumage by summer. **Flight:** Often flies low, dropping into vegetation; can take off directly from water. White triangular patch on inner upperwing; white axillaries. Prefers to hide rather than fly. **Hab:** Edges of well-vegetated lakes. **Voice:** Generally silent.

Subspp: Monotypic.

Adult m., summer CA/02

Adult m., winter CA/02

Adult f. AZ/03

Adult f. AZ/03

Ruddy Duck 1

Oxyura jamaicensis L 15"

Shape: Small, short-bodied, long-tailed, large-headed duck with a broad neck and proportionately fairly long, wide, deep-based bill. Bill length about 3/4 length of head. Sharply curved culmen blends with sloping forehead and fairly high rounded crown (sometimes peaked). Tail often held raised, but rarely fanned. **Ad:** M. (Apr.–Aug.) has black cap and nape contrasting with white cheek and pale blue bill; body reddish brown. Winter m. (Sep.–Mar.) grayish overall with black cap and nape; white cheek; grayish bill. F. dark brown overall except for whitish cheek with single dusky bar and white undertail coverts. **Juv/1st Winter:** (Aug.–Dec.) Much like ad. f.; m. gradually acquires ad. m. plumage by midwinter. **Flight:** Runs long distance over water for takeoff; very rapid wingbeats; rarely seen in flight. Wings dark with pale grayish axillaries. **Hab:** Open lakes in summer; also along coast in winter. **Voice:** A low grating *kit kit kitawaahn.*

Subspp: Monotypic.

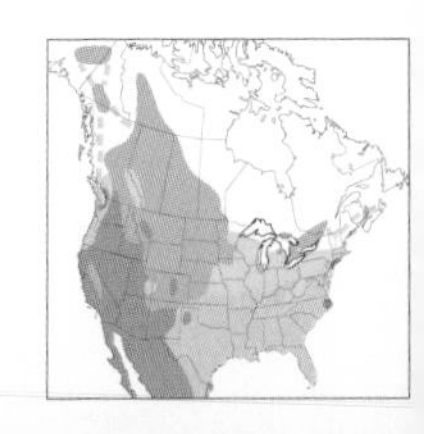

Chachalaca

Partridge

Grouse

Turkey

Pheasant

Quail

Plain Chachalaca 2

Ortalis vetula L 22"

Shape: Large, heavy-bodied, very long-tailed, small-headed bird with a relatively long neck and long legs. **Ad:** Appears brownish overall with a wash of reddish brown on belly and undertail coverts; slightly more grayish head. Tail indistinctly bluish green and tipped white; bill bluish; throat pouch bright red during breeding. **Juv:** Upperparts duller brown; tail tips with buffy wash (rather than white, as in adult). **Flight:** Large and heavy-bodied; long tail with white tips to outer feathers. **Hab:** Open tropical forests and thickets. **Voice:** Loud *chachalah-cah* is actually a duet of m. and f. calls overlapping; also loud *chak chak chak*.

Subspp: (1) •*mccalli*.

Chukar 2

Alectoris chukar L 14"

Game bird introduced from Asia. **Shape:** Midsized, stocky, potbellied game bird with short legs and short tail. **Ad:** Gray breast; bold black barring on whitish flanks; bold black band through eyes continues down across chest; bill and eye bright red. **Flight:** Bursts into flight gaining height, then glides. Reddish-brown outer tail feathers; plain gray back. **Hab:** Arid rocky slopes and canyons. **Voice:** Raspy clucking, *chukchukchukchukar, chukar, chukar;* also softer clucking sounds.

Subspp: (1) •*chukar*.

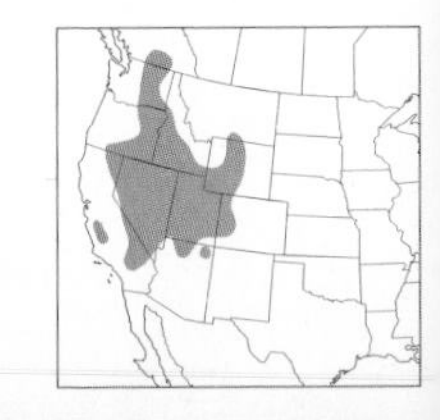

Adult NV/–

Himalayan Snowcock 2

Tetraogallus himalayensis L 28"

Asian game bird introduced into the Ruby Mts. of NV. **Shape:** Very large, heavy-bodied, relatively short-necked, and small-headed game bird with a moderately long tail. Large thick-based bill has strongly downcurved culmen. **Ad:** White sides of neck outlined by wide blackish-brown stripes; pale grayish breast with dark barring contrasts strongly with rest of dark brownish-gray body; wing coverts and back with short reddish-brown streaks; bill gray. **Flight:** Must run and flap to take off from level surface; usually takes off from high cliff and glides downward, then walks back up. **Hab:** Mountain ridges, cliffs, and meadows. **Voice:** A haunting rising wail, reminiscent of Common Loon, *ooowaaheeeooo;* also a rapid low *wawawawa.*

Subspp: (1) •*himalayensis.*

IDENTIFICATION TIPS

Game Birds are a varied group that includes grouse, ptarmigans, turkeys, pheasants, bob-whites, and quail, among others. Species vary greatly in size but in general have a large body, small head, stubby chickenlike bill, and proportionately short, broad, rounded wings. They spend most of their time on the ground feeding on plant material and some insects. The sexes generally look different.

Species ID: Note shape and length of tail, presence of head plumes or crests, and general size; also look for presence and color of eye combs and air sacs and color patterns of face, belly, and flanks.

Adult MT/–

Adult FIN/02

Gray Partridge 2

Perdix perdix L 13"

Game bird introduced from Asia. **Shape:** Fairly small, stocky, potbellied game bird with short legs, short tail, and short stubby bill. **Ad:** Pale to dark orange face and throat surrounded by gray crown, neck, and breast; broad brown barring on gray flanks; belly white with variable brownish patch on upper belly; patch can be large, small, or absent on either sex; patch tends to be larger in m.; f. may have darker brown streaking on scapulars and wing coverts. **Flight:** Bursts into flight to gain height, then glides and flaps to continue. Reddish-brown tail feathers. **Hab:** Open farmland and hedgerows. **Voice:** A rapid, high-pitched, scratchy *chiderit.*

Subspp: (1) •*perdix.*

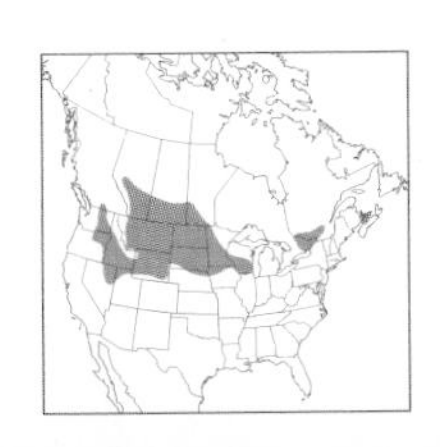

Adult m. with neck-ring, *Colchinus* Grp. MN/10

Adult m. "Green Pheasant," *Versicolor* Grp. JAP/04

Adult m. without neck-ring, *Colchinus* Grp. SCO/-

Adult f. CA/12

Ring-necked Pheasant 1

Phasianus colchicus L m. 33", f. 21"

Game bird introduced from Asia. **Shape:** Large, heavy-bodied, small-headed, long-legged game bird with a very long tail. M. tail almost as long as body; f. tail about ½ length of body. M. has thick neck, large head with ear tufts, and large bill; f. neck thinner, head and bill smaller, no ear tufts. **Ad:** M. red facial skin and wattles; pale yellow bill; iridescent greenish-purple neck usually with white collar; body varies from coppery to iridescent green (see Subspp.). F. cryptically light brown with variable darker brown spotting; dull reddish-brown wash on neck and upper breast; bill horn-brown. **Juv:** Similar to ad. f.; by 6–8 weeks begin to look like ad. of their sex. **Flight:** Rapid steep takeoff, then long glide with intermittent flaps; short wings and long pointed tail. **Hab:** Farmlands with woods edges or hedgerows, brushy marsh edges. **Voice:** A loud hoarse *skwok skwok* or *swogok* by m., often answered by f. with harsh *kia kia;* also *tukatuk tukatuk tukatuk,* trailing off and given when flushed.

Subspp: Of the 32 subspp. worldwide, several have been introduced into NAM. Repeated introductions, interbreeding of subspp., and specially bred and released birds have clouded subsp. distinctions. As a result, there are about 5 subsp. "groups" that usually can be distinguished in NAM but are not tied to specific geographic areas. **•*Colchicus* Group:** "Ring-necked Pheasant"; m. lacks white eyebrow, usually has white collar, buffy to reddish-brown lesser coverts. **•*Versicolor* Group:** "Green Pheasant"; m. has iridescent green body, no white collar, gray lesser coverts. **•*Chrysomelas* Group:** M. with broad white eyebrow, no white collar, white lesser coverts. **•*Mongolicus* Group:** M. has broad white eyebrow, wide white collar, white lesser coverts. **•*Torquatus* Group:** M. with little or no white eyebrow, narrow white collar (often partial), gray lesser coverts. **Hybrids:** Ruffed Grouse, Dusky Grouse, Sooty Grouse, Greater Prairie-Chicken.

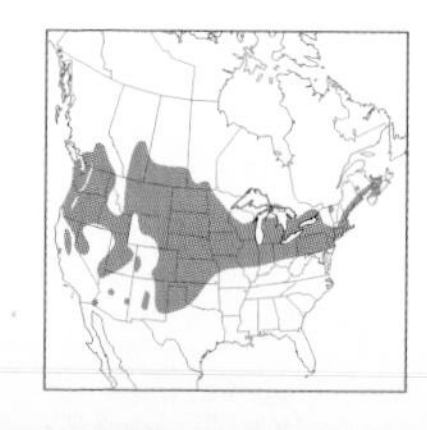

Adult m., red morph, *togata* NH/03

Adult m., gray morph USA/–

Adult m., red morph, *togata* NH/03

Adult f., gray morph, *incana* WY/02

Ruffed Grouse 1

Bonasa umbellus L 17"

Shape: Medium-sized, short-legged, heavy-bodied game bird with a broad, rounded, medium-length tail. Crest when raised creates triangular peak to otherwise rather squarish head; crest when lowered is inconspicuous. Ruff expanded and tail spread during displays. **Ad:** Flanks white with bold blackish barring; distinct whitish line along scapulars; tail with wide dark subterminal band. M. tail band usually complete; f. tail band usually broken in center. Two color morphs (defined by color on head, back, and tail) – red morph (looks reddish brown) and gray morph; intergrades also occur. Similar Sharp-tailed Grouse lacks bold dark barring on the flanks, instead has dark spotting or chevrons. **Juv:** (Jul.–Sep.) Similar to ad. but lacks dark tail band. **Flight:** Rapid takeoff, then long glide with intermittent flaps; short wings and broad rounded tail with dark subterminal band. **Hab:** Forests with mostly deciduous trees. **Voice:** A quiet *gweek gweek gweek;* a low-pitched accelerating throbbing created by the wings during m. display, when it spreads its tail and expands its ruff. 🎧**4**.

Subspp: (10) Sometimes divided into three groups. **Pacific Coastal Group:** Darkest overall. •*brunnescens* (s.w. BC), •*castanea* (w. WA–n.w. OR), •*sabini* (s.w. BC–n.w. CA). **Interior Western Group:** Generally palest overall. •*yukonensis* (w. AK–NT–n.w. SK), •*umbelloides* (BC–e. ID east to QC), •*phaia* (s.w. BC–n.e. OR–w. MT), •*incana* (s. SK–ND south to UT–CO). **Eastern Group:** Generally moderately dark. •*togata* (n. MN–NF south to s. ON–NH), •*monticola* (s.e. MI–w. PA south to n.w. AL), •*umbellus* (s.w. MN–w. KS to s.e. MI–AR, also w. NY–MA–e. MD). **Hybrids:** Ring-necked Pheasant, Spruce Grouse.

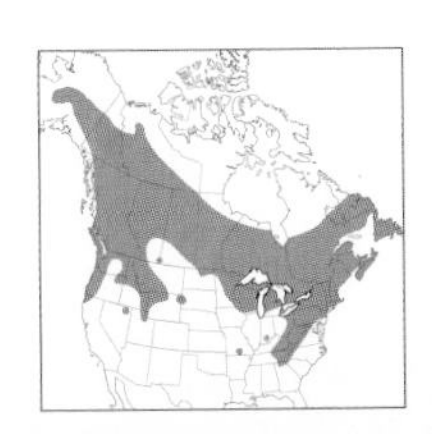

Adult m. CA/04

Adult f. CA/04

Ad. m. displaying CA/04

Adult m. CA/04

Adult m. displaying CO/04

Adult m. displaying CO/04

Greater Sage-Grouse 1

Centrocercus urophasianus L m. 28", f. 22"

Gather for group courtship displays in spring. **Shape:** Large, heavy-bodied, small-headed game bird with a relatively long pointed tail. During displays, m. chest sac inflated, sparse filoplumes on nape raised, yellow eye combs expanded, and tail feathers fanned. Similar m. Gunnison Sage-Grouse has much longer and denser filoplumes. **Ad:** Upperparts grayish brown with white speckles and streaks; belly black; thin buffy eyebrow behind eye. M. with breast whitish; f. breast mottled brown and white, like upperparts. Similar Gunnison has paler tail feathers with more contrasting light-and-dark barring. **Juv:** Similar to ad. **Flight:** Strong and rapid; long tail, black belly, white underwing coverts. **Hab:** Sagebrush flats. **Voice:** During display, m. gives 3 low-pitched coos; also dull percussive *clop clopal* produced by air sacs (m. Gunnison display calls end in up to 9 *clops*). F. gives 2–4-note *ka ka ka* coming to or leaving display ground; also quiet *pent* when traveling with her brood.

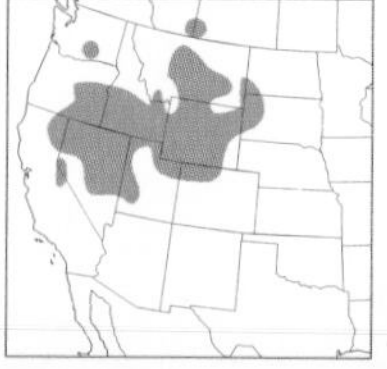

Subspp: Monotypic. **Hybrids:** Dusky Grouse, Sharp-tailed Grouse.

Gunnison Sage-Grouse 2

Centrocercus minimus L m. 22", f. 18"

Gather for group courtship displays in spring; highly specific range. **Shape:** Same shape as Greater Sage-Grouse but smaller overall; m. has much longer and more dense filoplumes on nape. **Ad:** Upperparts grayish brown with white speckles and streaks; belly black; thin buffy eyebrow behind eye. Tail feathers pale with distinct contrasting light-and-dark barring. M. with breast whitish; f. breast mottled brown and whitish like upperparts. Similar Greater Sage-Grouse has darker tail feathers with less distinct and less contrasting light-and-dark barring. **Juv:** Similar to ad. **Flight:** Strong and rapid; long tail, black belly, white underwing coverts. **Hab:** Sagebrush flats. **Voice:** Similar to Greater Sage-Grouse, but during display, m. gives 3 low-pitched coos that have slight rise and fall, and dull percussive *clops* produced by air sacs are usually in a series of 9 at a time. F. gives 2–4-note *ka ka ka* coming to or leaving display ground; also quiet *pent* when traveling with her brood.

Subspp: Monotypic.

Adult m., *franklinii* USA/–

Adult f., red morph, Canadian Grp. ME/0

Adult m., Canadian Grp. MB/06

Adult f., gray morph, *canadensis* AK/0

Spruce Grouse 2

Falcipennis canadensis L 15"

Shape: Medium-sized, short-legged, heavy-bodied game bird with short tail. Head rounded; neck broad-based. During displays, m. tail fanned, eye combs expanded. Similar Dusky and Sooty Grouse are comparatively long-tailed, longer-necked, with larger deeper-based bills. **Ad:** M. upperparts blackish to sooty or finely barred with gray, brown, and black (see Subspp.); throat black with thin white outline; foreneck black; rest of underparts boldly patterned black and white; eye combs red; tail all black or with reddish-brown (sometimes white) terminal band (see Subspp.). F. has red and gray morphs; cryptically colored above; always heavily barred with black on breast and belly, either over white (gray morph) or reddish-brown (red morph) background. Similar Sooty and Dusky Grouse males grayish with little or no patterning on underparts, female underparts with indistinct spotting (rather than heavy barring of f. Spruce Grouse). **Flight:** Good flier in woods habitat; short flights between trees. Once on ground, more likely to walk than fly. Rounded tail, dark above. **Hab:** Dense coniferous forests with some clearings. **Voice:** M. gives low grunting *ruhn ruhn;* also loud wing-clapping (in western birds only).

Subspp: (4) Divided into two groups. **Franklin's Group:** Brownish to blackish gray above, tail all black (or with thin white tip), uppertail coverts with broad white tips. •*franklinii* (cent. BC–s.cent. AB south to n. OR–n.w. WY), "Franklin's Grouse"; m. has all-black tail with no reddish brown on tip (sometimes with thin white tip), bold white tips to uppertail coverts and tertials, large white tips to flank feathers. •*isleibi* (Prince of Wales Is., AK) like *franklinii* but with longer tail, uppertail coverts with smaller white tips. **Canadian Group:** Blackish to sooty above, tail with reddish-brown tip, uppertail coverts with indistinct or no whitish to grayish tips. •*canadensis* (w. AK–cent. BC and east) m. has variable-width reddish-brown terminal tail band (rarely broadly white), small or no white tips to uppertail coverts and tertials, small white tips to flank feathers. •*canace* (s. MB–n. MN and east) m. similar to *canadensis*, but f. with more bright cinnamon barring on back. **Hybrids:** Ruffed Grouse, Willow Ptarmigan, Dusky Grouse.

Adult m., early summer AK/05

Adult f., summer, *maior** SIB/–

Adult m., late summer AK/–

Adult winter, *lagopus** FIN/03

Willow Ptarmigan 1

Lagopus lagopus L 15"

Shape: Medium-sized, heavy-bodied, short-tailed, short-legged game bird with a small head, fairly long neck, and stubby bill. Compared with other ptarmigans' bills, Willow's is deeper-based, slightly longer, and with more strongly curved culmen. **Ad. Summer:** Black tail (mostly hidden by uppertail coverts) and white wings all year. M. has bright reddish-brown head and neck in early summer; by late summer, flanks variably reddish brown, back reddish brown with dark barring; orange-red eye combs. F. is cryptically colored warm brown overall with fine barring on belly. Similar f. Rock Ptarmigan has shorter thinner-based bill with more gradually curved culmen. **Ad. Winter:** Both sexes all white with black tail. Similar f. Rock Ptarmigan differs in bill shape and, to some extent, its preference for rocky (not shrubby) habitat. **Juv:** (Jul.–Aug.) Grayish brown overall with white outer primaries (ad. has all-white wings). **Flight:** Rapid takeoff followed by glides; note black tail. **Hab:** Shrubby alpine areas in summer; lower elevations in winter. **Voice:** An extended, rapid, ducklike *gegegegegege,* loudest in the middle.

Subspp: (6) Subsp. differences often obscured by individual variation, intergradation, and complexities of 3 molts per year; thus subspp. are not described. Listed subspp.: *alexandrae, alascensis, albus, alleni, leucopterus, ungavus.* **L. l. lagopus* (EUR) and **L. l. maior* (SIB) used for good photographs. **Hybrids:** Spruce Grouse, Rock Ptarmigan.

Adult m., early summer, *nelsoni* AK/06

Adult m., winter, *rupestris* BC/05

Adult m., midsummer, *millaisi** SCO/04

Adult m. LAP/06

Adult f., summer, *rupestris* BC/05

Rock Ptarmigan 1

Lagopus muta L 14"

Shape: Medium-sized, heavy-bodied, short-tailed, short-legged game bird with a small head, fairly long neck, and stubby bill. Bill (like that of White-tailed Ptarmigan) shorter, thinner-based, and with less strongly curved culmen than that of Willow Ptarmigan. **Ad. Summer:** Black tail (mostly hidden by uppertail coverts) and white wings all year. Unlike other m. ptarmigans, m. Rock Ptarmigan delays its molt and retains its white winter plumage during spring courtship and into early summer; when the snow melts, it may dustbathe, making the plumage soiled white. In midsummer, m. molts and has brownish or blackish barred upperparts (m. Willow has reddish-brown upperparts) and white flanks blotched darker, contrasting with white belly. Another brief molt of contour feathers occurs in late summer before molt into white winter plumage. F. molts early in spring to cryptically barred brown; at this point, very similar to f. Willow Ptarmigan, except for smaller bill and, to some extent, rocky habitat preference. **Ad. Winter:** All white with black tail. M. has distinctive dark eyeline, red to orange eye combs; f. usually lacks or has suggestion of dark eyeline. **Flight:** Rapid takeoff followed by glides; note black tail. **Hab:** Tundra and barren rocky slopes. **Voice:** A low grating *r'r'r'r'r* like running a finger across a comb.

Subspp: (8) Subsp. differences often obscured by individual variation, intergradation, and the complexities of 3 molts per year; thus subspp. are not described. Listed subspp.: *atkhensis, captus, dixoni, evermanni, nelsoni, rupestris, townsendi, welchi.* **L. m. millaisi* (SCO) used for good photograph. **Hybrids:** Willow Ptarmigan.

Ad. m., early summer CO/06

Ad. f., early summer CA/08

Adult late summer CO/07

Adult winter USA/–

White-tailed Ptarmigan 2

Lagopus leucura L 12½"

Shape: Medium-sized, heavy-bodied, short-tailed, short-legged game bird with a small head, fairly long neck, and stubby bill. Bill (like that of Rock Ptarmigan) shorter, thinner-based, and with less strongly curved culmen than that of Willow Ptarmigan. **Ad. Summer:** White tail and wings all year. M. variable, mostly cryptically gray, brown, and white contrasting with white belly. F. similar but more finely patterned overall and with grayish belly; lacks warm browns of f. Willow and Rock Ptarmigans. In late summer (Jul.–Oct.) both sexes mostly grayish on head, breast, and back with some black and reddish-brown markings; underparts finely barred brown. **Ad. Winter:** Both sexes all white. **Flight:** Rapid takeoff followed by glides; generally walks but may fly to feeding grounds or to patrol territory. Note white tail. **Hab:** Rocky mountain tundra in summer; edge of tree line, often near shrub willow, in winter. **Voice:** A scratchy *zuk zeeek zeeek* and a lower *zuk zuk zuk*.

Subspp: Monotypic. **Hybrids:** Sharp-tailed Grouse.

Adult m. displaying WA/05

Adult f. CA/08

Sooty Grouse 2

Dendragapus fuliginosus L 19"

Shape: Large, heavy-bodied, long-tailed, and long-necked grouse with a large deep-based bill. Tail fairly rounded, individual tail feathers with fairly rounded tips, 18 tail feathers. **Ad:** Grayish overall with no strong contrasts, except whitish markings along flanks; northern populations generally darker than southern. M. eye combs generally yellow but can turn orange to red during courtship; air sacs (visible during display) yellow with surrounding feather bases narrowly white; dark gray tail has narrow to broad (see Subspp.) paler gray terminal band. See Dusky Grouse for comparison. F. generally dark grayish brown and cryptically colored; underparts mostly unmarked grayish. **Flight:** Rapid takeoff often followed by glide. **Hab:** Dry open forest or shrub/grassland habitats. **Voice:** M. gives hoots during courtship – usually 6, high-pitched, quite loud, and often given while perched in a tree and audible at relatively long distance; also a low *zuk zuk zuk*.

Subspp: (4) •*fuliginosus* (s.e. AK–s.w. YT–n.w. CA) and •*sitkensis* (s.e. AK–Q. Charlotte Is. BC) have narrow gray terminal tail band. •*sierrae* (cent. WA–cent. CA) and •*howardi* (s.cent. CA) have broad gray terminal tail band. **Hybrids:** Ring-necked Pheasant, Dusky Grouse.

Adult m. WY/06

Adult m. displaying WY/06

Adult f., *obscurus* CO/06

Dusky Grouse 2

Dendragapus obscurus L 19"

Shape: Large, heavy-bodied, comparatively long-tailed and long-necked grouse with a large deep-based bill. Tail fairly squared off, individual tail feathers fairly square-tipped, 18–20 tail feathers. Similar Spruce Grouse smaller with shorter tail, shorter thicker neck, smaller bill. Similar Sooty Grouse m. best separated by range, but also in shape has subtly more rounded tail, rounded tips to tail feathers, and 18 tail feathers. **Ad:** In general, grayish overall with no strong contrasts, except whitish markings along flanks; northern populations generally darker than southern. M. eye combs generally yellow but can turn orange to red during courtship; air sacs (visible during display) red with surrounding feather bases extensively white; dark gray tail may have broad to no gray terminal band (see Subspp.). Similar m. Sooty Grouse has yellowish air sacs with limited white at base of surrounding feathers and louder and different calls (see Voice). F. generally dark grayish brown and cryptically colored; underparts mostly unmarked grayish; use of shape and range can help distinguish from other f. grouse. **Flight:** Generally walks rather than flies; strong agile flier for short distances; rapid takeoff often followed by glide. **Hab:** Open forests, sage, scrub oak; winters near tree line. **Voice:** M. gives soft hoots during courtship – usually 5, low-pitched, and at relatively low volume, often given from ground and only detectable at short range; also a low *zuk zuk zuk*.

Subspp: (4) •*obscurus* (cent. WY south) and •*oreinus* (n.e. NV) usually have a broad gray terminal tail band. •*richardsonii* (YT–interior BC south to n.w. WY) and •*pallidus* (s. BC–n.e. OR) usually have little or no gray terminal tail band. **Hybrids:** Ring-necked Pheasant, Greater Sage-Grouse, Spruce Grouse, Sooty Grouse, Sharp-tailed Grouse.

Adult m., *columbianus* MT/06

Ad. m., *columbianus* USA/–

Adult f., *columbianus* MT/06

Sharp-tailed Grouse 2

Tympanuchus phasianellus L 17"

In spring groups of males display together. **Shape:** Medium-sized, compact, heavy-bodied grouse with a distinctive spikelike tail dominated by the 2 central feathers. Small crest, when raised, creates rather triangular head; legs comparatively long; neck short and thin. **Ad:** Above, brown variably spotted with white; below, buffy breast and belly with variable white spotting and brown chevrons (see Subspp.); bill grayish. M. has pink to violet air sacs, yellowish eye combs (seen during display), and dark-streaked crown. F. lacks large air sacs and eye combs and has paler, more barred crown. Similar Ruffed Grouse has bold dark barring (instead of spotting or chevrons) on flanks and broad tail; similar prairie-chickens have brown and whitish barring on all of underparts. **Flight:** Flights mostly short; rapid takeoff followed by long glides alternating with a few flaps. Note pale belly, pointed tail, white outer tail feathers. **Hab:** Open prairies with shrubs, regenerating clear-cuts, agricultural fields. **Voice:** A low-pitched rapid *owalow owalow* and a low cooed *oooo*.

Subspp: (2) •*phasianellus* (AK–n.w. SK–cent. QC); upperparts dark brown with white spotting. •*columbianus* (e. BC–n.cent. AB–s.e. MB–MI and south); upperparts pale brown with some darker mottling and less white spotting. **Hybrids:** Greater Sage-Grouse, White-tailed Ptarmigan, Dusky Grouse, Greater Prairie-Chicken.

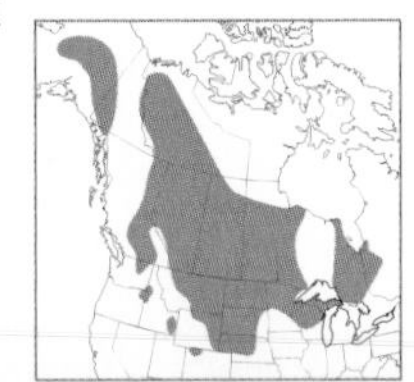

Adult m. NM/04

Adult m. displaying NM/04

Adult f. NM/04

Lesser Prairie-Chicken 2

Tympanuchus pallidicinctus L 16"

Shape: Shape identical to Greater Prairie-Chicken's, although marginally smaller. **Ad:** Nearly identical plumage to Greater Prairie-Chicken, except air sacs in m. dull red (rather than yellowish orange) and plumage overall with slightly paler, weaker, less even barring. Best distinguished by ranges, which do not overlap. **Flight:** Strong flier, using flaps and glides. **Hab:** Arid grasslands with sage or oak. **Voice:** A bubbling low-pitched *wuhla wuhla*.

Subspp: Monotypic. **Hybrids:** Greater Prairie-Chicken.

Adult m., *pinnatus* IL/04

Adult f., *pinnatus* IL/04

Adult m. displaying, *pinnatus* IL/04

Males, *pinnatus* IL/04

Greater Prairie-Chicken 2

Tympanuchus cupido L 17"

In spring groups of males display together. **Shape:** Medium-sized heavy-bodied grouse with a short rounded tail and long specialized feathers (pinnae) on nape that can be raised during displays (look like rabbit ears). M. also has inflatable air sacs and eye combs that can be enlarged. **Ad:** Bold, even, dark brown barring overall, especially on breast, flanks, and belly; pale buff unmarked chin. M. has dark unbarred tail, yellowish-orange air sacs, and large yellowish-orange eye combs. F. has barred tail, no air sacs, and small inconspicuous eye combs. Similar Lesser Prairie-Chicken has weaker paler brown and less even barring overall; Sharp-tailed Grouse has pale underparts with dark spot or chevrons and lacks pinnae of Prairie-Chicken. **Flight:** Strong flier, using flaps and glides. **Hab:** Prairie openings, agricultural fields. **Voice:** Low-pitched, musical, cooing *cooloo looo, cooloo looo,* and an exclamatory *whop whop.*

Subspp: (2) Very similar. •*attwateri* (s.e. TX), "Attwater's Prairie-Chicken"; slight reddish-brown wash to plumage, slightly smaller than *pinnatus.* •*pinnatus* (ND–MN south to OK–IL); more reddish brown on plumage, slightly larger. ▶ *T. c. cupido,* "Heath Hen," is extinct. **Hybrids:** Ring-necked Pheasant, Sharp-tailed Grouse, Lesser Prairie-Chicken.

Males, *silvestris* NH/04

Adult f., *intermedia* TX/04

Adult m. displaying, *intermedia* TX/04

Wild Turkey 1

Meleagris gallopavo L m. 46", f. 37"

Shape: Very large heavy-bodied game bird with long thick legs, long tail and neck, and proportionately very small head (smaller in f. than m.). Thin hairlike feathers on central breast (the "beard") typical of m., uncommon and smaller when present in f.; prominent flap of skin on throat (the "dewlap"), head bare and "warty," sometimes with sparse bristles. M. has prominent spur halfway up back of tarsus, which starts to appear on 1st-winter m. and continues to get larger over time. Displaying m. raises feathers and tail, becoming large and rounded with huge fan-shaped tail. In general, m. is larger, has proportionately larger head, has well-developed beard, and has spurs. **Ad:** Bare skin on head and neck bluish and pinkish (brighter on m.); body feathers iridescent metallic blue or bronze with dark tips in m. and brown or whitish tips in f. Exposed primaries contrastingly dull and strongly barred.

Flight: Large wings and huge body fairly shocking to see in flight. Flies up to roosts; glides down; longer-distance movements combine flaps and glides. **Hab:** Usually in open woods with nut-producing trees or pines with some clearings nearby. **Voice:** M. gives loud, explosive gobbling *skwaloogaloogalooga;* many other calls including *kwuk*s and rattles. 🎧**5**.

Subspp: (4) Introduction of subspp. and domestic turkeys into new areas complicates subspp. classification. •*silvestris* (s.e. NE–LA and east to NH–GA); large bird, short legs, long beard. •*merriami* (Great Plains and Rocky Mts., also introduced to ND–WY and west); large bird, average-length legs, comparatively short beard. •*intermedia* (KS south); medium to small bird. •*osceola* (cent. FL and south); small bird, long legs, long spurs, thin beard fibers.

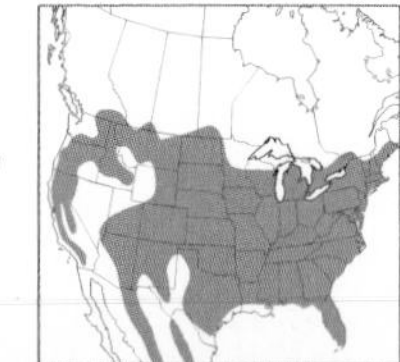

Adult CA/05

Adult, *pallida* NM/05

Mountain Quail 1

Oreortyx pictus L 11"

Shape: Relatively large short-tailed quail with distinctive long straight head plumes and bushy crest. Head a rounded triangle when crest is raised. **Ad:** Clear grayish breast and upper belly; dark reddish-brown throat and cheek with thin white border. Bold white barring on flanks. Sexes not reliably distinguished in field, but m. tends to have longer head plumes and grayer nape (f. browner) and crown in some populations. **Flight:** Usually runs to escape danger; will make short flights into trees. **Hab:** Chaparral, forests with shrub understory. **Voice:** A loud, penetrating, descending whistled *kwee-ark;* also series of short whistles.

Subspp: Monotypic. **Hybrids:** California Quail.

Adult, *castanogastris* TX/11

Scaled Quail 1

Callipepla squamata L 10"

Shape: Medium-sized short-tailed quail with a distinctive tufted crest. **Ad:** Strongly scaled appearance to nape and underparts; crest tipped white; relatively unmarked head and throat. Sexes very similar: m. may have unmarked buffy throat; f. may have grayer throat with fine dark streaks and slightly shorter crest. **Flight:** Prefers to run rather than fly; will make short flights. **Hab:** Arid shrublands. **Voice:** A percussive *chik chik;* loud whistled *whee-eet;* low *chuh-kar chuh-kar.*

Subspp: (3) •*castanogastris* (s. TX); belly dark buff, usually with dark reddish-brown central patch. •*pallida* (n.w. NM–w. TX and west); belly grayish buff with no dark patch. •*hargravei* (CO–n.w. NM–KS–w.cent. TX); belly whitish with no dark patch. **Hybrids:** California Quail, Gambel's Quail, Northern Bobwhite.

Adult m. CA/03

Adult f. CA/03

California Quail 1

Callipepla californica L 10"

Shape: Medium-sized relatively long-tailed quail with a short forward-arching head plume ("top-knot") that is shorter on female. Shape identical to that of Gambel's Quail. **Ad:** Black head plume; breast and belly yellowish with strongly scaled appearance; dull dark brown flanks with white streaks. M. chin black with white border; dull brown crown; pale forehead; large head plume; dark reddish-brown belly patch. F. head brownish, relatively unmarked; small head plume; no dark belly patch. Similar Gambel's Quail has a yellowish belly with no scaling and rich reddish-brown flanks. **Flight:** Prefers to run rather than fly; will make short flights. **Hab:** Open woodlands, shrubby areas, parks, suburbs, usually near water. **Voice:** A loud *sheeer* and a repeated *chi cah go, chi cah go;* also spitting call *pit pit pit.* 🎧**6**.

Subspp: Monotypic. **Hybrids:** Mountain Quail, Scaled Quail, Gambel's Quail, Northern Bobwhite.

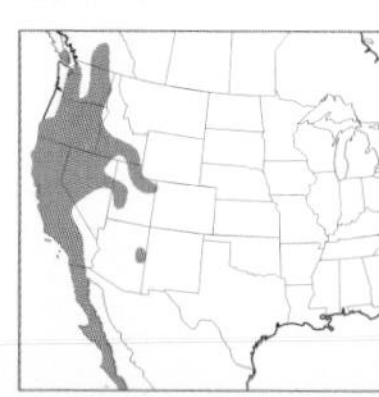

Adult m. AZ/01

Adult f. AZ/02

Gambel's Quail 1

Callipepla gambelii L 11"

Shape: Identical shape to that of California Quail. **Ad:** Black head plume; breast and belly yellowish with no scaling; reddish-brown flanks with white streaks. M. chin black with white border; rich reddish-brown crown; black forehead; large head plume; dark blackish belly patch. F. head grayish brown, relatively unmarked; small head plumes; thin reddish-brown streaking on belly. Similar California Quail has yellowish scaled belly and dull brown flanks. **Flight:** Much prefers to run rather than fly; will make short flights. **Hab:** Arid shrubby areas, riparian woodlands. **Voice:** Similar to California Quail's but more muted and with double note at end, *ga way gaga, ga way gaga;* also spitting call *pit pit pit.* 🎧**7**.

Subspp: Monotypic. **Hybrids:** Scaled Quail, California Quail.

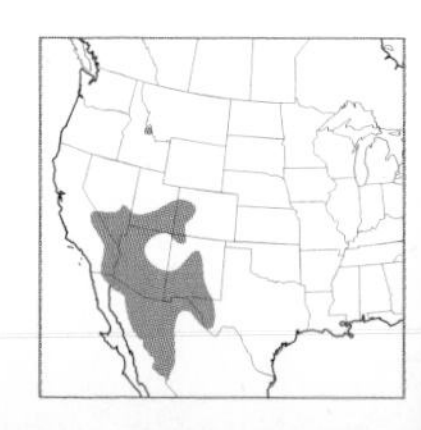

Adult m., *virginianus* NH/04

Adult, *texanus* TX/05

Ad. m., *ridgwayi* –/01

Adult f., *texanus* TX/11

Northern Bobwhite 1

Colinus virginianus L 9¾"

Shape: Very rounded, short-tailed, short-necked, compact quail. Slight bushy crest, when raised, creates rounded peak to head. **Ad:** Pale eyebrow and throat separated by broad dark eyeline that connects to dark collar. Flanks reddish brown with white streaks; belly whitish and scaled; upperparts cryptic. (See Subspp. for strikingly different "Masked Bobwhite" of the Southwest.) M. with contrasting white eyebrow and throat; f. with subtle warm buffy eyebrow and throat. **Flight:** Short flights; note plain gray wings and tail. **Hab:** Farmland, brushy fields, open woodland. **Voice:** A distinct whistled *hoot weeet,* sounding like "bob white"; also loud *koi-lee koi-lee* and soft *tu tu tu.* 🎧 **8.**

Subspp: (5) Introductions of subspp. into new areas have mixed subsp. characteristics and obscured ranges. Divided into two groups; main differences are between males. **Northern Group:** M. as described above. •*taylori* (SD–s.e. WY–n. TX–n.w. AR), •*texanus* (s.e. NM–s.w. TX), •*floridanus* (peninsular FL), •*virginianus* (rest of range in East). **Western Mexican Group:** M. with black head, thin to indistinct white eyeline, and unmarked reddish-brown underparts. •*ridgwayi* (s.e. AZ), "Masked Bobwhite." **Hybrids:** Scaled Quail, California Quail.

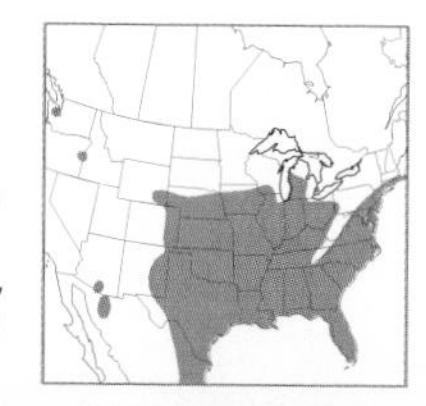

Adult m. AZ/08

Adult f. TX/05

Montezuma Quail 2

Cyrtonyx montezumae L 9"

Shape: Relatively small, rotund, short-tailed quail with a relatively stout bill and bushy crest, which when raised makes head appear somewhat helmeted. **Ad:** M. has distinctive white-and-black patterned face with reddish-brown crest; sides of breast and flanks black densely dotted with white; central breast and belly black. F. cryptically gray, brown, and reddish brown with fine paler streaking especially on back; indistinct facial pattern similar to that of male. **Flight:** Rapid takeoff followed by short glides; difficulty landing smoothly. **Hab:** Grassy open woodlands on semiarid mountain slopes, in canyons. **Voice:** M. display call a descending, slightly buzzy *djeeeeeen;* also descending whistle given by both sexes, and a quiet contact note like *whi whi.*

Subspp: (1) •*mearnsii.*

Loons

Grebes

Flamingo

Albatrosses

Petrels and Shearwaters

Storm-Petrels

Tropicbirds

Boobies and Gannet

Pelicans

Cormorants

Anhinga

Frigatebirds

Adult summer FIN/07

1st winter CA/11

Adult summer AK/06

1st winter NJ/01

Adult winter NJ/02

Adult winter NJ/01

Red-throated Loon 1

Gavia stellata L 21–25"

Shape: Small loon with a thin neck, small head, and thin bill. Bill with fairly straight culmen, angled bottom to lower mandible, and held tilted up. Neck often held straight off chest, creating 90-degree angle with water. **Ad. Summer:** Gray face and sides of neck; reddish-brown throat; thin white vertical streaks on nape; plain brownish back; black bill. **Ad. Winter:** Neck mostly white with sharply contrasting thin dark gray nape and crown; face paler than that of Common or Pacific Loon; back dark gray spotted with white; bill blackish. Similar Pacific Loon has thicker neck, bill held straight out (not tilted up), darker face, often a thin dark chin strap, and no spotting on back. **Juv/1st Winter:** (Aug.–Jan.) Like ad. winter but sides of neck grayish and blending to darker nape; back and scapular feathers with small white chevrons; bill grayish. **1st Summer:** Like ad. winter but with variable suggestions of ad. summer plumage on head and neck. **Flight:** Fast wingbeats and often low flight. Thin overall, small feet, thin neck and bill; head often held slightly down and then lifted up. **Hab:** Tundra lakes, arctic coasts in summer; coast, Great Lakes in winter. **Voice:** High-pitched, descending, plaintive wail, *weeeeaah,* given on breeding grounds; rarely heard on spring migration.

Subspp: Monotypic.

Adult summer, *arctica** ENG/06

Juv./1st winter, *arctica** ENG/12

1st summer, *viridigularis* CA/05

IDENTIFICATION TIPS

Loons are waterbirds seen at sea or on lakes; they are long-bodied, with fairly short necks and pointed spikelike bills. They may "snorkel" by holding their head under water to look for fish and then dive to catch them. They have different summer and winter plumages, but the sexes look alike.

Species ID: Note overall size, shape and size of bill, length and thickness of neck, and angle at which bill is generally held. Also look for color patterns on head, neck, and back; check bill color.

Grebes are waterbirds found on lakes during breeding and often on the ocean during winter. Most are proportionately short-bodied, and their bills vary greatly in shape from species to species. They have different summer and winter plumages, but the sexes look alike.

Species ID: Note overall size, shape of bill, and length of bill in relation to depth of head. Summer plumages of the various species are distinctive; winter plumages similar to each other. In winter note details of light and dark patterns on head and neck and whether light and dark areas are sharply demarcated or blend gradually.

Arctic Loon 2

Gavia arctica L 24–28"

Shape: Medium-sized to large loon. Bill moderately sized; often tilted slightly above horizontal. When at rest, sloping forehead angles to rather long flattened crown. Neck fairly long. Similar Pacific Loon holds bill horizontal and has shorter more rounded crown. **Ad. Summer:** Prominent white hip patch at waterline divides dark flanks from dark tail; may also show line of white along whole flank. Dark gray nape and blackish throat separated by thick, white, vertical stripes on side of neck (from a distance appears like a white patch) that connect with stripes on breast (separated from chest stripes by black in Pacific Loon); rows of bold white squarish spots on scapulars and back; bill black. **Ad. Winter:** Prominent white hip patch at waterline; may also show line of white along whole flank. Upperparts grayish brown; well-defined sharp border between white throat and rest of dark neck; eye usually within the dark area of upper head; bill gray with darker line on culmen. No dark chin strap across whitish throat. (See Pacific Loon for comparison.) **Juv/1st Winter:** (Aug.–Jan.) Like ad. winter but browner upperparts, with pale margins to back and scapular feathers creating finely scaled look. Note presence of white hip patch as in ad. **1st Summer:** Like juv. but with variable amount of ad. summer plumage on head and neck. **Flight:** Strong rapid wingbeats. Holds head and feet below center of body plane. Note large white hip patch and lack of black strap across white vent area (present in Pacific Loon). **Hab:** Tundra lakes in summer; coast in winter. **Voice:** Medium-pitched, rising and falling, resonant wail (on breeding grounds only), like *gahwaarrr;* sharply rising *weeeep, put weeeep, put;* a froglike grating *grrrt grrt.*

Subspp: (1) •*viridigularis.* **G. a. arctica* (w. Eurasia) used for good photograph. **Hybrids:** Common Loon, possibly Pacific Loon.

Adult summer AK/06

1st winter CA/12

Adult winter CA/12

Adult winter CA/11

Pacific Loon 1

Gavia pacifica L 22–26"

Shape: Medium-sized loon. Bill moderately sized; usually held horizontal. When at rest, head with well-rounded crown. Neck short, thick, and held near front of chest (not pressed back like Arctic Loon's). Similar Arctic Loon has more flattened crown and slightly larger bill usually held angled above horizontal. **Ad. Summer:** Pale gray nape and blackish throat separated by thin white stripes on side of neck (not conspicuous from a distance) which are divided from stripes on breast by black line; rows of bold white spots on scapulars and back; bill black. May show thin uneven white line along waterline (or none). Similar Arctic Loon has prominent white hip patch at waterline, darker nape, and more conspicuous white stripes on neck sides that connect to stripes on breast. **Ad. Winter:** Upperparts blackish; well-defined sharp border between white throat and rest of dark neck; eye usually within the black area of upper head; bill gray with dark line on culmen. Many individuals have a suggestion of a paler nape and show a thin dark chin strap. Similar Arctic Loon has prominent white hip patch at waterline; similar Common Loon has much larger bill, suggestion of paler markings on back, and whitish indentation into dark side of neck; similar Red-throated Loon has thinner neck, whiter face and neck, and fine white spotting on the back. **Juv/1st Winter:** (Aug.–Jan.) Like ad. winter but browner upperparts, with pale margins to back and scapular feathers creating finely scaled look. Note absence of white hip patch as in adult. **1st Summer:** Like juv. but with variable amount of ad. summer plumage on head and neck. **Flight:** Strong rapid wingbeats; usually low over water. Holds head and feet below center of body plane. Note presence of dark strap across white vent area (absent on Arctic Loon). **Hab:** Tundra lakes in summer; coast in winter. **Voice:** A low mooing wail (on breeding ground only), lower-pitched than call of Arctic Loon; a gargled *craw craw.*

Subspp: Monotypic. **Hybrids:** Common Loon, possibly Arctic Loon.

Adult summer MT/06

1st winter CA/11

Adult winter FL/01

Adult summer BC/08

1st winter FL/02

Adult winter CA/12

Common Loon 1

Gavia immer L 26–33"

Shape: Large, thick-necked, large-billed loon. Culmen slightly curved; lower mandible curved or slightly angled; bill usually held horizontal. Steep forehead angles to flat crown then angles to hindcrown; forecrown may be raised into a bump. Similar Yellow-billed Loon has straight culmen and tilts bill slightly up. **Ad. Summer:** Head and neck black with greenish tinge; bill blackish; band of thin vertical white stripes on side of neck, widest toward rear; back and scapulars finely checkered with white. Similar Yellow-billed Loon has pale yellow bill; band of vertical stripes on side of neck tapers to front and back, and white stripes are wider; back and scapulars broadly checkered with white. **Ad. Winter:** Head and back roughly concolor; back grayish brown with faint hint of paler checkering; wing coverts finely spotted with white (wear off by late winter). Midneck has whitish triangular indentation into black hindneck. Bill pale gray, usually with dark tip and culmen. Variable white around eye. Similar Yellow-billed Loon head and neck paler than back and may have dark auricular smudge; back grayish brown with faint hint of paler checkering; bill pale yellow to ivory with pale tip and culmen (may be dusky at base); much white around eye; less distinct division between dark and whitish areas on head and neck. **Juv/1st Summer:** (Aug.–Aug.) Like ad. winter but edges to back and scapulars create scaled look; no spots on wing coverts. **Flight:** Slow strong wingbeats; often flies very high. Large head, neck, and bill; large feet. **Hab:** Lakes in summer; coast in winter. **Voice:** A medium-pitched haunting wail, rising and falling *tuuweeeaarr;* a tremelous *wawawawa;* a repeated high-pitched *weeahweeep weeahweeep.* 🎧 **9.**

Subspp: Monotypic. **Hybrids:** Arctic Loon, Pacific Loon, Yellow-billed Loon.

Adult summer PA/05

Adult winter CA/04

1st winter TX/11

Yellow-billed Loon 2

Gavia adamsii L 28–34"

Shape: Like Common Loon but slightly larger; culmen straight and lower mandible angled up; bill tilted up. **Ad. Summer:** Like Common Loon but bill pale yellow; band of vertical stripes on side of neck tapers to front and back and has wider white stripes; back and scapulars broadly checkered with white. **Ad. Winter:** Head and neck paler than back, may have dark auricular smudge; back grayish brown with faint hint of paler checkering. Bill pale yellow to ivory with pale tip and culmen; may be dusky at base. Much white around eye. Less distinct division between dark and whitish areas on head and neck than Common Loon. See Common Loon for comparison. **Juv/1st Summer:** (Aug.–Aug.) Like ad. winter but pale edges to back and scapulars create scaled look; no spots on wing coverts. **Flight:** Slow strong wingbeats; often flies very high. Large head, neck, and bill; large feet. **Hab:** Tundra lakes, rivers in summer; coast, rarely inland lakes in winter. **Voice:** Sounds similar to Common Loon's but lower-pitched.

Subspp: Monotypic. **Hybrids:** Common Loon.

Adult summer, *bangsi* AZ/04

Adult winter, *brachypterus* TX/12

Least Grebe 2

Tachybaptus dominicus L 9"

Shape: Very small grebe with a comparatively short neck and short, thin, sharp-pointed bill. Rear squared off when tail is raised. **Ad. Summer:** Black cap, face, and upper throat; rest of upperparts dark gray; yellow to golden-yellow eye; whitish rear and belly; bill blackish. **Ad. Winter:** Like ad. summer but white throat and pale lower mandible. Similar Pied-billed Grebe has large dark eye with thin pale eye-ring and thick-based blunt-tipped bill. **Juv:** (Jun.–Oct.) Similar to ad. winter but sides of face with broad gray streaks. **Flight:** Rapid wingbeats; can take off more quickly than other grebes; runs on water for takeoff. Rarely seen in flight; has whitish secondaries and inner primaries, and feet trail behind body. **Hab:** Small ponds, lakes, rivers. **Voice:** A very high-pitched screeching *kreeet;* a chattering *chchchch.*

Subspp: (2) •*brachypterus* (s. TX, vag. to LA) darker and larger. •*bangsi* (vag. to CA, AZ) paler and smaller.

Adult summer MT/06

Adult winter ENG/02

Adult winter CA/12

Adult molting to summer CA/02

Horned Grebe 1

Podiceps auritus L 14"

Shape: Fairly small stocky grebe with a thin neck and relatively short, thin, straight bill. Forehead sloping and crown usually flattened with slight peak at rear; head roughly oblong (but can appear peaked and triangular like Eared Grebe's). Bill about ½ length of head. From front, head wider at top. Summer crest. Similar Eared Grebe has high peaked crown, relatively short triangular head, slightly upturned thinner bill, thinner neck; also, from the front, the head is evenly narrow. **Ad. Summer:** Black head with thick golden plumes from eye to nape; blackish back; neck and flanks reddish brown; eye deep red; lore reddish brown. **Ad. Winter:** Dark-capped look (vs. helmeted look of similar winter Eared Grebe); lower edge of black cap extends from bill base straight across base of eye and top of cheek, strongly contrasting with white cheek and white side of nape. Conspicuous whitish to indistinct dark gray supraloral dash; deep red eye connected to bill by thin dark loral line; bill gray with white tip. Whitish foreneck blends gradually with dark hindneck; back feathers with fine white margins. Late winter/early spring transitional plumage of Horned Grebe can have dusky auriculars and look similar to winter Eared Grebe, in which case note differences in head and bill shape (see Shape). Similar winter Eared Grebe has dark helmet with dark to dusky ear flap extending onto cheek and auriculars; its foreneck is usually dusky and its gray bill lacks a white tip; it can have an indistinct grayish supraloral dash; see also Shape. **Juv:** (Jul.–Oct.) Like ad. winter but face and nape mottled dusky; border between dark cap and face less distinct. Upperparts may have brownish wash and lack pale margins to feathers. Similar ad. winter Eared Grebe has high peaked crown, relatively short triangular head, slightly upturned thinner bill, thinner neck. **Flight:** Fast wingbeat; runs on water during takeoff; legs trail behind body. Note white secondaries. Body tilted slightly upward during flight. **Hab:** Marshy ponds, lakes in summer; coasts, some inland lakes in winter. **Voice:** A high-pitched scratchy *p'p'weeah p'p'weeah;* a rapid chattering.

Subspp: Monotypic. **Hybrids:** Eared Grebe.

Adult with young, summer, *holboellii* MT/06

Adult early winter, *grisegena** ENG/11

Adult winter, *holboellii* BC/02

Adult summer, *holboellii* AB/07

Red-necked Grebe 1

Podiceps grisegena L 18"

Shape: Medium-sized, thin-necked, large-headed grebe with a long relatively thick bill. Bill length about length of head; bill usually held angled down. Neat crest creates squared-off hindcrown (most prominent in summer). **Ad. Summer:** Black cap; whitish chin and cheek; reddish-brown neck with darker nape; bill dark with yellowish base; eye dark brown. **Ad. Winter:** Dusky overall without sharp contrasts; whitish on throat extends up behind dusky cheek and auricular; bill yellowish. Disting. from other grebes by bill shape; similar Red-throated Loon holds bill angled up, has contrasting white face and foreneck and spotted back. **Juv:** (Jul.–Oct.) Similar to ad. winter but with reddish-brown wash to neck and pale streaking on face. **Flight:** Fast wingbeat; runs on water during takeoff. White secondaries and white leading edge to inner wing; legs trail behind body. **Hab:** Lakes and ponds in summer; mostly coastal, also on Great Lakes in winter. **Voice:** A nasal whining *greeah;* a rapid chattering.

Subspp: (1) •*holboellii.* *P. g. grisegena* (EUR) is a potential vag. to n.e. NAM and is used for good photograph; it is smaller than *holboellii.*

Adult with young, summer MT/06

Adult winter CA/12

Adult winter CA/12

Juv. AZ/11

Eared Grebe 1

Podiceps nigricollis L 12"

Shape: Small grebe with a thin neck, small head, and relatively short, thin, slightly upturned bill. Steep forehead leads to peaked crown (highest point usually over eye). Head relatively short and roughly triangular, especially with summer crest. Bill about ½ length of head. From the front, head evenly narrow and about as narrow as neck. Similar Horned Grebe has more oblong head with highest peak usually at rear of crown, and from the front the head is wider at the top. **Ad. Summer:** Blackish head, neck, and back; brownish flanks; thin golden plumes spread out from behind eye over auriculars; eye deep red; bill and lore black. **Ad. Winter:** Dark helmeted look includes dark to dusky ear flaps that extend onto cheek and auriculars; foreneck usually dusky; back feathers all dark and without fine pale margins; bill all dark. Horned Grebe similar; see that account for comparison. **Juv:** (Jul.–Nov.) Like ad. winter but with reddish-brown wash on neck and duller eye. **Flight:** Fast wingbeat; runs on water during takeoff. Rarely seen in flight, but has white secondaries and legs trail behind body. **Hab:** Lakes and marshes in summer; coast and inland lakes in winter. **Voice:** Very high-pitched rising *weeee'pt weeee'pt* (on breeding grounds only); very high-pitched chatter.

Subspp: (1) •*californicus.* **Hybrids:** Horned Grebe.

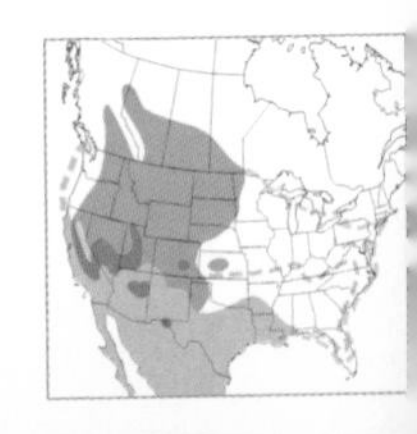

Adult summer UT/05

Adult summer CA/03

Adult winter CA/12

Adult winter CA/12

Western Grebe 1

Aechmophorus occidentalis L 25"

Shape: Large elegantly proportioned grebe with a long thin body, neck, and bill. Bill of m. almost as long as head; f. bill slightly shorter. Culmen straight, bottom of lower mandible angled up. **Ad. Summer:** Black cap and nape contrast strongly with bright white face and foreneck; black of cap extends below and surrounds red eye; black nape comparatively wide; bill dull greenish yellow with dark culmen; flanks grayish. Similar Clark's Grebe has black cap above eye, thinner dark nape, bright orangish-yellow bill, and more extensive whitish flanks; also see Voice. **Ad. Winter:** Like ad. summer but area surrounding eye dark gray. Some birds may have characteristics intermediate between winter Clark's and Western Grebe and cannot be identified. **Flight:** Fast wingbeat; runs on water during takeoff. Seen only occasionally in flight; legs trail behind body; head sometimes held below level of body; long neck and bill; indistinct white stripe on upperwing secondaries and inner primaries. **Hab:** Lakes and marshes in summer; coast and inland lakes in winter. **Voice:** High-pitched, harsh, 2-parted *k'dit k'deet*. 🎧**10**.

Subspp: Monotypic. **Hybrids:** Clark's Grebe.

Clark's Grebe 1

Aechmophorus clarkii L 25"

Shape: Large elegantly proportioned grebe with a long thin body, neck, and bill. Bill of m. almost as long as head; f. bill shorter. Culmen straight, bottom of lower mandible angled up. **Ad. Summer:** Black cap and nape contrast strongly with bright white face and foreneck; white of face extends above red eye; black nape very narrow; bill bright yellow to orangish yellow with well-defined dark culmen; flanks whitish; back gray. Similar Western Grebe has dark cap covering eye, wider dark nape, duller greenish-yellow bill, slightly grayer upperparts, and grayish flanks; see also Voice. **Ad. Winter:** Like ad. summer but area above eye dusky with white supraloral dash; area below eye white. Some birds may have characteristics intermediate between winter Clark's and Western and cannot be safely identified. Bill color and particularly flank color are best distinctions to use for birds at a distance. **Flight:** Fast wingbeat; runs on water during takeoff. Only occasionally seen in flight; legs trail behind body; head sometimes held below level of body; long neck and bill; distinct white stripe along upperwing secondaries and inner primaries. **Hab:** Lakes and marshes in summer; coast and some inland lakes in winter. **Voice:** A clear, high-pitched, whistled, single, slightly scratchy *peeet* (Western Grebe call usually 2-parted).

Subspp: Monotypic. **Hybrids:** Western Grebe.

Adult summer FL/03

Adult winter FL/01

Juv. CA/09

Adult BON/05

Imm. GAL/05

Adult GAL/10

Pied-billed Grebe 1

Podilymbus podiceps L 12½"

Shape: Small stocky grebe with a short, deep-based, blunt bill. **Ad. Summer:** Bill pale with wide dark median band; white eye-ring around dark eye; brownish overall except for blackish chin and crown; whitish rear. **Ad. Winter:** Like ad. summer but chin whitish and bill with indistinct or no dark band. **Juv:** (Apr.–Oct.) Like ad. winter but face with gray and light brown streaks. **Flight:** Fast wingbeat; runs on water for takeoff. Rarely seen in flight; has whitish terminal bar on secondaries, and feet trail behind body. **Hab:** Lakes, ponds in summer; also sheltered saltwater bays in winter. **Voice:** A low-pitched muffled *gawoh gawoh;* a giggling *heh heh heh heh.* 🎧**11**.

Subspp: (1) •*podiceps.*

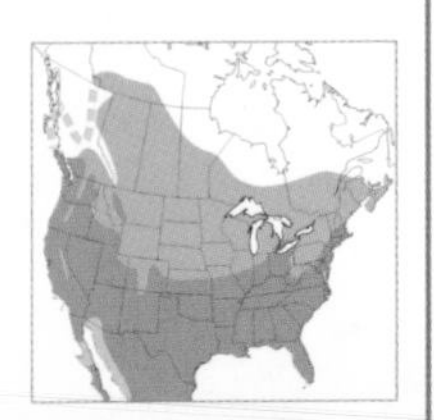

American Flamingo 3

Phoenicopterus ruber L 46"

Flamingos in FL and TX rare, likely represent wild birds; sightings away from these states are often escapees of other flamingo species. **Shape:** Very tall, long-legged, long-necked wader with distinctive, thick-based, somewhat triangular bill. Bill angled on top, widest in middle, and recurved. **Ad:** Dark pink overall except for black flight feathers (seen in flight) and black tip to bill. **Imm:** Grayish overall with dark streaking on wing coverts. Gradually becomes pale pink and then dark pink of ad. over 3–4 years. **Flight:** Flies with neck and legs extended. Black flight feathers contrast with reddish-pink wing coverts. **Hab:** Coastal bays, large areas of shallow water.

Subspp: Monotypic.

Adult TDC/01

Adult SAF/02

Yellow-nosed Albatross 4
Thalassarche chlororhynchos L 32"

Casual vag. to w. North Atl. and Gulf of Mexico. **Shape:** Long narrow wings, heavy body, and long, blunt, hooked bill typical of albatrosses. Compared with similar Black-browed Albatross, Yellow-nosed is smaller, more slender, with relatively long bill, long neck, and long tail. Similar imm. Northern Gannet has pointed bill and tail, smaller head and thinner neck, and body length equal to one wing length (body length clearly shorter than one wing length in Yellow-nosed Albatross). **Ad:** Bill black from distance; from close up has yellow ridge on culmen. Upperwings blackish with whitish outer primary shafts; underwings white with narrow black leading edge and fine black trailing edge. Head pale gray (in fresh plumage) to white; back blackish gray; rump white; tail gray. Similar ad. Black-browed Albatross has yellowish-orange bill with reddish tip and wider blackish margin to leading edge of underwing. **Juv:** Similar to ad. but head white; bill all dark; underwing often with wider dark leading edge. Similar juv. Black-browed Albatross has darker underwings and grayish nape and partial collar. **Flight:** Wings often bowed in flight; arcs and glides in heavier winds. **Hab:** Open ocean.

Subspp: (1) •*chlororhynchos.*

Adult, *salvini* NZE/12

Adult, *salvini* NZE/12

Shy Albatross 4
Thalassarche cauta L 37"

Casual vag. to n.e. Pacific from n. CA to WA. **Shape:** Long narrow wings, heavy body, and long, blunt, hooked bill typical of albatrosses. Compared with similar Laysan Albatross, Shy is larger with heavier body and longer thicker bill. **Ad:** Bill pale gray with yellow tip. Upperwings blackish gray with outer primary shafts whitish; underwings white with narrow black leading and trailing edges and distinctive black "spur," or "thumbmark," where leading edge of wing meets body. Head whitish, contrasting with gray back; back is palest toward head and blends into darker wings; rump white; tail gray. Similar Laysan Albatross is smaller with a darker mantle, heavily mottled underwings, smaller pinkish bill, and lacks black spur, or thumbmark. **Juv/Subadult:** Like ad. but bill grayish with dark tip; head all gray. **Flight:** Wings often bowed in flight; arcs and glides in heavier winds. **Hab:** Open ocean.

Subspp: (2) •*cauta* (vag. to WA), "White-capped Albatross," has all-white head. •*salvini* (vag. to CA, AK), "Salvin's Albatross," has gray head with white crown.

Adult NZE/12

Adult NZE/12

Adult ATP/12

Black-browed Albatross 5

Thalassarche melanophris L 35"

Accidental vag. to w. North Atlantic. **Shape:** Long narrow wings, heavy body, and long, blunt, hooked bill typical of albatrosses. Compared with similar Yellow-nosed Albatross, Black-browed is compact and heavy-bodied with a short thick neck and relatively thick bill. **Ad:** Bill yellowish orange with reddish tip. Upperwings and back blackish with outer primary shafts whitish; underwings white with narrow black trailing edge and broad black leading edge. Head white; back blackish gray; rump white; tail gray; eye with short blackish eyebrow. Similar imm. Northern Gannet has pointed bill and tail, smaller head and thinner neck. See Yellow-nosed Albatross for comparison. **Juv:** Similar to ad. but bill grayish with darker tip; underwings can be mostly dark with just a suggestion of lighter pattern of ad. **Subadult:** Similar to ad. but bill pale with blackish tip. **Flight:** Mostly soars on bowed wings; arcs and glides in heavier winds. **Hab:** Open ocean.

Subspp: (2) •*impavida* (NZE), •*melanophris* (Cape Horn–Antipodes Is.).

Adult ATP/12

Light-mantled Albatross 5

Phoebetria palpebrata L 35"

Accidental vag. off CA. **Shape:** Long narrow wings, heavy body, and long, blunt, hooked bill typical of albatrosses. Compared with other albatrosses, Light-mantled has a very long wedge-shaped tail. **Ad:** Distinctively grayish brown overall with a paler gray body and contrasting blackish head with conspicuous white eye crescents. Underwings all gray; upperwings gray with white shafts to primaries. **Juv:** Similar to ad. but browner overall; eye crescents less conspicuous. **Flight:** Distinctive long wedge-shaped tail. **Hab:** Open ocean.

Subspp: Monotypic.

Adult SAT/02

Imm. SAT/02

Imm. SAT/02

Imm. AUK/12

Juv./imm. NZE/–

Wandering Albatross 5

Diomedea exulans L 52"

Accidental vag. off West Coast. **Shape:** Long narrow wings, heavy body, and long, blunt, hooked bill typical of albatrosses. Compared with other albatrosses, Wandering is very large with very long wings and a thick deep-based bill. **Ad:** Large bill pink; head, body, and tail all white. Above, inner wings mostly white, contrasting with blackish outer wing. Below, wings all white except for black tips and thin black trailing edge. **Juv:** Brown body and tail; white face; pinkish bill. Underwings as in adult; upperwings mostly brown; mantle with some white scaling. **Imm:** Gradually acquires ad. white plumage over several years, starting with the belly and back and expanding out onto the wings and tail. **Flight:** Incredible flier with long glides on its long stiff wings. **Hab:** Open ocean.

Subspp: Classification still in flux.

IDENTIFICATION TIPS

Albatrosses are our largest seabirds, usually seen soaring over deep ocean water. They have long narrow wings, heavy bodies, and long hooked bills. Their plumage is mostly shades of brown, black, and white; there is no seasonal change due to molt, but worn birds can be paler; the sexes look alike. Some species take many years to reach adult plumage.

Species ID: Within their general body type, species vary subtly in overall size; length of bill and tail; and thickness of body. Note colors of bill; from above, look for color of tail, rump, and back and any light-colored patterns on wings. From below, note patterns of dark and light, particularly on wings.

Laysan Albatross 2

Phoebastria immutabilis L 32"

Shape: Long narrow wings, heavy body, and long, blunt, hooked bill typical of albatrosses. Compared with similar Shy Albatross, Laysan is smaller. **Ad:** Bill pinkish to yellowish with darker tip. Upperwings blackish with outer primary shafts whitish; underwings with narrow border of black surrounding a variable patchwork of white and black areas. Crown white; face gray; dark area around eye; black on back concolor with wings and extending slightly onto upper rump; rest of rump white; tail black; feet pink. **Juv:** Like ad. **Flight:** Wings often bowed in flight; arcs and glides in heavier winds. **Hab:** Open ocean.

Subspp: Monotypic. **Hybrids:** Black-footed Albatross.

Black-footed Albatross 1

Phoebastria nigripes L 32"

Shape: Long narrow wings, heavy body, and long, blunt, hooked bill typical of albatrosses. Comparatively, Black-footed is small and slender. **Ad:** Bill dark with variable amount of white feathers around base. Otherwise, mostly blackish brown with whitish primary shafts, whitish undertail (can appear mottled), and sometimes whitish uppertail coverts; dark legs and feet. Belly and underwings may become paler with age. **Juv:** Like ad. but white feathering at bill base less prominent; uppertail and undertail coverts blackish. Similar juv./subadult Short-tailed Albatross is much larger and has a proportionately much longer pink bill. **Flight:** Mostly soars on bowed wings; arcs and glides in heavier winds. **Hab:** Open ocean.

Subspp: Monotypic. **Hybrids:** Laysan Albatross.

Adult PAC/–

Imm. –/05

Juv. PAC/07

Imm. PAC/11

Imm. PAC/12

Imm. HI/–

Short-tailed Albatross 3

Phoebastria albatrus L 36"

Rare visitor to waters off AK and occasionally farther south along West Coast. **Shape:** Long narrow wings, heavy body, and long, blunt, hooked bill typical of albatrosses. Short-tailed is large, heavy-bodied, and has a large thick bill. **Ad:** Bill "bubblegum pink" with pale bluish-gray tip. Upperwings have distinctive whitish base where scapulars fan out from white back; dark humerals surrounded by white create black patch on inner wing; underwings white with narrow black border. Golden head and upper breast; back and rump white; tail black. Most NAM sightings are of juv. and subadult. **Juv:** All dark brown with some variable white around eye and on throat; bill pink with bluish tip. **Imm:** Gradually acquires ad. plumage over 10+ years; dark humeral patches and pale scapulars increasingly prominent. Similar juv. Black-footed Albatross is smaller with a proportionately shorter and dark bill. **Flight:** Mostly soars on bowed wings; arcs and glides in heavier winds. **Hab:** Open ocean, but historically, at least, occurred near shore.

Subspp: Monotypic.

Adult light morph, *glacialis* ATL/02

Adult intermediate morph, *glacialis* ATL/06

Adult dark morph, *glacialis* ATL/02

Adult light morph, *rodgersii* PAC/08

Adult dark morph, *rodgersii* PAC/06

Northern Fulmar 1

Fulmarus glacialis L 19"

Shape: Medium-sized, compact, heavy-bodied, broad-necked seabird with relatively short wings and short thick bill. Forehead steep and crown rounded. Shape somewhat similar to gulls but wings proportionately shorter, bill thicker and shorter, neck broader, and head smaller and more rounded. **Ad:** Polymorphic, with light to dark morphs and many intergrades; light morph most common in Atlantic, dark morph more common in Pacific. Bill pale yellow. Light morph upperwings and back gray with whitish patch at base of inner primaries; underwings white with dark margins; body white with dark smudge around eye; tail dark or light gray. Dark morph dark grayish brown overall (in Pacific) or bluish gray overall (in Atlantic); whitish patch at base of inner primaries on paler individuals. Similar Short-tailed Shearwater is slightly smaller with more pointed wings and tail and short, thin, dark bill. Similar Sooty Shearwater has more pointed wings and tail; fairly long, thin, dark bill; and pale silvery underwing coverts. Intermediate morph shows continuous gradation between light and dark morphs. Can be like light morph but with grayish wash to head and hindneck or with grayish body. **Flight:** Stiff rapid wingbeats alternate with glides; wing bent slightly at wrist. **Hab:** Ocean, northern coasts; more often seen from shore in Pacific than Atlantic. **Voice:** Feeding groups on ocean may give grunts and cackles.

Subspp: (2) •*rodgersii* (Pacific Ocean) are smaller with shorter wings, larger bill, show greater extremes in color in light and dark morphs, have dark tails in both morphs. •*glacialis* (w. Atlantic Ocean) are larger with larger wings, shorter bills, are mostly light morphs, have light tails in both morphs.

Adult ATL/–

Adult light morph ATL/–

Adult PAC/11

Adult PAC/11

Herald Petrel 3

Pterodroma arminjoniana L 15½"

Rare but regular visitor to Gulf Stream off southeast coast of U.S., late spring–early fall. **Shape:** Medium-sized, short-necked, long-tailed petrel with long narrow wings and short bill. **Ad:** Polymorphic; dark morph most commonly seen morph in NAM. All morphs have dark brownish upperparts (no whitish base to primaries); blackish underwings with large white patch at base of primaries sometimes extending onto greater coverts of primaries and secondaries; dark tail. Underparts of body reflect morph: dark morph all dark; light morph has dark head and upper breast, white belly; intermediate morph with dark head and neck and paler (but not white) belly. Dark morph similar to Sooty Shearwater, which differs by having a fairly long thin bill and short tail, and holds wings straight (not crooked); Sooty underwing also has paler coverts and dark primaries. **Flight:** Alternates flaps and glides; agile and fast. **Hab:** Deep ocean, mostly Gulf Stream.

Subspp: (1) •*arminjoniana.*

Murphy's Petrel 3

Pterodroma ultima L 16"

Rare offshore visitor, CA–WA. **Shape:** Fairly large, slender, broad-necked petrel with a short bill. **Ad:** Upperparts all dark gray with an indistinct blackish M across the wings; underparts all dark except for whitish throat and whitish flashes at the base of the underside of the primaries. **Flight:** Stiff-winged and agile; arcs and glides over waves; wings bent back at the wrist. **Hab:** Deep ocean.

Subspp: Monotypic.

IDENTIFICATION TIPS

Petrels and Shearwaters are large to medium-sized seabirds usually seen over deep ocean water, with some occurring near shore. They have generally short bodies, long narrow wings, and somewhat pointed tails; their bills are blunt, strongly hooked, and have tubelike nostrils on top. Their typical flight behavior alternates flaps with glides, and they often arc up and down over the waves. Their plumage does not vary seasonally, the sexes look alike, and some species have dark and light color morphs. Petrels (or gadfly petrels) in comparison to shearwaters are generally slightly smaller and stockier and have proportionately shorter thicker bills.

Petrel Species ID: Look for subtle differences in relative length of bill, tail, and wings; note carefully the patterns of dark and light on the underwings and on the head and neck. (See p. 92 for shearwater species ID.)

Adult NZE/12

Adult NZE/12

Mottled Petrel 2

Pterodroma inexpectata L 14"

Shape: Medium-sized, stocky, short-bodied, long-winged petrel. **Ad:** Upperparts gray with darker M across wings; underwings white, crossed from wrist to body by bold black bar (ulnar bar); underparts distinctive, with white throat and white undertail coverts contrasting with gray belly. **Flight:** Alternates flaps and glides; agile. **Hab:** Deep ocean.

Subspp: Monotypic.

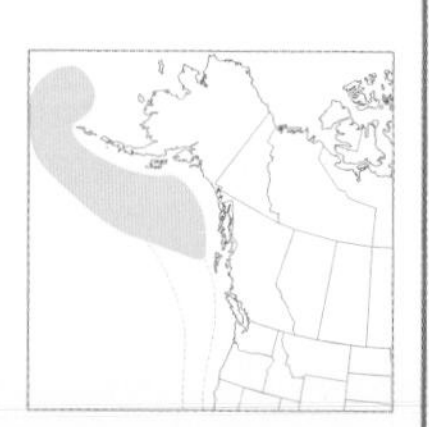

Adult ATL/09

Adult ATL/09

Adult ATL/05

Bermuda Petrel 3

Pterodroma cahow L 15"

Rare but regular visitor to Gulf Stream off Atlantic Coast, late spring–early fall (possibly year-round). **Shape:** Large, slender-bodied, narrow-winged petrel. **Ad:** Upperparts mostly dark brownish gray except for variable whitish U of uppertail coverts, white forehead, and blackish cap; indistinct darker M on upperwings may show in fresh plumage; underwings with dark flight feathers, white greater and median coverts, and dark ulnar bar crossing from wrist to body; throat and belly white. Similar Black-capped Petrel is slightly larger and heavier, often has a white collar across its nape (can be absent), and its white uppertail coverts combine with its white tail base to create a mostly white-tailed appearance. **Flight:** Aerobatic and agile, arcing high (almost to vertical) and gliding over waves. **Hab:** Deep ocean; in NAM only in Gulf Stream (to date).

Subspp: Monotypic.

Black-capped Petrel 2

Pterodroma hasitata L 16"

Shape: Large, heavy-bodied, thick-billed petrel. **Ad:** Upperwings and back dark brownish black with an indistinct darker M across wings (in fresh plumage); uppertail coverts and base of tail white, creating a mostly white-tailed look; tip of tail dark; throat and belly white; white forehead. In lighter variation black cap separated from dark back by broad white collar; underwings with dark flight feathers, white greater and median coverts, and dark ulnar bar crossing from wrist toward body. Darker variation lacks white collar, has more extensive dark on underwing, and less extensive white on uppertail coverts. Similar Great Shearwater is larger and heavier with a big dark smudge on the white belly, a dark forehead, a thin white U on the dark uppertail, and mottled axillaries. Similar Bermuda Petrel is slightly smaller, lacks a white collar across nape, and has a mostly black tail with a limited U of white on the rump. **Flight:** Aerobatic and agile, arcing high (almost to vertical) and gliding over waves. **Hab:** Deep ocean, mostly in Gulf Stream.

Subspp: Classification in a state of flux.

Hawaiian Petrel 4

Pterodroma sandwichensis L 17"

Casual visitor off Pacific Coast. **Shape:** Large, long-winged, fairly long-tailed petrel. **Ad:** Upperparts uniformly blackish; white forehead and throat; blackish partial hood extends down sides of neck; rest of underbody white. Underwings mostly white with outer wing and trailing edge outlined in black; also a thin black line angling back from wrist through secondary coverts. **Flight:** Aerobatic and agile; alternates quick flaps with arcs and glides. **Hab:** Deep ocean.

Subspp: Monotypic.

Adult ATL/05

Adult ATL/06

Adult NZE/02

Adult NZE/12

Fea's Petrel 3

Pterodroma feae L 14"

Rare but regular visitor to Gulf Stream from Cape Hatteras to Canadian maritimes, late spring–late summer. **Shape:** Medium-sized narrow-winged petrel with a short thick bill and pointed tail. **Ad:** Upperparts gray (sometimes with brownish wash), usually with dark M across wings; variable underwings usually mostly dark except for small white triangles near body on leading edge of inner wing. Gray on crown and nape extends down along sides of neck; chin and underbody bright white; tail and uppertail coverts pale gray; back darker gray. **Flight:** Aerobatic and agile; arcs and glides over waves. In strong winds, arcs high over horizon. **Hab:** Deep ocean.

Subspp: Monotypic. Classification still in flux.

Cook's Petrel 3

Pterodroma cookii L 10"

Rare visitor to Pacific Coast, spring–fall; has occurred in Salton Sea, CA. **Shape:** Small slender petrel with a relatively thin bill. **Ad:** Upperparts all gray with bold dark M across wings; head pale gray with darker eye patch; underparts all white with fine blackish margin to wings. Similar Stejneger's Petrel is darker above with less distinct darker M across wings and uniformly darker cap over head and eyes. **Flight:** Agile and erratic, often changing directions; in calm conditions, very buoyant flight. **Hab:** Deep ocean.

Subspp: Monotypic.

IDENTIFICATION TIPS

Shearwater Species ID: Note the colors of the bill and legs; look carefully at the patterns of dark and light on underwings, head and neck, and rump. Look for subtle differences in relative length and shape of bill, tail, and wings. Study the flight pattern: stiff vs. fluid wingbeats, more flapping than gliding.

Adult PAC/09

Adult PAC/09

Adult PAC/09

Streaked Shearwater 4

Calonectris leucomelas L 18"

Casual Asian vag. to CA coastal waters. **Shape:** Large, comparatively long-billed, long-tailed shearwater with long broad wings. **Ad:** Bill gray with a dark tip. Upperparts brownish with scaled appearance created by pale margins to feathers of back, rump, and wing coverts; underwing coverts mostly white, contrasting with dark primaries and secondaries. Pale head and nape variably streaked with brown; forehead, throat, and often foreface usually whitish; rest of underbody white. Similar Pink-footed Shearwater has pink bill with dark tip, uniformly dark back, dark undertail coverts, and a darker face and head. **Flight:** Long slow wingbeats; high arcing. **Hab:** Ocean. **Voice:** While feeding on ocean, may give cackling call.

Subspp: Monotypic.

Adult, *borealis* ATL/05

Adult, *borealis* ATL/08

Adult, *borealis* ATL/08

Cory's Shearwater 1

Calonectris diomedea L 18"

Shape: Large heavily built shearwater with relatively long broad wings and thick heavy bill. **Ad:** Bill yellow to ivory with dark tip (actually subterminal band). Upperparts warm brownish gray overall with variable whitish to buff crescentlike area on uppertail coverts and subtle darker M across wings when in fresh plumage; underwings mostly white with dark primaries and secondaries and thin line of dark marginal coverts along leading edge. Body whitish below with dusky sides to face, neck, and breast. Similar Great Shearwater has dark bill, dark back and upperwings separated from blackish cap by a white collar, and a narrow white U formed by the tips of the uppertail coverts. **Flight:** Wings distinctly bowed and bent back at wrist; long glides interrupted by short series of slow deep wingbeats. **Hab:** Ocean, occasionally seen from shore. **Voice:** Feeding groups on ocean may give bleating sound.

Subspp: (2) •*borealis* (regular e. Atlantic visitor to NAM) slightly larger, larger-billed, underwing with variable amount of white in all primaries except P10, which is all dark. •*diomedea* (Mediterranean visitor to w. Atl.), "Scopoli's Shearwater," slightly smaller, smaller-billed, underwing with white primary coverts blending with extensive white on undersides of primaries. ▶ Classification in flux; some consider these two different species.

Adult ATL/04

Adult PAC/10

Adult ATL/08

Adult PAC/09

Cape Verde Shearwater 5

Calonectris edwardsii L 17"

Accidental vag. that may occur with Cory's Shearwater. **Shape:** Like a slightly smaller slimmer version of Cory's Shearwater. In flight, narrower wings, proportionately longer tail than Cory's. Afloat, noticeably smaller with flatter crown, slimmer bill, thinner neck. **Ad:** Bill grayish with blackish tip. Upperparts dark grayish brown overall with white crescent-like area on uppertail coverts and subtle darker M across wings when in fresh plumage; underparts mostly white except for darker primaries and secondaries and thin line of dark marginal coverts along leading edge. Crown may be darker than back (creating a suggestion of a cap) and contrast strongly with white chin and throat. Similar Cory's has more concolor warm brownish upperparts that shade more to white chin and throat; also a paler bill. **Flight:** Wings distinctly bowed and bent back at wrist; long glides interrupted by short series of slow deep wingbeats. **Hab:** Ocean.

Subspp: Monotypic; formerly considered a subspecies of Cory's Shearwater.

Pink-footed Shearwater 1

Puffinus creatopus L 18"

Shape: Large heavy-bodied shearwater with long narrow wings and proportionately long outer wing. **Ad:** Bill pink with dark tip; feet pink. Upperparts uniformly grayish brown with blackish tail; underparts with whitish belly, dusky hood and undertail coverts, and dark tail; underwings centrally whitish with variable black around margins. Pale variation (typical) has whitish throat and more white on underwings; dark variation (rarer) has dark throat and white limited on underwings to mostly greater coverts. May show some white on upper secondary coverts when molting (as do most seabirds); scapulars may have pale margins, creating subtle scaled look. Similar Flesh-footed Shearwater is chocolate-brown overall. **Flight:** Long slow wingbeats; arcs and glides. Often angles wings back at wrist. **Hab:** Ocean; occasionally seen from shore. **Voice:** Feeding birds rarely give short nasal call.

Subspp: Monotypic.

Adult NZE/10

Adult PAC/10

Adult NZE/–

Adult dark morph AUS/01

Adult dark morph NZE/10

Adult light morph HI/11

Flesh-footed Shearwater 3

Puffinus carneipes L 17"

Shape: Medium-sized heavy-bodied shearwater with long narrow wings and proportionately long outer wing. Similar size and shape as Pink-footed Shearwater. **Ad:** Bill pinkish with dark tip; legs flesh-colored. Dark sooty brown overall with no lighter markings; underwings all dark (primaries may look silvery below in certain lights). Similar Pink-footed Shearwater usually has pale throat and belly (some Pink-footed are dark on throat); similar Sooty Shearwater has thinner all-dark bill and silvery underwing coverts; similar dark-morph Northern Fulmar has a short, thick, pale bill, a very stocky body, and underwings with contrastingly paler primaries. **Flight:** Long slow wingbeats; arcs and glides. Often angles wings back at wrist. **Hab:** Ocean.

Subspp: Monotypic.

Wedge-tailed Shearwater 4

Puffinus pacificus L 18"

Casual vag. to offshore CA; one inland record from Salton Sea, CA. **Shape:** Large, slender, long-tailed shearwater with a long thin bill. Tail projection almost twice that of head; tail tip wedge-shaped. **Ad:** Bill grayish with dark tip; legs and feet pinkish. Two morphs. Dark morph sooty brown overall with scaled look on scapulars and upperwing coverts, and slightly paler belly and underwing coverts. Light morph similar to above but underparts mostly whitish; body whitish with dusky sides of head, face, and dusky undertail coverts; underwings dark surrounding whitish median and greater coverts. Other similar dark shearwaters lack the Wedge-tailed's distinctive long thin bill and long wedge-shaped tail. **Flight:** Often flies low with wings angled back at the wrist; graceful. **Hab:** Ocean.

Subspp: Monotypic.

Adult ATL/08

Adult ATL/08

Adult ATL/08

Adult ATL/08

Great Shearwater 1

Puffinus gravis L 18"

Shape: Large shearwater with fairly long narrow wings and a fairly thin bill. **Ad:** Bill all dark. Sharply defined dark cap with white collar behind and white cheek below. Rest of upperparts dark grayish brown with scaled look on coverts, scapulars, and back, a thin U of white formed by white tips to uppertail coverts, and dark primaries; secondaries and their greater secondary coverts can look silvery when fresh. Underparts generally pale; underwings pale with mottled axillaries; body white with distinctive grayish smudge on lower belly (often hard to see) and grayish half collar near leading edge of wing. Similar Black-capped Petrel is smaller, has more extensive white near the base of the tail, cleaner white underparts (belly, axillaries), and tends to fly with wings bent at the wrist (most shearwaters fly with wings held straight out). **Flight:** Rapid wingbeats on stiff wings; arcs and glides. **Hab:** Ocean; rarely seen from shore. **Voice:** Feeding groups on ocean may give long whining call.

Subspp: Monotypic.

Adult PAC/08

Adult PAC/08

Buller's Shearwater 2

Puffinus bulleri L 16"

Shape: Medium-sized shearwater whose long body and long wings create elegant proportions; head and long neck about equal in length to long wedge-shaped tail; bill thin. **Ad:** Bill gray. Upperparts pale grayish with a distinct dark M across the wings and back (at a distance, pattern seen as contrasting dark and pale areas); underwings almost all white with only a narrow margin of black. Dark cap contrasts strongly with white cheek; underparts gleaming white except for black tail. **Flight:** Graceful, easy, measured flier with snappy wingbeats. **Hab:** Ocean.

Subspp: Monotypic.

Adult ATL/08

Adult ATL/08

Adult ATL/08

Sooty Shearwater 1

Puffinus griseus L 17"

Shape: Medium-sized, slender, narrow-winged shearwater with a long thin bill (thinnest in middle) and rather flattened crown. Afloat, has relatively long neck; bill is about 2/3 length of head; wing tips generally just reach tail tip. Similar Short-tailed Shearwater has higher more rounded crown, shorter evenly thin bill. **Ad:** Bill black; upperparts uniformly blackish brown; underparts with dark body (belly sometimes paler), dark flight feathers, and whitish underwing coverts, often brightest in the center of the outer wing. See Short-tailed Shearwater for complete comparison between these two similar species. **Flight:** Wingbeats (3–7) rapid with stiff wings, followed by glides of 3–5 seconds in a smoothly repeated pattern; high arcing in strong winds. **Hab:** Ocean; regularly seen from shore in Pacific, often in huge flocks; less commonly seen from shore in Atlantic.

Subspp: Monotypic.

Short-tailed Shearwater 2

Puffinus tenuirostris L 16½"

Shape: Medium-sized narrow-winged shearwater with a short thin bill and well-rounded crown. In flight, feet generally extend well beyond short tail. Afloat, appears to have short, thick neck; bill is about ½ length of head; wing tips usually extend well past tail tip. Very similar Sooty Shearwater has longer bill, longer, sloped crown, feet generally not projecting past tail, longer neck when afloat. **Ad:** Bill dark gray with slightly darker tip. Upperparts uniformly blackish brown; underwings either largely dusky or with variable whitish areas (some birds very similar to Sooty), often lightest on the median (and lesser) coverts of the inner wing. Underbody usually mostly dark except for paler chin; sometimes throat is pale and head appears darker and subtly helmeted. Similar Sooty usually has pale underwing and is palest on the median and greater coverts of the outer wing (primaries); Sooty generally much more common south of AK, except in late fall and winter, when both species can be expected equally; also see Shape comparison. **Flight:** Wingbeats very rapid with stiff wings and short glides, often with erratic changes in direction; arcing in strong winds. **Hab:** Ocean; occasionally seen from shore (regularly so in AK).

Subspp: Monotypic.

Manx Shearwater 2

Puffinus puffinus L 13½"

Shape: Small compact shearwater with relatively long extension of head and neck beyond wings. Medium-length thin bill; steep forehead and flattish crown. **Ad:** Bill black. Upperparts all blackish (may look brownish when worn), contrasting strongly with mostly white underparts (including undertail coverts). White on throat extends up behind auriculars in a white wedge (ear-surround) that separates head from grayish collar. Underwing mostly white surrounded by dark flight feathers and thin line of dark marginal coverts. Small white flank patch at trailing edge of wing often extends in a spur on sides of rump and is easily seen from above (as with Tree Swallow). Similar Audubon's Shearwater (in Atlantic) is smaller with relatively shorter wings, has dark undertail coverts, and has more extensive white on face; similar Black-vented Shearwater (in n. Pacific) is slightly larger, has black undertail coverts, has a dusky head and throat, and has black tips to axillaries that create a dark bar projecting into whitish inner wing. **Flight:** Wingbeats rapid on stiff wings with short glides; high arcing in strong winds. **Hab:** Ocean; occasionally seen from shore.

Subspp: Monotypic.

Black-vented Shearwater 2

Puffinus opisthomelas L 14"

Shape: Small, rather evenly proportioned, slim shearwater with a medium-length thin bill, sloping forehead, and rounded crown. **Ad:** Bill dark. Upperparts blackish brown with slightly paler head and nape. Underparts whitish and brightest on belly; pale to dusky throat, dark undertail coverts; underwing coverts white surrounded by dark flight feathers and dark marginal coverts; black tips to axillaries create dark bar projecting into whitish inner wing. Similar Manx Shearwater has more contrastingly white undertail coverts, a white ear-surround behind auriculars, and a more contrasting white throat. **Flight:** Wingbeats rapid on stiff wings with only short glides; high arcing in strong winds. **Hab:** Ocean near shore; regularly seen from shore.

Subspp: Monotypic.

Audubon's Shearwater 1

Puffinus lherminieri L 12"

Shape: Small slim shearwater with relatively short rounded wings, long bluntly wedge-shaped tail, and thin medium-length bill. **Ad:** Bill dark. Upperparts uniformly blackish brown. Underparts bright white with dark undertail coverts (occasionally mottled with whitish); underwing coverts white surrounded by dark flight feathers and dark marginal coverts; white throat and cheek fairly well defined from dark cap and nape. Feet and legs pink. Similar Manx Shearwater has longer bill, relatively longer more pointed wings and shorter tail, all-white undertail coverts, and darker face. **Flight:** Fluttery wingbeats rapid on stiff wings with only short glides; often stays low to water. **Hab:** Ocean.

Subspp: Monotypic.

Adult ATL/03

Adult ATL/08

Wilson's Storm-Petrel 1

Oceanites oceanicus L 7¼"

Shape: Small, short-winged, long-legged storm-petrel with a fairly square tail. Feet project beyond tail in flight. **Ad:** Upperparts sooty black with pale greater secondary coverts forming a diagonal bar across inner wing; uppertail coverts white, forming a broad white patch that wraps around onto undertail coverts and rear flanks. Underparts sooty black with extensive white on undertail coverts and rear flanks. Similar Leach's Storm-Petrel has notched tail, longer wings, feet that do not project beyond tail in flight, smudgy white uppertail coverts (often split), and dark undertail coverts and rear flanks. **Flight:** Alternates stiff fluttery wingbeats with short glides on straight wings. When feeding, hovers over water and patters feet on surface. **Hab:** Ocean; regularly seen from shore in Atlantic.

Subspp: (1) •*oceanicus*.

Adult ATL/–

Adult NC/08

White-faced Storm-Petrel 3

Pelagodroma marina L 7½"

Rare visitor to Atlantic Coast. **Shape:** Medium-sized long-legged storm-petrel with short, broad, rounded wings and a slightly rounded tail. Feet project well beyond tail in flight. **Ad:** Grayish crown, back, and wings (back and coverts may be contrastingly brownish); broad whitish rump; blackish tail; pale greater secondary coverts form short bar across inner upperwing. Face white with blackish bar from eye to auriculars; underbody white except for gray half collar that frames distinctive face pattern; underwings with dark flight feathers and white coverts. **Flight:** Rarely flaps; shallow stiff wingbeats alternate with very long glides. When feeding, seems to bounce about, kangaroo style, briefly touching its feet to the water. **Hab:** Open ocean.

Subspp: (2) •*eadesi* (Cape Verde Is., vag. to N. Atlantic) has whiter forehead and hindneck. •*hypoleuca* (Salvage Is., potential vag. to N. Atlantic) has darker forehead and hindneck.

Adult NC/05

Adult NC/05

Adult NC/05

European Storm-Petrel 4

Hydrobates pelagicus L 6½"

Casual vag. off Atlantic Coast from NS to NC. **Shape:** Small storm-petrel with a small head and short bill; wings are long and narrow; short legs do not project beyond tail in flight. **Ad:** Dark overall with bold white rump patch that extends slightly onto sides of rump. Upperwings all dark; underwings dark with distinctive white bar created by white greater coverts (best seen as bird hovers during feeding). **Flight:** In direct flight, wingbeats in bursts and wings kept mostly below plane of body. **Hab:** Open ocean.

Subspp: Monotypic.

Adult AUK/12

Adult INO/01

Black-bellied Storm-Petrel 5

Fregetta tropica L 8"

Accidental vag. off mid-Atlantic Coast. **Shape:** Medium-sized storm-petrel with long legs that project beyond tail in flight. **Ad:** White flanks separated by distinctive black line along belly connecting black head and chest with black undertail coverts and tail. Underwings with dark flight feathers and whitish greater and median coverts; upperwings mostly dark but with paler greater coverts creating a line on the inner wing. Rump white. **Flight:** Distinctive behavior of splashing into water with breast and pushing off with legs. **Hab:** Open ocean.

Subspp: Presumably •*tropica*.

IDENTIFICATION TIPS

Storm-Petrels are very small swallowlike seabirds that pick food off the ocean surface. Most have small bodies, square, forked, or pointed tails, short thin bills, and long thin legs. They have one plumage for all ages, and the sexes are alike.

Species ID: Note the shape of the tail tip, whether legs project beyond the tail, and the relative width and length of the wings. From above, look for the shape of any white patches on the rump, and any pattern of paleness on the wings. From below, look for any white on the underwings and belly and any pattern on the face. Each species also has a distinctive flight style.

Adult AK/09

Adult PAC/04

Fork-tailed Storm-Petrel 2

Oceanodroma furcata L 8½"

Shape: Relatively large, long-winged, short-legged storm-petrel with a forked tail. **Ad:** Pale gray overall (sometimes with darker belly, see Subspp.) with dark eye patch and dark bill. Variable darker gray M across upperwings; underwings with dark slate-gray coverts contrasting with paler flight feathers. **Flight:** Rapid shallow beats on wings often bent back at wrist; may patter feet over water surface. **Hab:** Ocean; rarely seen from shore.

Subspp: (2) Differences subtle. •*furcata* (Aleutian Is.–s.w. AK) is larger, paler overall; dark breast contrasts with whitish throat. •*plumbea* (s.e. AK–n.w. CA) is slightly smaller, darker overall; belly concolor with rest of underparts.

Adult ATL/04

Adult ENG/09

Leach's Storm-Petrel 1

Oceanodroma leucorhoa L 8"

Shape: Medium-sized, long-winged, short-legged storm-petrel with notched tail. Feet do not project beyond tail in flight. **Ad:** Upperparts blackish brown except for pale greater secondary coverts forming a long diagonal bar across inner wing; rump variably whitish to grayish (smudgy) to all brown; when whitish, somewhat triangular and typically divided by a central dark line. Underparts uniformly blackish brown. Similar Wilson's has short square tail, feet project beyond tail in flight, uppertail coverts extensively white, undertail coverts and rear flanks partly white. See also Flight. **Flight:** Wingbeats deep, deliberate (similar Wilson's wingbeats fluttery); flight path erratic, nighthawk-style (similar Ashy's flight direct, wingbeats deliberate, rarely rise above body plane; similar Black's flight direct, wingbeats deep, deliberate, rise well above body plane; similar Least very small, with direct flight, deep wingbeats). May arc and glide over waves. **Hab:** Open ocean.

Subspp: (4) Vary in size and proportion of white-rumped individuals. •*leucorhoa* (AK–cent. CA and N. Atlantic is.) largest of subspp., mostly white-rumped individuals. •*chapmani* (cent.–n. Baja Calif., MEX) with deeply notched tail, mostly dark-rumped individuals. •*socorroensis* (Isla Guadalupe, MEX) smallest, mostly dark-rumped individuals. •*cheimomnestes* (Isla Guadalupe, MEX) slightly larger than *socorroensis*, with more white-rumped individuals.

Ashy Storm-Petrel 2

Oceanodroma homochroa L 8"

Shape: Medium-sized, long-winged, short-legged storm-petrel with a relatively long forked tail. **Ad:** Dark overall with an ashy or brownish wash. Upperparts concolor except for pale greater secondary coverts forming a long diagonal bar across inner wing. Underparts concolor except for paler underwing coverts. **Flight:** Direct flight; wingbeats moderately fast, shallow, usually at or below the body plane, minimal gliding; flight path direct, often low. Similar Black Storm-Petrel typically raises wings above the body plane; similar Least Storm-Petrel is very small, has wedge-shaped tail and deep wingbeats. **Hab:** Ocean.

Subspp: Monotypic.

Band-rumped Storm-Petrel 2

Oceanodroma castro L 9"

Shape: Medium-sized, relatively long-winged, short-legged storm-petrel with a square to shallow-forked tail; feet do not project past tail in flight. **Ad:** Blackish brown overall. Upperparts blackish brown except for pale greater secondary coverts forming a long dull bar across inner wing and white U-shaped rump, which extends slightly onto undertail coverts. Underparts uniformly blackish brown. Similar Wilson's has proportionately shorter wings, longer legs that project past tail, and fluttery wingbeats; even more similar Leach's has proportionately longer, more deeply forked tail, usually divided rump patch, deep wingbeats, and erratic flight path. **Flight:** Wingbeats shallow, steady, with faster downstroke; flight path direct; glides on slightly bowed wings. Arcs and glides over waves like shearwater. **Hab:** Open ocean.

Subspp: Classification is still in flux. Based on vocalizations and molt schedule, there seem to be four distinct groups of Band-rumped Storm-Petrels (Grant's, Madeiran, Monteiro's, and Cape Verde), whether species or subspecies is still to be determined. The groups most likely seen offshore of NAM are Grant's and Cape Verde.

Adult –/12

Adult –/12

Adult GAL/07

Wedge-rumped Storm-Petrel 4

Oceanodroma tethys L 6½"

Casual vag. off CA. **Shape:** Small storm-petrel with long, narrow, pointed wings and short legs that do not project past tail in flight. Tail with only shallow notch. **Ad:** All black except for extensive white uppertail coverts and some white on undertail coverts. Uppertail coverts extend in a V nearly to tip of tail, creating an almost white-tailed look to the bird from certain angles. Undertail coverts less prominent. Upperwings have pale greater coverts to inner wing; underwings fairly uniformly dark. **Flight:** In direct flight, flies higher above water and with deeper wingbeats than many other storm-petrels. **Hab:** Open ocean.

Subspp: (2) •*tethys* (Galapagos Is.) larger; •*kelsalli* (off Peru) smaller. One record in NAM is believed to be *kelsalli*.

Adult PAC/03

Adult PAC/07

Black Storm-Petrel 2

Oceanodroma melania L 9"

Shape: Large, slim, long-winged storm-petrel with a long forked tail. **Ad:** Blackish overall except for pale greater secondary coverts forming a long diagonal bar across inner wing. Similar dark-rumped Leach's has bouncing flight path; similar Least is much smaller, with a wedge-shaped tail; similar Ashy slightly paler overall, with wingbeats generally at or below the body plane. **Flight:** Wingbeats deep, deliberate, spanning well above and below the body plane; flight path fairly direct. Arcs and glides over waves like shearwater. **Hab:** Ocean; rarely to occasionally seen from shore.

Subspp: Monotypic.

Adult PAC/08

Adult PAC/08

Least Storm-Petrel 3

Oceanodroma microsoma L 5¼"

Shape: Small, short-legged, proportionately rather average storm-petrel. Tail short with wedge-shaped tip. **Ad:** Blackish brown overall except for pale greater secondary coverts forming a diagonal bar across inner upperwing. Similar Black and Ashy Storm-Petrels much larger, with forked tails. **Flight:** Wingbeats surprisingly deep, deliberate for small bird; flight path direct. **Hab:** Ocean.

Subspp: Monotypic.

Adult, *catesbyi* NC/08

Juv., *catesbyi* ATL/07

White-tailed Tropicbird 3

Phaethon lepturus L 15" (not including streamers)

Shape: Slender-bodied tropicbird with long narrow wings and long tail streamers (as long as body on ad.), which can be missing. Moderate-sized bill; length clearly less than length of head (as long as head in Red-billed). **Ad:** Upperwings with black subterminal patch on outer 4–5 primaries (see Subspp.), cut off sharply at white primary coverts; also diagonal black stripe across inner wing. Underwings similar; translucent white tips to outer primaries (more obvious on Pacific subspp.). Bill orange, reddish orange, greenish yellow, or yellow depending on subsp. and breeding status (more reddish during breeding); thin black eyeline extends to auriculars; tail streamers white (can have yellowish or pinkish hue) with fine black shafts. **Juv:** Outer wing pattern like ad. but less distinct; upperparts barred with black; no tail streamers; eyelines do not connect on nape (do in juv. Red-billed); bill dull yellow. **Imm:** (2nd yr.) Like juv. but less barring on upperparts, short tail streamers, yellow bill. **Flight:** Faster wingbeat and more buoyant flight than other tropicbirds. **Hab:** Pelagic; nests on island cliffs.

Subspp: (2) •*catesbyi* (Atlantic, Gulf of Mexico) larger; outer primaries black with narrow white tips, bill orangish to orangish yellow. •*dorotheae* (Pacific) smaller; outer primaries black with broad white tips, bill greenish yellow.

Adult GAL/10

Adult GAL/10

Juv. ATL/05

Red-billed Tropicbird 3

Phaethon aethereus L 18" (not including streamers)

Shape: Heavy-bodied tropicbird with long narrow wings and long tail streamers (as long as or longer than body on ad.), which can be missing. Large heavy bill; length about as long as head (clearly shorter than head on White-tailed). **Ad:** Upperwing with black outer 5–8 primaries; black extends onto the primary coverts, ending in a rough point near wrist. Underwings mostly whitish with little or no translucent white tips on outermost primaries. Back, rump, and inner secondary coverts finely barred with black (appear gray from a distance). Bill red; wide black eyeline extends to nape; tail streamers white. **Juv:** Outer wing pattern like ad.; back more heavily barred; no tail streamers; tail with dark tip; dark eyelines meet at back of nape and encircle dark-spotted whitish crown; bill dull yellow to orange. **Imm:** (2nd yr.) Crown with little or no spotting; short white tail streamers. **Flight:** Stiff-winged and less buoyant than White-tailed. **Hab:** Pelagic; nests on island cliffs.

Subspp: (1) •*mesonauta.*

Adult HI/03

Adult CA/09

Red-tailed Tropicbird 4

Phaethon rubricauda L 18" (not including streamers)

Casual to offshore CA. **Shape:** Heavy-bodied tropicbird with long relatively broad wings and long tail streamers (as long as body on ad.), which can be missing. Very large bill tapers to fine point; length as long as head. **Ad:** Wings mostly white except for black centers on tertials. Body white overall, except for short black eyeline. Bill bright red; tail streamers whitish to pinkish. **Juv:** Upperparts like ad., but with thin black barring on back and inner secondary coverts and black spotting on crown and nape; no tail streamers; bill dull, changing to yellowish. **Imm:** (2nd yr.) Like juv. but lacks spotting on head and nape; bill turns orange to orangish red; tail streamers short and pinkish. **Flight:** Appears nearly all white; heavyset with broad wings and slow powerful wingbeats. **Hab:** Pelagic; nests on island cliffs.

Subspp: Monotypic.

IDENTIFICATION TIPS

Tropicbirds are medium-sized seabirds with long thin tail streamers on the adults; they catch their food by plunge-diving and readily rest on the water. There are only three species; they do not vary seasonally and the sexes are alike.

Species ID: Look for the pattern of black and white on the outer primaries and their coverts – this is distinctive for each of the three species in all ages. Also note the presence of black on back of adult birds. Note size, length, and color of tail streamers, also bill color and bill length in relation to head length.

Adult, *personata* LHI/08

Imm., *melanops** OMA/–

Adult, *dactylatra* FL/04

Juv., *personata* CLI/06

Masked Booby 3

Sula dactylatra L 32"

Shape: Pointed sharply on all extremities – head, tail, wings. Long slender body projects equally in front of and behind long relatively narrow wings. Thick-based long conical bill; fairly straight culmen continuous with short forehead and shallow crown. **Ad:** Bill yellowish or greenish yellow with black mask at base; body white; tail black. Upperwings black except for white median and lesser secondary coverts; underwings similar, but all coverts white. Feet gray, olive, or orangish yellow (see Subspp.). Similar white-morph ad. Red-footed Booby has white tail and tertials; similar ad. Northern Gannet has white secondaries, white tail, yellowish head, and grayish bill; similar advanced imm. Northern Gannet can have dark secondaries and look much like ad. Masked Booby but has whitish scapulars and substantial white in tail; similar rare ad. Nazca Booby (*S. granti*) has orange bill and often white central tail feathers (see Juv./Imm. also). **Juv/Imm:** Bill bluish gray with yellowish tip, gradually turning all yellow. Upperparts in juv. all brown except for broad, distinct, white collar; imm. gradually develops white from rump to central back; underparts white except for dark hood, tail, and flight feathers. Ad. plumage acquired in 3 yrs. Similar juv. Northern Gannet lacks white collar, has darkish underwing with brightest white on axillaries; similar rare juv./imm. Nazca Booby usually lacks the white collar or has a narrow less well-defined collar, often has a darker head than back, and base of bill gradually turns to orange. **Flight:** Often flies low to water; alternates flaps with glides; may join line formations; steep vertical dives after fish. **Hab:** Pelagic; usually seen far offshore; nests on islands.

Subspp: (2) •*dactylatra* (Atlantic, Gulf of Mexico) slightly smaller, bill smaller and yellow, legs and feet orangish to olive. •*personata* (Pacific) larger, bill larger and bright yellow, legs and feet bluish gray to greenish. **S. d. melanops* (w. Indian Ocean) used for good photograph.

Adult GAL/07

Juv. MEX/09

Adult GAL/07

Juv. GAL/07

Adult MEX/–

Blue-footed Booby 4

Sula nebouxii L 32"

Shape: Similar to Masked Booby (which see, under Shape) but slightly more slender and with slightly narrower wings. **Ad:** Bill grayish; feet blue to bluish gray. Head and neck streaked pale brown; belly white; tail whitish centrally with brown sides. Upperwings all dark brown; back dark brown with white barring and white rump patch (also small white patch on upper back); underwings mostly dark brown with white secondary coverts divided by a dark median covert bar. Similar juv./imm. Masked Booby has dark head with defined white collar, dark tail, and more white on underwing coverts; similar imm. and ad. Brown Booby has upperparts (including tail) all dark. **Juv/Imm:** Gray bill; dull blue feet; body dark brown overall except for contrasting whitish belly and whitish axillaries; wings and tail similar to ad. patterns. Ad. plumage acquired in 2 yrs. **Flight:** Often flies high over water; alternates flaps with glides. **Hab:** Marine; often seen from shore; nests on islands; U.S. records mostly from Salton Sea, CA.

Subspp: Monotypic.

Adult HI/03

Imm., *leucogaster* FL/05

Adult, *brewsteri* CLI/05

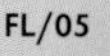

Brown Booby 3

Sula leucogaster L 30"

Shape: Similar to Masked Booby (which see, under Shape) but slightly smaller. **Ad:** Dark brown to grayish head and neck (see Subspp.) sharply demarcated from white belly and undertail coverts. Wings, back, and tail all dark brown; underwings mostly dark brown contrasting with white coverts of inner wing. F. bill pale yellow to pinkish with yellow facial skin and dark smudge before eye; m. bill pale yellow to greenish yellow with bluish facial skin; legs of both sexes pale yellowish; colors brightest during early breeding. **Juv/Imm:** Overall pattern like ad. but with whitish areas variably mottled with brown; some individuals can be almost wholly brown. Ad. plumage acquired in 2 yrs. Similar juv./imm. Red-footed Booby has all-dark underwings and a darker trailing edge to inner upperwings; it almost always shows dark band across breast, and imm. head is paler than back. **Flight:** Alternates flaps with glides. **Hab:** Pelagic; nests on islands.

Subspp: (3) •*leucogaster* (seen NB–TX) comparatively larger, yellow bill with bluish tip, brown head and nape slightly darker than back. •*brewsteri* (seen coastal s. CA, vag. to NV–AZ) comparatively smaller, yellow to white bill, head and nape paler than back, m. has whitish head. •*plotus* (possibly seen off CA) comparatively larger, yellow to greenish bill, upperparts concolor.

IDENTIFICATION TIPS

Boobies and the closely related Northern Gannet are large seabirds that are pointed at each extremity – they have long pointed bills with deep bases; long, narrow, pointed wings; and long pointed tails. They perform spectacular plunge-dives into the ocean to catch fish. They take several years to gradually attain adult plumage; they do not vary seasonally and the sexes are alike (except in Brown Booby).

Species ID: All members of this family are roughly the same shape, but note overall size, which varies. Note the color of the bill and head. From above, look for patterns of white and dark on the tail, rump, back, and secondaries. From below, look for patterns of light and dark on the inner wing.

Adult white morph HI/03

Adult brown morph GAL/05

Adult white morph HI/03

Imm. white morph HI/10

Adult white-tailed brown morph BON/–

Imm. brown morph GAL/05

Red-footed Booby 4

Sula sula L 28"

Casual vag. to offshore s. Atlantic and Gulf Coasts; also offshore cent.–s. CA. **Shape:** Similar to other boobies, with pointed head, tail, and wings. Differs in being significantly smaller and slimmer with proportionately longer tail; slightly convex culmen abuts a bulging forehead; somewhat high rounded crown. **Ad:** Several color morphs with many intermediates; in light of this, the 3 basic morphs listed below should be viewed as at the 2 extremes and the middle of a complex set of variations. In adults of all morphs, look for these constants: distinctive head and bill shape (see Shape); bluish-gray bill with pink facial skin; deep red or orangish-red legs and feet (leg and facial colors brightest just before breeding, then fade). In white morph, body mostly to all white; tail white, gray, or brown; primaries and secondaries black; tertials white (similar Masked Booby has black tertials, similar Northern Gannet has white secondaries); secondary coverts white; primary coverts on upperwing black, on underwing white with a dark central patch formed by the median coverts. White-tailed brown morph all brown except for white tail and variably white rump and rear belly. Brown morph all dark brown except for variably paler head and underbody. Similar juv. Brown Booby with dark head and underbody. **Juv/Imm:** Juv. mostly dark to lighter brown overall and usually shows a darker breastband in any morph; underwings all dark; legs and feet dull orangish yellow; bill all dark; head and bill shape distinctive (see Shape). Imm. more similar to ad. colors of respective morphs (but some white morphs may have brown back and wings); may have breastband to varying degree; bill pink with dark tip. **Flight:** Low over water; alternates flaps with glides. **Hab:** Pelagic; nests on islands.

Subspp: Monotypic. There are no consistent differences in tail color between Pacific and Atlantic populations.

Adult NC/03

2nd-yr. imm. QC/06

3rd-yr. imm. SCO/07

Juv. VA/11

Northern Gannet 1

Morus bassanus L 37"

Shape: Very large heavy-bodied seabird, sharply pointed on head, tail, wings. Long slender body projects equally in front of and behind long relatively narrow wings. Deep-based conical bill has relatively straight culmen abutting slightly steeper forehead and relatively high rounded crown; gape extends as a black line under cheek. **Ad:** Body all white except for golden wash over head. Above, white inner wing contrasts with all-black outer wing; below similar, but primary coverts white. Bill silvery; facial skin black; eye blue; feet blackish. **Juv/Imm:** At first, grayish brown spotted with white overall, except for white axillaries and uppertail coverts. Gradually acquires ad. plumage over 4–5 yrs., becoming white first on head and belly, then on back, wings, and tail (in that order). In late transitional stages, imm. bird may look like adult but have dark secondaries and thus look similar to ad. Masked Booby or ad. white-morph Red-footed Booby; but imm. Northern Gannet usually has both black and white on tail, while Masked Booby has black tail and white-morph Red-footed Booby has all-white tail. **Flight:** Flies at variety of heights; alternates flaps with glides; flight path often undulates over ocean waves; steeply dives for fish (American White Pelican feeds while on water); often in large flocks; may arc up and down in strong winds. **Hab:** Seacoasts and ocean waters; nests on coastal cliffs.

Subspp: Monotypic.

Adult prebreeding FL/03

Juv. ID/07

Adult postbreeding WY/07

Adult winter FL/01

Adult FL/02

American White Pelican 1

Pelecanus erythrorhynchos L 62"

Shape: Very large, heavy-bodied, long-winged bird with very long bill and large pouch typical of pelicans. Proportionately shorter bill and tail than Brown Pelican; bills of the 2 species about same length, but body of American White Pelican twice as heavy. **Ad. Summer:** Standing or afloat, appears all white with bright orange bill and legs. In flight, all white with black primaries and outer secondaries; wing coverts white except on upperwing, where primary coverts are black. Starting before migration north, breeding birds may get yellowish feathers on breast, crown, and upperwing coverts and grow a rounded keel on their culmen. By the time they are feeding young, crest feathers are replaced by shorter dusky feathers and keel is shed. **Ad. Winter:** Similar to postbreeding ad. summer but bill, pouch, and legs more dull yellowish than bright orange. **Juv:** (Jul.–Mar.) Like ad. winter but with dusky wings, neck, and crown; change to essentially ad. plumage by 1st spring. **Flight:** Graceful fliers, often forming long shifting line formations; groups often circle, rising high on thermals. **Hab:** Large inland lakes in summer; also along coasts in winter; nests on lake islands.

Subspp: Monotypic.

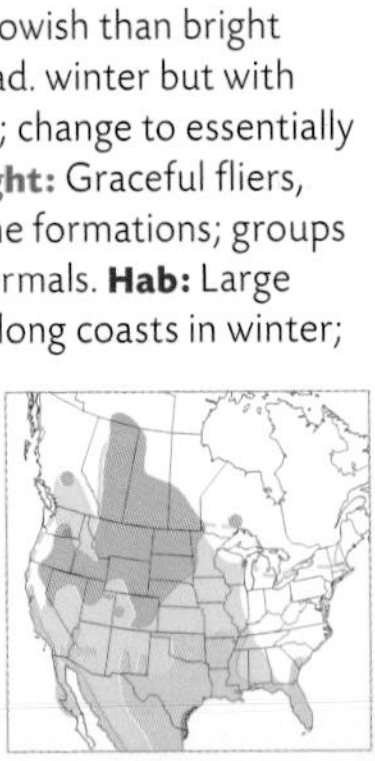

Adult breeding, *carolinensis* FL/02

Adult, *occidentalis* FL/02

Adult nonbreeding, *carolinensis* FL/01

Adult, *carolinensis* FL/02

Adult breeding, *carolinensis* FL/02

Juv., *carolinensis* FL/01

Juv./1st yr., *carolinensis* FL/01

Brown Pelican 1

Pelecanus occidentalis L 50"

Dives from air for fish. **Shape:** Very large, heavy-bodied, long-winged bird with very long bill and large pouch typical of pelicans. Proportionately longer bill and tail than those of American White Pelican. **Ad. Summer:** Body silvery gray except for blackish-brown belly. Upperwings with silvery inner wing, blackish outer wing; underwings mostly dark with slightly paler coverts. Head, neck, bill, pouch, and eye vary in color with season and breeding status. From courtship through incubation, head yellowish, nape ebony, foreneck white, eye whitish, pouch dark or with bright red base (see Subspp.), and bill darkish with pale base and reddish tip. Similar during nestling phase, but head whitish, eye dark, pouch all dark, bill pale. **Ad. Winter:** Head pale yellow, neck white, eye dark, pouch and bill dark. **Juv/1st Yr:** Mostly brownish gray overall with pale whitish belly. **2nd Yr:** Similar to 1st yr. but pale belly speckled with brown. Acquires ad. plumage in 3 yrs. **Flight:** Graceful flier; often coasts low over water; may create line formations; dives into water for food. Imm. has light underwing stripe formed by whitish greater coverts. **Hab:** Mostly coasts; nests on islands.

Subspp: (3) •*carolinensis* (coastal MD–TX) medium-sized, base of pouch dark; •*californicus* (coastal CA, accidental to Gulf of Mexico) largest, base of pouch bright red (Dec.–Jul.); •*occidentalis* (rare SAM, West Indies, seen s. FL) smallest, base of pouch dark, extensive blackish brown on belly.

Adult summer TX/04

Juv. TX/01

Adult prebreeding TX/04

Adult winter TX/12

Adult summer TX/12

Neotropic Cormorant 1

Phalacrocorax brasilianus L 25"

Shape: Small, slender, short-bodied, long-tailed cormorant with a thin neck and bill. Tail more than half length of rest of body (from base of tail to base of neck) – less than half in Double-crested. Back of gular pouch forms a horizontal V pointing into cheek (straight or slightly curved in Double-crested). **Ad. Summer:** Small, inconspicuous, yellowish to orangish-yellow gular pouch, rest of face and head dark (ad. Double-crested has orange supraloral area). V of gular pouch bordered by a thin line of white feathers, making the pattern more conspicuous; just before breeding, thin white plumes may grow on sides of face and neck; bill pale to dusky. **Ad. Winter:** Like summer but with no white plumes; white border to gular pouch less conspicuous or absent. **Juv:** Like ad. but browner above and paler on throat and breast; lacks white plumes and white border to gular pouch. **1st Winter:** Brownish head and back, glossy dark green back and rump, pale brown below, white border to gular pouch. **Flight:** Tail longer than that of other cormorants, thus projection behind wings about equal to that in front of wings (in other cormorants, neck and head longer than projection of tail and body behind wings); neck often kinked. **Hab:** Coasts, lakes; nests on islands.

Subspp: (1) •*mexicanus.*

IDENTIFICATION TIPS

Cormorants are large dark waterbirds with heavy bodies, longish necks, and large hooked bills. They dive from the surface and swim to catch fish. Their feathers are wetable and after diving they often hold their wings out as they dry. There are slight seasonal differences, but the sexes are alike.

Species ID: Note size, length, and thickness of bill and comparative length of tail. Also look for presence and shape of crest during breeding. Note subtle differences in the colors and pattern of facial skin at the base of the bill and on the gular pouch. Breeding birds grow thin white plumes on the sides of their head and/or neck.

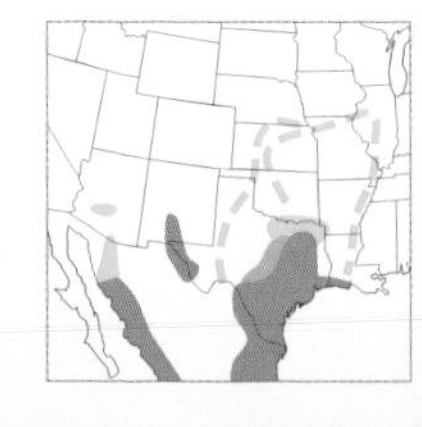

Adult FL/03

2nd yr. FL/02

1st yr. FL/01

Adult prebreeding, *albociliatus* CA/04

Adult drying wings FL/02

Adult prebreeding FL/03

Adult FL/02

Double-crested Cormorant 1

Phalacrocorax auritus L 33"

Shape: Large heavy-bodied cormorant with a thick neck and moderately thick deep-based bill, hooked at the tip (as in all cormorants). Overall, head fairly shallow – culmen curves slightly into sloping forehead and shallow somewhat flattened crown. Tail less than half length of rest of body (more than half in Neotropic). Spring to early summer ad. also has inconspicuous crests on either side of crown (whitish in some subspp.). **Ad:** Large gular pouch and supraloral stripe bright orange to yellowish orange; lore variably dusky. Upper mandible grayish (often with fine dark bars), lower mandible orange mottled with gray. **Juv/Imm:** Lower mandible and all facial skin orange. Underparts generally bicolored – buffy or mottled buffy upper breast contrasts with dark belly (opposite of juv. Great Cormorant); underparts brownish. Acquires ad. plumage in 2 yrs.: 1st yr. has all pale breast; 2nd yr. mottled breast. **Flight:** Strong deliberate flight often intermixed with occasional glides; neck kinked (not held straight); body usually held at slight upward angle; longer in front of wing than behind (about equal in Neotropic). Note extensive orange on face. **Hab:** Coasts, rivers, lakes; nests on islands.

Subspp: (4) •*auritus* (AB–NF south to s.e. CA–MA) fairly large, crests black and curled; •*floridanus* (TX–NC and south) smallest, crest black and curled; •*cincinatus* (s. AK) largest, crest white and straight; •*albociliatus* (coastal s. BC–CA) fairly small, crest variably white or black-and-white and straight.

Adult summer ENG/04

2nd yr. ME/02

1st yr. ENG/–

Adult ENG/01

Adult winter, drying wings ENG/01

1st yr. NJ/12

Great Cormorant 1

Phalacrocorax carbo L 36"

Shape: Large, heavy-bodied, relatively short-tailed cormorant with a thick neck and heavy deep-based bill. Head fairly wedge-shaped – culmen curves slightly into shallow forehead, flattened crown, and somewhat squared-off nape. **Ad. Summer:** Broad white border to back of gular pouch, starting behind eye and continuing under chin; gular pouch yellow to orange; lore all dusky (Double-crested has orange supraloral stripe); bill grayish, paler at base of lower mandible. In spring to early summer, also has whitish feathers on side of face and nape and white hip patches. **Ad. Winter:** Like summer but lacks extra white feathers on face and hip; in early winter, white border to gular pouch less conspicuous. **Juv/Imm:** Upperparts dark brown; underparts generally bicolored, with grayish-brown neck and upper breast contrasting with off-white belly (opposite of juv. Double-crested, which has pale breast and dark belly). White border to gular pouch limited to throat. Acquires ad. plumage in 2 yrs. **Flight:** Strong deliberate flight intermixed with occasional glides; neck slightly kinked, not held straight. Note white hip patch of breeding ad., white belly of juvenile. **Hab:** Coasts; nests on islands and remote cliffs.

Subspp: (1) •*carbo*.

Adult HI/03

Juv. CA/04

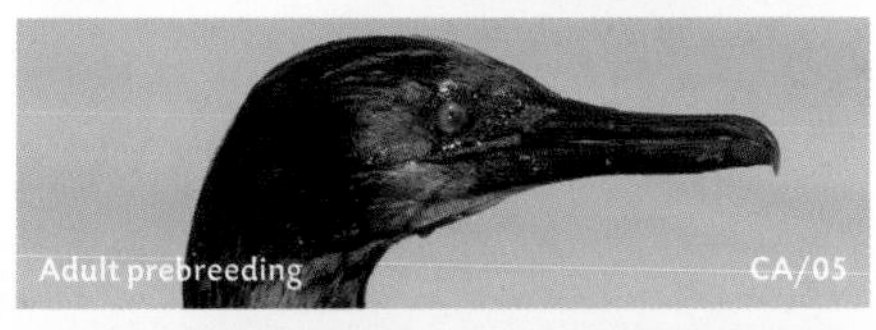

Adult prebreeding CA/05

Adult CA/03

Brandt's Cormorant 1

Phalacrocorax penicillatus L 34"

Shape: A large cormorant, similar in size and shape to Double-crested except tail about 15% shorter, head slightly shorter with a steeper forehead and more rounded crown. **Ad:** Head and bill all black with small buffy patch on throat. Body and wings black with bluish-green sheen. In Apr.–Jun. base of bill and gular skin become bright turquoise; white plumes grow off side of neck and to a lesser extent along scapulars. **Juv/Imm:** Brownish overall, except for buffy foreneck and breast blending onto darker belly. Acquires ad. plumage in 2 yrs. **Flight:** Usually flies with level body and straight neck (Double-crested has kinked neck and body angles up in flight). Wingbeats continuous, rarely gliding until landing. **Hab:** Rocky coasts; nests on islands.

Subspp: Monotypic.

Adult AK/06

Juv. AK/07

Adult late summer AK/07

Red-faced Cormorant 2

Phalacrocorax urile L 31"

Shape: Medium-sized, heavy-bodied, fairly compact cormorant with a relatively short thick neck. In fall and winter, crest on midcrown (can be raised or lowered); spring to summer, additional crest on nape. **Ad. Summer:** Body black with greenish to purplish iridescence; bright reddish facial skin from forehead to behind eyes; bill yellowish with dark tip and culmen and bluish base to lower mandible; gular pouch blue. White hip patches (spring and early summer). **Ad. Winter:** Similar to ad. summer but facial skin dull red; lower mandible mostly yellowish; no white plumes or white hip patches. **Juv/Imm:** Brownish overall (no paler underparts as in many other cormorants); little or no iridescence; lacks crests. Forehead feathered; facial skin and gular pouch pale yellowish to grayish pink; bill pale and slightly yellowish with dark tip (Pelagic has smaller, all-dark bill; also see Shape). Acquires ad. plumage in 2 yrs. **Flight:** Rapid wingbeats with few glides; neck usually straight; similar to Pelagic in flight but with thicker neck. Note broad white hip patch and red facial skin in spring and early summer. **Hab:** Rocky coasts; nests on cliffs.

Subspp: Monotypic.

Adult summer CA/02

Juv. BC/03

Adult winter AK/02

Adult summer CA/03

Pelagic Cormorant 1

Phalacrocorax pelagicus L 27"

Shape: Small slender cormorant with a long thin neck and small head. Bill short, thin, fairly even depth (pencil-like); feathered forehead rises steeply off bill to high, rounded, short crown. Feathering below bill often bulges out, making bill look stuck onto head. Crests on forecrown and nape in summer. Similar Red-faced has larger heavier body, thicker neck, larger head, thicker bill, bare forehead. Beware of some size variation in Pelagic. **Ad. Summer:** Body and wings black overall with greenish to purplish iridescence (Red-faced Cormorant has flat-toned brownish wings); white hip patches in early breeding season; black crests on forecrown and nape; gular pouch and limited facial skin at bill base dark red and inconspicuous; bill all dark. Similar Red-faced Cormorant has flat-toned brownish wings, crests on midcrown and nape, bill yellowish with dark tip and culmen. **Ad. Winter:** Like ad. summer but lacks crest on nape and white hip patches; facial skin duller and appears dark at any distance. **Juv/Imm:** Brownish overall with no paler areas on underparts; darkest of all juv./imm. cormorants; bill dark (pale yellowish in Red-faced); no iridescence; lacks crests. See also Shape. **Flight:** Moderate wingbeats with few glides; neck usually straight. Note small head and thin neck; in spring and summer, note broad white hip patch. **Hab:** Rocky coasts.

Subspp: Monotypic.

Adult m. FL/01

Juv./imm. FL/02

Adult m., breeding FL/03

Adult f. FL/02

Adult m. FL/03

Anhinga 1

Anhinga anhinga L 35"

Shape: Large, slender, long-tailed waterbird with a long snakelike neck and long, thin, sharply pointed bill. Broad wings often crooked and held out to sides, helping to dry feathers. Similar Double-crested Cormorant (which also holds wings out to dry) has shorter hooked bill and shorter tail. **Ad:** Bill mostly orangish; body, tail, and wings blackish; scapulars and greater secondary coverts silvery white; median and lesser secondary coverts spotted silvery white. M. has black head, neck, and upper breast; f. has dark-tipped crown and nape feathers and a buffy throat and upper breast strongly demarcated from black lower breast and belly. Just before breeding: whitish plumes grow on head and neck of both sexes; facial skin becomes emerald to turquoise; m. bill gets bright yellow; f. gets black patch at base of bill. **Juv/Imm:** Like ad. f. (with paler head, neck, and breast) but with brown wing and tail feathers and little or no silvery-white coverts or scapulars. Acquires ad. plumage in 2 yrs. **Flight:** Often alternates rapid flaps with short glides (accipiter-like); can soar very high on outstretched wings and slightly spread tail; may rise on thermals. When soaring, wing bases broadly join body, and neck and tail extensions beyond wings are about equal; in similar soaring Double-crested Cormorant, wing bases are pinched in before joining body, and neck and head are longer than tail. **Hab:** Lakes, ponds, slow-moving rivers, brackish bays. **Voice:** A series of low grating calls, trailing off at end, *e'e'e'e'e'e'e.*

Subspp: (1) •*leucogaster.*

Adult m. FL/02

Juv./1st yr. ATL/04

Adult m. displaying ANT/03

Imm./3rd–4th yr. ATL/04

Adult f. MEX/02

Magnificent Frigatebird 1

Fregata magnificens L 40"

Shape: Mostly aerial. Very long thin wings strongly bent at wrist form silhouetted M; long thin tail strongly forked for more than half its length; small head and long (twice length of head) strongly hooked bill. M. has inflatable red gular pouch, used during breeding displays. **Ad:** M. mostly all black except for gular pouch, which is red and conspicuous when inflated (breeding), orange and inconspicuous rest of year; feet gray to blackish; lacks pale diagonal bar across secondary upperwing coverts; axillaries all black. Similar m. Great Frigatebird has pinkish to reddish feet, a noticeable pale diagonal bar across secondary upperwing coverts, pale white to grayish-brown scalloping on axillaries. F. mostly black except for white breast; head and throat hooded with black, which extends down in a V onto white breast (making white breast patch look like a buttoned vest); blue orbital ring hard to see. Similar f. Great Frigatebird has a black hood with a gray throat that blends into white breast and a red orbital ring that is easier to see. **Juv:** (1st 1½–2 yrs.) White head and throat; diamond-shaped white breast patch; pale bluish feet; black axillaries. Similar juv. Great Frigatebird has cinnamon head and throat (head fading to whitish but throat with some cinnamon), oval white breast patch, pale pinkish feet. **Imm:** (Next 2 yrs.) White on belly more extensive and broadly connected to white throat and head; no cinnamon on head and throat; gradually becomes mottled with black on belly and head; bluish-gray feet. Similar imm. Great Frigatebird usually has some traces of cinnamon on head and/or throat; belly may become heavily mottled black while head is still light. **Flight:** Wings form strong M shape in flight; deeply forked tail; soars mostly, with rare wingbeats. **Hab:** Coasts, offshore islands.

Subspp: Monotypic.

Adult m. GAL/05

Adult m. HI/10

Adult f. HI/03

Juv. HI/03

Imm. –/01

Great Frigatebird 5

Fregata minor L 37"

Accidental vag. to offshore CA. **Shape:** Shape like Magnificent Frigatebird. **Ad:** M. with black underparts except for gular pouch, which is red and conspicuous when inflated (breeding), orange and inconspicuous rest of year; upperparts black except for pale brown diagonal bar across upperwing secondary coverts; axillaries tipped with white or grayish brown; pinkish to reddish feet. See Magnificent Frigatebird for comparison. F. blackish overall except for gray chin and throat, which blends into broadly whitish breast; orbital ring red (often seen). See Magnificent Frigatebird for comparison. **Juv:** (1st 1½–2 yrs.) Cinnamon head and throat (head may wear to whitish); isolated oval white breast patch; pale pinkish feet. See Magnificent Frigatebird for comparison. **Imm:** (Next 2 yrs.) White on belly more extensive and broadly connected to white throat and head; some traces of cinnamon on head and/or throat; belly may become heavily mottled black while head is still light. See Magnificent Frigatebird for comparison. **Flight:** Wings form strong M pattern in flight; deeply forked tail; soars mostly, with rare wingbeats. **Hab:** Coasts, offshore islands.

Subspp: Polytypic, but sightings off NAM not identified to subspecies.

Adult m. ALD/–

Imm. AUS/06

Adult f. AUS/–

Adult f. AUS/06

Lesser Frigatebird 5

Fregata ariel L 30"

Accidental vag. to either coast; has been seen also near Great Lakes. **Shape:** Mostly aerial. Very long thin wings strongly bent at wrist form silhouetted M; long thin tail strongly forked for more than half its length; small head and long (twice length of head) strongly hooked bill. M. has inflatable gular pouch, used during breeding displays. Lesser Frigatebird has wingspan comparable to an Osprey; Magnificent Frigatebird has wingspan comparable to a Bald Eagle. In all frigatebirds, males are smaller than females and size can vary between populations of the same species. **Ad:** In all ages and sexes, has distinctive white spur extending from edge of flanks into axillaries. M. all black with red gular pouch (usually inconspicuous) and prominent white spur extending from the upper edge of the flank into the axillaries; lacks any pale alar bars crossing the inner upperwing. Similar ad. m. Magnificent lacks white on axillaries; similar ad. m. Great lacks white on axillaries and has pale alar bars. F. has black head and white chest with black belly extending into chest in a V, but (unlike similar ad. f. Magnificent) white on chest extends in a spur onto axillaries. **Juv:** (1st 1½–2 yrs.) All black with variably cinnamon head, white breast that extends in spurs onto axillaries. **Imm:** (Next 2 yrs.) M. gradually acquires black head and breast; may have some white mottling on breast; still has white spurs on axillaries. F. gradually acquires black head and keeps white breast that extends as spurs onto axillaries. **Flight:** Wings form strong M pattern in flight; deeply forked tail; soars mostly, with rare wingbeats. **Hab:** Coasts, offshore islands.

Subspp: Polytypic.

Bitterns

Herons

Night-Herons

Ibises

Spoonbill

Storks

Adult TX/05

Juv. NJ/10

Adult FL/03

American Bittern 1

Botaurus lentiginosus L 25"

Shape: Large, relatively short-legged, heavy-bodied heron with fairly short thick neck, small head, and long, deep-based, sharply pointed bill. Bill often pointed toward vertical. Similar imm. night-herons have shorter, proportionately thicker-based, more abruptly pointed bills. **Ad:** Bill dull yellowish with dark culmen. Back heavily streaked tawny brown with fine black spotting; throat and breast whitish with warm reddish-brown streaking; strong black malar streak extends from bill to well down side of neck. Legs greenish yellow. **Juv:** (Jul.–Oct.) Like ad. but lacks black malar streak. **Flight:** Strong deliberate wingbeats; on upperwing, dark flight feathers contrast with lighter brown coverts. **Hab:** Marshes, wet meadows. **Voice:** A deeply resonant and repeated *gunk kerlunk;* a throaty *werk.* 🎧**12**.

Subspp: Monotypic.

Adult m. TX/05

Adult f. TX/04

Juv. FL/05

Adult m. TX/04

Least Bittern 1

Ixobrychus exilis L 13"

Shape: Very small, stocky, thick-necked heron with a proportionately large head and relatively long sharply pointed bill. Relatively short legs and large feet. **Ad:** Bill yellowish orange with dark culmen. Tawny face, neck sides, and wing coverts contrast with dark crown, back, and tail; throat and foreneck whitish with tawny streaks. M. has black crown, back, and tail; f. has reddish-brown crown, back, and tail (sometimes with purplish gloss). Dark morph ("Cory's Least Bittern") is dark reddish brown on face, neck, belly, and wing coverts; has occurred in some eastern states and provinces, but very rare. **Juv:** (Jul.–Oct.) Like ad. f. but crown paler, wing coverts pale buff with dark shafts. **Flight:** Fast and low; on upperwings, pale buffy secondary coverts contrast with dark flight feathers. **Hab:** Marshes with dense vegetation. **Voice:** Common call a raspy 3-note *wak wak wak;* a drawn-out *churrrr.*

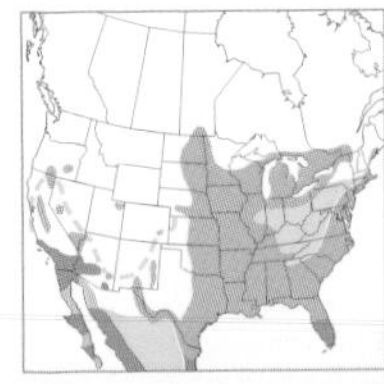

Subspp: (1) •*exilis.*

Adult FL/01

"Great White," *occidentalis* FL/02

"Wurdemann's" FL/02

Adult FL/01

1st yr. FL/01

Adult FL/03

Great Blue Heron 1

Ardea herodias L 46"

Shape: Very large, big-bodied, long-necked heron with a massive, long, deep-based bill. Bill thick for most of its length; abruptly tapered near tip on lower mandible; deep base blends into lines of head. Variable-length plumes off hindcrown (see Subspp.). **Ad:** Mostly grayish overall (can be all white, see subsp. *occidentalis*). Bill orangish with variably dusky culmen; facial skin blue. Head whitish with wide black lateral crown-stripes leading to short black plumes; neck sides pinkish gray; foreneck white with dark streaking; back and wing coverts grayish; legs dark (can be dull yellow, see subsp. *occidentalis*). **1st Yr:** Like ad. but with all-dark crown; dark upper mandible; extensive grayish streaking on neck. **Flight:** Long broad wings, usually somewhat bowed; slow graceful wingbeat; pale gray secondary coverts and back contrast with dark gray flight feathers, primary coverts, and tail. **Hab:** Marshes, wet fields, swamps, rivers, lakes. **Voice:** A deep, raspy *kraaaar.* 🎧**13**.

Subspp: (4) •*herodias* (s.e. BC–QC south to e. WA–SC); our smallest subsp. and as described above. •*fannini* (s.e. AK–Queen Charlotte Is. BC); medium-sized subsp., comparatively shorter bill and legs and longer tail, darker overall. •*wardi* (s.w. BC–s. CA–FL); fairly large subsp., occipital plumes long. •*occidentalis* (s. FL, possible vag. through East Coast states), "Great White Heron"; all white, large, with short head plumes (see additional descrip. below).

"Great White Heron" (*A. h. occidentalis*). **Shape:** Massive bill abruptly tapers near tip on lower mandible; short head plumes. Similar Great Egret has proportionately longer thinner neck, thinner bill that gradually tapers evenly on both mandibles, no head plumes. **Ad:** All-white plumage. Bill orangish with culmen slightly dusky; legs and feet dull yellow or horn; facial skin bluish. Similar Great Egret has black legs and yellow (to light green during breeding) facial skin (see also Shape).

"Wurdemann's Heron." Probably results from breeding between subspp. *occidentalis* and *wardi.* **Ad:** Similar to dark subspecies, but head mostly white and neck very pale.

Adult, *egretta* FL/01

Adult displaying, *egretta* FL/02

Adult, *egretta* FL/01

Adult breeding, *egretta* FL/03

Adult, *egretta* FL/02

Great Egret 1

Ardea alba L 39"

Shape: Large elegant heron with long thin legs, neck, and bill. Bill moderately deep-based and fairly evenly tapered on both mandibles; no head plumes. **Ad:** Plumage all white. Bill orangish yellow (see Subspp.); facial skin orange; legs and feet usually black (see Subspp.). **At Breeding:** Bill becomes orange with dark culmen, facial skin turns pale green; grows long plumes on back that extend past tail (never has head plumes). **Juv:** Like ad. **Flight:** Slow wingbeat; long folded neck projects well below body plane (in Snowy Egret, folded neck is more even with body plane). **Hab:** Marshes, wet fields, swamps, lakes. **Voice:** A low, short, rattly *kraaak*.

Subspp: (2) •*egretta* (NAM range) medium-sized, legs black, bill orangish yellow. •*modesta* (Asian vag. to w. AK islands) small, legs pinkish to yellowish to reddish (esp. Jan.–Jul.), bill mostly black, sometimes with paler base.

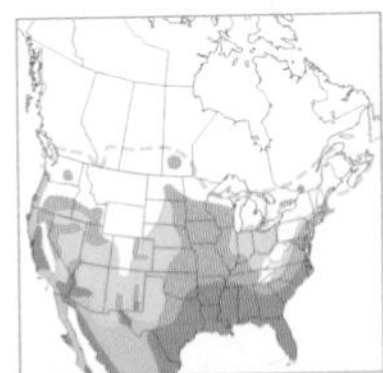

Adult SAF/01

Adult SAF/04

Little Egret 4

Egretta garzetta L 25"

Casual to Atlantic Coast. **Shape:** Medium-sized, fairly long-necked, long-legged heron with a very long fine-pointed bill. Bill noticeably longer than depth of relatively small head (bill length about 1½ x head length). In breeding plumage has 2 long head plumes. Similar Snowy Egret is shorter; has proportionately shorter neck, bill, and legs and thinner-based bill; during breeding has bushy head plumes and recurved back plumes; higher well-rounded crown; thinner legs; slightly slimmer overall. **Ad:** Plumage all white. Bill black; facial skin bluish gray (but variable, including bluish, greenish, grayish, pale yellow, whitish); legs black; feet greenish yellow and usually sharply demarcated from black legs (sometimes yellow extends up leg a short distance); eye pale. Similar Snowy has bright yellow facial skin, yellow eye, generally brighter yellow feet. **At Breeding:** Facial skin variable, can be red, orange, or yellow; feet become orange or red; grows 2 long head plumes. **Juv:** Like ad. but with pale base to bill; variably dark facial skin; variable yellow on legs, often extending up back of leg or just on tarsus. Bill may not be full adult length. **Flight:** Longer neck folds below body plane more than in Snowy Egret. **Hab:** Marshes, swamps, lakes. **Voice:** A harsh *shraaak.*

Subspp: (1) •*garzetta.* **Hybrids:** Western Reef-Heron, Snowy Egret.

Adult NH/08

Ad. with imm. Snowy Egret NH/08

Adult NH/08

Western Reef-Heron 5

Egretta gularis L 25"

Accidental to Atlantic Coast. **Shape:** Medium-sized long-necked heron with a very long, fairly deep-based, straight bill and long thick legs. Bill length 1½ x head length. During breeding season, 2–3 long plumes off hindcrown and stringy plumes on breast. Gular area extends halfway out on lower mandible. **Ad:** Dark gray overall except for white patch on chin and upper throat; this white area extends halfway out lower mandible. Bill and lore gray to blackish; eye pale; legs blackish with contrasting greenish-yellow feet; yellow may extend up onto leg with a clear demarcation or blend into the black of the upper leg. **Imm:** Like ad. but may be more brownish overall. **Flight:** Deliberate wingbeats. **Hab:** Ocean shores, marshes. **Voice:** A harsh *skwak.*

Subspp: (1) •*gularis.* **Hybrids:** Little Egret.

Adult FL/02

Juv. FL/01

Adult breeding TX/05

Adult FL/02

Snowy Egret 1

Egretta thula L 24"

Shape: Medium-sized, delicate, relatively short-necked heron with a thin-based spikelike bill. Bill moderately long, about length of head. Thin legs and bill add to delicate appearance. Crown relatively high and well rounded. During breeding, grows bushy plumes on crown and long recurved plumes on back. See Little Egret for comparison. **Ad:** Plumage all white. Bill black; facial skin bright yellow; legs black with yellow extending variably up back of tarsus; feet bright yellow; eye yellow. Similar imm. Little Blue Heron has two-toned bill (bluish base with darker tip) and legs all greenish yellow. **At Breeding:** Facial skin and feet become deep reddish orange; legs become all black. Grows long plumes on breast and recurved plumes on back; shorter plumes on head form shaggy crest. **Juv:** (Jul.–Apr.) Like ad. but legs extensively yellowish green, especially on back; feet dull yellow; facial skin can be dark just after fledging, then turn yellow. **Flight:** More rapid wingbeat than Great Egret; folded neck even with body plane (lower in Great Egret). **Hab:** Marshes, swamps, lakes, seashore. **Voice:** A low harsh *shraaak.*

Subspp: Monotypic. **Hybrids:** Little Egret, Little Blue Heron, Tricolored Heron, Cattle Egret.

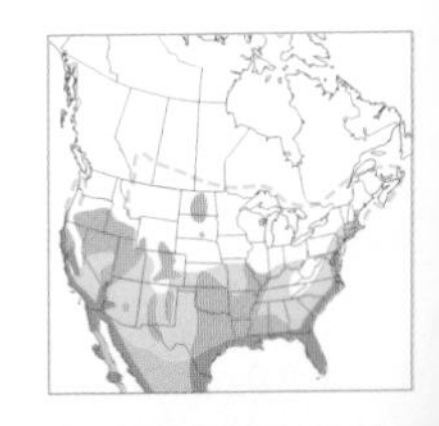

Adult FL/02

Juv. FL/02

Adult breeding FL/03

Adult FL/02

Little Blue Heron 1

Egretta caerulea L 24"

Shape: Medium-sized, fairly heavily built heron with a relatively short thick neck and a deep-based slightly downcurved bill. Neck is shorter than body. Typical feeding pose is neck at 45-degree angle and head pointed down at right angle to neck. **Ad:** Dark bluish-gray body with maroon head and neck. Bill bluish gray at base, dark at tip; facial skin grayish; legs and feet greenish yellow. **At Breeding:** Bill base and facial skin become cobalt blue; legs blackish. **Juv:** (Jun.–Apr.) Plumage all white, usually with dusky tips to primaries. Bill pale gray at base, dark at tip; facial skin pale greenish; legs and feet yellowish green. Hunts in same posture as adult. Similar Snowy Egret has thinner all-black bill, black legs with variable yellow up back, bright yellow feet. **1st Spring and Summer:** Like juv., but white plumage mottled with incoming ad. bluish-gray feathers. **Flight:** Fairly rapid wingbeats, agile. **Hab:** Marshes, swamps, lakes. **Voice:** A rather thin grating *shreeer.*

Subspp: Monotypic. **Hybrids:** Snowy Egret, Cattle Egret.

IDENTIFICATION TIPS

Bitterns, Herons, Egrets, and Night-Herons are medium to large, long-legged, long-necked wading birds with spikelike bills. They live mostly in marshes and shallow ponds and feed on a variety of small animals. Some marked changes in color of leg, bill, and facial skin occur right around breeding time; the sexes are alike.

- **Bitterns** are secretive birds of marshes and have comparatively broad necks and short legs. They are cryptically colored.
- **Herons and egrets** are graceful birds with very long necks and legs; they often feed in the open and are usually not hard to see.
- **Night-Herons** have comparatively short broad-based bills and fairly short legs. They generally feed at night and sleep secretively during the day.

Species ID: Length and shape of bill in relation to length of head is distinctive for most wading birds; also note length of neck and legs, colors of the bill, legs, and feet, and general coloring on the body. The apparent size of wading birds varies greatly based on whether the neck is extended or folded into the body.

Adult FL/01

Juv. FL/01

Adult breeding FL/03

Adult breeding FL/03

Adult FL/02

Tricolored Heron 1

Egretta tricolor L 26"

Shape: Medium-sized, long-necked, long-legged heron with a very long thin bill and relatively small head. Proportionately the longest-billed of all our herons; bill about twice length of head and very gradually tapered. **Ad:** Our only dark heron with a white belly; neck, back, and wings dark slate blue; breast and belly white; throat and foreneck whitish. Bill with dark upper mandible, orange lower mandible; facial skin orange; legs dull orange. **At Breeding:** Bill black at tip, cobalt blue at base and on adjacent facial skin; legs dark orange or reddish pink. Grows a few long purplish plumes on back and white plumes off hindcrown. **Juv:** (Jul.–Apr.) Like ad. but with rusty neck and rusty tips to upperwing coverts. **Flight:** Fold of long neck extends well below body plane; white belly and underwing coverts contrast with gray rest of body. **Hab:** Marshes, shores, mudflats, tidal creeks. **Voice:** A drawn-out harsh *creeer.*

Subspp: (1) *ruficollis.* **Hybrids:** Snowy Egret.

Adult dark morph FL/02

Adult white morph FL/03

Adult breeding, dark morph FL/03

Adult breeding, dark morph FL/03

Juv. dark morph FL/02

Adult breeding, dark morph FL/03

Reddish Egret 1

Egretta rufescens L 30"

Shape: Large, long-legged, long-necked heron with a fairly long deep-based bill. Bill about the length of head. Neck and breast distinctively shaggy in adult. **Ad:** Polymorphic. In all morphs and ages, legs dark gray to bluish gray and eye pale; ad. bill distinctive for species with pinkish base and clean-cut black tip (bill all dark for juv./1st yr.); ad. with fairly long shaggy plumes. Often feeds by making short dashes after fish, as if "dancing," but other herons also do this at times. Dark morph common; has reddish-brown head and neck; bluish-gray body and wings. Similar ad. Little Blue Heron smaller with proportionately shorter bill and legs, bluish bill base. White morph uncommon; all white; bill has distinctive pink base and clear-cut black tip. Intermediate morph like dark morph but with some white feathers, often on wings. **At Breeding:** In all morphs, bill turns bright, almost fluorescent pink at base, contrasting sharply with black tip; facial skin blue. **Juv/1st Yr:** All morphs have all-dark bill and dark facial skin. Dark morph pale gray overall, sometimes mottled darker. White morph all white with dark bill and dark facial skin. Large size and large deep-based bill helpful in distinguishing from similar herons. **Flight:** Large heron; ad. has two-toned bill with pinkish base easily seen. **Hab:** Mostly coastal wetlands, shores, lagoons. **Voice:** A grating *raaah.*

Subspp: Monotypic.

Adult, *ibis* FL/01

Adult breeding, *ibis* FL/02

Adult breeding, *ibis* FL/05

Adult breeding, *ibis* TX/04

Cattle Egret 1

Bubulcus ibis L 20"

Shape: Medium-sized, short-necked, short-legged, stocky heron with a short fairly deep-based bill with extensive feathering out onto lower mandible, and high rounded crown. **Ad:** White overall. Bill and facial skin yellowish orange; legs gray to blackish. **At Breeding:** Legs reddish; buffy wash on crown, breast, and back. **Juv:** (Jul.–Oct.) All white with dark legs and dark bill, which soon turns yellowish orange as on adult. **Flight:** All white with dull orangish bill; deep wingbeats. **Hab:** Open dry areas, lawns, fields, pastures with livestock. **Voice:** Generally silent away from breeding colony.

Subspp: (2) •*ibis* (all of NAM range) smaller with proportionately shorter bill and legs, buffy to orangish plumage (Feb.–Jun.) limited to crown, back, and breast. •*coromanda* (Asian vag. to s.w. AK) larger with proportionately longer legs and bill, orange plumage (Feb.–Jul.) on all of head, neck, breast, and back. **Hybrids:** Snowy Egret, Little Blue Heron.

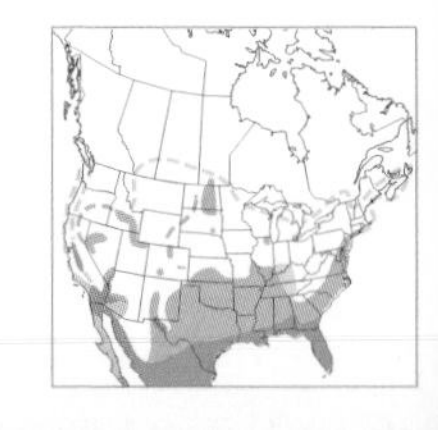

Adult, *virescens* TX/–

Juv., *virescens* USA/08

Adult breeding, *virescens* FL/03

Adult, *virescens* MEX/04

Green Heron 1

Butorides virescens L 19"

Shape: Small, short-legged, large-headed, flat-crowned, stocky heron with neck usually retracted. Bill proportionately long and thick-based. Bird is usually crouched when hunting and long thick neck only seen during a strike at food. **Ad:** Crown and back dark iridescent bluish green; neck and breast reddish brown with variable white streaking centrally; thin buff margins to upperwing coverts. Bill dark above, yellowish below; facial skin with greenish-yellow and black stripes; legs pale yellow. **At Breeding:** Facial skin and bill mostly dark; legs reddish orange. **Juv/1st Yr:** (Jul.–Mar.) Like ad. but with neck and breast more brownish and heavily streaked with white; upperwing coverts have pale dots at tips. **Flight:** Size and wing shape crowlike, but wings more cupped in flight. **Hab:** Shores and water edges with dense vegetation, salt marshes, streams. **Voice:** A sharp *keoow;* a repeated *keh keh keh.*

Subspp: (2) •*virescens* (cent. and e. NAM) smaller, no white on tips of flight feathers, back greener. •*anthonyi* (w. NAM) larger, fine white tips to some flight feathers, back grayer.

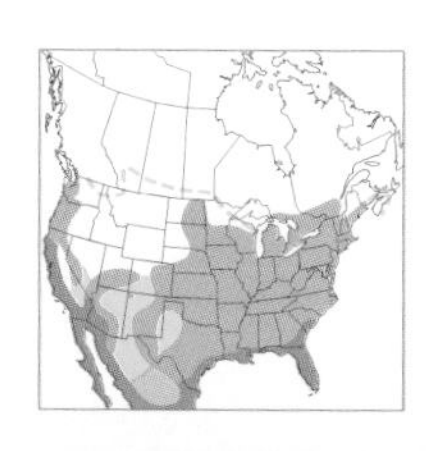

Adult FL/03

Juv. FL/02

2nd yr. FL/02

Adult TX/04

Black-crowned Night-Heron 1

Nycticorax nycticorax L 25"

Shape: Medium-sized, short-legged, stocky heron with a relatively short, deep-based, tapered bill. Head long with a shallow sloping forehead and flat crown; bottom of lower mandible mostly straight, culmen downcurved. Bill about as long as head. Similar Yellow-crowned has shorter head, more rounded crown, higher forehead, slightly shorter but thicker bill curved toward tip above and below, thinner neck, and slightly longer legs. **Ad:** Crown and back black with bluish iridescence; underparts pale gray; wings and tail darker gray. Bill blackish with paler base to lower mandible; legs yellowish. Thin white plumes on nape. **At Breeding:** Head plumes longer; legs reddish pink. **Juv/1st Yr:** Brown with whitish streaks on face, neck, breast; variable spotting on wing coverts and back; bill with greenish yellow at base of lower mandible. In all ages, can be distinguished from similar Yellow-crowned by absence of pale margins to greater coverts (Yellow-crowneds in all ages have pale margins to greater coverts). **2nd Yr:** Grayish brown overall with suggestion of darker cap of adult; white spotting on wings and back often gone or worn off. Acquires ad. plumage by 3rd yr. **Flight:** Gray upperparts with black back. Just part of toes project beyond tail (feet project completely beyond tail in Yellow-crowned). **Hab:** Freshwater streams, lakes, marshes; saltwater marshes. **Voice:** A single explosive *wok*. 🎧**14**.

Subspp: (1) •*nycticorax*. **Hybrids:** Tricolored Heron.

Adult, *violacea* FL/02

1st yr., *violacea* FL/02

2nd yr., *violacea* FL/02

1st yr., *violacea* FL/02

1st yr., *violacea* TX/04

Yellow-crowned Night-Heron 1

Nyctanassa violacea L 24"

Shape: Medium-sized, short-legged, stocky heron with a relatively short, deep-based, abruptly tapered bill. Head egg-shaped with a steep forehead and flat crown; bottom of lower mandible and culmen strongly curved toward tip; bill slightly shorter than head. Similar Black-crowned has longer, flatter head, more gradually tapered bill, mostly straight bottom edge to lower mandible, thicker neck. **Ad:** Smooth gray underparts; back and upperwing coverts darker gray with pale gray margins; head black except for whitish crown and cheek-stripe (sometimes tinged buff); bill all black; legs yellowish. **Juv/1st Yr:** Brownish gray with whitish streaks on face, neck, breast; variable spotting on back and wings; bill all dark. In all ages, can be distinguished from similar Black-crowned by presence of pale margins to greater coverts (Black-crowneds lack pale margins to greater coverts at all ages). **2nd Yr:** Dark grayish brown overall with suggestion of ad. darker blackish band through eye and across auriculars. Acquires ad. plumage by 3rd yr. **Flight:** Gray back and upperwing coverts; darker flight feathers. Feet project completely beyond tail (just part of toes in Black-crowned). **Hab:** Freshwater streams, lakes, marshes; saltwater marshes. **Voice:** A single *kwerk;* higher-pitched, a little softer, and hoarser than Black-crowned call. 🎧**15**.

Subspp: (2) •*violacea* (all of U.S. range) slightly paler overall, thinner bill; •*bancrofti* (seen s. AZ–s. CA) slightly darker overall, thicker bill.

Adult summer FL/04

Juv. UKM/10

Adult winter FL/02

Adult summer FL/04

Glossy Ibis 1
Plegadis falcinellus L 23"

Shape: Medium-sized somewhat short-necked wading bird with a very long, thin, downcurved bill. Similar in shape to White-faced Ibis. **Ad. Summer:** Reddish-brown head, neck, upper back, and underparts; rest of body glossy iridescent bronze-green. Eye brown and lore dark gray; pale blue lines above and below dark facial skin. Legs and bill grayish to pinkish gray. Similar White-faced Ibis has pale pinkish to reddish eye; pale pinkish to reddish facial skin often encircled by whitish line (this may be faint), including behind eye; legs in high breeding condition reddish. Intermediate eye and facial skin colors occur; these are probably hybrids. **Ad. Winter:** Head and neck finely streaked white and brown; rest of body metallic bronze-green. Eye brown and lore dark; pale blue lines remain above and below facial skin but thinner and duller than in summer. **Juv:** (Jul.–Nov.) Like ad. winter but browner with little iridescence; often with white streaking on the neck; may have wide dark band across middle of lower and/or upper mandible; bill may be pinkish in very young birds. By 1st fall/winter, may develop blue lines on facial skin; until then, indistinguishable from White-faced juv. **Flight:** Flies with neck extended; quick flaps alternated with glides. All dark. **Hab:** Freshwater and brackish marshes. **Voice:** A nasal *wahn* and *emp emp.*

Subspp: Monotypic. **Hybrids:** White-faced Ibis.

IDENTIFICATION TIPS

Ibises have long legs, long necks, and distinctive long, thin, downcurved bills. They live in marshes, shallow water, and fields, where they eat a wide variety of small animals. Some species undergo color changes seasonally, others just at breeding; the sexes are alike.

Adult summer UT/06

Juv. OR/08

Adult winter CA/02

Adult summer MT/06

White-faced Ibis 1

Plegadis chihi L 23"

Shape: Medium-sized somewhat short-necked wading bird with a very long, thin, downcurved bill. Similar in shape to Glossy Ibis. **Ad. Summer:** Reddish-brown head, neck, upper back, and underparts; rest of body glossy iridescent bronze-green. Eye deep red; facial skin pale pinkish to bright pinkish red; variable-width whitish band of feathering encircles pinkish facial skin (this may be faint), including behind eye. Legs pinkish (sometimes brightest at joint) to reddish (at high breeding condition); bill grayish to pinkish gray and paler at base than on Glossy Ibis. **Ad. Winter:** Head and neck finely streaked white and brown; rest of body metallic bronze-green. Eye and facial skin pinkish; mostly lacks pale white line around facial skin. Similar Glossy Ibis has brown eye and dark facial skin; pale blue lines do not meet behind eye. Intermediate eye and facial skin colors occur; these birds are probably hybrids. **Juv:** (Jul.–Nov.) Like ad. winter but browner with little iridescence; often with white streaking on the neck and a wide dark band in middle of bill; bill can be pinkish in very young birds. By 1st fall/winter, may develop reddish eye and pinkish facial skin; until then, indistinguishable from Glossy juv. **Flight:** Flies with neck extended; quick flaps alternated with glides. All dark. **Hab:** Freshwater and brackish marshes. **Voice:** A nasal *wahn* and *emp emp*.

Subspp: Monotypic. **Hybrids:** Glossy Ibis.

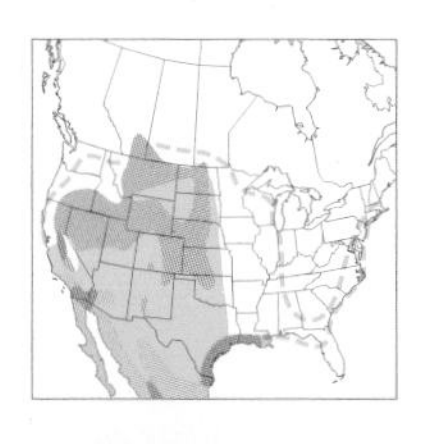

Adult FL/01

1st winter FL/01

Adult breeding FL/03

Adult breeding FL/03

1st spring FL/03

Adult FL/02

White Ibis 1

Eudocimus albus L 25"

Shape: Medium-sized, somewhat short-necked, heronlike bird with a very long, thin, downcurved bill. Gular sac enlarged during breeding (larger in females). **Ad:** Red bill, facial skin, and legs. Otherwise, all white except for tips of outer primaries, which are black with subtle iridescent sheen. **At Breeding:** Bill, facial skin, and legs turn bright red; bill tip turns black; bright red gular sac develops at base of bill. **1st Yr:** Reddish bill; pale legs. Brown back and wings; streaked brown head and neck; white belly. Ad. white feathers first appear on back and wings, creating mottled effect in 1st spring. **Flight:** Red downcurved bill, white belly. Flies with neck extended (herons retract neck in longer flights); quick flaps alternate with glides. **Hab:** Salt and freshwater lakes, marshes, swamps, tidal shores, mudflats, shores. **Voice:** A nasal honking *huunh.*

Subspp: Monotypic. **Hybrids:** Scarlet Ibis.

Adult FL/02

Adult FL/02

2nd yr. FL/02

1st yr. FL/02

Adult FL/03

Roseate Spoonbill 1

Platalea ajaja L 32"

Shape: Large, long-legged, fairly long-necked wading bird with a distinctive, flat, spoon-shaped bill. **Ad:** Bald green head; red eye; pale grayish bill with dark cracklike marks at base. Neck and upper back white; rest of plumage pink except for orange tail and dark carmine upperwing coverts and rump; legs deep pink. **2nd Yr:** Like ad. but head half-bald; little or no dark carmine on wings or rump; tail pink to orangish. **1st Yr:** Head, neck, and most of body white with pale pink wash over wings and belly; head feathered with white; eye red; bill grayish with pink edges; tail pink. **Flight:** Flies with neck outstretched; quick flaps alternated with glides. **Hab:** Salt water, fresh water, mangroves, coastal islands. **Voice:** Grunting *unh unh.*

Subspp: Monotypic.

Adult BRA/12

Adult VEN/01

Jabiru 4

Jabiru mycteria L 52"

Casual Cent. Am. vag. to TX, OK. **Shape:** Very large, massive-billed, big-bodied, long-legged wading bird. Distinctive bill somewhat boat-shaped and slightly upturned. **Ad:** Black bill, head, and upper neck; red lower foreneck; head and neck bald; rest of body and wings white. **Juv:** Dark downy feathers on head; body mostly gray; wing feathers mixed brown and white. **Flight:** Massive bill; all-white wings. **Hab:** Freshwater wetlands. **Voice:** Usually silent.

Subspp: Monotypic.

Adult FL/03

Imm. FL/03

Adult FL/03

Wood Stork 1

Mycteria americana L 40"

Shape: Very large, large-bodied, long-legged, relatively short-necked wading bird with massive deep-based bill slightly downcurved at the tip. **Ad:** Bald, scaly head and neck grayish and black; bill horn-colored; legs blackish with bright pink feet. Body white except for black flight feathers and tail. Wing coverts white except for upperwing primary coverts, which are black. **Imm:** Like ad. but head is feathered with grayish down and bill is dull yellow in 1st yr.; much like ad. after 1st yr., attains full ad. plumage by 4th yr. **Flight:** All white with black flight feathers and tail; much like American White Pelican, but all secondaries and tail are black (pelican has white inner secondaries and tail). Flocks often soar, turning in thermals, like pelicans, vultures, or raptors. **Hab:** Swamps, coastal shallows, ponds, ditches. **Voice:** Usually silent.

Subspp: Monotypic.

Vultures

Hawks

Kites

Eagles

Caracara

Falcons

Adult AZ/10

Adult AZ/05

Juv. AZ/05

California Condor 6
Gymnogyps californianus L 46"

Central CA; Grand Canyon, AZ–UT. Quickly dwindling condor populations in the mid-1900s led to a recovery program where all remaining wild birds were caught and kept in captivity for breeding and release into the wild. About 100 birds have been released so far, but they are still partially dependent upon food supplied by the recovery program. Because they are not naturally breeding and surviving on their own, they are considered "extinct" by the American Birding Association checklist. **Shape in Flight:** Extremely large, broad-winged, short-tailed, small-headed vulture. Tail slightly rounded when closed; tail length less than half width of wing. Wings angled slightly forward; held level or in slight dihedral. Up to 8 spread very long outer primaries create "multifingered" look. **Ad. in Flight:** Below, black overall except for white median and lesser coverts and white axillaries. Above, black overall except for thin white line created mostly by tips of greater coverts. Similar Turkey Vulture has black underwing coverts and silvery flight feathers; similar Golden Eagle has a feathered head and longer tail as an ad. and white patches at the base of the primaries as a juv./imm.; similar imm. Bald Eagle has a variable patchwork of white on underparts and flies with wings flat. **Shape Perched:** Fairly stocky with a bulbous head and prominent ruff; folded wings project past tail. **Ad. Perched:** Bare red to orange head and neck, black neck ruff, black over rest of body except for white wingbar on tips of greater coverts. **Imm:** Like ad. but white on wings grayish and less prominent; head and bill start blackish and gradually become like those of ad. over 4–5 yrs. **Flight:** Generally soars in steady manner (no tilting side to side like Turkey Vulture). **Hab:** Needs extensive wild areas for foraging. Grasslands, mountains, canyons. **Voice:** Hissing.

Subspp: Monotypic.

Adult FL/02

Adult FL/02

Juv. FL/06

Juv. FL/02

Black Vulture 1
Coragyps atratus L 25"

Shape in Flight: Comparatively short-tailed vulture with relatively short broad wings. Tail square when closed; tail length less than half width of wing. Wings angled forward and often at slight dihedral; up to 7 spread outer primaries create "multifingered" look. **Ad. in Flight:** All black with contrastingly pale bases to outer 6 primaries. **Shape Perched:** Stocky; comparatively long-necked and big-headed; folded wings extend just short of tail tip. **Ad. Perched:** Dark bill with pale tip; wrinkled bare head gray, looking like barrister's wig. **Juv:** (1st yr.) Head blackish with short, black, bristlelike hair. **Flight:** Soaring alternating with a few rapid shallow flaps; wings in slight dihedral or flat. Similar Turkey Vulture rarely flaps (and if so, slowly), often tilts side to side, and holds wings in a strong dihedral. **Hab:** Open country, dumps, urban areas.
Voice: Hissing.

Subspp: Monotypic.

Adult, *septentrionalis* FL/01

Adult, *septentrionalis* FL/03

Adult, *septentrionalis* FL/01

Juv., *septentrionalis* FL/02

Turkey Vulture 1

Cathartes aura L 26"

Shape in Flight: Comparatively long-tailed vulture with long broad wings. Tail rounded when closed; tail length greater than half width of wing. Wings held straight out with strong dihedral; up to 7 spread outer primaries create "multifingered" look. **Ad. in Flight:** Below, silvery flight feathers create a broad trailing edge to wing, contrasting with black underwing coverts; head reddish. Above, brownish overall with pale margins to wing coverts. **Shape Perched:** Small-headed; long folded wings create elongated look to body; folded wings project just past tall tip. **Ad. Perched:** Bare reddish head with pale bill; brownish overall with pale margins to wing coverts. **Juv:** (Jul.–Feb.) Like ad. but more blackish overall; head grayish and bill all grayish at first. Over 1st yr., head becomes pinkish and bill pale with dark tip; then like adult. Similar juv. Black Vulture's head larger and black; bill is black. **Flight:** Almost all soaring with frequent tilting side to side; wings in strong dihedral; occasional deep wingbeats. Similar Black Vulture alternates soaring with rapid shallow flaps and holds wings flat or in a slight dihedral. **Hab:** Open country, woods, dumps, urban areas. **Voice:** Hissing.

Subspp: (3) •*septentrionalis* (e. ON–e. TX and east) largest; extensive white tubercles around front of eye, lesser coverts with glossy blue centers. •*meridionalis* (w. BC–w. ON south to CA–cent. TX) medium-sized; a few white tubercles around front of eye, lesser coverts without glossy centers. •*aura* (s.e. CA–s. TX) smallest; a few white tubercles around front of eye, lesser coverts with glossy purple centers.

Adult m., *carolinensis* FL/03

Adult m., *carolinensis* FL/02

Adult f., *carolinensis* FL/02

Adult f., *carolinensis* FL/01

Juv., *carolinensis* TX/09

Osprey 1

Pandion haliaetus L 23"

Shape in Flight: Large long-winged hawk with a short fairly slender body. Wrists usually held up and forward, outer wing bent down and back, creating M-shaped silhouette from below or head-on. Wings fairly evenly broad, tapering slightly to tips; trailing edge of each a subtle S. Tail length ¾ wing width; wing length 3 x width. **Ad. in Flight:** Head white with broad black eye-stripe. Underparts white except for black greater secondary coverts and wrist patch (which form a black M). Paler inner primaries create light panel on outer wing. Flight feathers and tail finely barred (latter can appear reddish in some lights). **Shape Perched:** Appears small-headed, big-eyed, long-winged, and slender. Wing tips project well past tail. Steep forehead, rounded crown (often with a bushy nape), eye well forward; deep-based bill with long hook. **Ad. Perched:** White head with broad blackish eye-stripe; belly white, back and wings dark brown. Variable short streaking on breast forms "necklace," generally more prominent on females. **Juv:** (1st yr.) Like ad. but underwing secondary coverts often buffy, upperwing coverts and back feathers with pale margins (wear off by late fall), eye darker reddish orange (like ad. yellow by midfall), flight feathers with pale white tips. **Flight:** Wingbeats deep; hovers while hunting fish. **Hab:** Large lakes, rivers, coasts. **Voice:** Rising and descending whistled *ee ee ee ee ee ee;* loud, strongly rising, whistled *waaaeeeh;* a grating *grr grr grr grr.* 🎧 **16**.

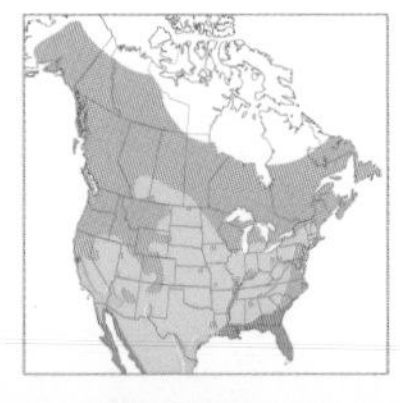

Subspp: (1) •*carolinensis.* ▶ *P.h. ridgwayi* (possible vag. to s. FL) is smaller, has little or no streaking on breast, more white on head.

IDENTIFICATION TIPS

Raptors—hawks, kites, eagles, and falcons—range in size from fairly small to large; they all have sharp talons and hooked beaks that are used to capture, kill, and tear apart their prey, which includes mammals, fish, birds, amphibians, reptiles, and insects. There are no seasonal differences in plumage, and for the majority there are only slight differences between the sexes, although females are almost always larger.

Adult m. TX/05

Juv. TX/10

Adult f. TX/05

Adult m. TX/05

Adult f. TX/05

Hook-billed Kite 3

Chondrohierax uncinatus L 18"

Shape in Flight: Medium-sized, long-tailed, slender-bodied hawk with short very broad wings pinched in at base. Wings often angled forward; tail with rounded corners. Tail length 7/8 wing width; wing length 1½ x width. **Ad. in Flight:** Head and neck dark; rest of underparts heavily barred; tail dark with 2 wide light bands; outer primaries translucent at base. M. underparts grayish; f. underparts reddish brown. Rare dark morphs have all-dark wings with some white spots on primaries; all-dark bodies with no barring. **Shape Perched:** Fairly large head; broad neck; long tail; short legs; wing tips reach just over halfway down tail. Sloping forehead and shallow crown; bill fairly thick and very long with long hook; short gape. **Ad. Perched:** White eye, orange to yellow eyebrow, gray bill with yellowish cere and lore. Light morph barred on belly; dark morph all dark below. F. has broad reddish-brown collar. **Juv:** (1st yr.) Similar to ad. f. but with white collar on hindneck; 3 pale bands on tail; underparts paler with variable barring and lacking strong reddish-brown tones. **Flight:** Wingbeats slow and floppy; wings flat during soar, slightly cupped during glide. **Hab:** Brushy woods, partially open areas. **Voice:** Very rapid, mellow, almost flickerlike *kehkehkehkeh*.

Subspp: (1) •*uncinatus*.

Adult FL/03

1st yr. FL/05

Adult FL/04

Adult FL/04

Swallow-tailed Kite 1

Elanoides forficatus L 22"

Shape in Flight: Distinctive silhouette with long deeply forked tail and long, narrow, sharply pointed wings. Tail length 2 x wing width; wing length 3 x width. **Ad. in Flight:** Below, white body and underwing coverts contrast strikingly with blackish flight feathers and tail. Above, mostly purplish blue with wide black band across leading edge of inner wing and upper back, contrasting with white head. **Shape Perched:** Small head and bill; short legs; long wing tips project just over 3/4 down long forked tail. **Ad. Perched:** White head and body; purplish-blue back, wings, and tail; eye dark reddish; bill black with pale bluish cere. **Juv:** (1st yr.) Similar to ad. but has shorter tail (1½ x greatest wing width); small white tips to flight and tail feathers; lacks any signs of wing molt. Buff wash on chest and head when very young. **Flight:** Sweeping flight as it hunts insects in air. **Hab:** Open areas, woodlands, swamps. **Voice:** Very high-pitched *ee ee ee;* very high-pitched *taaeee.*

Subspp: (1) •*forficatus.*

IDENTIFICATION TIPS

Kites are a varied group of birds that are only loosely related. They could be informally divided into two subgroups. The Swallow-tailed, Mississippi, and White-tailed Kites all have slender pointed wings and fairly long tails; they are buoyant fliers and catch insects on the wing as well as eat other small animals. The Hook-billed and Snail Kites have long, broad, rounded wings, much like many hawks; both feed mostly on snails. They are buoyant agile fliers.

Adult TX/04

Juv. CA/09

Adult CA/02

Adult CA/02

White-tailed Kite 1

Elanus leucurus L 15"

Shape in Flight: Medium-sized kite with long, narrow, pointed wings and a long thin tail. Trailing edge of wings fairly straight; closed tail slightly notched. Tail length slightly greater than wing width; wing length 3 x width. **Ad. in Flight:** Below, all white except for dark primaries and dark wrist patch; above, pale gray with blackish secondary coverts; white outer tail feathers. **Shape Perched:** Slender short-legged kite with a small, thin, strongly hooked bill; wing tips just reach tail tip. Head large with a steep forehead and well-rounded crown; large eye set well back on side of head; gape long, extending to below center of eye. **Ad. Perched:** Gray above; white below; dark secondary coverts; eye dark red with surrounding black eye patch; bill black. **Juv:** (May–Feb.) Like ad. but with buffy streaks or wash on breast, crown, and nape; eyes orangish yellow (ad. red); back and wing coverts tipped white. **Flight:** Hovers while hunting; may kite into strong winds; otherwise, wingbeats fairly deliberate and gull-like. Wings usually held in slight dihedral. **Hab:** Grasslands with scattered trees, marshes, near highways. **Voice:** Distinctive, rough, piercing *shrrrrr.*

Subspp: Monotypic.

Adult m. FL/–

Adult m. FL/–

Adult f. FL/05

Adult m. FL/04

Adult f. FL/02

Imm. FL/04

Juv. FL/04

Snail Kite 2

Rostrhamus sociabilis L 17"

Shape in Flight: A fairly large kite with long broad wings, fairly small head, and moderate-length tail. Wings of fairly even width. Tail length ¾ wing width; wing length just over 2 x width. **Ad. in Flight:** Below, generally dark with variable pale window at base of mostly outer primaries (most prominent on female); above, all dark except for broad white base to tail overlapped by white uppertail coverts. M. all-gray body; f. heavily streaked below on older adults. **Shape Perched:** Short-legged and fairly small-headed. Fairly long thin bill with long thin hook. Head with rounded crown; neck seems thick. Wings project just past tail. **Ad. Perched:** M. dark gray overall with yellow to orange legs and cere (bright red during breeding); f. with white throat (lacks paler cheek and eyebrow of juv./imm.), brown-and-white streaked breast, slight scaling on back, yellow legs and cere (orange during breeding), and red eye. **Imm:** May take 8–10 yrs. to get full ad. plumage; during this time there may be a variety of plumages midway between juv. and adult. **Juv:** (Jun.–Dec.) Like f. but with extensive tawny wash to head and breast; more extensive tawny scaling on back; more thinly streaked underparts; legs and cere dull yellow; eye brown (ad. red).

Flight: Deep wingbeats; soars on bowed wings. **Hab:** Freshwater marshes. **Voice:** A soft grating *gr'r'r'r'r.*

Subspp: (1) •*plumbeus.*

Adult m. CO/06

Adult m. CO/07

Adult f. CO/05

1st summer CO/07

Adult f. CO/07

Adult f. CO/08

Juv. OK/08

Mississippi Kite 1

Ictinia mississippiensis L 14"

Shape in Flight: Small kite with long tail and long, narrow, pointed wings. With wings extended, trailing edge is straight to slightly convex. Outermost primary considerably shorter than the next inner primary; tail square with sharp corners and, when closed, flares slightly at tip. In similarly shaped falcons, outer primary is just slightly shorter than the next and tail tapers to tip. Tail length = wing width; wing length slightly more than 3 x width. **Ad. in Flight:** Below, gray body and inner wings with darker outer wings and tail. Above, dark primaries and tail contrast with pale gray back and forewings; secondaries can be whitish or gray; head pale. M. head whitish gray, tail uniformly dark, undertail coverts all gray, secondaries above usually whitish; f. head gray, tail dark with pale smudges or lighter gray on outer feathers and dark terminal band, undertail coverts often mottled white and gray, secondaries above grayish or whitish. **Shape Perched:** Slender kite. Head egg-shaped; forehead sloped; crown well rounded at back. Wing tips project just past tail tip. **Ad. Perched:** Plain gray overall with paler head and dark wings and tail; red eye with surrounding black patch; bill black with dark gray cere. **1st Yr:** Similar to ad. f. but with banded tail and sometimes white mottling on breast and belly. **Juv:** (Jun.–Oct.) Underbody with broad brown streaking over buffy background; white eyebrow above dark eye patch; whitish chin. Below, coverts mottled reddish brown and white; variable white at base of flight feathers; tail dark or with up to 3 white bands. **Flight:** Light and aerobatic; does not hover. Wings held flat; often soars high on thermals. **Hab:** Open woodlands, wooded streams, swamps, well-vegetated residential areas, open areas with some trees. **Voice:** Very high-pitched, thin, whistled, and descending *tseeteeuu;* rapidly repeated, descending, short, whistled *tutututu.*

Subspp: Monotypic.

Adult m. MI/04

Juv., faded OH/05

Adult f., fresh plumage NM/11

Juv. TX/12

Northern Harrier 1

Circus cyaneus L 19"

Shape in Flight: Medium-sized long-tailed hawk with long narrow wings usually held in a strong dihedral. Thick neck and small head create a rather blunt projection in front of wing; long outer secondaries create shallow S-shaped trailing edge to each wing. Tail length equal to or slightly greater than wing width; wing length 2¼ x width (tail and wings proportionately longer in female). **Ad. in Flight:** All plumages have bold white uppertail coverts, forming conspicuous squarish patch on flying bird. Ad. m. is mostly pale gray with black tips to outer primaries and dark gray trailing edge to flight feathers (especially secondaries); darker gray head and neck can appear like a hood from below; gray tail with thin dark bars or none on some (possibly older males). F. brown above; streaked brown and buff below; flight feathers strongly barred; axillaries often dark; tail with several dark bands; dark brown facial disk and head can create hooded look. When soaring high overhead, can look similar to accipiters and some buteos; m. mostly pale gray beneath with dark wing tips; f. with dark inner wing. **Shape Perched:** Long slender hawk with rather small head and thick neck; crown flat; facial disk distinctive; bill small; gape short. Wing tips project about ⅔ down long tail; primary projection past tertials very long. **Ad. Perched:** Both sexes have facial disk slightly darker than rest of head; bill dark gray with greenish-yellow cere; eyes pale yellow. M. medium gray above and paler below; forehead whitish. F. mottled brown above, streaked brown and buff below; eyes surrounded by whitish "spectacles." **2nd Yr:** F. similar to ad. f. but with brown eyes; m. like ad. m. but darker gray above with wide barring on secondary coverts, often a rufous crown and facial disk, and faint darker streaking on the nape. **Juv:** (1st yr.) Similar to ad. f. but rich orangish buff below (gradually fades to pale buff by spring) with dark brown streaks primarily on flanks and upper breast (ad. f. has more streaking on belly); above, dark brown wings with cinnamon ulnar bar (ad. f. has grayish tones to flight feathers contrasting with dark brown coverts); eyes pale brown to pale gray

Adult m. CA/11

2nd-yr. m. CA/11

Adult f. TX/01

1st winter CO/01

at first but yellow by 1st winter. **Flight:** Typically hunts low over ground, wings in strong dihedral, and tilts side to side (like Turkey Vulture). Flies high during migration. **Hab:** Open fields, grasslands, prairies, marshes. **Voice:** Emphatic, scratchy, repeated *kree kree kree;* very high-pitched, thin, descending *wheeeuu.*

Subspp: (1) •*hudsonius.*

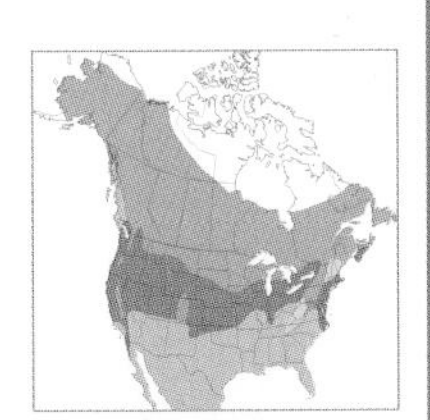

Adult IND/–

Imm. HI/02

Imm. HI/02

Steller's Sea-Eagle 4

Haliaeetus pelagicus L 36"

Casual vag. to AK. **Shape in Flight:** Very large, big-headed, long-tailed raptor with a distinctive sharply wedge-shaped tail. Wings long and broad with narrow base and wide wrist, creating S-shaped trailing edge on each wing. Tail length = wing width; wing length 2½ x width. **Ad. in Flight:** Mostly blackish with white leading edge to inner wing; white tail and undertail coverts; white leg feathers. Bill and feet orangish yellow. **Shape Perched:** Bill massive, deep, and long (bill much larger than Bald Eagle's); culmen continuous with shallow flattened crown. Broad-shouldered; wing tips reach only partially down long tail. **Ad. Perched:** All blackish except for conspicuous white "shoulder" and tail; huge bill orangish yellow. **Imm:** Dark overall with white mottling on axillaries and underwing coverts; tail whitish with dark terminal band. Superficially like imm. Bald Eagle, but Steller's tail shape and large bill distinctive. **Flight:** Wings held in a dihedral. **Hab:** Coasts, ocean islands.

Subspp: Monotypic. **Hybrids:** Possibly with Bald Eagle.

Adult AK/03

3rd yr. AK/03

2nd yr. AK/03

Adult FL/02

Juv., worn FL/01

4th yr. AK/03

Bald Eagle 1

Haliaeetus leucocephalus L 33"

Shape in Flight: Very large heavy-bodied raptor with long evenly broad wings. Wings usually held flat and straight out, like a dark 2 x 8 plank. Head and neck project substantially in front of wings, almost as far as well-rounded tail (Golden Eagle tail much longer than head projection). Tail length ¾ wing width; wing length 3½ x width. Juv. secondaries and tail feathers considerably longer than in other ages, creating a broader inner wing and longer tail. **Ad. in Flight:** All dark with white head and neck and whitish tail (tail can look orangish brown in some lights). **Shape Perched:** Large blocky body; broad shoulders continue in straight line down sides. Wing tips just short of tail tip in ad.; in imm. slightly longer. Long head; massive bill very long and deep; forehead short with flat shallow crown; eye well forward on head. **Ad. Perched:** White head, dark belly, white tail; bill and feet deep yellow.
▶ Distinguishing between imm. plumages of Bald Eagles can be difficult due to great variation in individual maturation; birds in flight give the best clues to age, and the evenness of the wings' trailing edge can be a help. **4th Yr:** More adultlike than any other imm. plumage, but can still have elements of imm. plumage. Perched: Bill and cere color varies from like 3rd yr. to all yellow like ad.; eye pale brown to pale gray to yellow; head whitish with thin dark (often indistinct) line behind eye; belly mostly dark but may have some white speckling. In flight: Trailing edge of wing even, with no protruding juv. secondaries; much like ad. with pale head and tail, and dark belly and underwings; head mostly whitish and body mostly dark with no contrasting dark bib; underwings (including axillaries) mostly dark; undertail white with dark terminal band and maybe dark outer edges. **3rd Yr:** In general darker overall than 2nd-yr. birds. Perched: Bill and cere variable in color, much like 2nd yr. (rarely all yellow like ad.);

Adult FL/02

3rd yr. AK/02

4th yr. WA/01

2nd yr. AK/02

Juv. AK/03

eye color also like 2nd yr. (pale brown or gray, rarely yellow); very similar to and hard to distinguish from perched 2nd yr.; in general has mostly dark underbody, but this is highly variable. In flight: Trailing edge to wings either even or with just 1–2 longer juv. secondaries projecting past the edge; less white on flight feathers than 2nd yr. **2nd Yr:** Perched: Variably pale and streaked tawny to whitish head and neck with dark auriculars (slightly Osprey-like); dark gray bill and cere, sometimes with thin yellow line in front of cere (lower base of upper mandible may be yellowish); pale brown or gray eye (rarely yellow); variably well-defined dark bib contrasts with streaked whitish to mostly whitish belly; tail whitish with dark border or terminal band. In flight: Trailing edge of secondaries very uneven, with new feathers shorter and having more rounded, darker tips; below, whitish to pale head and neck with dark auriculars; dark variably well-defined bib contrasts with mottled whitish belly; whitish axillaries, underwing coverts, and undertail. **Juv:** (1st yr.) Perched: Mostly dark brown head, neck, and bib; pale throat; black bill and cere; yellow gape; dark eye; belly dark brown (when fresh) to tawny to streaked white on flanks. In flight: Below, mostly dark brown with white axillaries and variably mottled white underwing coverts; 2–3 inner primaries are whitish, creating a translucent wedge in the outer wing; tail pale with thin dark terminal band; trailing edge of secondaries even. Above, paler coverts contrast with dark flight feathers. Bill and eye dark. Similar imm. Golden Eagle, in flight, mostly dark beneath with variable white at base of primaries; lacks white in axillaries. **Flight:** Usually seen soaring on straight flat wings; occasional wingbeats are deep with a distinct upstroke (Golden Eagle has even upstrokes and downstrokes). On migration may go miles without flapping. **Hab:** Coast, lakes, large rivers. **Voice:** Weak, high-pitched, short, whistled, variable *tweetwee twee twee;* a rising, drawn-out, whistled *ka'waaeee.* 🎧**17**.

Subspp: Monotypic. **Hybrids:** Possibly with Steller's Sea-Eagle.

Adult CO/12

3rd–4th yr. NAM/–

Adult NAM/–

2nd yr. NAM/12

Golden Eagle 1

Aquila chrysaetos L 30"

Shape in Flight: Very large, long-winged, comparatively long-tailed and small-headed eagle. Head projection past wings only ½ tail length (head and tail about equal projection on Bald Eagle). Inner secondaries shortest, creating slight pinched-in look (where wing joins body) and convex trailing edge to wing. Wings broad throughout and broadly rounded at tip. Tail length about ⅘ wing width; wing length slightly less than 3 x width. **Ad. in Flight:** Appears dark brown overall with golden nape. Below, flight feathers dark gray with faint darker barring and dark tips, creating dark trailing edge to wing; coverts may appear flecked white due to molt (not to be confused with extensive white on underwing coverts of imm. Bald Eagle); undertail coverts buffy or dark; tail dark with 0–2 gray bands. Above, median coverts pale brown, creating V across upperwings. **Shape Perched:** Broad-shouldered, big-bodied, and broad-tailed, with a relatively small head and large bill that is long and deep. Forehead short; crown flat, shallow, and long with a sloping nape. Long gape extends past center of eye. Wing tips just reach tail tip. **Ad. Perched:** Dark brown overall with paler wing coverts and golden nape. Bill with black tip, gray base, and yellow cere; eye pale yellow to dark brown. **4th Yr:** Much like ad. but may have some white at base of inner primaries. **3rd Yr:** Similar to 2nd yr., but trailing edge of wing uneven, due to mix of ad. and juv. secondaries. Tail usually has ad. central and outer feathers (mostly dark) with juv. feathers (dark with white base) in between. **2nd Yr:** Similar to juv., but central tail feathers replaced and may be mostly dark (adultlike) with inconspicuous grayish bands at base (in flight these darker feathers divide tail's white base in two), or juvenile tail feathers may be retained and worn and/or faded. More subtly, the inner 3–5 primaries are fresh and like adult's, with rounded tips (juv. primaries have more pointed tips). Juv. secondaries are all retained at this age. **Juv:** (1st yr.) Like ad., but tail variably white at

Juv. NAM/11

Adult NAM/12

Juv. NAM/04

4th yr./adult NAM/12

2nd yr. NAM/12

base (sometimes hidden below by long undertail coverts) with broad dark terminal band; also variable white patches at base of inner primaries and sometimes outer secondaries (always visible from below, usually from above; may be absent in some individuals). **Flight:** Soars with mild dihedral; during glides, inner wings in dihedral, outer wings leveled off. Wingbeats with equal upstrokes and downstrokes (higher upstroke than downstroke in Bald Eagle). **Hab:** Mountains, foothills, canyons, adjacent grasslands, deserts. **Voice:** Throaty, low-pitched, gull-like *kwow kwow kwow.*

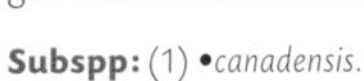

Subspp: (1) •*canadensis.*

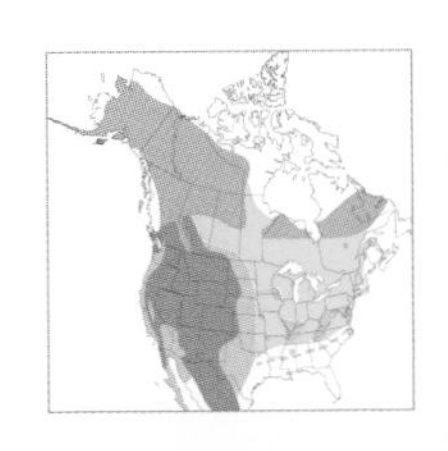

2nd yr. NAM/–

Juv. NAM/01

Adult FIN/03

Imm. (3rd–5th yr.) JAP/03

Adult JAP/12

Juv./imm. FIN/03

White-tailed Eagle 4

Haliaeetus albicilla L 30"

Casual vag. to AK and East Coast; has nested on Aleutian Is., AK. **Shape in Flight:** Large, broad-winged, relatively long-necked, small-headed eagle with a moderate-length wedge-shaped tail. **Ad. in Flight:** Brown overall with a clear white tail; head, neck, and upper breast paler brown than rest of body; bill yellow. **Shape Perched:** Appears fairly long-necked and small-headed; wing tips reach just short of tail tip. **Ad. Perched:** Milk-chocolate-brown head, neck, and upper breast is streaked whitish; dark chocolate body and wings; paler edging to wing coverts and back feathers creates scaled look; tail white; eye and bill yellowish. **Imm:** (3rd–5th yr.) Much like ad. in wing and body color, but tail white with dark terminal band; bill and eye yellow. **Juv/Imm:** (1st–2nd yr.) In flight has longer outer secondaries and longer less wedge-shaped tail. Body and wings variable dark brown (head darkest) with blackish tips to many feathers; bill gray, eye dark, lore pale. In flight, mostly dark with some pale edging to median coverts and axillaries; tail feathers light centrally with dark margins and tips creating a tail that appears mostly dark with a darker terminal band. In 2nd year may get some white blotching on back, wing coverts, and breast. **Flight:** Wings held flat in soar; may take short dips or rises in flight path; may soar for long periods. **Hab:** Coastal or large rivers and lakes.

Subspp: Monotypic.

Adult FIN/11

IDENTIFICATION TIPS

Eagles are the largest of the raptors and have very long, broad, rounded wings and heavy bodies. They do a great deal of soaring and feed on a wide variety of prey, including fish, birds, and mammals. Usually seen soaring.

Accipiters are named for their genus, although their common names end in "Hawk." They have relatively short, broad, rounded wings and comparatively long tails. They usually eat other birds caught in flight, but also eat insects, reptiles, and small mammals. They tend to alternate several flaps with glides when flying.

Adult MN/10

Juv. NV/10

Adult NV/10

Adult NV/10

Juv. NJ/10

Juv. MN/10

Northern Goshawk 1

Accipiter gentilis L 21"

Shape in Flight: Large heavy-bodied hawk with a long broad tail and fairly broad wings. Closed tail rounded at tip with a suggestion of corners. Outer wing more tapered (longer outer primaries) than on other accipiters (Sharp-shinned and Cooper's Hawks); tail broader; body heavier (may even look slightly potbellied; esp. f.). Tail length = wing width; wing length about 2¼ x width. **Ad. in Flight:** Bold whitish eyebrow distinctive for ad. accipiters and often visible in flight. Below, flight feathers and tail gray and heavily barred; underbody and underwing coverts pale gray; undertail coverts large and white. Upperparts dark gray with darker flight feathers and dark bars on tail. **Shape Perched:** Heavy-bodied, fairly long broad tail, sloping "shoulders," and a relatively small head. Bill deep-based; forehead sloping to fairly short shallow crown; eye well forward of center on side of head. Wing tips project less than ½ length of tail; short primary projection past tertials. **Ad. Perched:** Bold whitish eyebrow distinctive for ad. accipiters; bill gray with yellowish cere; eye yellow-orange to red. Below, pale gray with fine barring. **Juv:** (1st yr.) Fine brown streaking on head and neck; darker auriculars; whitish eyebrow, variably distinct; orangish-yellow eye. Back grayish brown with white feather margins and variable white spotting on scapulars; pale bars along side of closed wing are on median and greater secondary coverts. Tail with distinctively zigzag dark bands. Whitish below with heavy streaking on breast and belly (often on undertail coverts as well). Similar juv. Cooper's Hawk slimmer overall with longer, narrower, more rounded tail; also has largely straight bands on tail, and underparts with streaking finer and often densest on upper breast. **Flight:** Rapid flaps alternate with glides; a heavy hawk, more like a large buteo (especially Red-shouldered Hawk) than the other, smaller accipiters. **Hab:** Deep woods, mostly conifers. **Voice:** High-pitched, scratchy, repeated *kree kree kree*. Cooper's Hawk voice lower-pitched and more guttural.

Subspp: (1) •*atricapillus*.

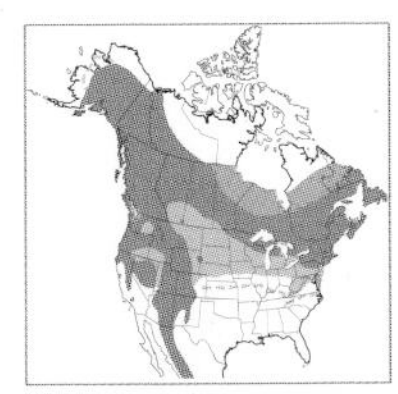

Adult UT/09

Juv. UT/09

Adult NJ/10

Adult NJ/10

Adult MN/09

Juv. MN/09

Sharp-shinned Hawk 1

Accipiter striatus L 11"

Shape in Flight: Small, long-tailed, small-headed hawk with short, broad, rounded wings. Closed tail squarish at corners, tip straight or slightly notched; small head projects just past forward-held wrists on leading edge of wing; long central secondaries create strong S shape to wing's trailing edge. Females considerably larger than males. Tail slightly longer than wing width; wing length about 2 x width. Similar Cooper's Hawk is medium-sized, longer-tailed with a well-rounded tip, larger-headed; comparatively large head and broad neck project well past straighter leading edge of wing; tail noticeably longer than widest portion of wing and often with a slightly wider white terminal band. **Ad. in Flight:** Below, tail and flight feathers whitish strongly barred black; tail with narrow white terminal band. Throat, chest, flanks, and underwing coverts barred reddish brown; large undertail coverts white. Similar Cooper's Hawk tends to have slightly wider white terminal tail band; wingbeats slightly slower and "countable." **Shape Perched:** Small slender hawk with a long somewhat squared tail. Head small; neck thin; forehead angles off culmen (creating small notch) to short rounded crown; eye just forward of center in rounded head creates wide-eyed look. Wing tips project less than ½ length of tail; short primary projection past tertials. Similar Cooper's Hawk has larger head, broader neck; line of culmen continuous with long flat crown; eye well forward of center and near shallow crown, creating fierce look. **Ad. Perched:** Crown and nape concolor with rest of upperparts (blue to blue-gray in m., gray in f.); breast and upper belly variably barred reddish brown (see Subspp.); bill gray with yellow cere; eye orange to deep red; cheeks bright reddish brown. **Juv:** (1st yr.) Head finely streaked dark brown (crown and nape darker); eye yellowish, changing to orange by spring; underparts white with variable streaking on breast and belly (can be distinctly barred on flanks); streaking often heavy and sloppy but can be thin and identical to that on juv. Cooper's Hawk. Upperparts brown and may have white spots along scapulars. Uppertail with wide, even, dark bands and thin white terminal band. **Flight:** Alternates quick flaps with short glides; wingbeats very fast, often too fast to count ("uncountable"); often buffeted about on wind due to small size; often gets in "dogfights" with its own and other species of hawks during migration. **Hab:** Woods and woods edges; regularly hunts at winter bird feeders. **Voice:** High-pitched, harsh, repeated *keu keu keu keu*. Cooper's Hawk voice lower-pitched and more guttural. 🎧**18**.

Subspp: (3) •*perobscurus* (s.w. AK–w. BC) ad. underparts dark reddish brown with indistinct whitish bars and spots, leg feathers unbarred dark reddish brown. •*suttoni* (s.e. AZ–s.w. NM) ad. underparts with sparse reddish-brown barring, leg feathers unbarred reddish brown. •*velox* (rest of USA and CAN range) ad. underparts with heavy reddish-brown patterning, leg feathers pale brown barred with white.

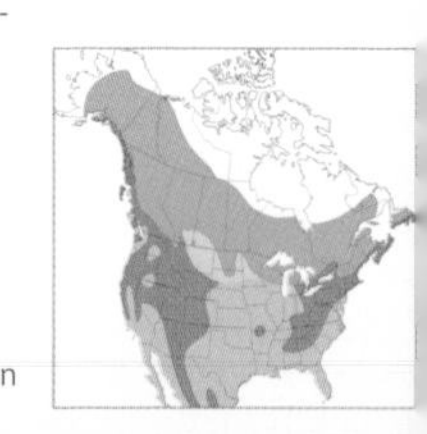

Adult NJ/10

Adult NJ/10

Juv. NJ/10

Juv. NJ/10

Adult PA/11

Juv. NJ/11

Cooper's Hawk 1

Accipiter cooperii L 16"

Shape in Flight: Medium-sized, long-tailed, large-headed hawk with short broad wings. Closed tail well rounded at tip; relatively large head and broad neck project well past straight leading edge of wing; long central secondaries create strong S shape on each wing's trailing edge. Females considerably larger than males. Tail length 1¼ wing width; wing length about 2 x width. See Sharp-shinned Hawk for comparison. **Ad. in Flight:** Below, tail and flight feathers whitish strongly barred black; tail usually with broad white terminal band (often quite worn by spring). Throat, chest, flanks, and underwing coverts barred reddish brown; large undertail coverts white and often held up onto the upper sides of the tail (creating a white-rumped appearance a little like a Northern Harrier's). **Shape Perched:** Medium-sized slender hawk with very long rounded tail. Head large; neck broad; line of culmen continuous with long flat crown; eye well forward of center and near shallow crown, creating fierce look. Wing tips project less than ½ length of tail; short primary projection past tertials. See Sharp-shinned for comparison. **Ad. Perched:** Blackish-gray crown contrasts with slate-gray or rufous nape and blue back of m. or gray to gray-brown back of f.; breast and upper belly barred reddish brown; bill gray with yellow cere; eye orange to deep red; cheeks brown. **Juv:** (1st yr.) Head finely streaked dark brown (crown and nape often darker, creating hooded look); regularly has variable whitish eyebrow; eye pale yellow, changing to bright yellow by 1st spring (eye orange in same-aged Sharp-shinned); auriculars and nape often warmer brown than in Sharp-shinned Hawk. Underparts white with thin dark streaking on breast and belly (often sparse or absent on belly); streaking on belly of similar juv. Sharp-shinned Hawk often has reddish-brown tone. Upperparts dark brown and may have large white spots on scapulars; uppertail with wide, even, dark bands with fine white edges; tail with wide white terminal band. See Northern Goshawk for comparison. **Flight:** Alternates quick flaps with short glides; wingbeats fast but "countable" (Sharp-shinned's often too fast to count); usually steady on wind due to large size; often gets in "dogfights" with its own and other species of hawks during migration. **Hab:** Woods, woods edges; regularly hunts at winter bird feeders. **Voice:** Midpitched, repeated, guttural, grating *gakgakgakgak;* also single *gak.* 🎧**19**.

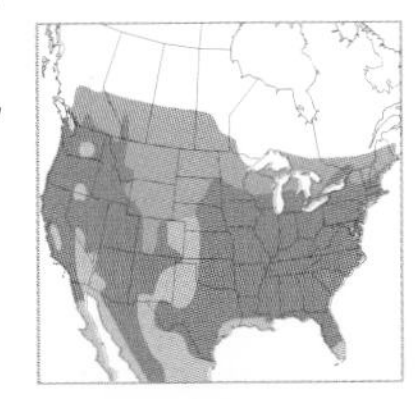

Subspp: Monotypic.

Adult TX/04

Adult TX/04

Juv. MEX/02

Juv. MEX/12

Common Black-Hawk 2

Buteogallus anthracinus L 21"

Shape in Flight: Large, small-headed, broad-winged hawk with a very short broad tail. Trailing edge of wing convex to very shallow S-shaped; wings held flat and often angled forward during soaring. Tail length ½ wing width; wing length about 2 x width. **Ad. in Flight:** Mostly black overall except for tail, which has a broad white basal band and fine white tip. Below, wing coverts and body black; flight feathers silvery and finely barred, with broad black trailing edge. Similar Zone-tailed Hawk has proportionately much longer wings and tail, a straighter trailing edge to wing, and usually holds wings in a dihedral as it tilts side to side like a Turkey Vulture. **Shape Perched:** Compact broad-chested hawk with a small head and broad neck. Long thick bill with short hook; legs long and tail short. Wing tips extend to just short of tail tip. **Ad. Perched:** All black; bill with black tip and very large yellow cere; eye brown; legs yellow. Fine white tip to black tail; broad basal white band mostly hidden. **Juv:** Below, flight feathers and tail heavily barred with thin, dark, wavy lines; tail with wider dark terminal band (wings with no dark trailing edge); underwing coverts and underbody coarsely streaked with dark brown. Streaked head has short white eyebrow and prominent dark malar streak that extends onto sides of neck. Bill black-tipped with dull yellow cere. **Flight:** Slow deep wingbeats; may hover or kite into wind briefly. Soars on flat wings with tail spread. **Hab:** Wooded streams, rivers, dry canyons. **Voice:** Rapidly repeated, rising, whistled *kwee kwee kwee* (almost Osprey-like).

Subspp: (1) •*anthracinus.*

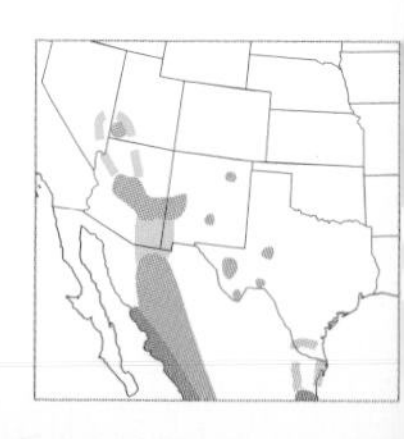

Adult TX/11

Adult TX/11

Juv. TX/12

Adult TX/05

Juv. AZ/12

Harris's Hawk 1

Parabuteo unicinctus L 20"

Shape in Flight: Large hawk with a long tail, fairly long broad wings, and relatively small head. Trailing wing edge shallow S-shaped. Tail length slightly less than wing width; wing length about 2¼ x width. **Ad. in Flight:** Above and below, wings and body blackish brown with dark reddish-brown wing coverts; tail broadly white at base, with broad black subterminal band and white tip; tail coverts white. **Shape Perched:** Slender, long-tailed, long-legged. Head relatively small with long, very deep bill; line of culmen continuous with sloping forehead and gently rounded crown; eye well forward of center. Wing tips project halfway down tail. **Ad. Perched:** Dark brown overall except for reddish-brown wing coverts and leggings; broad pale yellow cere and pale yellow legs; white undertail coverts and tail tip. **Juv:** (1st yr.) Above, dark brown with reddish-brown wing coverts. Below, finely barred tail and flight feathers contrast slightly with dark reddish-brown underwing coverts; tail coverts broadly white; inner primaries form a paler panel from below (much like that of a juv. Red-tailed Hawk). Underbody variably streaked dark brown and whitish. Leg feathers reddish brown or white and barred. Head variable, from mostly dark brown to having pale eyebrow and cheek. Large bill with gray tip and broad dull yellow cere. **Flight:** Alternates several flaps with long glides; wings bowed during soaring and glides. Flight usually low; may hover briefly. **Hab:** Desert scrub, mesquite woodlands; may hunt cooperatively in small groups. **Voice:** Harsh grating *chrrrrr.*

Subspp: (1) •*harrisi.*

Adult MEX/–

Adult CRI/01

Juv. MEX/12

Roadside Hawk 4

Buteo magnirostris L 14"

Casual Mex. vag. to s. TX. **Shape in Flight:** Medium-sized compact hawk of average proportions except for fairly long tail. Longer central secondaries create convex trailing edge to wing; wings held slightly forward. Tail length = wing width; wing length 2 x width. **Ad. in Flight:** Below, dark grayish head and upper breast; reddish-brown-barred belly; white undertail coverts; flight feathers finely barred with no dark trailing edge; tail with multiple thin black and whitish bars; underwing coverts pale reddish brown. Above, grayish brown overall with thin U of buff to white uppertail coverts with darker spots; evenly wide dark and light bands on tail. Similar Broad-winged Hawk's dark trailing edge to wing is straighter with a dark broad outline; similar Red-shouldered Hawk has finer white barring on tail and pale crescent-shaped window at base of outer primaries. **Shape Perched:** Slender upright hawk with a fairly long tail and short primary projection past tertials; wing tips extend slightly less than halfway down tail. Fairly large head with eye centrally placed; culmen of deep-based bill a continuous line with sloping forehead and shallow rounded crown. **Ad. Perched:** Throat and breast solid or indistinctly streaked grayish brown; belly barred reddish brown. Above, grayish brown; even-width light and dark tail bands. Bill blackish with yellowish cere; eye whitish to pale yellow; legs orangish yellow. **Juv:** (1st yr.) Breast with broad brown streaking; belly with irregular brown barring. Above, grayish brown with some white mottling; even-width tail bands gray and dark brown. Face with broad buffy eyebrow and buff surrounding bill; darker brown-streaked cheek and crown. Bill blackish with yellow cere; eye reddish brown; legs pale yellow. **Flight:** Rapid wingbeats alternate with short glides; rarely soars. **Hab:** Open country; often perches along roadsides. **Voice:** High-pitched, downslurred whistle, like *tseeer.*

Subspp: (1) •*griseocauda.*

IDENTIFICATION TIPS

Buteos (mostly in the genus *Buteo*) and their close relatives are medium-sized to large hawks with fairly long, broad, rounded wings and comparatively short to medium-length tails. They are often seen soaring with wings widely spread. They feed on mammals, reptiles, amphibians, birds, and insects. They do a great deal of soaring.

Adult AZ/05

Adult AZ/06

Juv. AZ/07

Juv. MEX/03

Gray Hawk 2

Buteo nitidus L 17"

Shape in Flight: Medium-sized, broad-winged, broad-tailed hawk with a strongly convex trailing edge to wings. Fairly large head and broad neck project well past relatively straight leading edge of wing. Tail length just short of wing width; wing length slightly more than 2 x width. **Ad. in Flight:** Below, mostly pale gray with flight feathers finely barred and fairly dark-tipped, creating a dark outline to wings' trailing edge. Tail black with white tip and 2 white bars. **Shape Perched:** Compact, sturdy-bodied, and somewhat long-tailed; wing tips project slightly less than halfway down tail. Bill fairly deep-based; short sloping forehead and gently rounded crown; neck broad; large eye placed forward of center on side of head. **Ad. Perched:** Pale gray overall with fine gray barring on underparts; tail black with 2 white bars. Bill black with yellow cere; eye brown. **Juv:** (1st yr.) Head whitish with dark crown, dark eyeline, and dark malar streak. Breast whitish and coarsely streaked dark brown; back brown with variable whitish mottling on coverts and scapulars; tail with multiple bars that narrow progressively toward tail base; white U on uppertail coverts; bill black with yellow cere; eye brown. Similar juv. Broad-winged Hawk's trailing edge to wing is straighter and more darkly outlined; also lacks bold white cheek and white U on the uppertail coverts of the juv. Gray Hawk; also can have all-white throat, while juv. Gray always has dark central streak on throat. **Flight:** Accipiter-like; bursts of flaps alternating with glides. Soars on flat wings. **Hab:** Wooded rivers in semiarid areas. **Voice:** High-pitched *sweee* followed by a mellow lower-pitched *kluweee kluweee.*

Subspp: (1) •*plagiata*. **Hybrids:** Red-shouldered Hawk.

Adult FL/02

Juv. MI/03

Adult MI/03

Juv. FL/03

Red-shouldered Hawk 1

Buteo lineatus L 17"

Shape in Flight: Medium-sized hawk with rather average proportions, although wings and tail can appear somewhat long. Wings often held slightly forward and wing tips often appear rather squared off. Long outer secondaries create bulge in central trailing edge of wing. Tail length = wing width; wing length about 2½ x width. **Ad. in Flight:** Below, flight feathers checkered black and white with broad dark band on trailing edge; tail black with narrow white tip and 1–3 narrow white bands (1 shows on closed tail, up to 3 on fanned tail); body and wing coverts reddish brown to pale buff; light whitish or grayish crescent-shaped window across outer wing due to translucent bases to outer primaries (seen in good light at all ages). Above, flight feathers also checkered, with some reddish brown on wing coverts. Similar ad. Broad-winged Hawk from below shows 1–2 broad white tail bands, has no banding on base of outer primaries, and wings with more pointed tips and straighter trailing edge; similar ad. Red-tailed Hawk from below has reddish unbanded tail, dark leading edge to inner wing, dark comma at tips of primary coverts, and a variable belly band. **Shape Perched:** Compact hawk with somewhat long tail; wing tips project ⅔ down tail. Small bill, steep forehead, well-rounded crown, and eye placed well forward of center. Short gape does not reach eye. **Ad. Perched:** Subspp. differ, especially in FL (see Subspp.). Generally wings appear strongly checkered black and white with some reddish brown on wing coverts; tail black with thin white bands. Belly barred reddish brown; breast solid or barred reddish brown. Black bill with yellow cere; dark brown eye. **Juv:** (1st yr.) Breast streaked

Adult, *extimus* FL/03

Juv., *elegans* CA/04

Adult, *lineatus* CT/12

Adult, *elegans* CA/11

Juv. FL/02

Juv. FL/02

brown; belly with brown spots or jumbled barring. Greater secondary coverts, seen on closed wing of perched bird, have 2–3 pale bars. In flight, pale buff translucent crescent at base of outer primaries from above and below. Upperparts brown mottled with white; upperwing coverts may have some reddish brown; tail with multiple dark thin bands. Eye pale brown; cere greenish yellow; bill black. Similar juv. Broad-winged Hawk perched has plain unbarred greater secondary coverts; in flight, has pale panel over all primaries, proportionately broader wings and shorter tail, as well as wings with more pointed tips and a straighter trailing edge. Similar juv. Red-tailed Hawk has pale window encompassing inner primaries, dark leading edge to inner wing, dark comma at tips of primary coverts, a dark belly band, and proportionately broader wings and shorter tail. **Flight:** Wings often pressed forward; bowed down during soars. Alternates a few rapid flaps with glides (accipiter-like). **Hab:** Woodlands, swamps. **Voice:** Wild, drawn-out, descending screech, like *kreeaaah kreeaaah;* shorter 2-part *kreega kreega.* **20**.

Subspp: (4) **Eastern Group:** •*lineatus* (s.e. OK–NC and north, can winter in FL) largest, throat and face contrastingly darker than breast, breast and belly barred pale reddish brown with many dark feather shafts. •*alleni* (OK–SC south to TX–cent. FL) medium-sized, throat and face not contrastingly darker than breast, breast and belly barred pale reddish brown with some dark feather shafts. •*extimus* (s. FL) smallest, throat and face paler than breast, breast and belly barred pale reddish brown with very few dark feather shafts. **Western Group:** •*elegans* (s.w. OR–s.w. CA, vag. to BC, AZ) fairly small, throat and face not contrastingly darker than breast, breast solid bright reddish brown, belly barred bright reddish brown, few or no dark feather shafts on underparts; juv. more adultlike than in Eastern Group. **Hybrids:** Gray Hawk.

Adult MN/09

Adult MI/04

Adult MI/04

Juv. MN/09

Broad-winged Hawk 1

Buteo platypterus L 15"

Shape in Flight: Fairly small compact hawk with tapered outer wing. Tail length 3/5 wing width; wing length about 2 x width. Trailing edge of wings convex during soar, straight to concave during glide. **Ad. in Flight:** Comes in 2 morphs; dark morph extremely rare (in East seen at a few hawk-watch sites, in West seen in Can. boreal forest and, most commonly, at hawk-watch sites). Light morph pale below with darker head and breast; wings pale with distinct dark trailing edge and tips; tail dark with dominant central white band and thin pale tip. Dark morph dark below, with all-dark body and wing coverts; flight feathers whitish except for wide black trailing edge; tail same as that of light morph. Similar ad. Red-shouldered Hawk has heavily checkered underwings, pale crescent-shaped window at base of outer primaries, proportionately narrower wings and longer tail, and wings with squarer tips and more S-shaped trailing edges. Similar ad. Red-tailed Hawk has reddish tail, dark leading edge to inner wing, dark comma at tips of primary coverts, often a dark belly band, and proportionately broader wings and shorter tail. **Shape Perched:** Small, rather slender, average-proportioned hawk; can look somewhat broad-chested. Wing tips project 2/3 down tail. Short steep forehead leads to shallow rounded crown. **Ad. Perched:** Light morph's head and back dark brown; breast and face reddish brown; belly barred reddish brown; undertail coverts whitish; bill black with yellow cere; eye pale orangish brown; wide pale central band on tail. Dark morph with

Juv. MN/09

Adult LA/04

Adult FL/04

Juv. TX/03

all-dark body. **Juv:** (1st yr.) Upperparts brownish gray with some whitish mottling; tail with multiple light and dark bands; underbody variably streaked brown centrally with heavier streaking and/or barring on flanks (juv. Red-shouldered usually more evenly streaked below); throat either all whitish or with central dark streak (Red-shouldered juv. has throat either all dark or with central dark streak). On perched bird, note unbarred greater secondary coverts (juv. Red-shouldered has barred greater secondary coverts); in flight, outer wing (with light from behind) has pale panel encompassing all of primaries. Cere and legs pale yellow; eye pale brown. Similar juv. Red-shouldered Hawk perched has pale barring on greater secondary coverts; in flight, pale crescent at base of outer primaries, proportionately narrower wings and longer tail, and spread wings with squarer tips and more S-shaped trailing edges. Similar juv. Red-tailed Hawk has pale panel encompassing inner primaries, dark leading edge to inner wing, dark comma at tips of primary coverts, a dark belly band, and proportionately broader wings and shorter tail. Dark-morph juv. Broad-winged mostly dark with thin or very thin tawny streaking on underbody and fine dark bands on pale tail from below. **Flight:** Rapid fairly stiff wingbeats; soars with wings flat; glides with wings slightly bowed. Gathers into large groups on thermals during migration. **Hab:** Woodlands. **Voice:** Extremely high-pitched, drawn-out, whistled *tsiseeeeaah.* 🎧 **21**.

Subspp: (1) •*platypterus.*

Adult light morph MEX/02

Adult dark morph MEX/02

Juv. light morph FL/12

Juv. dark morph FL/11

Short-tailed Hawk 2

Buteo brachyurus L 16"

Shape in Flight: Fairly small, compact, broad-winged hawk. During soar, wing tips taper to an abrupt point, trailing edge of wing convex, wings are pressed slightly forward, and tips bend noticeably upward. Tail length ¾ wing width; wing length about 2⅓ x width. **Ad. in Flight:** Dark and light morphs. Light morph pale below with darker flight feathers and tail, dark helmet, and dark trailing edge to wings and tail. Light morphs of all ages in AZ have strongly banded tails; those in FL have little or no tail banding. Dark morph dark below with blackish-brown body and wing coverts, paler flight feathers and tail, dark trailing edge to wings and tail. **Shape Perched:** (Rarely seen perched.) Fairly large-headed, small-bodied, broad-tailed, and long-legged. Bill short but deep-based; short forehead, flat crown; eye well forward on head. **Ad. Perched:** Light morph blackish brown above with striking white throat and underparts; touch of reddish brown on side of neck; white area surrounds bill base. Dark morph all blackish brown. Bill in both morphs black with yellow cere. **Juv:** (1st yr.) Light morph similar to ad. but with sparse streaking on sides of breast and some white on eyebrow and cheek. Dark morph like ad. but with some or much white streaking on belly. **Flight:** Mostly soars, often switching heights abruptly; may kite in a single spot for long periods. **Hab:** Swamps, woodlands bordering open areas.

Subspp: (1) •*fuliginosus.*

Adult light morph FL/02

Juv. light morph FL/04

Juv. dark morph FL/11

IDENTIFICATION TIPS

Hawk Species ID: Hawk identification can be quite challenging and requires using a wide variety of identification clues, including shape, plumage, and behavior. You also need to take slightly different approaches to identifying flying versus perched birds.

Flying Birds—Shape: Comparing wing length to wing width will help you get a sense of wing shape and proportions; comparing tail length to wing width can also give a sense of relative tail length among species. In this guide, when we compare something to wing width, we are always referring to the greatest width of the outspread wing (wings vary in width throughout their length). **Plumage:** Note the amount and width of barring on the tail. Note also the contrasting colors of underwing coverts versus flight feathers and any patterns of translucency (windows) on the underwing. Look for any patterns on the face or head. **Behavior:** Note speed of wingbeat and size of soaring circles; smaller birds have faster wingbeats and soar in tighter circles. Note any buffeting on the wind to help indicate size (smaller birds are more affected by the wind). Look for patterns of wingbeats: continuous flapping, intermittent flaps, alternating flaps and glides.

Perched Birds—Shape: Note where the tips of the folded wings fall relative to the length of the tail. Look at the size of the head, slope of the forehead, and shape of the crown; note the length of the bill and its depth at the base, also the placement of the eye on the head (forward or central). Judge the overall slenderness or heaviness of the body; note the length of the legs and the presence or absence of leg feathering. **Plumage:** Note any patterns on the head; look at the colors and streaking of the belly; look for tail barring; look for the colors of the bill, cere (the bare fleshy area surrounding the nostrils), and eye.

Adult light morph CA/04

Adult dark morph UT/05

Juv. light morph UT/08

2nd-yr. dark morph UT/06

Swainson's Hawk 1

Buteo swainsoni L 19"

Shape in Flight: Fairly large slender hawk with long narrow wings and fairly long tail. Outer wing tapers gradually to a point; wing's trailing edge is convex; wings are held in a shallow dihedral. Tail length = wing width; wing length about 2¾ x width. **Ad. in Flight:** Light, dark, and intermediate morphs occur, as well as a continuum of plumages in between. Light morph below has pale throat surrounded by dark brown to reddish-brown bib; rest of body and underwing coverts pale whitish to pale reddish brown contrasting with dark flight feathers; tail has multiple thin bands and slightly wider dark subterminal band. Above, narrow whitish U at tip of uppertail coverts. M. has gray head; f. has dark brown head. Intermediate morphs are primarily like light morph except the belly is reddish brown; most have the dark bib and sexual head variations of light morphs. A few males, since their belly is reddish brown and blends with the often reddish-brown bib, become all reddish brown on the underparts, except for the whiter undertail coverts. Dark morph has tail and wings similar to light morph's but body all dark brown to blackish except for pale buffy undertail coverts; rufous underwing coverts often have some darker mottling, so contrast with dark flight feathers is limited. Similar dark-morph ad. hawks have dark undertail coverts. Similar ad. Red-tailed Hawk has proportionately broader wings and shorter tail, dark bar along leading edge of inner wing, reddish tail. **Shape Perched:** Fairly heavy-bodied long-winged hawk. Wing tips usually extend just past tail tip; primary extension past tertials fairly long. Head small with well-rounded crown; eye fairly central on head; bill average but with short gape that just reaches forward edge of eye. **Ad. Perched:** Light morph with dark grayish to brown head, back, and wings, often with some paler edging to back and wing feathers; bill black, cere

Adult m., light morph MT/06

Adult dark morph UT/05

Adult f., light morph AZ/04

Juv. light morph CO/09

Juv. dark morph CO/09

yellow; white forehead and throat; reddish-brown to dark brown bib and neck sides; belly and flanks variably barred; m. with distinct gray head, reddish-brown breast; f. with browner head and chest. Dark morph all dark blackish brown except for pale buffy undertail coverts; tail light with fine barring and wider subterminal band. M. has gray head; f. has dark brown head. In similar Red-tailed Hawk, wings extend to tail tip or less, long gape reaches to below center of eye, head is larger and crown flatter. **2nd Yr:** Continues to have traces of juv. plumage but usually has wide dark subterminal tail band; many can look quite pale and worn with pale heads in spring and summer. Juv. and 2nd-yr. Swainson's Hawks' lack of a pale panel on the outer wing helps distinguish them from several other juv./imm. hawks that have them (such as Red-tailed). **Juv:** (1st yr.) In flight, light morph similar to ad. but lacks wider subterminal tail band in 1st yr., has streaked belly (unstreaked on palest birds), extended dark malar areas, and whitish throat and breast rather than complete dark bib; narrow whitish U at tip of uppertail coverts. Head variable, from whitish with dark malar streak to dark and streaked, with paler eyebrow, cheek, and forehead. Back feathers with pale edging and variable white mottling. Dark morph similar but darker overall; uppertail coverts mostly dark. **Flight:** Wingbeats strong and moderately deep; may hover or kite briefly during hunting; soars often and mostly during migration (large groups locally in fall), with wings in dihedral. **Hab:** Prairies, open land, agricultural fields. **Voice:** Very long, drawn-out, descending screech, *shreeeeaaah.* 🎧**22**.

Subspp: Monotypic. **Hybrids:** Rough-legged Hawk.

Adult TX/11

Adult TX/11

Juv., heavily marked TX/01

2nd yr. TX/11

White-tailed Hawk 2

Buteo albicaudatus L 20"

Shape in Flight: Large, long-winged, short-tailed hawk. Outer wing tapers to point; wing strongly pinches in at body, making trailing edge usually strongly convex. Tail length about ⅔ wing width; wing length about 2¼ x width. **Ad. in Flight:** Gray helmet and white throat; whitish belly and underwing coverts; flight feathers darker. Striking black subterminal band across white translucent tail. **Shape Perched:** Relatively compact fairly heavy-bodied hawk with very long wings. Wing tips project well past tail tip; primary projection past tertials very long. Head average size; bill long and deep-based; gape extends to below mideye. **Ad. Perched:** Medium gray head and back; white throat and underbody; white tail with dark subterminal band; variably reddish brown on scapulars and lesser secondary coverts. Cere pale bluish green to yellow. **2nd Yr:** Similar to 1st yr. but with dark subterminal band to otherwise finely barred tail; reddish brown on scapulars. **Juv:** (1st yr.) Below, body variably mottled to all dark with contrasting white breast patch; flight feathers and tail with fine dense barring; underwing coverts lightly to heavily mottled with dark; head mostly dark with variably white auriculars; cere pale green. Above, white U created by uppertail coverts. Similar juv. Red-tailed Hawk has pale flight feathers, pale panel encompassing inner primaries below, pale panel that encompasses all of primaries from above, more broadly rounded wing tips, shorter wings that do not project past tail on perched bird. Similar juv. Swainson's Hawk has dark tail, proportionately narrower wings and longer tail, and wings broadly join body (not strongly pinched in as in White-tailed). **Flight:** Wingbeats slow and heavy; wings held in strong dihedral during soaring; may hover or kite during hunting. **Hab:** Open and partially open grasslands, coastal savanna, marshes. **Voice:** Short harsh *chawee chawee.*

Subspp: (1) • *hypospodius.*

Adult AZ/06

Adult NM/06

Juv. NM/07

Juv. NM/07

Zone-tailed Hawk 2

Buteo albonotatus L 20"

Shape in Flight: Large long-winged hawk. Wings evenly broad to tip; wings' trailing edges usually slightly convex; head appears small; wings held in strong to moderate dihedral. Tail length ¾ wing width; wing length 3 x width. **Ad. in Flight:** Below, black body and wing coverts; flight feathers silvery with fine dense barring and wide dark trailing edge; tail black and, when spread, shows 2 white bands on m. and 3 white bands on f.; bill pale at base with yellow cere; legs and feet yellow; forehead white. Tilts side to side in gusty winds. Above, all blackish gray with grayish tail bands. Similar Turkey Vulture lacks bands on tail and flight feathers; has smaller, bald, reddish head; similar Common Black-Hawk has proportionately much shorter and broader wings, shorter tail, soars on flat wings, and does not tilt side to side. **Shape Perched:** Compact fairly small-headed hawk; perches vertically. Long wings project just past tail; primary projection past tertials long. Bill short but deep-based; head small and rounded; gape short; eye well forward. **Ad. Perched:** All black with pale-based dark-tipped bill and yellow cere; white forehead; eye dark with whitish orbital ring; tail black with broad white central band. **Juv:** (1st yr.) Like ad. but tail with fine narrow bands and wide dark terminal band. **Flight:** Wingbeats deep; holds wings in strong to moderate dihedral and may tilt side to side like Turkey Vulture. **Hab:** Open country near canyons or cliffs. **Voice:** Rising and falling, harsh, wailing *shareeaah.*

Subspp: Monotypic.

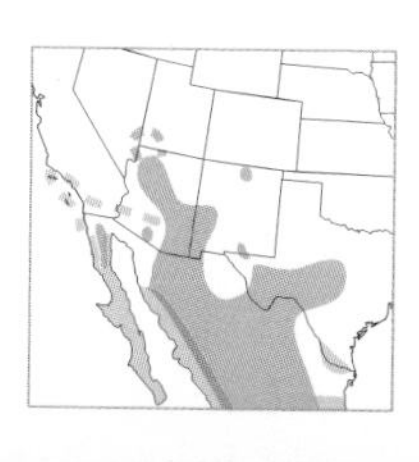

Adult, *borealis* AZ/02

Juv., *borealis* NH/10

Adult, *borealis* CT/10

Juv., *borealis* CT/10

Red-tailed Hawk 1

Buteo jamaicensis L 20"

Shape in Flight: Large, heavy-bodied, broad-winged, relatively short-tailed hawk. Tail length 3/4 wing width; wing length 2 1/3 x width. Longer outer secondaries give "bulging biceps" look to wing width; wing's trailing edge markedly convex or even slightly angled where secondaries meet primaries. Wings held in moderate to slight dihedral or wide shallow U with primary tips curved up. **Shape Perched:** Heavy-bodied and broad-chested tapering abruptly to tail. Adult wing tip in relation to tail length varies with subspecies: wing tips fall noticeably short of tail tip in *borealis* and *harlani*; reach tail tip in *calurus, fuertesi*, and *alascensis*; can reach tail tip in *umbrinus.* Primary projection past tertials fairly long. Bill fairly short but deep-based, so that line of culmen is fairly continuous with shallow flattish crown; head overall quite rounded (not elongated like that of Ferruginous Hawk); gape extends to below center of eye.

Color Morphs: There are light, intermediate, and dark morphs of Red-taileds, depending on the subspecies, and an absolute cline in plumage between all morphs. The most widespread subspecies of Red-taileds are *borealis* (throughout the East) and *calurus* (in most of the West). *Borealis* has only a light morph; *calurus* is 80–90% light morph. Three other subspp., which have limited ranges, have only light morphs: *umbrinus* (FL), *fuertesi* (Southwest), and *alascensis* (n.w. coast). The only intermediate and dark morphs are in 10–20% of *calurus* individuals and in the uncommon "Harlan's Red-tail" (*harlani*), which breeds from cent. AK to YT and winters from e. WA to n.e. CA–IA–LA. This is an important perspective when trying to comprehend color morphs of Red-taileds.

In general, light and intermediate morphs can be identified to species by a dark bar (the patagial bar) on leading edge of inner underwing; dark-morph adults by tail patterns; and juveniles by a pale trapezoidal window encompassing, on the upper surface, the inner primaries and primary coverts and, on the underside, the inner primaries. Descriptions below are first for light morphs; intermediate and dark morphs and subspecies distinctions for adults are dealt with in the Subspp. section. Separating subspecies juveniles is subtle and beyond the scope of this guide.

Light Morphs—Ad. in Flight: Almost all ad. have some reddish brown on the tail, with thin to medium-width dark subterminal band and thin

Ad. light morph, *borealis*, "Krider's" TX/01

Adult, *fuertesi*, "Fuertes's" TX/02

Juv., *borealis* NJ/11

Juv., *borealis* CO/11

Adult, *borealis* CT/01

Adult, *borealis* CT/01

Juv. light morph, *borealis*, "Krider's" LA/02

white terminal band. All ad. with pale underwing coverts (80–90% of all Red-tails) have a dark patagial bar on the leading edge of the inner underwing (also visible on most intermediate morphs; true dark morphs are all dark on underwing coverts). All ad. have dark trailing edge to pale wing. Many ad. have light streaking across the belly, creating a subtle belly band. Almost all light morphs have a somewhat hooded look, with many having white throats. **Ad. Perched:** Reddish-brown tail; suggestion of a belly band (in many individuals); white mottling on scapulars, forming a rough V on the back; often white mottling on upperwing coverts. Young ad. has pale eye; older ad. dark eye. Bill dark with yellowish cere; grayish lores. **Juv:** (1st yr.) In flight, distinctive pale panel above and below encompassing inner primaries. Tail light brown with numerous, even-width, fine dark bands above and below. Trailing edge to wings pale (dark in ad.); pale eye (dark in older ad.); dark coarsely streaked belly band. Juv. generally appears to have narrower wings and longer tail than adult. **Flight:** Wings usually held in a slight dihedral; wingbeats deep and usually slow; may kite into the wind; often soars on thermals. **Hab:** Generally open country with some trees, farmland, rural land. **Voice:** High-pitched, very harsh, scratchy *shreeaaah;* short, 2-part, whistled *klooeee klooeee.* 🎧**23**.

Subspp: (6) The three main subspp. are *calurus* and *harlani* in the West and *borealis* in the East. Two other subspp., *umbrinus* and *fuertesi,* are largely resident and not often seen outside their limited ranges. Subsp. *alascensis* breeds in AK and BC; most depart in winter to ranges still unknown. All subspecies intergrade broadly with neighboring subspp. They are divided into three groups.

Eastern Group: Light morphs only; underparts and underwings mostly light with few to some dark markings; ad. tail reddish brown. •*borealis* (s.w. AB–e. TX and east), "Eastern Red-tail." Light morph only. Above, brown head and upperparts with marked white mottling on scapulars; uppertail coverts white. Tail all reddish brown with few or no bands; throat pale (can be streaked, rarely all dark); body pale with very little reddish-brown wash; dark belly band; unmarked leg feathers. Birds from e. CAN can be more heavily marked. "Krider's Red-tail"—a scarce, very pale morph of *borealis* (found mostly in Great Plains region)—typically has a whitish head, is strongly whitish on central scapulars; white underwing coverts have little or no marking except for reddish-brown patagial bar, tail is whitish for basal 2/3 and pale reddish brown on outer 1/3.•*fuertesi* (s.e. CA–e. UT–s.e. TX), "Fuertes's Red-tail." Light morph only. Similar to *borealis* but with little or no belly band (just a little barring on sides) and gener-

Adult light morph, *calurus* UT/11

Adult light morph, *calurus* NM/01

Adult rufous morph, *calurus* CA/01

Adult rufous morph, *calurus* TX/11

Adult dark morph, *calurus* CO/10

Adult dark morph, *calurus* TX/11

ally a white throat; underwing coverts white or buffy with prominent brown patagial bar; tail pale reddish brown with little or no dark subterminal band. •*umbrinus* (peninsular FL), "Florida Red-tail." Light morph only. Very similar to light-morph *calurus;* generally larger than other subspp.

Western Group: Light, intermediate, and dark morphs; light- and intermediate-morph underparts and underwings with extensive dark marking; ad. tail reddish brown on all but *harlani*. •*calurus* (cent. AK–w.cent. MB south through CA–NM), "Western Red-tail." This is the most variable subsp., with light (most common), intermediate, and dark morphs. Light-morph ad. brown above, with extensive white mottling on scapulars. Below, throat dark or streaked, body and underwing coverts usually washed pale reddish brown, with darker streaks and barring (especially on belly); variably streaked belly band; tail all pale reddish brown, often with fine dark bars (about 50% have a single band on tail); leg feathers spotted or barred. Intermediate-morph ad. similar to light morph (with dark patagial bar) but with dark reddish-brown breast, belly, and legs; minimal white on scapulars. Dark-morph ad. with body and underwing coverts all dark below; tail dark reddish brown with fine barring and dark subterminal band; no white on scapulars. •*alascensis* (coastal, s.e. AK–s.w. BC), "Alaskan Red-tail." Light morph only. Very similar to light- or

Adult light morph, *harlani*, "Harlan's" CO/11

Adult light morph, *harlani*, "Harlan's" CO/11

Adult intermediate morph, *harlani*, "Harlan's" CA/10

Ad. dark morph, *harlani*, "Harlan's" CO/11

Juv. dark morph, *harlani*, "Harlan's" CO/11

Juv. intermediate morph, *harlani*, "Harlan's" CO/11

intermediate-morph *calurus*, and some consider it the same subsp.; the two can be difficult to distinguish in the field. Richly reddish brown below, with wide dark subterminal band on tail.

Harlan's Group: Mostly dark morphs, tail whitish to grayish. •*harlani* (cent. AK–YT–n. BC; winters e. WA–n.e. CA–IA–LA), "Harlan's Red-tail." This subsp. is largely black and white with little or no reddish brown on tail. There are light (uncommon), intermediate (common), and dark (uncommon) morphs. Light-morph ad. has substantial white on face surrounding eye (white "spectacles"); mostly white below, with black patagial bar and variable black streaking on belly and underwing coverts; tail whitish and translucent with thin, irregular, black subterminal band. Intermediate-morph ad. has dark brownish-black body and wing coverts with variable whitish mottling; patagial bar does not stand out. Dark-morph ad. has brownish-black body and wing coverts with no white mottling; whitish flight feathers with fine blackish barring and very wide blackish trailing edge; tail grayish with some fine banding and sometimes a dark subterminal bar. Similar Rough-legged Hawk has a wide, neat, dark subterminal tail band and, on the underwing, dark patches at the wrist.

Hybrids: Rough-legged Hawk.

Adult light morph CO/06

Juv. light morph CA/12

Adult dark morph CA/01

Juv. dark morph UT/11

Ferruginous Hawk 1

Buteo regalis L 23"

Shape in Flight: Large long-winged hawk with a moderate-length tail. Outer wing often tapered to blunt point; trailing edge of wing slightly convex or gently S-shaped; wings held in a slight to moderate dihedral. Tail length just under wing width; wing length almost 3 x width. **Ad. in Flight:** There is a cline in plumages from intermediate to dark morph, but not between light and intermediate morphs. Light morph (common): Below, body and underwings mostly white with variable reddish-brown barring or mottling on coverts and belly; dark rusty brown feathering on legs forms a V that contrasts with pale body and pale (sometimes washed reddish) unmarked tail; dark comma formed by tips of primary coverts; trailing edge of wing has thin darkish subterminal band; pale primaries with distinctively small black tips (gray in juv.). Above, pale translucent base of primaries contrasts with dark rest of wing; back and wing coverts variably reddish brown. Intermediate morph (uncommon): Underparts, including underwing coverts, are richly reddish brown (some individuals with white streaking or mottling on breast). Upperparts, including uppertail coverts, are edged with reddish brown. Dark morph (rare): Below, body and wing coverts all dark chocolate-brown (with no reddish-brown tones), contrasting with pale flight feathers and whitish unmarked tail; variable narrow white comma in primary coverts. Above, all dark brown except for pale translucent base of primaries. **Shape Perched:** Robust, large-chested, broad-shouldered

Adult light morph MT/07

Juv. light morph CO/12

Adult dark morph CO/02

Juv. dark morph NAM/02

hawk with a relatively large head and broad powerful neck. Wing tips almost reach tail tip. Large deep-based bill; long, flat, shallow crown; head overall rather elongated; gape extends under eye; legs fully feathered. **Ad. Perched:** In all plumages, bill black, cere yellow, and long gape yellow. Light morph's back and wing coverts reddish brown; eyeline dark; underparts mostly whitish to variably mottled brown; leg feathers contrastingly reddish brown; tail from below whitish and unmarked. Intermediate-morph underparts richly reddish brown, sometimes with whitish mottling or streaking on breast; head gray on m., brown on f. Dark morph dark brown overall with no paler mottling or tones of reddish brown; tail whitish and unmarked from below. **Juv:** (1st yr.) In all plumages, bill black, cere and long gape yellow. Light morph with distinct dark eyeline (lacking in similar light-morph juv. Red-tails). Below, pale and relatively unmarked overall except for small dark gray tips to primary coverts; lacks patagial bar of juv. Red-tail; tail pale, with dark wide barring on outer 2/3 contrasting with unbarred base; trailing edge to wing has no dark barring; feathering on legs pale to washed tawny. Above, pale base to primaries contrasts with rest of dark upperwing; pale rump contrasts with darker back and tail. Dark and intermediate morphs much like ad. but with outer tail barred darker with 3–4 bands. **Flight:** Wings held in moderate to slight dihedral during soar; wingbeats fairly shallow and stiff. **Hab:** Semiarid open land and grasslands. **Voice:** Raspy, whistled, strongly descending *peeeaaah.*

Subspp: Monotypic.

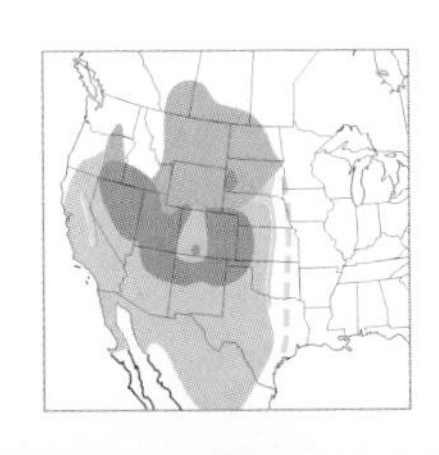

Adult m., light morph, *sanctijohannis* UT/11

Adult f., light morph, *sanctijohannis* UT/11

Adult dark morph, brown type, *sanctijohannis* UT/11

Adult dark morph, black type, *sanctijohannis* MN/10

Rough-legged Hawk 1

Buteo lagopus L 21"

Shape in Flight: Large, heavyset, long-winged, fairly long-tailed hawk. Outer wing mostly even width and rounded; trailing edge of wing slightly convex or gently S-shaped. Tail length about ⅘ wing width; wing length 3 x width. **Ad. in Flight:** Light and dark morphs, with a cline of plumages in between. Light morph (common): Below, blackish primary coverts create squarish wrist patch (carpal patch) just beyond bend in wing which contrasts with variably mottled secondary coverts and pale flight feathers; broad dark trailing edge to pale flight feathers; tail with a wide dark subterminal band; often a pale U on variably dark chest. M. has barred flanks; f. has solid dark (unbarred) flanks. Above, white base of tail contrasts with rest of dark upperparts. Dark morph (uncommon): Comes in "brown" type (m. and f.) or "black" type (m. only). Below, dark brown or black body and underwings ("brown" type often has paler breast); wide dark trailing edge to pale flight feathers; tail with broad black subterminal band and 0–4 thin whitish to gray inner bands (both sexes share most tail patterns in either morph). All-black birds are always m.; brown birds can be either sex. On brownish dark morphs, the black carpal patch may be apparent (hidden on blackish dark morphs) because of contrast with the reddish underwing coverts. Above, base of tail dark and concolor with rest of upperparts. **Shape Perched:** Heavyset, broad-necked, with relatively small head. Wing tips reach or just exceed tip of tail. Bill small and stubby; gape short; forehead steep; crown high, short, and sharply rounded; eye well forward on head. Feathered legs and noticeably small feet. **Ad. Perched:** In all adults, note distinctive small pale "mask" created by the combined whitish forehead and front half of the lore; most obvious on light-morph males and both sexes of dark morphs. Light morph has pale head with a finely streaked crown and nape; dark eyeline; blackish bill with yellow cere; dark back heavily mottled with gray; tail with dark subterminal band. F. with solid dark flanks; m. with barred flanks. Dark morph mostly dark brown or sooty black with some paler mottling on back, sometimes a paler head; tail has wide dark subterminal band and 0–4 thinner inner bands; note distinctive white "mask." **Juv:** (1st yr.) Trailing edge of wings grayish (black in ad.). Light morph:

Juv. light morph, *sanctijohannis* AZ/–

Adult m., light morph, *sanctijohannis* UT/11

Adult f., light morph, *sanctijohannis* CO/12

Juv. dark morph, *sanctijohannis* CO/–

Adult dark morph, *sanctijohannis* CO/11

Juv. light morph, *sanctijohannis* CO/02

Below, dark carpal patches contrast with paler rest of underwing; tail with white base and grayish tip; dark belly patch. Above, base of primaries pale but coverts dark; base of tail white, contrasting with rest of dark upperparts. Dark morph: Distinctive small pale "mask." Below, dark carpal patches contrast with variable dark mottling on underwing coverts; tail with whitish base, grayish tip. Above, base of tail dark and concolor with rest of upperparts. Takes up to 2 yrs. to attain ad. plumage. **Flight:** Wings held in a mild dihedral during soaring; during glides, inner wings in dihedral, outer wings leveled off. While hunting, flaps loosely or hovers, often weaving back and forth. **Hab:** Tundra in summer; open country, often with scattered trees, in winter. **Voice:** Clear, descending, sometimes quavering *keeaaah*.

Juv. dark morph, *sanctijohannis* NAM/07

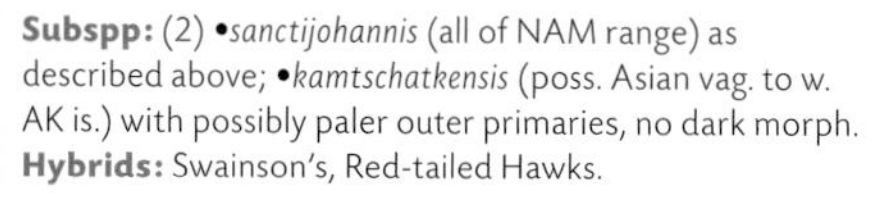
Subspp: (2) •*sanctijohannis* (all of NAM range) as described above; •*kamtschatkensis* (poss. Asian vag. to w. AK is.) with possibly paler outer primaries, no dark morph. **Hybrids:** Swainson's, Red-tailed Hawks.

Adult TX/02

Adult TX/12

Crested Caracara 1

Caracara cheriway L 23"

Shape in Flight: Large, long-necked, long-tailed raptor with long narrow wings. Projection of head and bill past wings extensive. Tail length = wing width, approx.; wing length slightly less than 3 x width. Wings evenly broad with well-rounded tips. Bill very long and as deep as head. **Ad. in Flight:** Whitish wing tips, tail base, and neck contrast with rest of blackish body and wings. Whitish tail has broad black terminal band and inconspicuous finer bands interiorly. Cere and lores orange to yellow; bill pale blue. **Shape Perched:** Slender, long-tailed, long-legged, long-necked. Massive bill; culmen continuous with line of shallow flat crown; crest often squared off at nape. **Ad. Perched:** Blackish cap and nape; whitish cheeks; bill blue, cere and lores orange to yellow. Neck and breast white; breast, lower neck, and back finely barred darker. Rest of body and wings blackish. Legs yellow. **Juv:** (1st yr.) Like ad., but dark areas brown (not black), neck and face buffy, legs grayish, and bill pale with horn-colored to pinkish cere and lores. Upperwing coverts tipped pale; chest streaked rather than barred. In 2nd yr., like juv., but breast barred as in adult. **Flight:** Wings often bowed down during glides; wingbeats even and rowing. Flight path often low. **Hab:** Grasslands, mesquite, livestock fields, dumps. **Voice:** Soft, ratchety, rapid *chchchchch*.

Subspp: Monotypic.

Juv. TX/12

Adult m. EUR/05

Adult m. OMA/05

Eurasian Kestrel 4

Falco tinnunculus L 13½"

Casual Eurasian vag. to w. AK; accidental to either coast. **Shape in Flight:** Medium-sized long-tailed falcon with outer wing strongly tapered and slightly blunt-tipped. Wings often pressed forward at wrist, almost matching projection of head. Tail length about 1½ x wing width; wing length about 3 x width. **Ad. in Flight:** Below, wings whitish with fine even barring; body buffy with dense steaking and/or spotting on the breast. Above, reddish-brown back and secondary coverts contrast with dark outer wing (primaries and their coverts). M. back and inner wings reddish brown with scattered darker spots; tail below grayish with 2 dark subterminal bars (bars may appear joined); tail above grayish with single subterminal bar. F. back and inner wings heavily barred brown and reddish brown; tail pale reddish brown above and below, with fine barring and wider dark subterminal bar. Similar American Kestrel smaller with proportionately narrower wings; has 2 dark cheek bars on each side of face; paler underparts; m. with bluish-gray upperwing coverts; f. with reddish-brown upperwing coverts. **Shape Perched:** Long-winged, long-tailed, short-legged look. Bill short and stubby; forehead steep and crown well rounded. Appears neckless. Wing tips extend just short of tail tip; primary projection past tertials very long. **Ad. Perched:** M. has reddish-brown back and inner wings with sparse black spotting; breast buffy with dense spotting; one dark moustache streak below eye; head grayish. F. heavily barred brown and reddish brown on back and inner wings; one dark moustache streak below eye; head buffy with dark streaking. **Flight:** Continuous with little gliding; often hovers with tail bent down and fanned while hunting. **Hab:** Open fields and marshes near trees.

Subspp: (1) •*tinnunculus*.

Adult f. OMA/01

IDENTIFICATION TIPS

Falcons have pointed wings and comparatively long tails. They are agile fliers and feed on birds, insects, and small mammals. Compared with accipiters, they tend to beat their wings more continuously when flying.

Adult m. UT/09

Adult m. FL/01

Adult m. UT/09

Adult f. UT/10

American Kestrel 1

Falco sparverius L 9"

Shape in Flight: Small long-tailed falcon with long, narrow, pointed, and strongly tapered wings. Trailing edge of extended wing usually slightly convex or straight. Tail length greater than wing width; wing length 2½ x width. **Ad. in Flight:** Underwings whitish with fine barring throughout; 2 dark vertical stripes on side of whitish face; back reddish brown with fine dark barring. M. with blue-gray upperwing coverts; reddish-brown tail with wide dark subterminal bar, and white outer tail feathers usually with wide dark bars; trailing edge of wings has subterminal row of white translucent dots; belly spotted black. F. all reddish brown above with fine dark barring; belly streaked reddish brown; tail reddish brown with multiple fine dark bars, last often widest; trailing edge of wings has subterminal row of buffy spots. Similar Merlin has one moustachial bar and (except for subsp. *richardsonii*) is darker overall and lacks reddish-brown or bluish-gray tones to upperparts. Similar Eurasian Kestrel is larger with proportionately broader wings and has one cheek bar; m. and f. have reddish-brown inner wings and sharply black outer wings. **Shape Perched:** Fairly compact, proportionately large-headed, and long-tailed; wings extend just over halfway to tail tip;

Male UT/09

Adult m. FL/01

Female NV/10

Adult f. FL/01

primary projection past tertials long. Sexes about equal in size (in our other falcons, f. noticeably larger). **Ad. Perched:** Typically pumps tail while perched. Two dark vertical bars on whitish face; lightly barred reddish-brown back. M. with blue-gray wings and crown; underparts whitish with dark spotting. F. with reddish-brown wings and crown; underparts whitish streaked brown. **Juv:** (1st fall) Similar to ad. of respective sex; juv. m. may have fine dark streaking on upper breast. **Flight:** Agile and buoyant; outer wings often swept back, creating sickle shape or "drawn bow" look; often hovers while hunting. **Hab:** Wide variety of open habitats, including urban areas. **Voice:** High-pitched quickly repeated *kleekleekleeklee*. 🎧**24**.

Subspp: (3) •*sparverius* (most of NAM range) large with proportionately small bill, m. belly heavily spotted; •*paulus* (s. LA–cent. SC and south) smaller with proportionately large bill, m. belly largely unmarked; •*peninsularis* (s. CA–s. AZ) like *paulus* but smaller and with virtually no marking on back or belly.

Adult m., *columbarius* NJ/10

Adult f., *columbarius* NJ/10

Adult f., *columbarius* NJ/10

Adult m., *richardsonii* CO/02

Merlin 1

Falco columbarius L 10"

Shape in Flight: Small compact falcon with a relatively short tail and moderately long strongly tapered wings. Head broad; neck short. Tail length slightly greater than wing width; wing length $2\frac{1}{2}$ x width. **Ad. in Flight:** Generally appears dark overall, but subspp. vary (see Subspp.). Below, heavily streaked breast and belly with paler throat and undertail coverts; underwings generally dark with pale spotting; tail dark with pale bands. Above, back and wings solid brown (f.) or dull bluish gray (m.). Faint cheek bar on each side of face. Similar American Kestrel has 2 cheek bars on each side, is generally pale underneath, has reddish-brown tail. **Shape Perched:** Compact sturdy-bodied falcon with a fairly big head and long tail. Wing tips extend $\frac{2}{3}$ length of tail; primary projection past tertials moderately long. Short stubby bill; steep forehead; rather squared-off crown. **Ad. Perched:** Back, wing coverts, and crown concolor and unmarked bluish gray (m.) or brownish (f.) (differences subtle). Underparts heavily streaked on whitish to buffy background; throat pale. Faint cheek bar on paler cheek; thin white eyebrow in some subspp. **Juv:** (1st yr.) Like ad. female; may be distinguished from ad. f. in late summer and fall by lack of molt in wings. **Flight:** Fast, direct, with rapid powerful wingbeats and occasional glides. Wings usually held flat. **Hab:** Forest edges, farmlands, open woods, prairies, marshes, beaches. **Voice:** Loud, rapid, fairly high-pitched *keekeekeekee*.

Adult m., *suckleyi* AK/09

Adult f., *columbarius* CA/04

Subspp: (3) •*columbarius* (AK–OR east to NF–ME), "Boreal Merlin" or "Taiga Merlin"; underparts with dark brown streaking and reddish-brown leg feathers, underwings dark brown with pale spotting, tail dark with 3–4 narrow pale tawny or gray bands, whitish eyebrow. •*richardsonii* (AB–MB south to WY), "Richardson's Merlin" or "Prairie Merlin"; underparts paler with thin brown streaking and whitish to very pale reddish-brown leg feathers, underwings pale grayish with paler spotting, tail dark with 3–4 broad pale tawny or gray bands, whitish eyebrow. •*suckleyi* (coastal s.e. AK–n.w. WA), "Black Merlin"; very dark overall, underparts heavily streaked with dark brown and leg feathers reddish brown, underwings very dark with little or no pale spotting, tail dark with 0–2 faint partial bands, little or no whitish eyebrow.

Adult MOR/09

Adult JAP/09

Adult OMA/09

Eurasian Hobby 4

Falco subbuteo L 12½"

Casual Eurasian vag. to w. AK; accidental to WA. **Shape in Flight:** Medium-sized long-tailed falcon with long narrow wings. Outer wing long and gradually tapered to thin point; tail squarish. Tail length substantially greater than wing width; wing length 3 x width. **Ad. in Flight:** Head has dark mask and strong vertical cheek bar. Below, dark with heavy barring on wings and heavy streaking on breast and belly; throat whitish; undertail coverts and leg feathers reddish brown; above, uniformly dark gray. Ad. tail plain above, many fine dark bars below. Similar Merlin lacks reddish-brown undertail coverts, has barred tail above, has pale cheek bar and no dark mask; similar Peregrine Falcon lacks reddish-brown undertail coverts and is much larger and bulkier. **Shape Perched:** Slim, long-winged, long-tailed falcon; wings reach tail tip and primary projection past tertials extremely long. Relatively large head with a steep forehead, rather flattened crown, and sloping nape; substantial bill for a falcon. **Ad. Perched:** Head with gray crown, fine whitish eyebrow, black mask, and strong black vertical cheek bar contrasting with white neck and throat. Heavily streaked breast; reddish-brown undertail coverts and "leggings." **Juv:** (1st yr.) Similar to ad. but pale tips to primaries and coverts create scaled look on wings; underparts buffy, lack reddish brown of adult. **Flight:** Fast, powerful, and often low; outer wing often bent back and finely pointed. **Hab:** Variable; farmland, marshes, forests, river edges.

Subspp: (1) •*subbuteo.*

Adult light intermediate morph AK/–

Juv. gray morph AK/08

Adult AK/06

Adult gray morph ID/01

Gyrfalcon 2

Falco rusticolus L m. 20", f. 25"

Shape in Flight: Very large, heavy-bodied, broad-necked falcon; tail very broad and moderately long; wings fairly long and broad-based for a falcon; outer wing tapers to blunt tip. Tail length = wing width; wing length 2⅔ x width. F. substantially larger than m. **Ad. in Flight:** Light, gray (most common and widespread), and dark morphs as well as intermediates. Light morph: Below, body and wing coverts mostly white with some fine spotting; flight feathers and tail white with fine dark barring; primary tips blackish. Gray morph: Below, body whitish, heavily barred and spotted darker; cheek bar evident; underwing coverts whitish with dark barring, making them slightly darker than indistinctly barred grayish primaries and secondaries; primaries tipped blackish; tail finely barred. Dark morph: Below, body and underwing coverts dark brown with some pale spotting and/or streaking; flight feathers contrastingly pale gray with darker tips to primaries; tail finely barred. **Shape Perched:** Heavy-bodied, with broad chest and sometimes hunched back; legs fully feathered and short, adding to compact muscular shape. Head small, rounded, with relatively long deep-based bill. Tail moderately long; wing tips extend ⅔ down tail; primary projection past tertials fairly short. **Ad. Perched:** Light morph white overall with blackish spotting

Adult light morph CO/-

Juv. gray morph CO/12

Adult light intermediate morph AK/06

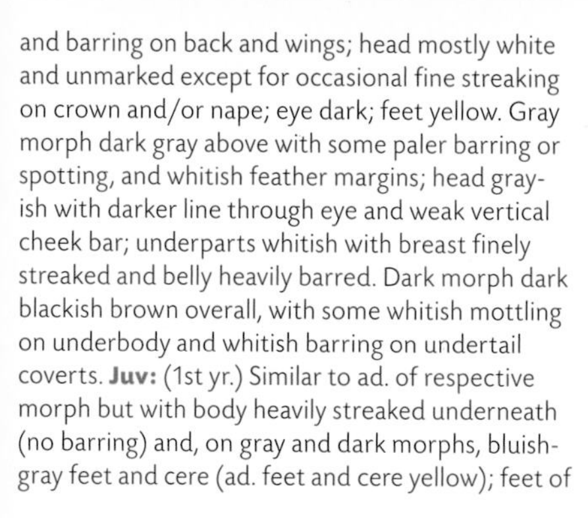

and barring on back and wings; head mostly white and unmarked except for occasional fine streaking on crown and/or nape; eye dark; feet yellow. Gray morph dark gray above with some paler barring or spotting, and whitish feather margins; head grayish with darker line through eye and weak vertical cheek bar; underparts whitish with breast finely streaked and belly heavily barred. Dark morph dark blackish brown overall, with some whitish mottling on underbody and whitish barring on undertail coverts. **Juv:** (1st yr.) Similar to ad. of respective morph but with body heavily streaked underneath (no barring) and, on gray and dark morphs, bluish-gray feet and cere (ad. feet and cere yellow); feet of white-morph juv. are pinkish. May take 2–3 yrs. to attain full ad. plumage. **Flight:** Slow, powerful, deep wingbeats. **Hab:** Arctic tundra, often near coasts, in summer; open country, often near water, in winter. **Voice:** Fairly weak, high-pitched, percussive *k't't't k't't't;* also louder grating call.

Subspp: Monotypic. **Hybrids:** Peregrine Falcon.

Adult, "Eastern type" NJ/10

Adult, *tundrius* NJ/–

Adult, "Eastern type" NJ/10

Juv., *tundrius* NJ/10

Peregrine Falcon 1

Falco peregrinus L 16"

Shape in Flight: Large falcon with a moderate-length tail and moderately long wings; proportionately long outer wing tapers to a point. Tail length = wing width; wing length 2½ x width. Closed tail is slightly tapered. **Ad. in Flight:** Below, wings and body uniformly barred gray except for whitish breast and throat (see Subspp.); wide black cheek bar often visible (see Subspp.). Above, generally uniformly smooth bluish gray except for slightly paler rump and uppertail coverts, which contrast with darker (see Subspp.) finely barred tail. In general m. is more bluish, f. more brownish. Similar Prairie Falcon is paler, has thinner cheek bars, dark axillaries, and, in ad., tail paler than back and rump; similar light-morph Gyrfalcon concolor grayish above, lacks pale rump and dark tail of Peregrine, underwings usually two-toned with coverts darker than flight feathers, cheek bar thin and indistinct. **Shape Perched:** Moderately heavy-bodied, broad at the chest, tapering evenly to the tail tip. Wings project just to tail tip; primary extension past tertials fairly long. Bill deep-based; crown well rounded to sloping nape. **Ad. Perched:** Dark crown, face, and cheek bar create helmeted look; throat and breast whitish to buffy and clear or streaked; belly and undertail coverts variably barred. Bill black-tipped with bluish base and yellow cere. See Subspp. for variations.

Adult, *anatum* CA/01

Adult, "Eastern type" CT/01

Juv., *pealei* WA/03

Juv., *anatum* NM/08

Juv: (10–14 mos.) Similar to ad. but brownish overall. Below, body pale to rich buff with dark brown streaking (ad. is barred). **Flight:** Fast and seemingly effortless; slices through air with great speed and grace; wingbeats moderately deep; soars and glides on flat wings. **Hab:** Open country near cliffs, urban areas, coast. **Voice:** Loud, fairly high-pitched, harsh *shreeshreeshreeshree*.

Subspp: (3) •*pealei* (coastal from Aleutian Is. to w. WA, largely nonmigratory, winters along West Coast), "Peale's Peregrine"; darkest subsp., wide cheek bar, dark gray hood generally concolor with dark gray back, breast white with distinct streaks or spots, underparts heavily streaked, forehead and lore black or whitish. •*tundrius* (n.w. AK–n. NF, migratory, winters s. CA coast and Gulf and Atlantic coasts), "Arctic Peregrine" or "Tundra Peregrine"; palest subsp., narrow cheek bar, blackish hood contrasts with pale gray back, breast whitish and unmarked, underparts less heavily marked, forehead and lore usually white. •*anatum* (most of NAM range, migratory, winters in lower 48 states), "American Peregrine"; moderately dark subsp., moderately wide cheek bar, dark gray hood contrasts slightly with bluish-gray back, breast cinnamon with few or no dark streaks or spots, underparts moderately barred, forehead and lore usually all black. Reintroduced birds, especially in the East, may combine traits of several subspecies as well as the European subspp. *peregrinus* and *brookei*. They are sometimes referred to as "Eastern type." **Hybrids:** Prairie Falcon, Gyrfalcon.

Adult WY/06

Adult UT/05

Adult WY/06

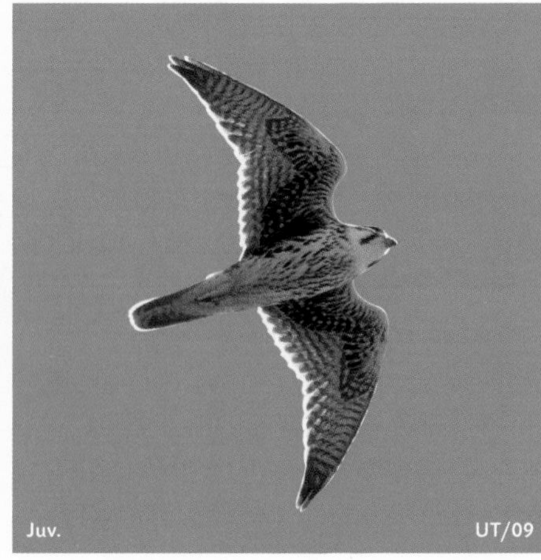
Juv. UT/09

Prairie Falcon 1

Falco mexicanus L 16"

Shape in Flight: Large falcon with a long tail and moderately long broad wings; proportionately long outer wing tapers to a blunt point. Closed tail is slightly tapered to a rounded tip. Tail length slightly greater than wing width; wing length 2½ x width. **Ad. in Flight:** Below, dark brown greater coverts and axillaries contrast with rest of pale brown barred wings and tail; throat whitish; belly finely spotted with brown. Above, uniformly barred brown with tail paler than back and rump. Similar Peregrine Falcon has wide cheek bar, underwings and axillaries darker and concolor (no contrastingly darker axillaries), and tail concolor with back. **Shape Perched:** Moderately heavy-bodied, long-tailed, fairly big-headed; somewhat slender-looking. Bill deep-based; short forehead leads to shallow

Adult UT/05

Adult NAM/06

Juv. UT/11

flattish crown. Wings extend ¾ to tail tip; primary extension past tertials fairly long. **Ad. Perched:** Variably barred brown back, wings, and tail. Head with brown crown, white eyebrow, and thin brown cheek bar; white on cheek extends narrowly up behind eye. Underparts whitish; belly lightly spotted with brown. **Juv:** (1st yr.) Similar to ad. but heavily streaked on breast and belly. Cere and orbital skin pale gray, turning yellow by winter. Above, tail concolor with rest of upperparts (on ad. it is slightly paler). **Flight:** Wingbeats shallow, stiff, and usually mostly below the body plane; often flies low when hunting; gliding is common; soars and glides on flat wings. **Hab:** Plains, grasslands, marshes, other open country. **Voice:** High-pitched series of flat calls, *kwee kwee kwee kwee.*

Subspp: Monotypic. **Hybrids:** Peregrine Falcon.

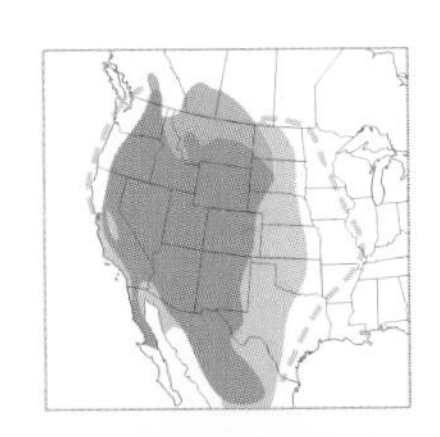

Adult TX/05

Adult VEN/01

Juv. TX/09

Juv. TX/09

Aplomado Falcon 3

Falco femoralis L 16"

Rare resident of s. TX. **Shape in Flight:** Large falcon with a very long very thin tail; wings moderately long, and outer wing gradually tapered to a fine point. Head and neck project substantially past wings' leading edge, but wrists often pressed forward by about the same amount. Tail length 1½ x wing width; wing length 2½ x width. **Ad. in Flight:** Below, relatively colorful: body three-colored with whitish throat and breast, blackish belly band (narrowed centrally), pale buffy lower belly and undertail coverts. Wings mostly dark with fine pale spotting, except for buffy primary coverts and leading edge to inner wing and broad white trailing edge to secondaries; axillaries blackish. Tail dark with thin gray bars; head strongly patterned (see Ad. Perched). Above, dark gray overall. M. has unstreaked white breast; f. similar but with fine streaking on breast. **Shape Perched:** Fairly slender and long-tailed with a substantial head and neck. Bill comparatively large and deep-based; line of culmen blends smoothly with sloping forehead and flat crown. Wings extend ⅔ down tail; primary projection past tertials long. **Ad. Perched:** Distinctive head has black crown, wide white eyebrow, black mask and vertical cheek bar contrasting with white cheek and throat. Breast white, contrasting with black belly; lower belly and undertail coverts pale buffy brown. **Juv:** (1st yr.) Similar to ad. but upperparts brownish; eyebrow, cheek, and throat buffy; breast heavily streaked; tail with more numerous thin white bars. **Flight:** Agile, with deep wingbeats; may hover while hunting. **Hab:** Grasslands with yuccas. **Voice:** High-pitched rapid *kikikikiki*.

Subspp: (1) •*septentrionalis*.

Rails

Gallinules

Coot

Limpkin

Cranes

Adult TX/04

Adult OK/10

Yellow Rail 2

Coturnicops noveboracensis L 7¼"

Shape: Very small, compact, round-bodied rail with a very short tail and short somewhat deep-based bill. **Ad:** Back blackish brown with short, broad, buffy streaks and crossed by fine white lines; underparts buffy, palest on belly. Head with dark crown and diffuse dark eyeline; eyebrow and rest of face buffy. Bill usually blackish except during breeding, when m. bill becomes yellow; m. bill may be slightly paler than f. bill at all times. Similar juv. Sora larger and lacks buffy streaking on back. **Juv:** Similar to ad. but duller overall and with fine white spotting on crown and nape. **Flight:** In short flights, body held at an angle with feet dangling beneath; stealthy and rarely flushes. White secondaries show in flight; below, axillaries also white. **Hab:** Wet meadows, coastal or inland marshes.

Voice: Dry percussive sounds in regular rhythm of 2 slow, 3 fast: *tik tik tiktiktik, tik tik tiktiktik*.

Subspp: (1) •*noveboracensis*.

Adult, *jamaicensis* TX/04

Adult, *jamaicensis* TX/04

Black Rail 2

Laterallus jamaicensis L 6"

Shape: Tiny compact rail with a short pointed tail and relatively short slender bill. **Ad:** Grayish black overall with variable reddish brown on back, nape, and shoulders (see Subspp.); variable fine white spotting over back, wings, and tail. Eye deep red; bill black. F. has pale gray to white throat; m. has pale gray to medium gray throat. Similar chicks of Sora and Virginia Rail are also black, but they are somewhat fluffy, have pale bills, and lack the white spotting on the back and reddish-brown nape of the Black Rail. **Juv:** (May–Aug.) Similar to ad. but eye dusky to reddish brown. **Flight:** In short flights, body held at an angle with feet dangling beneath; all-dark wings; stealthy and rarely flushes. **Hab:** Salt and freshwater marshes, wet meadows. **Voice:** Distinctive, with 2 percussive sounds followed by a low mellow sound, much like the written *kikipooo, kikipooo; grrr grrr grrr* calls often during aggression; also sharp bark.

Subspp: (2) •*jamaicensis* (e. and cent. NAM) larger, with comparatively large thick bill; reddish brown on nape and shoulders, crown grayish to grayish brown. •*coturniculus* (CA–s.w. AZ), "California Black Rail," smaller, with thinner bill; extensive reddish brown on back, nape, and shoulders, crown dark brown.

Adult SCO/–

Adult summer FL/02

Adult winter GA/04

Juv. BC/09

Corn Crake 4

Crex crex L 11"

Casual Eurasian vag. to east coast of NAM, mostly fall. **Shape:** Fairly large, slightly heavy-bodied, short-necked, short-tailed bird with a short, deep-based, conical bill. Often quite upright in stance. **Ad:** Back feathers warm reddish brown with blackish centers; belly barred reddish brown; bill pinkish. M. crown and broad eye-stripe reddish brown, rest of face and throat gray. F. more reddish brown overall on face and throat. **Juv:** Like ad. f. **Flight:** Above, reddish-brown inner wings. Legs trail behind tail. **Hab:** Wet meadows, lush fields, marshy lake edges. **Voice:** Display call a repeated, rising, hoarse *ereep ereep ereep,* given mostly at night.

Subspp: Monotypic.

Sora 1

Porzana carolina L 8¾"

Shape: Small, stocky, chickenlike rail with a short triangular tail, often flipped up, and short, deep-based, conical bill. Legs short and thick. **Ad. Summer:** Bright yellow bill abuts black mask on foreface. Upperparts dark brown and black with fine white streaking. Face and breast gray; throat black; flanks dark with whitish barring; undertail whitish. **Ad. Winter:** Like summer, but black on throat may be hidden by pale gray tips to fresh feathers. **Juv:** (Jun.–Mar.) Like ad. but with buffy face and breast, white chin, no mask, and brownish bill. **Flight:** Short flights; body held at an angle with feet dangling beneath. **Hab:** Salt and freshwater marshes, wet meadows. **Voice:** Sharply rising *kewweeet kewweeet;* long, sharply descending, and slowing series of toots, *keekeekakakokokuku,* like running your fingernail over the teeth of a comb; a single high-pitched *keeek.* 🎧**25**.

Subspp: Monotypic.

Adult, Western Grp. CA/03

Adult, *saturatus* TX/04

Adult, Western Grp. CA/09

Clapper Rail 1

Rallus longirostris L 14½"

Shape: Large heavy-bodied rail with a short triangular tail, relatively long thick neck, and long, deep-based, slightly downcurved bill. Legs short and thick. Generally smaller than King Rail, but larger m. Clapper overlaps size of smaller f. King. **Ad:** Brightly colored western subspp. share many features with King Rail but are geographically separated (See Subspp. for descrip.). Duller eastern subspp., which overlap range of King Rail, are described here. Back feathers with brownish centers and wide grayish margins; auriculars primarily gray; upperwing coverts brown to dull reddish brown; chin whitish and not strongly contrasting with rest of throat; chest and belly gray to reddish brown, sometimes with a grayish breast-band (see Subspp.); flanks grayish with thin whitish barring. Eye deep red; bill yellowish to orangish with dark culmen. Similar King Rail has back feathers with blackish centers and wide warm brown to reddish-brown margins; auricular mostly brownish or reddish brown; upperwing coverts and underparts bright reddish brown; white chin contrasts sharply with reddish-brown throat; flanks blackish with crisp white barring; eye deep red; bill yellowish to orangish with dark culmen. Disting. from King Rail with difficulty along Gulf Coast, where some Clapper subspp. resemble King and where the two can interbreed. **Juv:** (Jun.–Sep.) Grayish to blackish gray on back and flanks; paler on breast and throat; eye dark (red in ad.); dark bill. **Flight:** Short flights with body held at an angle and feet dangling beneath; longer flights with body and feet level; shallow rapid wingbeats. **Hab:** Saltwater and brackish marshes;

Adult, *crepitans* FL/01

Adult, *crepitans* NJ/05

Juv., *crepitans* VA/–

sometimes freshwater marshes in winter. **Voice:** Dry, fairly rapidly repeated, percussive *kek kek kek kek;* sometimes fastest in center of series and slowing down at ends, at other times all fairly fast. Also a more resonant, fairly rapid, grunting *jupe jupe jupe jupe.* Both calls at fastest delivery are about 4–6 sounds per second. Similar King Rail calls are slightly lower-pitched and slightly slower, with fastest delivery averaging about 2 sounds per second; caution should be used, since the calls of the two species broadly overlap. Clapper also gives a *kek kek kek krrrr.* **26**.

Subspp: (8) Divided into two groups. **Western North American Group:** Large; dark centers to back feathers, rich reddish-brown underparts and auriculars (much like King Rail), more contrasting white chin (also like King Rail). •*obsoletus* (San Francisco Bay), "California Clapper Rail," flanks grayish with indistinct paler bars; •*levipes* (Santa Barbara and south), "Light-footed Clapper Rail," flanks brown with white bars; •*yumanensis* (s.e. CA–s.w. AZ), "Yuma Clapper Rail," flanks grayish with white bars. **Eastern North American and Caribbean Group:** Small; underparts vary from grayest in north to pale reddish brown in south, auriculars grayish. •*crepitans* (MA–NC, winters on s.e. coast), "Northern Clapper Rail," very gray above and below; •*waynei* (s. NC–n.e. FL, winters Gulf Coast), "Wayne Clapper Rail," dark above, pale cinnamon below, gray breastband; •*scotti* (coastal FL to panhandle), "Florida Clapper Rail," darkest of all upperparts, cinnamon below; •*insularum* (FL Keys to Boca Grande), "Mangrove Clapper Rail," like *waynei* but smaller, shorter bill, more gray above, gray breastband; •*saturatus* (n.w. FL–TX), "Louisiana Clapper Rail," reddish-brown underparts (similar to King), gray feather margins above. **Hybrids:** King Rail.

Adult TX/05

Adult TX/04

Adult NJ/10

King Rail 1

Rallus elegans L 15½"

Shape: Identical in shape to Clapper Rail. **Ad:** Back feathers with blackish centers and wide warm to reddish-brown margins; auricular mostly brownish or reddish brown; upperwing coverts and underparts bright reddish brown; white chin contrasts sharply with reddish-brown throat; flanks blackish with crisp white barring. Eye deep red; bill yellowish to orangish with dark culmen. Similar Clapper Rail has back feathers with brownish centers and wide grayish margins; auriculars primarily gray; upperwing coverts brown to dull reddish brown; chin whitish and not strongly contrasting with rest of throat; chest and belly gray to reddish brown, sometimes with a grayish breastband; flanks grayish with thin whitish barring; eye deep red; bill yellowish to orangish with dark culmen. **Juv:** (Jun.–Sep.) Grayish to blackish gray on back and flanks; paler on breast and throat; eye dark (red in ad.); dark bill. **Flight:** Short flights with body held at an angle and feet dangling beneath; longer flights with level body and feet and shallow rapid wingbeats. **Hab:** Freshwater marshes, ditches, wet fields; sometimes brackish marshes. **Voice:** Dry, fairly deliberate, repeated, percussive *kek kek kek kek;* usually delivered fairly evenly throughout a series. Also a more resonant, grunting *jupe jupe jupe jupe*. Both calls at fastest delivery are about 2 sounds per second. See Clapper Rail, Voice, for comparison. King also gives a *kek kek kek krrrr.* 🎧**27**.

Subspp: (1) •*elegans*. **Hybrids:** Clapper Rail.

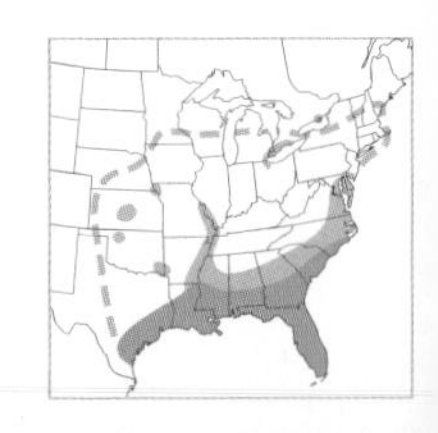

Adult AZ/06

Adult CT/04

Adult CT/05

Virginia Rail 1

Rallus limicola L 9½"

Shape: Medium-sized compact rail with a very short triangular tail, relatively short thick neck, and proportionately long, deep-based, slightly downcurved bill. Legs short and thick. Similarly plumaged King Rail has longer neck, larger tail, proportionately shorter bill, longer body. **Ad:** Much like small version of King Rail in plumage but with much grayer face, deeper-red eye, and more reddish base of bill. See Shape. **Juv:** Dark and blotchy overall with paler throat and belly. Best identified by shape. **Flight:** Short flights; body held at an angle with feet dangling beneath. From above, flight feathers dark, coverts reddish brown. **Hab:** Freshwater and brackish marshes; may winter in salt marshes as well. **Voice:** Dry, percussive, mostly paired sounds like *kik kidik kidik kidik*. Other calls include a more resonant *jupe jupe jupe* and a *kek kek kek krrrrrt*, which are very similar to the corresponding calls of King and Clapper Rails, but may be slightly higher-pitched. 🎧**28**.

Subspp: (1) •*limicola*.

Adult FL/02

1st winter FL/03

Adult FL/03

Purple Gallinule 1

Porphyrio martinica L 13"

Shape: Fairly large chickenlike bird with long thick legs and very long toes. Bill deep-based, conical, slightly downcurved at tip; extends onto forehead as a frontal shield. Wings long, primary projection past tertials long; tail short and triangular, often flipped up. **Ad:** Iridescent green and turquoise above; deep blue and violet below; bill bright red with yellow tip and pale blue frontal shield; legs and toes bright yellow. **1st Winter:** (Sep.–Apr.) Gradually molts into ad. plumage; looks like juv. with blotches of violet, blue, and green; bill still dark with pale yellow tip. **Juv:** (Jun.–Aug.) Brownish upperparts with a variable greenish wash; buffy face and underparts, palest on belly; all-white undertail coverts; bill variably brownish and yellow. **Flight:** Short flights with legs dangling beneath; long flights with legs trailing behind. Note long wings, yellow legs, blue upperwing coverts. **Hab:** Freshwater marshes, lake edges. **Voice:** Series of low-pitched, muffled, chickenlike clucks of fairly even spacing or slightly speeding up, *buk buk buk bugaak buk;* high-pitched *keeek.*

Subspp: Monotypic.

Adult summer FL/04

Juv. FL/03

Adult winter FL/03

Common Moorhen 1

Gallinula chloropus L 14"

Shape: Fairly large chickenlike bird with a relatively short neck, long thick legs, and fairly long toes. Bill deep-based, conical, slightly downcurved; extends onto forehead as a frontal shield. Wings relatively short, with primary projection past tertials short (about as long as bill); tail short and triangular, often flipped up. **Ad. Summer:** Dark grayish black overall with a dark brown wash over wings and lower back; bill bright red with yellow tip; frontal shield also bright red. Undertail pure white on each side, black in center; a few white feathers along upper flanks. Legs yellow, with patch of red above tarsus. **Ad. Winter:** Like summer but paler gray on chin, throat, and underparts; bill base and frontal shield dull brown; bill tip dull yellow. **Juv:** (Jul.–Feb.) Brownish gray overall with brown wings; face, chin and underparts paler; whitish flank marks and white undertail divided by dark line as in ad. Bill dull brown at base, yellowish at tip. Similar juv. American Coot has longer neck, longer primary projection past tertials, and lobed toes and lacks white line along upper flanks. **Flight:** Short flights with legs dangling beneath; long flights direct with legs trailing behind. Note dark trailing edge to secondaries (Am. Coot has white trailing edge). **Hab:** Freshwater ponds, marshes, lakes. **Voice:** Rapid churring followed by a series of double or single toots that slow down and seem to "run out of steam," *churrrrdoot-doot doot doot doot dahr dahrr* (sometimes descending in pitch a little like Sora, but never as markedly as Sora, and Sora lacks churring start); short grating *chrrrt;* short high-pitched *keeek.* 🎧**29**.

Subspp: (1) •*cachinnans.*

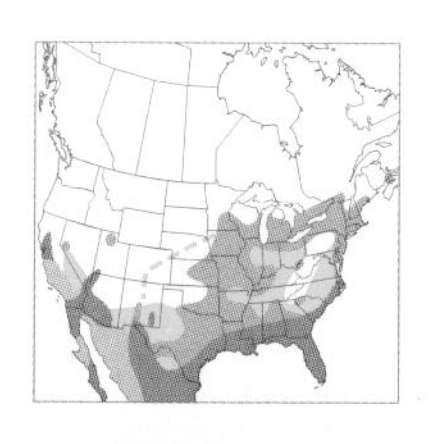

Adult FL/05

Juv. CA/07

Adult FL/02

Adult FL/02

Juv. CA/06

American Coot 1

Fulica americana L 15½"

Shape: Heavy-bodied bird with relatively long thick neck, fairly small head, and very deep-based, straight, conical bill. Short wings with very short primary projection past tertials (shorter than bill). Toes with lobed projections on the sides of each segment. **Ad:** Slate-gray body and wings; black head; white bill with dark partial ring near tip and maroon rounded area on frontal shield (latter may be absent); dark red eye. Wide white streaks on either side of dark undertail coverts. **Juv:** (Jul.–Oct.) Grayish-brown back, wings, and tail; paler beneath; head pale gray; chin and throat whitish; bill dusky. Similar juv. Common Moorhen has longer primary projection past tertials, shorter neck, toes with no lobes, and white line along upper flanks. **Flight:** Runs along water to get airborne, then flight direct and strong. Note white trailing edge to secondaries. **Hab:** Marshy lakes in summer; also along coast in winter. **Voice:** Short, low-pitched, grating *grrrt grrrt* or *grrit grrit;* short toots, as on a party horn, *toot toot;* 2-part toots, *kitoot kitoot,* or grating calls, *grrdit grrdit*. 🎧**30**.

Subspp: (1) •*americana.*

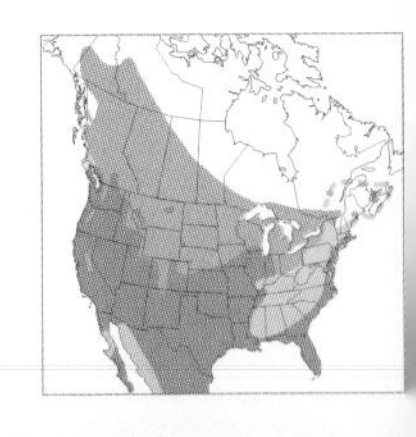

Adult, *pictus* FL/02

Adult, *pictus* FL/02

Adult, *pictus* FL/02

Limpkin 2

Aramus guarauna L 26"

Shape: Large heronlike bird with a relatively large body and short neck; long heavy bill slightly downcurved and of even thickness until abrupt point. Long broad wings, when folded, add to the impression of a large body. **Ad:** Brown overall with white triangular marks densest on head and neck and scattered over back, wings, and breast. Bill orangish at base, dusky at tip; legs dark. **Juv:** (Jun.–Apr.) Like ad. but slightly paler brown, and white marks on wings and back thinner (more like streaks than triangles). **Flight:** Long broad wings; flies with quick upstroke (as in cranes), slower downstroke; neck angled slightly down and bill held down at 30-degree angle. **Hab:** Freshwater swamps, marshes. **Voice:** Loud, squealing, descending scream, *screeaah.*

Subspp: (1) •*pictus.* ▶ *A. g. dolosus* (possible vag. from MEX) has less white streaking on back, broader white streaking on wings.

Adult NM/10

Adult NM/10

Juv./1st yr. NM/12

Sandhill Crane 1

Grus canadensis L 41–46"

Shape: Tall, graceful, heronlike bird with long legs, long neck, and a long slender body. Characteristic "bustle" at tail is created by longer decurved inner secondary coverts and tertials that extend over tail. **Ad:** Pale gray overall; bright red to rosy skin from midcrown to base of bill; eye yellowish. Variable rusty staining from midneck down created by bird rubbing iron-rich soil on plumage; this is most pronounced in summer, does not exist in winter when birds are in fresh plumage; only occurs in regions where suitable soil exists. **Juv/1st Yr:** (Jun.–2nd Sep.) Similar to ad. but feathers of crown, nape, back, and wings tipped cinnamon; forecrown to bill feathered and dark cinnamon. **Flight:** May run to take off; quick upstroke, slow downstroke; flies with neck and feet extended. **Hab:** Grasslands in summer; marshes, agricultural fields in winter. **Voice:** Mellow, rolling, fairly low-pitched, repeated *kukukuroo;* juv. "Lesser Sandhill Cranes" give high *treee treee* call (Jul.–Apr.). **31**.

Subspp: (4) Subspecies in North smallest, in South largest. •*canadensis* (AK–NU, winters s. CA–TX), "Lesser Sandhill Crane," is the smallest. The other three larger subspp. are collectively called "Greater Sandhill Cranes." They include •*tabida* (s.e. AK islands–ON south to n. CA–OH, winters CA–FL), "Canadian Sandhill Crane"; •*pratensis* (s.e. GA–cent. FL), "Florida Sandhill Crane"; and •*pulla* (s. MS), "Mississippi Sandhill Crane," which has strong contrast between paler cheek and darker rest of head and neck. Last two are nonmigratory; the others migratory. **Hybrids:** Common Crane.

Adult SPA/–

Juv. SPA/–

Common Crane 4
Grus grus L 44–50"

Casual Eurasian vag. to West and Midwest, fall–spring; records farther east may be escapees. **Shape:** Tall, graceful, heronlike bird with long legs, long neck, and a long slender body. Characteristic "bustle" at tail is created by longer decurved inner secondary coverts and tertials that extend over tail. **Ad:** Grayish overall with black neck and throat; bare crown black in front and red behind, cheek white, chin black; yellowish bill. Head and neck may be stained with cinnamon wash. **Juv:** Reddish-brown head and neck; darker chin and throat; brownish-gray body. **Flight:** Pale gray coverts; black flight feathers. **Hab:** Prairie, marshes, fields. **Voice:** Loud *krooo* or *kroooah.*

Subspp: Monotypic. **Hybrids:** Sandhill Crane.

Adult TX/03

Juv. WI/10

Adult TX/03

Whooping Crane 2
Grus americana L 55"

An endangered species that is highly protected on its breeding and wintering grounds; there are attempts to establish new breeding colonies. **Shape:** Tall, graceful, heronlike bird with long legs, long neck, and a long slender body. Characteristic "bustle" at tail is created by longer decurved inner secondary coverts and tertials that extend over tail. **Ad:** White overall with black primaries and their greater coverts (hidden when at rest); crown and broad malar streak deep red; front of face black; small gray patch on nape. **Juv/1st Yr:** Buffy head and upper neck; buffy tips to feathers of body and wings. **Flight:** Long, broad wings with slow wingbeats. Note black primaries, their greater coverts, and alula on otherwise white wing. **Hab:** Freshwater marshes in summer; saltwater marshes, fields in winter. **Voice:** Common call a 2-part *keeekaaa,* actually a duet between m. and f. (f. call the higher part).

Subspp: Monotypic.

Lapwing and Plovers

Oystercatchers

Stilt and Avocet

Jacana

Sandpipers, Phalaropes, and Allies

Adult m., summer FIN/04

Adult m., summer ENG/04

Adult winter NET/02

Adult summer ENG/08

Juv. ENG/08

Northern Lapwing 4

Vanellus vanellus L 12½"

Casual Eurasian vag. to East Coast; rarely inland. **Shape:** Large, heavy-bellied, long-crested plover with a relatively short thin bill. Head squarish; outer wing broad, well rounded. **Ad. Summer:** Dark back and wings (iridescent close up); black breast contrasts with white belly and light face; undertail coverts rusty. M. has black foreface and throat, very long crest; f. throat mottled white, crest shorter. **Ad. Winter:** Buffy face; black markings around eye; white chin and throat; buffy tips to scapulars and wing coverts. **Juv:** Like ad. winter but crest very short (about lore length); less dark on face; more extensive buffy edges to upperpart feathers; grayish legs. **Flight:** Wingbeat jerky, with slow deep strokes. Dark primaries white-tipped; underwing secondary coverts white; tail white with black central band (squarish on closed tail). **Hab:** Moist short-grass fields. **Voice:** Flight call *payyeet payyeet.*

Subspp: Monotypic.

IDENTIFICATION TIPS

Shorebirds are a large and varied group of birds often seen feeding in large groups in the shallow water of shorelines, marshes, and mudflats but also in meadows and agricultural fields. Most are cryptic, with pale underparts and darker upperparts; they are generally graceful in proportions, both standing and in flight, with long legs, usually long thin bills, and fairly narrow pointed wings. Most breed in the far north and winter along southern beaches; thus the majority of people see them during migration, which occurs throughout the Midwest as well as along both coasts.

Species ID: Shape is a key feature of shorebird identification. Note the size of the bird and the length of the bill compared to the depth of the head—is it shorter than, equal to, or longer than the head? Note the length of the neck and the shape of the bill. In some of the plovers and peeps (an informal term for our small sandpipers), it is important to see the position of the wing tips in relation to the tip of the tail—do they fall short of, just reach, or extend past the tail tip? With experience, you can identify many shorebirds by shape alone. Color differences between many shorebirds are subtle, but look closely at the patterns on the head in addition to everywhere else on the bird.

Adult m., summer AK/06

Juv. MN/09

Adult winter FL/01

Adult winter FL/01

Adult, spring transition FL/03

Adult winter CA/09

Black-bellied Plover 1

Pluvialis squatarola L 11½"

Shape: Large, deep-chested, thick-necked plover with a relatively large head and long, thick, blunt-tipped bill. Primaries extend just beyond tail tip; primary projection past tertials shorter than bill. **Ad. Summer:** Black face, throat, and belly bordered by white on crown and neck; vent and undertail coverts white. M. has all-black face, breast, belly; back checkered mostly dark gray and white; crown mostly whitish. F. with variable whitish mottling on face, breast, belly; back checkered mostly brown and white; crown mostly dark. **Ad. Winter:** Upperpart feathers brownish gray with margins pale or pale-spotted; brown crown contrasts slightly with indistinct whitish eyebrow; breast whitish with diffuse brown spotting; belly and undertail coverts white. **Juv:** Like ad. winter but upperparts distinctly checkered gray and white with crisp white feather fringes (may have golden wash in fresh plumage); breast and flanks more distinctly and coarsely streaked than adult's. **Flight:** Below, distinctive black axillaries; above, distinctive white rump and whitish tail; bold white wing-stripe on inner primaries. **Hab:** Tundra in summer; mudflats, shores in winter; fields and pastures in migration. **Voice:** Plaintive whistled *peeahhweet* or *pahweeah* or *tseee.* 🎧**32**.

Subspp: Monotypic.

Adult m., summer NOR/06

Adult winter ENG/02

Juv. ENG/09

Winter UKM/03

Winter UKM/11

European Golden-Plover 4

Pluvialis apricaria L 10½"

Casual Eurasian vag. to NF and n. Atlantic Coast, mostly in spring. **Shape:** Large, very round-bellied, short-legged plover with a thick neck, relatively small head, and short, gradually tapered, fine-pointed bill. Primaries fairly long and extend well past tertials and just past tail; 3–4 primary tips are exposed. Appears smaller-billed, shorter-legged, and slightly stockier than American or Pacific Golden-Plover. **Ad. Summer:** Has the most gold in the smallest dots of all golden-plovers at all stages. M. similar to Pacific Golden-Plover, with white extending down sides of breast, along flanks, to mostly white undertail coverts; however, white expands more on breast, leaving a narrower area of black in the center than on most Pacifics. Crown, nape, and upperparts heavily spotted with tiny dots of gold. F. similar but may have more white on underparts and paler face; geographic variation in this species can have some males just as light, thus sexing is difficult. **Ad. Winter:** Extensively and finely speckled with gold over dark crown, nape, back, and wings; breast finely streaked with gold and black; belly whitish with gold mottling on flanks. **Juv:** Very similar to ad. winter but in general more strongly marked underparts (may be hard to age in the field). **Flight:** Distinguished from other golden-plovers by white axillaries and underwing coverts (gray on Pacific and American) and white wing-stripe on inner primaries (lacking on American and Pacific). **Hab:** Tundra in summer; plowed fields, tidal flats in winter. **Voice:** High-pitched *cheerwee cheerwee;* flight call a simple whistled *tuuu.*

Subspp: (1) •*albifrons.*

Adult m., summer AK/06

Juv. NJ/09

Adult, fall transition NJ/08

Adult summer YT/05

Adult winter BOL/03

Adult winter USA/09

American Golden-Plover 1

Pluvialis dominica L 10½"

Shape: Large plover, smaller-headed, thinner-necked, slimmer overall than Black-bellied Plover; compared with Pacific Golden-Plover, is slightly larger and heavier with proportionately slightly shorter legs and bill and deeper belly. Primaries long (longer than Pacific's) and extend well past tertials and well past tail; 4–5 primary tips are exposed (last 2 primaries nearly equal in length). Extension of primaries past tertials 1–1½ x bill length; extension of primaries past tail ½–1 x bill length. Legs with no or very little extension past tail in flight. **Ad. Summer:** Similar to Pacific Golden-Plover but less golden, with more white spotting, and more grayish on back at all stages. M. has black face, throat, and central breast with broad white patches on sides of breast; belly and undertail coverts all black; upperparts blackish heavily dotted with bold gold and some white spots and notches. F. similar but with variable flecks of white on face, flanks, and undertail and less colorful upperparts. **Ad. Winter:** Head, neck, breast, and upperparts grayish brown with whitish to gray spotting on back. Dark gray streaked crown and ear patch contrast strongly with white eyebrow; face otherwise grayish (washed golden on Pacific); underparts grayish with smudgy lighter spotting. See Shape. **Juv:** Like ad. winter but spotting on upperparts gold to whitish with gold-spotted rump; breast spotting finer and more distinct. **Flight:** Below, underwing and axillary grayish; above, upperparts and inner wings all grayish brown; no wing-stripe. **Hab:** Rocky drier tundra in summer; open areas with sparse vegetation, plowed fields, turf farms, drier mudflats in winter. **Voice:** Very high-pitched scratchy *tsoocheeet* or rapid *queedleet.*

Subspp: Monotypic.

Adult m., summer AK/06

Juv. HI/03

Adult winter HI/01

Pacific Golden-Plover 2

Pluvialis fulva L 10¼"

Shape: Large plover, smaller-headed, thinner-necked, and slimmer overall than Black-bellied Plover; compared with American Golden-Plover, has proportionately slightly longer legs and bill and deeper chest. Primaries short (shorter than American's) and extend just past tertials and slightly past tail; 1–3 primary tips are exposed (last 2 primaries nearly equal in length). Extension of primaries past tertials is less than bill length. Legs may extend noticeably past tail in flight. **Ad. Summer:** Similar to American Golden-Plover but generally with more golden on back at all stages. M. has black face, central throat, and belly; relatively thin white line on sides of breast continues along flanks to mostly whitish undertail coverts; upperparts blackish heavily dotted with gold and some white. F. similar but with variable flecks of white on black areas of underparts. **Ad. Winter:** Head, neck, breast, and upperparts grayish brown heavily spotted with gold. Dark brown-and-gold streaked crown contrasts slightly with pale golden eyebrow; face otherwise washed with gold (washed gray on American); often with dark spot behind eye; underparts grayish with smudgy gold spotting. See Shape. **Juv:** Similar to ad. winter but more finely and extensively spotted with gold above and below; coarsely streaked upper breast contrasts with paler buff unmarked belly; gold face and eyebrow contrast little with darker crown. **Flight:** Below, underwing and axillary grayish; above, upperparts and inner wings all grayish brown with golden wash; no wing-stripe. **Hab:** Moist tundra in summer; fields, beaches, tidal flats in winter. **Voice:** *Pee-puweee* or simple shrill *pweee.*

Subspp: Monotypic.

Adult summer, *mongolus* HKO/04

Juv. JAP/08

Adult winter NZE/05

Lesser Sand-Plover 3

Charadrius mongolus L 7½"

Asian species; rare visitor and casual breeder to w. AK is.; casual on West Coast; accidental on Gulf and East coasts and Great Lakes. **Shape:** Small, relatively long-legged, fairly large-billed plover. Bill longer than on Semipalmated Plover, but shorter than on Wilson's Plover. **Ad. Summer:** Distinctive rusty chest and nape contrast with white throat and belly; dark mask from lore through eye. Sexes similar; m. averages more extensive and brighter rust on breast and blackish mask; f. often more buffy on breast with variable brownish mask. Legs greenish to dark gray. **Ad. Winter:** Dark grayish-brown upperparts with no white collar (as exists in all of our other small plovers); underparts white; dark extensions of incomplete breastband onto breast sides (may join for complete band); white on forehead continues back to form thin white eyebrow. **Juv:** (Aug.–Nov.) Like ad. winter but light margins to back and wing coverts create faint scaled pattern; as in ad., no white collar; light buffy wash on breast and face. **Flight:** Thin white wing-stripe on inner primaries; tail mostly dark with fine white edge; white sides to rump. **Hab:** Tundra in summer; mudflats, beaches in winter. **Voice:** Rapid *kwid'd'dip* or *kwidip.*

Subspp: (2) **Mongolus Group:** •*stegmanni,* •*mongolus.*

Adult m., summer CA/03

Adult winter FL/01

Adult m., summer FL/03

Adult summer TX/04

Snowy Plover 1

Charadrius alexandrinus L 6½"

Shape: Small, compact, long-legged, proportionately large-headed plover with a fairly long distinctively thin bill; steep forehead angles to gently rounded crown. Similar Piping Plover has shorter thicker bill. **Ad. Summer:** Dark brown to pale grayish-brown upperparts; white collar. On m., incomplete breastband, ear patch, and forehead bar are black; on f., dark brown. Legs gray or horn-colored. **Ad. Winter:** Like ad. summer but black or dark brown areas paler and concolor with or only slightly darker than upperparts. East Gulf Coast birds paler than West Coast birds in all seasons. **Juv:** (Jul.–Oct.) Similar to ad. winter but with pale margins to back and wing feathers, creating subtle scaled look. **Flight:** Distinct wing-stripes; tail dark-centered with white sides. **Hab:** Beaches, salt and alkaline flats in summer; beaches, pond flats in winter. **Voice:** Rising, soft *chuuweeep* or short *jr'r'rt*.

Subspp: (1) •*nivosus*.

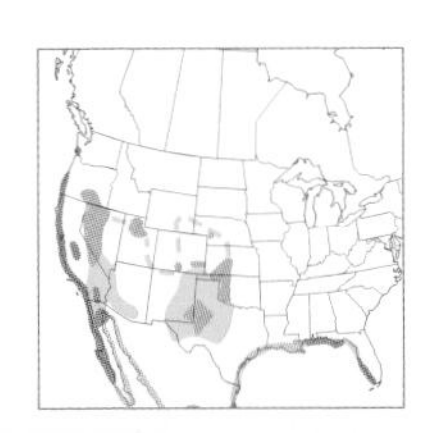

IDENTIFICATION TIPS

Plovers are small to medium-sized shorebirds with short thick bills, rounded heads, proportionately large eyes, and moderate-length legs; they often look fairly deep-chested. They typically feed by looking for food, rather than probing with their bills (which is more common in other shorebirds), and they usually alternate short runs with abrupt stops as they hunt.

Adult m., summer LA/04

Adult winter FL/02

Adult summer FL/03

Adult MEX/05

Adult winter FL/03

Adult summer FL/05

Wilson's Plover 1

Charadrius wilsonia L 7¾"

Shape: Fairly small, attenuated, large-headed plover with a long thick bill. Bill looks disproportionately large; steep forehead leads to long fairly flat crown. Often runs or "skulks" in horizontal posture. **Ad. Summer:** Upperparts dark brown; underparts white except for complete breastband, which can appear thick or narrow depending on posture of the bird. Complete neckband, forecrown, and lores on m. black; on f. brown and neckband may be incomplete. May be some rust at back of eyebrow and behind cheek; legs dull pinkish gray. **Ad. Winter:** Both sexes similar to summer f.; breastband may be browner, thinner, and incomplete. **Juv:** (Jun.–Oct.) Like ad. winter but neckband is paler and more indistinct; back and wing feathers with pale margins, creating scaled effect. **Flight:** Large bill; distinct white wing-stripe; dark central tip to tail. **Hab:** Beaches with sparse vegetation in summer; beaches, mudflats in winter. **Voice:** Short *chidit* or loud, rising, whistled *chweeep*.

Subspp: Monotypic.

Adult OH/05

Adult NV/05

Adult OH/05

Adult IL/04

Juv. NY/08

Killdeer 1

Charadrius vociferus L 10½"

Shape: Large, slim, long-tailed plover with a long thin bill. Relatively small head and broad neck. The 2 black neckbands and the rusty rump and distinctive tail pattern (seen when flying or displaying) are diagnostic in all plumages. **Ad:** Upperparts dark grayish brown except for rusty uppertail coverts and tail base; 2 complete black breastbands; bright red eye-ring. May have scattered rusty feathers on upperparts. **Juv:** Like adult but with fine buff edges to mantle and wing covert feathers. **Flight:** Long tail with rufous rump distinctive; strong white wing-stripes. **Hab:** Open barren or short-grass fields, mudflats. **Voice:** Loud very high-pitched *kideeer* or *kidideeer* or *deeah* or repeated *kyit kyit* or rolling trill. 🎧**33**.

Subspp: (1) •*vociferus*.

Adult m., summer TX/05

Adult FL/02

Adult winter FL/01

Adult m., summer USA/08

Juv. NH/09

Adult m., summer NY/08

Semipalmated Plover 1

Charadrius semipalmatus L 7¼"

Shape: Small plover with a relatively small head, rounded crown, and short stubby bill. Piping Plover has proportionately shorter legs, more rounded body, and less tapered rear. **Ad. Summer:** Upperparts dark brown; underparts white except for black neckband. Forecrown, eye patch, and neckband deep black and more extensive in male; browner and less extensive in f. Legs are dull yellow to orange; bill with orange base and black tip. **Ad. Winter:** Similar to ad. summer but areas of black are brown; neckband complete or incomplete; bill all dark or with some orange at base. **Juv:** (Jul.–Oct.) Like ad. winter but with fine pale margins to wing and crown feathers. **Flight:** White wing-stripe, tail centrally dark with a thin white margin. **Hab:** Barren arctic areas in summer; beaches, mudflats, fields in winter. **Voice:** Fairly harsh emphatic *chu-weeet* or harsh, short, repeated *cheet*.

Subspp: Monotypic.

Adult m., summer NOR/–

Juv. POR/10

Adult f., summer ENG/06

Winter CYP/11

Adult m., summer UKM/10

Common Ringed Plover 2

Charadrius hiaticula L 7½"

Shape: Almost identical to Semipalmated Plover, but with a longer thinner-based bill. **Ad. Summer:** Most reliably distinguished from Semipalmated Plover by voice; there are also several subtle plumage clues that, together with voice, can back up an identification. Dark areas on face and breastband generally more extensively black; lacks obvious orbital ring; broad dark lores; white forehead patch often extends back toward eye in a point and touches eye; extensive bright white eyebrow extends behind eye (Semipalmated Plover summer m. has either no white eyebrow or only a faint line; f. Semipalmated has a white eyebrow); white on throat extends just to bottom of gape (usually extends just above gape in Semipalmated). **Ad. Winter:** Very similar to Semipalmated Plover; see Shape, Voice. **Juv:** Like ad. winter but with fine buffy margins to back and wing feathers, creating a scaled look. **Flight:** Like Semipalmated Plover but with more distinct wingstripe. **Hab:** Barren arctic areas in summer; beaches, mudflats, fields in winter. **Voice:** Soft rising *pooeeet* (softer than Semipalmated Plover's emphatic *chuweeet*) or simple mellow *peet* or repeated *tooweeah tooweeah*.

Subspp: Monotypic.

Adult summer, *melodus* FL/03

Adult summer, *circumcinctus* TX/04

Adult winter FL/02

Adult summer MI/06

Piping Plover 2

Charadrius melodus L 7¼"

Shape: Rounder and less attenuated body than Semipalmated Plover, with relatively small rounded head and short stubby bill; legs proportionately shorter than Semipalmated Plover's and most other plovers'. Snowy Plover has noticeably longer and thinner bill. **Ad. Summer:** Upperparts pale gray; underparts white; dark forecrown bar and breastband (complete or incomplete) are black on m., reduced and brownish black on f.; bill orange at base, dark at tip; legs almost always dark orange. **Ad. Winter:** Dark areas replaced by pale gray; breastband reduced to grayish patches on sides of breast; bill all dark or with some orange at base. **Juv:** (May–Sep.) Like ad. winter but with back and wing coverts tipped with buff; all-dark bill. **Flight:** Pale back and inner wing contrast with darker outer wing and tip of tail. **Hab:** Sandy beaches, river bars, alkaline flats in West in summer; beaches, mudflats in winter. **Voice:** Clear mellow *peetoo* or repeated *pit pit* or plaintive *turwee*.

Subspp: (2) Vary in darkness of dark areas. •*melodus* (Atlantic Coast) has paler back, paler dark facial areas, breastband usually incomplete; •*circumcinctus* (Midwest) has darker back and face, breastband complete in adults, incomplete in young birds. Often black flecks in lores.

Adult summer CO/06

Juv. CA/11

Adult winter CA/12

Adults CA/01

Mountain Plover 2

Charadrius montanus L 9"

Shape: Large, deep-chested, long-legged plover with a relatively thin neck and thin bill. **Ad. Summer:** Pale, relatively unmarked plover that blends in with the barren fields it often inhabits. Pale grayish brown above, with rust feather fringes; white below (sometimes with buffy brown on breast sides); blackish lore and forecrown bar; legs gray. **Ad. Winter:** Similar to ad. summer but with pale lores, no dark mark on forecrown, and wide pale brown breast sides. **Juv:** (May–Feb.) Like ad. winter but darker brown above and with dark smudge before eye; when fresh, buffy margins to back and wing feathers, and head and breast may be washed rich buff. Dark smudge before eye. **Flight:** Dark subterminal band on tail; strong white wing-stripe; white on sides of rump. **Hab:** Open barren or short-grass areas, plowed fields. **Voice:** Harsh *grrrt* or soft *pweet*.

Subspp: Monotypic.

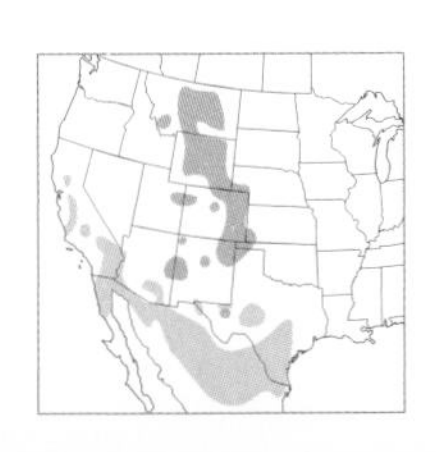

Adult summer FIN/06

Adult winter CA/02

Juv. -/-

Eurasian Dotterel 4

Charadrius morinellus L 8¼"

Eurasian species that is a rare breeder in w. AK and casual vag. to West Coast. **Shape:** Large, compact, rotund plover with a relatively long neck, small head, short bill. **Ad. Summer:** In all plumages, distinctive broad whitish eyebrows extend down behind eyes onto nape. Reddish-brown breast and flanks separated from grayish throat by thin white band; black belly; yellowish legs. F. more brightly colored than male. **Ad. Winter:** Similar to summer but more muted with grayish crown, brownish breast, and white belly; hint of summer breastband; white eyebrow and as in summer. **Juv:** (Jun.–Sep.) Eyebrow buffy; breastband faint; buffy below; grayish above with scaled effect from buffy feather margins. **Flight:** Dark outer wing; brownish inner wing; no wing-stripe. **Hab:** Stony tundra, barren ground in summer; dry, barren, sandy or gravelly areas in winter. **Voice:** Whistled *weeet weetweet* or soft *jjert*.

Subspp: Monotypic.

Adult CA/10

Juv. CA/10

Adult CA/02

Black Oystercatcher 1

Haematopus bachmani L 17½"

Shape: Large heavily built shorebird with very long deep (but narrow) bill and very thick relatively short legs. **Ad:** Appears all black; only head is pure black; rest of body and wings tinged brown; bill bright red; legs pale pink; eye yellow with red orbital ring. May be variable amount of white mottling on underparts, tail coverts, and white on inner wing; birds in southern portion of range more likely to have some white (possibly the result of hybridization with American Oystercatcher in MEX). **Juv/1st Yr:** Like ad. but eye dark, bill orangish with dark tip. **Flight:** All dark with red bill and yellow eye. **Hab:** Rocky coast; only rarely on mudflats. **Voice:** Loud whistled *wheeet* either singly or repeated in a series.

Subspp: Monotypic. **Hybrids:** American Oystercatcher.

Adult, *palliatus* FL/02

Imm., *palliatus* TX/02

Adult (r.) and 2 juvs., *palliatus* NJ/–

American Oystercatcher 1

Haematopus palliatus L 18½"

Shape: Large heavily built shorebird with very long deep (but narrow) bill and very thick relatively short legs. **Ad:** Black head and neck; dark brown back and wings; bright white below. Legs pale pink; yellow eye with red orbital ring. F. slightly larger and longer-billed than male. **Imm:** (2nd yr.) Similar to ad., with yellow eye, pinkish legs, and orange bill, but bill still has substantial duskiness near tip (ad. bill all orange). Head may be grayish black (black in ad.). **Juv:** (1st yr.) Similar to ad. but with brownish head and neck, pale fringes to wing coverts; dark eye; bill mostly dark with dull orange base; legs grayish pink. **Flight:** Broad white diagonal stripe through inner wing; white uppertail coverts; mostly white underwing coverts. **Hab:** Coastal beaches, mudflats, salt marshes. **Voice:** Whistled *peeet,* drawn-out whistled *wheeeah,* or *wheeeah* followed by a rolling trill.

Subspp: (2) Geographically separate. •*palliatus* (Gulf and Atlantic coasts) has white wing-stripe that continues onto inner 4 primaries, white uppertail coverts, brownish back; •*frazari* (s. CA) has white wing-stripe on secondaries and their greater coverts that does not extend onto primaries, dusky uppertail coverts, dusky back. **Hybrids:** Black Oystercatcher.

Adult m. FL/03

Juv. CA/08

Adult f. FL/03

Adult m. CA/05

Black-necked Stilt 1

Himantopus mexicanus L 14"

Shape: Large very slender shorebird with distinctive long thin legs and needlelike bill. **Ad:** Blackish upperparts; bright white underparts; pink legs; white patch over eye. M. back bluish black; f. back brown. **Juv:** (Jun.–Jan.) Brownish above; pale margins to back and wing feathers; legs dull pink and bill pinkish at base at first. **Flight:** Upperparts black with white tail and wedge up lower back; underparts white with black wings; legs conspicuously trailing behind. **Hab:** Shallow water of marshes, ponds, fields, impoundments. **Voice:** Short, repeated, barklike *kwit*.

Subspp: (1) •*mexicanus*. **Hybrids:** American Avocet.

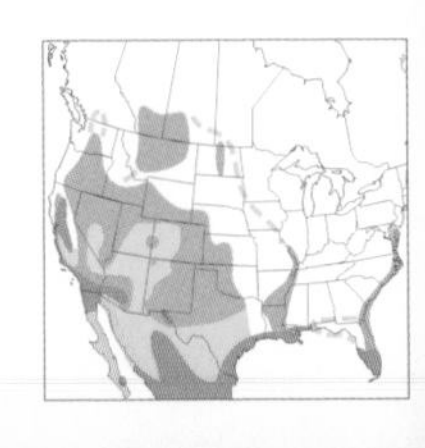

Adult m., summer FL/03

Adult m., summer TX/04

Adult f., summer TX/04

Adult f., winter CA/11

Adult m., winter FL/02

American Avocet 1

Recurvirostra americana L 18"

Shape: Large, long-necked, long-legged shorebird with distinctively needle-thin upcurved bill. **Ad. Summer:** Bold black-and-white pattern on back and wings in all seasons; buffy cinnamon head and neck; white underparts; legs pale bluish. M. bill straighter, slightly longer, and slightly upcurved at tip. **Ad. Winter:** Like ad. summer, but head and neck grayish white. **Juv:** (Jun.–Oct.) Like ad. winter but with pale cinnamon wash to neck and brownish wings. **Flight:** Bold black-and-white patterns on back; bold white stripe on inner wing. Below, inner wing white, outer wing black. **Hab:** Alkaline lakes, estuaries, shallow water in summer; beaches, salt marshes, impoundments in winter. **Voice:** High-pitched, harsh, repeated *chleep*.

Subspp: Monotypic. **Hybrids:** Black-necked Stilt.

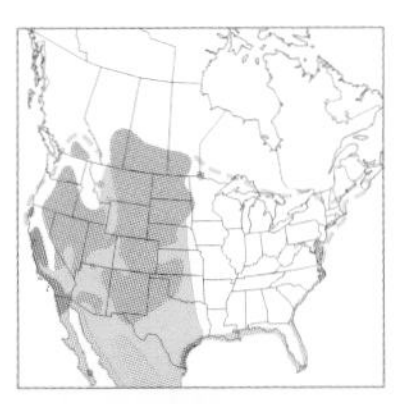

Adult BEL/10

Juv. CRI/02

Adult BEL/10

Northern Jacana 4

Jacana spinosa L 9½"

Casual Mexican vagrant to s. TX. **Shape:** Similar to gallinule–heavy body, thick bill, frontal shield, and extremely long legs and toes. **Ad:** Chestnut body; glossy black head and neck; bright yellow bill with pale blue base; yellow frontal shield. Male 30% smaller than female. **Juv/1st Yr:** Brownish upperparts; bold whitish eyebrow; broad dark eye-stripe connects to nape behind eye; bill pale yellow with small pale frontal shield. Gradually acquires ad. plumage over 1st yr. **Flight:** Striking yellow flight feathers seen in flight and when stretching; narrow chestnut wing coverts. **Hab:** Vegetated freshwater marshes. **Voice:** Harsh repeated *jit jit jit* (rail-like quality).

Subspp: Monotypic.

Adult summer HKO/05

Juv. JAP/08

Adult winter NZE/02

Adult FIN/07

Terek Sandpiper 3

Xenus cinereus L 9"

Rare Asian vag. to w. AK, casual farther south. **Shape:** Medium-sized, stocky, short-legged shorebird with a distinctive long, thin, upturned bill. Bill length about 1½ x head length. **Ad. Summer:** Gray above with fine dark streaks on back and wing coverts; wider dark streaking on some scapulars forms dark stripe; head and breast finely streaked; rest of underparts white. Bill black, sometimes with paler brownish to orangish base; legs greenish yellow to orange. Very active when feeding, running over rocky shorelines and beaches; may roost with slightly similar Gray-tailed Tattler. **Ad. Winter:** Like ad. summer but with less streaking overall; fairly prominent white supraloral dash. **Juv:** Like ad. winter but slightly browner with thin pale margins to scapulars, coverts, and tertials. May have thin dark scapular stripe. **Flight:** White tips to secondaries create broad white trailing edge to inner wing; tail gray, finely edged with white. **Hab:** Boreal forests near lakes or rivers in summer; coastal mudflats in winter. **Voice:** Soft, repeated, whistled *peepeepeet*.

Subspp: Monotypic.

Adult summer ENG/08

Juv. ENG/08

Adult winter OMA/01

Adult OMA/03

Common Sandpiper 3

Actitis hypoleucos L 8"

Rare Asian vagrant to w. AK. **Shape:** Like Spotted Sandpiper but tail slightly longer; tail extension past primaries equals lore plus eye (in Spotted, tail extension about equal to lore). **Ad. Summer:** Bobs tail but not as incessantly as Spotted Sandpiper. Clear white underparts with sides of breast washed with brown; conspicuous white notch just in front of wings; brown upperparts; back and scapular feathers with thin dark bars at tips. Bill dark gray; legs dull yellow. **Ad. Winter:** Like ad. summer. Distinguished from similar winter Spotted Sandpiper by all-dark bill and dark dots along tertials (Spotted has orangish lower mandible and generally unmarked tertials). See also Shape, Voice, Flight. **Juv:** (Jun.–Oct.) Very little contrast between patterning of wing coverts and back (strong contrast in Spotted); brownish wash on sides of breast can be streaked or barred (pale and plain in Spotted); tertials with dark dots on margins (usually lacking on Spotted). **Flight:** Bursts of stiff shallow wingbeats followed by glides, usually low over water. Strong white wing-stripe continues through secondaries to base of wing (fades out in midsecondaries on Spotted). **Hab:** Open areas near water edges. **Voice:** Very high-pitched *tseeseeseesee* (higher and more sibilant than Spotted Sandpiper).

Subspp: Monotypic. **Hybrids:** Green Sandpiper.

Adult summer NY/05

Juv CA/09

Adult winter FL/02

Adult winter FL/02

Spotted Sandpiper 1

Actitis macularius L 7½"

Shape: Small, horizontally oriented, short-legged sandpiper that continuously bobs its rear while walking or standing. **Ad. Summer:** Upperparts dark brown with sparse black barring; underparts bright white with dark spots; conspicuous white notch just in front of wing. Bill about length of head and orangish to yellowish with dark tip; legs pinkish. Only likely confusion is with rare Common Sandpiper or Solitary Sandpiper (which has similar call but lacks white notch before wing). **Ad. Winter:** Plain slaty brown above with some pale fringing on the wing coverts; white below; white shoulder notch. Occasionally a few dark spots or fine streaks on undertail coverts, less commonly on belly. Bill mostly dark; legs yellowish to greenish yellow. **Juv:** (Jul.–midwinter) Like ad. winter but extensive black-and-buff barring on wing coverts and pale fringes on tertials and scapulars. Bill base may have pinkish tone, rest of bill dark. **Flight:** Low over water with bursts of stiff, shallow wingbeats alternating with short glides. Short white wing-stripe starts in primaries and fades out in midsecondaries (continues through secondaries on Common Sandpiper). **Hab:** Open areas near water edges, including pond and stream edges, mudflats, and rocky shores. **Voice:** Loud whistled *peetweet* or rapid series of *peet* calls or rolling intro to *peet* calls.

Subspp: Monotypic

Adult summer FIN/06

Adult winter OMA/11

Juv. ENG/08

Green Sandpiper 4

Tringa ochropus L 8¾"

Casual Eurasian vag. to w. AK. **Shape:** Much like Solitary Sandpiper, but less attenuated rear body. **Ad. Summer:** Dark brown upperparts with fine white spotting on back and wings; bold white eye-ring; breast streaked brown; rest of underparts white; rump white; tail white with 3–4 broad blackish bars. Similar Solitary Sandpiper has dark rump; whitish tail centrally dark with fine dark barring on sides (see Flight). Green Sandpiper may also bob rear (like Spotted Sandpiper) rather than bob head (like Solitary). **Ad. Winter:** Like ad. summer but with less distinct streaking on head and breast and finer dots on back and wings. **Juv:** Like ad. winter, but more heavily spotted above, especially on tertials. **Flight:** Tail white with 3–4 bold dark bars near tip (Solitary tail dark centrally and barred on sides). **Hab:** Northern forest bogs, marshes in summer; inland lakes, ponds in winter. **Voice:** Very high-pitched *seetweet*.

Subspp: Monotypic. **Hybrids:** Common Sandpiper.

IDENTIFICATION TIPS

Tringids are named for the largest genus in the group of shorebirds that contains members of the genera *Tringa, Xenus*, and *Actitus*. They are mostly medium-sized gracefully proportioned shorebirds with small heads, long thin bills, fairly long necks, long thin legs, and long, narrow, pointed wings. They often feed in shallow water, where they pick their food off the surface or strike out for small fish.

Adult summer, *solitaria* NH/05

Adult BC/08

Adult winter VEN/02

Adult summer USA/11

Juv., *cinnamomea* CA/09

Solitary Sandpiper 1

Tringa solitaria L 8½"

Shape: Medium-sized shorebird with a slim body, long bill, short neck, relatively short legs, and long wings; distinctively long horizontal look. **Ad. Summer:** Repeatedly bobs front of body. Dark brown back with fine white notches to feather margins; streaked head and breast; white belly; white eye-ring; dark lore. Yellowish-green legs. In migration, most often alone or in groups of 2–3. Similar Spotted Sandpiper has white notch before wing; similar Lesser Yellowlegs has proportionately much longer legs. **Ad. Winter:** Like ad. summer but with less distinct streaking on head and breast and finer dots on back and wings. **Juv:** (Jun.–Nov.) Like ad. winter but seen in late summer. Generally warmer brown overall with fine dots evenly distributed over back and wing coverts; may be brownish wash over head and breast, with variably distinct streaking. Feathers fresh and not worn like adults'. **Flight:** Wings very dark above and below; tail centrally dark with heavy dark barring on sides. Flies short distances with wings mostly below body, a little like Spotted Sandpiper but slower and more labored. **Hab:** Wet areas in spruce forests in summer; small inland pools, freshwater marshes, muddy river or lake edges in winter. **Voice:** High-pitched emphatic *peetweet* or *peetweetweet* (higher and more emphatic than Spotted), or short harsh *chit*.

Subspp: (2) •*solitaria* (s.e. YT–NF south to s.e. BC–QC) is smaller, has dark brown back, solid dark lore; juv. with white spots on upperparts. •*cinnamomea* (AK–n.w. BC–n.e. MB) is larger, has slightly lighter brown back, less distinct streaked lore; juv. with cinnamon spots on upperparts. Winter ranges may overlap.

Adult summer HKO/05

Juv. JAP/-

Adult winter NZE/11

Adult JAP/05

Gray-tailed Tattler 3

Tringa brevipes L 10½"

Rare visitor to w. AK, vag. farther south along West Coast. **Shape:** Like Wandering Tattler except for subtle differences. In most cases, wings extend just to tail tip (in Wandering Tattler they usually extend well past tail tip). **Ad. Summer:** Slightly paler gray upperparts than Wandering Tattler; pale supraloral dashes (more obvious than on Wandering) meet on forehead above bill; barring limited on flanks and only faint or none on undertail coverts, thus more white on underparts than in Wandering. Uppertail coverts barred with black (seen if bird preens or stretches). **Ad. Winter:** Like ad. summer but with pale gray wash replacing barring on underparts. Distinguished from Wandering Tattler by white rather than gray flanks, more obvious supraloral dashes meeting on forehead, long nasal slit, and call. **Juv:** (Jul.–Sep.) Like ad. winter but with scapulars, tertials, and wing coverts narrowly fringed, dotted, or notched with white (more extensive than in Wandering). **Flight:** Very similar to Wandering Tattler, but uppertail coverts barred with black. **Hab:** Mountain streams or lakes in summer; rocky coasts, lakeshores, mudflats in winter. **Voice:** Best disting. from Wandering by voice. Call a rising *tooweet.*

Subspp: Monotypic.

Adult summer CA/04

Juv. CA/10

Adult winter GAL/05

Adult HI/03

Wandering Tattler 1

Tringa incana L 11"

Shape: Medium-sized stocky shorebird with a short neck, short thick legs, and bill slightly longer than length of head. **Ad. Summer:** Bobs body often. Smooth dark gray above; streaked gray on throat; extensively barred gray on breast, flanks, undertail coverts; legs bright yellow. Dark lores; white supraloral dashes that do not meet over forehead; bold white eye crescents. Only confused with rare Asian Gray-tailed Tattler (which see for differences). **Ad. Winter:** Like ad. summer but barring and streaking on underparts replaced by smooth gray wash on breast and flanks; central belly and undertail white; legs dull yellow. **Juv:** (Jul.–Sep.) Like ad. winter but with pale margins to scapulars and tertials, creating scaled look. **Flight:** Compact and all dark above. **Hab:** Mountain streams or lakes in summer; rocky coasts, jetties, beaches (sometimes), mudflats (rarely) in winter. **Voice:** Best distinguished from Gray-tailed Tattler by calls. Call a rapid, high-pitched, rolling *hidididi* of variable length, or *toowideedee* all on one pitch (rising in Gray-tailed Tattler).

Subspp: Monotypic.

Adult m., summer NOR/07

Adult winter OMA/10

Adult f., summer FIN/05

Juv. CYP/09

Spotted Redshank 4

Tringa erythropus L 12½"

Casual Eurasian visitor to both coasts; accidental inland. **Shape:** Large fairly long-necked shorebird with long legs and long thin bill. Bill length almost 2 x head length; bill droops very slightly at tip. **Ad. Summer:** Black to reddish-black legs; red on basal half of lower mandible (all plumages). M. all black with white notching on margins of back and wing feathers. F. similar, but browner and more heavily marked with white notching above and white barring below. Both sexes strikingly blotchy as they molt into and out of paler winter plumage. **Ad. Winter:** Pale gray above; white below; bright reddish-orange legs. Scapulars and wing feathers edged pale with fine white notching; lore dark with contrasting broad white supraloral line. **Juv:** (Jun.–Sep.) Brownish overall, with spotted upperparts, paler streaked head and neck, and barred flanks. Distinctive bill with red base to lower mandible. Dark lore and white supraloral line as in ad. winter; legs dark orange. Similar Greater Yellowlegs lacks reddish base to bill. **Flight:** White wedge up back, dark evenly barred tail and rump; no wing-stripe; toes extend fully beyond tail. **Hab:** Tundra in summer; freshwater wetlands, coastal bays in winter. **Voice:** Loud *chuweet,* much like Semipalmated Plover.

Subspp: Monotypic.

Adult summer FIN/05

Juv. ENG/10

Winter ENG/10

Adult JAP/–

Common Greenshank 3

Tringa nebularia L 13½"

Rare Eurasian vagrant to AK, CA, and e. CAN. **Shape:** Superficially similar to Greater Yellowlegs but with proportionately shorter legs and larger head; shorter, thicker neck; slightly thicker bill that appears more upturned. **Ad. Summer:** Legs gray, grayish green, or dull yellow (on Greater Yellowlegs, bright yellow to orange); basal half of bill gray in all seasons (basal ⅓ gray in winter Greater Yellowlegs), tip is dark. Head, neck, and breast finely streaked with black; belly white; flanks slightly streaked (barred in Yellowlegs). Back and scapular feathers mixed black and dark gray with white margins notched with black. **Ad. Winter:** Like summer but face and foreneck broadly white and unstreaked (streaked on Yellowlegs); bird appears clearly pale-headed. Bill as in summer. **Juv:** (Jul.–Sep.) Like ad. winter but feathers of back and scapulars washed brown and with wider pale margins, creating strongly scaled look. Scapular feathers pointed (rounded in ad.). **Flight:** White tail with only faint barring; white rump and distinctive white wedge up back (lacking in Yellowlegs); dark wings with no wing-stripe. **Hab:** Openings in boreal forest in summer; shallow water of tidal flats and wetlands in winter. **Voice:** Series of 5 or more high-pitched harsh whistles, *tew tew tew tew tew.*

Subspp: Monotypic.

Adult summer, *semipalmata* NJ/05

Adult summer, *inornata* FL/03

Adult summer, *semipalmata* ME/07

Adult summer, *inornata* ND/06

Willet 1

Tringa semipalmata L 12½–14½"

Two subspecies that are fairly distinct in structure and plumage. During breeding, they are geographically separate; in migration, both subspp. may be found along the East and Gulf coasts. In winter, Eastern Willet (*T. s. semipalmata*) is not in North America; the Western Willet (*T. s. inornata*) can be found along the Pacific, Atlantic, and Gulf coasts. The two subspecies are described separately here.

"Western Willet"

Tringa semipalmata inornata

Shape: Large fairly graceful shorebird with an attenuated body, long fairly thick legs, and long, fairly thin, fine-pointed bill. Bill length about 1½ x head length. Bill sometimes subtly upturned. **Ad. Summer:** Upperparts pale gray with little to some darker barring; flanks whitish with moderate dark barring (E. Willet flanks buffy with heavy dark barring); face and crown concolor pale gray except for paler supraloral area; bill often all dark, contrasting with pale head. Legs gray. **Ad. Winter:** Upperparts smooth pale gray overall; underparts whitish. Legs gray; bill dark with pale base. **Juv:** Pale buffy-gray upperparts with crisp buff fringes and dark subterminal markings; little or no contrast between scapulars and coverts nor between crown and auricular. **Flight:** Distinctive broad white stripe extends from tips of secondaries through center of dark primaries; tail grayish; rump white. **Hab:** Grassy wetlands, dry meadows in summer; beaches, tidal flats in winter. **Voice:** Song and calls of W. Willet generally lower-pitched, raspier, and slower than those of E. Willet. Song a ringing repeated *paweel weeel willet;* also repeated harsh *kew* or harsh *chidit* or plaintive downslurred *cheurr.*

"Eastern Willet"

Tringa semipalmata semipalmata

Shape: Medium-sized heavily built shorebird with fairly thick neck, fairly short thick legs, and fairly long, thick, abruptly pointed bill. Bill may droop slightly at tip. Bill length slightly more than head

Adult winter, *inornata* FL/02

Juv., *inornata* CA/07

Adult winter, *inornata* CA/12

Adult summer, *semipalmata* NJ/06

length. **Ad. Summer:** Upperparts dark brown with darker barring; flanks buffy with heavy dark barring; dark brown crown contrasts with paler face and whitish supraloral area; bill grayish, often with pinkish base and not strongly contrasting with head and face. Legs gray with pinkish cast. **Juv:** Brown back and dark-centered scapulars contrast with paler wing coverts; dark crown contrasts with paler supraloral stripe and paler auricular. Primaries with little or no extension beyond tail. **Flight:** Distinctive white stripe extends from tips of secondaries through center of dark primaries; tail grayish; rump white. **Hab:** Salt marshes, beaches in summer; does not winter in NAM. **Voice:** Song and calls of E. Willet generally higher-pitched, less raspy, and more rapid than those of W. Willet. Song a ringing repeated *pawil willet;* also repeated harsh *kew* or harsh *chidit* or plaintive downslurred *cheurr.* **34**.

Adult winter, *inornata* FL/02

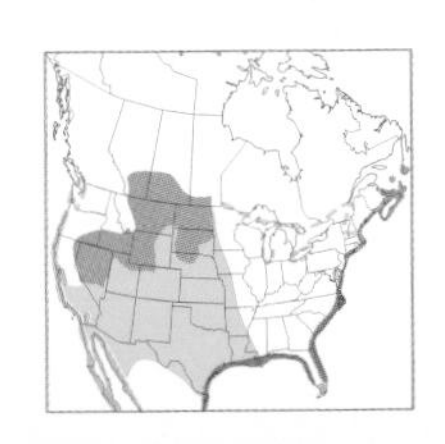

Adult summer MB/07

Juv. NY/07

Adult winter FL/02

Adult YT/05

Lesser Yellowlegs 1

Tringa flavipes L 10½"

Shape: Medium-sized gracefully proportioned sandpiper with long legs, slim body, long thin neck, and medium-length thin bill. Bill of Lesser Yellowlegs about the length of head and always straight; bill of Greater Yellowlegs about 1½ x length of head and either slightly upcurved or straight. **Ad. Summer:** Legs bright yellow or orange. Plumage similar to Greater's except with little or no barring on belly (Greater with barred belly). **Ad. Winter:** Plumage similar to Greater's except head often with slightly darker crown, creating a capped effect; bill all black (usually gray base on Greater); upperparts more finely spotted; flanks less barred. **Juv:** (Jul.–Sep.) Evenly brownish-gray to dark gray back and wings with well-spaced large whitish dots; head and neck more faintly streaked than ad. summer. Bill dark with yellow base; occasionally all dark. **Flight:** Dark back and long pointed wings with no wing-stripe; tail white with fine barring; rump white, contrasting with dark back. **Hab:** Open boreal forest in summer; shallow water of tidal flats and wetlands in winter. **Voice:** Usually 2 harsh whistled notes, *teeteer*, or repeated and spaced *teer* calls; mellower and less emphatic than calls of Greater Yellowlegs. 🎧**35**.

Greater (front) and Lesser Yellowlegs FL/03

Subspp: Monotypic.

Adult summer FL/04

Juv. BC/07

Adult winter FL/02

Juv. CA/08

Greater Yellowlegs 1

Tringa melanoleuca L 14"

Shape: Large sandpiper with long legs, slim body, long thin neck, and long thin bill; contours of neck and chest often appear angular (smoother in Lesser Yellowlegs). Bill of Greater Yellowlegs about 1½ x length of head and either slightly upcurved or straight; bill of Lesser about the length of head and always straight. **Ad. Summer:** Legs bright yellow or orange; head and neck strongly streaked with black; upper belly and flanks strongly barred black; back and wings mottled gray and black and heavily marked with white spotting and chevrons; bill all dark. Similar Lesser Yellowlegs usually lacks barring on belly; see Shape to distinguish. **Ad. Winter:** Like ad. summer but head and neck less streaked; belly and flanks mostly white with little barring; back and wings brownish gray and feathers notched with white dots; bill dark with paler gray at base for about ⅓ of length. **Juv:** (Jun.–Sep.) Brownish upperparts (without black spotting of ad.) with buffy to whitish dots along feather edges, creating finely spotted look; bill usually with gray base (nearly half; ad. bill all dark in summer). Similar juv. Lesser Yellowlegs often has yellow base to bill. **Flight:** Dark back and long pointed wings with no wing-stripe; tail white with fine barring; rump white contrasting with dark back. **Hab:** Openings in boreal forest in summer; shallow water of tidal flats and wetlands in winter. **Voice:** Usually a series of 3–4 harsh loud whistles, *teeteeteer, teeteeteeteer,* or repeated *teer* notes, or repeated mellow *kluwee.* Lesser Yellowlegs usually only 2 notes, *teeteer*, and mellower. 🎧**36**.

Subspp: Monotypic.

Adult summer CYP/05

Juv. FIN/08

Adult winter OMA/12

Adult summer CYP/07

Wood Sandpiper 2

Tringa glareola L 8"

Fairly regular visitor to w. AK; vagrant down West Coast and along Northeast Coast. **Shape:** Much like Solitary Sandpiper but with slightly longer legs. Primary extension past tertials very short, less than 1/4 bill length (almost length of bill in similar Lesser Yellowlegs). **Ad. Summer:** In all plumages, broad whitish eyebrow (from forehead to nape) contrasts with dark caplike crown and dark eye-stripe. Upperparts dark grayish brown with bold white notches along feather edges (similar Lesser Yellowlegs has finer spotting); head and neck finely streaked; breast and flanks lightly barred; legs yellowish green. Bill with paler base (all black and longer and thinner in Lesser Yellowlegs). May bob front or rear body. **Ad. Winter:** Like ad. summer but with muted markings; spotting smaller on upperparts; neck and breast indistinctly streaked and washed with brownish gray. Bill with paler base. **Juv:** (Jul.–Aug.) Like ad. summer but back dark brown with finer buffy spotting. **Flight:** Dark wings and back; no wingstripe; rump white; white tail with narrow barring. Toes extend fully beyond tail. **Hab:** Bogs, marshes in summer; inland wetlands in winter. **Voice:** High-pitched mellow *dewdewdew* or slowly repeated *jit*.

Subspp: Monotypic.

Adult summer POR/04

Juv. JAP/09

Adult, trans. to summer HKO/04

Adult NOR/07

Adult winter ENG/02

Adults, summer CYP/08

Common Redshank 5

Tringa totanus L 11"

Accidental Eurasian vag. to n.e. NAM (records from NF). **Shape:** Medium-sized shorebird with long legs, long straight bill, and rather compact deep-chested body. Bill length about 1¼ x head length. **Ad. Summer:** Orange-red legs and base of both mandibles; thin white eye-ring; pale supraloral dash. Upperparts dark brown with darker flecking; head and underparts heavily streaked; some barring on flanks. **Ad. Winter:** Upperparts, head, and breast smooth gray-brown; dark barring on flanks and undertail coverts; thin eye-ring on plain head. Bill and legs as in summer. Back and scapulars have fine pale margins with tiny white dots. **Juv:** (Jun.–Sep.) Upperparts dark brown; feathers with whitish margins and large whitish notches, creating a scaled effect; breast and throat finely streaked. Bill paler at base (may have no red), and legs may be more orange than red. White eye-ring; dark lore. **Flight:** Distinctive broad white trailing edge to secondaries and inner primaries; white wedge up back; tail white with fine black barring overall. **Hab:** Grassy wetlands in salt or fresh water in summer; coastal mudflats in winter. **Voice:** Series of midpitched soft whistles, *k'weet k'weet k'weet*.

Subspp: (1) •*robusta*.

Adult MT/06

Juv. NJ/09

Adult NJ/06

Upland Sandpiper 1

Bartramia longicauda L 10½"

Shape: Large, heavy-bodied, long-tailed, and long-winged shorebird with long skinny neck, relatively small head, large dark eyes, and upright stance. Bill short, about length of head; long tail projects well past wing tips. **Ad:** Plain pale buffy face; head and neck streaked brown; breast and flanks with dark brown chevrons. Back and wing feathers dark brown with pale fringes and dark bars. Bill yellow with dark tip and upper culmen; legs yellowish. **Juv:** (Jun.–Oct.) Like ad., but scapulars and wing coverts round-tipped and with buffy margins, creating scaled look. **Flight:** Alternates stiff-winged flight with long glides. Below, brownish with barred underwings and pale belly. Bill short; tail long. **Hab:** Wide-open areas with low vegetation, including prairies, other grasslands, airports, blueberry barrens, in summer; grasslands, cultivated fields in winter. **Voice:** Quiet, ascending, guttural, rolling trill precedes slow "wolf whistle," *g'g'g'g'g waaeeet wheeeoo*, or short, fast, churring *jibibib*.

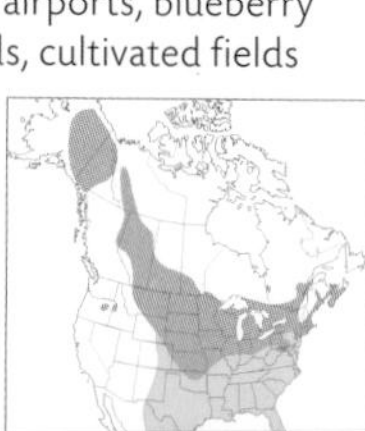

Subspp: Monotypic.

Adult, worn CA/09

Juv. –/–

Juv. CA/09

Little Curlew 5

Numenius minutus L 11"

Accidental Asian vag. to w. AK and West Coast, spring or fall. **Shape:** Similar size and proportions to Upland Sandpiper, but with shorter tail and longer, tapered, downcurved bill. Bill clearly longer than head; wings project just to tail tip. **Ad:** Dark crown (with pale central stripe) contrasts with beige eyebrow; eyeline thin and broken, restarts under eye and continues back to nape; head and neck streaked brown; flanks barred brown; back and wing feathers dark brown with pale margins. Head, neck, sides of breast, and flanks variably washed with golden buff. Bill dark with pinkish base to lower mandible. **Juv:** (Jul.–Oct.) Like adult; best distinguished in fall when plumage is fresh (adults' will be worn). **Flight:** Compact curlew with short bill and short tail; toes project just past tail; underwing brownish. **Hab:** Open grasslands in summer; agricultural fields, beaches, mudflats, marshes in winter. **Voice:** High-pitched, rapid, ascending series, *teeteeteetee*.

Subspp: Monotypic.

Adult, *hudsonicus* FL/02

Adult, *variegatus* RUS/06

Juv., *hudsonicus* CA/09

Adult, *hudsonicus* CA/02

Adult, *phaeopus* POR/04

Whimbrel 1

Numenius phaeopus L 16"

Shape: Large heavy-bodied shorebird with moderate-length legs, relatively short neck, and long downcurved bill. Bill length 1½–2 x head length. **Ad:** Generally grayish brown overall with no warmer reddish tones as seen in similar Long-billed Curlew. Dark brown lateral crown-stripes, separated by pale median stripe, create dark "cap," which contrasts with pale eyebrow; usually has wide dark brown eyeline. All-dark bill has pink base to lower mandible in winter. **Juv:** (Jul.–Mar.) Like ad., but scapulars and tertials darker than coverts and boldly notched with white along margins; bill may be shorter than adult's (about 1¼ x head length). **Flight:** Underwings and belly buffy with no reddish tones; any white below limited to central belly; upperparts concolor or grayish brown. **Hab:** Tundra in summer; tidal flats, meadows in winter. **Voice:** Midpitched rapid series mostly on one pitch, *teeteeteeteetee*.

Subspp: (3) Differences best seen in flight. •*hudsonicus* (throughout NAM range) has brown back, rump, and tail; •*phaeopus* (EUR vag. to Atlantic Coast) has white wedge up back, white rump, brown tail barred white at base; •*variegatus* (Siberian vag. to w. AK–CA) has white wedge up back, darkish rump, dark tail slightly barred with white at base.

IDENTIFICATION TIPS

Curlews are medium-sized to large shorebirds with long to very long, thin, downcurved bills; most are fairly heavy-bodied and relatively short-legged. Most are brownish overall. They feed by probing into mud, sand, or grasses with their long downcurved bills. The group includes Upland Sandpiper (the only example with a short straight bill) and Whimbrel, in addition to those with "curlew" in their name.

Adult HI/02

Adult HI/03

Juv. HI/10

Adult LHI/–

Adult RUS/06

Adult RUS/08

Bristle-thighed Curlew 2

Numenius tahitiensis L 16"

Local and uncommon breeder in w. AK. **Shape:** Similar in size and proportions to Whimbrel. Bristle-like feather shafts project off body at top of legs (only seen close up). **Ad:** In general very similar to Whimbrel but more rich buff on rump, tail, and flanks; flanks less barred; streaking on breast ends more abruptly on belly; lower mandible mostly pinkish. On fresh-plumaged birds (Nov.–Mar.) upperparts dark brown with warm buff spotting; worn birds less colorful. See Flight. **Juv:** Similar to winter adults. **Flight:** Distinctive, clear, warm buffy rump contrasts with dark brown back (Whimbrel upperparts all dark grayish brown); tail same color as rump but with thin dark brown bars. **Hab:** Tundra with shrubs in summer; variable, including shorelines, meadows in winter. **Voice:** Flight call like attention-getting human whistle, a sharp *keeooweet.*

Subspp: Monotypic.

Far Eastern Curlew 4

Numenius madagascariensis L 25"

Casual Asian visitor to w. AK spring–fall; accidental in BC. **Shape:** Very similar to Long-billed Curlew but larger and with little or no primary projection past tertials (primary projection on Long-billed and Eurasian Curlews equals lore + eye). **Ad:** Very similar to Long-billed Curlew but with cool brown tones overall (Long-billed rich cinnamon tones overall). See Flight. **Juv:** Very similar to juv. Long-billed but with cool brown tones overall (Long-billed rich cinnamon tones overall). **Flight:** Underwings pale grayish brown with faint broad barring (Long-billed underwings strongly cinnamon with faint thin barring); rump and back dark (Eurasian Curlew has white rump and white wedge up back; white underwing coverts). **Hab:** Open bogs and swamps in summer; beaches, wetlands in winter. **Voice:** A mellow whistled *keeurweee.*

Subspp: Monotypic.

Adult ENG/02

Adult WAL/11

Adult WAL/11

Juv. FIN/07

Eurasian Curlew 4

Numenius arquata L 22"

Casual Eurasian visitor to n. Atlantic Coast, mostly fall–winter. **Shape:** Very similar to Long-billed Curlew but slightly larger; as in Long-billed, primary projection past tertials equals lore + eye (Far Eastern Curlew has little or no primary projection past tertials). In flight, toes project halfway beyond tail. **Ad:** Very similar to Long-billed Curlew but belly more heavily streaked over white background and hind belly and undertail coverts white and unstreaked (Long-billed has buffy to cinnamon underparts that are largely unstreaked). See Flight.

Juv: (Jul.–Sep.) Similar to ad. but with feathers of upperparts more broadly edged buff; bill may be shorter by $\frac{1}{2}$. **Flight:** White rump and white wedge up back; white rear belly and undertail coverts; pale lightly marked underwing coverts (Long-billed underwings cinnamon, Far Eastern underwings grayish brown). **Hab:** Open bogs and swamps in summer; beaches, mudflats, wetlands in winter. **Voice:** Mellow whistled *kurree* or low-pitched, initially ascending, mellow, whistled *kwarkwarkwar kweeu kweeu kweeu*.

Subspp: (1) •*arquata*.

Adult CA/09

Juv. CA/08

Adult FL/02

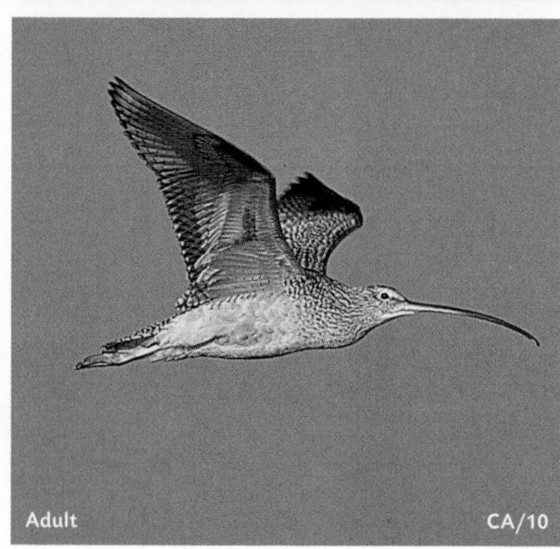
Adult CA/10

Long-billed Curlew 1

Numenius americanus L 20–26"

Shape: Very large shorebird with extremely long downcurved bill. Bill 2½ x head length in ad. males and juveniles and 3½ x head length in ad. females. **Ad:** Buffy cinnamon look overall, especially on wings, breast, and belly; face usually rather plain, with little or no contrasting eyeline or eyebrow; darker streaking on crown creates suggestion of shallow cap, but this is never as clearly dark as on Whimbrel, nor is there a pale central stripe. Aside from bill, can be similar to Marbled Godwit, but Long-billed Curlew has gray (not black) legs and little or no projection of dark primaries past tertials (Marbled Godwit's dark primaries project noticeably past its tertials). **Juv:** (May–Sep.) Wing coverts buff with dark central streaks, rather than barred as in adult. Breast less streaked or plain. Bill length at the shorter extreme of range (about 2½ x head length), still longer than adult Whimbrel's (1½–2 x head length). **Flight:** Above, secondaries and inner primaries cinnamon with dark notches; below, lesser and median coverts dark cinnamon, rest of underwing pale. **Hab:** Grasslands in summer; mudflats, beaches, pastures, agricultural fields in winter. **Voice:** Slowly repeated, single, long whistle, *curweee,* or more rapid series of rising and falling whistles, or harsh 2-part whistle, *chuleeet.*

Subspp: Monotypic.

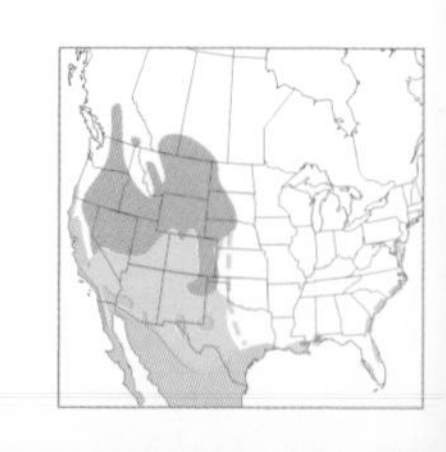

Adult m., summer MB/06

Juv ON/–

Adult f., trans. to summer TX/05

Adult NJ/09

Adult, trans. to winter CA/08

Adult, trans. to winter MA/08

Hudsonian Godwit 1

Limosa haemastica L 15½"

Shape: Relatively short-necked and short-legged godwit with a slightly upcurved bill; primary projection past tertials longer than lore. **Ad. summer:** M. with finely streaked whitish to buffy neck; thin dark barring on dark reddish-brown breast and belly. F. similar but paler reddish brown below, often with some white. **Ad. Winter:** Generally plain gray above and whitish below. **Juv:** (Jul.–Oct.) Brownish gray above, whitish below; dark inner tips and shafts of scapulars and dark shafts of pale-fringed wing coverts create fine dotted effect on the scapulars and streaked effect on coverts. Bill colored for only basal third (ad. bill colored for half or more). **Flight:** Distinctive. Above, black tail, white uppertail coverts, narrow white wing-stripe that fades out in secondaries; below, black median and lesser underwing coverts; toes project only partially beyond tail. **Hab:** Wet meadows, taiga pools in summer; mudflats in winter. **Voice:** High-pitched rapid *weebeep* or spaced *weet weet*.

Subspp: Monotypic.

Adult m., summer JAP/–

Juv. JAP/09

Adult winter ENG/11

Adult winter MOR/09

Black-tailed Godwit 3

Limosa limosa L 16½"

Rare Eurasian vag. to w. AK and Atlantic Coast, fall–spring (see Subspp.). **Shape:** Relatively long-necked long-legged godwit with a very long bill that is straight to barely upturned or barely downturned (see Subspp.); primary projection past tertials equal to or shorter than lore (clearly longer than lore in Hudsonian Godwit). **Ad. Summer:** M. with unstreaked reddish-brown neck; variable dark barring on variably reddish-brown lower breast, belly, and flanks; whitish on rest of belly and undertail coverts. F. similar but paler-colored with less barring on underparts; bill longer. Bill orangish to pinkish on basal two-thirds, dark on rest. **Ad. Winter:** Grayish above with thin paler margins to wing coverts; gray wash over neck, breast, and flanks; whitish belly. **Juv:** (Jul.–Oct.) Head, neck, and breast buffy; belly white; scapulars and wing coverts dark with buffy margins, creating a scaled effect. **Flight:** Distinctive. Above, black tail, white rump, broad white wingstripe that extends all the way to body (fades out in secondaries on Hudsonian); below, white underwing coverts with dark flight feathers (Hudsonian underwing coverts black); feet project completely beyond tail. **Hab:** Wet meadows in summer; mudflats, wetlands in winter. **Voice:** Harsh repeated *chuweechuweechuwee*.

Adults NY/08

Juv. JAP/09

Subspp: (3) Distinguished with difficulty due to complex molt, sex, and age differences, regional variation; ad. summer m. most distinctive. •*islandica* (vag. to East Coast, winters w. EUR) ad. summer m. has most extensive and darkest reddish brown, extending to flanks and belly; heavily barred underparts include belly; bill straight or subtly upturned. •*limosa* (vag. to East Coast, winters Africa) ad. summer m. has less extensive and paler reddish brown, limited to neck and upper breast; lightly barred underparts (little on belly); bill straight or subtly upturned. •*melanuroides* (vag. to w. AK from Asia) ad. summer m. has dull reddish brown limited to neck and upper breast; bill shorter and straight or subtly drooped at tip.

IDENTIFICATION TIPS

Godwits are large shorebirds with long straight to slightly upcurved bills and fairly long legs. They feed by probing into mud, sand, or short grasses.

Adult m., summer, *baueri* AK/06

Juv., *baueri* CA/11

Adult f., summer, *baueri* AK/06

Adult summer, *baueri* AK/05

Adult winter, *lapponica* ENG/12

Juv., *lapponica* UK/10

Bar-tailed Godwit 2

Limosa lapponica L 16"

Shape: Relatively compact, short-necked, short-legged godwit with a long slightly upcurved bill; primary projection past tertials longer than lore. **Ad. Summer:** M. with rich reddish-brown neck, breast, and belly, with fine streaking only on nape and sides of breast; above, dark brown with darker mottling and white notches. F. varies below from almost all whitish to heavily barred darker on breast and dark chevrons along flanks; above, grayish-brown feathers with pale margins and darker centers create a striped effect (similar to winter plumage); face and neck finely streaked. Bill of m. and f. mostly all dark; f. bill longer. **Ad. Winter:** Below, belly whitish, neck and upper breast finely streaked, dark chevrons along flanks; above, grayish-brown feathers with pale margins, creating striped effect. Basal half of bill pinkish. **Juv:** (Jul.–Nov.) Heavily spotted above with buff and white; belly whitish; face and neck finely streaked; neck and breast variably washed with buff. Similar juv. Hudsonian Godwit lacks contrasting spotting above and streaking on face. **Flight:** Above, grayish brown with darker primary coverts and whitish to barred brown-and-white rump (see Subspp.); below, underwings uniformly pale. **Hab:** Open tundra and wetlands in summer; mudflats and wetlands in winter. **Voice:** Loud repeated *kaweeah* or *kawee,* or forceful *keekeet.*

Subspp: (2) Distinguished in flight. •*lapponica* (Eurasian vag. to Atlantic Coast, spring–fall) has contrasting whitish wedge from rump up back; •*baueri* (breeds w. AK, Asian vag. to Pacific or Atlantic coasts spring or fall) has barred brown-and-white rump, back concolor with rest of upperparts.

Adult summer, *fedoa* MT/06

Adult summer, *fedoa* MT/06

Adult winter, *fedoa* FL/02

Adult summer, *fedoa* MT/06

Marbled Godwit 1

Limosa fedoa L 18"

Shape: The largest godwit, with relatively long legs and long slightly upturned bill; primary projection past tertials shorter than lore. **Ad. Summer:** Dark brown above with whitish and gold barring and spotting; cinnamon below with fine dark barring; head and neck streaked; blackish legs. **Ad. Winter:** Above, dark with cinnamon spotting; below, all pale cinnamon with only slight barring on flanks. **Juv:** (Jun.–Sep.) Like ad. winter but with little or no barring on flanks and less heavily marked wing coverts. **Flight:** Distinctive. Wings cinnamon above and below, richest color on underwing coverts (distinguished from similar Long-billed Curlew by upcurved bill and blackish rather than gray legs). **Hab:** Prairies in summer; mudflats and beaches in winter. **Voice:** Repeated harsh *kweeet,* or 2-part *kweecut,* or repeated *kaweeya.*

Subspp: (2) •*beringiae* (s.w. AK) is slightly heavier with slightly shorter wings and legs than •*fedoa* (rest of range).

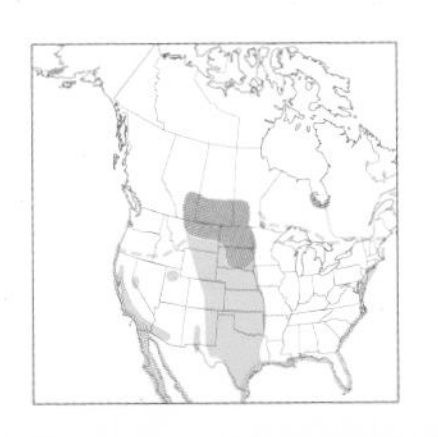

Adult m., summer AK/06

Adult winter FL/02

Adult f., summer AK/06

Juv. NJ/09

Adult summer NJ/05

Ruddy Turnstone 1

Arenaria interpres L 9"

Shape: Medium-sized, small-headed, deep-chested shorebird with short legs and short wedge-shaped bill. Appears front-heavy and walks with distinctive jerky motion. **Ad. Summer:** Wide black loops on breast sides enclose white marks; head white with complex black markings; back with broad areas of reddish brown and black; belly white; orange legs. F. similar to m. but with less rufous on upperparts and less distinct facial pattern and fine black streaks in white facial areas. Walks along water edge flipping over debris in search of food. **Ad. Winter:** Like summer, but grayish-brown head and upperparts and loops on breast brownish and less well defined. **Juv:** (Jul.–Oct.) Like ad. winter, but back feathers and wing coverts edged pale, creating scaled look. Similar juv. Black Turnstone has more extensive dark brown on head and chest, and border between chest and belly is a straight line rather than loops. **Flight:** Bold white stripes on either side and center of back; also bold white wing-stripes. **Hab:** Tundra and rocky coasts in summer; shorelines in winter. **Voice:** High thin *tseeah*, or harsh chattering *ch'ch'ch'chrt,* or single *churt,* or *chewki chewki.*

Subspp: Monotypic.

Adult summer CA/05

Juv./1st winter CA/09

Adult winter CA/01

Adult winter BC/02

Black Turnstone 1

Arenaria melanocephala L 9"

Shape: Medium-sized, small-headed, stocky shorebird with short legs and short wedge-shaped bill. Appears front-heavy. **Ad. Summer:** Blackish brown overall with a white belly and faint white mottling on breast; small white oval across loral area; suggestion of white eyebrow; dark blackish-orange legs. Walks along water edge flipping over debris in search of food. **Ad. Winter:** Sooty brown overall with white belly; head, breast, and back without white markings of summer. Legs dull blackish orange. **Juv:** (Jul.–Oct.) Similar to ad. winter, but scapulars round-tipped (pointed in ad.) and more brownish than adult. Similar juv. Ruddy Turnstone's chest is less extensive dark brown and border with belly is in 2 loops. **Flight:** Bold white stripes on either side and center of back; also bold white wingstripes. **Hab:** Grassy wetlands in tundra in summer; rocky shorelines, beaches, mudflats in winter. **Voice:** Slow *pit too wit* or rapid mellow chattering.

Subspp: Monotypic.

Adult summer CA/04

Juv./1st winter CA/10

Adult winter CA/11

Adult CA/07

Surfbird 1

Aphriza virgata L 10"

Shape: Medium-sized deep-chested shorebird with short thick legs, relatively small head, and short blunt bill (noticeably shorter than depth of head). **Ad. Summer:** Gray above with scapulars variably reddish brown; head and upper breast streaked; belly and flanks white with gray chevrons; legs yellow; bill dark with yellow to orange base of lower mandible in all plumages. **Ad. Winter:** Similar to ad. summer, but upperparts and breast smooth gray; belly white with a few dark chevrons. **Juv:** (Jul.–Oct.) Similar to ad. winter but with head finely streaked, breast finely barred, and upper belly finely dotted; wing coverts with thin pale margins and thin dark subterminal bands. **Flight:** Dark gray above with boldly contrasting white rump and wing-stripe. **Hab:** Rocky mountain tundra in summer; rocky coast in winter. **Voice:** Two-part repeated *wayuh wayuh wayuh,* or a quiet *if if if* or chattering while feeding among rocks.

Subspp: Monotypic.

Adult summer RUS/07

Adult winter AUS/03

Juv. JAP/09

Great Knot 4

Calidris tenuirostris L 11"

Casual Asian vag. to w. AK and accidentally farther south, spring or fall. **Shape:** Large heavy-bodied shorebird with relatively short legs, short neck, and small head. Fairly long bill with slight droop to tip and length equal to or slightly greater than head length. Similar proportions to a Red Knot but larger, longer-legged, and with longer, more strongly tapered bill that droops slightly at the tip. **Ad. Summer:** Similar to Surfbird. Dark brownish gray above with reddish-brown scapulars; wing coverts dark-centered with pale margins, creating scaled effect; head and neck strongly streaked gray; breast heavily mottled with black; belly white with dark spotting along flanks. **Ad. Winter:** Similar to ad. summer, but no reddish brown on scapulars; breast with fine gray streaks and spots. Distinguished from similar winter Red Knot by scaled effect on back and wings (smooth unpatterned gray on Red Knot) and streaks along flanks (Red Knot has barred flanks). **Juv:** (Jul.–Sep.) Similar to ad. winter, but wing coverts and scapulars strongly margined with white and with dark subterminal "anchors." **Flight:** White rump contrasts with brownish-gray upperparts; faint wing-stripe. **Hab:** Barren uplands in summer; beaches, mudflats in winter. **Voice:** Mostly silent.

Subspp: Monotypic.

Adult summer NJ/05

Juv. CA/09

Adult summer TX/05

Adult summer NJ/–

Adult winter FL/02

Adults, winter NJ/02

Red Knot 1

Calidris canutus L 9½"

Shape: Medium-sized deep-chested shorebird with a short neck, short legs, and relatively small head. Bill about the length of the head or slightly shorter; straight and rather blunt. Bird often appears hunchbacked. **Ad. Summer:** Reddish-brown or orangish face, throat, and upper belly; gray crown and nape with fine dark streaking; upperparts vary from silver with a few dark bars to heavily marked with reddish brown, black, and white (see Subspp.); legs blackish. **Ad. Winter:** Appears smooth gray on back and wings, although feather shafts are dark; head and breast finely streaked; paler whitish eyebrow; flanks variably barred; legs yellowish green. Primary projection past tertials as long as lore (on winter dowitchers with bills hidden, primary projection much shorter than lore). **Juv:** (Jul.–Oct.) Similar to ad. winter, but light buff wash on breast, and feathers of back and wing coverts have fine white margins and dark subterminal lines, creating scaled appearance. **Flight:** Upperparts grayish brown with lighter rump and darker primary coverts; moderate wing-stripe. **Hab:** Tundra in summer; beaches, mudflats in winter. **Voice:** Mostly quiet; flight call a soft *tuweet*.

Subspp: Monotypic.

Adult summer FIN/06

Juv. NH/10

Adult winter FL/01

Adults, summer NJ/05

Juv. NH/09

Sanderling 1

Calidris alba L 7½"

Shape: Comparatively heavy-bodied, broad-necked, and large-headed sandpiper. Bill short, tubular, straight; feet lack hind toe (distinctive for this species); prominent primary projection past tertials (obvious in winter). **Ad. Summer:** Head, neck, and breast variably reddish brown with fine dark streaking; feathers of back and wing coverts variably reddish brown with dark subterminal tips and whitish margins; clear division between darker bib and white rest of underparts; legs black. Some individuals have little to no reddish brown on upperparts and instead have silvery look with black streaks. Familiar "wave chaser" of beaches. **Ad. Winter:** Palest of our winter peeps. White face and underparts; pale silver-gray upperparts. On perched bird, black lesser coverts often visible and look like a black "shoulder" (actually wrist). **Juv:** (Jul.–Nov.) Similar to ad. winter, but upperparts darker and checkered black and white; crown streaked darker; lore and auricular dark. When fresh, may have buffy wash on breast and wing coverts. **Flight:** Strong white wing-stripe through dark wing pattern—wing-stripe bordered in front by black primary and lesser secondary coverts and in back by broad black tips to primaries and secondaries; rump white with dark central stripe. **Hab:** Barren coastal tundra in summer; beaches in winter. **Voice:** Irregularly repeated short *pit* or scratchy *cheet,* which given together by a flock create a chattering sound.

Subspp: Monotypic.

Adult summer AK/06

Juv. NH/09

Adult summer TX/05

Juv. ME/09

Adult winter BAH/03

Juv. NH/09

Semipalmated Sandpiper 1

Calidris pusilla L 6¼"

Shape: Small sandpiper with moderate proportions. Bill generally short, straight, and tubular with a blunt or slightly bulbous tip; on average, bills finer-tipped and longer in eastern birds, blunter and shorter in western birds. In eastern U.S. there is an overlap in bill shape and length with m. Western Sandpipers. Wing tips just reach or fall just short of tail tip; primary extension past tertials about length of lore. Some webbing between middle and outer toes. **Ad. Summer:** Dark legs. Brownish above, with variable rufous on head, back, and scapulars. Spotting or streaking on sides of breast but variable to none on flanks; when present it is thin dashes, not chevrons as on Western Sandpiper. **Ad. Winter:** Completed plumage rarely seen, since birds winter outside U.S. and do not molt completely into winter plumage until on wintering grounds. Plumage very similar to Western Sandpiper's—generally unmarked grayish brown above; white underparts except for possibly some faint dusky streaks on sides of breast (flanks unmarked); eastern birds with long bills may be very difficult to separate in the field from Western Sandpipers (except Westerns are generally further along in molt). **Juv:** (Jul.–Dec.) Extremely variable in plumage, from bright rufous to dull grayish brown. Buffy wash on breast; pale margins to upperparts feathers create strong finely scaled effect; scaly appearance on back more uniform than on juv. Western. Back and head buffy to grayish brown with little or no reddish brown; when reddish brown is present on back, it is confined to feather margins. Dark crown contrasts with paler gray nape (crown and nape concolor grayish on juv. Western). **Flight:** Weak wing-stripe; dark-centered rump and tail. **Hab:** Tundra in summer; mudflats, beaches in winter. **Voice:** Short *chirt* or higher-pitched scratchy *cheet,* sometimes given in a rapid series.

Subspp: Monotypic.

Adult summer AK/07

Adult, trans. to summer FL/03

Adult winter FL/01

1st winter FL/01

Juv. CA/09

Adult winter FL/02

Juv. CA/08

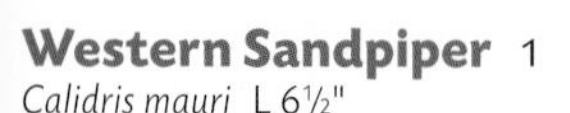

Western Sandpiper 1

Calidris mauri L 6½"

Shape: Small, rather heavy-set, front-heavy sandpiper with a relatively large head, broad neck, and fairly long, broad-based bill that tapers to fine point, slightly drooping at tip. Wing tips reach or fall just short of tail tip; primary extension past tertials about length of lore. As in all small sandpipers, bill length varies, usually longer in females, shorter in males. Some webbing between middle and outer toes. **Ad. Summer:** Dark legs. Upperparts generally grayish brown with rich reddish-brown scapulars, crown, and auriculars (amount of reddish brown varies); chevron-shaped spotting on breast and flanks. **Ad. Winter:** Upperparts grayish and rather unmarked; white on face, throat, and belly with a little fine streaking on sides of breast. Appears more white below than many other small sandpipers (similar winter Dunlin has grayish-brown chest). Generally changes to winter plumage earlier in fall than Semipalmated Sandpipers. **Juv:** (Jul.–Oct.) Pale margins to upperparts feathers create strong scaled effect; scaly appearance on back and wings less uniform than on juv. Semipalmated Sandpiper. Bold, crisp, reddish-brown margins to upper scapulars and sometimes on mantle (amount on scapulars varies, but always present); lower scapulars have subterminal dark "anchors." Underparts white except for some fine streaking on sides of breast; buffy wash across breast when fresh. Eyebrow whitish and fairly indistinct (whiter and more prominent on juv. Semipalmated Sandpiper); head paler overall than on juv. Semipalmated. **Flight:** Weak wingstripe; dark-centered rump and tail. **Hab:** Tundra in summer; mudflats, beaches in winter. **Voice:** High-pitched *cheet,* often with several syllables in a rolling manner, like a rapid *cheeveet* or *cheeveveet.*

Subspp: Monotypic.

Adult summer JAP/07

Juv. JAP/08

Adult summer CT/07

Juv. JAP/–

Adult winter NZE/09

Adults, summer HKO/05

Red-necked Stint 3

Calidris ruficollis L 6¼"

Regular migrant and rare breeder to w. AK; casual on both coasts and accidental inland, spring–summer. **Shape:** Small attenuated sandpiper with a short, straight, deep-based, and strongly tapered bill (more strongly tapered than similar Semipalmated Sandpiper's). Forehead steep (sloping in Western or Semipalmated Sandpipers) and angling to a flattish crown, creating a somewhat blocky look to head. Wing tips extend just past tail tip; primary projection past tertials as long as lore (shorter than lore in Western and Semipalmated Sandpipers). Legs relatively short (shorter than those of Little Stint). No webbing between toes. **Ad. Summer:** Legs dark. Mantle and scapulars with bold reddish-brown margins; tertials are grayish with paler edging and little or no reddish brown and are contrastingly duller than rest of upperparts (in Little Stint, tertials are darker-centered with reddish-brown edges and blend with rest of upperparts). From June to mid-July, ad. has dark to pale reddish-brown head, throat, and upper breast with a white chin, though some f. may have whitish throat and upper breast (Little Stint has white throat and white central breast); dark streaking on breast forms a necklace below area of reddish brown (spotting throughout reddish-brown area on breast of Little Stint). In late summer, reddish brown may be worn away, resulting in just a pale wash across the breast and a whitish, sometimes subtly split eyebrow; in these cases, look for streaking on the breast below the reddish-brown area. Flanks unmarked or with a few fine streaks (as in Semipalmated and Western). **Ad. Winter:** Upperparts grayish brown; feathers with dark shafts (not dark-centered as in Little Stint). Little or no streaking on flanks. **Juv:** (Jul.–Oct.) Grayish-buff wash to head and breast; scapulars dark-centered; upper scapulars fringed with rich reddish brown, contrasting with duller lower scapulars; wing coverts and tertials fringed with grayish buff. Note long primary projection past tertials. **Flight:** Weak wing-stripe; dark-centered rump and tail. **Hab:** Tundra in summer; mudflats, shores, beaches in winter. **Voice:** Scratchy high-pitched *jeet,* similar to Western Sandpiper.

Subspp: Monotypic.

Adult summer CYP/05

Adult winter UAE/01

Adult summer NOR/07

Juv. ENG/09

Adult winter OMA/12

Juv. CYP/08

Little Stint 4

Calidris minuta L 6"

Casual Eurasian vag. to both coasts, accidental inland, mostly in summer, rarer in spring and fall. **Shape:** Small fairly compact sandpiper with a short, straight (sometimes slightly drooped), deep-based, strongly tapered, and fine-pointed bill. Forehead more sloping and bill slightly longer and more fine-pointed than on Red-necked Stint. Wing tips project just past tail tip, and primary projection past tertials is as long as lore. Legs slightly longer than Red-necked Stint's; toes with no webbing. **Ad. Summer:** Legs dark. Upperpart feathers strongly edged with reddish brown, including wing coverts and tertials (tertials dull-centered and pale-edged in Red-necked Stint). Variable orangish brown on head and sides of breast; throat and central breast white. Dark spotting on sides of breast but not on flanks; when orangish brown present on breast, spotting contained within it (streaking below reddish-brown area on Red-necked Stint). **Ad. Winter:** Upperparts grayish brown; similar to Red-necked Stint, but some feathers dark-centered (all just thin dark shafts on Red-necked Stint). Shape subtly different (see Shape). **Juv:** (Jul.–Oct.) Scapulars, wing coverts, and tertials extensively black-centered and bordered by reddish brown and white, creating scaled appearance, and not contrasting with each other (as in Red-necked Stint, where coverts are more grayish, with black restricted to arrow-shaped shaft streak). Tertials edged reddish brown (no reddish brown in Red-necked Stint tertials). **Flight:** Weak wing-stripe; dark-centered rump and tail. **Hab:** Tundra in summer; mudflats, shores in winter. **Voice:** Very high-pitched, very short, repeated *peet.*

Subspp: Monotypic. **Hybrids:** Temminck's Stint.

Adult summer ENG/05

1st winter CYP/11

Adult summer FIN/07

Juv. CYP/09

1st winter POR/01

Adult NOR/06

Temminck's Stint 3

Calidris temminckii L 6¼"

Eurasian vag. to w. AK islands, rare farther south, spring or fall. **Shape:** Small, short-legged, attenuated sandpiper with a thin fine-pointed bill that droops slightly at tip. Wing tips extend to tail tip; practically no primary projection past long tertials. **Ad. Summer:** Legs pale greenish yellow to yellow. Rather plain brown upperparts with scattered black and reddish-brown markings on scapulars and coverts. Fine brown streaking on throat and breast creates bibbed effect. Face with no eyebrow but fairly conspicuous eye-ring. Belly and flanks white with no streaking. Similar Least Sandpiper has a pale eyebrow and lacks the conspicuous eye-ring. **Ad. Winter:** Plainest of all stints. Relatively unmarked grayish-brown back, head, and breast; pale throat and chin and white belly separated by well-demarcated breastband. **Juv:** (Jul.–Sep.) Similar to ad. winter; has unique stint feather pattern on back—grayish-brown feathers with fine dark subterminal margins and pale buff margins; subtle scaled look. **Flight:** Pure white outer tail feathers (other stints have grayish tail feathers); dark central tail feathers; moderate wing-stripe. Tends to tower up after being flushed, rather than just flying off. **Hab:** Bogs and marshes in summer; freshwater wetlands in winter. **Voice:** Fast rolling series of short high-pitched *peet* calls, like *pidididdeet*.

Subspp: Monotypic. **Hybrids:** Little Stint.

Adult summer HKO/04

Adult winter SIN/12

Adult summer HKO/04

Juv. JAP/09

Adult summer, worn JAP/09

Juv. JAP/09

Long-toed Stint 3

Calidris subminuta L 6"

Eurasian vag. to w. AK islands, rarer on rest of West Coast, spring or fall. **Shape:** Small, long-legged, and long-necked sandpiper with a small head and thin finely pointed bill that appears to droop slightly at tip. Wing tips project to tail tip; almost no primary projection past tertials. Often adopts upright stance. **Ad. Summer:** Legs pale greenish yellow to yellow. Reddish-brown crown, back, wing coverts, and tertials; breast finely streaked dark brown; head and breast can be washed with bright reddish brown; belly and flanks white and unmarked. Distinguished from Least Sandpiper by more heavily streaked crown and back; dark forehead (Least's pale eyebrows usually meet on forehead, over bill); base of lower mandible often pale. Broad reddish-brown margins to scapulars and tertials last through summer; whitish edges to wing coverts. **Ad. Winter:** Large dark centers and pale margins to scapulars and wing coverts create strongly scaled appearance (more muted on winter Least). **Juv:** (Jul.–Sep.) Bold pale eyebrows do not meet over bill; rufous-fringed scapulars and tertials contrast with paler-fringed and grayer-toned wing coverts. **Flight:** Toes project well beyond tail; tail pale, dark centrally. **Hab:** Northern bogs in summer; freshwater wetlands in winter. **Voice:** Relatively low-pitched *churp;* Least Sandpiper call much higher.

Subspp: Monotypic.

Adult summer CA/05

Juv. NY/08

Adult summer NJ/05

Juv. MA/08

Adult winter, trans. to summer FL/02

Adult winter FL/03

Adult FL/02

Least Sandpiper 1

Calidris minutilla L 6"

Shape: Very small, small-headed, comparatively wide-eyed, thin-necked sandpiper with a thin fine-pointed bill that droops slightly at tip. Wing tips project to tail tip; practically no primary projection past tertials. Back often hunched and legs angled sharply forward during feeding. **Ad. Summer:** Our only common pale-legged (legs usually yellowish) small sandpiper. Feathers on back and wings dark brown with rufous or white margins; breast with fine black streaks over buffy wash, contrasting with rest of white underparts; thin white lines (braces) down each side of mantle; white tips to scapulars. Dull pale eyebrows meet over bill. Evenly streaked head and breast create hooded appearance. **Ad. Winter:** Dull brown on head, back, wings (appear somewhat scaled), and breast; brown breast forms bib against rest of white underparts. **Juv:** (Jul.–Sep.) Like ad. summer but more brightly colored with rufous above; breast less streaked and with buffy wash. **Flight:** Moderate wing-stripe; tail dark centrally. Dark flight feathers distinctive among all peeps. **Hab:** Northern bogs in summer; mudflats, muddy shorelines in winter. **Voice:** High-pitched, drawn-out, slightly ascending *kreeet* or 2-syllable *kreeyeet;* also a short *peet* and chattering.

Subspp: Monotypic.

Adult, trans. to summer TX/05

Juv. NH/11

Adult summer TX/05

Adult winter CHI/01

Juv. NJ/05

Adult summer NE/05

White-rumped Sandpiper 1

Calidris fuscicollis L 7½"

Shape: Medium-sized, deep-chested, strongly attenuated, short-legged sandpiper with a tapered fine-pointed bill that droops slightly at tip. Wing tips project just past tail tip (see also Baird's Sandpiper); primary projection past tertials longer than lore. **Ad. Summer:** Legs dark; base of lower mandible often orangish brown (in all plumages). Back and wing feathers dark-centered with reddish-brown to whitish edges; dark streaks on head and breast extend onto flanks, often as chevrons; strong whitish eyebrow; variable reddish brown on crown and auriculars. All-white rump. Similar Baird's Sandpiper has all-dark bill, no reddish brown on back and wings, dull eyebrow, no streaking on flanks, and dark central rump feathers. **Ad. Winter:** Grayish back and wings with dark feather shafts; often orangish-brown base to bill; grayish streaking on head and breast extends along flanks; white rump. Similar Baird's has all-dark bill, no streaking on flanks, and dark central rump feathers. **Juv:** (Jul.–Nov.) Neatly scaled look to back and wings; scapulars with bright reddish-brown to white margins; thin white lines on either side of back; fine streaking on head and breast extends along flanks; usually an orangish-brown base to bill; rusty crown and auriculars; white rump; conspicuous white eyebrow. **Flight:** Distinctive all-whitish rump seen when bird flies or preens. **Hab:** Tundra in summer; short-grass areas, mudflats in winter. **Voice:** Very short, quiet, insectlike *tsik* given singly or in series.

Subspp: Monotypic. **Hybrids:** Dunlin, Buff-breasted Sandpiper.

Adult summer TX/05

Juv. AZ/08

Adult, trans. to summer TX/04

Juv. NJ/09

Adult winter TDF/12

Baird's Sandpiper 1

Calidris bairdii L 7½"

Shape: Medium-sized, short-legged, strongly attenuated sandpiper with a straight, tapered, fine-pointed bill (White-rumped Sandpiper's bill slightly longer, heavier, less fine-pointed, droops at tip). Wing tips project noticeably past tail tip and are often crossed; primary projection past tertials longer than lore. **Ad. Summer:** Dark legs; bill all dark. Grayish-brown upperparts; dark-centered scapulars with silver edges; back and wings with little or no reddish brown; fine streaking on head and breast does not extend onto flanks; rusty auricular patch; dark lore with intruding whitish supraloral dot; dull eyebrow. **Ad. Winter:** Upperparts grayish with dark feather shafts and white feather fringes; head and breast finely streaked and with buffy wash; belly white; flanks unstreaked. **Juv:** (Jul.–Nov.) Upperparts pale grayish brown and scaly; no red on scapulars or white lines on back; eyebrow faint (whiter and more obvious on juv. White-rumped); dark lore with intruding whitish supraloral dot; flanks unstreaked; white throat contrasts with streaked buffy breastband. **Flight:** Narrow white wing-stripe; rump dark centrally. **Hab:** Tundra in summer; shorelines (inland), wet meadows in winter. **Voice:** Call a rather low, raspy, reedy *kreeet*.

Subspp: Monotypic.

Adult summer, worn AK/09

Juv. CA/10

Adult summer TX/05

Juv. CA/10

Adult winter BOL/03

Juv. CA/09

Pectoral Sandpiper 1

Calidris melanotos L 8¾"

Shape: Medium-sized, deep-chested, short-legged, small-headed shorebird. Bill about as long as head, strongly tapered, slightly drooping at tip. **Ad. Summer:** Bill dark with paler brownish or orangish base; legs yellowish. Upperparts, head, and breast brownish; dense streaking on breast ends abruptly at white belly and may come to a small point in breast center. Reddish-brown margins to scapulars; light buff margins to wing coverts. Dark streaked crown often with reddish brown. M. larger with darker and much more extensive breast streaking than female. **Ad. Winter:** Like ad. summer but duller upperparts without reddish brown. **Juv:** (Jul.–Nov.) Like ad., but usually with brighter reddish-brown margins to crown, scapulars, and tertials, creating a scaly look to upperparts; breast with finer streaks and washed with pale buff; white lines on mantle and scapulars. **Flight:** Faint wing-stripe, dark central tail feathers. **Hab:** Tundra in summer; marshes, wet meadows, mudflats in winter. **Voice:** Relatively low-pitched, short, burry trill, like *jreeef,* all on one pitch; unusual flight song a deep cooing *oooah oooah.*

Subspp: Monotypic. **Hybrids:** Curlew Sandpiper (hybrid called "Cox's Sandpiper").

Adult summer HKO/04

Juv. CA/11

Adult winter AUS/01

Adult winter AUS/12

Sharp-tailed Sandpiper 3

Calidris acuminata L 8½"

Rare fall Eurasian mig. to w. AK and Pacific Coast; casual in spring inland and along East Coast. **Shape:** Like Pectoral Sandpiper but with slightly shorter bill. **Ad. Summer:** Like Pectoral Sandpiper but less paleness at bill base, less abrupt division between breast streaking and clear belly, and more pronounced head pattern—darker rufous cap, brighter whitish eyebrow especially behind eye, and white eye-ring. Also has bold chevrons on flanks, belly, and undertail coverts. **Ad. Winter:** Like summer ad. but upperparts duller; underparts whitish with buffy wash and fine dotting on breast. **Juv:** (Jul.–Nov.) Distinctive rich buffy wash across breast, with fine streaking restricted to sides; head markings like ad. but more vivid, with bright reddish-brown crown. Becomes duller over winter. Fine dark streaks on undertail coverts. Similar juv. Pectoral Sandpiper has breast with less buff and streaking all across it; brown crown; duller eye-ring and eyebrow; longer bill. **Flight:** Faint wing-stripe, dark central tail feathers. **Hab:** Tundra in summer; marshes and mudflats in winter. **Voice:** Fairly high-pitched mellow *weeet* or *weeetweeet*.

Subspp: Monotypic. **Hybrids:** Curlew Sandpiper (hybrid called "Cooper's Sandpiper").

Adult summer –/–

Adult winter NJ/02

1st winter NH/11

Adult winter NJ/12

Adult winter UKM/11

Purple Sandpiper 1

Calidris maritima L 9"

Shape: Medium-sized, fairly rotund, short-legged sandpiper with a strongly tapered fine-pointed bill that droops slightly at tip. Bill as long as or slightly longer than length of head. **Ad. Summer:** Dark bill with orangish base; legs orangish to yellowish. Feathers of back and wings dark brown with fine paler margins; head and neck finely streaked brown; breast heavily spotted; rest of underparts sparsely streaked or spotted. **Ad. Winter:** Head, neck, and upper breast a smooth brownish gray; rest of breast and underparts sparsely spotted or streaked. Scapulars gray with paler gray margins and variable purplish shafts; wing coverts gray with white margins that wear thinner over winter. **Juv:** (Jul.–Sep.) Like summer ad., but wing feathers all with bold white margins, creating scaled appearance. First winter, still has bold white margins to wing coverts. **Flight:** Narrow white wing-stripe. Dark tail and central rump. **Hab:** Open tundra in summer; rocky coasts in winter. **Voice:** Midpitched smooth *kweet* or *kweekweet;* when rapidly repeated becomes chattering.

Subspp: Monotypic. **Hybrids:** Dunlin.

Adult summer, *tschuktschorum* AK/06

Adult winter, *tschuktschorum* AK/03

Adult summer, *ptilocnemis* AK/06

Adult winter, *ptilocnemis* AK/03

Adult summer, *couesi* AK/06

Juv., *ptilocnemis* AK/08

Rock Sandpiper 2

Calidris ptilocnemis L 9"

Shape: Medium-sized, fairly rotund, short-legged sandpiper with a strongly tapered fine-pointed bill that droops slightly at tip. Bill averages slightly shorter than length of head. **Ad. Summer:** Head boldly marked with streaked chestnut crown, pale eyebrow, and prominent dark ear patch. Upperparts dark with variable reddish brown on back, scapulars, and crown. Underparts white with fine streaking and variable mottled black patch on lower breast. Bill and legs black. Similar Dunlin has longer bill and black belly patch that extends beyond legs. **Ad. Winter:** Medium to pale gray upperparts; head and neck smooth gray; flanks and belly variably spotted or streaked with gray (see Subspp.). Bill slightly greenish yellow at base; legs greenish yellow. **Juv:** (Jun.–Sep.) Like summer ad., but back and wing feathers strongly margined paler, creating scaled appearance; bill yellowish at base. **Flight:** Wing-stripe narrow to broad and tail dark to light (see Subspp.); dark central stripe on rump. **Hab:** Tundra in summer; rocky shores, sometimes sand- or mudflats in winter. **Voice:** Repeated rolling *jurwee jurwee jurwee* or short hoarse *djreet.*

Subspp: (4) Geographically distinct; plumage differences slight. •*ptilocnemis* (Pribilof and St. Matthew Is.): In summer, dusky belly patch usually mottled with white, back feathers with few or no white tips; in winter, upperparts pale gray, flanks with little or no streaking; has widest wing-stripe, most white on tail, and whitest underwing of all subspp. •*tschuktschorum* (AK from Bristol Bay to Seward Peninsula, also St. Lawrence and Nunivak Is.): In summer, black belly patch usually unmottled, back feathers with white tips; in winter, upperparts fairly dark and flanks with fairly extensive streaking. •*couesi* (Aleutian Is., AK Peninsula, Shumagin and Kodiak archipelagos): In summer, dusky belly patch small or absent, back feathers with white tips; in winter, upperparts dark gray, flanks with extensive streaking. •*quarta* (vag. to w. AK): In summer, dusky belly patch small and mottled with white, back feathers with few or no white tips; in winter, upperparts medium gray, flanks with extensive streaking.

Adult summer, *pacifica* AK/06

1st winter CA/01

Adult summer, *hudsonia* TX/05

Juv. NJ/09

Adult winter FL/03

Winter FL/03

Dunlin 1

Calidris alpina L 8½"

Shape: Medium-sized fairly deep-chested shorebird with a relatively small head, short neck, and long bill. Bill about 1½ x head length, tapered to a fine point, and drooping at the tip. Wing tips fall just short of tail tip; primary projection past tertials shorter than lore. **Ad. Summer:** Large solid black patch on belly, extending beyond legs; head, neck, and breast whitish with fine dark streaks; crown with some rufous and dark streaking. Back and scapulars mostly reddish brown; feathers with darker centers and variable lighter tips. Legs dark gray to black. Feeds in "sewing machine" fashion like dowitchers, or picks at surface like peeps. Similar Rock Sandpiper has shorter bill and black patch limited to lower breast. **Ad. Winter:** Breast and upperparts plain grayish brown with darker shafts to some feathers; chin and belly white. Resembles dowitcher when feeding or sleeping, but mostly white undertail and black legs distinctive. Similar Western Sandpiper has shorter bill and more extensive white on breast, throat, and face. **Juv:** (Jul.–early Sep.) Like summer ad., but with dark splotches on breast and more brightly margined feathers on back and scapulars, creating scaled effect. Molt into winter plumage mostly on breeding grounds. **Flight:** Broad white wing-stripe; dark central tail feathers. Often flies in large flocks with synchronous movement. **Hab:** Wet tundra in summer; mudflats, beaches in winter. **Voice:** Fairly drawn-out *jjeeep;* also rapid chattering series of *jee* calls.

Subspp: (5) Plumage differences slight. •*hudsonia* (NU–n. ON) has fine streaking on flanks and undertail coverts. •*pacifica* (w. AK) has little or no streaking on flanks or undertail coverts. •*articola* (breeds on AK north slope) has little or no streaking on flanks or undertail coverts. •*arctica* (GRN vag. to n.e. CAN–MA) has black belly patch mottled with white; unlike the above North American subspp., which molt to winter plumage on the breeding ground, *arctica* molts after fall migration and so is seen in worn breeding plumage during fall migration. •*alpina* (EUR vag. to MA–SC) has black belly patch with little or no mottling. **Hybrids:** White-rumped Sandpiper, Purple Sandpiper.

Adult summer CYP/05

Juv. ENG/08

1st summer POR/05

Adults, summer POR/04

Adult winter TX/05

Curlew Sandpiper 3

Calidris ferruginea L 8½"

Rare Asian vag. to East and West coasts; casual inland. **Shape:** Like Dunlin but with slightly longer legs and neck and a finer-tipped more evenly downcurved bill. Wing tips extend beyond tail tip; primary projection past tertials about as long as lore (Dunlin's wing tips fall short of tail tip, primary projection shorter than lore). **Ad. Summer:** M. dark brick-red head, neck, and underparts except for white undertail coverts with dark spots; white eye-ring. F. finely mottled black and white with traces of reddish brown on face, scapulars, and belly; whitish eyebrow. Dark legs. **Ad. Winter:** Unmarked gray above. Head, sides of breast, and neck finely streaked with gray and white; central breast usually clear; prominent whitish eyebrow (indistinct in similar Dunlin) contrasts with dark crown. Dark legs. Similar Stilt Sandpiper has longer and yellowish legs. **Juv:** (Jul.–Nov.) Upperparts grayish brown with fine buff margins to feathers, creating scaled appearance. Breast with buffy wash and some fine darker streaking along sides. Dark legs. Similar Stilt Sandpiper has longer and yellowish legs. **Flight:** Bold white wing-stripe; white rump; gray tail. Toes project just past tail. **Hab:** Wet tundra in summer; mudflats, beaches, shorelines in winter. **Voice:** Rapid burst of *ch'd'deet ch'd'deet* given on breeding grounds.

Subspp: Monotypic. **Hybrids:** Sharp-tailed Sandpiper (hybrid called "Cooper's Sandpiper"), Pectoral Sandpiper (hybrid called "Cox's Sandpiper").

Adult summer MB/06

Adult winter TX/03

Adult summer TX/05

Juv. QC/09

Adult winter CA/01

Adult TX/04

Stilt Sandpiper 1

Calidris himantopus L 8½"

Shape: Medium-sized, slender, long-necked, long-legged shorebird. Bill noticeably longer than head, not strongly tapered, slightly drooping at tip. **Ad. Summer:** Reddish-brown lore, auriculars, and crown contrast with distinct whitish eyebrow. Head and neck streaked; breast and belly strongly barred. Black-and-white scapulars contrast with dull grayish wing coverts; legs greenish yellow. Feeds in water with rapid "stitching" motion, like dowitchers. Similar Short-billed Dowitcher has much longer and straight bill, heavier-looking body, dark brown lore and crown, pale eyebrow. **Ad. Winter:** Plain grayish brown above with distinct pale eyebrow contrasting with dark lore and crown; whitish below with fine streaks on breast. Undertail coverts usually spotted (seen when bird feeds). First-winter birds have bold white margins to wing coverts and white-centered tail feathers with grayish margins (no barring as on similar yellowlegs). **Juv:** (Jul.–Sep.) Brownish crown and eyeline contrast with whitish eyebrow. Back and wing feathers generally dark-centered with contrasting pale margins of white, buff, or occasionally reddish brown, creating strong scaled effect; back feathers quickly molted to plain gray. Head to flanks streaked; breast washed with buff at first. **Flight:** Plain dark upperwing; whitish rump (can be some mottling); pale gray tail in ad. (tail feathers with white centers and dark margins on juv./1st winter). Toes project completely beyond tail. **Hab:** Wet tundra in summer; shallow pools, lakes, marshes, wet meadows in winter. **Voice:** Midpitched *jeew* or rising *jooeet* or harsh *chrit*.

Subspp: Monotypic.

Adult summer RUS/-

Adult winter THA/01

Juv. JAP/-

Spoon-billed Sandpiper 4
Eurynorhynchus pygmeus L 6"

Casual Asian vag. along Pacific Coast. **Shape:** Small, long-winged, fairly long-legged sandpiper with a distinctive bill that is long, thick, broad, and widely flared at tip; flare may be hard to see from side, but bill still distinctively long and thick. Wing tips extend considerably beyond tail tip; primary projection past tertials as long as lore. **Ad. Summer:** Bright reddish-brown face and neck with spotting below on upper breast, similar to Red-necked Stint; feathers of wings and back with much reddish brown, some black internal markings, and white tips. Bill shape distinctive. **Ad. Winter:** Gray above with broad white feather margins; face extensively white; underparts white. Bill shape distinctive. **Juv:** (Jul.–Oct.) Extensive reddish brown on crown, back, and wings; face and underparts white with dark eye patch, dark streaking on sides of breast, and buffy wash across breast; thin white lines on either side of back. **Flight:** Dark wings with faint stripe; rump and tail dark. **Hab:** Coastal tundra in summer; mudflats in winter. **Voice:** Soft *wheet*.

Subspp: Monotypic.

Adult summer, *falcinellus* ENG/09

Adult, trans. to summer, *sibirica* AUS/03

Winter, *sibirica* AUS/03

Juv., *falcinellus* FIN/08

Broad-billed Sandpiper 4

Limicola falcinellus L 7"

Casual Eurasian vag. to w. AK, accidental on East Coast. **Shape:** Small, deep-bellied, short-legged, long-winged shorebird with a fairly long, broad (from in front or above), deep-based bill that droops abruptly near tip. Wing tips extend beyond tail tip; primary projection past tertials equal to lore. Somewhat similar to Dunlin, but smaller, head proportionately larger, bill proportionately shorter and equals length of head; Dunlin's bill longer than head length (in all but subsp. *arctica*), wing tips just short of tail tip. **Ad. Summer:** Legs dull olive to blackish. Distinctive dark crown with thin white lateral stripes in all plumages; whitish eyebrow; dark line through eye from lore to auricular. Dark grayish to brownish above; back and wing feathers dark-centered, edged pale gray to reddish brown; head and breast whitish with fine dark streaking; flanks heavily streaked. **Ad. Winter:** Streaked crown; feathers of back and wings gray and edged paler gray; underparts white with variable light streaking on breast. **Juv:** (Jul.–Sep.) Streaked crown and shape of bill distinctive. Back and wing feathers dark-centered with reddish-brown to whitish margins; white lines on either side of back; head and neck finely streaked; breast may have buffy wash. **Flight:** Fine wing-stripe; white rump dark centrally. **Hab:** Tundra in summer; mudflats in winter. **Voice:** Drawn-out buzzy *jjreeet*.

Subspp: (2) •*falcinellus* (possible vag. to Atlantic Coast) summer ad. has less reddish brown above and dark brown when worn; •*sibirica* (vag. to w. AK islands) summer ad. has more reddish brown above, buff wash across breast.

Adult summer, *hendersoni* MB/06

Adult summer, *griseus* NJ/06

Adult summer, *caurinus* CA/05

1st winter FL/02

Short-billed Dowitcher 1

Limnodromus griseus L 11"

Shape: Medium-sized plump shorebird with a very long straight bill. Bill length 1½ to almost 2 x head length. Wing tips usually extend to tail tip (subspp. *griseus* and *hendersoni*) or slightly beyond tail tip (subsp. *caurinus*). Bill length variable and can overlap with Long-billed Dowitcher's, but short-billed Short-bills and long-billed Long-bills are distinct. **Ad. Summer:** This plumage generally distinct for each subsp. and from Long-billed Dowitcher. Identification is complicated by migrating birds being in various stages of molt and because some individuals within each subsp. vary greatly (particularly *griseus*). Combine all plumage clues for best identification; voice is best clue to species. In general, Short-billed has paler back and scapulars with wide reddish-brown edges and buff tips (Long-billed scapulars have narrow dark rufous edges with white tips); central tail feathers with white and black barring (Long-billed central tail feathers often with reddish-brown and dark barring); neck sparsely dotted (except for subsp. *griseus*, which has heavily spotted neck) and breast sides heavily spotted (Long-billed neck with dense dark spotting, dark barring on sides of breasts); belly either all orangish (*hendersoni*) or with substantial white on belly and vent (*griseus*) or with white mostly on vent (*caurinus*). Long-billed belly mostly all dark reddish brown, except in fresh plumage, when belly feathers may have white tips. Short-billeds migrate south ½–1 month earlier than Long-billeds, with some adults first reaching lower 48 states around July 1. Juvs. also migrate earlier, first Short-billeds in late July, first Long-billeds in early September. **Ad. Winter:** Short-billed best told from Long-billed Dowitcher by its mellow *tututu* call (Long-billed's call is a sharp higher-pitched *teeek*) and its preference for saltwater and coastal, rather than freshwater and inland, locations. Nevertheless, there are subtle plumage clues. Short-billed has less extensive and less demarcated gray on breast, often paler and in spots (rather than even wash as in Long-billed); mantle feathers evenly gray (Long-billed's have darker centers). Winter

Adult winter FL/01

Juv., *caurinus* CA/08

Adult winter, *caurinus* CA/01

Adult, trans. to summer, *caurinus* CA/04

Short-billed subspp. have subtle differences: *griseus* is typically darker with darker central shafts and limited pale fringing on upperparts and extensive barring or spotting on flanks; *hendersoni* has pale-centered feathers on upperparts, crisp white fringes to wing coverts, and mostly spotting on flanks; *caurinus* is most like Long-billed Dowitcher, with dark upperparts, no white fringing to wing coverts, and an extensively dark hooded appearance. **Juv:** (Jun.–Oct.) Fresh boldly edged feathers of juv. very different from worn adult plumage in late summer and fall. Disting. from Long-billed juv. by the wide gold margins to back and wing feathers, pale buff or rust internal markings of tertials, warm buffy wash to breast, and darker reddish-brown and black streaked crown. Long-billed juv. lacks all of these. Juvs. of both species do not fully molt into winter plumage until Oct.–Nov. **Flight:** Dowitchers similar—dark above with paler tail; white wedge up back. **Hab:** Boreal forests, tundra in summer; coastal saltwater and brackish areas in winter (much more common than Long-billed on larger tidal mudflats). **Voice:** Rapid *tututu* or *tutututu* can be variably pitched but often fairly low-pitched and mellow; distinct from high-pitched *teeek* or *keeek* of Long-billed Dowitcher.

Subspp: (3) Distinct as summer adults (described below), somewhat similar in winter (see text above) and juv. plumages. •*griseus* (s.e. NU–cent. QC–w. LB, migrates along Atlantic Coast) has reddish-brown wash on throat and breast, belly with much white; throat and breast heavily spotted and flanks heavily barred. •*hendersoni* (n.w. BC–e. YT–s. NU–n. ON, migrates through Midwest and along Atlantic Coast) has rich reddish brown on neck, chest, and underparts; no spotting on throat; light spotting on sides of breast; light barring on flanks. •*caurinus* (s. AK–s.w. YT and n.w. BC, migrates along Pacific Coast) is similar to *griseus* but generally mostly orange underparts with whitish vent and less dark marking on underparts; variable white lower belly and vent; wing tips usually project just past tail in this subsp.

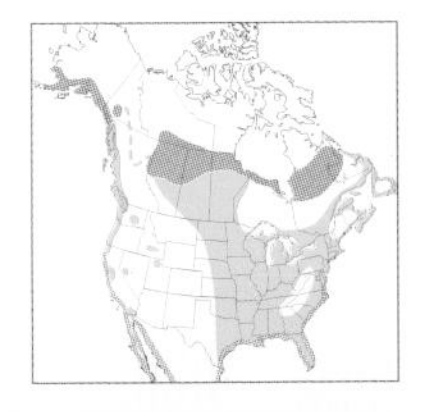

Adult summer CA/04

1st winter AZ/–

Adult summer TX/05

Adult winter TX/11

Long-billed Dowitcher 1

Limnodromus scolopaceus L 11½"

Shape: Medium-sized plump shorebird with a very long, straight, tapered bill. Bill length 1½ to slightly more than 2 x head length. Bill length variable and can overlap with Short-billed Dowitcher's, but short-billed Short-bills and long-billed Long-bills are distinct. **Ad. Summer:** Breeding grounds generally distinct from very similar Short-billed Dowitcher's. Identification complicated because migrating birds are in various stages of molt and because individuals vary greatly within species. Combine all plumage and shape clues for best identification; voice is best clue to distinguish species. See Short-billed for comparison. In general, Long-billed has dark back and scapulars with narrow reddish-brown edges and white tips; reddish-brown and dark barring on central tail feathers; dense dark spotting on neck; dark barring on sides of breast, with distinctive white fringes; mostly all-dark reddish-brown belly, except in fresh plumage, when belly feathers may have white tips. Long-billeds migrate south ½–1 month later than Short-billeds, with adults reaching lower 48 states around mid-Jul. to Aug. 1. **Ad. Winter:** Long-billed best told in winter from very similar Short-billed Dowitcher by its sharp *teeek* call (Short-billed's call is a mellow *tututu*) and preference for inland freshwater, rather than coastal

Juv. CA/09

Juv. CA/09

Adult CA/08

saltwater, locations, although it can be seen on typically smaller tidal mudflats. Nevertheless, there are subtle plumage clues. Long-billed has more extensive and demarcated gray on breast, usually as an even wash; gray mantle feathers have darker centers. **Juv:** (Jun.–Nov.) Fresh boldly edged feathers of juv. very different from worn adult plumage in late summer and fall. Told from Short-billed juv. by its dark relatively unmarked back with rust fringes; tertials with narrow paler margins and no paler internal markings; dull pinkish wash to breast; slightly lighter reddish-brown and black streaked crown. See Short-billed Dowitcher juv. for comparison. Juveniles of both species do not fully molt into winter plumage until Oct.–Nov. **Flight:** Both dowitchers similar—generally dark above with a paler tail and white wedge up the back. White lesser underwing coverts often distinctive. **Hab:** Coastal tundra in summer; inland freshwater lakes and impoundments in winter. **Voice:** High-pitched *teeek;* can be single or in a series like *teeteeteeteek,* always higher-pitched than Short-billed Dowitcher's call.

Subspp: Monotypic.

Adult m., summer UKM/–

Adult m., summer UKM/–

Adult m., summer UKM/–

Adult f., summer SAK/02

Ruff 3

Philomachus pugnax L m. 11", f. 9"

Rare visitor throughout NAM, most common on coasts. **Shape:** Large deep-bellied shorebird with a proportionately small head. Bill about the length of head and drooping slightly at tip. Long loose scapulars may ruffle up in breeze. Size varies greatly between sexes; m. larger and deep-bellied (about the size of Greater Yellowlegs); f. smaller and more slender (about size of Lesser Yellowlegs). **Ad. Summer:** Plumage varies greatly between sexes and individuals. In both sexes the tertials are boldly barred. M. variable with ear tufts and neck ruffs in chestnut, white, or black; face bare with wattles and warts that can be orange, green, or yellow; bill during breeding may be orange, yellow, or pink. F. less variable, being generally brownish above and paler below; head and neck plain brown to grayish; back and wing feathers generally dark-centered or boldly barred black and reddish brown with thin pale margins; underparts whitish with variable black mottling; bill all dark or with orangish base. About 1% of males have plumage like that of females and cannot be distinguished except by size, which is intermediate between females and more common

Adult f., winter HI/03

Juv. CA/09

Adult m., winter, variant ENG/11

Adult CYP/07

larger males. Active when feeding, often running while picking up insects off surfaces. **Ad. Winter:** Sexes similar. Grayish brown above with pale feather fringes; often a band of white feathering around bill base; bill all dark or with orange to pinkish base (m.); underparts variably whitish with variable blurred spotting or streaking on sides; legs orangish yellow. Some individuals may have mostly whitish head and neck (often m.). **Juv:** (Jul.–Nov.) Head, neck, and breast smooth plain warm buff to darker chestnut; feathers of back and wings dark-centered with wide buffy to rufous margins, creating bold scaled effect. Distinctive short eyeline only behind eye. Bill dark, but may have some brown to orange at base; legs dull greenish to yellowish. **Flight:** Thin white wing-stripe; conspicuous white oval on either side of uppertail coverts; toes trail partially past tail. Rather languid flier with slower, more measured wingbeats than many other shorebirds. **Hab:** Wet tundra, meadows, marshes in summer; wet fields, shallow water, mudflats in winter. **Voice:** Nasal low-pitched grunt, like *gegent,* or higher-pitched *jeejeejeet.*

Subspp: Monotypic.

Adult summer AK/06

Juv. CA/09

Adults, summer, displaying NE/05

Juv. NY/09

Juv. BC/08

Buff-breasted Sandpiper 1

Tryngites subruficollis L 8¼"

Shape: Medium-sized deep-bellied sandpiper with a thin neck, small head, and short bill. Bill thin, fine-pointed, clearly shorter than depth of head. **Ad:** In all plumages, plain buffy face, neck, and underparts. Crown streaked darker; back and wing feathers dark with buffy margins, creating scaled effect. Legs yellow. **Juv:** (Jul.–Oct.) Like ad., but margins of back and wing feathers whitish, underparts paler buff with whitish belly. **Flight:** Uniformly buffy brown above; below, wing coverts mostly white with primaries and secondaries dark. **Hab:** Drier tundra, wetlands in summer; short-grass habitats, margins of wetlands, upper beaches in winter and migration. **Voice:** Ascending *creeit* or soft *jert.*

Subspp: Monotypic. **Hybrids:** White-rumped, Baird's Sandpipers.

Adult MT/06

Adult MN/05

Adult MN/05

Wilson's Snipe 1

Gallinago delicata L 10½"

Shape: Medium-sized, short-legged, deep-bellied shorebird with a broad neck and very long, straight, thick bill. Bill about 2 x head length. **Ad:** Dark back with conspicuous white or pale buff long stripes; head boldly patterned with dark and light stripes; dark crown continues over forehead; brown barring on whitish flanks. **Flight:** After flushes usually flies off in zigzag path. Disting. from American Woodcock by pointed wings, white stripes on back, orange tail, grayish underwings, and overall brownish, rather than buffy, underparts. Wilson's and Common Snipes have same pattern of dark and light barring on underwing coverts, but are subtly different–Wilson's with narrower barring of white and dark gray, Common paler overall with wider barring of silver and gray. In flight display, bird circles high in sky, then dives down; accompanied by a whistling sound made by air passing through outer tail feathers. **Hab:** Wet meadows, pond edges, ditches, and marshes. **Voice:** During breeding, loud, high-pitched, measured, repeated *pika pika pika* calls or breathy rapid series of low-pitched *weu weu weu* sounds (created by air passing through tail feathers in display flight); also series of scratchy *cheet* calls or harsh *skyeep* in alarm.

Subspp: Monotypic.

Adult HKO/01

Adult BOR/03

Adult CYP/09

Common Snipe 3
Gallinago gallinago L 10½"

Rare but regular Eurasian visitor to w. AK islands. **Shape:** Like Wilson's Snipe. **Ad:** Very similar to Wilson's; on average may be buffier overall and flanks less heavily barred. Best clues are wings, seen in flight. May not be able to be identified except in hand. **Flight:** Wider white trailing edge to secondaries than Wilson's. Wilson's and Common Snipes have same pattern of dark and light barring on underwing coverts, but are subtly different—Wilson's with narrower barring of white and dark gray, Common paler overall with wider barring of silver and dark gray. In flight display, bird circles high in sky, then dives down; accompanied by a whistling sound made by air passing through outer tail feathers. **Hab:** Wet meadows, pond edges, and marshes. **Voice:** Harsh *skyeep* in alarm (like Wilson's Snipe) or low-pitched, vibrating, crescendoing noise made in display flight (unlike *weu weu weu* of Wilson's Snipe).

Subspp: (1) •*gallinago.*

Adult OH/05

Adult NJ/02

American Woodcock 1
Scolopax minor L 11"

Shape: Medium-sized, short-legged, short-tailed, deep-bellied shorebird with a short deep neck, relatively large head, large eyes, and very long straight bill. Bill 2 x length of head. **Ad:** Above, back centrally blackish with wide silvery bars on each side; scapulars with subterminal black marks and buffy tips. Below, unmarked pale cinnamon. Hindcrown black with buffy transverse bars; forehead gray; bill brownish at base, darker at tip. **Flight:** Flushes up from ground. During courtship, performs circling upward flight with rapid drop, accompanied by twittering sounds made by air passing through outer primaries. Cinnamon belly and underwing coverts. Disting. from Wilson's Snipe in flight by short rounded wings, deep belly, and dark tail with reddish-brown uppertail coverts. **Hab:** Woods edges, open fields in summer; wet woods, thickets, and fields in winter. **Voice:** Loud, nasal, buzzy *zeeent* or *peeent* when on the ground; extended high-pitched twittering and chirping sounds during flight display. 37.

Subspp: Monotypic.

Adult m., summer UT/07

Adult f., summer ID/05

Adult winter CA/04

Juv., trans. to winter CA/08

Juv. CA/08

Adult summer BC/06

Juv. NY/08

Wilson's Phalarope 1

Phalaropus tricolor L 9¼"

Shape: Large-bodied, small-headed, relatively long-necked shorebird with a fairly long, very thin, needlelike bill. Bill equal to or slightly longer than head length (noticeably shorter than head length in Red-necked and Red Phalaropes). Sloping forehead. Largest of our phalaropes. Seen both swimming and spinning in water and dashing about mudflats near water's edge. **Ad. Summer:** Dark upperparts with no paler streaking (similar Red-necked with buff streaks); whitish underparts; bold dark streak through eye continues down side of neck; cinnamon wash on throat; white chin; white line on nape. As in all phalaropes, ad. f. more colorful than ad. male. F. with gray crown, gray and reddish-brown back; bolder face and neck colors. M. with dark brown crown, dark brown back; duller face and neck colors. Legs black. **Ad. Winter:** Pale gray above and white below with little marking; thin pale gray eyestripe starts behind eye and continues down neck; legs paler, often yellowish. **1st Winter:** Like ad. winter but with pale buffy to whitish (Sep. onward) margins to tertials and wing coverts. **Juv:** (Jun.–Aug.) Back and wing feathers have dark brown centers, buffy margins; buffy wash across breast at first, but this soon fades; eyebrow broad and whitish; eyeline thin and brown. Quickly molts into winter ad. plumage on breeding grounds and while flying south, molting back and scapulars first; at this stage, looks much like winter ad. but with dark juv. wing feathers contrasting with new gray scapulars. Legs yellowish. **Flight:** Plain dark upperwing; mostly pale whitish unmarked underwing; white rump and grayish tail. Toes project past tail. **Hab:** Grassy wetlands in summer; mudflats, shallow water in winter. **Voice:** Low-pitched repeated *wherp wherp wherp* or higher repeated *wep wep wep*.

Subspp: Monotypic.

Adult m., summer AK/06

Juv. CA/09

Adult f., summer AK/06

Juv. NY/08

Adult winter CA/11

Juv. CA/08

Red-necked Phalarope 1

Phalaropus lobatus L 7"

Shape: Small slender shorebird with a fairly short, thin, needlelike bill. Bill noticeably shorter than head length (about as long as head in Wilson's Phalarope). Upper neck relatively thin; neck fairly long (Red Phalarope with proportionately shorter thicker neck); steep forehead. The smallest phalarope. **Ad. Summer:** Dark back with long buffy streaks; dark grayish or brown sides of breast and flanks. Head and neck generally dark except for white throat and white patch over eye. F. with dark head, white throat, chestnut collar. M. similar but duller—dark crown, whitish or chestnut eyebrow, dark cheek patch, white throat. M. varies from almost as dark as f. to much paler. Legs gray to black. **Ad. Winter:** Head white with gray to blackish crown (mostly hindcrown) and dark patch behind eye; back dark gray with variable paler streaking (similar winter Red Phalarope has unstreaked back, shorter thicker neck, thicker-based bill). **Juv:** (Jul.–Oct.) Back blackish gray with bold streaks—buff early, then whitish; head similar to ad. winter with dark eye patch, but whole crown dark; face and underparts dark grayish early, then mostly white. Does not begin molt from juv. plumage until Nov. (Red Phalarope juv. begins molt on breeding grounds in Aug.). **Flight:** Upperwings dark with broad white wing-stripe; underwings whitish on coverts, darker on flight feathers and leading edge (most boldly patterned underwing of the phalaropes). Rump and tail grayish, dark centrally. Feet do not project beyond tail. **Hab:** Tundra ponds in summer; ocean, bays, lakes in winter. **Voice:** Short high-pitched *chit* calls or high-pitched squeaky twittering.

Subspp: Monotypic.

Adult m., summer AK/06

Adult f., summer NOR/06

Adult winter CA/11

Juv., late stage PA/09

Juv., trans. to winter NJ/08

Adult f., summer CA/05

Adult winter CA/12

Red Phalarope 1

Phalaropus fulicarius L 7¾"

Shape: Compact phalarope with a short thick neck and a short relatively thick bill. Bill noticeably shorter than head length (thicker than Red-necked and Wilson's Phalaropes', shorter than Wilson's). Steep forehead. Rides higher in water than Red-necked Phalarope, often with longer tail cocked upward. **Ad. Summer:** Bill yellowish with dark tip. Unmarked reddish-brown neck and underparts; dark back and wings with buff streaking; dark crown and forehead; white patch over eye and auriculars. F. rich brick-red below, with black crown, boldly streaked back, and bold white cheek; m. duller reddish brown below, streaked crown, and thinner buff streaks on back. **Ad. Winter:** Pale gray unstreaked back and slightly thicker bill and neck distinguish it from similar Red-necked Phalarope. Black bill (sometimes with yellow base); black eye patch. **1st Winter:** Similar to ad. winter but with darker crown and dark tertials. **Juv:** (Jul.–Nov.) Very similar to juv. Red-necked Phalarope but breast and flanks buffy rather than gray at first; bill thicker (sometimes with paler base); underwing in flight less patterned. Begins molt to 1st-winter plumage on breeding grounds, and rarely seen outside breeding grounds without plain gray back of 1st-winter plumage. **Flight:** Bold white wing-stripe, central dark line through rump, which is reddish in summer, whitish in winter. Underwings mostly white (Red-necked has darker flight feathers, Wilson's is mostly pale). **Hab:** Tundra ponds in summer; mostly ocean in winter. Our most pelagic phalarope; spends most of year out at sea, only rarely seen inland on lakes. **Voice:** Buzzy *jreeep* or short high *jip* or *kip*.

Subspp: Monotypic.

Gulls

Terns

Skimmer

Skuas

Jaegers

Dovekie

Murres

Guillemots

Murrelets

Auklets

Puffins

IDENTIFICATION TIPS

Gulls are medium-sized to large birds strongly associated with water. They have large bodies, medium-length necks, and from thin to fairly thick bills. Their legs are fairly short and their feet webbed. The colors of adult plumage are mostly white, gray, and black. All North American gulls take 2–4 years to acquire full adult plumage; during early immature stages of most species, they have substantial brown on their body and/or wings. Gulls feed on a wide range of foods in varied ways, but they generally do not dive to catch fish like their close relatives the terns.

Species ID: Many gulls look much alike, and their identification is a challenge. Overall body size and proportions and the shape of the bill and head are subtle differences (also affected by sex) but can be very useful. In large gulls, males are quite a bit larger than females and often have a more strongly sloping forehead, flatter crown, longer thicker bill, and more snoutlike appearance where the feathers meet the base of the bill; the smaller female often has a steeper forehead, more rounded crown, and smaller bill.

Shape: For each species we give a rough proportion (on a standing bird) of the distance from the front of the chest to the legs versus the distance from the legs to the tip of the folded wings. This ratio is fairly constant within a species but differs significantly between certain species. Larger gulls tend to have a more equal distance for the two measurements and look more stocky (like Herring Gull); smaller and more long-winged gulls tend to have a longer leg-to-wing-tip distance and look more sleek or streamlined (like Laughing Gull). This ratio can help to determine the size of a gull when there is no basis for comparison with another bird. Another aspect of shape in gulls involves the gonys and gonydeal angle. The gonys is where two plates of the lower mandible meet near the tip of the bill. On some gulls the gonys is expanded into what is sometimes called the gonydeal angle, and the absence, presence, or degree of expansion can be useful in identifying some gulls.

Colors: In identifying adults, always look at the colors on the bill and legs, the shade of color on the back and wings, and the patterns of black and white on the wing tips (often seen best in flight). In some species, adults have black hoods in summer and patterns of gray shading on the head in winter, and these can be helpful.

Wing Patterns: Wing patterns, above and below, are often crucial to identification and are usually best seen when the bird is flying; be sure to look at the top of the wing as the bird banks or on the wing downstroke. In a few species the underwing is distinctively dark. Two terms are useful in looking at gull wings: "mirror" and "ulnar bar." A mirror is a white dot in the dark tip of the outer primaries; the ulnar bar is a dark band that goes diagonally across the secondary coverts of the upperwing on some species (usually immature stages). The term "window" is sometimes used in wing descriptions and refers to a paler area of a wing usually noticed when the light is behind the wing. In some cases, there is a need to refer to the patterns on particular primary feathers. Gulls have 10 primaries and they are numbered starting with the innermost; thus the outermost primary is the 10th. In the text these are referred to as P1, P2, etc., to P10.

Aging: For the large 4-year gulls, there are some helpful tips to determining their age using the colors on the tail, rump, and bill. These tips are generalizations and will only give you a rough approximation of age; this is because, within species, 4-year gulls can vary greatly in the relative speed of their development into new plumages. In 3-year and 4-year gulls, the plumage just before full adult plumage is usually very similar to adult.

General Pattern of Aging for Most Large 4-Year Gulls

1st yr. tail and rump dark
bill largely dark

2nd yr. tail dark, rump light
bill dark with pale base

3rd yr. tail mostly white, rump white
bill pale with dark subterminal ring

4th yr. tail and rump white
bill pale with red and/or black marks

Adult summer, *tridactyla* NOR/07

Adult summer -/-

Adult winter, *pollicaris* CA/03

Adult winter, *pollicaris* CA/03

Black-legged Kittiwake 1

Rissa tridactyla L 17½"

Three-year gull. **Shape:** Medium-sized short-legged gull with a usually somewhat upright stance. Bill relatively short, thin, with little or no gonydeal angle and culmen gently curved to tip. Forehead fairly steep and crown well rounded; eye proportionately large. Chest-to-leg vs. leg-to-wing-tip ratio about 1:2½. Similar Red-legged Kittiwake has proportionately shorter more abruptly pointed bill and shorter wings. **Ad. Summer:** Bill all yellow; legs black (rarely yellow or pinkish); eye dark. Medium gray back; outer primaries with black ends and little or no white on tips. **In flight,** gray inner wing, paler outer wing; outer primaries sharply tipped with black (no white mirrors). Short squarish tail slightly notched. Below, wings all whitish. **Ad. Winter:** Similar to ad. summer but with dark ear patch and blackish gray from nape to hindcrown. **2nd Winter:** Like ad. winter, but legs may be dark brownish or grayish;

1st winter, *pollicaris* CA/02

1st winter, *tridactyla* ENG/02

Juvs., *tridactyla* NF/03

Juv., *pollicaris* CA/10

bill may have darker tip and cutting edge. **In flight,** like ad. winter but with some black extending onto outer primary coverts. **1st Winter:** Like juv. but with bill brownish; less extensive dark markings in wings. Dark collar reduced or absent by summer. **In flight,** like juv. **Juv:** Bill blackish; legs dark; eye dark. Dark collar and ear patch; dark around eye. Lesser coverts mostly black, as are inner greater coverts and some tertials. **In flight,** dark outer primaries and inner wing coverts create strong M pattern across upperwings; pale inner primaries and secondaries. Tail with narrow black band. **Hab:** Ocean shores, sea cliffs, estuaries, coastal freshwater lakes. **Voice:** Greeting ceremony call (at nest) a *geeeeeh geeeeeeh;* alarm call an *oh oh oh oh.*

Subspp: (2) Differences slight. •*tridactyla* (n. Atlantic coast) is slightly smaller, with smaller bill, less black on primaries. •*pollicaris* (n. Pacific coast) is slightly larger, with larger bill, more black on primaries.

Adult summer AK/07

Adult summer RUS/06

Adult winter AK/08

2nd winter RUS/12

1st summer RUS/06

1st summer RUS/06

Red-legged Kittiwake 2

Rissa brevirostris L 16"

Three-year gull. **Shape:** Medium-sized short-legged gull with long narrow wings and usually a somewhat upright stance. Bill very short, thin, slightly hooked, with no gonydeal angle, and abruptly curved to tip. Forehead steep and crown high and slightly flattened or rounded; eye proportionately large. Chest-to-leg vs. leg-to-wing-tip ratio about 1:3. Similar Black-legged Kittiwake has longer more finely pointed bill and longer wings. **Ad. Summer:** All-yellow bill; bright red legs; dark eye. Dark gray back and wings with black wing tips. **In flight,** long narrow wings evenly gray all the way out to well-defined triangular black tips. Short square tail slightly notched. Below, gray outer wing contrasts with white inner wing coverts. Stiff-winged flight more like that of terns than of other gulls. **Ad. Winter:** Like ad. summer but with dark earspot and gray smudge around eye. **2nd Winter:** Like ad. winter. **In flight,** black on wing tips may extend onto primary coverts. **1st Winter:** Like juv. but with smaller earspot and little or no dark collar. Legs become brownish yellow to reddish and bill yellowish at base by 1st summer. **Juv:** Bill blackish; legs blackish; eye dark. Dark earspot and thin dark collar. Gray mantle and wings; little or no dark bar on wing coverts. **In flight,** broad white trailing portion of wing, deepest at bend, suggests Sabine's Gull. Some dark on secondary coverts; tail all white (only gull with white tail as juv.). **Hab:** Ocean shores, sea cliffs, estuaries, coastal freshwater lakes. **Voice:** Greeting ceremony call a *keeyika keeyika;* alarm call a repeated *aooh aooh;* sounds are higher-pitched than those of Black-legged Kittiwake.

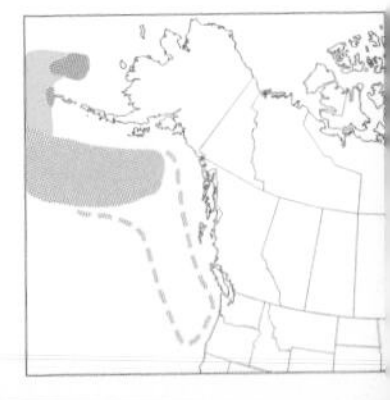

Subspp: Monotypic.

Adult summer WAL/02

Adult summer NOR/06

Juv. NF/09

Juv. NS/01

Ivory Gull 3

Pagophila eburnea L 18"

Two-year gull. **Shape:** Medium-sized, short-legged, deep-chested gull with a proportionately large head and short thin bill. Steep forehead; fairly high and rounded crown. Chest-to-leg vs. leg-to-wing-tip ratio 1:2. **Ad:** Bill bluish gray to grayish green with yellow tip; legs black; eye dark. All white. **In flight,** all white with dark yellow-tipped bill. Adult has essentially the same plumage all year. **Juv:** (1st yr.) Juv. plumage kept through winter and into 1st summer. Like ad. but with variably sooty foreface; bill often darker, with less distinct paler tip. Upperwing feathers variably tipped dark, creating fine dotted lines across wings and back and dotted trailing edge to wings. **In flight,** white except for dark spotting along tips of coverts and flight feathers. Tail with narrow subterminal band. **Hab:** Sea ice, beaches, open ocean. **Voice:** Long-call a harsh, descending, screamlike *keeeeer keeeeer;* alarm call a short, descending, high-pitched *tseeo.*

Subspp: Monotypic.

Adult summer CA/06

Adult summer NAM/08

1st summer NJ/09

1st summer NJ/09

Juv. NJ/09

Juv. VA/09

Sabine's Gull 1

Xema sabini L 13½"

Two-year gull. **Shape:** Small, short-legged, short-tailed gull with a short, moderately thin, blunt-tipped bill. Wings relatively long and pointed; tail tip, when closed, is slightly concave (can have a slightly forked look). Chest-to-leg vs. leg-to-wing-tip ratio 1:2. **Ad. Summer:** Bill black with yellow tip; legs variable—whitish, pinkish, gray, or black. Unique dark gray hood with thin black band at base; dark gray mantle; black outer primaries with large white tips. **In flight,** distinctive wing pattern at all ages composed of three colors: black wedge of outer primaries and their coverts; white midwing triangle with its apex at the wrist; gray inner triangle of secondary coverts. Below, pattern similar to above but muted and subtle. **Ad. Winter:** Winter plumage generally not seen in NAM. Like ad. summer but with grayish half-hood on back of head; bill tip yellow. **1st Summer:** Complete molt in late winter produces this plumage. Like ad. winter, but primaries with small or no white tips and bill pale-tipped or all black. **Juv:** Grayish brown on half-hood, nape, mantle, wing coverts, and tertials; grayish brown also extends down sides of breast. Back and wing coverts tipped paler, creating scaled look. Bill all black or black with pale base to lower mandible. Primaries blackish with thin white tips. **In flight,** tail shape and upperwing pattern distinctive like ad., but inner triangle on secondary coverts grayish brown and scaled instead of smooth gray; also, tail with black terminal band, narrower at the ends. **Hab:** Open ocean, tundra, estuaries, beaches. **Voice:** Long-call a *kyeeer kyeeer kyeeer kyuh kyuh kyuh;* alarm call a rapid grating *krkrkrkrkr.*

Subspp: Monotypic.

Adult summer MB/06

Adult ENG/–

Adult winter UKM/03

1st summer RUS/06

1st winter JAP/01

1st winter SCO/02

Ross's Gull 3

Rhodostethia rosea L 13"

Two-year gull. **Shape:** Small, deep-bellied, short-necked, short-legged gull with a very short thin bill. Steep forehead and high rounded crown create dovelike look. Wings relatively long; central tail feathers longest, creating wedge shape. Chest-to-leg vs. leg-to-wing-tip ratio 1:2½. **Ad. Summer:** Bill black; legs red; eye dark. Distinctive narrow black collar encircles white head. Back and wings medium gray; body and head often with pinkish blush. **In flight,** upperwings medium gray; underwings darker gray with wide white trailing edge to secondaries and inner primaries and dark wing tip. Similar Little Gull's upperwings pale gray, underwings dark gray with complete white trailing edge and broad white wing tip. **Ad. Winter:** Like ad. summer but lacks black collar; head and neck gray like back; small dark earspot, dark shading around eye. **1st Winter:** Similar to ad. winter but with black primary tips and, on upperwing, black lesser and median coverts and tertials. Legs pinkish. **In flight,** strong M pattern on upperwings formed by dark outer primaries and their coverts and dark median secondary coverts. Broad white trailing edge to wing, widest at bend, visible above and below; underwings otherwise gray; dark central tip to wedge-shaped tail. **Juv:** Sooty-brown face, crown, neck, and sides of breast; whitish underparts. Back and wing coverts blackish with broad buff fringes; legs pinkish. **In flight,** similar to 1st winter but back dark and collar brown. **Hab:** Ocean shores, estuaries, coastal freshwater marshes. **Voice:** Higher-pitched and more like voice of terns than of other gulls; a clipped *adac adac adac;* alarm call a dry rapid series like *kikikiki.*

Subspp: Monotypic.

Adult summer ME/05

Adult summer NH/05

Adult winter FL/02

Adult winter FL/03

Bonaparte's Gull 1

Chroicocephalus philadelphia L 13½"

Two-year gull. **Shape:** Small narrow-winged gull whose rounded dovelike head and short, thin, straight bill create a delicate and graceful appearance. Short-legged and fairly deep-bellied; chest-to-leg vs. leg-to-wing-tip ratio 1:2½. Floats high on water, long wings and tail angled up. Similar rare Black-headed Gull (which is sometimes found in flocks of Bonaparte's) is larger, has a longer slightly downcurved bill, and has proportionately longer legs and broader wings. **Ad. Summer:** Bill black; legs reddish pink. Black hood (does not cover lower nape); light gray mantle and wings. **In flight,** obvious translucent white leading edge of wings, from above and below, created by white outer primaries and their associated coverts; outer primaries also tipped black. Similar Black-headed has a red bill and dark underwing. **Ad. Winter:** Hood replaced

1st winter FL/02

1st winter CA/04

Juv. OH/08

by dark earspot and 2 faint gray "headbands." Gray from mantle extends onto nape and sides of breast. Legs pink. **In flight,** as in summer. Similar Black-headed has a red bill and dark underwing. **1st Winter:** Perched bird like ad. winter but with dark wing coverts and dark centers to tertials (rather than all gray). **In flight,** dark outermost primaries and dark brownish ulnar bar suggest M; wings also have distinct thin dark trailing edge. Tail with narrow black subterminal band. **Juv:** Warm brown crown, nape, back, and scapulars; back and scapulars pale-edged, creating scaled look. Dark brown earspot; median coverts dark brown while greater coverts gray. Legs pinkish; bill black. **In flight,** strong M pattern from above; from below, mostly pale. **Hab:** Estuaries, lakes, marshes, plowed fields; nests in conifers around boreal lakes and marshes. **Voice:** Long-call a buzzy descending *gaagaagaagagaga;* alarm call a buzzy *geh geh* or *geh geh geh.*

Subspp: Monotypic.

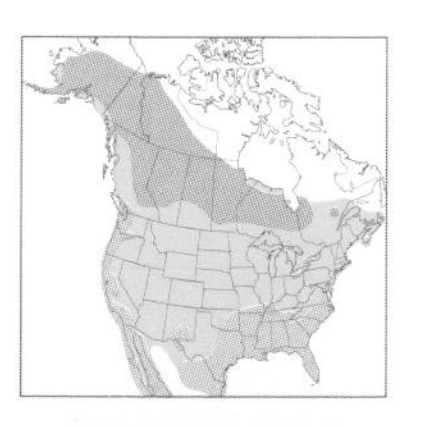

Adult summer, *ridibundus* FRA/03

Adult summer, *ridibundus* UKM/–

Adult winter, *ridibundus* NET/02

Adult winter IL/02

Black-headed Gull 3

Chroicocephalus ridibundus L 15½"

Two-year gull. **Shape:** Medium-sized graceful gull with a small dovelike head and fairly short thin bill. Chest-to-leg vs. leg-to-wing-tip ratio 1:2½. Similar to Bonaparte's Gull but larger; bill longer and slightly downcurved (straight in Bonaparte's); legs proportionately longer and wings broader. **Ad. Summer:** Bill red with variably dark tip; legs red. Dark brown hood (often appears black and does not cover nape); pale gray mantle. **In flight,** similar to Bonaparte's Gull above, with white leading edge of wings formed by white outer primaries and associated coverts; outer primaries also tipped with black. Below, outer primaries white (similar to Bonaparte's), but adjacent primaries grayish black (pale gray to whitish in Bonaparte's). **Ad. Winter:** Like ad. summer, but dark brown hood replaced by smudgy earspot and, often, vertical bars behind eye that nearly meet on crown. Legs still red, bill red with dark tip (bill black on Bonaparte's). **1st Winter:** Bill reddish pink with dark tip; legs reddish pink. Perched bird has gray mantle, dark wing coverts, dark-centered tertials. **In flight,** upperwings with

1st winter, *ridibundus* UKM/–

1st winter, *ridibundus* UKM/–

Juv., *ridibundus* ENG/08

Juv., *ridibundus* NAM/–

broad dark ulnar bar and wide dark trailing edge, widest at bend in wing; median and greater coverts both warm brown (greater coverts gray in 1st-winter Bonaparte's). Underwings similar to adult's, with outer primaries white and inner primaries grayish black. Tail with narrow subterminal band. **Juv:** Bill pinkish with dark tip; legs pinkish. Warm brown on mantle, scapulars, and wing coverts, with feathers edged paler, creating scaled look; median and greater coverts both warm brown (greater coverts gray in juv. Bonaparte's). Dark crown and mantle separated by white collar. **In flight,** like 1st winter but mantle and scapulars brown. **Hab:** Ocean shores, estuaries, lakes, rivers, plowed fields. **Voice:** Alarm call *kek kek kek kek.*

Subspp: (2) •*ridibundus* (n.e. NAM) smaller, with smaller bill, more rounded head, shorter wings, shorter legs. •*sibiricus* (Asian visitor to AK) larger, with heavier bill, flatter and shallower crown, longer legs and wings. **Hybrids:** Ring-billed, Laughing Gulls.

Adult summer ENG/05

Adult summer FIN/06

2nd winter OH/03

Adult winter OH/03

Little Gull 3

Hydrocoloeus minutus L 11½"

Two–three-year gull. **Shape:** Smallest of all gulls. Deep-chested, short-tailed, short-legged gull with a very short, thin, finely pointed bill. Steep forehead and high rounded crown give head a dovelike look. Adult has rounded wing tips (similar Bonaparte's Gull has longer more pointed wings). Chest-to-leg vs. leg-to-wing-tip ratio 1:2. **Ad. Summer:** Bill black; legs red. Pale gray mantle and extensive black hood; wing tips pale gray and white, with no black. **In flight,** wings distinctively all dark below and all light above. Upperwings pale gray with white trailing edge; underwings all dark with broad white trailing edge. Quick wingbeats; erratic flight path when it chases aerial insects. **Ad. Winter:** Like ad. summer, but hood replaced by dark earspot and small variably dark cap. Legs reddish pink. **2nd Winter:** Like ad. winter, but upper wing tips black with white tips; legs pinkish. **In flight,** upperwings pale

1st winter/summer ME/06

1st summer ENG/05

Juv. FIN/07

Juv. FIN/07

gray with black on wing tips; underwings grayish. **1st Winter:** Head and mantle like ad. winter, but greater coverts, tertials, and primary tips blackish brown; legs pink. Distinctive dark cap and earspot (cap lacking in similar Bonaparte's). **In flight,** upperwing with wide blackish ulnar bar, dark gray flight feathers; underwings white with dark tips on outer primaries. Tail has narrow terminal bar, widest in the center; bar may not extend over outer tail feathers. **Juv:** Distinctive dark cap, earspot, and dark half collar that extends onto sides of breast; mantle blackish. **In flight,** like 1st winter, but upperwings more extensively dark, with bold dark M pattern; underwings mostly pale with dark tips to primaries. **Hab:** Ocean shores, estuaries, lakes, rivers. **Voice:** Long-call like *kaay kaay keke keke keke kyoo kyoo;* alarm call a staccato *tk tk tk tk.*

Subspp: Monotypic.

Adult summer FL/03

Adult summer FL/03

Adult winter FL/02

Adult winter FL/02

Laughing Gull 1

Leucophaeus atricilla L 16"

Three-year gull. **Shape:** Medium-sized, slender-bodied, long-winged gull with a strongly sloping forehead and relatively long thick bill that appears to droop slightly, enhanced by the gradual curve of the upper mandible. Looks strongly attenuated; chest-to-leg vs. leg-to-wing-tip ratio 1:3. **In flight,** wings narrow; outer wing gradually tapers to sharp point. Similar Franklin's Gull is slightly smaller with shorter wings and shorter straight bill, steep forehead, high rounded crown. **Ad. Summer:** Bill dark red; legs dark (rarely, blotched with orange). Black hood; dark gray mantle and wings; small white tips to primaries (may wear off in summer). Thin white eye crescents do not meet behind eye. **In flight,** outer primaries extensively black, above and below; tail all white. **Ad. Winter:** Like summer but bill black; partial gray hood, sometimes looking like headphones. **2nd Winter:** No white tips on primaries. **In flight,** like ad., but black on primaries extends onto adjacent coverts toward wrist; may be traces of black band on tail. **1st Winter:** Extensive grayish brown on head, nape, and breast; foreface whitish; belly white. Back dark gray; upperwing coverts pale brown; flight feathers dark brown. **In flight,** broad dark tail band (tail almost all black) contrasts with white rump; primaries all dark; secondaries dark with white tips, forming a trailing dark subterminal band. From below, patterned, usually with diagnostic darkish axillaries. Similar 1st-winter Franklin's has mostly all-white underwing, thinner dark central subterminal tail band, white nape, white breast. **Juv:** Warm brown extends evenly over upperparts and onto breast and flanks; feathers of back and wing coverts with wide pale margins, creating strongly scaled look. **In flight,** broad black tail band extends completely across tail. Head appears mostly brown. Underwings mottled light and dark and have dark carpal bar. Similar juv. Franklin's Gull has white face, chest, and flanks and narrow black central tail band. **Hab:** Ocean shores, estuaries, lakes, rivers, dumps, parking lots. **Voice:** Long-call a few high-pitched notes followed by shorter notes, like *keeeaah keeaaah kya kya kya kya;* alarm call a laughlike *gagagaga.* 🎧**38**.

Subspp: (2) •*megalopterus* (all of NAM range) is larger with shorter bill; wing tips have smaller black area and larger white tips. •*atricilla* (possible West Indies visitor to s. FL). **Hybrids:** Black-headed, Ring-billed Gulls.

2nd winter FL/02

2nd winter FL/01

1st winter FL/01

1st winter FL/02

Juv. NJ/09

Juv. NJ/08

Adult with orange legs FL/03

Adult summer MT/06

Adult summer ND/06

Adult winter CHI/12

Adult winter CA/11

Franklin's Gull 1

Leucophaeus pipixcan L 14½"

Three-year gull (2 complete or nearly complete molts each year). **Shape:** Fairly small, deep-bellied, long-winged, short-legged gull with a fairly short, thin, straight bill. Rather small rounded head; steep forehead and high rounded crown. Thick white eye crescents can look like goggles. Chest-to-leg vs. leg-to-wing-tip ratio 1:2½. Similar Laughing Gull has longer, thicker, slightly downcurved bill; longer more sloping forehead; longer thinner wings; and thinner eye crescents. **Ad. Summer:** Bill dark red; legs dark red. Black hood; dark gray mantle and wings; prominent white spots on tips of primaries. Pink blush on underparts. **In flight,** white band separates gray inner wing from black tip. White tail has unique pale gray central feathers. **Ad. Winter:** Bill black, often with small red tip; legs black. Dark half-hood or "shawl" is prominent and contrasts with white forehead. Primaries have large white tips (wings molted twice each year). **In flight,** as in summer. **1st Summer/2nd Winter:** Like ad. winter, but white tips of primaries may be smaller; white band between black and gray of upper primaries usually

1st winter FL/01

1st winter PER/11

Laughing (l.) and Franklin's, adult summer TX/04

absent. Similar to Laughing Gull, but shape can be helpful to distinguish. **1st Winter:** Like ad. winter except for wings and tail. Wings dark gray with black tips (no white separating them); variable brown on coverts. Tail pale gray with thin subterminal band on all but outermost 2 feathers of each side. Breast and belly white, nape whitish; underwings mostly clear white. 1st-winter Laughing has gray nape, breast, and flanks; underwings with dark axillaries; and tail band complete to tail edges. **Juv:** Warm brown upperparts with back and wing covert feathers strongly edged paler, creating scaled look. Blackish half-hood with contrasting white face; white belly, flanks, breast, and chin. **In flight,** note white face contrasting with dark hood, narrow black tail band that does not extend onto white outer tail feathers. This plumage kept for very short time, with molt beginning in July. **Hab:** Ocean shores, estuaries, lakes, rivers, dumps, marshes. **Voice:** Long-call starts with distinctive short note, like *kah keeaah keeaah kah kah kah;* alarm call a staccato *kekekeke.*

Subspp: Monotypic. **Hybrids:** Ring-billed Gull.

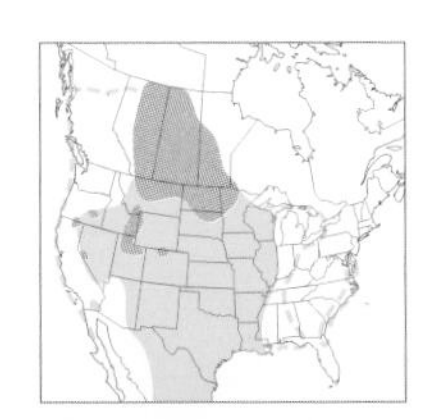

Adult summer JAP/02

Adult winter VT/10

Adult winter JAP/02

Adult winter VT/10

Black-tailed Gull 4

Larus crassirostris L 18"

Four-year gull. Casual vag. from Asia that can occur just about anywhere; most records near coasts. **Shape:** Medium-sized, long-winged, short-tailed, and relatively short-legged gull with a slender chest and belly. Wing tips project well past tail tip. Head relatively small and crown well rounded. Bill long and thin with weak gonydeal angle. Chest-to-leg vs. leg-to-wing-tip ratio about 1:2¼. **Ad. Summer:** Bill yellow with black subterminal band and red tip; legs yellow; eye pale. Dark gray back and wings with thin tertial crescent. **In flight,** dark gray back, tail with white base and wide black band surrounded by a thin margin of white. **Ad. Winter:** Like ad. summer but with dark streaking on back of head and sides of neck. **3rd Winter:** Similar to ad., but outer primary coverts blackish (gray in ad.); little or no white on tips of outer primaries; bill base may be pale yellowish gray. **2nd Winter:** Bill pale grayish at base with well-defined black tip; legs pale grayish yellow; eye fairly dark. Dark gray back and median coverts; other coverts and tertials brown. Foreface generally whitish, but rest of head, breast, and flanks pale chocolate brown; white crescents above and below eye. **In flight,** from above, dark outer wing and dark secondaries that contrast with rest of pale inner wing; tail blackish with no white surrounding edge. **1st Winter:** Bill pinkish with well-defined black tip; legs pale pinkish; eye dark. Very dark brown overall except for paler foreface and contrastingly whitish uppertail coverts. Faint or no white eye crescents. **In flight,** dark bird with fairly uniformly dark upper wing. Pure white rump and undertail coverts (California Gull has mottled or barred undertail coverts); all-black tail. **Juv:** Like 1st winter but with some barring on the rump. **Hab:** Ocean shores, estuaries, lakes, rivers. **Voice:** Similar to American Herring Gull's but slightly higher-pitched.

Subspp: Monotypic.

2nd winter JAP/02

2nd winter JAP/10

1st winter JAP/09

1st winter JAP/11

Juv. JAP/09

Adult summer CA/03

Adult summer CA/–

Adult winter CA/12

Adult winter CA/12

Heermann's Gull 1

Larus heermanni L 19"

Four-year gull. **Shape:** Medium-sized, slender-bodied, long-winged gull with a rounded crown and fairly thin straight bill with a slight gonydeal angle making it bulbous-tipped. Long primary projection past tail. Chest-to-leg vs. leg-to-wing-tip ratio about 1:2½. **Ad. Summer:** The only largely gray gull with a red bill in NAM. Bright red bill with small black tip; blackish legs (a helpful clue for all ages); dark eye. White head; gray body with darker gray back and wings; broad white tertial crescent. **In flight,** upperwings with blackish primaries and secondaries, gray coverts; pale gray rump contrasts with all-dark tail; secondaries with narrow white trailing edge. **Ad. Winter:** Similar to ad. summer, but with fine gray streaking and/or mottling on head and duller bill color. **3rd Winter:** Very similar to ad. winter, but bill dull red with broader black tip; head smooth or mottled gray (rarely streaked); thin white tertial crescent. **In flight,** uniformly grayish bird with thin white trailing edge to secondaries, black tail, and red bill with black tip. **2nd Winter:** Bill orangish red at base, dark at tip; smooth brown overall with darker brownish head; thin tertial crescent; thin pale eye crescents. **In flight,** dark overall; light gray rump contrasts with blackish tail; secondaries and tail narrowly tipped pale gray or white (white tips may quickly wear off). **1st Winter:** Bill pinkish at base with dark tip; dark brown overall; wing coverts with pale fringes and worn look; no marked tertial crescent; legs blackish. **In flight,** all dark grayish brown with black tail that is evenly rounded; primaries and secondaries darkest parts of wings; bill pale with dark tip; no white tips to secondaries or tail. **Juv:** Like 1st winter, but feathers fresh and with fine pale margins, creating subtle scaled look. Bill all dark. **Hab:** Coastal areas, harbors. **Voice:** No known long-call; alarm call *caw ca ca ca*.

Subspp: Monotypic. **Hybrids:** California Gull.

3rd winter CA/02

3rd winter CA/11

2nd winter CA/03

2nd winter CA/08

1st winter CA/11

1st winter CA/01

Juv. CA/08

Adult summer, *brachyrhynchus* AK/05

Adult summer, *brachyrhynchus* AK/05

Adult winter, *brachyrhynchus* CA/01

Adult winter, *brachyrhynchus* CA/02

Mew Gull 1

Larus canus L 16½"

Three-year gull. **Shape:** Medium-sized, very long-winged, small-headed gull with a short thin bill. Rather steep forehead and high rounded crown along with short thin bill create somewhat dovelike look. Chest-to-leg vs. leg-to-wing-tip ratio almost 1:3. **Ad. Summer:** Bill dull greenish yellow; legs yellow; eye mostly darkish (small percentage may have light eyes). Medium gray back; broad white tertial crescent (Ring-billed Gull has pale gray back and little or no tertial crescent, black ring on bill); primaries with black ends, white tips. **In flight,** medium gray wings and back; limited black on outer primaries; large mirrors on P9 and P10; on PP5–8 white dots behind the black ends. Broad white trailing edge to secondaries. **Ad. Winter:** Like ad. summer but with grayish-brown streaking and mottling on crown, nape, and neck and onto sides of breast; bill may have faint subterminal ring; legs more greenish. **2nd Winter:** Like ad. winter, but bill with dark tip; eye dark; no white tips to primaries; tertials may have dark smudging; tail has partial to complete thin subterminal band; wing coverts may have some brown. **In flight,** like ad. winter, but upperwing with black on outer primary coverts; may have black marks on tail; mirror on P10 only. **1st Winter:** Bill with yellowish-gray base and dark tip; legs pinkish; eye dark. Medium gray back; dark tertials; face mostly brown; lower neck and rest of underparts mottled brown with dark belly smudge. **In flight,** dark outer primaries and somewhat paler inner primaries create modest window effect; dark secondaries contrast with paler secondary coverts; underwings dark (unlike Ring-billed). Rump barred brown and does not contrast strongly with back; tail mostly dark brown. **Juv:** Bill mostly all dark; legs pink; eye dark. Fairly smooth buffy brown overall with paler edges of back and wing feathers, creating scaled look. **In flight,** tail mostly dark; rump barred brown and whitish, not contrasting strongly with back (whitish rump of Ring-billed Gull contrasts with darker back). **Hab:** Ocean shores, estuaries, lakes, rivers, plowed fields. **Voice:** Long-call *ka ka ka keeaah keeeaah keeeaah kewkewkewkewkewkew;* alarm call a repeated *kwew kwew kwew.*

Subspp: (3) Classification still in flux; these 3 subspp. occurring in NAM may be 2–3 separate species. The above description is for •*brachyrhynchus,* the only NAM breeding subsp. The other 2 subspp. are vagrants; following are the ways they differ from *brachyrhynchus.*

L. c. canus (vag. to Atlantic Coast from EUR), "Common Gull." **Shape:** More like Ring-billed, with slightly longer thicker bill; may be slightly heavier overall. **Ad:** Mantle not quite as dark as that of *L. c. kamtschatschensis* or *brachyrhynchus.* More black on wing tip, esp. P8 and P9;

2nd winter, *brachyrhynchus* CA/11

2nd winter, *brachyrhynchus* CA/02

1st winter, *brachyrhynchus* CA/12

1st winter, *brachyrhynchus* BC/10

Juv., *brachyrhynchus* BC/09

Juv., *canus* FIN/08

few or no white dots between gray bases and black tips on outer primaries; eye averages less dark than that of *kamtschatschensis*. **1st Winter:** Whiter head, underparts, and underwings; back paler gray. **Juv:** Whiter head (esp. foreface), whiter breast with brown mottling and whitish belly (rather than smooth buffy brown throughout); underwing secondaries form dark bar on wing's trailing edge (rather than all pale brown); tail with narrow blackish subterminal band and whitish base (rather than tail almost all brown); rump more contrastingly whitish (rather than brownish).

L. c. kamtschatschensis (Asian vag. to w. AK and possibly farther south), "Kamchatka Gull." **Shape:** Larger, with sloping forehead and slightly thicker longer bill. **Ad:** Eye usually pale. Like *canus,* lacks obvious pale subterminal spots on black wing tip. **Juv:** Whiter head (esp. foreface); whiter breast and belly with heavy dark mottling (rather than smooth buffy brown throughout); underwing secondaries form dark bar on wing's trailing edge (rather than all pale brown); tail with broad blackish subterminal band and whitish base (rather than tail mostly brown).

Hybrids: Ring-billed, Black-headed Gulls.

Adult summer FL/03

Adult winter FL/12

Adult winter FL/02

Ring-billed Gull 1

Larus delawarensis L 18½"

Three-year gull. **Shape:** Medium-sized gull with rather average proportions; a good gull to know well and use as a comparison for the shape of other species. Thick blunt-tipped bill; sloping forehead to fairly rounded crown; fairly long head. Chest-to-leg vs. leg-to-wing-tip ratio about 1:2. Similar Mew Gull has shorter, thinner, more gradually tapered bill; smaller well-rounded head; steep forehead and high rounded crown; longer wings; no ring on bill. **Ad. Summer:** Bill yellow with dark subterminal band; legs yellow; eye pale. Pale gray back and wings; outer primaries with black ends and white tips; tips may start to wear off in summer. A small percentage of spring adults have a pink cast to body plumage. **In flight,** broad black ends to outer primaries; large mirror on P10, smaller one on P9; tail all white. **Ad. Winter:** Like ad. summer but with fine dark streaking on head; sometimes duller yellow bill and legs; large white tips on primaries. **2nd Winter:** Like ad. winter but bill duller with broader band; eye usually pale; small or no white tips to primaries; may be traces of dark on tertials; legs duller yellow. **In flight,** black on outer primaries extends onto their coverts toward wrist; one small mirror on P10; usually some dark marks on tail. **1st Winter:** Bill pinkish with dark end and small pink tip; legs pinkish; eye dark. Pale gray back; brown wing coverts; dark brown tertials; brown streaking or mottling on head and underparts. **In flight,** back gray; rump whitish; tail with variably-sized dark subterminal band that can cover most of tail or be distinct and slightly thinner. Outer primaries and their coverts dark; inner primaries pale gray, creating a pale window at base of outer wing; dark secondaries contrast with pale-edged brown coverts. **Juv:** Like 1st winter but back feathers brownish with pale margins, creating scaled look. **In flight,** like 1st winter but back scalloped brown. Juv. plumage sometimes retained into winter. **Hab:** Ocean shores, estuaries, lakes, rivers, dumps, parking lots. **Voice:** Long-call *goh keeeah keeeah kyakyakyakya;* alarm call a loud *kakakakaka.* 🎧**39**.

Subspp: Monotypic. **Hybrids:** Black-headed, Franklin's, Laughing, California, Mew Gulls.

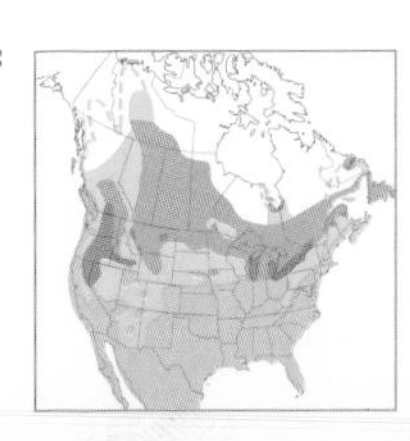

2nd winter NH/09

2nd winter NH/10

1st winter NH/09

1st winter FL/02

Juv. NH/09

Juv. NH/09

Adult summer, *wymani* CA/03

Adult summer CA/03

Adult winter CA/12

Adult winter, *occidentalis* CA/03

Western Gull 1

Larus occidentalis L 25"

Four-year gull. **Shape:** A large, deep-bellied, long-legged gull with fairly short broad wings and short primary projection past the tail (often shorter than the bill). Somewhat rounded crown; bill noticeably thick with a strong gonydeal angle, making it look bulbous-tipped. Chest-to-leg vs. leg-to-wing-tip ratio 1:2. **Ad. Summer:** Bill yellow with red gonydeal spot; legs pink at all ages (may have yellowish tinge near start of breeding); eye yellow to brownish yellow (see Subspp.). Dark gray back and wings (see Subspp.); outer primaries with black ends and white tips. Broad white band of tertial crescent often continues along the secondaries as a white line ("skirt") at the edge of the dark greater coverts. **In flight,** dark gray above with broad white trailing edge; outer primaries black with small white tips (may wear off by late summer); single mirror on P10, rarely smaller mirror on P9. Below, primaries and secondaries dark gray (contrasting with white coverts), creating very dark tip and trailing edge to wing. **Ad. Winter:** Like ad. summer. May have some light streaking on head and nape. **3rd Winter:** Similar to ad., but bill with black tip; small or no white tips to primaries; traces of brown on wing coverts; remnants of black on tail. **In flight,** like ad. but with traces of black on tail and no mirrors on primaries. Often, no white tips to primaries. **2nd Winter:** Bill with pale base, dark subterminal band, pale tip; eye dark, becoming ad. color over winter. Back and median coverts adult gray to variable extent; head and body whitish with some darker streaking on head and collar; tertials and rest of upperwing coverts dark brown. **In flight,** dark tail; white rump; back gray; body whitish; broad white trailing edge to secondaries. **1st Winter:** Bill all black (sometimes a paler area at base of lower mandible, particularly in late winter). Head and body all sooty brown, more evenly colored and darker than similar-aged Herring Gull. **In flight,** similar to juv. **Juv:** Bill all black; eye dark. Body and wings dark sooty brown overall, with a darker mask through face and eyes. Remains dark through winter. (Similar juv. Herring Gull paler brown and more finely mottled or checkered overall; head and breast whiten through winter.) **In flight,** wings very dark above and below, with no pale wing window; secondaries have narrow white trailing edge; tail all dark; light brown rump contrasts with rest of dark brown upperparts. **Hab:** Coastal areas, coastal freshwater lakes, harbors, sea cliffs. **Voice:** Long-call *kwooh kwooh kwooh kleeah kleeah kya kya kya;* alarm call *kehkehkeh.* Higher-pitched than similar Yellow-footed Gull sounds. 🎧**40**.

3rd winter, *wymani* CA/11

3rd winter CA/04

2nd winter CA/02

2nd winter CA/03

1st winter CA/01

1st winter CA/10

Juv. CA/12

Subspp: (2) •*occidentalis* (cent. CA north) larger; paler gray upperparts, often darker eye in comparative ages. •*wymani* (cent. CA south) smaller; darker gray mantle, paler eye in comparative ages. ▶ Some intergradation occurs where breeding subspp. overlap. **Hybrids:** Glaucous-winged Gull. Hybridization so common that during census counts birds are often not distinguished to species.

Adult summer CA/02

Adult summer NT/07

Adult winter CA/12

Adult winter BC/02

California Gull 1

Larus californicus L 21"

Four-year gull. **Shape:** Medium-sized, short-legged, slender-bellied gull with a moderately sloping forehead and fairly high rounded crown. Bill relatively long with a small gonydeal angle, giving it an even thickness throughout. Wings relatively long and narrow. Midway in size between smaller Ring-billed and larger Herring Gulls. Chest-to-leg vs. leg-to-wing-tip ratio about 1:2½. **Ad. Summer:** Bill yellow with black-and-red spot near tip (black spot may be reduced or absent in summer); legs yellow; eye dark at all ages. Medium gray back; primaries with black ends and white tips; wide tertial crescent, small scapular crescent. **In flight,** medium gray back; extensive black across outer ⅓ of upperwing (outer 3 primaries all black to coverts; outer 1–2 all black to coverts on Herring Gull); large white mirror on P10, smaller mirror on P9. Broad white trailing edge to secondaries and inner primaries. On underwings, gray secondaries and primaries contrast slightly with white underwing coverts. **Ad. Winter:** Like ad. summer, but black on bill more extensive. Forehead and face generally unstreaked, but nape and collar heavily streaked with grayish brown, creating a scarflike or slightly hooded effect. Legs more grayish green. **3rd Winter:** Similar to ad., but no large white tips to primaries, bill and legs greenish gray to dull yellow; bill has black subterminal band and sometimes red on gonys; some brown on greater coverts. Tertial and scapular crescents obvious. Legs grayish green to greenish yellow. **In flight,** black on wing tips extends onto outer primary coverts; variable traces of black on otherwise white tail; mirrors small or absent. **2nd Winter:** Bill pale gray or dull bluish gray with dark subterminal band; legs distinctively grayish blue; eye dark. Medium gray back and median coverts; other coverts and tertials brownish. Body mainly pale with variable brown flecking, strongest on collar. Similar 1st-winter Ring-billed has shorter bill with pink base, pinkish legs, lightly streaked nape, black band on tail. Similar 2nd-winter Herring is larger, has pale iris, pinkish legs; only traces of black on tail. **In flight,** back and inner wing medium gray; dark secondaries and outer primaries; small paler window on inner primaries; tail all black and rump white. **1st Winter:** Bill pinkish-based with dark tip sharply delineated from fall on (in Herring Gull, dark tip blends variably to pink base until late winter). Body mostly brown, sometimes paler on breast and foreface; gray shadings to some back feathers. **In flight,** from above, 2 darker bars on inner wing formed by dark secondaries and dark bases to greater coverts (the latter lacking in Herring Gull). Tail dark; rump brownish like back. **Juv:** Bill all black; legs pale pink. Variably pale to quite dark overall; body generally brown with breast and belly paler, sometimes with cinnamon wash; back can appear checkered. Can be very similar to juv. Herring Gull, but smaller and migrates to West

3rd winter CA/04

3rd winter CA/03

2nd winter CA/01

2nd winter CA/02

1st winter CA/03

1st winter CA/09

Juv. CA/08

Coast in late July, while juv. Herring arrives in same areas in early Oct., when California Gull no longer in full juv. plumage. **In flight,** wing dark brown with darker primaries and secondaries. Greater and median coverts may have whitish tips, forming 2 thin wingbars extending the length of the wing. Tail dark and rump strongly barred with brown. **Hab:** Ocean shores, estuaries, lakes, rivers, dumps, parking lots. **Voice:** Long-call *koh koh keeeah keeeah kyahkyahkyahkyah;* alarm call a descending *keeeow.*

Subspp: (2) Differences subtle. •*californicus* (cent. CA–s. MT–CO) is slightly smaller and with darker gray mantle. •*albertaensis* (NT–n.e. MT–ND) is larger with paler gray mantle—approaching paleness of Ring-billed and Herring Gulls. **Hybrids:** Herring, Ring-billed Gulls.

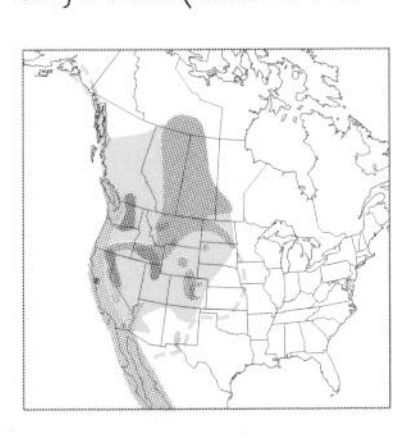

Adult summer, *smithsonianus* MB/06

Adult summer, *smithsonianus* NH/07

Adult winter, *smithsonianus* NH/12

Herring Gull 1

Larus argentatus L 22–27"

Four-year gull. **Shape:** Large, long-legged, long-winged, deep-bellied gull. Bill fairly long with a moderate to obvious gonydeal angle. Forehead often strongly sloping, crown often rather flat, but both variable. Common and widespread gull and one of the most variable in size, structure, and immature plumages. Chest-to-leg vs. leg-to-wing-tip ratio about 1:2½. **Ad. Summer:** Bill yellow with red gonydeal spot; legs pink in all plumages; iris pale yellow, orbital ring yellowish orange. Pale gray back; primaries with black ends, white tips. Aging easiest when tail and rump seen in flight. **In flight,** tail and rump all white. Moderate amount of black at tips of wing seen above and below (only outer 2 primaries all black to coverts). Evenly pale underwings. Large mirror on P10, smaller mirror on P9 (only on eastern birds). **Ad. Winter:** Like ad. summer but dense streaking and/or mottling on head and neck. May be slight black marks at bill tip along with red spot. **3rd Winter:** Similar to ad. but with traces of brown on greater coverts and tertials. Eye very pale and surrounded by dark shading, giving bird an eerie look. Bill pinkish gray with thin dark subterminal band. **In flight,** tail mostly white with traces of black band. Otherwise, similar to adult but black of wing tip extends onto primary coverts. **2nd Winter:** Bill pale pinkish gray with sharply demarcated dark subterminal band. Back varies from pale gray (similar to adult) to still mostly brown. Body brownish, tertials dark. Eye first dark, then becomes pale over winter. **In flight,** tail mostly black, contrasting with white rump. Wings brown and gray with a large paler window on inner primaries seen from above and below. Secondaries, as well as outer primaries and their coverts, tend to be the darkest portions of wing. **1st Winter:** Bill all dark at first, becoming variably pinkish gray at base with a generally poorly defined darker tip. Bird variable, from all dark brown to pale brown with a paler whitish head and neck, and variations in between. In flight, tail all dark; rump mottled brown and white and not contrasting substantially with back. Wings brown with a large paler window on inner primaries. Dark secondaries form a dark trailing edge to inner wing that contrasts with brown greater coverts. **Juv:** Bill all dark; body all dark brown with fine pale mottling and pale edges to primary tips. **In flight,** all dark brown except for pale window on inner primaries. **Hab:** Ocean shores, estuaries, lakes, rivers, dumps, parking lots, fields. **Voice:** Long-call *kyoh kyoh kyoh kleeah kleeah kia kia kia;* alarm call a *keh keh keh.* Higher-pitched than similar Great Black-backed Gull sounds. 🎧**41**.

Subspp: (4) Taxonomy in flux; these 4 subspecies may be 2–3 species. •*smithsonianus* (all NAM range), "American Herring Gull," described above; other subspp. described in comparison. •*vegae* (w. AK is. and probable vag. south along Pacific Coast), "Vega Gull"; 1st winter with pale base to tail; ad. larger, upperparts darker gray, wing tips with more extensive black, eye pale yellow to brownish, orbital ring more reddish. •*argentatus* (EUR vag. to n.e. NAM), "European Herring Gull"; 1st winter with pale base to tail; ad. upperparts darker gray, wing tips less extensively black, eye and orbital ring similar to *smithsonianus.* •*argenteus* (EUR vag. to n.e. NAM), "European Herring Gull"; 1st winter with pale base to tail; ad. smaller, upperparts paler, otherwise like *smithsonianus.* **Hybrids:** Glaucous, Great Black-backed, California, Kelp, Glaucous-winged, Lesser Black-backed Gulls.

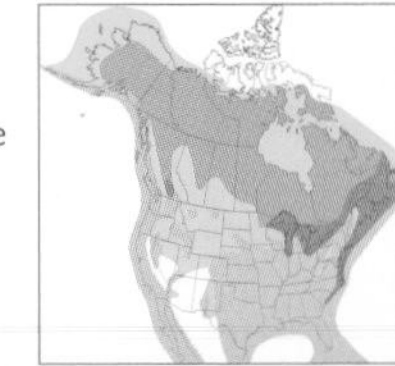

3rd winter, early stage, *smithsonianus* NH/09

3rd winter, *smithsonianus* –/01

2nd winter, *smithsonianus* NH/10

2nd winter, late stage, *smithsonianus* FL/01

1st winter, *smithsonianus* FL/02

1st winter, *smithsonianus* CA/09

Juv., *smithsonianus* NH/09

Juv., *smithsonianus* NH/10

Adult summer FRA/02

Adult summer POR/04

Adult winter MOR/09

Yellow-legged Gull 4

Larus michahellis L 25"

Four-year gull. Casual southern European vag. to East Coast. **Shape:** Large, stocky, deep-chested gull. Bill relatively deep and blunt with a fairly marked gonydeal angle. Chest-to-leg vs. leg-to-wing-tip ratio about 1:2½. **Ad. Summer:** Bill bright yellow with large red gonydeal spot, which may extend onto upper bill (smaller and limited to lower mandible on Herring); legs yellow (about 1% of Herring Gulls have yellow legs in late winter and spring); eye yellow with red orbital ring (Herring orbital ring orangish). Back slightly darker gray than Herring Gull's. **In flight,** large mirror on P10; smaller mirror on P9. Similar to Herring Gull, but wing tip with less white and more black. Below, gray primaries and secondaries contrast slightly with white underwing coverts (Herring underwing is generally all light). Difficult to distinguish from Herring Gull and Herring Gull x Lesser Black-backed hybrids (although all of these have extensive streaking on head and neck throughout winter). **Ad. Winter:** Like summer but with limited streaking around eyes and on central crown; this is gone by late fall, due to earlier molt schedule than our native gulls, leaving head and body all white through most of winter (Herring Gull and Lesser Black-backed are streaked on head and neck throughout winter). May have some black near bill tip; legs grayish yellow. **3rd Winter:** Like ad. winter, but bill with more extensive dark marks near tip; less white on tips of primaries. Eye pale. **In flight,** like ad. summer but with black on wing tip extending onto primary coverts; some dark markings on outer portion of white tail. **2nd Winter:** Bill mostly dark with some paler areas at base and a pale tip; legs pale and tinged yellow; eye dark. Generally has adult gray on back, median coverts, and some tertials; extensively dark around eye. **In flight,** wide dark tail band, white rump, and gray mantle. **1st Winter:** Similar to juv. but with a whiter body. **Juv:** Bill all dark; legs pinkish; eye dark. Generally brown with paler foreface, central belly, and undertail coverts; darkest streaking around the eye. Back feathers edged paler, creating a strong scaled effect. In plumage, more like juv. Lesser Black-backed than juv. Herring. **In flight,** generally all-dark wings with little suggestion of a paler window on inner primaries. Tail with broad dark band and white base; rump and uppertail coverts mostly white. **Hab:** Ocean shores, estuaries, lakes, rivers, dumps, parking lots, fields. **Voice:** Similar to voice of Lesser Black-backed and Herring Gulls; lower-pitched than Herring Gull's.

Subspp: (2) Two subspp. in Europe that may arrive in NAM. Difficulty of distinguishing from Herring Gulls, let alone between the subspp., has made it hard to know the

3rd winter POR/04

3rd winter ITA/11

2nd winter MOR/09

2nd winter POR/05

1st winter MOR/09

1st winter MOR/09

subsp. designation of each NAM occurrence. Description above is for •*michahellis* (western Europe) which also can have mirrors on only P10 or both P10 and P9. •*atlantis* (Azores) is more compact with shorter wings and legs and stouter bill; also darker overall, showing more streaking on head and neck of ad. winter, but this is lost by early winter due to earlier molt schedule. **Hybrids:** Lesser Black-backed Gull.

Adult summer MB/06

Adult winter CA/12

Adult winter CA/02

Adult winter BC/02

Thayer's Gull 2

Larus thayeri L 23"

Four-year gull. **Shape:** Fairly large, short-legged, deep-chested gull with a fairly steep forehead and high rounded crown. Eye is proportionately large. Bill is relatively short with a slight gonydeal angle. Overall, this is a stocky gull with a rather sweet look. Chest-to-leg vs. leg-to-wing-tip ratio about 1:2. Shape similar to Iceland Gull. ▶Thayer's Gull and the *kumlieni* subspecies of Iceland Gull are very similar in plumage at all ages, except for ad. color of mantle and pattern of primary tips, which can be highly variable and overlap in appearance. Some suggest that Iceland and Thayer's Gulls are a continuum of one species, with Thayer's being at the darkest extreme of mantle color and having the most extensive dark patterning on the primary tips, the Iceland Gull subsp. *glaucoides* at the lightest extreme of mantle color and having the least extensive dark patterning on the primary tips, and Iceland Gull subsp. *kumlieni* in between. Therefore, in the field it is not always possible to identify to the species level a gull with intermediate characteristics, especially in juv./1st-winter birds. **Ad. Summer:** Bill greenish yellow with red spot; legs reddish pink; eye usually dark (about 90% of time), but can range to pale yellow; purple to reddish orbital ring. Outer primaries pale gray in inner webs, dark gray to black on outer webs, creating striped effect when spread. In general, differs from Herring Gull by having a slightly darker back, darker pink legs, darker eye, more rounded head, shorter bill, and less black on outer primaries from above and especially from below. In general, differs from Iceland Gull by having a darker mantle and more dark markings on outer primaries. **In flight,** from above, outer webs of outer primaries dark gray to black, inner webs pale gray, creating striped effect on outer wing. From below, wing all pale and translucent, with dark tips only to outer primaries. **Ad. Winter:** Similar to ad. summer but with variable grayish-brown streaking and/or smudging on head and neck; bill usually greenish yellow with red spot. **3rd Winter:** Like ad. winter, but bill is pale horn with a thin dark subterminal band; may be some pale brown on wing coverts. **In flight,** primary coverts may have dark marking, and remnants of dark tail band still clearly evident. **2nd Winter:** Bill pale pink at base with dark tip. Paler overall than 1st winter; back and sometimes median coverts becoming adult gray, more so over winter; rest of wings pale brown. Primaries darker than rest of body, but brownish black (not blackish as in imm. Herring Gull). **In flight,** striped pattern of outer primaries in brown, not black (as ad.); paler inner primaries create wing window; tail with broad brown band, contrasting with paler lightly barred rump. **1st Winter:** Very similar to juv. since juv. plumage not molted until late winter. Generally remains all dark, not like 1st-winter Herring, which often develops a pale head and neck. Often dark smudging around eye, usually lacking in 1st-winter Iceland Gull. Bill dark, developing some limited paleness at base (beginning in late Jan.–Feb.), but generally not as extensive as on Herring Gull of same age. **In flight,** like juvenile. **Juv:** Bill all

3rd winter CA/02

3rd winter/summer MB/06

2nd winter ON/10

2nd winter CA/01

Juv./1st winter CA/02

Juv./1st winter CA/10

Juv. CA/12

dark; legs pinkish; eye dark. Overall fairly uniformly tan-brown with fine paler mottling. Primaries darker than rest of body, but brown with pale chevrons at tips (not black and with minimal chevrons as in juv./1st-winter Herring Gull). Tertials medium brown and often blend with back and wings (not strongly darker as in Herring). **In flight,** from above, outer primaries darker than inner ones; darkness is on only outer webs, creating a striped effect when primaries are spread. Dark secondaries create dark trailing edge to wing. Below, flight feathers appear evenly pale, the pattern of the tips seen from above not seen below. Tail has extensive dark brown terminal band, contrasting with lighter barred brown rump. Similar juv./1st-winter Iceland is uniformly paler overall, lacking the contrastingly darker tail, secondaries, and outer primaries of juv./1st-winter Thayer's. **Hab:** Coastal areas, estuaries, sea cliffs, dumps, parking lots. **Voice:** Long-call *kweee kweee kyoow kyoow kyoow;* alarm call a *gagagaga.*

Subspp: Monotypic. **Hybrids:** Possibly Iceland Gull subsp. *kumlieni*, Glaucous-winged Gull.

Adult summer, *kumlieni* NF/02

Adult winter, *kumlieni* ME/01

Adult winter, *kumlieni* NH/01

Adult winter, *kumlieni* MA/02

Iceland Gull 2

Larus glaucoides L 22"

Four-year gull. **Shape:** Fairly large, short-legged, deep-bellied gull with a fairly steep forehead and high rounded crown. Eye is proportionately large and well below crown. Bill is relatively short with a slight gonydeal angle. Overall, this is a large stocky gull with a rather sweet look. Chest-to-leg vs. leg-to-wing-tip ratio about 1:2. Shape like Thayer's Gull. ▶ Darker Iceland Gulls can overlap in characteristics with lighter Thayer's Gulls and are not always separable. Lighter Iceland Gulls can be confused with Glaucous Gulls; structural characteristics best for making distinction. **Ad. Summer:** Bill yellow with red gonydeal spot; eye creamy (rarely dark) with purple to reddish orbital ring; legs dark pink. Pale gray back. Outer primaries variable but with no black; usually have dark gray regions on outer webs near tips but may have none. When darker gray is present, it can be as little as just a dark line on P10 to extensive gray on P6–P10; P10 tends to have an extensive white tip (on some darker birds this may be a mirror). **In flight,** pale gray back and wings; outer webs of outer primaries sometimes with darker gray markings seen from above and more faintly from below; broad white trailing edge to whole wing. **Ad. Winter:** Like ad. summer, but head and neck variably streaked or mottled with brown and bill may be greenish yellow at base. **3rd Winter:** Like ad. winter, but bill pale greenish yellow with dark subterminal band or spot; some pale brown on lesser and greater coverts and tertials; primaries can be slightly darker. **In flight,** like ad. but with brown tones to wing coverts; tail generally white but sometimes with a pale grayish band. **2nd Winter:** Bill variable; generally pale pinkish with a poorly defined darker tip, more rarely pink with well-defined dark tip (as in Glaucous, but Glaucous larger overall); dark eye may get paler over winter. Generally paler than 1st winter and with gray mottling (which increases over winter) on otherwise white mantle. Grayish-brown spotting on wing coverts and tertials like 1st winter. Primary tips are grayish brown, often darker than in 1st winter. **In flight,** generally pale brown overall, with outer webs on outer primaries darker gray-brown. Grayish-brown tail band contrasts with white base of tail and white rump. **1st Winter:** Bill all dark or with slightly pale base (later in season), eye dark. Body and wings vary from grayish brown overall to mostly white; in either case, back and wings are evenly spotted with pale brown; tail faintly to strongly mottled with brown; tertials mottled brown. Primaries usually concolor with body color. **In flight,** evenly pale overall; tail may have slightly darker band or be concolor with rest of upperparts.

3rd winter, *kumlieni* NF/02

2nd winter, *kumlieni* NF/02

2nd winter, *kumlieni* NF/02

2nd winter, *kumlieni* NC/12

Juv./1st winter, *kumlieni* NF/02

Juv./1st winter, *kumlieni* NF/02

Juv: Like 1st winter but slightly darker overall. **Hab:** Coastal areas, sea cliffs, harbors, beaches. **Voice:** Long-call *kweee kweee kyoow kyoow kyoow;* alarm call a *gagagaga*.

Subspp: (2) Classification in flux; the account above describes subsp. *kumlieni.* These two subspp., along with Thayer's Gull, may be part of a continuum of one species, with *glaucoides* at the lightest end, *kumlieni* in the middle, and Thayer's Gull at the darker end. •*glaucoides* (GRN, rare vag. to e. CAN) ad. generally has light gray mantle, lightest primary tips (to all white), least patterning on primary tips (often none), and shortest bill. •*kumlieni* (n.e. CAN, winters on northeast coast and Great Lakes), "Kumlien's Gull," ad. generally has light to medium gray mantle, gray primary tips (sometimes appear mostly white), some patterning on primary tips, and slightly longer bill. **Hybrids:** Thayer's Gull.

Adult summer FL/02

Adult winter FL/01

Adult summer FL/02

Lesser Black-backed Gull 2

Larus fuscus L 21"

Four-year gull. **Shape:** Medium-sized, slim-bodied, long-winged, short-legged gull with a fairly rounded head. Bill relatively short and thick with a slight gonydeal angle. Slightly smaller than Herring Gull; much smaller than Great Black-backed Gull. Chest-to-leg vs. leg-to-wing-tip ratio 1:2½; although this ratio is the same as for Herring Gull, the Lesser Black-backed's slim body makes it look more attenuated and "all wing." **Ad. Summer:** Bill deep yellow with large red gonydeal spot; legs bright yellow; eye pale yellow. Dark gray back and wings; subtly contrasting black primaries. **In flight,** from above, black on outer primaries contrasts with rest of dark gray wing; small mirror on P10 and sometimes a smaller mirror on P9; narrow white trailing edge to wing. Below, gray primaries and secondaries contrast with white coverts. Legs bright yellow. **Ad. Winter:** Like ad. summer but heavily streaked on head and neck, darkest around eyes. **3rd Winter:** Like ad. winter, but bill dull yellow with dark subterminal band, sometimes with some red present near tip; legs dull yellow or pinkish yellow. Primaries may be lightly tipped with white. **In flight,** similar to adult, but black on primaries more extensive and less well defined; some black streaking on tail. May be small mirror on P10, usually no mirror on P9. **2nd Winter:** Bill black with paler base; legs pinkish yellow; eye dark at first, changing to yellow over winter. Back mostly dark gray of adult (occasionally with just some gray); wing coverts variably dark gray to brown. Heavily streaked head and more lightly streaked breast and flanks; streaking fades over winter. **In flight,** as in 1st winter, but dark tail band may be narrower and back is usually dark gray. **1st Winter:** Bill usually all black well into 1st spring (Herring Gull bill usually turns pale at base by midwinter), rarely paler at base; legs pinkish; eye dark. Back and wings generally checkered dark brown and white; noticeably paler head, breast, and belly with distinct dark markings on sides of breast. Note smaller size and more attenuated look in relation to any nearby Herring or Great Black-backed Gulls. **In flight,** whitish rump contrasts with dark back and broad dark terminal tail band. Above, outer wing uniformly dark with no paler window (Herring Gull has pale window); dark greater coverts and dark secondaries form 2 dark bars on trailing edge of inner wing (Herring and Great Black-backed Gulls have only dark secondaries). **Juv:** Bill all black; legs pink. Back feathers dark brown with paler margins, creating scaled effect. Head and underparts streaked paler brown. **In flight,** like 1st winter. **Hab:** Ocean shores, estuaries, lakes, rivers, dumps, parking lots, fields. **Voice:** Similar to Herring Gull's but lower-pitched.

Subspp: (2) •*graellsii* (w. EUR visitor to e. NAM, rarely w. NAM) our most common subsp.; back is slate gray. •*intermedius* (w. EUR vag. to e. NAM) much less common in NAM; back is darker, nearly black. **Hybrids:** Yellow-legged, Herring Gulls.

3rd winter FL/02

3rd winter FL/02

2nd winter FL/01

2nd winter POR/11

1st winter FL/03

1st winter FL/02

Juv. NF/10

Juv. MOR/09

Adult summer JAP/09

Adult summer SIB/06

Adult winter RUS/11

Adult winter CA/01

Slaty-backed Gull 3

Larus schistisagus L 25"

Four-year gull. Rare Asian visitor to w. AK and, very rarely, farther south throughout NAM. **Shape:** Large, deep-chested, long-necked gull with a moderate-sized bill with little or no gonydeal angle. Chest-to-leg vs. leg-to-wing-tip ratio 1:2. **Ad. Summer:** Bill yellow with red gonydeal spot; legs deep pink (all ages); iris pale yellow. Slate-gray back and wings; white body. Broad white tertial and scapular crescent and white "skirt" of broad secondaries visible below coverts of closed wing. The only dark-backed gull likely to be seen in AK; on West Coast, ad. only confused with Western Gull, which is slightly lighter-backed, even in the darker southern subsp. *wymani,* and has a larger bill. **In flight,** dark blackish-gray wings and back; broad white trailing edge to wings; outer primaries with dark outer webs, lighter inner webs, creating striped effect on wing tips. Also, usually an extra line of subterminal white dots on the outer primaries (4–8) creates an uneven so-called "string of pearls" seen from above and even more clearly from below (this pattern evident from 2nd winter on). Below, gray primaries and secondaries contrast with white underwing coverts. Large mirror on P10, smaller mirror on P9. **Ad. Winter:** Like ad. summer but with dark smudging around eye, giving bird a "stern" look (as in Lesser Black-backed); streaking on head, nape, and sides of breast. **3rd Winter:** Like ad. winter, but bill pale with dark subterminal band; some brown shading on wing coverts and sometimes belly; small white tips to primaries. **In flight,** like ad. but with variable dark tail band; "string of pearls" less pronounced. **2nd Winter:** Bill with pale base, dark tip and cutting edge; eye pale. Dark gray back of ad. gradually acquired over winter; usually complete by spring; wing coverts pale brown (can fade to very pale); tertials darker brown; tail and outer primaries brown, fading over winter. **In flight,** dark gray back; pale brown wings (darkest at tip); white rump; dark brown tail. **1st Winter:** Bill at first all black, then acquires pale base by summer. Considerable fading between 1st winter and 1st summer: head and rump at first dark, then fade to whitish; tail fades to dull brown but is still all dark, except for some lightness at very base; lighter inner webs of outer primaries fade, making striped pattern more obvious; inner primaries fade, creating wing windows as on Herring Gull; inner wing fades with only a suggestion of dark secondary bar. **Juv:** Black bill; deep pink legs; dark eye. Pale brown overall, with dark wing tips and dark brown tertials. Muted barring or marking on wings and back. Differs from Herring Gull in having little or no barring on the greater coverts and tertials

3rd winter AK/06

3rd winter RUS/12

2nd winter JAP/02

2nd summer RUS/06

1st winter JAP/02

Juv. JAP/09

and having brown instead of black wing tips. **In flight,** darker outer webs of outer primaries create a striped effect; wing fairly uniformly dark, without marked paler window on inner primaries; secondaries dark. In contrast, Herring Gulls have more obvious pale window on inner primaries; outer primaries all dark. **Hab:** Ocean shores, estuaries, lakes, rivers, dumps, parking lots, fields. **Voice:** Long-call similar to Western Gull's but lower-pitched and slower.

Subspp: Monotypic. **Hybrids:** Glaucous-winged Gull, Herring Gull (subsp. *vegae*).

Adult summer CA/04

Adult summer CA/03

Adult winter CA/02

Adult winter CA/12

Glaucous-winged Gull 1

Larus glaucescens L 26"

Four-year gull. **Shape:** Large deep-bellied gull; bill stout with strong gonydeal angle, making it look bulbous-tipped; eye proportionately small; wings short and broad. Chest-to-leg vs. leg-to-wing-tip ratio about 1:2. **Ad. Summer:** Bill yellow with red gonydeal spot; legs pink to purplish pink; eyes dark (rarely light). Pale gray above, with ends of primaries about concolor with back and wings; primaries tipped white. Gray of mantle slightly darker than that of Glaucous Gull. Broad white tertial crescent and white visible "skirt" below wing coverts due to long white-tipped secondaries. **In flight,** wings uniformly gray above (rather than white-tipped as in Glaucous Gull); below, ends of outer primaries paler than their bases. Large white mirror on P10 and less well defined subterminal white dots on rest of primaries create "string of pearls" effect. **Ad. Winter:** Like ad. summer but with diffuse brownish-gray mottling on head and neck. **3rd Winter:** Like ad. winter, but bill pinkish to yellowish with dark subterminal band, tail with extensive dark streaking, and wing coverts with slight brown tinge. **In flight,** like ad. winter but with grayish tail band. **2nd Winter:** Bill mostly dark, with paler base increasing over winter. Back mostly pale gray; coverts and tertials brownish, only faintly marked, and concolor with primaries. **In flight,** back pale grayish and unmarked; wings pale brown with paler flight feathers; paler rump contrasts with diffuse grayish-brown tail. **1st Winter:** Bill all black. Uniformly grayish to grayish brown overall, including wing tips; relatively unmarked but may appear finely marked with white above. By 1st summer, birds fade to very pale and base of bill may start to lighten. **In flight,** above, whitish with variable brown spotting on back and wings, with a slightly darker tail (occasionally a subtly darker secondary bar). Below, evenly grayish brown, except for flight feathers, which appear paler and translucent in contrast to rest of wing. **Juv:** Similar to 1st winter but slightly darker overall. **Hab:** Coastal areas, estuaries, harbors, beaches, lakes, dumps. **Voice:** Long-call *chaaoo chaaoo kleeaah kleeaah kya kya kya;* alarm call *kaka* or *kakaka.*

Subspp: Monotypic. **Hybrids:** Western, Herring, Glaucous, Slaty-backed Gulls. Glaucous-winged x Western Gull hybrids are abundant in Puget Sound area and are the most commonly seen hybrids in California in winter. Glaucous-winged x Herring Gull hybrids are common in Cook Inlet (Anchorage, AK) region and frequent in CA in winter. Hybrids can represent all intergradations between parent species, making identification difficult or impossible. In many cases, one can only make an educated guess that an individual is a mixture of two species.

3rd winter CA/11

3rd winter RUS/12

2nd winter RUS/06

2nd winter RUS/12

1st winter CA/12

1st winter CA/02

Juv./1st winter CA/12

Adult summer AK/06

Adult summer NF/02

Adult winter AK/03

Adult winter RUS/12

Glaucous Gull 1

Larus hyperboreus L 27"

Four-year gull. **Shape:** Very large broad-winged gull. Bill long with little or no gonydeal angle, making it look straight and of even depth (not deep and bulbous-tipped like Glaucous-winged's and Western's); crown somewhat flattened and long; eye comparatively small. Chest-to-leg vs. leg-to-wing-tip ratio about 1:2. **Ad. Summer:** Bill yellow with red gonydeal spot; legs pink (at all ages); eye yellow. Wing tips paler than rest of wing in all ages. Back pale gray, wing tips all white (thus no mirrors or lighter tips to primaries as in other gulls) and blending with white tertial crescent. **In flight,** all pale gray above, with paler unmarked white wing tips. Below, uniformly white. **Ad. Winter:** Like ad. summer except with variable brown streaking on head and neck; can be extensive, creating a brownish hooded look, or minimal, limited to crown and nape. In 4th winter, may still have some dark on bill. **3rd Winter:** Like ad. winter, but bill pale yellow with dark subterminal band; some beige on wing coverts. **In flight,** like ad. winter, but note pale bill with dark subterminal band and some beige on wing coverts above and below. **2nd Winter:** Like 1st winter, but extreme tip of bill may be pale; dark eye turns pale over winter. Increasingly pale gray on mantle over winter. **In flight,** like 1st winter. May be hard to distinguish between 1st and 2nd winter. **1st Winter:** Bill pale with well-defined and extensive black tip (unlike Glaucous-winged or Iceland Gulls); eye dark; legs pink. Body color variable, from pale mottled tan to light brown overall with lighter wing tips to almost all white; generally gets whiter over winter and into summer. In either case, brown flecks or bars always present on wing and undertail coverts and scapulars. **In flight,** mottled pale beige above; below, primaries and secondaries unmarked, translucent, and paler than rest of wing. Note bicolored bill. **Juv:** Like 1st winter but darker. Note dark-tipped pink bill (except all dark when very young), which distinguishes this species from most other juvenile gulls, which have all-dark bills. **Hab:** Coastal areas, estuaries, harbors, beaches, lakes, dumps. **Voice:** Long-call *gaao gaao keeeah keeeah kya kya kya;* alarm call *kekeke.*

Subspp: (3) •*hyperboreus* (n. CAN) large, with medium gray mantle, long thin bill. •*barrovianus* (w. and n. AK) medium-sized, with darkest mantle. •*pallidissimus* (w. AK islands) larger, with pale gray mantle, relatively short thick bill. **Hybrids:** Herring, Glaucous-winged, Great Black-backed Gulls.

3rd winter MB/06

3rd winter RUS/12

2nd winter NF/02

2nd winter RUS/12

Juv./1st winter CA/12

1st winter CA/02

Juv. NH/12

Adult summer USA/03

Adult winter USA/03

Adult winter NF/02

Adult winter NJ/01

Great Black-backed Gull 1

Larus marinus L 25–31"

Four-year gull. **Shape:** Very large, deep-chested, deep-bellied gull. Head long with flattened crown; eye proportionately small; bill massive, with moderate gonydeal angle creating a bulbous tip. Legs relatively long and thick. Chest-to-leg vs. leg-to-wing-tip ratio 1:2. World's largest gull. **Ad. Summer:** Bill pale yellow with red gonydeal spot (sometimes additional dusky spot); legs pale pinkish (at all ages); eye pale yellow to dark. Back and wings dark blackish gray; primary ends black with white tips. **In flight,** broad dark blackish-gray wings with black outer primaries; broad white trailing edge to secondaries; large white mirror on P10 contiguous with mirror on P9. Below, gray primaries and secondaries contrast with white coverts. **Ad. Winter:** Almost identical to ad. summer; minimal streaking on head and neck; bill often more muted yellow. **3rd Winter:** Like ad., but bill pale yellow with subterminal black and red markings (red especially toward spring); legs pink; eye pale. Back and median coverts dark gray, rest of wing dark brown. Body white with some faint brown streaking on nape. **In flight,** lesser coverts of wings brownish; some patterning on underwing caused by darker tips to underwing coverts; tail with variable remnants of dark band of earlier plumages; mirrors on PP9–10. **2nd Winter:** Like 1st winter, but bill pale pinkish at base with dark tip and dark along cutting edge; eye dark; greater coverts buffier and more finely barred than 1st winter. May be a few dark gray markings on back, suggesting adult color, but these are more common in 2nd-summer birds. **In flight,** like 1st winter, but may have small mirror on P10. **1st Winter:** Like juv., but back feathers barred or with darker tips, muting the checkering of the juvenal plumage; head whiter than juv.; some paleness to base of bill (juv. bill all dark). **Juv:** Bill black with pale extreme tip; legs pinkish; eye dark. Head mostly white; body whitish with streaking along sides of chest and flanks; back and wings checkered dark brown and white, giving "salt and pepper" look. **In flight,** tail with black terminal band and black-and-white mottling on rest of tail; rump white. Above, dark secondaries contrast with lighter greater coverts (Lesser Black-backed has dark secondaries and dark secondary greater coverts). **Hab:** Ocean shores, estuaries, lakes, rivers, dumps, fields. **Voice:** Long-call low-pitched *kwooh kwooh kwooh kleeah kleeah kya kya kya;* alarm call *keh keh keh.* Lower-pitched than similar Herring Gull sounds.

Subspp: Monotypic. **Hybrids:** Herring and Glaucous Gulls.

3rd winter NH/09

3rd winter –/–

2nd winter NH/09

2nd winter FL/01

1st winter FL/01

1st winter NH/10

Juv. NH/09

Adult CHI/12

3rd winter TX/01

Adult CHI/12

Kelp Gull 4

Larus dominicanus L 24"

Four-year gull of Southern Hemisphere; has bred in Gulf of Mexico, off LA, casual elsewhere. **Shape:** Large, slim-bellied, broad-winged gull with a relatively rounded crown (especially f.) and a short thick bill with a strong gonydeal angle. Chest-to-leg vs. leg-to-wing-tip ratio about 1:2. **Ad:** Adult has roughly same plumage all year. Bill yellow with red gonydeal spot; legs dull greenish to pale yellowish (Great Black-backed Gull has pink legs; Yellow-footed has bright yellow legs); eye varies from pale yellow to dark brown. Very dark black back shows little contrast with dark primary tips. **In flight,** one small mirror on P10; broad white trailing edge to secondaries and inner primaries. Below, dark flight feathers contrast with white underwing coverts. **3rd Winter:** Like ad., but bill pale with variable dark subterminal band; variable dark spotting on white head and chest; variable traces of black on tail; some brown on upperwing coverts. **In flight,** similar to ad. except for variable traces of black on tail and some brown on upperwing coverts; may lack mirror on P10. **2nd Winter:** Bill with pale base and dark tip. Back with variable amount of brown and ad. gray; head and body whitish with moderate streaking on neck and breast; belly spotted with brown; greater coverts contrastingly dark brown and unpatterned; tertials dark brown with lighter tips. **In flight,** tail all dark with fine white border; rump all white or slightly barred; wings dark brown with fine white trailing edge. **1st Winter:** Bill with pale base, dark tip and cutting edge. Grayish back with darker feathers interspersed; head and body whitish with slightly darker face mask; breast and belly mottled with pale brown dots. **Juv:** Bill all dark; legs dark; eye dark. Very dark brown all over, with dark relatively unbarred greater coverts and tertials; dark pale-fringed feathers on back and on lesser and median coverts create scaled look. **In flight,** all dark brown except for whitish lightly barred rump; dark tail finely edged with white; bill all dark. **Hab:** Ocean shores, estuaries, lakes, rivers, dumps, parking lots, fields. **Voice:** Similar to Herring Gull's but lower-pitched.

Subspp: (1) •*dominicanus.* **Hybrids:** Herring Gull.

2nd winter ARG/11

2nd winter NZE/11

1st winter NZE/12

1st winter CHI/02

Juv. NZE/01

Juv. CHI/11

Adult MEX/12

Adult MEX/–

2nd winter MEX/01

1st winter MEX/01

Juv./1st winter MEX/01

Yellow-footed Gull 2

Larus livens L 25"

Three-year gull. Uncommon gull of s. CA coast but common in Salton Sea, most numerous in summer and early fall. **Shape:** Similar shape to southern subsp. of Western Gull, *wymani,* but with a slightly thicker bill. ▶This is a large gull that takes only 3 yrs. to mature (other large gulls take 4 yrs.). Breeding and molt cycle 3–4 months ahead of our other large gulls, thus terms "summer" and "winter" in this case are relative. Adult has roughly same plumage all year. **Ad:** Bill bright yellow with red gonydeal spot; legs bright yellow; eye yellow. Back dark gray; wide white tertial crescent; broad white edges of secondaries show as white "skirt" below greater coverts. **In flight,** from above, broad dark gray wings with wide white trailing edge to secondaries; yellow legs; small mirror on P10, occasionally one on P9. Below, extensively dark primaries and secondaries contrast with white wing coverts. **2nd Winter:** Similar to ad., but bill pale with dark well-defined subterminal band and pale extreme tip; legs yellowish (can be tinged pink); some brown on wing coverts; some light streaking on face and collar. **In flight,** wings dark gray with no mirrors; may be traces of dark on tail. **1st Winter:** Like juv., but bill dark with usually bright yellow base to lower mandible; legs pinkish (can be tinged yellow); some adult gray on back (increasing over winter, extensive by summer); head and underparts paler. **In flight,** similar to juv. **Juv:** Bill all black; feet and legs pink; eye dark. Dark grayish-brown back and wings; pale head with darker mask over eye and auricular. Streaked brown breast and flanks, but distinctively whitish on belly and undertail coverts. **In flight,** wings fairly uniformly dark above and below; white rump contrasts with black tail and dark back. **Hab:** Coastal areas, estuaries, fields. **Voice:** Long-call *kyoh kyoh kyoh kleeah kleeah kleeah kya kya kya;* alarm call *kekeke.* Lower-pitched than similar Western Gull sounds.

Subspp: Monotypic.

Adult FL/05

1st summer FL/04

Adult ANT/–

Adult FL/04

Brown Noddy 2

Anous stolidus L 15"

Breeds on Dry Tortugas; vag. to Gulf and Atlantic coasts. **Shape:** Medium-sized short-legged tern with a fairly long head and thick neck; bill fairly long (equal to or slightly shorter than length of head), fairly deep-based, with slightly curved culmen. Tail appears pointed when closed. Similar Black Noddy is slightly smaller and has a smaller rounded head; thinner and longer neck; and long, very thin, straight bill (longer than head). **Ad:** Dark brown overall except for blackish primaries; white forehead blends smoothly into gray nape; black primaries contrast with brown tertials and back. **In flight,** secondary upperwing coverts slightly paler and browner than rest of blackish wing; underwing coverts appear paler than body (wings all dark in ad. Black Noddy). **1st Summer:** Like ad., but forehead variable—brownish to smudgy brown to mostly white; crown and nape brown; may have a thin whitish supraloral line; upperwing coverts often pale from wear. **Juv:** (Jun.–Mar.) Like 1st summer but back and wing coverts fringed pale; forehead ashy but crown and hindneck dark brown; often a thin white supraloral line. **Hab:** Open ocean, ocean islands. **Voice:** Barking sounds like *ark ark ark;* juv. gives high-pitched *tseee.*

Subspp: (1) •*stolidus.*

Adult FL/04

Adult –/03

Juv., *melanogenys** HI/04

Adult summer RUS/06

Adult summer AK/06

Black Noddy 3
Anous minutus L 13"

Rare vag. to Dry Tortugas; accidental to TX coast. **Shape:** Medium-sized short-legged tern with a small rounded head and thin neck; bill long (longer than length of head), thin-based, with fairly straight culmen. Tail appears pointed when closed. See Brown Noddy for comparison. **Ad:** Blackish overall (dark brown when worn); extensive white on forehead contrasts with gray hindcrown and nape, with more abrupt transition than the even blending on Brown Noddy. **In flight,** wings appear all dark. **Juv/1st Summer:** (1st yr.) Like ad. but with faint buffy fringes to feathers of back and wing coverts; distinct white forehead and forecrown end fairly abruptly at brownish nape. In 1st summer, can appear brownish and often has paler brownish worn upperwing coverts. **Hab:** Open ocean, ocean islands. **Voice:** A harsh *garrr;* juv. gives high-pitched *sweee.*

Subspp: (1) •*americanus.* **A. m. melanogenys* (HI) used for good photograph.

Aleutian Tern 2
Onychoprion aleuticus L 12"

Shape: Medium-sized, fairly short-billed, short-legged tern. **Ad. Summer:** Bill and legs black. Dark cap with embedded white forehead; medium gray body and wings contrast with white cheek and tail. **In flight,** white rump and tail contrast with gray back and belly. Below, dark secondaries create distinctive dark trailing edge to inner wing; outer primaries broadly darker at tips; outermost primary with dark inner web. **Ad. Winter:** Does not winter in NAM. Underparts white, collar white. Black cap reduced to a thin black mask from eye to eye around nape. **In flight,** like ad. summer but leading edge of inner wing may be dark. **Juv:** (Aug.–Sep.) Bill dark with paler base to lower mandible; legs orangish. Brownish cap; dark-centered back and tertial feathers edged with reddish orange; cinnamon collar and sides of breast. **In flight,** dark brownish back above; dark gray lesser and median coverts contrast with medium gray rest of wing. **Hab:** Sparsely vegetated coastal islands, estuaries, freshwater marshes. **Voice:** A thin clear whistle, a grating *grrr,* and a short *chit.*

Subspp: Monotypic.

Adult, *fuscatus* FL/–

Adult, *fuscatus* FL/04

Juv., *serratus** AUS/12

Adult, *fuscatus* FL/04

Imm., *fuscatus* NC/08

Sooty Tern 2

Onychoprion fuscatus L 16"

Shape: Medium-sized, slender-bodied, long-billed, long-legged tern with a fairly long shallow crown. Tail long and deeply forked; projects beyond wings in summer, shorter than wings in winter. **Ad. Summer:** Bill and legs black. Black cap with an embedded white forehead that extends to eye. Dark loral area narrows down to bill (similar Bridled Tern has even-width loral line). Strongly black and white. **In flight,** primaries all blackish below (Bridled has whitish bases to underside of primaries); above, tail with thin white edges that are barely visible from a distance (Bridled has extensive white on outer tail feathers). Flight strong and steady. **Ad. Winter:** Very similar to ad. summer, but with mottled nape and broken loral line. **Imm:** Similar to ad., but with some dark spotting on belly and/or flanks for up to 5 years. **Juv:** (Jun.–Dec.) Legs and bill black. Dark overall except for whitish rear belly and undertail coverts; white tips to back and wing covert feathers create spotted look. **In flight,** like ad. but underwing coverts and axillaries variably spotted blackish (all white in ad.). Tail shorter but still strongly forked. **Hab:** Open ocean, tropical islands, sparse rocky or beach areas. **Voice:** Loud *widawayk;* alarm call a low harsh *wup.*

Subspp: (2) •*fuscatus* (Atlantic is., FL–TX) slightly smaller, mostly white underparts; •*crissalis* (e. Pacific is., accidental to s. CA) slightly larger, underparts tinged gray. **O. f. serratus* (AUS) used for good photograph.

Adults, *recognita* FL/05

Juv., *recognita* NC/09

Adult, *anaethetus** HKO/07

Bridled Tern 2

Onychoprion anaethetus L 14"

Shape: Medium-sized, slender-bodied, long-billed tern with a fairly long, shallow crown. Tail long and deeply forked; projects beyond wings in summer, shorter than wings in winter. **Ad. Summer:** Bill and legs black. Black cap with an embedded white forehead that extends well past eye. Dark loral area evenly wide down to bill (similar Sooty Tern's loral line narrows to bill). Generally tricolored—black cap, brownish-gray back, white underparts (Sooty Tern is all black and white). **In flight,** black cap contrasts with grayish back; white collar usually separates cap from back (black cap and black back concolor and connected on Sooty). Below, dark primaries have whitish bases (all dark in Sooty); above, tail gray with white (can be extensive) on outer feathers. Often seen sitting on flotsam well offshore (similar Sooty not seen doing this). Flight buoyant and erratic. **Ad. Winter:** Like ad. summer, but with dark markings on head variably paler—loral area may be whitish, crown may be streaked. **1st Winter:** Like ad. summer, but head may be white with just a hint of gray on the hindcrown; back and wing coverts may become paler with wear. **Juv:** (Jun.–Oct.) Pale margins to feathers of back and wing coverts create scaled look; crown streaked gray and white. **In flight,** wings from below as in adult. **Hab:** Open ocean, small ocean islands, sparse rocky or beach areas. **Voice:** A loud *greer;* alarm call a repeated *wap,* often rising in pitch.

Subspp: (2) •*recognita* (Atlantic and Gulf of Mex.) smaller; underparts white in summer, upperparts brownish gray. •*nelsoni* (vag. to s. CA and AZ) larger; underparts tinged gray in summer, upperparts sooty. **O. a. anaethetus* (Taiwan to Indonesia and Philippines) used for good photograph.

Adult summer CA/08

Adult summer TX/04

Adult winter TX/09

Juv. CT/07

Juv. CA/08

Least Tern 1

Sternula antillarum L 9"

Shape: Aptly named, this is our smallest tern. Bill proportionately long and thin. Head flattened on crown and squared off in back. **Ad. Summer:** Bill yellowish with variable small black tip (rarely all yellow); legs reddish orange. Crown black with embedded white triangular patch on forehead. Outer 2 primaries and their coverts black and seen as thin black line on perched birds. **In flight,** generally pale overall except for distinctive black outer 2 primaries and their coverts, most obvious from above; tail gray with white outermost feathers. **Ad. Winter:** Seen in late summer, just before birds leave U.S. and Canada in fall. Bill black; legs grayish yellow to blackish. White forehead sometimes variably speckled with black; black half-hood extends to eyes. **In flight,** dark lesser coverts create line along leading edge of wing, sometimes seen on perched birds. Outer 2 primaries dark gray after molt in midwinter, but soon become black with wear; yellow bill acquired by late winter. **1st Summer:** Like ad. winter, with black bill all summer. Cap may be reduced to a thin black mask as on juvenile. **Juv:** (Jul.–Sep.) Bill dark with pale base to lower mandible; legs variably orangish. Thin black mask from eyes back to nape; dark subterminal bars on back create scaled look; dark carpal bar shows on perched bird. **In flight,** dark outer primaries and dark leading edge to inner wing. **Hab:** Sparsely vegetated sandy or muddy areas along rivers and coasts; bays, estuaries. **Voice:** A short *kidik* often given by m. carrying food; alarm calls include a rising *chreeet*, a clipped *kit kit kit*, and a deeper *croog*.

Subspp: (3) Distinctions minor, mostly geographic. •*antillarum* (Atlantic and Gulf coasts) has comparatively longer and thicker bill. •*athalassos* (Mississippi watershed, Rio Grande) has comparatively short bill. •*browni* (CA) has comparatively thin bill.

Adult summer GRE/05

Adult summer TX/03

Adult winter TX/11

Juv./1st winter NY/08

Juv. NJ/07

Juv. NJ/07

Gull-billed Tern 1

Gelochelidon nilotica L 13½"

Shape: Medium-sized comparatively long-legged tern with a rounded crown and nape and a short, deep-based, tapered bill. No projecting crest. Long finely pointed outer wing; relatively short tail with shallow notch. **Ad. Summer:** Bill black; legs brownish orange to black; cap and nape black; cap broadly joins bill base. Pale gray upperparts; white underparts. **In flight,** uniformly pale gray above; white below; ends of outer primaries moderately tipped with black. Flight is deliberate and gull-like. Does not plunge-dive after fish like most other terns; prefers to pick up insects and other food items from water's surface, mudflats, and fields. Generally does not gather with other terns or gulls. **Ad. Winter:** Like ad. summer, but head white with small variable black mask behind eye (never as large as on winter Forster's Tern). Molt of black cap occurs over whole cap, creating a black-and-white peppered look in midmolt, rather than starting in front like many other terns. **1st Winter:** Like ad. winter, but tertials with dark markings and primary tips more extensively dark. Bill may be slightly shorter than on adult. **In flight,** like winter adult, but outer primaries increasingly dark into 1st summer. **Juv:** (Jul.–Sep.) Like 1st winter, but with variable pale base to bill and buffy wash over crown, back, and lesser and median coverts. **Hab:** Coastal marshes, sandy beaches, ocean inlets, dredged spoil islands, mudflats. **Voice:** Upslurred *kaweeek;* alarm call a staccato series like *ah ah ah ah ah.*

Subspp: (1) •*aranea.* **Hybrids:** Forster's Tern.

Adult summer CA/04

Adult summer CA/04

Adult winter FL/03

1st winter TX/04

Juv. OH/09

Juv. CA/07

Caspian Tern 1

Hydroprogne caspia L 21"

Shape: Large, short-necked, fairly deep-bellied tern with a thick bill that tapers abruptly to tip (proportionately shorter and more abruptly tapered than in similar Royal Tern). Head with little or no shaggy crest (appears to have a crew cut). **Ad. Summer:** Bill deep red (faint dusky subterminal band and small yellowish tip); legs black (may be mottled or all orange to yellow). Cap all black for extended period (late spring to early fall); white gap between cap and gape. Pale gray back and wings; white underparts. **In flight,** underwings distinctively dark for outer third; upperwings with dark gray trailing edge to outer primaries. Wings broad and flight deliberate. **Ad. Winter:** Like ad. summer, but area of cap variably speckled with black and white; this speckling extends to gape. Bill dull red with slightly broader dark subterminal band. **In flight,** like ad. summer, but outer primaries may be darker from above. **1st Winter:** Like ad. winter but primary tips darker. **In flight,** like ad., but darker secondaries form dark trailing edge to inner wing; also, outer primaries and their coverts are dark. **Juv:** (Jul.–Oct.) Bill shorter, more orange, and with darker subterminal band than in ad.; legs at first brownish yellow, turning black by late fall. Speckled black-and-white cap has buffy wash overall. Feathers of back and wing coverts with dark gray subterminal V's. **Hab:** Coastal beaches, salt marshes, islands; interior river islands and freshwater and salt lakes. **Voice:** Drawn-out raspy *schaaar;* young birds give high-pitched *whiiiew.*

Subspp: Monotypic.

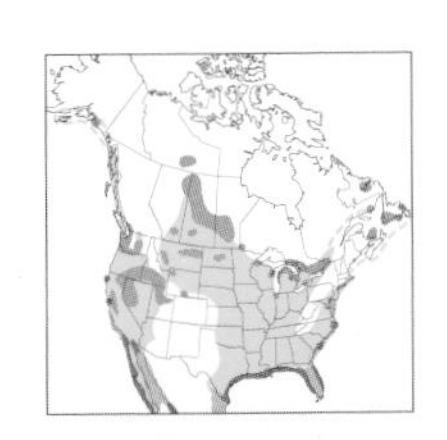

Adult summer ND/06

Adult summer ND/06

Adult winter TX/09

Adult, trans. to winter MB/08

Black Tern 1

Chlidonias niger L 9½"

Shape: Small, thin-billed, very short-tailed tern. Wings relatively broad; tail slightly notched. Bill slightly longer than visible portion of legs. **Ad. Summer:** Bill black; legs black (sometimes with reddish to pinkish tinge). Black head and underbody contrast with white rear belly and undertail coverts; gray back, wings, and tail. **In flight,** from below, black body contrasts with white undertail coverts; underwing coverts light gray; secondaries and primaries subtly darker gray. Above, wings dark gray; tail and back concolor gray. (White-winged Tern has pale gray upperwings, black underwing coverts, contrasting white tail and rump.) **Ad. Winter:** Underparts all white; white forehead and black cap with "earmuffs"; back, wings, and tail concolor gray (lesser coverts can be darker, especially with wear). **In flight,** wings like ad. summer, but body white below with distinctive dark spur on sides of upper breast. **Summer to Winter:** Starts body molt before migrating south, so transitional plumages with mottling on head and body are commonly seen in late summer and fall; head is molted first, followed by breast and belly. **1st Winter/Summer:** Birds seen in spring with winterlike plumage are probably 1st-yr. birds. This plumage kept all summer and gradually molted into ad. winter by fall. Differences from ad. winter are seen in flight on the upperwing, where darker lesser coverts and secondaries create dark edges to the inner wing and the dark outer primaries contrast with the fresher paler inner primaries. **Juv:** (Jul.–Nov.) Similar to ad. winter, but bill black with reddish base; legs pinkish; back brownish with variable darker barring; wing coverts with beige margins; distinctive "spur" on sides of breast (in all winter plumages also). **In flight,** similar to ad. winter. **Hab:** Freshwater marshes, prairies, ponds, plowed fields, lakes; coastal in winter. **Voice:** A *kewdik kewdikdik;* alarm call a high-pitched repeated *kik.*

Subspp: (1) •*surinamensis.* **Hybrids:** White-winged Tern.

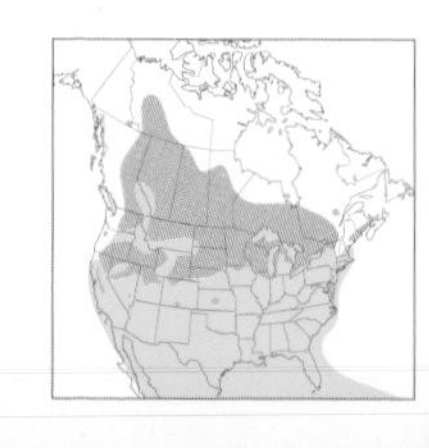

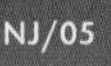
1st winter/summer NJ/05

Juv. CA/10

Juv. TX/09

IDENTIFICATION TIPS

Terns are small to large sleeker relatives of gulls; they have pointed bills, long thin wings, and, in many cases, long forked tails. Most terns are gray above, white below, with a black cap in summer; some are mostly dark. Most terns mature in 2 years, and their immature plumage is similar to ad. winter plumage but often with dark-centered tertials; juveniles usually have some buff or brown on their backs and heads at first. Most terns feed through agile diving into water to catch fish; some feed also on aerial insects.

Species ID: The size of the bird and the size and shape of the bill are good starting points with terns. Also note the colors of the bill and legs and the pattern of colors on the head.

Juv. TX/09

Adult summer GRE/05

Adult summer HKO/05

Adult winter HKO/09

Adult winter NAM/04

Juv. –/09

Juv. POL/07

White-winged Tern 4

Chlidonias leucopterus L 9½"

Casual vag. from EUR to either coast. **Shape:** Like Black Tern, but shorter bill, larger head, and thicker neck create a slightly stockier look overall. Bill slightly shorter than legs (the opposite in similar Black Tern). **Ad. Summer:** Bill black; legs red. Black head, back, and belly contrast with white rear belly and undertail coverts; white wing coverts contrast strongly with black body and back. Similar to Black Tern and best disting. in flight by underwing patterns and tail. **In flight,** distinctive black underwing coverts contrast with silvery flight feathers below. Above, pale gray wings and white tail contrast with black back; outer 1–4 primaries become contrastingly darker over summer. **Ad. Winter:** Bill black; legs reddish. Body white; wings and back pale gray; head white with small streaked gray cap and isolated black earspot. **In flight,** gray wings and back contrast with white rump and head above. Below, lacks dark "spur" on sides of breast (distinctive on winter Black Tern); variable dark edge to greater coverts may create thin black line through gray underwing. **1st Summer:** Like ad. winter, but may lack black line through underwing and may have some brownish darker juvenal feathers on the back and among tertials. **Juv/1st Winter:** Head with dark cap and "earmuffs" (much like juv. Black Tern). Back brownish (darker than in Black Tern) and contrasts strongly with rest of bird. Back feathers and wing coverts with beige margins, creating scaled look. In fall, white supercilium separates ear patch from cap. **In flight,** white rump that contrasts with dark back and pale gray tail above. Below, no distinct dark "spur" on sides of upper breast. Tail pale gray, with white outer web on outermost feathers. **Hab:** Freshwater marshes, prairies, ponds, plowed fields, lakes; coastal in winter. **Voice:** Similar to Black Tern's but harsher and lower-pitched.

Subspp: Monotypic. **Hybrids:** Black Tern, Whiskered Tern (*Chlidonius hybrida*).

Adult summer ME/06

Adult summer NJ/06

Adult winter NJ/09

Juv. USA/–

Roseate Tern 2

Sterna dougallii L 12½"

Shape: One of 4 similar medium-sized terns with no crests (with Forster's, Common, Arctic). Bill comparatively longer and thinner than in Common, Arctic, and Forster's Terns; bill shape accentuates relatively elongated head; long forked tail extends substantially beyond wing tips on perched bird; wings shorter than on Common, Arctic, and Forster's. **Ad. Summer:** Bill mostly all black; in early breeding develops red at base, which increases to almost half bill when chicks fledge; changes back to black in Aug.; legs reddish orange. Overall, our palest smaller tern; back and wings very pale gray; tail and underparts all white; subtle pinkish wash over breast and belly in early breeding (soon fades); outer 2–4 primaries blackish and visible on perched bird. **In flight,** very pale all over; white rump blends with whitish tail and very pale silver-gray back and upperwings; no dark trailing edge to wing from below; outer 2–4 primaries dark from above (wear darker over summer). Wingbeats stiff and comparatively rapid. **Ad. Winter:** Like ad. summer, but bill all black; legs dark; forehead white; may have faint dark bar on lesser coverts; subtle pinkish wash on breast. **1st Winter/Summer:** Like ad. winter; black legs. **In flight,** similar to ad. summer, but with dark secondaries and lesser coverts creating dark leading and trailing edge to inner wing. **Juv:** (Jul.–Aug.) Bill and legs black; feathers of back and wing coverts with dark subterminal V's, creating scaled look (juv. Common and Arctic Terns lack these dark markings); dark brownish cap wears lighter by fall; darkish carpal bar. **Hab:** Rocky coastal islands, barrier beaches, estuaries. **Voice:** A sharp *keeek,* a *chivik chivik,* and alarm calls like *kekekeke* and longer *keeep.*

Subspp: (1) •*dougallii.*
Hybrids: Common, Arctic Terns.

Adult summer, *hirundo* NH/07

Adult summer, *hirundo* NH/07

Adult summer, *hirundo* NH/07

Adult summer, *hirundo* NH/07

Common Tern 1

Sterna hirundo L 12"

Shape: One of 4 similar medium-sized terns with no crest (with Forster's, Arctic, Roseate). Bill moderately long (shorter than in Forster's and Roseate but longer than in Arctic); crown relatively high, slightly elongated, and well rounded; moderately long forked tail does not extend beyond wing tips on perched bird. **Ad. Summer:** Bill orangish red with dark tip (culmen may be mostly black in early breeding), becoming mostly red by midsummer (rarely all red); legs red (brightest during breeding). Back and wings light to medium gray; belly and lower breast pale gray; sides of face, neck, and upper breast whitish; cap black. **In flight,** wings white below, with a broad dark trailing edge to outer primaries and dark inner web of outermost primary. Above, outer primaries darker than rest of wing, increasingly so from about midsummer into fall. In spring, 1 or 2 central primaries may be darker and contrast with newer innermost primaries, creating a thin dark wedge on the midwing. White rump contrasts with gray back; tail white with dark outer webs to outermost feathers. Wingbeats smooth and measured. Similar Arctic Tern has narrow dark trailing edge to underside of outer primaries; primaries and secondaries translucent; from above, wing evenly gray; wingbeats spaced and snappy. **Ad. Winter:** Almost all Common Terns leave our area for winter. Winter

Trans. to adult winter, *hirundo* TX/09

1st summer, *hirundo* NH/08

Juv., *hirundo* ENG/07

Juv., *hirundo* NY/07

adult has shorter tail, black bill, white belly, white forehead, black from eye to eye around nape (Forster's has just black eye patches), dark legs (see Subspp.). Dark carpal bar may be seen on perched birds and on leading edge of inner wing in flying birds. **1st Summer:** Like ad. winter, with dark half-hood that extends from eye to eye around nape; white forehead; black bill (which can be red at base); white belly; dark carpal bar. **In flight,** dark outer primaries and their coverts; dark secondaries and lesser coverts. **Juv:** (Jul.–Nov.) Bill with black upper mandible, orangish-red to pinkish base on lower mandible; legs orangish. Black half-hood; brown forehead wears to white; back feathers with dark subterminal bars, brownish turning to gray; underparts white. Dark carpal bar shows on perched bird. **In flight,** leading and trailing edges of upper inner wing dark; outer wing becomes darker with age. Below, like adult. **Hab:** Coastal beaches and islands, salt or freshwater (less commonly) marshes, coastal inlets. **Voice:** Extended harsh *keeeeh,* shorter harsh *djah djah,* and repeated *kip kip kip.* 🎧**42**.

Subspp: (2) •*hirundo* (throughout NAM range) as described above; •*longipennis* (Siberian vag. to w. AK) slightly darker gray below, bill all black (Mar.–Aug.), legs reddish brown to black. **Hybrids:** Roseate, Arctic Terns.

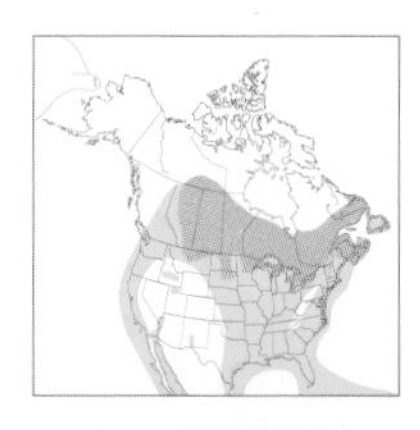

Adult summer NH/07

Adult winter –/03

1st summer NJ/06

Juv. ENG/06

Adult summer NH/07

Adult summer NH/07

1st summer NJ/06

Arctic Tern 1

Sterna paradisaea L 12"

Shape: One of 4 similar medium-sized terns without crests (with Forster's, Common, Roseate). Bill relatively short and thin; legs very short (shorter than bill); head with a well-rounded crown; tail usually extends well beyond tips of wings on perched bird (but can be even with it). **In flight,** wings are relatively long and thin and placed well forward on short body, and neck is short, making bird in flight seem short in front and long behind. **Ad. Summer:** Bill deep red, sometimes with dark shading at tip; legs deep red. Pale gray back and wings; pale gray on breast and belly usually extends up sides of neck, leaving a contrasting white stripe across cheek just below dark cap. **In flight,** from below, wings white with a fine black trailing edge to outer primaries; primaries and secondaries translucent. Above, wings evenly pale gray. White rump contrasts with gray back; tail white with darker outer webs to outer feathers. Wingbeats spaced and snappy. See Common Tern for comparison. **Ad. Winter:** Almost all Arctic Terns leave our area before winter and before molting into winter plumage, and molt into summer plumage before returning. Plumage like ad. summer, but bill and legs black; forehead and underparts white; may have slight dark carpal bar. **1st Summer:** Like ad. winter; dark half-hood extends from eye to eye around nape; white forehead; black bill and legs; white belly; faintly darker carpal bar. **In flight,** wing evenly light except for darker leading edge to inner wing (weaker than in Common Tern); secondaries white and lighter than rest of wing (unlike 1st-yr. Common Tern). **Juv:** (Jul.–Dec.) Brown on forehead wears to white; brown scaling on back wears to gray; black half-hood; white underparts. Bill black with orange to red base; legs variably reddish. Faintly darker carpal bar shows on perched bird. **Hab:** Open areas in tundra, usually near water, coastal beaches, marshes, meadows, glacial moraines. **Voice:** Fairly high-pitched, harsh, downslurred *djeeeah;* repeated *jip jip.*

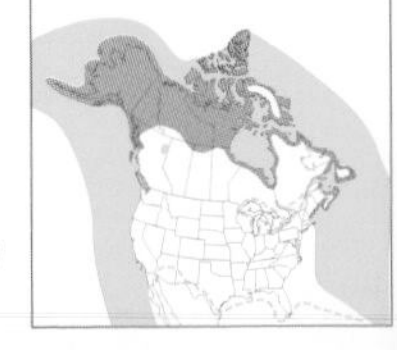

Subspp: Monotypic. **Hybrids:** Roseate, Forster's, Common Terns.

Adult summer FL/03

Adult summer AB/07

Adult winter NY/09

Adult winter FL/03

1st winter FL/02

Juv. NJ/07

Forster's Tern 1

Sterna forsteri L 13"

Shape: One of 4 similar medium-sized terns without crests (with Common, Arctic, Roseate). Bill relatively long and thicker than in Common, Arctic, and Roseate; legs longer than in Roseate and Common, much longer than in Arctic; head somewhat elongated, with fairly flattened crown; in summer, tail extends beyond wing tips on perched bird (in winter, may be shorter). **In flight,** head and neck extend well beyond wings; wings relatively short and broad; tail forked. **Ad. Summer:** Bill orangish red with substantial dark tip; legs orange to reddish orange. Back and wings pale gray; underparts all white (not grayish as in Common and Arctic); black cap. **In flight,** from below, dark trailing edge of outer primaries broad, ill-defined (grayer than on Common; broader than Arctic); little or no translucence in wings. Above, whitish outer primaries and secondaries paler than rest of wing. Primaries all pale in spring; over summer, outer primaries darken with wear, starting at tips. Tail grayish with dark inner web and white outer web to outermost feather (Common and Arctic have black outer web to outermost tail feather). **Ad. Winter:** Distinctive separate black mask over each eye (not connected by solid black around nape as in other similar terns); nape variably gray to speckled with black, but eye masks always darker and distinct. Underparts white; bill black; legs reddish brown. **In flight,** like ad. summer. **1st Winter:** Like ad. winter, but dark smudges on tertials. **Juv:** (Jul.–Nov.) Like ad. winter, with distinctive black eye masks; back and coverts brownish wearing to light gray; no dark carpal bar. **Hab:** Fresh- and saltwater marshes, beaches, estuaries, lakes, rivers. **Voice:** Fairly short harsh *djar djar;* also longer *jeeer jeeer.* 🎧**43**.

Subspp: Monotypic. **Hybrids:** Gull-billed, Arctic Terns.

Adult summer FL/03

Adult winter FL/02

1st winter FL/02

Juv. NJ/09

Adult summer TX/05

Adult winter FL/01

1st winter FL/02

Royal Tern 1

Thalasseus maximus L 18"

Shape: Large, slender-bodied, shaggy-crested tern with a long, fairly thin, deep-based bill and long flat-crowned head. Bill is straight (sometimes downcurved in Elegant Tern) and about ¾ length of head (often equal to or longer than head in Elegant Tern); bill is less deep-based and less abruptly tapered than in Caspian Tern; not as thin-based and evenly tapered as in Elegant Tern. Long narrow wings; tail moderately forked. **Ad. Summer:** Bill deepest orange to orange-red in early breeding, paler orange for rest of year (never darker at tip as in Caspian); legs black (sometimes mottled with orange). Cap and crest briefly all black during early breeding, but soon change to streaked crest with a white forehead and midcrown. Pale gray back and wings; white underparts. **In flight,** dark gray trailing edge to outer primaries below (not all-dark outer underwing like Caspian); above, outer primaries usually dark. **Ad. Winter:** Like ad. summer, but black cap reduced to shaggy loop around back of head; dark eye is separate and isolated from remains of cap. **2nd Winter:** Head and cap as ad. winter, tertials faintly smudged sooty (ad. tertials all pale gray). **In flight,** faint darker secondaries form a subtly darker gray trailing edge to inner wing; tail may have traces of gray. **1st Winter:** Head and cap as ad. winter, but tertials heavily smudged with dark. **In flight,** from above, dark secondaries, dark line of lesser coverts on front of wing, and variably dark outer wing; tail marked with gray. **Juv:** (Jul.–Nov.) Like 1st winter, but partial dark cap may connect with dark eye; greater and median coverts dark-centered with pale margins. **In flight,** 3 dark bars on inner wing created by dark secondaries, greater coverts, and lesser coverts; tail marked with gray. **Hab:** Coastal beaches, river mouths. **Voice:** Fairly low-pitched harsh *djaar* or 2-part *djarit.*

Subspp: (1) •*maxima.*

Adult summer FL/04

Adult winter FL/02

1st winter FL/03

Juv. UKM/08

Adult summer TX/04

Adult winter FL/02

1st winter –/11

Sandwich Tern 1

Thalasseus sandvicensis L 15"

Shape: Medium-sized crested tern with a very long, thin, evenly tapered bill. Bill longer than head and considerably longer than visible portion of legs. Crown relatively long, flat, and squared off to nape. Long narrow wings; tail moderately forked. **Ad. Summer:** Bill black with variable yellow tip (the "mustard on the sandwich"; see Subspp.), which can be difficult to see; legs black. Full black crown and crest during early breeding (sometimes forecrown spotted white), soon changing to white forehead in midsummer; back and wings pale gray; underparts white (may have pinkish wash). Our only crested tern with a black bill (Gull-billed Tern lacks crest and has noticeably thick bill). **In flight,** note bill colors; below, wings with substantial dark trailing edge to outer primaries. **Ad. Winter:** Like ad. summer, but forehead and crown mostly white (this plumage can be acquired as early as June). **1st Winter:** Like ad. winter, but bill may have less yellow at tip or be all black; tertials with dark centers or smudging. **In flight,** dark secondaries above; below, dark trailing edge of outer primaries narrower than in ad.; outer tail feathers gray, central tail white. **Juv:** (Jul.–Sep.) Bill shorter than ad., may be paler along gape, and has little or no yellow at tip. Feathers of back and wing coverts marked with dark V's at their tips; tertials variably mottled with dark. **In flight,** inner upperwing and back extensively marked with dark. **Hab:** Flat coastal islands or spoil islands near shore, beaches. **Voice:** Short harsh *djit djit* or longer *djaarit.*

Subspp: (2) •*acuflavida* (all NAM range) has black bill with yellow tip, slightly paler back, shorter crest. •*eurygnatha* (vag. from Caribbean, SAM), "Cayenne Tern," has all-yellow to yellowish-orange bill (sometimes dark marks mid-bill), slightly darker gray back, longer crest. **Hybrids:** Elegant Tern (hybrids may occur on West Coast).

Adult summer CA/06

Adult summer CA/06

1st winter CA/09

1st winter CA/10

Juv. CA/06

Elegant Tern 1

Thalasseus elegans L 16½"

Shape: Fairly large, slender-bodied, short-legged tern with a very long, thin-based, evenly tapered bill that may be slightly downcurved (longer and more downcurved in m. than f.). Head is relatively small with a fairly short rounded crown (Royal Tern has long flat crown). Crest very long, bushy, and continues down nape (only on crown in Royal). Bill about length of head and longer than visible portion of legs. Wings long and thin; tail moderately forked. **Ad. Summer:** Bill generally orange-red to yellow at base, blending to pale yellow to orange at tip (usually paler than in Royal Tern); legs black (may be mottled or solid orange). Black cap and crest; pale gray wings and back; white underparts, often with pink wash to underparts. **In flight,** primaries above pale to dark; below, outer primaries broadly tipped dark. **Ad. Winter:** White forehead; black on crest extends forward to usually include eye (eye clearly isolated by white in similar Royal Tern). **In flight,** like ad. summer. **1st Winter:** Like ad. winter, but tertials may be smudged with gray. **In flight,** from above, dark secondaries and dark lesser coverts create dark front and trailing edge to inner wing; outer primaries very dark. **Juv:** (Jul.–Oct.) Feathers of back and wing coverts brownish gray with pale edges, creating a scalloped effect; bill shorter and often paler than in ad.; legs range from yellow and red to black. **Hab:** Low, sandy or muddy, sparsely vegetated islands; spoil islands and dikes; estuaries. **Voice:** Adult gives harsh *keereek;* juv. gives a short trill.

Subspp: Monotypic. **Hybrids:** Sandwich Tern.

Adult, trans. to summer CA/01

Adult winter FL/01

1st winter FL/01

Juv., worn CA/11

Adult winter FL/02

Adult winter FL/01

Adult winter FL/01

Black Skimmer 1

Rhynchops niger L 18"

Shape: Large, long-winged, short-tailed, short-legged ternlike bird with an unmistakable bill that is long, deep-based, gradually tapered, and with the lower mandible much longer and protruding; bill looks thick from side but is razor-thin in cross section. Unique behavior of flying low over water skimming lower mandible through the water. Male bill longer and deeper than female's. **Ad. Summer:** Black cap, nape, back, and wings contrast with white forehead and underparts, giving the bird a "tuxedo" look. Bill with red base, black tip; legs orange-red. **In flight,** wings often held above body as bird skims its lower bill through water surface. Above, wings black with white trailing edge to secondaries and inner primaries; tail white with black central feathers. Below, white inner wing shades to dark outer wing. **Ad. Winter:** Similar to ad. summer, but nape is whitish, separating black crown from black back; back and wing feathers may look brownish from fading. **1st Winter/Summer:** Like ad. winter, but cap grayish; wing coverts in early winter with pale margins, creating scaled look; in late winter, mix of brown and black wing coverts; by 1st summer wing coverts considerably worn. **Juv:** (Jul.–Dec.) Shorter bill with orangish base shading to dark tip; legs orangish. Crown buffy with fine dark streaking; feathers of back and wing coverts dark-centered with broad buffy margins and browner and more checkered-looking overall than other ages. **Hab:** Primarily coastal beaches and sandbars with sparse vegetation; some inland lakes, like Salton Sea, CA. **Voice:** Short, low-pitched, barking sounds, like *vark vark.*

Subspp: (1) •*niger.*

Adult ENG/06

Adult ENG/07

Adult VA/02

Adult ENG/06

Great Skua 3

Stercorarius skua L 23"

Shape: Very large, heavy-bodied, broad-necked, short-tailed seabird with broad abruptly tapered wings. Head large; bill short and thick. About Herring Gull size. In flight, has hunchbacked, heavy-bodied look with short broad tail. **Ad:** Appears dark with large white crescent at base of primaries creating a white flash above and below. Variably reddish brown, dark brown, or pale brown overall with coarse paler buffy or cinnamon streaking on back and scapulars; finer pale streaking on neck; head generally darker than rest of body (except on some paler individuals); face dark with paler auricular. Upperwing coverts with pale spots; underwing coverts tipped reddish brown. **Juv:** (Sep.–Feb.) Dark reddish-brown body with darker head; dark wings with warm brown covert tips above and below; brownish barring on back; white crescent at base of primaries may be less extensive than on adult; tail coverts dark and unbarred. Similar juv. and imm. Pomarine Jaegers (dark and light morphs) have distinctive pale barring on tail coverts, subtle pale barring on underwings; white wing crescents often smaller; wings proportionately longer and thinner. **Flight:** Steady powerful wingbeats, gull-like but with slightly more flexible wing. **Hab:** Ocean. **Voice:** Silent at sea.

Subspp: Monotypic.

IDENTIFICATION TIPS

Skuas and Jaegers are large oceanic birds with pointed wings and fairly thin strongly hooked bills. Skuas have short squared-off tails and are mostly dark; jaegers have distinctive long central tail feathers as breeding adults and are either all dark (dark morphs) or dark above with light collar and belly (light morphs).

Species ID: Identification is challenging because the species look similar and immature plumages can last up to 3 years. Note size and shape of bill, body, and wings, and on jaegers the shape of central tail feathers. Try to see the extent of white at base of primaries; on adult light-morph jaegers, look for the shape of their dark cap, color of the bill, and degree and quality of shading on flanks and breastband. Along the coast, Pomarine and Parasitic Jaegers are the ones most likely to be seen from shore during migration (Parasitic most often); in interior North America, Long-tailed and Parasitic Jaegers are the species most often seen.

1st yr. NC/08

Adult intermediate morph AAR/02

Adult light morph SOR/12

Adult dark morph AAR/01

South Polar Skua 2

Stercorarius maccormicki L 21"

Shape: Very large, heavy-bodied, broad-necked, short-tailed seabird with broad abruptly tapered wings. Head large; bill short and thick. About Herring Gull size. In flight, has hunchbacked, heavy-bodied look with short broad tail. **Ad:** Light, intermediate, and dark morphs. In all morphs has uniformly grayish to grayish-brown dark wings, lacking the contrastingly paler and warmer brown coverts of the similar Great Skua. Large white crescents at bases of primaries create white flashes from above and below. Head generally concolor with underbody, although face may be darker; nape generally paler. Blackish underwing coverts. Light morph has pale grayish body, dark back and wings with whitish feather tips. Dark morph dark overall with paler nape and little or no brownish or reddish-brown tones (as seen on Great Skua). **Juv:** (Apr.–Jul.) Slate gray overall with darker wings; pale edges to upperwing coverts; bill with pale base and dark tip. Similar juv. and imm. Pomarine Jaegers (light and dark morphs) have distinctive pale barring on their tail coverts and subtle paler barring on their underwings; white wing crescents often smaller; wings proportionately longer and thinner. **Flight:** Steady powerful wingbeats, gull-like. **Hab:** Ocean. **Voice:** Silent at sea.

Subspp: Monotypic. **Hybrids:** Brown Skua.

Adult summer, light morph CA/10

Adult winter, light morph CA/10

Adult summer, dark morph RUS/06

2nd-yr. light morph NC/05

Pomarine Jaeger 1

Stercorarius pomarinus L 21"

Shape: Large, deep-bellied, deep-chested, broad-necked jaeger with broad inner wing and short abruptly tapered outer wing. Rounded forehead leads to long flattened crown; bill relatively long, thick, and markedly hooked. Tail (not counting streamers) broad and shorter than greatest wing width. Ad. and 3rd-yr. streamers (central tail feathers) long, broad, with rounded tips and twisted 90 degrees, looking spatulate from side. Even when short, they are broader and more rounded than those of Parasitic Jaeger. Largest jaeger. **In All Plumages:** Above, white shafts to outer 3–8 primaries; below, white base to outer primaries creates distinct white crescent; smaller second white "comma" often created by whitish base to primary greater coverts. Similar ad. Parasitic has single white crescent on underwing; ad. Long-tailed Jaeger lacks white patch on underwing. **Ad. Summer:** Bill pale-based, dark-tipped. Concolor back and wings. In flight, large white crescent on primaries above and often 2 white patches below (rare in Parasitic). Light and dark morphs; light morphs most common (up to 90%). Light morph with black cap extending to malar area and base of upper mandible, creating unique helmeted appearance. Collar yellowish white. Breastband complete, wide, and heavily mottled (probably f.) or incomplete, spotted, or absent (probably m.). Belly white with dark mottling or barring on flanks. Dark morph all dark with no barring on back or wings; bicolored bill. **Ad. Winter:** Like ad. summer but streamers molted and shorter as they grow back; bill grayish with slightly paler base to lower mandible. Light morph with black-and-white barring on uppertail and undertail coverts. Similar to imm. but underwings not barred. **Imm:** (2–3 yrs.) 1st summer similar to juv. but worn. 2nd and 3rd years like winter ad. but with pale barring on underwings; 2nd year has shorter tail streamers than 3rd yr. or adult. **Juv:** (Aug.–Feb.) Dull brown overall with thin white barring on underwing, undertail, and uppertail coverts; bill grayish with dark tip; lacks dark cap. Body and head vary from all dark brown (dark morph) to medium brown (intermediate morph) to pale brown body with whitish head (light morph). Streamers shortest of all juv. jaegers, rounded, and barely project past tail. **Flight:** Powerful, direct, often low over water; constant flapping; little arcing or gliding. Wingbeats slower than those of Parasitic. **Hab:** Usually seen 2–50 miles off coast. **Migration:** South from mid-Jul. to mid-Nov.; north from mid-Apr. through May. **Voice:** Two-part *whichew* sometimes followed by quavering *wahwahwahwah,* mostly on breeding grounds, sometimes at sea.

Subspp: Monotypic.

Adult summer, light morph AK/07

Juv./imm. CA/08

Adult summer, light morph CA/08

Adult summer, dark morph ENG/06

Parasitic Jaeger 1

Stercorarius parasiticus L 19"

Jaeger most often seen from shore along coasts. **Shape:** Medium-sized slender-bodied jaeger with moderately broad inner wing and gently tapered outer wing. Tail (not counting streamers) narrow and about equal in length to greatest wing width. Thin mildly hooked bill. Ad. streamers long, sharp-pointed, and narrow. Disting. from Pomarine Jaeger by more slender body, smaller head, thinner less-hooked bill, narrower inner wing, more gently tapered outer wing, and narrow pointed streamers. **In All Plumages:** Above, white shafts to outer 3–8 primaries; below, white base to outer primaries creates distinct white crescent (generally no double areas of white on underwing as seen on Pomarine). **Ad. Summer:** All-black bill. Concolor back and wings. Dark cap; forehead at base of bill narrowly whitish, forming pale spot. Light and dark morphs. Light morph with dark brownish-black cap that extends just over eye (not dipping strongly into malar area as on Pomarine). Collar yellowish; breastband, usually present, is a smooth gray wash; belly white; flanks unmarked (similar Pomarine has mottled breastband and flanks). Dark morph dark grayish brown overall; faintly darker cap and breastband may be visible. **Ad. Winter:** Like ad. summer but uppertail and undertail coverts barred black and white; variable barring on belly but none on underwing coverts. **Juv:** (Aug.–Mar.) Light to dark morphs. Rusty tinge to plumage overall, especially on underwing, leading edge of inner wing, edging to feathers of upperwing, and collar, which usually contrasts with darker head (other jaegers lack rusty tones). Single white patch at base of underwing primaries. Buffy barring on uppertail and undertail coverts. Streamers pointed and project just past rest of tail (see Shape). Pale and intermediate morphs have dark tail with whitish base. **Flight:** Swift and direct; much arcing and gliding. Chases of other birds can be long and aerobatic; shrug preening (a quick shake in flight) can be common. **Hab:** On migration may enter rivers and bays looking for gulls and terns to pirate. Most frequently encountered jaeger within 1–2 miles of shore. **Migration:** Southward Jul.–early Oct..; northward Apr.–May. **Voice:** Wailing *waaheeeya;* 2-part *keeya.* Silent away from breeding grounds.

Subspp: Monotypic.

Adult summer AK/06

Adult summer RUS/07

Adult summer, streamers growing RUS/07

1st summer ME/06

Juv. MA/08

Long-tailed Jaeger 1

Stercorarius longicaudus L 15"

Shape: Relatively small slender-bodied jaeger with long, narrow, finely pointed wings. Head rather short with rounded crown; bill short and thick. Tail (not including streamers) narrow and longer than greatest wing width. Ad. streamers (central tail feathers) very long, narrow, pointed, and flexible (can be broken or missing). **In All Plumages:** On upperwing, only outer 2–3 primaries have white shafts. **Ad. Summer:** Only occurs in light morph. Underwing all dark (no white crescent on primary bases as in other jaegers). Upperparts two-toned: smooth gray back and secondary coverts contrast with black trailing edge to flight feathers and primary coverts. Broad yellowish collar; breast white; belly and flanks grayish; black cap extends straight back from bill (not dipping into malar area as on Pomarine Jaeger) and extends over forehead to base of upper mandible (not stopping short of bill as on Parasitic). Bill all black; legs pale (dark on other ad. jaegers). **Ad. Winter:** Like ad. summer but tail streamers molted and shorter as they grow back; undertail and uppertail coverts barred black and white; gray wash on sides of neck may suggest breastband. **Juv:** (Aug.–Mar.) Occurs in light, intermediate, and dark (least common) morphs. Shows distinct whitish fringes, but not black bars, on back and all wing and tail coverts; boldest fringing of all juv. jaegers; more obvious on light morphs; from distance, fringing usually appears whitish (but never rusty as in Parasitic). Streamers shorter than in adult and blunt-tipped, but longest of all juv. jaegers. Limited white on upperwing (1–2 primaries), primary bases white on underwing. Light morphs have pale head and whitish belly; intermediate morphs pale grayish head and variable whitish belly; dark morphs dark gray to sooty brown overall. Other species of juv. jaegers are never as pale. In general, this is only juvenile jaeger that can have a whitish belly. Bill two-toned; pale with dark tip. **Flight:** Ternlike, aerobatic, graceful flight. **Hab:** Stays farthest out to sea of all jaegers during migration, usually 20–50 miles offshore (can be closer in or farther off). **Migration:** Along West Coast, south mid-Jul. through mid-Oct.; north mid-Apr. through mid-Jun. Flies along continental shelf. Occasionally inland at smaller bodies of water. **Voice:** Sharp *keek;* short *weeeah.*

Subspp: Monotypic.

Adult summer RUS/07

Adult winter MA/03

Adult winter ENG/11

Adult winter ENG/11

Dovekie 2

Alle alle L 8"

Shape: Very small, short-tailed, heavy-bodied seabird with broad neck, proportionately large head, and very short, stubby, conical bill. **Ad. Summer:** Black upperparts, head, and upper breast sharply contrast with rest of white underparts; thin white streaking on scapulars; white tips of secondaries form a white line across wing. **Ad. Winter:** Like ad. summer but chin and breast white; black cap that loops down over eye; black partial collar extends onto sides of breast. **Juv:** (Jul.–Oct.) Most individuals like ad. summer, but with black areas more flat and brownish; some individuals may have whitish throats and appear intermediate between ad. summer and winter. **Flight:** Fast with blurred wingbeats; underwings dark. **Hab:** Coastal cliffs in summer; at sea in winter. **Voice:** High-pitched *kik kik kik;* drawn-out, harsh, screamed *jjreeeah.*

Subspp: Monotypic.

Adult summer ME/08

Adult winter NJ/01

Juv./1st winter NC/04

Adult summer ME/08

Razorbill 1

Alca torda L 17"

Shape: Large heavy-bodied seabird with a large, deep, blunt-tipped bill (flattened laterally); adult has 1–3 shallow vertical grooves at bill tip. Disting. from murres by deep blunt-tipped bill and long tail that projects well past tips of folded wings. **Ad. Summer:** Black head, neck, and upperparts contrast sharply with clean white underparts. Bill black with thin white vertical bar near tip; another thin white stripe along lore. **Ad. Winter:** Like ad. summer but with white on throat that extends up behind eye; suggestion of dark line behind eye; little or no white line on lore. **Juv/1st Winter:** Similar to ad. winter but bill smaller and with no white vertical bar. **Flight:** Fast flight with rapid wingbeats. **Hab:** Coastal cliffs in summer; at sea in winter. **Voice:** Very low-pitched snoring *zzrooor.*

Subspp: (1) •*torda.* **Hybrids:** Common Murre.

Adult summer, *aalge*, bridled form ME/06

Adult summer, *aalge*, bridled form ME/–

Adult winter, *californica* AK/03

Adult summer, *californica* CA/07

Adult summer, *aalge*, bridled form ME/08

Adult winter, *californica* CA/10

Common Murre 1

Uria aalge L 17"

Shape: Large fairly slender-bodied seabird. Bill relatively long, thin, sharply pointed, and with culmen gently curved to tip. Feathering on culmen extends ⅓ length of gape; exposed culmen same length as lore. Forehead angles off culmen to short rounded crown. Comparatively short tail projects just past tips of folded wings (longer tail projection in Razorbill). Similar Thick-billed Murre's feathering on culmen extends ½ length of gape, exposed culmen ¾ length of lore, and forehead is almost continuous with angle of culmen, creating Roman-nosed look; also crown can be somewhat flattened. **Ad. Summer:** All-blackish-brown head, neck, and upperparts contrast sharply with rest of white underparts (flanks with indistinct dusky streaking). White breast meets blackish-brown throat in a rounded inverted U; bill black (rarely with subtle white line on gape). Some Atlantic birds (subsp. *aalge*) may have thin white eye-ring and thin white line off back of eye ("bridled" form). Similar Thick-billed Murre's white breast meets black throat in an inverted V and, during breeding, it has a thin white line on gape. **Ad. Winter:** Black crown, nape, collar, face, and line off back of eye; rest of head and underparts whitish. **Juv/1st Yr:** Much like ad. winter but bill shorter at first. **Flight:** Fast flight with rapid wingbeats. **Hab:** Coastal cliffs in summer; at sea in winter. **Voice:** Low drawn-out growls, like *grrraaahr.*

Subspp: (2) •*californica* (AK–cent. CA) has upperparts with brownish wash, generally no bridled form. •*aalge* (n.e. NAM) has upperparts more blackish brown, and a minority of individuals are bridled form. **Hybrids:** Thick-billed Murre, Razorbill.

Adults, summer, *arra* AK/–

Adult summer, *arra* AK/07

Adult winter, *arra* CA/10

Adult summer, *arra* AK/08

Thick-billed Murre 1

Uria lomvia L 17"

Shape: Large fairly slender-bodied seabird. Bill relatively long and deep-based, with culmen abruptly curved to tip. Extensive feathering on culmen extends ½ length of gape; exposed culmen ¾ length of lore. Forehead almost continuous with angle of culmen, creating Roman-nosed look, and crown can be somewhat flattened. See Common Murre for comparison. Comparatively short tail projects just past tips of folded wings (longer projection in Razorbill). **Ad. Summer:** Head, neck, and upperparts blackish, contrasting sharply with rest of white underparts (flanks unstreaked). White breast extends up black throat in an inverted V, especially when neck is extended; bill with thin white line along gape during breeding. See Common Murre for comparison. **Ad. Winter:** Like ad. summer but with white throat. **Juv/1st Yr:** Much like ad. winter but bill shorter at first. **Flight:** Fast flight with rapid wingbeats. **Hab:** Coastal cliffs in summer; at sea in winter. **Voice:** Low, drawn-out, buzzing *brzzzz.*

Subspp: (2) •*lomvia* (n.e. NAM) smaller; ad. winter with browner head and face, narrower white tips to secondaries. •*arra* (AK–BC–NT) larger; blacker head and face, longer more gradually tapered bill, broader white tips to secondaries. **Hybrids:** Common Murre.

IDENTIFICATION TIPS

Alcids, which include Dovekie, murres, guillemots, murrelets, auklets, and puffins, spend most of their time at sea and breed on remote islands and fairly remote mainland habitats. They are in general heavy-bodied birds with short wings and tails and fly with rapid wingbeats. They range in size from very small to large, and their plumages are mostly black and white or brownish. Many have breeding and nonbreeding plumages, and the sexes look alike. Alcids use their wings to propel them during dives. Shape and size are extremely useful identifiers for all species of alcids, especially the shape of bill and head.

Adult summer, *grylle* EUR/05

1st summer, *grylle* ME/06

Juv./1st winter, *grylle* NJ/12

Adult summer, *grylle* ME/08

Black Guillemot 1

Cepphus grylle L 13″

Shape: Medium-sized, short-tailed, small-headed seabird with a moderately long spikelike bill (bill longer than lore). Body looks rounded, head slender and pointed. Forehead gently sloping off culmen, creating a shallow angle. Similar Pigeon Guillemot has shorter bill, steeper forehead, more stocky look, dark underwings. **Ad. Summer:** Mostly black with large white oval patch on secondary coverts of upperwing; underwings have white coverts with trailing blackish flight feathers. Feet and legs red. **Ad. Winter:** Wings as ad. summer. Body grayish with variable dark speckling on head, crown, and around eye; variable dark barring on back and rump; variable dark mottling on flanks and sides of breast. Northern subsp. with mostly white body. **Juv/1st Winter:** Like ad. winter but with dark barring on secondary upperwing coverts and white tips to primary greater coverts. May also be more heavily marked with dark than adult. **1st Summer:** Like ad. summer but with some black barring on upperwing coverts (resembling Pigeon Guillemot). **Flight:** Usually low over water; rapid wingbeats; note underwing colors mentioned above. **Hab:** Coastal waters and cliffs. **Voice:** Very high-pitched, faint, squeaking *ssseet ssseet.*

Subspp: (2) •*mandtii* (n. AK–n. LB–n. MB–ON) body mostly white in winter. •*grylle* (cent. LB–MA) as described above; body whitish with variable darker markings in winter.

Adult summer CA/07

Juv./1st winter AK/02

1st summer CA/08

Juv./1st winter AK/08

Adult summer CA/07

Pigeon Guillemot 1

Cepphus columba L 13½"

Shape: Medium-sized, short-tailed, small-headed seabird with a fairly short spikelike bill. Bill length about equal to lore. Bill abuts fairly steep forehead, creating obvious angle. Body looks rounded, head compact and rounded. Similar Black Guillemot has longer bill, longer head, less steep angle where forehead meets bill, more attenuated look. **Ad. Summer:** Mostly black with large white oval patch on upperwing secondary coverts penetrated by a dark wedge formed by dark bases to greater coverts. Underwings are fairly concolor gray. Feet and legs red. **Ad. Winter:** Wings as ad. summer. Body whitish with variable dark speckling on head, crown, and around eye; variable dark barring on back and rump; and variable dark mottling on flanks and sides of breast. **Juv/1st Winter:** Similar to ad. winter but with much duskier head, sides of breast, and flanks. **1st Summer:** Like ad. summer but with multiple dark bars on white upperwing coverts. **Flight:** Usually low over water; rapid wingbeats; gray concolor underwings (Black Guillemot has two-toned underwings). **Hab:** Coastal waters and cliffs. **Voice:** Very high-pitched soft *seep seep seep*.

Subspp: (1) •*columba*.

Adult summer WA/06

Adult winter AK/02

Juv. BC/08

Adult summer CA/07

Juv./1st fall ENG/11

Juv./1st fall KY/10

Marbled Murrelet 1

Brachyramphus marmoratus L 9¾"

Shape: Small stocky body, relatively large head, and thick neck typical of murrelets. Comparatively short thin bill. **Ad. Summer:** Fairly uniformly dark brown overall with reddish-brown tones; crown and upper face may appear darker, suggesting a cap. **Ad. Winter:** Generally black above, white below, with white scapular line. Black on head and nape meets white throat and neck in an uneven line, looping below eye, and greatly indenting on nape, creating a white wedge on the side of the neck. Black on shoulder extends onto sides of breast. Similar Long-billed Murrelet has more even line where white meets black on face, and black on shoulder extends less onto sides of breast. **Juv:** (Aug.–Oct.) Similar to ad. winter but with dusky scapulars, breast, and flanks. **Flight:** Rapid flight with fast wingbeats; underwings all dark (Long-billed underwing has whitish greater coverts). **Hab:** Coasts and, during nesting, a bit inland; nests on ground or in trees. **Voice:** High-pitched downslurred *tseeer tseeer.*

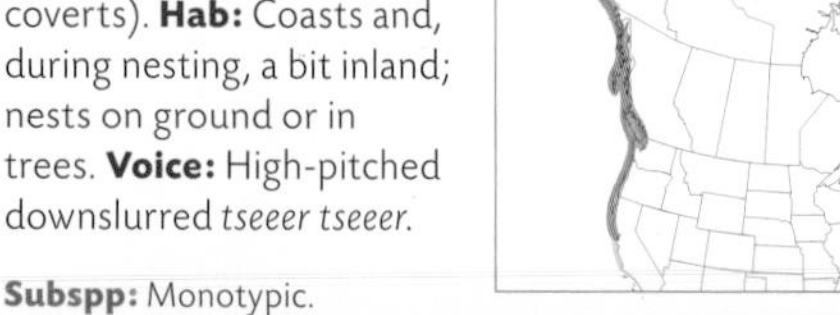

Subspp: Monotypic.

Long-billed Murrelet 3

Brachyramphus perdix L 10"

Rare Siberian vag. possible anywhere in NAM late summer to early winter. **Shape:** Small stocky body, relatively large head, and thick neck typical of murrelets. Comparatively long thick bill. **Ad. Summer:** Fairly uniformly dark brown with a paler throat. **Ad. Winter:** Generally black above, white below, with white scapular line. Black on head and nape meet white throat and neck in fairly smooth continuous line from base of bill to just below eye and down side of neck; black on shoulder does not extend onto sides of breast; faint pale patches on either side of wide dark nape. Similar Marbled Murrelet has much more uneven line of white on face where it meets black, looping well below eye, and greatly indenting on nape; also, black on shoulders usually extends more onto sides of breast. **Juv:** (Aug.–Nov.) Similar to ad. winter but with pale edges to back and head feathers, creating barred look. **Flight:** Rapid flight with fast wingbeats; underwings with whitish greater coverts (Marbled has all-dark underwings). **Hab:** Coasts; vagrants on inland lakes.

Subspp: Monotypic.

Adult summer AK/06

Adult winter BC/–

Adult summer AK/06

Kittlitz's Murrelet 2

Brachyramphus brevirostris L 9"

Shape: Small stocky body, relatively large head, and thick neck typical of murrelets. Comparatively small, with a very short blunt-tipped bill. **Ad. Summer:** Light brown speckling overall (slightly darker on crown) except for dark wings and whitish belly and undertail coverts. **Ad. Winter:** Generally black above, white below, with white scapular line. Shallow black cap; white face surrounding dark eye; narrow dark nape. Black collar usually complete. **Juv:** (Aug.–Oct.) Similar to ad. winter but with dusky face, scapulars, breast, and flanks. **Flight:** Rapid flight with fast wingbeats; underwings are all dark; white tail sides; black collar. **Hab:** Coasts and slightly inland in summer; at sea in winter. **Voice:** Low-pitched nasal *eeer.*

Subspp: Monotypic.

Adult, *scrippsi* CA/08

Adult, *scrippsi* CA/08

Adult, *scrippsi* CA/08

Xantus's Murrelet 2

Synthliboramphus hypoleucus L 9¾"

Shape: Like other murrelets but smaller, with a fairly short fairly thick-based bill and somewhat slender head. Swims low, flanks often hidden. **Ad:** All black above; white below with dusky flanks. Rare southern subsp., *hypoleucus,* has distinctive broad white area extending into black cap before and above eye. More common subsp., *scrippsi,* lacks broad white area around eye (but usually has thin white crescents above and below eye). Similar Craveri's Murrelet has dark partial collar, relatively long thin bill; black on cap continues very slightly onto chin; also see Flight. **Juv:** (Aug.–Oct.) Like ad. but with fine dark barring on flanks. **Flight:** Underwing two-toned with white coverts and black flight feathers (Craveri's underwing mostly dark). **Hab:** Coastal islands in summer; at sea in winter. **Voice:** High-pitched peeping.

Subspp: (2) Differ in face pattern—see Ad., above. •*hypoleucus* (breeds Guadalupe Is. to San Benito Is., MEX, disperses to WA in fall); •*scrippsi* (breeds Channel Is. CA to San Benito Is., disperses to s. BC in fall).

Adult summer AK/03

Juv. CA/10

Adults, winter WA/12

Ancient Murrelet 2

Synthliboramphus antiquus L 10"

Shape: Small stocky body, relatively large head, and thick neck typical of murrelets. Comparatively blocky-headed and broad-necked, with a short deep-based bill. **Ad. Summer:** Bill dark-based but conspicuously pale-tipped. Black hood with diffusely streaked white eyebrow from over eye to nape. Back gray; white cheek bordered by thin black partial collar finely barred with white. Underparts white with flanks mottled dusky. **Ad. Winter:** (Held briefly in fall–early winter) Chin grayish, throat white; eyebrow and barring on collar less distinct. **Juv:** (Aug.–Oct.) Similar to ad. winter but chin white, flanks grayish brown, back darker. **Flight:** Fast with rapid wingbeats. Underwing coverts mostly white; flanks gray; white wedge on cheek. **Hab:** Coast and coastal islands in summer; at sea in winter. **Voice:** High-pitched twittering.

Subspp: Monotypic.

Adult CA/03

Adult CA/03

Craveri's Murrelet 3

Synthliboramphus craveri L 9½"

Shape: Small stocky body, relatively large head, and thick neck typical of murrelets. Comparatively small, with a fairly long thin bill and somewhat slender head. Swims very low in water, flanks often hidden. **Ad:** All brownish black above with no white on scapulars; white below with dusky flanks. Narrow dark partial collar extends onto sides of breast; black on cap continues very slightly onto chin. Similar Xantus's Murrelet lacks partial collar on sides of breast, has an all-white chin, and has a short thick-based bill; also see Flight. **Juv:** (Aug.–Oct.) Like ad. but with fine dark barring on flanks. **Flight:** Fast with rapid wingbeats. Underwing mostly dark (some paler areas on coverts), while Xantus's has white coverts and black flight feathers. **Hab:** Coast and coastal islands in summer; at sea in winter. **Voice:** High-pitched raspy trill.

Subspp: Monotypic.

Adult CA/04

Adult CA/04

Cassin's Auklet 1

Ptychoramphus aleuticus L 9"

Shape: Small, stocky, short-necked seabird with a sharp-pointed deep-based bill. Culmen fairly straight, bottom of lower mandible angles sharply up to tip. **Ad:** Dark gray overall with paler belly and white undertail coverts. Pale spot at base of lower mandible and just above whitish eye visible at close range. **Juv/Imm:** Like ad. but eye dark; eyes may take up to 3 yrs. to become fully whitish. **Flight:** Fast with rapid wingbeat; mostly dark with pale belly. **Hab:** Coasts in summer; at sea in winter. **Voice:** Midpitched 2-part *chet-eeer chet-eeer.*

Subspp: Monotypic.

Adult summer AK/06

Adult, late winter CA/04

Adult summer AK/07

Parakeet Auklet 2

Aethia psittacula L 10"

Shape: Small, stocky, well-rounded seabird with a short, blunt, bulbous bill and high rounded crown. **Ad. Summer:** Blackish head and upperparts with black throat variably mottled with white; white breast and belly with variable dark mottling on breast; whitish eye with thin white line off back; bill orangish red. **Ad. Winter:** Like ad. summer but bill duller red, throat white. **Juv:** Similar to ad. winter but white eyeline less distinct. **Flight:** Broader wings and slower wingbeat than other auklets. White belly and orangish-red bill. **Hab:** Coastal in summer; at sea in winter. **Voice:** Drawn-out rattling trill.

Subspp: Monotypic.

Adults, summer AK/06

Adult, trans. to winter AK/07

Least Auklet 2

Aethia pusilla L 6¼"

Shape: Very small, very compact, large-headed, short-necked seabird with a short, blunt, deep-based bill and high rounded crown. **Ad. Summer:** Dark gray head and upperparts with variable white on scapulars. White eye; fine white facial plumes on lores, forehead, and behind white eye; white chin. Underparts highly variable from all blackish to all white; most commonly whitish with grayish mottling. Bill has blackish base with small red tip; small knob grows on culmen base during breeding. **Ad. Winter:** Similar to ad. summer but lacks most white facial plumes, bill duller, and underparts all white. **Juv/1st Yr:** Like ad. winter; juv. bill black; 1st summer may have dark or mottled throat and worn paler primaries; eye gray. **Flight:** Agile and fast; often forms large flocks with coordinated movements (similar to shorebirds). **Hab:** Ocean islands. **Voice:** Harsh short sounds, like *djee djee djee.*

Subspp: Monotypic.

Adult summer RUS/07

Adult summer AK/06

Whiskered Auklet 2

Aethia pygmaea L 7½"

Shape: Very small compact seabird with a moderate-sized head, fairly long neck, and short deep-based bill. Thin crest plumes curl forward over forehead; white plumes project over head from loral area. **Ad. Summer:** Dark gray overall with slightly paler gray belly; V of white facial plumes ("whiskers") off lores; white eye with thin white plumes behind eye. Bill orangish red. **Ad. Winter:** Like summer but lacks vertical plumes off lores; crest plumes may be shorter. **Juv/1st Winter:** Grayish overall with no head plumes; grayish eye; small dark bill; pale gray undertail coverts. **Flight:** Fast with rapid wingbeats. White whiskers and red bill of ad. visible. **Hab:** Ocean islands. **Voice:** Downslurred *meeew;* also chattering.

Subspp: Monotypic.

Adult summer AK/06

1st yr. AK/08

Adult winter AK/02

Adult summer AK/06

Crested Auklet 2

Aethia cristatella L 10½"

Shape: Relatively large long seabird with a moderate-sized head, somewhat flattened crown, fairly long neck, and short deep-based bill. Long bushy crest on forehead curls forward over bill. Wide plates on bill during breeding make it slightly larger and create broad "grin" on sides. M. bill deeper than long with a curved culmen and hooked tip; f. bill longer than deep with a fairly straight culmen and no hook. **Ad. Summer:** Dark blackish gray overall with only slightly paler underparts. Orange bill and added side plates create wide "grin"; eye white with thin white line behind. **Ad. Winter:** Similar to ad. summer but crest plumes and eyeline shorter and bill smaller (lacks plates) and dull grayish to blackish brown. **1st Yr:** Like ad. winter but bill blackish; crest nonexistent to short; eye gray. **Flight:** Long pointed wings; strong direct flight. May form flocks with coordinated movements (similar to shorebirds). **Hab:** Ocean islands. **Voice:** Mellow *weeuh weeuh;* also chattering.

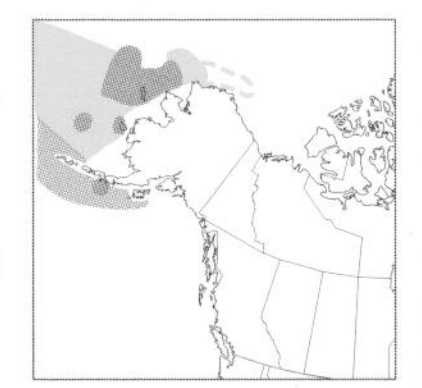

Subspp: Monotypic.

Adult summer CA/02

Adult winter CA/10

Juv./1st yr. BC/08

Adult summer CA/04

Rhinoceros Auklet 1

Cerorhinca monocerata L 14"

Shape: Large, heavy-bodied, broad-necked seabird with a large heavy bill and flattened crown. Culmen strongly downcurved; bottom of lower mandible angled to tip. Ad. has small whitish horn projecting vertically from base of upper mandible in summer, only remnant of horn in winter. **Ad. Summer:** Upperparts sooty brown; underparts slightly paler grayish brown, palest on lower belly and undertail coverts. Bill dull orangish to darkish with dull orangish base. Thin white plumes off gape and from behind eye. Eye varies from brown to pale buffy. **Ad. Winter:** Like ad. summer but horn lacking or small; bill darker and more brownish at tip; no white plumes on face. **Juv/1st Winter:** Like ad. winter but bill all dark; white lower breast and belly with a suggestion of darker barring; eye dark. **Flight:** Fast with rapid wingbeats; flies mostly low over water; dark overall, dark underwings, pale lower belly. **Hab:** Ocean islands. **Voice:** Nasal snoring *waaarh* or *wooorh.*

Subspp: Monotypic.

Adult summer ME/08

2nd summer ME/08

Adult winter VA/02

Adult summer ME/08

Adult summer ME/08

Atlantic Puffin 1

Fratercula arctica L 13″

Shape: Medium-sized, stocky, big-headed seabird with a massive, abruptly pointed, extremely deep-based bill. In summer, horny plates grow over bill, near eye, and at gape; these are shed soon after breeding, leaving bill more constricted at base through winter. Deep grooves on outer bill somewhat reflective of age: 2 yrs. old = 1 groove; 4 yrs. old = 2 grooves; greater than 4 yrs. old = 2–4 grooves. **Ad. Summer:** Black upperparts and collar; white face and underparts. Bill orange, yellow, and bluish gray; feet bright reddish orange. White (or gray in winter) facial feathers extend onto chin and throat (on similar Horned Puffin white facial feathers stop at gape). **Ad. Winter:** Starting after breeding, bill plates are shed and bill colors are dull; white face becomes dark gray; feet and legs dull orange. **Juv/1st Yr:** Similar to ad. winter but bill smaller and portions of face may be darker. Takes 2 yrs. to become adult. **Flight:** Fast with rapid wingbeats; underwings dark; adult bill obvious. **Hab:** Coastal islands in summer; at sea in winter. **Voice:** Like a small, low-pitched chain saw, *rrrrrrrah raaar.*

Subspp: Monotypic.

Adult summer AK/07

3rd yr. CA/09

Adult winter CA/02

Adult summer AK/07

Tufted Puffin 1

Fratercula cirrhata L 15"

Shape: Medium-sized, heavy-bodied, large-headed seabird with massive deep-based bill and, in summer, long crest feathers from both sides of crown reaching down the nape. In summer, horny plates grow over bill; these are shed soon after breeding, leaving bill more constricted at base through winter. **Ad. Summer:** All-black body with white mask leading to golden crest feathers. Bill with orangish-red tip and orangish base. Legs and feet bright orange. **Ad. Winter:** Like ad. summer but face grayish brown, golden crest variably mostly lost, orange plates on bill shed, leaving blackish base and dull orange tip. **Juv/1st Yr:** Like ad. winter but eye dark at first, then pale; bill smaller and blackish at first, then orangish; no crest; legs and feet blackish. Grooves on upper bill indicate age: 1 groove = 1–2 yrs.; 2 grooves = 3 yrs.; 3+ grooves = adult. **Flight:** Strong and direct; all-dark wings and underbody. **Hab:** Coastal islands and cliffs in summer; at sea in winter. **Voice:** Short, spaced, harsh sounds, like *chuk . . . chuk*.

Subspp: Monotypic.

Adult summer AK/07

Adult winter CA/05

Adult summer AK/07

Horned Puffin 1

Fratercula corniculata L 15"

Shape: Medium-sized, stocky, big-headed seabird with a massive, abruptly pointed, extremely deep-based bill. In summer, horny plates grow over bill, up from eye ("horns"), and at gape; these are shed soon after breeding, leaving bill more constricted at base through winter. **Ad. Summer:** Black upperparts and collar; white face and underparts. Bill with yellow base and orange tip; feet bright reddish orange. White (or gray in winter) facial feathers extend only to gape (on similar Atlantic Puffin white facial feathers continue onto chin). **Ad. Winter:** Starting after breeding, bill plates are shed and bill colors are dull; white face becomes gray; feet and legs dull orange. **Juv/1st Yr:** Similar to ad. winter but bill smaller and portions of face may be darker. **Flight:** Fast with rapid wingbeats; underwings dark; bill obvious. **Hab:** Coastal islands in summer; at sea in winter. **Voice:** Low-pitched sequence of snores, *waah waah waah wooor.*

Subspp: Monotypic.

Pigeons and Doves

Parakeets and Parrot

Cuckoos and Anis

Owls

Nightjars

Adult NH/06

Adult NH/06

Rock Pigeon 1

Columba livia L 12½"

Shape: Heavy-bodied, broad-shouldered, short-tailed pigeon with a relatively short neck and short stubby bill. Folded wings fall just short of tail tip, but primary extension past tertials very long (longer than tail length). **Ad:** Due to domestic breeding, colors variable, from all white to all black. Most often dark gray head; iridescent neck and breast; pale gray back; 2 dark wingbars formed by dark tips to secondaries and dark bases to greater coverts and tertials. Sexes similar, but f. may have less iridescence on neck and breast. **Flight:** Long pointed wings. White rump; gray tail with dark terminal band. Underwings whitish (except on darkest individuals). **Hab:** Cities, parks, farms, bridges, cliffs. **Voice:** Low-pitched gurgling *cucucurooo.*

Subspp: (1) •*livia.* **Hybrids:** Band-tailed Pigeon.

Adult f. FL/04

Adult m. BAH/04

White-crowned Pigeon 2

Patagioenas leucocephala L 13½"

Shape: Heavy-bodied, broad-shouldered, moderately long-tailed pigeon with a long thin neck and flattened crown. Tips of folded wings just reach end of uppertail coverts; primary projection past tertials moderate (shorter than exposed portion of tail). **Ad:** M. blackish gray overall with striking white crown; lower nape barred with iridescence; iris whitish; bill with reddish base and whitish tip. F. similar but with brownish wash to gray body; duller whitish crown; reduced iridescence on nape. Not all individuals can be reliably sexed by plumage. **Juv:** (May–Nov.) All dark brown with fine paler margins to feathers of back and wing coverts; eye dark. **Flight:** Long pointed wings; broad square tail. Appears all black with noticeable white crown; dark wing linings. **Hab:** Woods, mangroves, wooded parks. **Voice:** Cooing *woocuk wooo.*

Subspp: Monotypic.

Adult MEX/03

Adult MEX/05

Red-billed Pigeon 2

Patagioenas flavirostris L 14½"

Shape: Similar shape to White-crowned Pigeon (see for descrip.), but slightly larger with slightly shorter primary projection past tertials. **Ad:** M. with dark reddish-purple head, neck, and breast; grayish over rest of body except for pinkish-purple patch of lesser coverts; red to orange eye; dull red legs and feet; bill red with pale tip. F. similar but with grayer crown; browner head and breast; duller lesser coverts. **Juv:** (May–Sep.) Brownish gray overall with little or no purplish or reddish brown; greater coverts with pale margins; eye dark. **Flight:** Long fairly pointed wings; broad square tail. Note reddish-purple head, neck, and breast; dark wing linings. **Hab:** Wooded rivers, bottomlands. **Voice:** Low-pitched cooing *ooaahoo woik cucucooo.*

Subspp: (1) •*flavirostris.*

Juv., *fasciata* AZ/06

Adult m., *monilis* CA/06

Adult, *monilis* BC/09

Band-tailed Pigeon 1

Patagioenas fasciata L 14½"

Shape: Heavy-bodied, broad-shouldered, moderately long-tailed pigeon. Tips of folded wings just reach end of uppertail coverts; long primary projection past tertials about as long as exposed portion of tail. **Ad:** M. gray overall with pinkish to purplish cast to head; thin white half collar on upper nape with extensive iridescent flecking below, extending nearly to wing bend; bill yellowish with black tip; legs and feet yellow. F. similar but with brownish cast to head; thinner collar; iridescence limited to midnape. Not all birds reliably sexed. **Juv:** (All yr.) Brownish gray overall with no markings on nape; back and covert feathers with thin paler margins. **Flight:** Long pointed wings; broad square tail. Broad pale gray terminal band on tail; dark blackish flight feathers contrast slightly with paler gray coverts. **Hab:** Pine forests inland, oak and mixed forests along coast, gardens, parks. **Voice:** Low-pitched cooing *hooah wooo;* raspy *djeeeah.*

Subspp: (2) •*monilis* (s. AK–CA) large, more richly colored; •*fasciata* (UT–CO to NM–AZ) smaller, paler. **Hybrids:** Rock Pigeon.

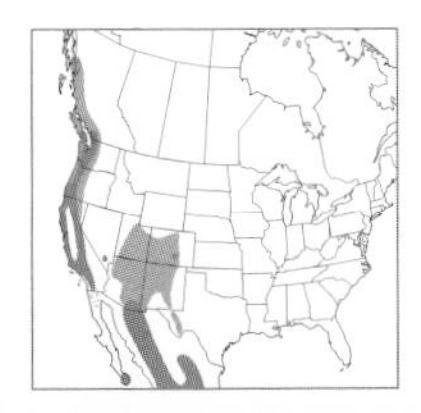

Adult HKO/02

Adult SKO/08

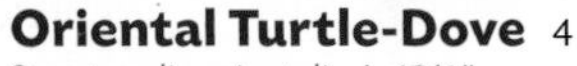

Oriental Turtle-Dove 4

Streptopelia orientalis L 13½"

Casual to w. AK islands; accidental on W. Coast. **Shape:** Large, fairly heavy, long-tailed dove with a fairly short neck, small head, and high rounded crown. Primary projection past tertials fairly long (about as long as exposed tail). Similar Spotted Dove has very short primary projection. **Ad:** Grayish crown and back; purplish breast and beige belly; striking, broad, buffy to brown margins to upperwing coverts and tertials create distinctive scaled look to wings; black patches on neck sides have rows of bluish dots. M. with gray crown and back; f. with brownish wash over crown and back. **Juv:** Like ad. but without prominent patches on sides of neck; margins of tertials and upperwing coverts pale buff and thin. **Flight:** Dark tail with white terminal band. **Hab:** Suburbs, open woods. **Voice:** *Keh-keh cooo cooo.*

Subspp: (1) •*orientalis.* **Hybrids:** Eurasian Collared-Dove.

Adult CA/11

Adult HKO/09

Spotted Dove 2

Streptopelia chinensis L 12"

Asian species introduced into s. CA in early 1900s; population now in decline. **Shape:** Large, stocky, long-tailed dove with a fairly long thin neck and small rounded head. Primary tips do not reach tips of uppertail coverts; primary projection past tertials very short (less than ½ length of exposed tail). **Ad:** Gray crown; reddish-brown face, neck, and breast; broad black collar with white spots; back and wing feathers with thin pale margins. Legs pinkish; bill dark. **Juv:** (All yr.) Similar to ad. but lacks spotted collar. **Flight:** Relatively short, rounded, dark wings; long dark tail with white tips to outer feathers. **Hab:** Parks, suburbs. **Voice:** Fairly loud *cuweee cooo.*

Subspp: (1) •*chinensis.*

Adult FL/02

Adult FL/02

Adult FL/02

Adult FL/02

Eurasian Collared-Dove 1

Streptopelia decaocto L 13"

Shape: Large, fairly heavy, long-tailed dove with a fairly short neck, small head, and somewhat flattened crown. Primary projection past tertials fairly long (about as long as exposed tail); tail square-tipped. **Ad:** Very pale tan overall except for contrasting blackish primaries; black half collar surrounded by white on lower nape; grayish undertail coverts; eye dark red. **Juv:** (Feb.–Nov.) Similar to ad. but feathers of back and wing coverts with pale margins; little or no black collar (until 3 mos. old); eye dark brown. **Flight:** Tail long with wide white band on all but central feathers; blackish primaries contrast with rest of grayish wing. **Hab:** Parks, suburbs. **Voice:** Repeated cooing *cucu cooo;* a kazoolike buzz in display flight. 🎧**44**.

IDENTIFICATION TIPS

Pigeons and Doves share the characteristics of fairly large deep-chested bodies; rather small heads; short, thin, blunt bills; and short legs. Most eat seeds and leaves while walking on the ground. They are all excellent and graceful fliers.

Species ID: Each genus has a fairly distinctive shape that can be useful in identifying silhouettes. The colors of the bill, head, and neck vary considerably and are helpful distinctions; the amount and shape of spotting on the wings is also critical in distinguishing several species.

▶ The Ringed Turtle-Dove is a domesticated version of the African Collared-Dove, *Streptopelia roseogrisea,* and a common caged bird in NAM. Escapees occur in almost any state but rarely have longtime self-sustaining populations; thus they are not on the ABA checklist and not included in this guide. Ringed Turtle-Doves are similar in appearance to Eurasian Collared-Doves but smaller and paler overall, especially on the wings and undertail coverts. Some believe they interbreed with Eurasian Collared-Doves, possibly blurring visual distinctions between the two.

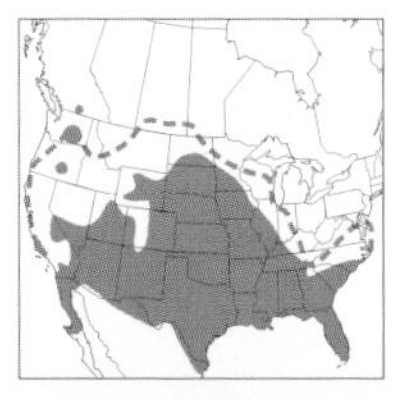

Subspp: Monotypic. **Hybrids:** Oriental Turtle-Dove; possibly Mourning Dove.

Adult, *mearnsi* AZ/09

Juv., *mearnsi* AZ/07

Adult, *mearnsi* AZ/07

White-winged Dove 1

Zenaida asiatica L 11½"

Shape: Medium-sized, somewhat deep-chested, long-tailed dove with a long, thin, slightly downcurved bill (longer than lore). Moderate primary projection past tertials. **Ad:** Unmarked pale brown overall with paler grayish-brown belly; distinctive broad white streak along edge of folded wing; small black patch just under cheek; eye reddish orange with surrounding blue skin. Similar Mourning Dove has dark spots on wing coverts and lacks white edge to folded wing. **Juv:** (Mar.–Oct.) Similar to ad. but eye is dark; bill may have pale base; areas of rusty brown wash over wings and throat. **Flight:** Bold white band across midwing, white tips to all but central feathers of rounded tail. **Hab:** Arid areas, towns, parks, suburbs. **Voice:** *Coo-coo cuh coooo.* 🎧**45**.

Subspp: (2) •*mearnsi* (s. CA–s.w. NM, vag. to east) large, paler overall; •*asiatica* (s.e. CO–FL, vag. throughout East) smaller, darker overall.

Adult, *zenaida** VIS/01

Adult, *aurita* BAR/11

Zenaida Dove 5

Zenaida aurita L 10½"

Accidental West Indies vag. to FL. **Shape:** Medium-sized somewhat deep-chested dove with a small head and thin fairly long bill. Similar to Mourning Dove but tail squared off and relatively short. **Ad:** Warm brown tone overall; thin white patch along rear edge of folded wing (white tips of secondaries); a few black marks on wing coverts. M. has extensive iridescence on sides of neck; f. less so. Similar Mourning Dove warm brown to grayish brown, has many black spots on wing coverts, and has a long, thin, pointed tail. **Juv:** (All yr.) Like ad., but white margins to feathers of back and wing coverts create scaled effect. **Flight:** White tips to all but central pair of feathers on rounded tail; white trailing edge to secondaries and inner primaries. **Hab:** Open woodlands. **Voice:** Like Mourning Dove, *hoowaoo hoo hoo hoo.*

Subspp: (1) •*aurita*. **Z. a. zenaida* (BAH and Greater Antilles) used for good photograph.

Adult m., *carolinensis* FL/02

Adult f., *carolinensis* FL/02

Juv., *carolinensis* FL/03

Adult, *carolinensis* FL/02

Adult, *carolinensis* FL/02

Mourning Dove 1

Zenaida macroura L 12"

Shape: Medium-sized, somewhat deep-chested, very long-tailed dove with a small head and thin fairly long bill. Can appear quite slim-bodied when fully alert. Central tail feathers long and pointed; rest of tail strongly graduated; some molting birds may have temporarily lost their long central tail feathers. **Ad:** Warm brown to grayish brown overall with black ovals on tertials and wing coverts; small black crescent just under auricular; variable iridescence on sides of neck. M. has bluish-gray crown, pinkish breast, iridescence on sides of neck; f. has brownish crown and breast and less iridescence; differences sometimes subtle and sexes not always reliably distinguished by plumage alone. **Juv:** (Jan.–Dec.) Brown overall with light margins to back and wing feathers, creating scaled effect; dark mark under auricular longer than in adult. **Flight:** Typically makes whistling sound at takeoff; long pointed and graduated tail with white-tipped outer feathers fanned to each side. **Hab:** Extremely varied, from deserts to woods to parks. **Voice:** Low-pitched *hoowaoo hoo hoo hoo;* also single *hoowaoo.* 🎧 **46**.

Subspp: (3) •*marginella* (MI–cent. TX and west) large, fairly pale overall; •*carolinensis* (MI–cent. TX and east) medium-sized, fairly dark overall; •*macroura* (FL Keys) smallest, fairly dark overall. **Hybrids:** Possibly Eurasian Collared-Dove.

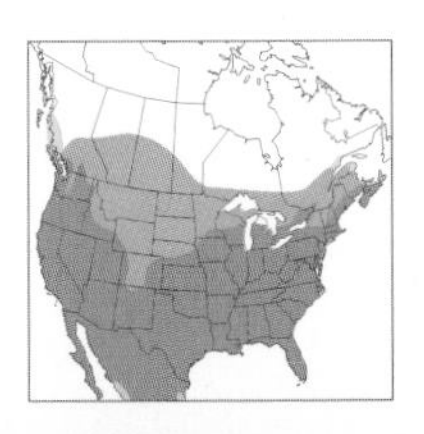

Adult TX/04

Adult TX/06

Adult m., *rufipennis* PAN/07

Adult TX/09

Adult f., *rufipennis* TRI/03

Adult m., *eluta* CA/10

Inca Dove 1

Columbina inca L 8¼"

Shape: Small, deep-bellied, long-tailed dove with a short neck, small head, and short bill. Wings very short; primary extension past tertials negligible. **Ad:** Pale gray to sandy overall with almost all feathers tipped dark brown, creating overall scaled look (sometimes less on breast); bill dark; feet and legs pinkish. No reliable plumage criteria for sexing. **Juv:** (Mar.–Nov.) Similar to ad. but less distinct barring overall. **Flight:** Above, reddish-brown primaries and their coverts; central tail brown, flanked by dark brown, outer tail feathers white. Below, wing reddish brown with darker-tipped coverts. Long tail. **Hab:** Arid lands, gardens, parks, suburbs. **Voice:** Two-part *waoo hoo;* also repeated *wuh wuh*.

Subspp: Monotypic.

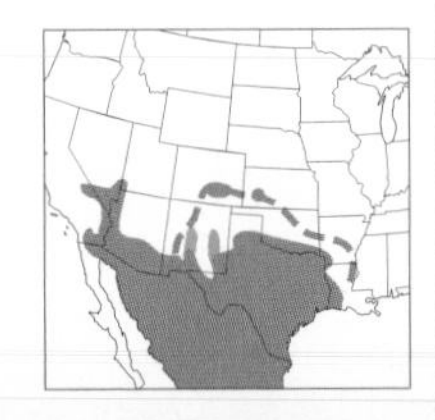

Ruddy Ground-Dove 3

Columbina talpacoti L 6¾"

Shape: Similar to Common Ground-Dove but a little heavier and tail and bill slightly longer. **Ad:** Dull gray-brown to deep reddish brown overall (see Subspp.), black bars on wing coverts and tertials, thinner dark dashes on scapulars; bill dark with gray base; no scaling on head or body. M. deep reddish brown to reddish brown overall (see Subspp.), except for pale face and bluish-gray crown; f. varies from dull gray-brown to reddish brown (see Subspp.) with paler head, usually thin white margins to greater coverts and tertials. Similar Common Ground-Dove lacks dark marks on scapulars; has rounded spots on wing coverts; scaling on crown and chest; red, pink, or orange bill base; no white margins to coverts and tertials. **Flight:** Short rounded wings; fairly short tail. Above, m. mostly reddish brown; in f., reddish-brown primaries contrast with pale brown body and secondary coverts. Below, wings reddish brown with mostly dark coverts. **Hab:** Open ground near brush or woods. Often with Inca Dove or Common Ground-Dove. **Voice:** Two-part *hoo-wa hoo-wa*.

Subspp: (2) •*eluta* (rare from CA–NM) m. reddish brown, f. pale grayish brown; •*rufipennis* (vag. to s. TX) m. deep reddish brown, f. dark brown.

Adult m., *pallescens* CA/03

Adult m., *passerina* FL/02

Adult f., *pallescens* CA/01

Adult f., *passerina* FL/02

Juv., late stage TX/12

Adult m. (front) and f., *passerina* FL/02

Common Ground-Dove 1

Columbina passerina L 6½"

Shape: Very small, deep-bellied, short-tailed dove with short neck and relatively large head. Wings short. **Ad:** Pale sandy to grayish brown overall (see Subspp.) with dark spots (somewhat oval) on wing coverts and tertials, but not on scapulars; bill pink, red, or orange with dark tip; variable scaled appearance on crown and breast (see Subspp.), none on back. M. with gray crown, rosy breast and secondary coverts; spots on wing coverts iridescent bluish black. F. more concolor sandy to brown, spots on wing coverts dark brown to coppery brown. Similar Inca Dove has long tail and extensive scaling. Similar f. Ruddy Ground-Dove lacks scaling on crown and breast; has dark dashes on scapulars and dark bars on wing coverts (in both cases black); gray-based bill; whitish margins to coverts and tertials (also see Flight). **Juv:** (Mar.–Nov.) Like ad. f., but feathers of wing and back with whitish margins; fewer or no black marks on scapulars; bill often mostly dark. **Flight:** Short rounded wings; short tail. Wings bright reddish brown; underwing coverts concolor with reddish-brown flight feathers (coverts blackish in Ruddy Ground-Dove); tail corners tipped white. **Hab:** Open ground near brush or woods; sandy areas. **Voice:** Low-pitched upslurred *ooowah ooowah*.

Subspp: (2) •*pallescens* (cent. TX and west) pale sandy brown overall; f. scaling may be restricted to head and throat; m. with pale rosy cast. •*passerina* (e. TX and east) grayish brown overall; f. scaling extensive on head and breast; m. with deep rosy cast.

Adult TX/04

Adults TX/04

White-tipped Dove 2

Leptotila verreauxi L 11½"

Shape: Medium-sized, heavy-bodied, deep-chested dove with a stout neck and moderate-sized head and bill. Tail relatively short; primary projection past tertials short. **Ad:** Plain and unmarked; grayish-brown back and wings contrast with whitish belly and undertail coverts; head and neck smooth pale gray with wash of purple iridescence on crown and nape; iris pale yellow to orange; orbital ring red (sometimes blue). Sexes similar; m. may be more purple on head and breast. **Juv:** (May–Oct.) Like ad., but pale edges to scapulars and wing coverts create scaled look; eye dark; no iridescence on head. **Flight:** Above, mostly dark brown, with white tips to outer tail feathers. Below, whitish belly; reddish-brown median and lesser coverts contrast with rest of grayish wing. **Hab:** Riverside vegetation, woodlands, citrus groves, well-vegetated suburbs. **Voice:** Like blowing across a bottle; low-pitched 2-part *hoo-hooo.*

Subspp: (1) •*angelica.*

Adult CUB/03

Key West Quail-Dove 4

Geotrygon chrysia L 12"

Casual West Indies vag. to s. FL. **Shape:** Medium-sized, heavy-bodied, deep-chested dove with a small head. Tail relatively short, moderate primary projection past tertials. **Ad:** Distinctive bold white stripe across cheek; brown back and wings with wash of purple iridescence on back and coverts; crown and nape iridescent greenish; throat through belly pale grayish white; bill deep red at base, dark tip. Sexes similar, but f. duller with reddish tones and less iridescence. **Juv:** (Apr.–Aug.) Like ad. f. but lacks all iridescence; scapulars and wing coverts with paler margins, creating scaled look. **Flight:** Short tail and wings; reddish wings with paler underwing coverts. **Hab:** Open woodlands, scrub. **Voice:** Low-pitched slightly downslurred cooing, *hoooowoh.*

Subspp: Monotypic.

Adult m. TRI/–

Adult f. TRI/–

Ruddy Quail-Dove 5

Geotrygon montana L 9½"

Accidental West Indies and Central American vag. to s. FL and s. TX. **Shape:** Relatively small, short-bodied, deep-chested dove with a short tail, thick neck, and small head. Bill relatively long. **Ad:** Brown above, buffy below; head brown with variably distinct pale buff stripe across cheek; iridescent purplish wash over back and wing coverts; pale chin and throat; bill variably reddish with dark tip; pale eye with reddish orbital ring. M. reddish brown above; f. duller brown above, cheek-stripe less distinct. **Juv:** Like ad. f., but pale margins of back and wing feathers create scaled look. **Flight:** Short tail and wings; concolor reddish underwing. **Hab:** Moist woods. **Voice:** Low-pitched simple *coo,* becoming quieter at end.

Subspp: (1) •*montana.*

Adults AUS/10

Budgerigar 3

Melopsittacus undulatus L 7"

Australian species with populations established from escaped caged birds; found in small numbers mostly along the Gulf Coast. **Shape:** Very small parakeet with a long tail and relatively small stubby bill. **Ad:** Head, back, and wings yellow with fine black barring from hindcrown and face over back and wings; underparts unmarked lime-green except for throat, which is yellow with 4 black dots and bluish malar spots. Tail with central feathers blue, outer feathers yellow with black tips. Sexes alike. **Juv:** Similar to ad., but throat without black dots and crown and forehead barred. **Flight:** Long thin central tail feathers, graduated outer tail; short blunt head. **Hab:** Urban yards with mixed vegetation. **Voice:** High-pitched chirps.

Subspp: Monotypic.

Adult FL/02

Adult with nesting material FL/02

Monk Parakeet 2
Myiopsitta monachus L 11½"

Introduced from captive populations; small colonies have occupied many East Coast states as well as IL and OR. **Shape:** Medium-sized parakeet with a large head, broad neck, fairly large bill, and long tail. **Ad:** Upperparts mostly unmarked bright green; crown, face, breast, and upper belly pale gray; belly and undertail yellowish green; flight feathers bluish; bill orangish pink. Sexes alike. **Juv:** Forehead washed green. **Flight:** Long tail and dark bluish wings. **Hab:** Urban parks with palms and other trees. **Voice:** Loud harsh squawks and softer chirps.

Subspp: Most individuals probably •*monachus;* some may be •*cotorra.*

Adult TX/12

Green Parakeet 2
Aratinga holochlora L 13"

Mexican species introduced into s. TX. **Shape:** A fairly large parakeet with a large head, large bill, and long tail. **Ad:** Bright green overall, slightly paler on belly, sometimes with a few scattered orange feathers. Bill buffy; wide gray orbital ring. Sexes alike. **Juv:** Similar to adult; usually all green underneath with no traces of orange. **Flight:** Fairly uniformly green with flight feathers slightly darker. **Hab:** Urban areas with mixed vegetation. **Voice:** Harsh calls and screeches.

Subspp: It is unclear which subsp. occurs in our area.

Adult FL/04

White-winged Parakeet 2

Brotogeris versicolurus L 8¾"

South American species established through escaped caged birds; seen in CA and FL. **Shape:** A fairly small parakeet with a large head and bill and moderate-length tail. **Ad:** Green overall except for yellow upper greater coverts and white secondaries and inner primaries (the latter mostly hidden on perched bird); outer primaries bluish. Sexes alike. **Flight:** White inner flight feathers obvious in flight. **Hab:** Urban areas with mixed vegetation. **Voice:** A harsh *shree shree* and various quieter chitters.

Subspp: Monotypic.

Adults TX/–

Red-crowned Parrot 2

Amazona viridigenalis L 13"

Mexican species established from escaped caged birds; seen in CA, TX, and FL. **Shape:** A medium-sized parrot with a large head and bill and a short tail. **Ad:** Green overall except for red forehead or crown and bluish barring on nape. Tail has broad yellow terminal band; red bases to outer secondaries may be seen on perched bird. M. has more red on crown than f. **Juv:** Similar to ad. female. **Flight:** Bright red on secondaries seen from above; yellow band on tail tip. **Hab:** Urban areas with palms and other mixed vegetation. **Voice:** A descending whistle and variety of harsh calls.

Subspp: Monotypic.

Adult m. CYP/04

Adult f., gray morph HUN/05

Adult m. FIN/06

Adult f., hepatic morph CYP/04

Adult f., hepatic morph EST/06

Common Cuckoo 3

Cuculus canorus L 13"

Rare visitor to w. AK islands; casual to mainland AK; accidental to MA. **Shape:** Fairly large, long-tailed, very long-winged, short-bodied bird with a rounded crown and short, pointed, slightly downcurved bill. Primary projection past tertials extremely long (like a swallow or swift). **Ad:** M. with gray upperparts; unbarred grayish throat and upper breast; belly whitish with fine dark barring; vent and undertail coverts white and variably barred. Bill dark with yellowish base to lower mandible; orbital ring yellow. Similar Oriental Cuckoo m. has buffy, sometimes barred, vent and undertail coverts; underwing coverts more blackish, contrasting more with paler flight feathers; upperparts average darker gray. F. comes in two morphs. Gray-morph female is like ad. m. but with barred buffy throat and breast, white vent and undertail coverts. (Similar Oriental Cuckoo f. gray morph has pale buffy vent and undertail coverts.) Hepatic-morph female has upperparts densely barred with reddish brown; throat, breast, and belly white with fine dark barring; rump and uppertail coverts largely unbarred or with thin barring when present. (Similar Oriental Cuckoo f. hepatic morph has thick dense barring on rump and uppertail coverts.) **Juv:** Similar to hepatic-morph f. but with white tips to back and wing feathers, creating thin white spotting or barring on upperparts; also white patch on nape. **Flight:** Long pointed wings create falconlike look. Below, dark primaries strongly barred with white. **Hab:** Variable, open lands, tundra. **Voice:** M. gives 2-part, well-spaced *cook-coh,* much like proverbial cuckoo clock; f. call a loud bubbling trill.

Subspp: (1) •*canorus* (Eurasian vag. to AK).

Adult m. PHI/03

Adult f., hepatic morph SIN/11

Adult f., hepatic morph SIN/11

Adult f., gray morph CHI/07

Oriental Cuckoo 4

Cuculus optatus L 12½"

Casual visitor to w. AK islands. **Shape:** Identical to Common Cuckoo; may have slightly shorter and thicker bill. **Ad:** M. and gray-morph f. similar to Common Cuckoo but with buffy (instead of white) undertail coverts; underwing coverts more solid black and contrasting more with paler, less-barred flight feathers. Hepatic-morph f. like similar Common Cuckoo but with thick heavy barring on rump and uppertail coverts (rather than thinly barred or unmarked rump). **Juv:** Similar to hepatic-morph f. but with white tips to back and wing feathers, creating thin white spotting or barring on upperparts. **Flight:** Long pointed wings create falconlike look. **Hab:** Variable, open lands, woods. **Voice:** M. gives series of low-pitched, muffled, 2-part coos, *coo-coo, coo-coo, coo-coo;* f. call a thrushlike *weeb weeb weeb.*

Subspp: Monotypic.

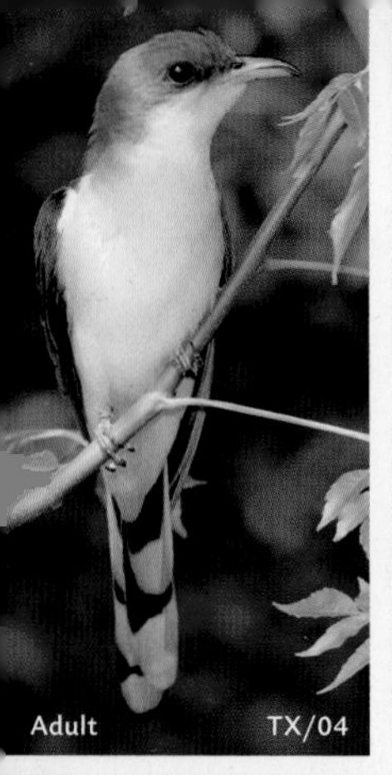
Adult TX/04

Juv./1st winter TX/04

Adult TX/04

Yellow-billed Cuckoo 1

Coccyzus americanus L 12"

Shape: Long thin bird with a very long, narrow, rounded tail and long, heavy, slightly downcurved bill. Similar Black-billed Cuckoo has slightly thinner bill. **Ad:** Bill deep yellow except for dark culmen; orbital ring gray (yellow in 1st winter); sharp line of contrast between brownish cap and white cheek and throat. Below, tail blackish with bold white tips to outer feathers. Primaries reddish brown, seen on folded wing and in flight. See Black-billed Cuckoo for comparison. **Juv/1st Winter:** (Jun.–May) Like ad. but orbital ring pale yellow; bill dusky for about 1st 2 months, then yellow like adult's. Distinguished from juv. Black-billed by tail pattern, orbital ring color, bill shape, and lack of pale tips to feathers of back. **Flight:** Very long tail; fairly rounded wings. Often flies low between areas of cover. Upperwing brown with reddish-brown wash on primaries; underwing reddish brown with whitish coverts. **Hab:** Trees, brush, woods edges. **Voice:** Series of dry percussive notes like *kuk kuk kuk kuk,* often slowing and ending with *kawlp kawlp kawlp* or *k'dawlp k'dawlp* sounds; also low-pitched, resonant, downslurred, well-spaced *keeyo keeyo keeyo.*

Subspp: (2) Differ slightly in size. •*occidentalis* (CO–w. TX and west) slightly larger; •*americanus* (rest of range). **Hybrids:** Black-billed Cuckoo.

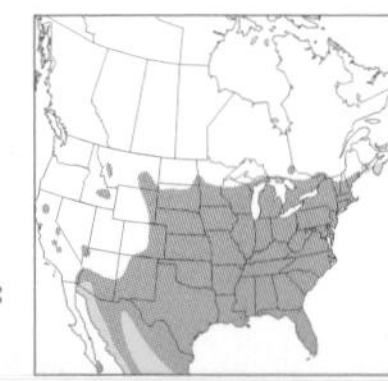

Adult FL/03

Adult FL/03

Adult FL/03

Mangrove Cuckoo 2

Coccyzus minor L 12¼"

Shape: Long thin bird with a very long, narrow, rounded tail and long, heavy, slightly downcurved bill. **Ad:** Upper mandible black, lower mandible deep yellow with dark tip; orbital ring yellowish; grayish brown above with dark mask from lore through auricular; pale cinnamon wash to underparts deepest on undertail coverts. Below, tail blackish with bold white tips to outer feathers. Primaries dull brown. **Juv/1st Winter:** Like ad. but with no black mask; orbital ring grayish (through winter); feathers of upperparts with pale or cinnamon tips. Disting. from Yellow-billed juv. by gray orbital ring, pale cinnamon wash to underparts, extent of yellow on bill, pale tips to wing and back feathers. **Flight:** Very long tail; fairly rounded wings. Often flies low between areas of cover. Wings brown above with no reddish on primaries; wings below brownish with whitish coverts. **Hab:** Mangroves, hardwood hummocks. **Voice:** Low, guttural, long series like *ga gaga gagagaga* followed by swallowed guttural notes like *gaaw gaaw gaaw gaaw.*

Subspp: Monotypic.

Adult USA/07

Juv. NJ/10

Adult –/06

Black-billed Cuckoo 1

Coccyzus erythropthalmus L 11¾"

Shape: Long thin bird with a very long, narrow, rounded tail and fairly long downcurved bill. Similar Yellow-billed Cuckoo has slightly thicker bill. **Ad:** Bill black above, gray below with dark tip; orbital ring red (may be yellowish in winter); brownish cap often blends into white cheek and throat. Below, tail gray with small whitish tips to outer feathers. Primaries brown without reddish-brown tones. Similar Yellow-billed has sharp line of contrast between cap and cheek, dark tail with bold white spots below, gray orbital ring, and reddish brown on primaries. **Juv/1st Winter:** (Jun.–May) Like ad. but with pale buffy to dusky throat and undertail coverts; whitish breast; orbital ring greenish to dusky; feathers of back and wing coverts tipped pale. **Flight:** Very long tail; fairly rounded wings. Often flies low between areas of cover. Upperwing brown without strong reddish tones; underwing grayish brown with whitish coverts. **Hab:** Trees, brush, woods edges. **Voice:** Series of low-pitched, resonant, monotonous *cu cu cu* or *cu cu cu cu;* also a sharp *kdawlp kdawlp* similar to Yellow-billed Cuckoo but higher-pitched; series of short, high-pitched, downslurred screeches like *cheew cheew.*

Subspp: Monotypic. **Hybrids:** Yellow-billed Cuckoo.

Adult AZ/02

Adult NM/01

Greater Roadrunner 1

Geococcyx californianus L 23"

Shape: Large very long-tailed bird with relatively long neck and thick legs. Heavy bill is deep-based, pointed, and slightly hooked. Bushy crest can be raised or sleeked; tail often cocked up. **Ad:** Streaked brown and white overall; bare skin behind eye bluish, can also have red and white (colors more intense during breeding); long tail feathers dark brown with pale edges. **Juv:** Like adult. **Flight:** Usually runs along ground with head and tail outstretched; flies for short distances, mostly gliding with wings and tail spread. **Hab:** Arid areas with sparse brush. **Voice:** Deep, mellow, well-spaced cooing *wooh wooh wooh whoa whoa whoa* (trails off at end).

Subspp: Monotypic.

Adult FL/–

Juv./Imm. CUB/03

Adult FL/05

Adult VEN/12

Smooth-billed Ani 3

Crotophaga ani L 14½"

Shape: Long-tailed, heavy-bodied, thick-necked bird with a short, deep, blunt bill. Culmen strongly and abruptly curved; lower mandible thin and fairly straight. Sides of bill smooth; lower mandible cutting edge has abrupt downward bend toward tip. Ad. has keel-like ridge on culmen that rises higher than crown. Rather shaggy appearance, with tail often held at odd angles. **Ad:** All glossy black with variable brownish margins to head and neck feathers, bluish margins to back and breast feathers. Distinguished from Groove-billed Ani with difficulty; use calls, range (but imm. Groove-billeds most likely to be vagrants), shape of lower mandible, presence of keel on adult, lack of any grooves (but juv. Groove-billed may also have few grooves). **Juv/Imm:** (Apr.–2nd Aug.) Body feathers dull black to mixed with some gloss; a few brownish wing feathers may be mixed with black ones; bill smaller, with no keel on culmen at first; keel gradually acquired over 1st year; tail shorter. **Flight:** Flight appears labored and floppy; flaps followed by glides; proportions not unlike a large grackle's. **Hab:** Open grassy areas with trees and brush, often near water. **Voice:** Strongly upslurred *hoooeeek hoooeeek*.

Subspp: Monotypic.

Adult, *sulcirostris* TX/12

Adult, *sulcirostris* FL/04

Adult, *sulcirostris* TX/06

Juv./Imm., *sulcirostris* TX/–

Groove-billed Ani 2

Crotophaga sulcirostris L 13½"

Shape: Long-tailed, heavy-bodied, thick-necked bird with short, deep, blunt bill. Culmen strongly and abruptly curved; lower mandible thinner and straighter; sides of bill variably grooved (more so with age); lower mandible has weak gonydeal angle and cutting edge has gradual downward bend toward tip. **Ad:** All glossy black with bluish to dusky margins on head, back, and breast feathers. Distinguished from Smooth-billed Ani with difficulty; use calls, range (imm. Groove-billeds most likely to be vagrants), shape of lower mandible, presence of any grooves on bill. **Juv/Imm:** (Apr.–2nd Aug.) Body feathers dull black to mixed with some gloss; a few brownish wing feathers may be mixed with black ones; bill smaller with few grooves at first; grooves gradually acquired over 1st year. Imm. the most likely age for Groove-billed to be vagrant. **Flight:** Flight appears labored and floppy; flaps followed by glides; proportions not unlike a large grackle's. **Hab:** Brushy areas, often near water. **Voice:** Short *cheurp* or *cheurpeek;* also short *ooeek,* similar to Smooth-billed but shorter.

Subspp: (2) •*sulcirostris* (s. TX, vag. MN–VA and south); body feathers black with bright iridescent bluish-green edges. •*palidula* (vag. AZ–CA); body feathers dusky with dull iridescent bluish-green edges.

Adult m. AZ/02

Adult f. AZ/01

Adult, *frontalis* AZ/05

Adult, *frontalis* AZ/05

Barn Owl 1

Tyto alba L 16"

Shape: Medium-sized proportionately large-headed owl with a small tapered body and long legs. Large facial disk strongly heart-shaped. Long wings extend well beyond short tail. **Ad:** White facial disk with dark eyes. Generally warm beige above with variable darker gray markings; paler below with fine dark spotting. M. usually paler, with underparts white to less than half beige with a few to some dark spots; areas adjacent to sides of facial disk usually white. F. darker below, usually more than half beige with many dark spots; areas adjacent to sides of facial disk usually beige. **Juv:** (Any month) Whitish body; beige wings; strong heart-shaped facial disk. **Flight:** Slow deep wingbeats; sometimes hovers briefly. Long wings; short tail; large head. **Hab:** Open country, grasslands, marshes, farmlands. **Voice:** Harsh raspy screech, like *sshrreeet.* 🎧**47**.

Subspp: (1) •*pratincola.*

Flammulated Owl 2

Otus flammeolus L $6^{3}/_{4}$"

Shape: Very small short-bodied owl with a proportionately large head and short rounded ear tufts, which are usually flattened against the head. Stocky but tapered body not much wider than head. **Ad:** Our only small owl with dark eyes; distinct facial disk outlined by thin dark ruff. Cryptically colored like the screech-owls, but smaller; largely buffy scapulars create distinct V on back. Legs feathered; toes bare. Some birds more reddish brown, but there is no distinct red morph as in Eastern Screech-Owl. **Juv:** (Jun.–Sep.) Barred with gray overall; dark eyes distinctive for our small owls. **Flight:** Quick agile flight; may hover briefly. **Hab:** Open dry coniferous forests at higher elevations; aspens. **Voice:** Two-part introductory call followed by single hoot, *huduh hoo, hudu hoo;* also just single hoots.

Subspp: (2) Differences slight, much individual variation. •*idahoensis* (s. CA–w. ID and north) often paler overall, indistinct facial disk, narrow streaking to underparts. •*frontalis* (rest of range) often darker overall, more distinct facial disk, wide streaking on underparts.

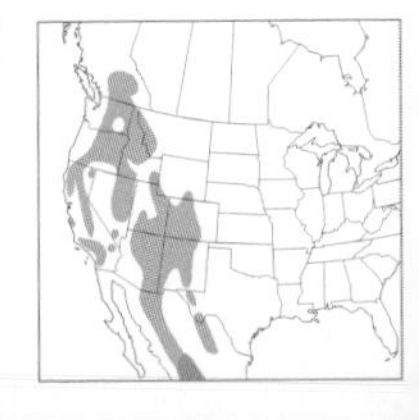

Adult AZ/04

Adult CA/05

Adult AZ/02

Adult, Southern Grp. AZ/04

Juv., Northern Grp. BC/06

Whiskered Screech-Owl 2

Megascops trichopsis L 7¼"

Shape: Similar in shape to other screech-owls but smaller and with proportionately smaller feet. **Ad:** Golden to orangish eyes; bill yellowish green with pale tip; legs and toes entirely feathered. Similar Eastern and Western Screech-Owls' toes unfeathered near tips; also see Voice. **Juv:** (May–Oct.) Note smaller feet, feathered toes, and orangish eyes when compared to Western Screech-Owl. **Flight:** Agile; rapid wingbeats; may glide or hover briefly. **Hab:** Wooded canyons, oak or oak-conifer forests. **Voice:** Series of short hoots given in "Morse code" rhythms, often 2 and then 1, *hoodoo hoo, hoodoo hoo hoo hoo.*

Subspp: (1) •*aspersus.*

Western Screech-Owl 1

Megascops kennicottii L 8½"

Shape: Like Eastern Screech-Owl. **Ad:** Eyes yellow; bill dark gray or blackish with lighter tip. Facial disk outlined in dark, especially on lower half. Underparts with thin dark streaking and variable finer cross-barring. Legs feathered; toes feathered on basal half, bare near tips. Monochromatic, birds in Northwest tend to be browner. Similar Eastern Screech-Owl has all-pale bill; also see Voice. **Juv:** (May–Oct.) Grayish like adults; yellow eyes; dark bill; underparts mostly barred. **Flight:** Wingbeats even and rapid. **Hab:** Varied, from suburban areas to woods. Needs natural cavity or birdhouse for nesting. **Voice:** Series of short high-pitched hoots, speeding up at end, *hoo hoo hoo hoohoohoo.*

Subspp: (6) Two groups. **Northern Group:** Darker, varies from large to small. •*kennicottii* (coastal AK–n.w. CA), •*bendirei* (s.w. CA–WY and north). **Southern Group:** Paler, medium to small. •*aikeni* (s.e. CA–s.e. CO to cent. AZ–s. NM), •*yumanensis* (s. NV–w. AZ), •*gilmani* (s.w. AZ), •*suttoni* (s.w. NM–s.w. TX). **Hybrids:** Eastern Screech-Owl.

Adult red morph, *asio* OH/11

Adult, gray morph, *asio* OH/11

Ad. red morph, *floridanus* FL/03

Ad. intermediate morph, *floridanus* FL/02

Juv., *asio* CT/05

Eastern Screech-Owl 1

Megascops asio L 8½"

Shape: Small, fairly broad-shouldered, stocky owl with relatively long ear tufts that can be held flat against head. Proportionately large feet. **Ad:** Eyes yellow; bill pale yellowish green or pale gray with a lighter tip. Facial disk prominently outlined in dark, especially on lower half. Underparts with thin dark streaking and variable finer cross-barring. Legs feathered; toes feathered on their basal half, bare near tips. Polychromatic, with red or gray morphs and intermediate brownish plumages. Red morph more common in the Southeast, rarer farther north and west (see Subspp.). Similar Western Screech-Owl bill is dark gray or blackish with a paler tip; also see Voice. **Juv:** (May–Oct.) Polychromatic like adults; yellow eyes, pale bill; underparts mostly barred. **Flight:** Wingbeats even and rapid. **Hab:** Varied, from suburban areas to woods. Needs natural cavity or birdhouse for nesting. **Voice:** Rising and falling screech, like *schreeeoooo;* a rapid series of monotone mellow notes given evenly and on one pitch. 🎧**48**.

Subspp: (5) Divided into two groups. **Western Group:** Medium-sized to large, fairly dull plumage, only about 10% red morphs. •*maxwelliae* (OK–CO and north), •*hasbroucki* (KS–OK south to n. TX–AR), •*mccallii* (s. TX); *mccallii* has no red morphs and gives monotone whistle only. **Eastern Group:** Medium-sized to small, brightly colored, 35–75% red morphs. •*asio* (cent. MS–GA and north), •*floridanus* (s. AR–FL). **Hybrids:** Western Screech-Owl.

IDENTIFICATION TIPS

Owls are large-bodied, large-headed, short-necked birds with both eyes looking forward on their face, giving them a unique appearance among our birds. Their large talons are used to catch and kill prey, mostly at night. Rounded facial disks formed by special feathers help them detect sounds of prey.

Species ID: Most owls can be identified by shape alone, for they have fairly distinct proportions of head, body, and tail. Look then at the pattern of the facial disk and feathering within it. Many owls are heard more than seen (due to their nocturnal activity), so it is good to familiarize yourself with their hoots and other sounds.

Adult, Pac. Coastal Grp. CA/04

Juvs., late stage, Pac. Coastal Grp. CA/04

Adult, virginianus NY/03

Juv., early stage, virginianus FL/02

Great Horned Owl 1

Bubo virginianus L 22"

Shape: Very large, rectangular-headed, large-bodied owl with long widely spaced ear tufts. Shoulders and body slightly wider than head; often appears short-necked. **Ad:** Eyes yellow; broad reddish-brown to gray facial disk outlined by thin dark ruffs; white bib on throat; dark barring on belly and undertail coverts. Overall color varies from pale gray to dark brownish gray; palest gray birds, when ear tufts are flattened, can resemble the darkest f. Snowy Owls (but see Shape). Legs and feet feathered. **Juv:** (Mar.–Oct.) Plumage with dark bars below; white bib on throat like adult; colors vary like adults (see Subspp.); legs and feet feathered. **Flight:** Wingbeats measured, with wings stiff and held below horizontal. **Hab:** Adaptable to many habitats, including woods, farmlands, deserts. **Voice:** Deep low-pitched hoots often grouped in 1's, 2's, or 3's, *hoohoohoo hoohoo hooo* or *hoo hoohoohoo hoo hoo.* 🎧**49**.

Subspp: (9) Divided into three groups. **Pacific Coastal Group:** Small, dark overall, with grayish feet. •*saturatus* (s.e. AK–n.w. CA), •*pacificus* (n.w. and s. CA–n.w. NV). **Interior Western Group:** Large, pale, with whitish feet. •*algistus* (n.w. AK), •*lagophonus* (cent. AK–YT south to n.e. OR–n.w. MT); •*subarcticus* (NT–n.w. CA south to w. KS) can look very white; •*pallescens* (s.e. CA–s. UT south to s. TX). **Eastern Group:** Medium-sized, brownish, with pale feet. •*scalariventris* (n.e. MB–s.e. QC), •*heterocnemis* (n. QC–NF), •*virginianus* (MN–NS and south).

Adult m. –/02

1st-yr. f. BC/06

1st-yr. m. OH/01

Snowy Owl 2
Bubo scandiacus L 23"

Shape: Very large, fairly small-headed, large-bodied owl with a semicircular domed head. No ear tufts; relatively large feet. When perched on ground, body can appear like right triangle with back the hypotenuse. Often seen perched on ground, dune tops, posts, buildings. **Ad:** Eyes yellow; no defined facial disk; face and throat white; rest of body variably barred with dark. Legs and feet heavily feathered. M. generally white overall with suggestion of grayish barring; f. heavily barred overall. 1st-yr. birds are usually the darkest of their sex (thus, 1st-yr. f. darkest). **Juv:** (Jul.–Oct.) Downy and grayish with white flight feathers. **Flight:** Deliberate wingbeats with quick upstroke and slow deep downstroke; glides. Note white underwings. **Hab:** Open tundra, barrens, short-grass areas, coastal dunes. **Voice:** Very high-pitched finchlike *tseeah* and *tseetseetsee;* also low-pitched croaking *kwohk kwohk*. Mostly silent away from nest.

Subspp: Monotypic.

Adult MN/01

Adult ON/12

Adult MN/01

Northern Hawk Owl 2
Surnia ulula L 16"

Shape: Medium-sized, broad-bodied owl with distinctive, long, narrow tail projecting well past wing tips. Rounded head; no ear tufts. **Ad:** Eyes yellow; gray facial disk well defined by broad dark sides; forehead and crown with dense white spotting; underparts finely barred. **Juv:** (May–Oct.) Downy; gray with suggestion of ad. facial pattern. **Flight:** Agile fast flight; may hover; pointed wings and long tail. **Hab:** Coniferous or mixed woods near clearings. **Voice:** Long rapid series of mellow toots often rising near end, *wuwuwuwuwawawawiwiwi;* also an upslurred screech, *sshhreeech*. Mostly silent away from nest.

Subspp: (1) •*caparoch*.

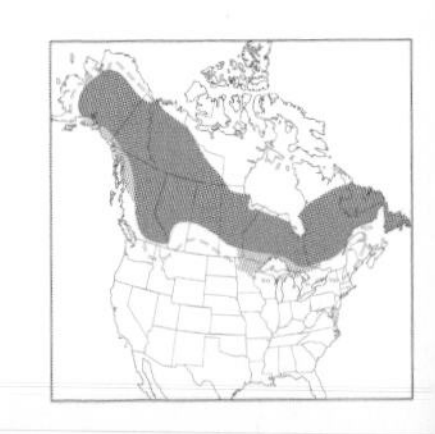

Adult BC/01

Adult BC/01

Adult AZ/05

Adult MEX/05

Northern Pygmy-Owl 2

Glaucidium gnoma L 6¾"

Shape: Small rounded owl with a short fairly narrow tail projecting well past short wings (like the handle of a cutting board). Head rounded; no ear tufts. **Ad:** Yellow eyes; no defined facial disk. Belly whitish with broad brown streaks; forehead, crown, and nape with dense white spots; rest of upperparts sparsely spotted; 2 black marks on lower nape. Tail dark, barred with white. M. tends to be grayish; f. brown to reddish brown. **Juv:** (May–Sep.) Brownish with suggestion of ad. patterns. **Flight:** Bursts of wingbeats alternating with closed-wing glides creates undulating flight path. Long narrow tail; short wings. **Hab:** Mixed forests, canyon woodlands. **Voice:** Midpitched, well-spaced, whistled hoots, *hooot, hooot, hooot;* also has continuous hooting in a short trill. Coastal subspecies' hoots tend to be higher-pitched and more widely spaced than those of interior subspecies. Calls of subsp. *gnoma* are different, being mostly paired hoots delivered in a more rapid and long series.

Subspp: (5) •*grinnelli* (coastal AK–cent. coastal CA) comparatively large; brownish-gray upperparts with reddish-brown spotting, underparts streaked brown; 3–4 indistinct pale bars on tail past uppertail coverts. •*swarthi* (Vancouver Is., BC) like *grinnelli* but medium-sized; dark brown underparts with pale spotting. •*californicum* (BC–AB south to s. CA) large; brownish-gray upperparts with pale spotting, underparts streaked black; 4–5 distinct pale bars on tail past uppertail coverts. •*pinicola* (UT–WY south to NV–NM) like *californicum* but large; pale gray upperparts with pale spotting. •*gnoma* (s.e. AZ), "Mountain Pygmy-Owl," small; brownish-gray upperparts, underparts streaked brown; 4–5 distinct pale bars on tail past uppertail coverts (may be a separate species based on song difference; see Voice).

Ferruginous Pygmy-Owl 3

Glaucidium brasilianum L 6¾"

Shape: Like Northern Pygmy-Owl. **Ad:** Like Northern Pygmy-Owl, but forehead, crown, and nape streaked with white; tail barred dark brown and paler reddish brown. Similar Northern Pygmy-Owl has spotted forehead, crown, and nape; tail barred dark and white. **Juv:** (May–Sep.) Brownish with suggestion of ad. patterns. **Flight:** Bursts of wingbeats alternating with closed-wing glides creates undulating flight path. Long narrow tail; short wings. **Hab:** Streamside woods, oak-mesquite forests, cactus desert. **Voice:** Series of high-pitched notes, a little like a small dog barking, *sheek sheek sheek;* rapid high-pitched trill.

Subspp: (1) •*cactorum.*

Adult, *whitneyi* CA/04

Adult, *whitneyi* AZ/06

Adult, *hypugaea* CA/01

Adult, *floridana* FL/02

Adult, *whitneyi* AZ/04

Adult, *floridana* FL/02

Elf Owl 2

Micrathene whitneyi L 5¾"

Shape: Our smallest owl. Tiny with a relatively small rounded head and a short tail; shoulders broader than head. No ear tufts. **Ad:** Yellow eyes. Weakly defined buffy facial disk; back grayish brown with buffy mottling; variable cinnamon streaking on breast (see Subspp.). **Juv:** Gray with fine barring on breast; very small. **Flight:** Direct rapid flight. Note rather pointed wings and short tail. **Hab:** Deserts with saguaro cacti, wooded streams, pine-oak woods. **Voice:** Short series of barking sounds, loudest in middle, *yowchechechechow;* high-pitched descending scream, *tcheer.*

Subspp: (2) •*whitneyi* (s.w. TX and west) has brownish tones to upperparts, face and underparts with much cinnamon. •*idonea* (s. TX) has grayish upperparts, face and underparts with little cinnamon.

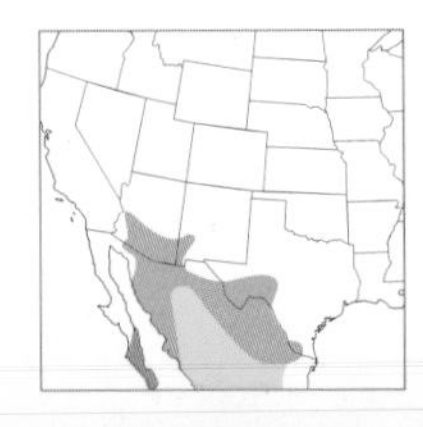

Burrowing Owl 1

Athene cunicularia L 9½"

Shape: Small, slim, long-legged owl with a relatively small oval head. Long wings project well past short tail. **Ad:** Yellow eyes; facial disk roughly outlined in white. Large white chin; horizontal whitish eyebrows connect broadly above bill. Pale spots on warm brown upperparts; variable spotting on breast; broad dark barring on belly. F. usually darker than m. **Juv:** (Feb.–Sep.) Similar to ad. but with plain whitish unbarred belly; legs and toes sparsely feathered. **Flight:** Mildly undulating flight; can hover. **Hab:** Open plains, grasslands, desert scrub. **Voice:** Midpitched dovelike *kwakoooo, kwakooo;* also high-pitched *cheecheechee.*

Subspp: (2) •*floridana* (FL) small, dark with whitish spotting; •*hypugaea* (rest of range in West) large, pale with buffy spotting.

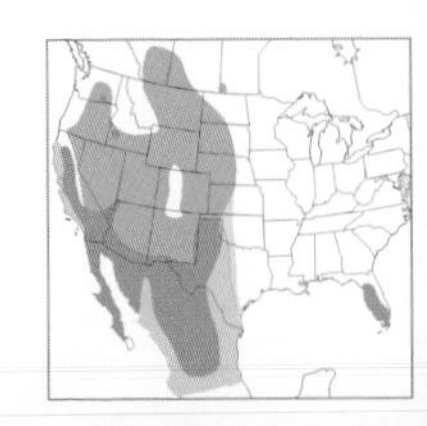

Adult, *huachucae* AZ/08

Adult, *huachucae* AZ/04

Adult, *caurina* OR/–

Spotted Owl 2

Strix occidentalis L 18"

Shape: Medium-sized, large-headed, broad-bodied owl with relatively short tail and short wings. Large apple-shaped facial disk; no ear tufts. **Ad:** Dark eyes; brownish facial disk outlined by thin dark ruff. Upperparts dark brown with whitish spotting; underparts brown with whitish spotting on throat, chest, and belly. Bill pale yellowish. Legs and feet fully feathered. Similar Barred Owl has barring on throat and chest, streaking on belly. **Juv:** (May–Oct.) Downy; back with narrow barring (wide barring on juv. Barred Owl). **Flight:** Agile with quick wingbeats and glides. Wings short and broad. **Hab:** Old-growth forests in North, canyon woodlands in South. **Voice:** Irregular series of midpitched raspy hoots, sometimes ending with longer inflected hoot, like *hoohoo, hoo, hoo, ahooowa*; a rising whistle.

Subspp: (3) Distinct ranges. •*caurina* (coastal BC–cent. CA) dark brown, breast spotting small and indistinct; •*occidentalis* (e. CA and coastal s. CA) medium brown, breast spotting medium-sized and indistinct; •*huachucae* (s. UT–s. CO and south) pale with large white dots on breast. **Hybrids:** Barred Owl.

Adult, *varia* ME/01

Adult, *varia* ME/01

Adult, *georgica* FL/02

Barred Owl 1

Strix varia L 21"

Shape: Large, large-headed, broad-bodied owl with relatively short tail and short wings. Large apple-shaped facial disk; no ear tufts. **Ad:** Dark eyes; brownish facial disk outlined by thin dark ruff. Upperparts dark brown with whitish spotting. Underparts with white barring on brownish throat and chest; brown streaking on pale belly. Bill pale yellowish. Legs and feet fully feathered. Similar Spotted Owl has white spotting on underparts. **Juv:** (May–Oct.) Back with wide bars. Legs feathered; toes mostly bare (see Subspp.). **Flight:** Agile with quick wingbeats and glides. Wings short and broad. **Hab:** Woods, wooded swamps. **Voice:** Series of midpitched hoots sounding like "Who cooks for you?"–*hoo hoo hoohooo;* sometimes with introductory *oh ah ah ah hoo hoo hoohoo;* also strong downslurred scream, *eeeooow.* 🎧**50**.

Subspp: (3) Differences minor. •*helveola* (s. TX) pale brown overall, toes unfeathered; •*georgica* (e. TX–NC and south) dark brown overall, toes unfeathered; •*varia* (rest of range) dark brown overall, toes feathered at base. **Hybrids:** Spotted Owl.

Adult MN/01

Adult, *tuftsi* CA/09

Adult TX/04

Adult QC/02

Adult, *tuftsi* CA/01

Great Gray Owl 2
Strix nebulosa L 27"

Shape: Very large, large-headed, long-tailed owl with a very large facial disk and proportionately small close-set eyes. **Ad:** Eyes yellow; facial disk gray with fine concentric black barring, outlined by a thin dark ruff; white markings on either side of dark chin create thin "bow tie" look. Underparts pale gray with darker gray streaking. Legs and feet fully feathered. **Juv:** (May–Sep.) Pale yellowish bill. **Flight:** Deep slow wingbeats; can hover. Large head, fairly long tail. Above, buffy bases to primaries; below, dark underwing. **Hab:** Northern forests, woods openings. **Voice:** Series of slow, very low-pitched, even-spaced hoots, *wooo wooo wooo wooo.*

Subspp: (1) •*nebulosa.*

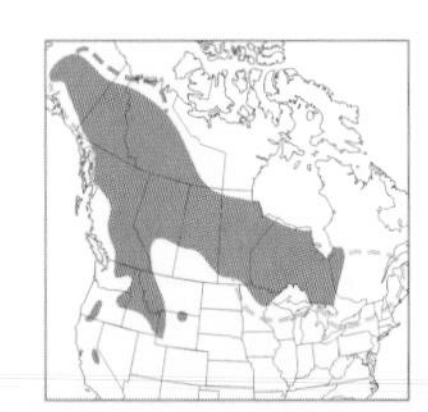

Long-eared Owl 2
Asio otus L 15"

Shape: Medium-sized, slender-bodied owl with long close-set ear tufts; facial disk split and shaped like quarters of an apple. Long wings project well past tail **Ad:** Eyes yellow to yellowish orange; facial disk well defined, split in half, rich buff to whitish outlined with thin dark ruff. Chest with blurry blackish streaks; belly with fine streaks and cross-barring. Legs and toes fully feathered rich buff. **Juv:** (Apr.–Sep.) Dark bill; facial disk 2 dark halves. **Flight:** Long glides alternate with rapid wingbeats; may hover over prey before attacking. Below, wings buffy with extensive dark barring on primary tips. **Hab:** Woods and groves near fields and marshes. **Voice:** Series of midpitched, evenly spaced hoots, *hooo, hooo, hooo;* also catlike descending scream, *yeeeowwwwah.*

Subspp: (2) •*tuftsi* (SK–w. TX and west) small, pale overall, distinct tail bars; •*wilsonianus* (MB–s. TX and east) large, dark overall, indistinct tail bars.

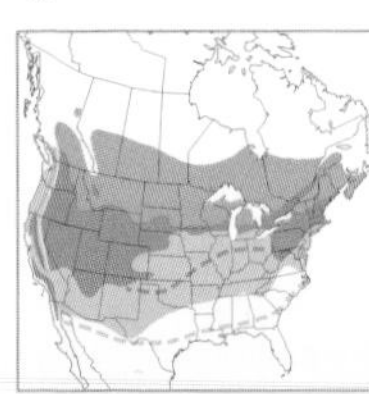

Adult, *flammeus* UT/06

Adult, *flammeus* NH/12

Adult, *flammeus* NH/12

Adult, *flammeus* CA/02

Short-eared Owl 1

Asio flammeus L 15"

Shape: Medium-sized, long-winged, slender-bodied owl with very short close-set ear tufts (not usually seen) and a rounded facial disk. Long wings project just past tail. **Ad:** Eyes yellow with dark masks to the side; facial disk fairly well defined with a thin white ruff, broken at the top. Underparts buff to whitish, with heavy streaking on chest thinning out over belly. F. darker overall than m. Legs and toes fully feathered. **Juv:** (May–Sep.) Facial area dark and not split as in juv. Long-eared Owl. **Flight:** Slow stiff wingbeats and glides; may hover. Note long narrow wings, broad buffy bases to upperwing primaries, minimal dark barring on primary tips, pale whitish underwings. **Hab:** Open fields, marshes, dunes, grasslands. **Voice:** Series of several short high-pitched barks, like *chek chek chek chek.*

Subspp: (2) •*flammeus* (all of NAM range) paler overall, toes fully feathered, uppertail coverts contrastingly paler than back; •*domingensis* (vag. to FL from Cuba) darker overall, toes sparsely feathered, uppertail coverts concolor with back, dark brown patch behind ruff on either side of head.

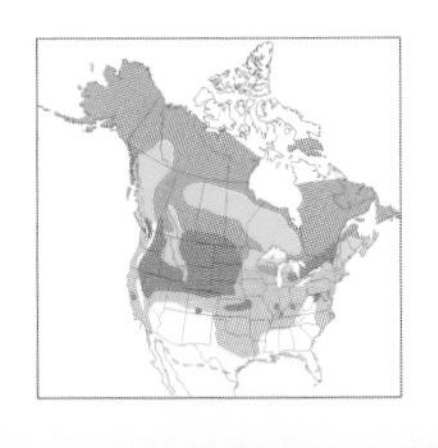

Adult, *richardsoni* MN/01

Adult, *richardsoni* ON/01

Boreal Owl 2

Aegolius funereus L 10"

Shape: Fairly small, large-headed, broad-shouldered owl with a short body that tapers abruptly to rear—shaped like a small shield. No ear tufts. **Ad:** Yellow eyes; facial disk divided into two halves by V-like extension of dark forehead; black lines on sides of forehead connect to eyes. Forehead is dark with dense white spots. Underparts whitish with grayish-brown irregular spots, sometimes forming jagged streaks. Bill pale horn. Similar Northern Saw-whet Owl has dark forehead with white streaks. **Juv:** (Jun.–Sep.) All dark grayish brown with a slightly paler belly; bill pale. **Flight:** Rapid wingbeats, often below horizontal. Short broad wings. **Hab:** Mixed coniferous-deciduous woods. **Voice:** Slight ascending series of mellow sounds, like *wowowowawawa* (like a winnowing snipe); also a descending *skew*.

Subspp: (2) •*richardsoni* (all of NAM range) dark brownish overall, underparts brownish streaked white; •*magnus* (vag. to AK islands) pale grayish overall, underparts whitish streaked pale brown.

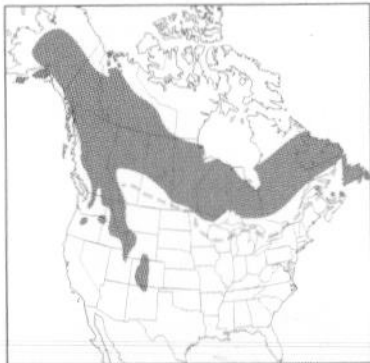

Adult, *acadicus* OH/–

Adult, *acadicus* MN/10

Northern Saw-whet Owl 2

Aegolius acadicus L 8"

Shape: Small, large-headed, broad-shouldered owl with a short body that tapers abruptly to rear—shaped like a small shield. No ear tufts. Broad white V between eyes creates wide-eyed expression. **Ad:** Yellow eyes; facial disk divided into 2 halves by white V-like extension of forehead. Forehead is dark with fine pale streaking. Underparts whitish with broad warm reddish-brown streaking. Bill dark gray. Similar Boreal Owl has dark forehead with white spots. **Juv:** (May–Sep.) All chocolate-brown above with a bright reddish-brown belly; white V over eyes; dark gray bill. **Flight:** Rapid wingbeats, often below horizontal. Short broad wings. **Hab:** Mixed coniferous-deciduous woods. **Voice:** Fairly shrill, drawn-out, high-pitched sound, *chooot, chooot;* similar shorter sound given in a faster series of about 2 per second, like *choot choot choot;* also a rising screech.

Subspp: (2) •*brooksi* (Queen Charlotte Is. BC) has dark facial disks, underparts buffy dark grayish brown with thin paler streaking. •*acadicus* (rest of range) has lighter facial disks, underparts brown to reddish brown with heavy paler streaking.

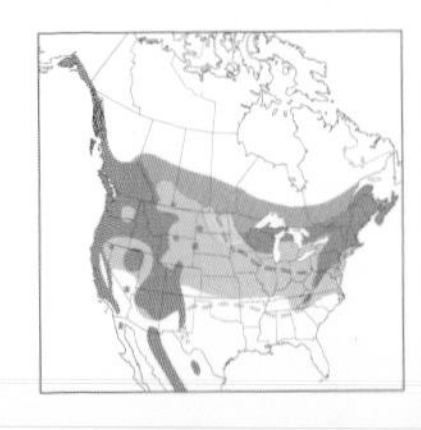

Adult TX/04

Adult –/09

Common Pauraque 2

Nyctidromus albicollis L 11"

Shape: Large nightjar. In flight, wings moderate length, well rounded; tail long, rounded; head large and blunt. On perched bird, wing tips project only halfway down tail; head large; primary projection past tertials fairly long. **Ad:** In flight, dark grayish above. Below, buffy belly and underwings; broad white or buffy primary bar usually wider toward trailing edge; central tail feathers dark, outer feathers variably white or buff. M. has wide white primary bar, white throat patch, and extensive white on tail; f. has smaller buffy primary bar, brownish throat patch, and reduced white on tail. Perched bird cryptic; scapulars grayish brown with black pointed marks at their tips and distinct whitish to buffy margins which form pale broken lines on sides of back; primary bar barely visible under tertials. **Juv:** (May–Sep.) Like ad. f. but with smaller buffy primary bar. **Flight:** Usually below 10 ft. off ground; active at dusk. May use same perch between short flights. **Hab:** Brushy woodlands, forest edges, open fields, roadsides. **Voice:** Downslurred, slightly buzzy, well-spaced *vweeer, vweeer*; also *puh puh puh paweeer.*

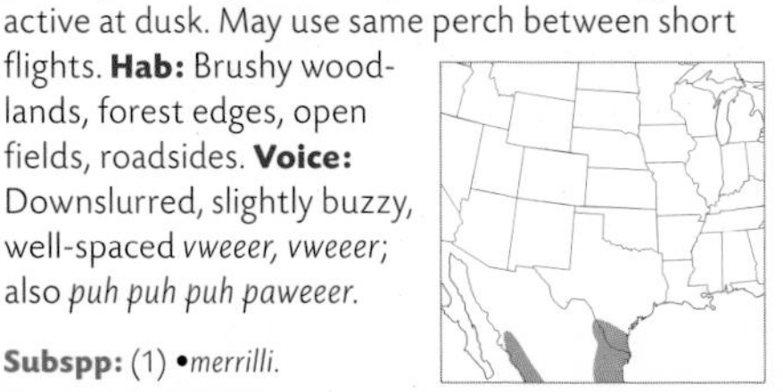

Subspp: (1) •*merrilli.*

Adult FL/05

Adult BAH/04

Adult m. BAH/05

Antillean Nighthawk 2

Chordeiles gundlachii L 8½"

Breeds on FL Keys; rare to s. FL; accidental to LA and NC. **Shape:** In flight, just like Common Nighthawk, but slightly smaller; wing tips slightly more blunt. On perched bird, wing tips just reach tail tip (slightly longer in Common Nighthawk). **Ad:** In flight, primary bar may be slightly more mottled than in Common Nighthawk. Perched, tertials may be contrastingly paler than back. Best distinguished by voice. **Juv:** As in Common Nighthawk. **Flight:** Wingbeats are quick flicks; alternates beats with much gliding; wings held in V during glide. Tends to fly high over ground (above 40 ft.); agile. **Hab:** Open areas, airports, urban areas. **Voice:** Short, buzzy, rapid series like *beebeebeebeep*; also 3-part *killy-ka-dik.*

Subspp: Monotypic.

Adult m. CA/04

Adult f. TX/–

Adult m. CA/06

Adult f. CA/06

Lesser Nighthawk 1

Chordeiles acutipennis 9"

Shape: In flight, long, thin, blunt-tipped wings (P10, outermost primary, usually shorter than P9) often pressed forward at wrist, bent back at tips; tail long, thin, slightly notched; outer third of outer wing has pale primary bar that often gets narrower toward trailing edge. On perched bird, wing tips just reach tail tip. Similar Common Nighthawk has more pointed wings (P10 usually longest primary), primary bar nearer center of outer wing and, on perched bird, wing tips longer than tail. **Ad:** In flight, wings spotted brown and buff; underparts largely buffy with finer dark barring. M. has white throat, white subterminal tail band, and white primary bar; f. has buffy throat, no white tail band, and pale buffy smaller primary bar. Molts on breeding grounds, thus some flight feathers missing in late summer (Common and Antillean molt on wintering grounds). Perched bird cryptically colored; on folded wing, pale primary patch located at or just beyond tips of tertials (short of tertial tips on Common Nighthawk); buffy spots on primaries behind pale bar. Buffy primary bar of f. harder to see than white bar of m. **Juv:** (Jun.–Sep.) Primaries broadly tipped buff and primary bar buffy and less distinct.

Body more buff overall; lacks bold pattern of ad. on scapulars. **Flight:** Wingbeats are quick flicks; alternates beats with much gliding; wings held in V during glide. Tends to fly low over ground (10–20 ft.); agile. **Hab:** Desert grasslands, fields. **Voice:** Very long, low-pitched, mellow trill (10–60 sec. or more) given from ground or near ground during breeding; low trill can also be given in flight; a bleating call, often during chases.

Subspp: (1) •*texensis.*

IDENTIFICATION TIPS

Nighthawks and Nightjars have short bodies, large heads, thick necks, and proportionately long wings. Their bills are short and stubby, but their mouths are large and bordered on the sides by long, stiff, hairlike structures that help them collect insects in the air. They feed mostly at night and sit quietly, camouflaged, during the day.

Species ID: Many have whitish bands across the outer wing or whitish patterns on the tail which can help with identification. Proportions of tail and wings in flight are also useful. In perched birds shape will be helpful again, particularly relative size of the head and length of the folded wings in relation to the tip of the tail.

Adult m. TX/05

Adult f. TX/05

Adult m. TX/05

Adult f. TX/05

Common Nighthawk 1

Chordeiles minor L 10"

Shape: In flight, long, thin, pointed wings (P10, outermost primary, usually longer than P9) often pressed forward at wrist, bent back at tips; tail long, thin, slightly notched; pale primary bar near center of outer wing usually gets wider toward trailing edge. On perched bird, wing tips slightly longer than tail. Similar Lesser Nighthawk has blunt-tipped wings and primary bar in outer third of outer wing; on perched bird, wing tips just reach tail tip. **Ad:** In flight, wings variably dark sooty gray to brownish, with paler barring on inner wing (no barring on outer wing); underparts variably whitish to reddish brown with fine dark barring. M. has white throat, white subterminal tail band, and white primary bar; f. has buffy throat, smaller whitish primary bar, and lacks white tail band. Perched bird cryptically colored; on folded wing, white primary patch located just short of tertial tip; lacks brownish spots on primaries. (See Lesser Nighthawk for comparison.) **Juv:** (Jul.–Sep.) Paler than adult; indistinct throat patch; lacks bold pattern of ad. on scapulars. **Flight:** Wingbeats are quick flicks; alternates beats with much gliding; wings held in V during glide. Tends to fly high over ground (above 40 ft.); agile. **Hab:** Forests, plains, urban areas. **Voice:** Buzzy *beeent* in flight (both sexes); wing-created booming sound at bottom of m. display flight. 🎧**51**.

Subspp: (7) Divided into three groups. **Eastern Group:** Dark with mostly blackish tones. •*minor* (AK–BC east to QC–NC); •*chapmani* (s.e. KS–NC and south). **Southwestern/Great Basin Group:** Pale with mostly buffy to grayish tones. •*sennetti* (e. MT–s. SK east to MN–IA); •*howelli* (w. KS–n. TX west to UT–WY); •*aserriensis* (cent. TX and south); •*henryi* (w. TX–AZ) has buffy spots at base of primaries (like Lesser Nighthawk). **Western Group:** Medium dark with mostly grayish tones. •*hesperis* (s. SK–s. MB south to n. CO) has dark grayish tones.

Adult AZ/05

Adult AZ/05

Common Poorwill 1

Phalaenoptilus nuttallii L 8"

Shape: Small stocky nightjar. In flight, short-bodied, short-tailed, fairly large-headed; wings broad and rounded. On perched bird, wing tips just reach tip of short tail. **Ad:** Grayish and brownish morphs expressed in color of upperparts. In flight, grayish or brownish back and upperwing coverts; no primary bar; flight feathers rich buff with dark barring above and below; undertail coverts and underwing coverts also rich buff; pale corners to tail. Sexes similar; m. tail tends to have extensive white edges; f. tail tends to have smaller buffy corners. **Juv:** (Apr.–Sep.) Like ad. f. but with small or no black dots on scapulars. **Flight:** Wingbeats deep and floppy; often alternates flaps with glides; may fly straight up to get insect and then glide down. **Hab:** Dry shrubby areas, rocky hillsides and canyons, grassy plains, open pines. Often seen in roads at dawn and dusk. **Voice:** Whistled *poor will chut,* the last note harsh and soft, and may not be heard from distance.

Subspp: (4) •*californicus* (s. OR–s.w. CA) and •*nuttallii* (e.cent. CA–TX and north) both relatively large, dark above, grayish below with dark barring; •*adustus* (s. AZ) medium-sized, paler above, pale below with grayish barring; •*hueyi* (s.e. CA–s.w. AZ) relatively large, pale above, fairly pale below.

Adult TX/04

Adult TX/04

Chuck-will's-widow 1

Caprimulgus carolinensis L 12"

Shape: Large nightjar. In flight, tail and wings both long and rounded; head blunt. On perched bird, primaries reach about halfway down tail (longer in nighthawks, shorter in Common Pauraque); fairly long primary projection past tertials; proportionately large-headed (almost half length of body) with long flat crown. **Ad:** Varies from overall grayish to warm brown. In flight, no primary bar; generally warm brown below with dark barring on wings. In 2nd-yr. and older birds, m. distinguished by white tail with buffy terminal band; f. and 1st-yr. birds above have buffy corners to otherwise brown barred tail. On perched birds look for pale scapulars with black dots near tips. **Juv:** (Apr.–Sep.) Like ad. f. but paler overall and with small or no black dots on scapulars. **Flight:** Flaps alternated with glides; generally flies about 10–60 ft. above ground when feeding; often uses same perch after flights. **Hab:** Woods with openings, forest and stream edges. **Voice:** Repeated *chut will widow, chut will widow,* introductory note harsh and soft, *will* and *wid* loudest. In Whip-poor-will, the accent is on the *will.*

Subspp: Monotypic.

Adult MEX/10

Adult MEX/10

Buff-collared Nightjar 3

Caprimulgus ridgwayi L 8¾"

Rare Mexican species that breeds in s.e. AZ; accidental to s.cent. AZ and s.w. CA. **Shape:** Medium-sized nightjar with fairly short tail and rounded wings. On perched bird, wing tips reach about ⅔ down short tail; primary projection past tertials fairly short. **Ad:** In flight, no primary bar; above, primaries spotted with brown; below, wings barred brown; complete narrow buffy collar around neck. M. tail white below, white corners above; f. with buffy tail corners. On perched bird, note complete buffy collar. **Juv:** (Apr.–Sep.) Like ad. f. but with small or no black dots on scapulars. **Flight:** Flaps alternated with glides; agile; often uses same perch after flights. **Hab:** Arid areas with shrubs and small trees. **Voice:** Ascending and quickening *tu tu tu titititeeteeteedeelip.*

Subspp: (1) •*ridgwayi.*

Adult m., *vociferus* NJ/05

Adult m., *arizonae* AZ/05

Whip-poor-will 1

Caprimulgus vociferus L 9¾"

Shape: Medium-sized nightjar with fairly short tail and fairly short rounded wings. On perched bird, wings project about ⅔ down tail; fairly long primary projection past tertials; head moderately large (about ⅓ length of body). **Ad:** Occurs in gray or brown morphs. In flight, no primary bar. M. tail below tipped white (see Subspp.); above, broadly white on either side of dark central feathers. F. tail above with buffy corners, below reddish brown. Perched, note broad dark median crown-stripe. **Juv:** (Apr.–Sep.) Like ad. f. but paler overall and with small or no black dots on scapulars. **Flight:** Wingbeats deep and floppy; often alternates flaps with glides; may fly straight up to get insect and then glide back down to perch. **Hab:** Dry open woods. **Voice:** Repeated *whip poor will*, accent on 3rd syllable (see Subspp.).

Subspp: (2) Distinct ranges. •*vociferus* (SK–n. TX and east) small with extensive white on m. tail. •*arizonae* (s. CA–w. TX) larger with reduced white on m. tail; song lower-pitched, slower, and slightly raspier than that of *vociferus* (sounds like "purple rrip"); may be a separate species.

Swifts

Hummingbirds

Trogons

Kingfishers

Adult CA/05

Juv. CA/07

Adult, *bouchellii** CRI/11

Adult, *mexicana* GUA/–

Black Swift 2

Cypseloides niger L 7¼"

Shape: Large swift with long, broad-based, but finely pointed wings. Wings usually swept back in sickle shape. Tail fairly long; head comparatively small; ratio of head-neck length to tail length about 1:2½. Adult m. tail slightly notched; f. and juv. tail roughly squared off. **Ad:** Blackish overall except for some fine whitish margins to feathers of chin and forehead and a short fine whitish eyebrow. **Juv/1st Winter:** (Jul.–Apr.) Like ad. f. but with whitish tips to body feathers (esp. lower belly and undertail) and wing coverts, creating scaled appearance visible at close range. **Flight:** Bursts of wingbeats followed by veering glides; usually flies very high. Slower wingbeats than other swifts. **Hab:** Steep cliffs, often near sea or waterfall. **Voice:** Low twittering *chips*.

Subspp: (1) •*borealis.*

White-collared Swift 4

Streptoprocne zonaris L 8½"

Casual neotropical vag. possible anywhere in NAM. **Shape:** Very large swift with long fairly even-width wings. Tail long and tapered with very small notch; head comparatively small; ratio of head-neck length to tail length about 1:2½. **Ad:** All blackish except for complete white collar; may have paler grayish or brownish forehead. **Juv:** Blackish with whitish collar, most well defined on nape; pale margins to feathers of body and wings. **Flight:** Glides most of the time; wingbeats in rapid bursts. **Hab:** Usually seen with other swifts. **Voice:** High-pitched sibilant *tseeet tseeet tseeet*. Mostly quiet away from nest site.

Subspp: (2) •*mexicana* (vag. from MEX); •*pallidifrons* (vag. from CUB). * *S. z. bouchellii* (NIC–PAN) used for good photograph.

IDENTIFICATION TIPS

Swifts spend most of the day airborne and cling to vertical surfaces at night. They eat aerial insects and have particularly long primaries. Their wings often appear thin and sickle-shaped.

Species ID: Look at the proportion of head and neck in front of the wing to body and tail behind the wing, for this varies substantially among species. Although many swifts are all dark, some have white markings or lighter shadings that are distinctive and useful.

Adult IL/05

Adult, *vauxi* CA/09

Adult IL/05

Adult, *vauxi* CA/09

Chimney Swift 1

Chaetura pelagica L $5\frac{1}{4}$"

Shape: Medium-sized swift with moderately long rather even-width wings. Wings usually only slightly swept back. Tail short and tapered; head relatively large; ratio of head-neck length to tail length about 1:$1\frac{1}{2}$. **Ad:** Dark grayish brown overall with slightly paler chin and throat and often upper breast. **Juv:** Like ad. except has fresh wing feathers when adults are molting theirs. **Flight:** Wingbeats in rapid bursts; glides most of the time. Flies both near ground and extremely high. **Hab:** Varied. Rural or urban where there are chimneys (more rarely, hollow trees) in which to nest or roost. **Voice:** High-pitched repeated *tseer tseer tseer tseer;* also high-pitched chittering. **52**.

Subspp: Monotypic.

Vaux's Swift 1

Chaetura vauxi L $4\frac{3}{4}$"

Shape: Medium-sized swift; slightly smaller than Chimney Swift but with very similar relative proportions—body length roughly equals wing length; ratio of head-neck length to tail length about 1: $1\frac{1}{2}$. **Ad:** Grayish brown overall with paler throat and breast (Chimney not as pale below, but differences can be subtle); paler rump contrasts with darker rest of upperparts (Chimney has less-contrasting upperparts). **Juv:** Like ad. **Flight:** Wingbeats in rapid bursts; glides most of the time. Flies both near to ground and extremely high. **Hab:** Mature and old-growth forest with large hollow trees for nesting and roosting; occasionally uses chimneys. **Voice:** Similar to Chimney Swift but higher-pitched and faster; repeated high-pitched *tsit tsit tsit;* also a chittering *tsittery tseeet*.

Subspp: (2) •*vauxi* (all NAM range) slightly paler, upperparts dull brownish, thin whitish eyebrow, dusky underparts. •*tamaulipensis* (Mex. vag. to AZ) slightly darker, upperparts glossy black, little or no eyebrow, underparts dark gray.

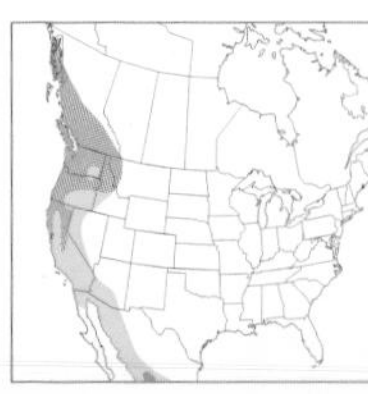

Adult THA/12

Fork-tailed Swift 4

Apus pacificus L 7¾"

Casual Asian vag. to w. AK is. **Shape:** Large, long-winged, long-tailed swift. Thin wings fine-pointed; thin tail deeply forked (although fork can be closed in flight, making tail look just pointed). Ratio of head-neck length to tail length about 1:2½. **Ad:** Upperparts all sooty except for broad white rump; underparts blackish with paler fringe to body feathers, creating scaled look, and throat paler gray. Underwing coverts tipped with white, creating thin dotted bars. **Flight:** See Shape. **Hab:** Open areas. **Voice:** High-pitched 2-part *tsee-eee,* mostly at nest site.

Subspp: (1) •*pacificus.*

Adult CA/05

Adult CA/05

Adult CA/01

White-throated Swift 1

Aeronautes saxatalis L 6½"

Shape: Large swift with long thin wings, pinched in at the base. Relatively short tail and large head; ratio of head-neck length to tail length about 1:2. Forked tail usually held closed. **Ad:** Striking black-and-white plumage (but may appear mostly dark at a distance or in poor light). Upperparts black; white throat and central breast; white hip patches; white narrow trailing edge to secondaries. **Juv:** (Jul.–Oct.) Like ad. but upperparts brownish; head and nape paler. **Flight:** Flight may be higher, faster, and with more gliding than Chimney or Vaux's Swifts'. **Hab:** Varied; needs natural or human-made crevices in which to nest, such as cliff, outcrops, bridges. **Voice:** Rapid series, high-pitched *tsetetetete.*

Subspp: (1) •*saxatalis.*

Adult m., *gracilirostris** HON/–

Juv./imm. TX/01

Green-breasted Mango 4

Anthracothorax prevostii L 4⅔"

Mexican species that is a casual vag. from s. TX to NC. **Shape:** Large hummingbird with long broad tail slightly notched or rounded. Bill fairly long (about 1¼ x head length), noticeably thick, and decurved. **Ad:** M. dark iridescent green overall with black to bluish throat and purple tail; undertail coverts blackish. F. iridescent green above; throat and belly white with vertical dark central line, black on throat, green on chest and belly; tail with violet base, black subterminal band, white tip. **Juv:** (Jun.–Sep.) Similar to ad. f. but with reddish brown variably on sides of throat, breast, and belly. **Flight:** Purple tail (with white tip in f. and imm.). **Hab:** Open land with some trees, feeders. **Voice:** Calls include high-pitched *tsip* and a buzzy *dzzzt*.

Subspp: (1) Most likely •*prevostii*. **A. p. gracilirostris* (El Salvador–HON–cent. CRI) used for good photograph.

Adult m., *cabanidis** CRI/07

Juv., *crissaliis** PER/07

Green Violetear 3

Colibri thalassinus L 4½"

Rare Mexican vag. to s. and cent. TX but also accidental anywhere to s. CAN. **Shape:** Large hummingbird with long, broad, often slightly notched tail. Bill about length of head and gently decurved. **Ad:** Dark iridescent green overall; violet-blue auriculars and chest patch; tail blue with broad blackish subterminal band. M. with extensive violet-blue areas; f. overall duller, more golden or bronze-green on crown, less violet-blue on auriculars and chest. **Juv:** (Jun.–Sep.) More bronzy green overall with gray underparts; pale fringes to crown feathers when fresh. **Flight:** Blue tail with black band. **Hab:** Woods edges, feeders. **Voice:** Calls include a short dry rattle and 2-part *chitsee;* song is 1–3-part metallic chip given frequently throughout the day.

Subspp: (1) Most likely •*thalasinnus*. **C. t. cabanidis* (CRI and PAN) and *C. t. crissalis* (Andes, PER–ARG) used for good photographs.

Adult m. AZ/07

Adult m. AZ/05

Adult f. AZ/07

Adult f. AZ/05

Broad-billed Hummingbird 2

Cynanthus latirostris L 3¾"

Shape: Medium-sized hummingbird with a fairly long tail, forked (m.) or notched (f.). Bill long (about 1⅓ x head length), slightly decurved, broad-based. **Ad:** M. dark green overall with deep blue throat; bill red-based, black-tipped; undertail coverts whitish; tail bluish black. F. upperparts pale iridescent green with gray feather margins; underparts grayish with green flanks; postocular eye-stripe (occasionally just a spot) is white, short, and borders top of dark gray auricular; bill dark red at base of lower mandible; small white tips to outer tail feathers. **Juv:** (Mar.–Aug.) Similar head and bill to ad. f. but crown feathers tipped buff. M. underparts grayish with variable blue on throat and green flecks on belly; no white on tail tips. F. much like ad. f., has buffy margins on fresh crown feathers. Complete molt into ad. plumage occurs Aug.–Feb. **Flight:** Often bobs and spreads tail during feeding. M. has strongly forked tail with bluish iridescence and no white tips. **Hab:** Streamsides and canyons of arid areas, feeders. **Voice:** Calls include sharp *tchit* or *tchidit;* song an extended, rough, chattered *tchededede.*

Subspp: (1) •*magicus.* **Hybrids:** Violet-crowned, Magnificent Hummingbirds.

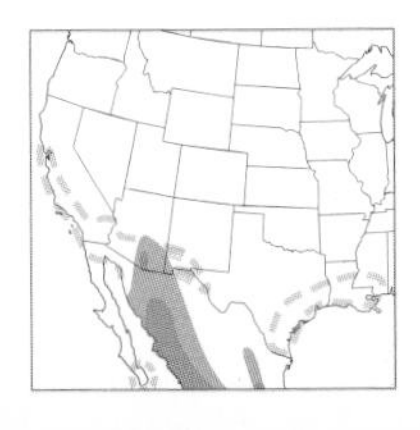

Adult m. AZ/08

Adult f. AZ/05

Adult m. MEX/09

Adult f. AZ/05

White-eared Hummingbird 3

Hylocharis leucotis L 3¾"

Rare Mexican visitor to s.e. AZ and s.w. NM; vag. to NM, TX, CO, and MS. **Shape:** Medium-sized stocky hummingbird with a high rounded crown and squared-off tail. Bill medium-sized (length of head) and fairly straight. **Ad:** M. bill red-based, black-tipped; head and throat appear blackish with long white postocular stripe that continues down side of neck; underparts greenish with white central chest and belly; grayish undertail coverts. F. bill red-based, black-tipped, or red just on base of lower mandible; throat pale, auricular black, crown dark; long white postocular stripe; underparts grayish heavily flecked with green; back and rump dark green; whitish tips to outer tail feathers. **Juv:** (May–Aug.) Like ad. f., but juv. m. with some dark feathers on throat. **Flight:** Broad white postocular stripe, whitish central belly stripe and undertail coverts. **Hab:** Mountain canyons, feeders. **Voice:** Call a sharp *tsik;* song a series of rapid chips and longer rising *dzeeek* notes.

Subspp: (1) •*borealis.* **Hybrids:** Broad-tailed Hummingbird.

IDENTIFICATION TIPS

Hummingbirds are the little iridescent green jewels that visit flower gardens and readily come to feeders with sugar water. Their long bills are designed to reach way into tubular flowers, where they lick up the nectar; but these birds are also good at catching aerial insects, a key source of their protein. In most cases the sexes are different in appearance.

Species ID: Most adult male hummingbirds have distinctively shaped and colored gorgets (throat patches) that can shine brilliantly in the right light but the rest of the time often appear blackish. Many of the females are greenish on the upperside and whitish on the underside, and they can be difficult to distinguish. Some helpful clues are the length and shape of the bill, and where the tips of the folded wings fall relative to the tip of the tail; also look at the overall size and shape of the body, the color of the flanks, and the presence of any streaking or patterning on the throat. Many juvenile hummingbirds of both sexes look like the adult female, but they often have fine paler edges to their crown feathers, making them look minutely scaled.

Adult m. MEX/–

Adult m. MEX/09

Xantus's Hummingbird 5

Hylocharis xantusii L 3½"

Accidental Mexican vag. to West. **Shape:** Very similar shape to White-eared Hummingbird, although slightly smaller. **Ad:** Similar to m. and f. White-eared but underparts all warm buff and tail reddish brown. **Juv:** (Aug.–Mar.) Like ad. f. but with buff margins to fresh crown feathers. M. with variable dark spotting on throat; f. paler buff below. Complete molt into ad. plumage occurs Mar.–Sep. **Flight:** White eyebrow, buffy belly and undertail coverts, reddish-brown tail. **Hab:** Arid woodlands, residential gardens. **Voice:** Calls include a sharp *tsik* or series of harsh chips; song a mix of harsh chips and chittering.

Subspp: Monotypic.

Adult MEX/09

Adult AZ/08

Berylline Hummingbird 3

Amazilia beryllina L 3¾"

Rare summer visitor from MEX to Southwest. **Shape:** Medium-sized hummingbird with a moderate-length tail; bill medium-sized (about length of head) and fairly straight. **Ad:** Dark overall; throat and breast dark green; belly grayish brown to buffy; extensive reddish brown at base of wing feathers (seen on perched and flying birds); purplish tail and uppertail coverts. Sexes similar but f. may be slightly paler overall and have some whitish flecking on throat and chin. Similar Buff-bellied Hummingbird (ranges do not overlap) has dusky wings; extensive red at base of bill; and reddish-brown tail and uppertail coverts. **Juv:** (May–Aug.) Like ad. f. but underparts mostly buffy. **Flight:** Note reddish brown on wings. **Hab:** Mountain canyons. **Voice:** Very short *tsit* or *tsidit*.

Subspp: (1) •*viola*. **Hybrids:** Magnificent Hummingbird.

Adult TX/11

Adult LA/12

Adult TX/11

Adult LA/12

Buff-bellied Hummingbird 2

Amazilia yucatanensis L 4"

Shape: Medium-sized hummingbird with a moderate-length tail; bill medium-sized (about length of head) and fairly straight. **Ad:** Dark overall; throat and breast dark green; belly buffy; wings dusky; bill with extensive red on base; tail and uppertail coverts reddish brown. Sexes similar, but f. may be slightly paler overall and have some whitish flecking on throat and chin. Similar Berylline Hummingbird (ranges do not overlap) has extensive reddish brown at wing bases, red on only lower mandible (sometimes hard to see), and purplish tail and uppertail coverts. **Juv:** (May–Aug.) Like ad. but upperpart feathers edged with buff; throat and breast buffy or buffy mixed with gray or green. **Flight:** Note lack of reddish brown in wings; presence of reddish brown in tail. **Hab:** Open woods, gardens, feeders. **Voice:** Calls include single and multiple bursts of short *djit* sounds.

Subspp: (1) •*chalconota.*

Adult AZ/07

Adult AZ/08

Adult AZ/08

Violet-crowned Hummingbird 2

Amazilia violiceps L 4¼"

Shape: Fairly large, slender, small-headed hummingbird with a fairly broad and squared-off to slightly notched tail. Bill fairly long (about 1¼ x head length) and straight. **Ad:** Bill bright red with dark tip. Crown and auriculars bluish violet; rest of upperparts brownish bronze; underparts mostly snowy white with dusky smudging on flanks. Sexes alike. **Juv:** (Feb.–Aug.) Like ad. but upperpart feathers edged buff, with little or no violet; bill dull red at base, shading to dark tip. **Flight:** Note red bill, bright white underside, lack of green iridescence. **Hab:** Streamside deciduous woods, usually with sycamores, feeders. **Voice:** Calls include explosive short bursts of squeaky chips; song is several downslurred, high-pitched whistles, like *tseeoo*.

Subspp: (1) •*ellioti*. **Hybrids:** Broad-billed, Magnificent Hummingbirds.

Adult AZ/08

Juv. AZ/07

Plain-capped Starthroat 4

Heliomaster constantii L 4¾"

Casual Mexican visitor to s. AZ. **Shape:** Large attenuated hummingbird with a squarish tail. Bill very long (1⅔ x head length) and relatively straight. **Ad:** Sexes alike. Distinctive head with whitish postocular stripe, dark auricular, wide whitish malar streak, and dark throat. Upperparts dull bronzish green with long whitish patch on lower back; white tufts on lower flanks (not always seen on perched bird). Underparts gray except for blackish throat, which may show red iridescence on lower portion. Outer tail feathers tipped white. **Juv:** Like ad. but feathers of upperparts fringed buffy when fresh; throat grayish with little or no iridescence. **Flight:** Note long bill, white rump and flank patches, white-tipped tail. **Hab:** Open dry woods, deserts, hillsides, feeders. **Voice:** Loud single *tseeek;* song a variable series of chips and chitters.

Subspp: (1) •*pinicola*.

Adult m., *bessophilus* AZ/04

Adult m., *bessophilus* AZ/05

Adult f., *bessophilus* AZ/06

Adult f., *bessophilus* AZ/06

Blue-throated Hummingbird 2

Lampornis clemenciae L 5"

Shape: Very large broad-tailed hummingbird; bill medium-sized (about length of head) and straight. **Ad:** Greenish head and upper back blend to blackish rump; grayish below. Even white postocular stripe and thin white malar streak; broad white tips to large dark tail. M. throat deep blue (often looks just dark); f. throat gray. Similar f. Magnificent Hummingbird has long bill, jagged postocular stripe, and small white tips to tail. **Juv:** (May–Nov.) Like ad. but feathers of upperparts edged with gray; postocular stripe less distinct. Juv. m. with less blue on throat than ad. male. **Flight:** Large size; large white spots at corners of dark tail. **Hab:** Wooded canyon streams, feeders. **Voice:** Calls include high-pitched *tseeet* (single or in series); song similar but in a series. Magnificent gives a harsh *chik*.

Subspp: (2) •*bessophilus* (AZ–NM) has medium-length thin bill; underparts pale gray. •*phasmorus* (w. TX) has shorter, wider bill; underparts dark gray. **Hybrids:** Magnificent, Black-chinned, Anna's, and Costa's Hummingbirds.

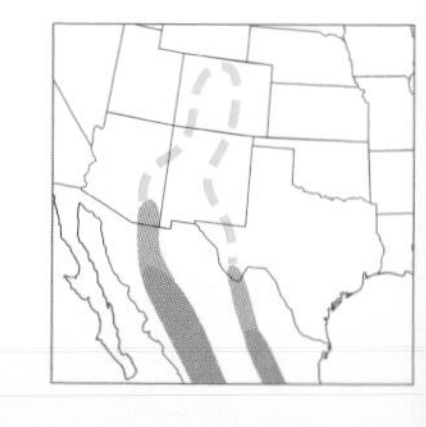

Adult m. AZ/04

Adult m. AZ/04

Adult f. AZ/04

Adult f. AZ/04

Magnificent Hummingbird 2

Eugenes fulgens L 5"

Shape: Very large, fairly slender, and elongated hummingbird; bill long (about 1⅓ x head length) and straight. Appears big-headed and long-necked; crest on male often raised. M. tail notched, f. tail mostly squared off. **Ad:** Both sexes have bright white postocular dot. M. often appears almost black, iridescence only rarely showing to full extent; crown and forehead iridescent purple; throat bright green to turquoise; belly blackish; undertail coverts dusky; tail dark. F. has variable jagged whitish line trailing off postocular spot and thin whitish malar streak; underparts gray heavily spotted with darker gray or green; upperparts iridescent green; corners of greenish tail narrowly tipped white or pale gray. Similar f. Blue-throated Hummingbird has medium-length bill, even postocular stripe, broad white tips to tail. **Juv:** (Jun.–Sep.) Head like ad. f., with jagged postocular line and slight whitish malar streak. Feathers of upperparts fringed with buff. Juv. m. with green throat patch, notched tail (sometimes with subtly paler tips); juv. f. with evenly spotted gray throat, square tail with small white outer tips. **Flight:** Note large size and long bill. **Hab:** Open woods, canyons, feeders. **Voice:** Calls include a harsh *chik* and an ascending series of whistles like *see see see sidideet;* song a complex mix of harsh notes and *tseeps*.

Subspp: (1). •*fulgens*. **Hybrids:** Broad-billed, Berylline, Violet-crowned, and Blue-throated Hummingbirds.

Adult m. AZ/08

Adult m. AZ/09

Adult f. AZ/09

Adult f. TX/03

Lucifer Hummingbird 2

Calothorax lucifer L 3¾"

Shape: Small hummingbird with a long tail that is forked but usually mostly closed to a narrow point. Bill long (about 1¼ x head length) and downcurved. On perched bird, m. tail projects well past wing tips by almost length of bill, f. tail just beyond wing tips. Overall, appears large-headed and often perches in hunched posture. **Ad:** M. has rosy purple gorget that is long and broad, with a scalloped lower edge and slightly elongated corners. F. with buffy flanks, breastband, and postocular stripe; auriculars dusky; dark lore; outer tail feathers with reddish-brown base, broad black subterminal band, white tips. **Juv:** (May–Oct.) Like ad. f. but feathers of upperparts fringed buffy when fresh; juv. m. with forked tail and some iridescent spots on throat. Complete molt into ad. plumage occurs Jul.–Feb. **Flight:** Note male's long forked tail, female's buffy breastband and reddish-brown tail base. **Hab:** Desert canyons, rocky hillsides. **Voice:** Calls include a fairly low *chet* and short bursts like *tsitsitsit*.

Subspp: Monotypic. **Hybrids:** Black-chinned Hummingbird.

Adult m. NH/06

Adult m. TX/04

Adult m. MN/07

Adult f. TX/04

Adult f. MN/07

Ruby-throated Hummingbird 1

Archilochus colubris L 3½"

Shape: Small, slender, elongated hummingbird with a relatively long tail and defined neck. Bill medium-sized (length of head) and straight. On perched bird, m. wing tips fall well short of tail tip; f. wing tips just short of tail tip. **Ad:** M. iridescent green above; below, dark open vest of green and grayish buff; ruby-red gorget (scalloped lower border) with broad white collar below; black patch from chin to below and behind eye. F. iridescent green above; underparts whitish with slight green or buff to flanks; may have a few dusky (rarely red) spots on throat; generally holds tail still while flying; white tips to outer tail feathers. F. primaries on perched bird taper gently to tip and have a gradual curve over their length; similar Black-chinned Hummingbird f. primaries on perched bird wide to tip and curve at their tip. **Juv:** (May–Feb.) Like ad. f., but juv. m. starts to develop red spots on throat by Aug. Juv. m. tail has white tips but is slightly more forked than juv. f. tail. Complete molt into ad. plumage occurs Oct.–Mar. **Flight:** Generally holds tail in plane of body and quivers it; may pump tail in wind or while maneuvering, but not habitually as Black-chinned does. **Hab:** Woodland edges and clearings, gardens, feeders. **Voice:** Calls include repeated *tseeet ch'ch'ch'ch* and single soft *chew;* song a long series of *chits*, occurs at dawn.

Subspp: Monotypic. **Hybrids:** Black-chinned Hummingbird.

Adult m. AZ/07

Adult m. AZ/04

Adult f. AZ/04

Adult f. AZ/04

Black-chinned Hummingbird 1

Archilochus alexandri L 3½″

Shape: Small, slender, elongated hummingbird with a relatively long tail and defined neck. Bill long (about 1¼ x head length) and straight. On perched bird, wing tips fall just short of tail tip. **Ad:** M. iridescent dull green above; below, dark open vest of grayish green; black chin and violet gorget (fairly straight lower border) with broad white collar below. F. iridescent dull green above, grayish green on crown; underparts whitish with slight green or buff to flanks; may have a few dusky spots on throat; constantly flips tail up and down while flying; white tips to outer tail feathers. F. primaries on perched bird wide to tip and curve at their tip. **Juv:** (May–Oct.) Like ad. f., but feathers of head, nape, and back with extensive buffy edges. Juv. m. usually has noticeable streaking on throat and 1–2 black gorget feathers. Complete molt into ad. plumage occurs Nov.–Apr. **Flight:** Habitually pumps tail up and down. **Hab:** Canyons, streamsides, feeders. **Voice:** Calls include rapid *ch'ch'ch'chit* or *tseee chchchchit*, also a soft *chew.*

Subspp: Monotypic. **Hybrids:** Blue-throated, Lucifer, Ruby-throated, Anna's, Costa's, Broad-tailed, and Allen's Hummingbirds.

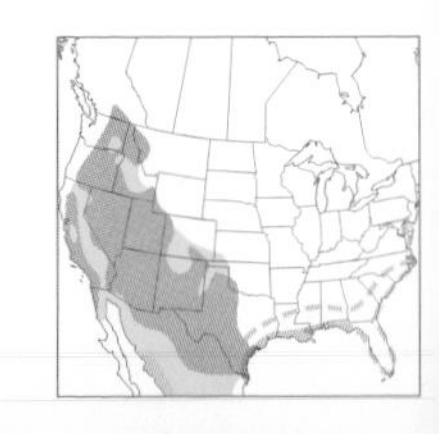

Adult m. CA/03

Adult m. AZ/02

Adult f. CA/03

Adult f. AZ/08

Anna's Hummingbird 1

Calypte anna L 3¾"

Shape: Medium-sized, stocky hummingbird with a long tail and compact large-headed look. Bill relatively short (shorter than head length) and fairly straight. On perched bird, m. wing tips well short of tail tip, f. wing tips reach or fall just short of tail tip. **Ad:** M. with rose to reddish-orange iridescence on gorget and crown, appearing like a helmet; gorget flares slightly at corners. Upperparts iridescent green; underparts grayish with green scaling, heaviest on flanks, lighter on belly and breast. F. upperparts iridescent green; underparts grayish with heavy green scaling on flanks; variable small rosy pink iridescent patch in center of throat. **Juv:** (Jan.–Aug.) Like ad. f. but upperpart feathers with buff edging; juv. f. with little or no dusky scaling or rose iridescence on throat; juv. m. with blotches of rose iridescence on throat. Complete molt into ad. plumage occurs Apr.–Jan. **Flight:** Tail held in body plane and quivered in flight. **Hab:** Varied; canyons, chaparral, open woodlands, gardens, feeders. **Voice:** Short, spaced, 1–2-syllable harsh calls, *chit, chit, chidit, chit;* m. song a varied and extended series of harsh, high-pitched, scratchy sounds.

Subspp: Monotypic. **Hybrids:** Black-chinned, Costa's, Calliope, Allen's Hummingbirds.

Adult m. CA/06

Adult m. CA/03

Adult f. AZ/01

Adult f. CA/04

Costa's Hummingbird 1

Calypte costae L 3¼"

Shape: Small, rather potbellied, large-headed hummingbird. Bill medium length (about head length) and straight or slightly decurved. **Ad:** M. iridescent purple gorget and crown; very long corners of gorget cross broad white collar; green scaled open vest on flanks and belly. F. has iridescent green upperparts; throat pale and generally unmarked (some older f. may develop small violet throat patch); rest of underparts whitish, sometimes with greenish wash on flanks; pale gray auricular. **Juv:** (Mar.–Sep.) Like ad. f. but upperpart feathers with buff edging; juv. m. with blotches of purple iridescence on throat and crown. Complete molt into ad. plumage occurs Jun.–Nov. **Flight:** Tail held in body plane, often flipped and spread. **Hab:** Desert scrub, chaparral, some open dry pine woodlands. **Voice:** High-pitched *tsik tsik;* slightly longer buzzy or squealing calls; song of m. very long, very high-pitched, ascending then descending *tsuiiieeer.*

Subspp: Monotypic. **Hybrids:** Blue-throated, Black-chinned, Anna's, Calliope, and Broad-tailed Hummingbirds.

Adult m. AZ/06

Adult m. MT/06

Adult f. AZ/08

Adult f. CA/07

Calliope Hummingbird 1

Stellula calliope L 3¼"

Shape: Small, rather potbellied, large-headed hummingbird. Tail short, and wing tips extend beyond tail when bird is perched. Bill short (less than head length) and fairly straight. **Ad:** M. iridescent green above; streaked reddish-purple gorget extends over shoulders; underparts whitish with green spots and buffy wash on flanks. F. iridescent green above; whitish below, with buffy flanks sometimes extending into faint buffy collar; throat lightly streaked with spots; tail has very limited or no reddish brown at base and looks blackish; loral line broken in middle. **Juv:** (Jun.–Oct.) Like ad. f. but upperpart feathers edged with buff; juv. m. may have some purple iridescence on throat. **Flight:** Tail generally held still in body plane. **Hab:** Mountains, early-succession scrub habitats, feeders. **Voice:** Short *tsip tsip tsidip;* high-pitched buzzy chitter; m. wing whir.

Subspp: Monotypic. **Hybrids:** Anna's, Costa's, Broad-tailed, and Rufous Hummingbirds.

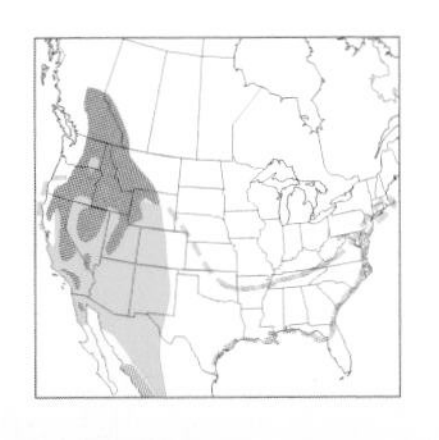

Adult m. AZ/08

Adult m. CA/03

Adult females NM/08

Adult f. CA/04

Rufous Hummingbird 1

Selasphorus rufus L 3½"

Shape: Small stocky hummingbird with a relatively long rounded tail. Bill short (less than head length) and fairly straight. On perched bird, wing tips fall well short of tail tip. **Ad:** M. upperparts usually mostly orangish brown with some variable green flecking on back and crown; gorget iridescent orange-red and slightly flared at corners; underparts mostly orangish brown with white chest and variably white central belly. One to two percent of males have extensive green on back and are difficult to distinguish from m. Allen's Hummingbird. F. iridescent green above; below, orangish-brown flanks, white forecollar and central belly; throat finely streaked with bronze or green dots, often with irregular central orange-red blotch; tail with extensive orangish brown at base. Flat horizontal lower edge to lores (f. Calliope Hummingbird has uneven lower edge). Nearly identical in the field to f. Allen's and cannot be distinguished except by hard-to-see shapes of individual tail feathers. In f. Rufous, next-to-central tail feathers slightly notched on inner web, and outer tail feathers nearly as broad as adjacent ones. In similar f. Allen's, next-to-central tail feathers generally unnotched, and outermost tail feather narrower than in Rufous, but difference slight. **Juv:** (Jun.–Nov.) Like ad. f., but juv. f. may have whitish throat with few markings; juv. m. may have more heavily marked throat with larger iridescent reddish spots. Complete molt into ad. plumage occurs Sep.–Mar. **Flight:** Note extensive orangish-brown body and tail of ad. male. **Hab:** Variable; open woods, mountain meadows, shrubby areas, gardens, parks, feeders. **Voice:** High-pitched, short, harsh *chit, chitit, chit* interspersed with loud *dzeeet* calls; high trill created by m. wings in flight; chittering call at bottom of m. dive display.

Subspp: Monotypic. **Hybrids:** Anna's, Calliope, possibly Allen's Hummingbirds.

Adult m. CA/05

Adult m. CA/01

Adult f. CA/04

Adult f. CA/06

Allen's Hummingbird 1

Selasphorus sasin L 3½"

Shape: Small stocky hummingbird with a relatively long rounded tail. Bill short (less than head length) and fairly straight. On perched bird, wing tips fall well short of tail tip. **Ad:** M. body mostly orangish brown with dull green crown and variably iridescent green back; gorget iridescent orange-red and slightly flared at corners; white chest. (See Rufous Hummingbird for comparison.) F. iridescent green above; below, orangish-brown flanks, white forecollar and central belly; throat finely streaked with bronze or green dots, often with irregular central orange-red blotch; tail with extensive orangish brown at base. Nearly identical in the field to f. Rufous and cannot be distinguished except by hard-to-see shapes of individual tail feathers. In f. Allen's, next-to-central tail feathers generally unnotched, and outermost tail feather narrower than in Rufous, but differences slight. In similar f. Rufous, next-to-central tail feathers slightly notched on inner web, and outer tail feathers nearly as broad as adjacent ones. **Juv:** (Mar.–Aug.; see Subspp.) Like ad. f., but juv. f. may have whitish throat with few markings; juv. m. may have more heavily marked throat with larger iridescent reddish spots. Complete molt into ad. plumage occurs Aug.–Feb. **Flight:** Note extensive orangish-brown body and tail of ad. male. **Hab:** Variable; open woods, shrubby areas, gardens, parks. **Voice:** High-pitched, short, harsh *chit, chitit, chit* interspersed with loud *dzeeet* calls; high trill created by m. wings in flight; long *tseeeek* call at bottom of m. dive display.

Subspp: (2) Differ slightly in size and flank color. •*sedentarius* (Channel Is. and s.w. coastal CA) has slightly longer bill and wings, f. often with extensive green spotting on flanks. •*sasin* (rest of range) has slightly shorter bill, little or no green spotting on flanks. **Hybrids:** Black-chinned, Anna's, and possibly Rufous Hummingbirds.

Adult m. AZ/07

Adult m. CA/05

Adult f. MEX/10

Adult f. AZ/04

Broad-tailed Hummingbird 1

Selasphorus platycercus L 3¾"

Shape: Medium-sized rather slender hummingbird with a long very broad tail (when spread, almost as wide as extended wings); m. tail squarish; f. tail rounded. Bill medium length (about length of head) and fairly straight. On perched bird, wing tips fall well short of tail tip. **Ad:** M. upperparts iridescent green; gorget rosy red with very fine white line where chin meets bill; area below and behind eye grayish (not black like similar m. Ruby-throated Hummingbird); underparts whitish with broad white collar and open vest of gray and green on flanks. Male (except when in molt) creates metallic trill with wings when moving from one place to another (not while hovering to feed). F. very similar to f. Calliope Hummingbird, with buffy flanks, finely streaked throat, and some reddish brown on tail base, but lores are dark with no white over gape (also see all elements of Shape). **Juv:** (Jun.–Nov.) Like ad. f. but upperpart feathers with buff edging; juv. m. may have some iridescence on throat. **Flight:** Tail generally held still in body plane. **Hab:** Open woodlands near shrubs and meadows. **Voice:** M. wing whir is a repeated, fairly soft, short trill, like *djeee djeee djeee;* also gives short *tsip,* multisyllable *chiterdit,* and drawn-out *dzeeet*.

Subspp: (1) •*platycercus.* **Hybrids:** White-eared, Black-chinned, Costa's, and Calliope Hummingbirds.

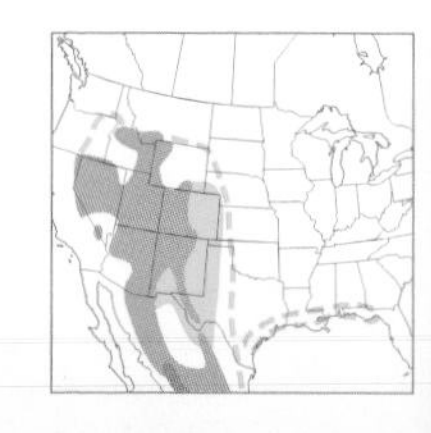

Adult m., *goldmani* AZ/05

Adult m., *goldmani* AZ/05

Adult m. AZ/12

Adult f., *goldmani* AZ/05

Juv. m., *goldmani* AZ/08

Adult f. AZ/01

Elegant Trogon 2

Trogon elegans L 12½"

Shape: Large rather heavy-bodied bird, long square-tipped tail, short deep-based bill. Often appears hunchbacked. **Ad:** Orangish-yellow bill. M. with bright red belly, thin white breastband, iridescent green chest; upperparts green with gray wing coverts and bronze tail. F. has gray head and upper breast, whitish belly, red undertail coverts; upperparts grayish with warm brown tail; broad white vertical dash behind eye. **Juv:** (Jun.–Nov.) Like ad. f., but lacks red undertail coverts and has large white spots on upperwing coverts and tertials. Juv. m. has some green on back; juv. f. lacks green on back. **Flight:** Takes short flights to pick off insects or fruit. Note long tail with white barring underneath. **Hab:** Higher-elevation streamside woodlands. **Voice:** Loud, hollow, low-pitched notes like *kwo kwo kwo kwo*, or *choy choy*, or *churrit churrit churrit;* a longer, harsh, downslurred *chuweer.*

Subspp: (2) •*goldmani* (AZ–NM) slightly larger, m. red paler, f. grayish; •*ambiguus* (vag. to TX) smaller, m. red dark, f. brownish.

Eared Quetzal 4

Euptilotis neoxenus L 14"

Casual Mexican visitor to s.e. and cent. AZ. **Shape:** Large heavy-bodied bird with a long very broad tail, relatively small head, and short bill. Somewhat pigeonlike proportions and size. **Ad:** Bright red belly and undertail coverts; iridescent green upperwing coverts; outer tail feathers broadly tipped with white; dark gray bill. M. with iridescent green head and breast; f. with dark gray head and breast. **Juv:** (Jun.–Sep.) Brownish head and breast. **Flight:** Large wide tail broadly tipped white on corners; red belly and undertail coverts. **Hab:** Mountain canyon woodlands. **Voice:** Whistled, high-pitched, repeated *peedeedeet;* a 2-part *pseeet-chuk* or single *pseeet.*

Subspp: Monotypic.

Adult m. MEX/12

Adult f. ARG/–

Adult m. MEX/12

Ringed Kingfisher 2

Megaceryle torquatus L 16"

Shape: Large, stout-bodied, big-headed bird with a bushy crest and long, massive, sharp-pointed bill. **Ad:** Bluish head and upperparts; white collar; tail barred black and white. M. with brick-red breast and belly; white undertail coverts. F. with bluish breast; thin white breastband; brick-red belly and undertail coverts. **Juv:** (Jun.–Dec.) Like ad. m. but feathers of upperparts tipped with white, creating spotted appearance. **Flight:** Relatively long wings with white spots on flight feathers. M. underwing coverts whitish; f. underwing coverts concolor with red belly. **Hab:** Large rivers with steep banks. **Voice:** Rapid rattling series of harsh notes, *ch'ch'ch'ch'ch,* slower than Belted Kingfisher (countable).

Subspp: (1) •*torquatus.*

Adult m. BRA/08

Adult f. TX/12

Green Kingfisher 2

Chloroceryle americana L 8¾"

Shape: Fairly small, slim, big-headed bird with a long sharp-pointed bill. Crown flat and long with small inconspicuous crest at hindcrown. **Ad:** Emerald-green head and upperparts; broad white collar; white belly and undertail coverts. M. with reddish-brown breast; f. with 2 mottled green breastbands. **Juv:** (Jun.–Apr.) Like ad. but often with small buffy spots on crown, back, and upperwing coverts. Juv. m. breast mix of reddish brown, green, and white; juv. f. breast brownish or washed with buff. **Flight:** White underwing coverts, white outer tail feathers spotted with green. Flies fast and low over water; rarely hovers. **Hab:** Rivers and lakes with shrubby edges. **Voice:** Short series of dry insectlike *tiks.*

Subspp: (2) •*hachisukai* (s.w. TX, vag. to AZ) upperparts are green with bronze overtones; white lines on forehead extend above eye. •*septentrionalis* (s. TX) upperparts are all green; white on forehead limited, usually does not extend above eye.

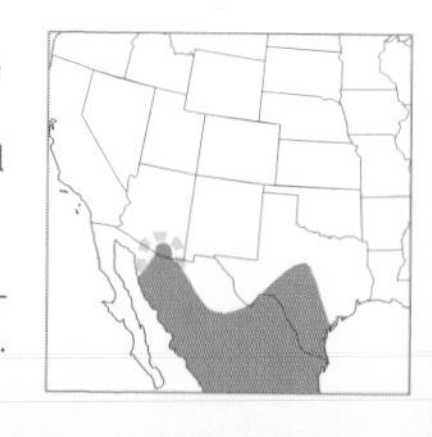

Adult m. TX/–

Adult NY/–

Adult f. MN/09

Adult f. FL/02

Juv. OH/06

Belted Kingfisher 1

Megaceryle alcyon L 13"

Shape: Medium-sized, sturdy-bodied, big-headed bird with a bushy crest and long, massive, sharp-pointed bill. **Ad:** Bluish-gray head and upperparts; broad white collar; bluish-gray breastband. M. has white unmarked belly; f. has white belly with reddish-brown belly band and flanks. **Juv:** (Jun.–Dec.) Like ad. but with reddish-brown feathers in breastband. Reddish brown on flanks but not belly indicates juv. m.; complete belly band indicates f.; partial belly band is inconclusive. **Flight:** Whitish underwing coverts; above, whitish base of primaries creates white patch on outer wing. **Hab:** Near water; lakes, rivers, coasts. **Voice:** Very rapid rattling series of harsh notes, *chchchchchch.* 🎧**53**.

Subspp: Monotypic.

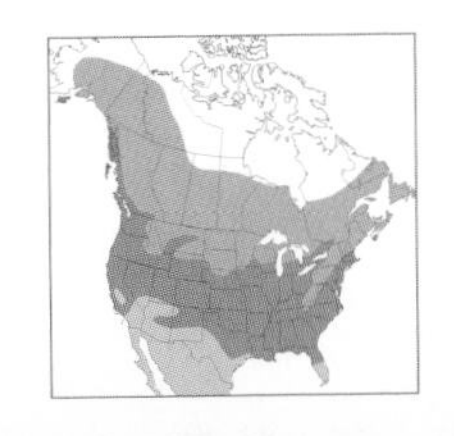

Woodpeckers

Sapsuckers

Flickers

Adult OR/07

Juv. OR/08

Lewis's Woodpecker 1

Melanerpes lewis L 10¾"

Shape: Large slender woodpecker with relatively long tail and small head. **Ad:** Black crown, back, wings, and tail with hint of iridescent green in certain lights; wide grayish collar extends across breast; belly washed pinkish red; face dark red. Sexes alike. **Juv/1st Yr:** (Jun.–Feb.) Head, breast, and flanks mostly dark brown, gradually changing to ad. colors by Feb.; lacks gray collar on nape. **Flight:** Steady crowlike wingbeats on broad wings. **Hab:** Open country with scattered trees. **Voice:** Short, soft, squeaky *chiv* (m.) or *chivit* (f.); lower, raspy *churr churr* (m.); descending series of chattering notes; drum is a roll, sometimes with a few slower hits at the end (does not drum often).

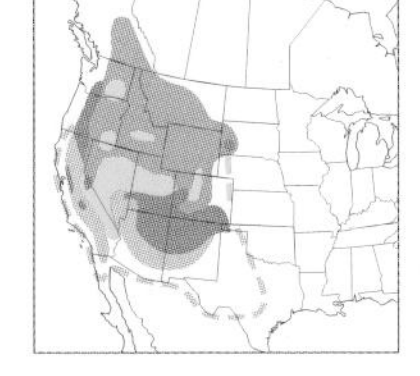

Subspp: Monotypic.

IDENTIFICATION TIPS

Woodpeckers typically perch vertically on the sides of trees or cacti using their stiff tail feathers as a brace. They use chisel-like bills to excavate nest holes and go after insects in wood. Their loud rhythmic drumming is a communication before and during breeding; chipping away wood has a more erratic rhythm. In most cases the sexes look different, the male often having slightly more red on head.

Species ID: Note size of the bird and size and shape of the bill. Then look at colors and patterns on the head (including the exact location of any red) and patterns of barring or spotting on back and wings.

Adult TX/04

Juv., trans. to 2nd yr. OK/11

Juv. TX/07

Red-headed Woodpecker 1

Melanerpes erythrocephalus L 9¼"

Shape: Medium-sized sturdily built woodpecker with a large deep-based bill. **Ad:** Striking crimson-red hood; black upper back; white lower back. In all ages, has distinctive large white patch consisting of secondaries, tertials, and rump; obvious on perched or flying birds. Sexes alike. **Juv/1st Yr:** (Jun.–2nd Sep.) Head grayish brown, gradually molting to dull red by 1st spring/summer; lighter margins on back feathers create scalloped look; 1–2 dark bars across white secondaries and tertials (may still be seen on 2nd- and 3rd-yr. birds); underparts white. **Flight:** Broad white band on trailing half of inner wings continues across white rump. **Hab:** Deciduous woods, open country with trees. **Voice:** Slightly rising *queeek;* raspy *churrr;* drum is a rapid short roll.

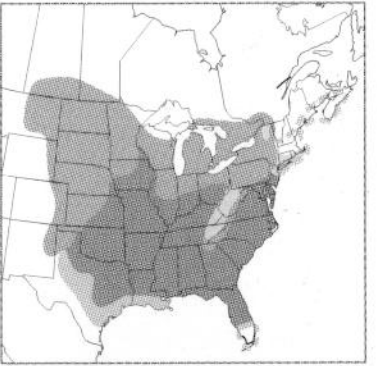

Subspp: Monotypic.

Adult m., *bairdi* CA/06

Adult f., *formicivorus* AZ/05

Adult m. CA/04

Adult f. AZ/05

Acorn Woodpecker 1

Melanerpes formicivorus L 9"

Shape: Medium-sized sleek woodpecker with a relatively small head, broad neck, and short bill. **Ad:** Distinctive white eye surrounded by black mask on face; red crown; white forehead and cheek; glossy black back. White below with black chest and black streaking on upper belly and flanks. M. has red crown touching white forehead; f. has red hind-crown and black midcrown touching white fore-head. **Juv:** (May–Oct.) Like ad. m. but iris dark; red on crown orangish; throat can be pinkish. **1st Yr:** (Oct.–Sep.) Like ad. but dull blackish-brown flight feathers contrast with shiny black wing coverts and back. **Flight:** White belly; white rump; white patches at base of primaries. **Hab:** Woods with oaks. **Voice:** An active, social, and very vocal wood-pecker. Loud *queeyer;* raspy *churrrit churrrit-chuk; wakka wakka wakka;* a chatter; drum is a short even roll of 10–20 beats.

Subspp: (2) •*bairdi* (s. WA–s. CA) is large-billed with broad chestband; •*formicivorus* (AZ–w. TX) is smaller-billed with narrow chestband.

Gila Woodpecker 1

Melanerpes uropygialis L 9¼"

Shape: Medium-sized sleek woodpecker with a relatively long neck and long thick bill. **Ad:** Body tan; wings, back, and rump finely barred black and white; central and outermost tail feathers barred black and white; rest of tail black. M. has red crown, tan nape; f. all-tan head. **Juv:** (Apr.–Sep.) Like ad. but black-and-white barring on back and wings indistinct. M. with mottled red crown; f. with none or 1–2 red feathers on crown. **Flight:** Barred black and white above, including rump and central tail feathers; white patch at base of primaries. **Hab:** Arid areas with large trees or cacti. **Voice:** High-pitched flat *keeah keeah keeah; churr;* drum is a simple roll.

Subspp: (1) •*uropygialis.* **Hybrids:** Golden-fronted Woodpecker.

Adult m. TX/02

Adult f. TX/11

Golden-fronted Woodpecker 1

Melanerpes aurifrons L 9½"

Shape: Medium-sized sleek woodpecker with a relatively long neck and long thick bill. **Ad:** Body tan; wings and back finely barred black and white; rump white; central tail feathers black; nape and nasal tufts gold to orange. M. has red crown; f. tan crown. **Juv:** (Apr.–Sep.) Like ad., but indistinct barring on back, fine dark streaking on breast and crown; yellow tinge on nasal tufts and nape. M. has a few red feathers on crown; f. has a few to none. **Flight:** Barred black-and-white back and wings; white rump; black tail. **Hab:** Variable; open woodlands to brushlands. **Voice:** A rolling *churrr;* a repeated *kek kek kek kek;* a modulated *dsuka dsuka;* drum is a simple roll with 1–4 single taps before or after.

Subspp: (1) •*aurifrons.* **Hybrids:** Red-bellied and Gila Woodpeckers.

Adult m. FL/01

Adult f. FL/02

Red-bellied Woodpecker 1

Melanerpes carolinus L 9¼"

Shape: Medium-sized sleek woodpecker with a relatively long neck and long thick bill. **Ad:** Body tan; wings and back finely barred black and white; rump white with a few black markings; central tail feathers variable, from mostly black to white with black bars; nape red (rarely yellow); nasal tufts orangish red. Variable reddish wash on belly (often hard to see). M. has red midcrown; f. tan midcrown. **Juv:** (May–Sep.) Like ad., but head dusky with no red or yellow on nape or nasal tufts and indistinct barring or spotting on back. M. usually has a few red feathers on crown; f. may have a few to none. **Flight:** Central tail and rump white mottled with black; white patch at base of primaries. **Hab:** Variable; mixed hardwood forests, bottomland woods, residential areas, parks. **Voice:** A bubbling *churrr;* a repeated, harsh *chew chew chew,* increasing in speed to a *chew chew chechechechurrr;* drum is a simple roll of 15–20 beats. 🎧**54**.

Subspp: Monotypic. **Hybrids:** Golden-fronted Woodpecker.

Adult m. CO/07
Adult f. CA/06

Adult m. IL/03

Adult f. CT/06

Juv. m. CA/07

Juv. f. OR/08

Juv. NH/09

Williamson's Sapsucker 1

Sphyrapicus thyroideus L 9"

Shape: Medium-sized woodpecker with a medium-length bill. **Ad:** M. has mostly black upperparts with a bold white wing patch; head black with white moustache and eye-stripe. Broad black breastband separates red throat from yellowish belly. F. with brown relatively unmarked head; rest of upperparts brownish black with fine pale barring; flanks finely barred gray and blackish; breast with a central black patch; belly yellow. **Juv:** (Jun.–Sep.) Like respective adults but m. has white throat and f. lacks dark central breast patch. **Flight:** Extensive white rump; m. with white wing patch; f. wing all brown. **Hab:** Coniferous forests at mid to high elevations; also mixed woodlands. **Voice:** Short repeated *jit jit jit jit;* low hissing *shhh shhh;* a harsh *cheeer;* drumming is a short roll followed by several shorter rolls.

Subspp: Monotypic. **Hybrids:** Red-naped Sapsucker.

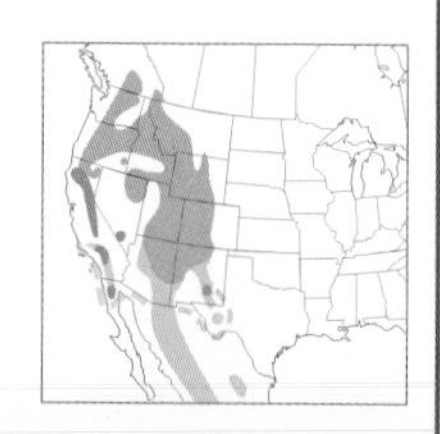

Yellow-bellied Sapsucker 1

Sphyrapicus varius L 8½"

Shape: Medium-sized woodpecker with a medium-length bill. **Ad:** Bold, long, white wing patch; black breastband; dark back evenly and extensively mottled with white, buff, or gold. Nape white and black (rarely red). M. has red throat completely bordered by black; red crown. F. has white throat completely bordered by black; mostly red crown. Rarely, f. may have all-black crown and/or a few red feathers on her throat. Similar Red-naped Sapsucker has red nape (usually), back mostly black with limited white mottling often concentrated into 2 rows down either side, and throat usually all red with no black border at rear or red and white. **Juv/1st Yr:** (Jun.–2nd Sep.) Retains juv. plumage into Dec. and may not acquire full ad. plumage until following fall. Thus, any sapsucker with substantial juvenal plumage Oct.–Jun. is most likely Yellow-bellied. Disting. from other juv. sapsuckers (Jul.–Sep.) by fine pale streaking on brown crown; other spp. have uniformly dark crowns. **Flight:** Whitish rump and white shoulder patch. **Hab:** Early succession forests. **Voice:** Downslurred *jeeer,* a shrieking *shreech shreech;* drum an uneven rhythm suggestive of Morse code, often slower at end.

Subspp: Monotypic. **Hybrids:** Red-naped and Red-breasted Sapsuckers.

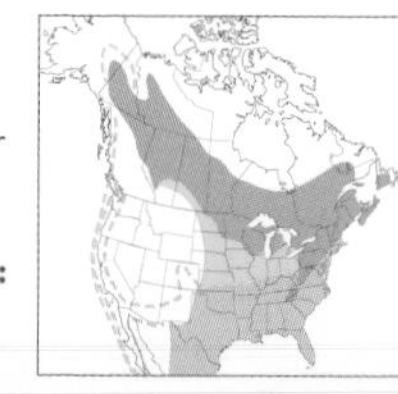

Adult m. BC/06

Adult f. CO/06

Adult, *daggetti* CA/06

Juv. CA/07

Adult, *ruber* NV/11

Red-naped Sapsucker 1

Sphyrapicus nuchalis L 8½"

Shape: Medium-sized woodpecker with a medium-length bill. **Ad:** Bold white wing patch; black breast-band; back mostly black with limited white mottling often concentrated into 2 rows down either side. Nape red and black (rarely white); crown red. M. has red throat that extends over black border of malar streak. F. has white chin (sometimes with just a few red feathers) and red throat completely bordered by black. Similar Yellow-bellied Sapsucker lacks red on nape, has extensive white mottling on white back, and has either red throat outlined with black or all-white throat. **Juv:** (Jun.–Sep.) Quickly molts on breeding ground into ad. plumage. Uniformly dark brown head and brownish wash to breast; white moustache. Generally darker than Yellow-bellied juv. and paler than Red-breasted juv. **Flight:** Whitish rump and white shoulder patch. **Hab:** Deciduous and mixed woodlands, aspens. **Voice:** High-pitched downslurred *cheeer;* drum is a few rapid taps followed by slower taps (similar to drum of Yellow-bellied and Red-breasted).

Subspp: Monotypic. **Hybrids:** Yellow-bellied, Red-breasted, and Williamson's Sapsuckers.

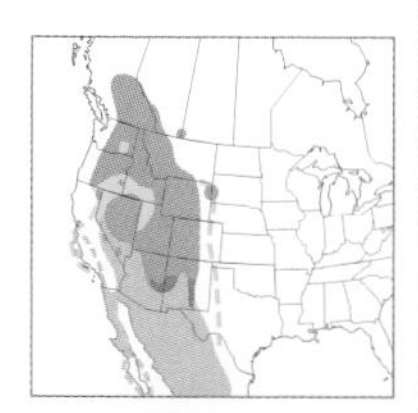

Red-breasted Sapsucker 1

Sphyrapicus ruber L 8½"

Shape: Medium-sized woodpecker with a medium-length bill. **Ad:** Bold white wing patch; extensive deep red on head and breast; white on face limited and variable (see Subspp.); back with variable white spotting (see Subspp.). Sexes cannot be reliably distinguished. **Juv:** (May–Sep.) Molts quickly on breeding ground into ad. plumage. Uniformly dark brown head and chest; white moustache. Generally darker than Red-naped or Yellow-bellied juvs. **Flight:** White central rump feathers; bold white shoulder patch; red head. **Hab:** Variable; mixed woodlands, coniferous forests, bottomlands. **Voice:** High-pitched downslurred *cheeer;* drum is a few rapid taps followed by slower taps.

Subspp: (2) •*ruber* (s. AK–n. OR) large; black back sparsely spotted white to yellowish; red breast clearly defined from dark yellow belly. •*daggetti* (s. OR–s. CA) small; black back densely spotted white; red breast blends to light yellow belly. **Hybrids:** Yellow-bellied and Red-naped Sapsuckers.

Adult m., *japonicus** JAP/02 Adult f., *japonicus** JAP/06

Adult m. AZ/04

Adult f. AZ/05

Juv. m., *major** FIN/07

Adult m. CA/05

Great Spotted Woodpecker 4

Dendrocopos major L 9"

Casual Eurasian vag. to w. AK. **Shape:** Medium-sized woodpecker with a medium-length fairly deep-based bill. **Ad:** Black cap; bold white auricular extends to surround eye; long white scapular line, reminiscent of sapsuckers; bright red undertail coverts; long black spur extends onto sides of white breast. M. has red nape patch; f. lacks red nape patch. **Juv:** Like ad. but undertail coverts tinged pinkish; m. crown red; f. crown black. **Flight:** Bold white scapulars on either side of black back. **Hab:** Woods. **Voice:** Rapid squeaky *kweekweekweekwee;* loud irregular *teek*s; drum is a rapid roll, subtly faster in middle or end.

Subspp: (1) •*kamtschaticus.* *D. m. major (n. EUR) and *D. m. japonicus* (KOR and JAP) used for good photographs.

Ladder-backed Woodpecker 1

Picoides scalaris L 7¼"

Shape: Small woodpecker with fairly long bill. Slightly smaller than similar Nuttall's Woodpecker, but with proportionately larger bill and indistinct nasal tufts, which blend into forehead line. **Ad:** "Zebra-backed." White barring extends all the way up black back; thin black eye-stripe and thin black malar streak connect behind white cheek; outer tail feathers have 3–4 black bars; often buffy wash to underparts. M. has red crown that extends variably to above eye; f. has black crown (rarely with a few red feathers). **Juv:** (May–Sep.) Like ad. but duller; variable white and red spotting on crown; m. usually with more red on crown than female. **Flight:** Finely barred back and wings. **Hab:** Creeks and washes of arid areas. **Voice:** Short high-pitched *cheek;* a lower *kweeek;* a high-pitched descending whinny, like *chichichichurrr;* drum is a very rapid roll (often high-pitched). Calls and drum similar to Downy Woodpecker's.

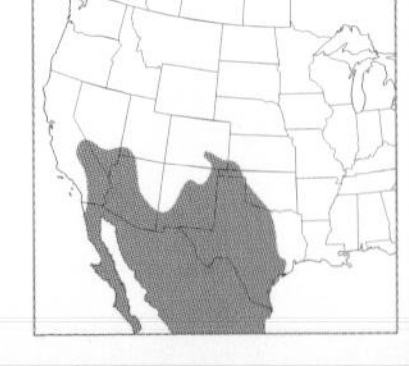

Subspp: (1) •*cactophilus.* **Hybrids:** Nuttall's and Hairy Woodpeckers.

Adult m. CA/05

Adult f. CA/11

Nuttall's Woodpecker 1

Picoides nuttallii L 7½"

Shape: Small woodpecker with short bill. Slightly larger and shorter-billed than similar Ladder-backed Woodpecker. Nasal tufts mounded and distinct. **Ad:** "Zebra-backed." White barring on black back stops before nape, leaving variably wide unbarred upper back; wide black eye-stripe and thin black malar streak connect behind white cheek; outer tail feathers have 1–2 black bars; white underparts. M. has red hindcrown—red ends before reaching eye; f. has black crown (rarely with a few red feathers). **Juv:** (May–Sep.) Like ad. but duller; variable white and red spotting on crown; m. usually with more red on crown than female. **Flight:** Finely barred back and wings. **Hab:** Mostly oak and riparian woodlands. **Voice:** Short 2-part *chivit;* a lower *kweeek;* raspy slightly descending whinny; drum is a rapid extended roll (often high-pitched).

Subspp: Monotypic. **Hybrids:** Ladder-backed and Downy Woodpeckers.

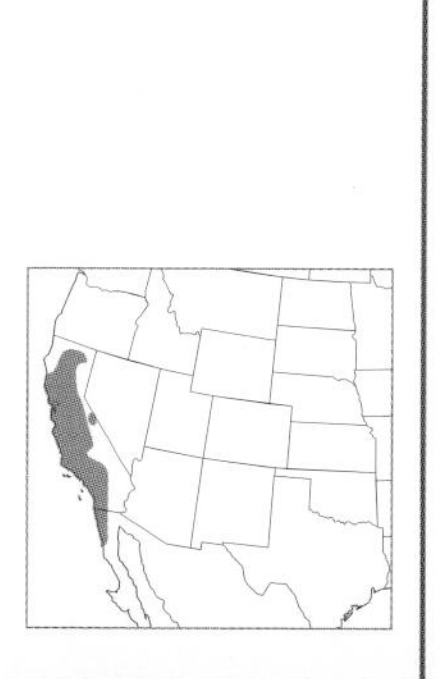

Adult m. AZ/04

Adult f. AZ/05

Arizona Woodpecker 2

Picoides arizonae L 7½"

Shape: Small woodpecker with a relatively large head, medium-length bill, and defined neck. **Ad:** Our only woodpecker with an all-dark-brown back. Large brown ear patch surrounded by extensive white. Below, whitish heavily spotted with brown; spots arranged into barring on flanks, lower belly, and undertail coverts. Outer tail feathers white with many dark bars. M. has red nape patch; f. no red nape patch. **Juv:** (May–Sep.) Like ad. but red on central crown; more in m. than f. **Flight:** Relatively dark overall except for large white patch on sides of nape. **Hab:** Oak or pine woodlands, often in canyons. **Voice:** A loud *teeek;* a raspy downslurred whinny; drum is a simple rapid roll.

Subspp: (1) •*arizonae.*

Adult m., Coastal Pac. Grp. CA/11

Adult m., Int. Western Grp. CO/07

Adult m., Eastern Grp. NH/12

Adult f., Coastal Pac. Grp. CA/11

Adult m., Int. Western Grp. CO/07

Adult f., Eastern Grp. TX/05

Downy Woodpecker 1

Picoides pubescens L $6\frac{3}{4}$"

Shape: Small, short-billed woodpecker with proportionately large head and short broad neck. Bill less than ½ depth of head; nasal tufts proportionately large. In similar Hairy Woodpecker, bill usually greater than ½ depth of head; nasal tufts proportionately smaller. **Ad:** White to dusky central back and underparts; black wings variably spotted or barred white to dusky; usually has dark bars on outer tail feathers; usually lacks extension of malar streak into a spur or point on side of breast. M. has red nape patch; f. no red nape patch (rarely just a few red feathers). Similar Hairy Woodpecker usually lacks dark bars on outer tail feathers and has dark spur onto side of breast (see also Shape). **Juv:** (May–Sep.) Like ad. but with shorter bill; forehead dotted with white. M. usually has red on central crown extending over area above eyes; f. may have some red on hindcrown. **Flight:** Small size; white back; black rump. **Hab:** Variable; deciduous woodlands, often riparian, gardens, feeders. **Voice:** A loud *teeek*; a descending whinny; a loud repeated 3–5 *kweeeks*; drum is a simple rapid roll. **55**.

Subspp: (7) Divided into three groups. **Eastern Group:** Medium-sized, heavily spotted wings, spotting and underparts whitish to tinged brown. •*pubescens* (s.e. KS–s.e. VA and south), •*medianus* (n.cent. AK–n.e. KS and east). **Interior Western Group:** Large, reduced spotting on wings, spotting and underparts pale gray to white. •*glacialis* (s.cent. AK); •*leucurus* (s.e. AK–w. NE south to AZ–NM) may lack barring on tail. **Coastal Pacific Group:** Small, reduced spotting on wings, spotting and underparts grayish to brownish. •*fumidus* (coastal s. BC–WA), •*gairdnerii* (coastal OR–n. CA), •*turati* (n.cent. WA–s. CA). **Hybrids:** Nuttall's Woodpecker.

Adult m., Coastal Pac. Grp. CA/11

Adult m., Int. Western Grp. CO/07

Juv. m., Coastal Pac. Grp. OR/08

Adult f., Coastal Pac. Grp. CA/06

Adult f., Eastern Grp. NH/07

Juv. m., Eastern Grp. NH/06

Hairy Woodpecker 1

Picoides villosus L 9¼"

Shape: Medium-sized woodpecker with a long bill (usually more than ½ depth of head). Compared to similar Downy Woodpecker, Hairy has thinner body, more oblong head, defined neck, longer bill, proportionately smaller nasal tufts. **Ad:** White to brownish central back and underparts; black wings variably spotted or barred with white to dusky; usually has white unbarred outer tail feathers (except in NF and Northwest); malar streak usually extends as a pointed spur onto side of breast. M. has red nape patch; f. no red nape patch (rarely, a few red feathers). Similar Downy Woodpecker has white outer tail feathers usually barred with black and usually lacks prominent black spur on side of breast (see also Shape). **Juv:** (Apr.–Sep.) Like ad. but flanks may have dusky wash (sometimes with grayish dots); forehead dotted with white. M. usually has red (rarely yellow or pink) on central crown; f. has some or no red on hindcrown. **Flight:** Medium-sized; whitish back; black rump. **Hab:** Mature deciduous and/or coniferous forests, residential areas, feeders. **Voice:** Loud sharp *teeek;* repeated *wiki wiki wiki;* descending harsh whinny; drum is a simple rapid roll. **56**.

Subspp: (10) Divided into four groups. **Coastal Pacific Group (CPG), Interior Western Group (IWG), and Mexican Group (MG):** Underparts drab white to dull cinnamon or sooty brown, wing coverts with no to few white spots (except *hyloscopus*). •*sitkensis* (CPG; coastal s. AK–cent. BC), •*hyloscopus* (CPG; n.cent. CA–s. CA), and •*orius* (IWG; s. BC–s.e. CA to s.w. UT–s.w. TX) have all-white outer tail feathers; •*harrisi* (CPG; coastal c. BC–n. CA), •*picoideus* (CPG; Q. Charlotte Is. BC), and •*icastis* (MG; s. NM–s. AZ) have outer tail feathers usually marked with black; •*septentrionalis* (IWG; s.cent. AK–QC south to n. AZ). **Eastern Group:** Underparts white to pale gray, wing coverts heavily spotted white (except *terraenovae*). •*villosus* (cent. ND–n. TX east to NS–n. VA) and •*audubonii* (s. IL–s. VA south) have all-white outer tail feathers; •*terraenovae* (NF) outer tail feathers usually marked with black. **Hybrids:** Ladder-backed Woodpecker.

Adult m. FL/03

Adult f. LA/05

Red-cockaded Woodpecker 2

Picoides borealis L 8½"

Shape: Medium-sized, long-tailed woodpecker with a medium-length bill and well-defined neck. **Ad:** Large white cheek patch; back and wings black with white spotting and barring; underparts whitish with scattered dark spotting, especially on breast and flanks. M. with inconspicuous, very thin, red line at lower edge of hindcrown; f. with no red. **Juv:** (Apr.–Sep.) Like ad. but with red confined to central crown. **Flight:** Dark upperparts; pale belly; large white cheek patch. **Hab:** Mature longleaf pine forests. **Voice:** Short raspy *kweek; jee jee jee jee;* squeaky calls; drum is a simple rapid roll.

Subspp: Monotypic.

Adult m., *albolarvatus* OR/08

Adult f., *albolarvatus* OR/08

White-headed Woodpecker 1

Picoides albolarvatus L 9¼"

Shape: Medium-sized woodpecker with a fairly rounded head, slightly defined neck, and deep-based chisel-like bill. **Ad:** White head; body all black; wings black with white streak along folded edge of primaries. M. with red nape; f. no red. **Juv:** (Jun.–Sep.) Like ad. f. but with some to many red feathers on central crown. **Flight:** All black except for white head and white patch at base of primaries. **Hab:** Montane coniferous woods. **Voice:** Rapid 2-part *jeejeet;* a *kweek kweek kweek;* drum is a simple rapid roll.

Subspp: (2) •*albolarvatus* (s. BC–cent. CA) has smaller bill; •*gravirostris* (s. CA).

Adult m., *dorsalis* CO/07

Adult f., *dorsalis* UT/07

Adult m., *bacatus* ME/07

American Three-toed Woodpecker 2

Picoides dorsalis L $8^{3}/_{4}$"

Shape: Medium-sized woodpecker with a relatively large head, thin neck, and long bill. Only three toes per foot. **Ad:** Black back with variable white barring (see Subspp.). Head mostly dark with thin white eyestripe and slightly wider white moustachial stripe. Blackish nasal tufts; some black barring on white outer tail feathers. Underparts white with dark barring on sides (Downy and Hairy Woodpeckers lack dark barring on underparts). M. crown flecked with white and central yellow patch; f. crown flecked with white and no yellow patch. **Juv:** (Jun.–Sep.) Like ad. f., with a few to no yellow feathers on crown; underparts buffy with brownish spots on flanks. **Flight:** Dark above; pale belly with barred flanks. **Hab:** Boreal and high montane forests. **Voice:** A short *jeeek;* a long rattlelike call; drum is a simple roll, subtly speeding up and trailing off at end.

Subspp: (3) •*fasciatus* (Northwest) has strongly flecked crown, extensive white barring on back; •*dorsalis* (Rocky Mts.) has less flecking on crown, more extensive white on back; •*bacatus* (eastern portion of range) has very little white flecking on crown, very little white on back.

Adult m. CA/07

Adult f. NH/04

Black-backed Woodpecker 2

Picoides arcticus L $9^{1}/_{2}$"

Shape: Medium-sized woodpecker with a relatively large head, defined neck, and long deep-based bill. Only three toes per foot. **Ad:** Glossy bluish-black cap, back, and wings; broad white moustachial stripe starts above bill and continues down side of neck; little or no thin stripe behind eye; black nasal tufts; barred flanks; little or no black barring on outer tail feathers. M. with yellow patch on crown, no white flecking; f. all-black crown. **Juv:** (Jun.–Sep.) Like ad. but upperparts dull black; crown with few to small patch of yellow feathers and no white flecking. **Flight:** Strongly black above; pale belly, strongly barred flanks. **Hab:** Boreal and montane forests. **Voice:** A flat *kuk;* short *jik jik;* a lower *kweek;* drum is a simple roll, subtly speeding up and trailing off at end.

Subspp: Monotypic.

Adult m., *pileatus* FL/02

Adult f., *pileatus* FL/02

Adult f., *abieticola* CT/01

Adult f., *pileatus* FL/02

Adult f., *pileatus* FL/02

Pileated Woodpecker 1

Dryocopus pileatus L 16½"

Shape: Very large, long-necked, long-tailed woodpecker with a large crest. Bill massive. **Ad:** Brownish-black back, wings, and tail except for small white patch at base of primaries; bright crimson-red crest; thin white eyebrow; wide black eye-stripe; white line from above bill to down neck; pale eye (m. always, f. usually). Upper mandible slate gray; lower mandible pale yellow or horn-colored with darker extreme tip. M. has red moustachial stripe (not always obvious), red forehead; f. has black moustachial stripe; black to brownish forehead (occasionally slightly speckled). **Juv:** (May–Sep.) Like ad. m. and f. but duller black; crest averages pinker; dark eye. **Flight:** Below, wings with bright white leading half (wing coverts and bases of primaries), wide black trailing edge (outer primaries and secondaries). Above, wings black with broad white crescent across base of primaries. **Hab:** Mature forests; younger forests with large dead and fallen trees. **Voice:** Loud *kekekeke;* drum a low-pitched roll that slows and tapers off at end. 🎧**57**.

Subspp: (2) •*abieticola* (cent. CA–MD and north) large with long bill; •*pileatus* (s. IL–cent. PA and south) smaller with shorter bill.

Adult m. (colorized image) LA/–

Adult f. LA/–

Adult m. LA/–

Adult f. LA/–

Ivory-billed Woodpecker 6

Campephilus principalis L 19½"

Thought to be extinct until possible sightings occurred in 2004; however, as of the writing of this guide, no uncontested conclusive evidence of its existence has been provided. Archival photographs used. **Shape:** Extremely large, long-necked, long-tailed woodpecker; bill massive. **Ad:** Face mostly black with pale yellow eye and ivory bill (Pileated chin white; eyeline white); body black with thin white stripe starting at cheek and continuing down through scapulars to secondaries. Upperwing coverts glossy black; outer portions of inner primaries, secondaries, and tertials all white, creating large white patch across lower back of perched bird. M. has red hindcrest; f. black crest. **Juv:** (May–Sep.) Like ad. f. but black areas less glossy; crest shorter and black; all primaries with white tips, decreasing in size to the outermost. **Flight:** Above and below, large white trailing half of inner wing extending slightly onto outer wing; below, second, narrower band of white on leading edge of wing formed by white lesser and median coverts. **Hab:** Mature hardwoods, bottomlands. **Voice:** A sharp nasal *kent,* sounding like a toot from a small toy horn; instead of drumming does a fast, resonant double knock, *tadut*.

Subspp: (1) •*principalis.*

Adult m., Yellow-shafted Grp. FL/04

Adult f., Yellow-shafted Grp. FL/04

Adult m., Yellow-shafted Grp. FL/02

Adult f., Yellow-shafted Grp. NH/09

Adult m., Red-shafted Grp. CA/07

Adult f., Red-shafted Grp. CA/05

Juv. m., Red-shafted Grp. UT/06

Northern Flicker 1

Colaptes auratus L 12½"

Shape: Large woodpecker with a relatively long thick neck and long slightly downcurved bill. **Ad:** Pale brown back with thin black barring; black breast patch; bold black spotting on buffy belly; has either a gray or a red nape (see Subspp.). M. has either red or black moustachial stripe; f. lacks moustachial stripe (see Subspp.). **Juv:** (May–Sep.) M. of all subspecies have orange moustache and often a reddish wash to crown; juv. f. much like ad. f. **Flight:** Bold white rump; above, wings brown; below, wings and tail reddish or bright yellow (see Subspp.). Long streamlined body. **Hab:** Open woodlands, suburbs. **Voice:** Loud single *keeoh;* soft *woika woika woika;* a *kekekeke;* drum a simple roll, often slightly muffled. **58**.

Subspp: (5) In two groups. **Yellow-shafted Group:** Red nape; gray crown; tan face; yellow under wings and tail; black moustache on male. •*luteus* (cent. AK–NF south to n. TX–w. NC) larger; •*auratus* (e. TX–s.e. VA and south) smaller. **Red-shafted Group:** Gray nape; brown crown; gray face; reddish under wings and tail; red moustache on male. •*cafer* (coastal s. AK–n. CA) upperparts brown; •*nanus* (s.w. TX) upperparts pale grayish brown; •*colaris* (rest of West) pale brown. **Hybrids:** Gilded Flicker.

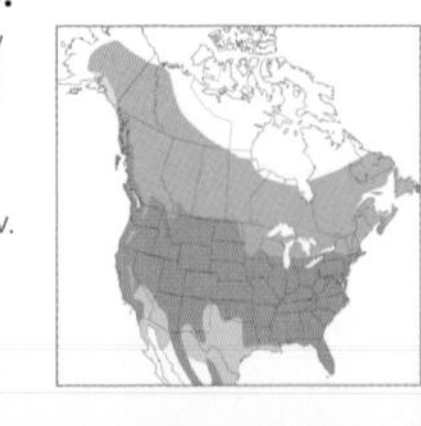

Adult m. AZ/01

Adult f. AZ/04

Juv. m. AZ/05

Gilded Flicker 2

Colaptes chrysoides L 11"

Shape: Large sturdy woodpecker with a relatively long thick neck and long slightly downcurved bill. **Ad:** Pale brown back with thin black barring; black breast patch; bold black spotting on buffy belly; wings and tail strongly yellow below. Distinguished from Northern Flicker by cinnamon crown and nape; thicker black breast patch; thinner dark bars on back. M. has red moustachial stripe; f. face plain bluish gray with no red stripe. **Juv:** (Apr.–Sep.) Like ad. but m. usually has orange moustache and reddish-tinged crown; f. much like ad. f. **Flight:** Bold white rump; above, wings brown; below, wings and tail bright yellow. Long streamlined body. **Hab:** Deserts with large cacti; riparian woodlands. **Voice:** Similar to Northern Flicker's. Loud single *keeoh;* soft *woika woika woika;* a *kekekeke;* drum a simple roll, often slightly muffled.

Subspp: (1) •*mearnsi.* **Hybrids:** Northern Flicker.

Flycatchers

Pewees

Phoebes

Kingbirds

Juv., *ridgwayi* AZ/05

Adult, worn, *ridgwayi* AZ/08

Adult, worn, *ridgwayi* AZ/08

Adult TX/05

Adult CA/05

Adult TX/05

Northern Beardless-Tyrannulet 2

Camptostoma imberbe L 4½"

Shape: Very small, relatively large-headed, broad-necked flycatcher with a moderate-length tail. Bill short, strongly curved culmen (somewhat vireolike). Steep forehead blends to upsloping crown and crest (bushy when raised) that angles down to fairly straight nape. **Ad:** Body pale grayish overall, paler on throat and belly (which can have yellowish cast); wings and tail darker. May show thin dark eyeline and paler eyebrow; variably distinct pale eye-ring. Bill with dark culmen and tip; orangish base to lower mandible. With fresh feathers (fall): 2 broad grayish-brown wingbars; tertials and secondaries with pale gray margins. **Juv:** (May–Sep.) Upperparts brownish gray; underparts buffy with pale yellow cast to belly; wingbars buffy. **Flight:** Underwing coverts pale; may hover while gleaning insects. **Hab:** Streamside woods. **Voice:** High-pitched, clear whistles, usually descending, like *peeeah peeeah peeeah;* or mixed with short clipped introductory notes, like *pit pit peee peeeah.*

Subspp: (2) Geographically separate; differences slight. •*ridgwayi* (s. AZ); •*imberbe* (s. TX).

Olive-sided Flycatcher 1

Contopus cooperi L 7½"

Shape: Large, long-bodied, short-tailed flycatcher with a big head and broad neck. Bill relatively long, broad-based, deep-based, and spikelike. Tail slightly notched. Crest when raised forms rounded peak. Long primary projection past tertials about equal to projection of tail past primary tips. **Ad:** Dark brownish to olive-gray head and upperparts; pale yellow to whitish throat, central breast, and belly; sides of breast and flanks strongly to mildly streaked darker, almost meeting on the breast like a buttoned vest; pale undertail coverts variably tipped dark. White patches on either side of lower back sometimes visible. Tertials edged with white. Upper mandible dark; lower mandible variable from almost all dark to orange-based to mostly orange. Wingbars distinct in 1st yr., indistinct thereafter. **Juv:** (Jun.–Nov.) Like ad. but with thin buffy wingbars. **Flight:** Fast, direct, deep wingbeats; short tail; pointed wings. **Hab:** Openings in coniferous or deciduous woods. Typically perches on dead branches at very top of trees. **Voice:** Loud whistled 3-part song—a clipped first note, higher-pitched emphatic note, and final downslurred note, much like "Quick, three beers"; also lower-pitched *pip* given singly or in a series.

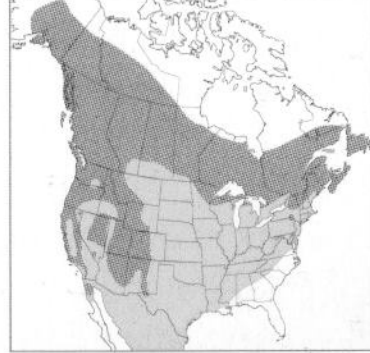

Subspp: (2) Differences weak. •*marjorinus* (s. CA) slightly larger with larger bill; •*cooperi* (rest of range).

Adult TX/01

Adult CA/05

Adult CA/05

Western Wood-Pewee 1

Contopus sordidulus L 6½"

Shape: Medium-sized, large-headed, broad-necked, short-legged flycatcher with a moderate-length tail. Long primary projection past tertials about equal to longest tertial. Rounded crest peaks at hindcrown, then angles down straight nape. Bill moderately long, broad-based. Tail slightly notched. Similar Empidonax flycatchers are smaller, usually lack peak at hindcrown, and have primary extension past tertials much shorter than longest tertial. **Ad:** Upperparts grayish brown; sides of breast and flanks grayish brown (creating a vestlike look); chin, central breast, and lower belly paler gray; undertail coverts pale gray with variably darker gray centers. Two pale wingbars, the lower wingbar often brighter and more prominent; tertial edges pale. Bill black above; lower mandible dull orange with variable darker tip. Partial eye-ring behind eye fairly indistinct. Similar Eastern Wood-Pewee best distinguished by voice and range but also tends to have brighter wingbars (equal in intensity), paler underparts, and brighter and more extensive orange on lower mandible. **Juv:** (Jun.–Nov.) Upperparts more brownish than in ad.; wingbars and tertial edges cinnamon. Not separable from Eastern Wood-Pewee in field. **Flight:** Direct with fast wingbeats on long flights. **Hab:** Woods, canyons. **Voice:** Song a *t'pah didip bzeeer,* the last note buzzy and downslurred; also single downslurred *bzeeer* and *k'zeee*. Song of similar Eastern Wood-Pewee all clear whistles with no buzzy quality.

Subspp: (2) Differences weak. •*saturatus* (s.e. AK–cent. OR) has concolor crown and back, whitish breast and belly. •*veliei* (rest of range) has crown darker than back; breast and belly may be washed with pale yellow. **Hybrids:** Greater Pewee, Willow Flycatcher.

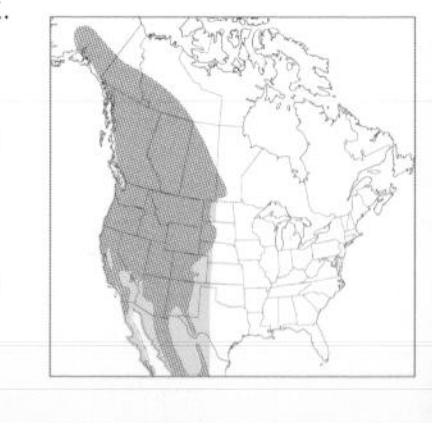

Adult TX/05

Adult TX/04

Adult MI/05

Eastern Wood-Pewee 1

Contopus virens L 6½"

Shape: Identical to Western Wood-Pewee: medium-sized, large-headed, broad-necked, short-legged flycatcher with a moderate-length tail; long primary projection past tertials about equal to longest tertial; rounded crest peaks at hindcrown, then angles down straight nape; bill moderately long, broad-based; tail slightly notched. Similar Eastern Phoebe has longer tail (which is frequently bobbed) and shorter primary projection past tertials: Phoebe primary projection = ½ length of longest tertial; Wood-Pewee primary projection = longest tertial. Similar Empidonax flycatchers are smaller, lack peak at hindcrown, and have primary extension past tertials much shorter than longest tertial. **Ad:** Similar to Western Wood-Pewee; use song and range as primary ways to distinguish in field. Secondary characteristics of plumage subtle and variable. They include these generalizations about Eastern: may have greenish tinge to upperparts or breast; duskiness on sides of breast often does not extend onto flanks; lower mandible generally brighter orange; underparts may be slightly paler than Western's. **Juv:** (Jun.–Nov.) Identical to Western Wood-Pewee in field. **Flight:** Direct with fast wingbeats on long flights. **Hab:** Woods. **Voice:** Song clear, strongly modulated whistles like *peeahweee peeeuuu,* the last note strongly downslurred and drawn out; also *peee widip* and rising *poooeee;* calls include a dry *chipit*. Song of similar Western Wood-Pewee has distinctly buzzy quality and is less drawn out.

Subspp: Monotypic.

Adult BEL/02

Adult AZ/06

Greater Pewee 2

Contopus pertinax L 8"

Shape: Similar shape to Olive-sided Flycatcher. Distinguished by proportionately shorter primary projection and proportionately longer tail (primary projection past tertials much shorter than projection of tail past primary tips). Also crest when raised forms sharp ragged peak (full and rounded peak in Olive-sided). **Ad:** Head and upperparts dark olive-brown; paler smooth gray throat, breast, flanks; belly and unmarked undertail coverts whitish, sometimes with a yellowish wash. Upper mandible black; lower mandible deep orange. Indistinct gray to whitish wingbars. No white patches on either side of rump (can be hidden in Olive-sided Flycatcher). **Juv:** (Jun.–Sep.) Upperparts with brownish tones; wingbars buffy; belly and undertail coverts may have yellow wash. **Flight:** Strong and direct; note fairly long notched tail and lack of contrasts on underparts. **Hab:** Open pine woodlands. **Voice:** Song a languid clearly whistled *hosay mareeeah;* introductory notes include a short *kweedee* or *kweedeeup;* also a rolling ascending *prrrt* and short *kip.*

Subspp: (1) •*pertinax.* **Hybrids:** Western Wood-Pewee.

IDENTIFICATION TIPS

Flycatchers are a large and fairly diverse family of birds that in the U.S. and Canada generally share the habit of waiting on exposed perches as they look for insects and then making short flights out to catch them. They tend to perch upright, not horizontally. They range from small to medium-sized, and most have dark upperparts in tones of brown, gray, or olive and paler underparts that are whitish, dusky, yellow, or olive. Being similarly colored can make them hard to identify; but their "songs" and calls, heard mostly during breeding, are often distinct and a great aid. In most cases the sexes are alike.

Species ID: Flycatchers can be divided into several genera by looking carefully at size, behavior, shape, and overall color. Assigning a bird to one of these genera is a good first step in flycatcher ID. Suggestions on how to approach identifying the species within two of the main groups of flycatchers, pewees and empids, are below. Identification tips for phoebes and Myiarchus flycatchers are on page 469; tips for kingbirds (genus *Tyrannus*) are on page 477. In addition to these groups, there are several individual species of flycatchers that are quite distinctive.

- **Pewees** are medium-sized flycatchers that are generally dark to dusky. Perched birds all have long primary projection past the tertials, and their crest when raised makes their crown a triangular peak. They tend to sit rather vertically on high open perches as they look for insects. The species can be identified by shape, color of underparts, and songs and calls.
- **Empids** are a large group of small flycatchers in the genus *Empidonax.* They may have a small peak at the back of their crown, their bodies are rather compact and stocky, and most have fairly prominent paler eye-rings. The visual differences between species are subtle and include both shape and colors. Their calls and songs, however, can be quite distinctive and one of the surest ways to identify them. In identifying species, aspects of shape to look for are the length and width of the bill, the length of the primary projection past the tertials, and the length and width of the tail. For colors, look at the tones of the back and subtle tonal variations on the throat, breast, and belly; also look for the extent of black on the tip of the orangish lower mandible.

Adult MN/05

Adult MN/05

Adult MN/06

Adult MN/06

Yellow-bellied Flycatcher 1

Empidonax flaviventris L 5½"

Shape: Small, stocky, big-headed empid (member of *Empidonax* genus) with a fairly short broad-based bill, moderately long primary projection past tertials (⅔ length of longest tertial); fairly short thin tail. Crown usually rounded, but can have small peak on hindcrown when alert. Large head and rounded chest create rather front-heavy look. **Ad:** Upperparts olive-green with slightly darker head; underparts from throat through central breast to belly yellow to washed yellow; dusky sides of breast and flanks; yellow throat blends fairly evenly into olive face; lower mandible all orangish to yellowish to pinkish; eye-ring pale yellow and may widen slightly at rear. Yellowish to whitish wingbars and tertial edges contrast strongly with blackish wings. Prebasic molt on wintering ground, so late summer/fall ad. has worn plumage, grayer above, grayish white below, still with gradual blend of throat to head color. Actively flicks wings and tail. **Juv/1st Winter:** (Jun.–Feb.) Like fresher spring ad. but with buffy wingbars; throat can be whitish; plumage fresh in fall (when adult plumage is worn). **Flight:** Slightly undulating; also short out-and-return feeding flights. **Hab:** Boreal conifer forests. **Voice:** Song a raspy *chevik* (similar to Least Flycatcher's, but lower-pitched and less clipped); also a rising whistled *chuweee* and a short downslurred *pew*.

Subspp: Monotypic.

Adult OH/05

Adult OH/05

Adult TX/06

Adult TX/06

Acadian Flycatcher 1

Empidonax virescens L 5¾"

Shape: Fairly large empid with a long (longer than lore) broad-based bill, long primary projection past tertials (equals longest tertial), and broad, medium-length, usually parallel-edged tail. Hindcrown often slightly peaked. On average, largest bill and longest wings among empids. **Ad:** Upperparts relatively bright greenish olive; underparts from throat to midbelly whitish gray; hindbelly to undertail variably washed yellow; whitish throat blends with pale malar area; lower mandible all orangish yellow (rarely with dusky tip); eye-ring usually distinct and whitish. Yellowish to whitish wingbars and tertial edges contrast strongly with blackish wings. Prebasic molt on breeding ground, so late summer/fall ad. has fresh plumage with relatively bright colors. Flicks wings and tail less actively than most other empids. **Juv/1st Winter:** (May–Feb.) Juv. upperparts distinctive, with buff edges to feathers creating scaled effect; wide buffy wingbars; whitish underparts, sometimes with yellow wash. 1st winter loses scaled look and is more like ad. but with buffy wingbars. **Flight:** Fast and direct; often hovers as it feeds. **Hab:** Wooded river floodplains, wooded streamside ravines. **Voice:** Song a clipped, buzzy, explosive *pizza!;* also a buzzy *zeet,* a soft *pipipipi,* and a short *pew.*

Subspp: Monotypic.

Adult MN/06

Adult MI/05

Adult MI/05

Adult OH/05

Least Flycatcher 1

Empidonax minimus L 5¼"

Shape: Small, compact, large-headed empid with short medium-width bill, moderate to short primary projection past tertials (½–⅔ length of longest tertial), and short, narrow, slightly notched tail often pinched in at base. Crown usually round; slightly peaked hindcrown when alert. **Ad:** Upperparts grayish brown with olive wash to back; underparts mostly grayish with yellow tinge when fresh; throat whitish, contrasting with darker head; lower mandible mostly orangish yellow; eye-ring distinct, whitish, conspicuous, widest at rear. White wingbars and tertial edges contrast strongly with blackish wings. Prebasic molt occurs mostly on wintering grounds; worn adults migrate in August. Very active, changing perches often, often flicking wings and tail. **Juv/1st Winter:** (Jun.–Feb.) Like ad. but with buffy wingbars and tertial edges. **Flight:** Fast, direct; short out-and-return flights. **Hab:** Edge habitats with taller vegetation. **Voice:** Song a high-pitched, clipped, repeated *chebek* (similar Yellow-bellied Flycatcher song lower-pitched and less clipped); also rolling *prrt* and sharp *whit*.

Subspp: Monotypic.

Adult MI/06

Adult MI/06

Adult MN/06

Adult OH/05

Alder Flycatcher 1

Empidonax alnorum L 5¾"

Shape: Fairly large empid with a moderately long wide-based bill, fairly long primary projection past tertials (¾ length of longest tertial), and broad, medium-length, parallel-edged tail. Crown rounded with slight peak on hindcrown when alert. Identical structure to eastern subspp. of Willow Flycatcher. **Ad:** Best identified by voice. Upperparts concolor dull greenish olive; white throat contrasts strongly with darker face; lower mandible usually all yellow-orange but sometimes with dusky tip; eye-ring narrow, whitish, distinct to rarely absent; lore pale. Dull whitish wingbars and tertial edgings contrast moderately with blackish wings. Prebasic molt occurs on wintering grounds, so late summer/fall ad. looks worn, with colors muted, wingbars worn, eye-ring indistinct. See Willow Flycatcher for comparison. Flicks wings and tail less actively than most other empids. **Juv/1st Winter:** (Jun.–Feb.) Wingbars buffy; colors brighter and feathers less worn than in ad. **Flight:** Short out-and-return flights during feeding. **Hab:** Wetland shrubs. **Voice:** Song a burry rolling *jreebeeah* or *freebeer!* with the emphasis on the last part (song of similar Willow Flycatcher is shorter, more clipped, with emphasis on first part, like *itsbeer*); also buzzy *zreeer,* abruptly upslurred *jweeep,* and short sharp *jip.*

Subspp: Monotypic.

Adult, Western CA/05

Adult, Western CA/05

Adult, Eastern OH/07

Adult, Eastern OH/07

Juv., Western CA/08

Willow Flycatcher 1

Empidonax traillii L 5¾"

Shape: Structure identical to Alder Flycatcher (see for description) except for western subspp. of Willow, which have shorter primary extension past tertials (see Subspp.). **Ad:** Best identified by voice. Nearly identical plumage to Alder Flycatcher (see for descrip.). Differences subtle—Willow often has upperparts with browner tones, eye-ring less distinct (lacking in many), wingbars slightly less contrasting with wing. Differences least between Alder and eastern subspp. of Willow; more apparent between Alder and western subspp. of Willow. Willow and Alder Flycatchers not clearly distinguished are sometimes called "Trail's" Flycatcher, their former name when they were one species. Prebasic molt occurs on wintering grounds, so late summer/fall ad. looks worn, with colors muted, wingbars worn, eye-ring indistinct. Flicks wings and tail less actively than most other empids, often just after landing. **Juv/1st Winter:** (Jun.–Feb.) Wingbars buffy; colors brighter and feathers less worn than in ad. **Flight:** Short out-and-return flights during feeding. **Hab:** Wetland shrubs. **Voice:** Song a short explosive *fitzbew* or *itsbeer*, with the emphasis on the first part (lacks the rolling burry first part and emphasized second part of Alder Flycatcher's song); also buzzy upslurred *jjjrip* (often given with song) and a soft *whit* (call of Alder a sharp *jip*).

Subspp: (5) Slight structural and color differences between eastern and western subspp. **Eastern:** •*campestris* (s. AB–cent. CO to s. QC–n. AR) and •*traillii* (rest of range to east) like Alder Flycatcher in structure and color. **Western:** •*brewsteri* (w. BC–s.w. CA) and •*adastus* (e. BC–s.w. AB to e. CA–cent. CO) have slightly darker brown upperparts than other subspp.; •*extimus* (s. NV–cent. CO to s. CA–s.w. TX) is palest subsp. All western subspp. have short primary projection past tertials and gray-buff tertial edges (easterns have white tertial edges). **Hybrids:** Western Wood-Pewee.

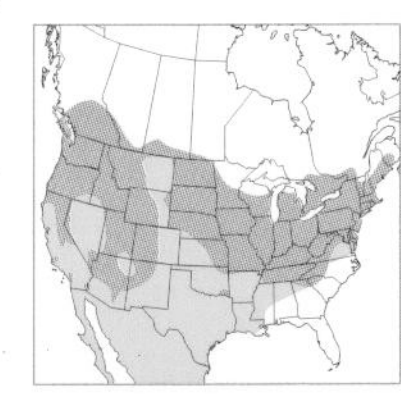

Adult CA/04

Adult CA/05

Adult CA/04

Adult CA/05

Hammond's Flycatcher 1

Empidonax hammondii L 5½"

Shape: Small fairly big-headed empid with a fairly short narrow bill, fairly long primary projection past tertials (⅔ length of longest tertial), and moderate-length, narrow, slightly notched tail often pinched in at base. Crown rounded or with rounded peak at hindcrown. **Ad:** Upperparts grayish with olive wash on back; underparts grayish from throat through breast; belly to undertail coverts paler and whitish, usually with a yellow wash; grayish throat concolor with grayish face; lower mandible mostly dark, some yellow at base; eye-ring white and wider behind eye. Buffy to grayish wingbars and tertial edges form subdued contrast with dark grayish wing. Prebasic molt occurs on breeding grounds, so late summer/fall ad. has fresh plumage with yellowish wash below and buffy wingbars, like juv. Actively flicks wings and tail. **Juv/1st Winter:** (Jun.–Feb.) Brownish wash to upperparts; buffy wingbars. **Flight:** Fast, direct; short out-and-return flights. **Hab:** Coniferous or mixed deciduous and coniferous forests with little or no understory. **Voice:** Song three 2-part calls: a sibilant *tseebik,* a raspy *jurrik,* and a descending *tsee-jurr,* given in varying combinations and order; also sharp *peep* (Dusky, Least, and Gray Flycatcher calls a short *whit*).

Subspp: Monotypic. **Hybrids:** Dusky Flycatcher.

Adult TX/06

Adult AZ/03

Juv. CA/08

Juv. CA/08

Gray Flycatcher 1

Empidonax wrightii L 6"

Shape: Large fairly slender empid with a fairly long narrow bill, short primary extension past tertials (less than ½ length of longest tertial), and long narrow tail. Crown usually rounded. **Ad:** Only empid to bob tail downward (often just after landing), like phoebe. Pale overall. Upperparts pale gray with tinge of olive on back; underparts pale; throat grayish and not contrasting with rest of gray head; lower mandible pinkish orange with well-defined dark tip; whitish eye-ring indistinct against pale gray head. Whitish wingbars and tertial edges form weak contrast with light gray wing; whitish outer web to outermost tail feathers. Prebasic molt occurs on wintering grounds, so late summer/fall ad. has worn plumage with little contrast. Bobs tail down; may occasionally flick wings. **Juv/1st Winter:** (Jun.–Feb.) Slight olive tinge to back; may have pale yellow wash to belly; buffy wash to wingbars. **Flight:** May take flights to ground to feed; out-and-return flights. **Hab:** Open coniferous woods with sagebrush; pinyon-junipers. **Voice:** Song a raspy *jibek;* also a whistled downslurred *pseer,* raspy upslurred *tseyeet,* and short *whit*.

Subspp: Monotypic. **Hybrids:** Dusky Flycatcher.

Adult AZ/–

Juv. CA/08

Adult CA/08

Juv. CA/09

Dusky Flycatcher 1

Empidonax oberholseri L 5¾"

Shape: Medium-sized empid with a medium-length narrow bill, moderate primary extension past tertials (about ½ length of longest tertial), and fairly long narrow tail. Crown rounded or with rounded peak at hindcrown when alert. **Ad:** Upperparts dull with grayish head and slightly darker olive-washed gray back; white throat tinged grayish, may contrast with gray head; lower mandible with orangish base shading to dark tip; whitish eye-ring distinct against grayish face; often a prominently pale loral area. Grayish wingbars and tertial edges form weak contrast with dark gray wings. Prebasic molt occurs on wintering grounds, so late summer/fall ad. has worn plumage with little contrast. One of the less active wing and tail flickers among empids. Early-winter birds in Southwest are most colorful, with belly washed with yellow, gray breast, olive back, and buffy wingbars. **Juv:** (May–Sep.) Similar to ad. but wingbars buffy. **Flight:** Out-and-return flights. **Hab:** Open woods, or open areas with strong shrub layer. **Voice:** Song a raspy *jibek;* also a whistled upslurred *tseeyeet,* raspy *d'jeet,* and short *whit*.

Subspp: Monotypic. **Hybrids:** Hammond's, Gray Flycatchers.

Adult AZ/06

Adult AZ/04

Adult AZ/04

Buff-breasted Flycatcher 2

Empidonax fulvifrons L 5"

Shape: Very small finely featured empid with a medium-length medium-width bill, moderate primary extension past tertials (about ⅔ length of longest tertial), and fairly short, narrow, often notched tail. Crown usually rounded. **Ad:** Upperparts brownish with pale head; underparts distinctively buffy; lower mandible all yellowish orange; whitish eye-ring may be wider at rear and slightly pointed and does not contrast strongly with pale brown head. Dull whitish wingbars and tertial edges do not form strong contrast with blackish wings. Prebasic molt occurs on breeding grounds, so late summer/fall ad. has bright buff below and whitish wingbars. May pump tail occasionally, but does not actively flick wings and tail. **Juv:** (Jun.–Sep.) Brownish upperparts; buffy wingbars. **Flight:** Out-and-return flights. **Hab:** Open pine woods with grasses, canyons. **Voice:** Song a repeated *chivik chiveer;* also sharp *pit,* rolling *prrew.*

Subspp: (1) •*pygmaeus.*

Adult CA/05

Adult CA/05

Adult CA/07

Adult CA/07

Pacific-slope Flycatcher 1

Empidonax difficilis L 5½"

Shape: Small fairly large-headed empid with a medium-length broad-based bill; moderate primary extension past tertials (about ⅔ length of longest tertial), and a moderate-length fairly narrow tail. Hindcrown often with rounded peak. **Ad:** Upperparts greenish olive to brownish olive; yellowish wash on throat and belly; lower mandible all bright orange (occasionally with small dark tip); dull yellowish to whitish almond-shaped eye-ring distinctively wider (and often pointed) at back, sometimes broken at top. Dull yellowish wingbars and tertial edges form weak contrast with dark wings. Prebasic molt occurs on wintering grounds, so late summer/fall ad. has worn plumage, grayer above, whitish below, wingbars thin and whitish. Actively flicks wings and tail. Virtually identical to Cordilleran Flycatcher; best distinguished by breeding range and voice (see Voice). If outside of range and silent, often called "Western" Flycatcher (an earlier name for when the two were considered one species). **Juv:** (May–Oct.) Brownish wash overall, buffy wingbars. **Flight:** Out-and-return flights. **Hab:** Moist lowland and foothill forests. **Voice:** Song very high-pitched, 3 parts—a whistled *tseeyaweet,* a percussive *p'tik,* and a short *tsit*. M. contact call—a single note, a strongly upslurred *peeyeee*—is most distinct identification call (all other calls and song very similar to Cordilleran Flycatcher's); f. contact call a sharp *tsip*.

Subspp: (3) Differences weak, mostly in size; geographically distinct. •*difficilis* (AK–CA) medium-sized, short-billed; •*insulicola* (Channel Is. CA) medium-sized, long bill; •*cineritius* (Baja CA vag. to AZ) small, long bill.

Adult AZ/06

Adult AZ/08

Adult AZ/06

Adult, worn AZ/08

Cordilleran Flycatcher 1

Empidonax occidentalis L 5½"

Shape: Like Pacific-slope Flycatcher (see for descrip.). **Ad:** Indistinguishable by plumage or proportions from Pacific-slope Flycatcher, although Cordilleran averages slightly greener above (see Pacific-slope account for all descriptions). Best distinguished by breeding range and voice (see Voice); only calling males can be reliably distinguished. **Juv:** (May–Oct.) Brownish wash overall, buffy wingbars.

Flight: Out-and-return flights. **Hab:** Moist lowland forests. **Voice:** Song (m. only) very high-pitched, 3 parts—a whistled *tseeyaweet,* a percussive *p'tik,* and a short *tsit*. M. contact call with 2 syllables, *pit-peeet,* is most distinct identification call (all other calls and song very similar to Pacific-slope Flycatcher's); f. contact call a sharp *tsip*.

Subspp: (1) •*hellmayri*.

Adult CA/04

Adult CA/02

Black Phoebe 1

Sayornis nigricans L 7"

Shape: Fairly short-bodied, long-tailed, big-headed flycatcher with a thin finely pointed bill. Forehead blends with sloping crown that is often peaked at rear. **Ad:** Black overall, with white belly and undertail coverts; tertials and secondaries edged with white; greater coverts with grayish tips; outer web of outermost tail feathers white nearly to tip. Actively bobs tail like all phoebes. **Juv:** (Apr.–Sep.) Like ad. but wingbars reddish brown. **Flight:** Direct without undulation; may hover briefly. Note white belly and dark breast; underwing coverts pale. **Hab:** Shrubs and trees near fresh water. **Voice:** Song 2 high-pitched, repeated, whistled phrases, like *tsit see tsitseew,* the last part downslurred; also a single downslurred *tseeew* and a rather flat *chip.* 🎧**59**.

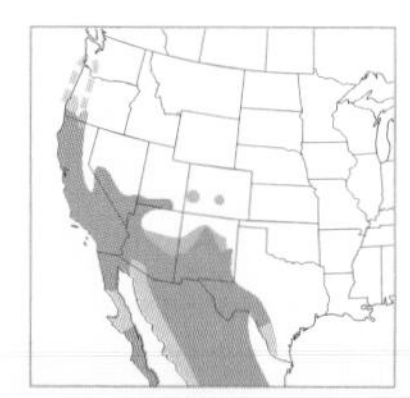

Subspp: (1) •*semiatra.* **Hybrids:** Eastern Phoebe.

Adult OK/04

Adult FL/03

Adult TX/01

Eastern Phoebe 1

Sayornis phoebe L 7"

Shape: Fairly short-bodied, long-tailed, big-headed flycatcher with a thin finely pointed bill. Steep forehead blends with sloping crown that is often peaked at rear. Short primary projection past tertials about ½ length of longest tertial (in similar wood-pewees long primary projection is equal to longest tertial). **Ad:** Upperparts brownish gray; underparts whitish with dusky sides of breast; variable pale yellow wash over throat and/or belly (most common on fresh plumage of fall/winter ad.); lower mandible all black. Tertials and secondaries edged with white; greater coverts with grayish tips; outer web of outermost tail feathers white from base to halfway down feather. Actively bobs tail downward. **Juv:** (Jun.–Aug.) Upperparts brownish; indistinct buffy wingbars; pale yellowish belly and undertail coverts. **Flight:** Steady wingbeats, direct without undulation; may hover briefly. Note white or pale yellow belly; pale underwing coverts. **Hab:** Breeds in shrubs and trees near fresh water. **Voice:** Song 2 slightly buzzy, high-pitched, whistled phrases, like *sweesi-dip sweejjeer;* also a fairly sweet, slightly descending *chirp*, a 2-part *t'keet,* and a chattering call. 🎧**60**.

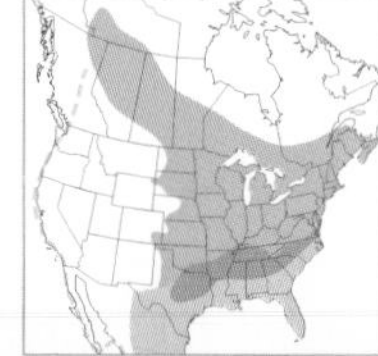

Subspp: Monotypic. **Hybrids:** Black Phoebe.

Adult AZ/03

Adult CA/01

Adult CA/02

Say's Phoebe 1

Sayornis saya L 7½"

Shape: Similar to Black Phoebe but slightly larger overall, with proportionately smaller head and larger bill. Sloping forehead blends with rather flat shallow crown, sometimes with a slight peak at hind-crown. **Ad:** Upperparts pale gray; face and crown darker grayish brown; lore slate gray; below, throat and breast pale gray blending evenly to variably reddish-brown belly and undertail coverts. Median and greater coverts with paler gray tips and margins, creating diffuse indistinct wingbars. Actively flares tail; other phoebes bob tail. **Juv:** (Apr.–Aug.) Distinct buffy wingbars on median and greater coverts. **Flight:** Steady wingbeats, direct without undulation; may hover briefly. Note orangish-brown belly and underwing coverts. **Hab:** Dry open country with perches; sagebrush. **Voice:** Song 2 whistled phrases, the first clear and downslurred, the second buzzy and rising, like *p't'peer p'd'jjeet* (sometimes just the first phrase repeated); also a plaintive downslurred *peeeuu*.

Subspp: (3) Differ slightly. •*yukonensis* (OR–AK) has smallest bill, darkest upperparts, deepest reddish-brown belly. •*quiescens* (s.e. CA–s.w. AZ) has largest bill, palest upperparts, pale tan belly. •*saya* (rest of range) has medium bill, upperparts medium gray, tan belly.

IDENTIFICATION TIPS

Phoebe Species ID: Phoebes are medium-sized flycatchers. Two species, Black Phoebe and Eastern Phoebe, have the habit of repeatedly bobbing their long tails. They tend to perch low and are often found near water. The differences in species are obvious in the color of their underparts.

Myiarchus Species ID: Myiarchus flycatchers are a small group of fairly large flycatchers that belong to the genus *Myiarchus*. They have large heads, bushy crests, fairly heavy bills, sleek bodies, and long tails. They are brownish gray above with grayish breasts and yellowish bellies and varying amounts of reddish brown on their primaries and tails. Species can be identified by overall size, comparative size of bill and head, the vividness of and contrast between their gray breast and yellow belly, and the extent and subtle patterns of reddish brown on their undertail.

Adult m., *flammeus* CA/04

Adult f., *flammeus* CA/12

1st-yr. m., *flammeus* AZ/06

1st-yr. f. TX/04

Vermilion Flycatcher 1

Pyrocephalus rubinus L 6"

Shape: Fairly small, short-tailed, big-headed flycatcher with a medium-length, thin, sharply pointed bill. M. crest, when raised, creates peaked or squared-off hindcrown; f. crown shallow and gently rounded. Tail slightly notched and often wagged or bobbed. **Ad:** M. with brilliant red crown and underparts separated by dark mask from bill through eye to nape; rest of upperparts grayish brown. In fresh plumage (fall) whitish tips to median and greater coverts form 2 thin wingbars (worn off by summer). F. brownish gray above; below, throat to upper belly whitish with fine brown streaking; lower belly, flanks, and undertail coverts washed with pinkish red; fine white wingbars wear away by summer (similar Say's Phoebe larger and lacks streaks on breast). **Juv:** (Apr.–Sep.) Like ad. f. but with no pink on belly or undertail coverts. **1st Yr:** M. similar to ad. m. but crown variably mixed red and brown, underparts variably mottled red and whitish. F. like ad. f. but lower belly and undertail coverts pale yellow. **Flight:** Long flights direct, no undulation; feeding flights out-and-return or longer erratic chases of insects; may hover briefly. **Hab:** Open farmlands, grasslands with shrubs; often near water. **Voice:** Song a variable series of repeated short sounds (often speeding up) with an explosive ending, like *pi pi pi pi p't'deeah;* also a sharp *peeet.*

Subspp: (2) Geographically separate. •*flammeus* (w. TX and west) ad. m. grayish brown above, orangish red below (sometimes with pale mottling); •*mexicanus* (s.cent. TX) ad. m. dark brown above, all deep red below.

Adult, *olivascens* AZ/05

Adult, *olivascens* AZ/05

Adult, *olivascens* AZ/06

Dusky-capped Flycatcher 2

Myiarchus tuberculifer L 7"

Shape: Slim body, big head, bushy crest, and fairly large bill typical of Myiarchus flycatchers. Comparatively, Dusky-capped is small and noticeably large-headed. **Ad:** Pale gray throat and breast contrast with fairly bright lemon-yellow belly. Little or no reddish brown in tail, except finely on outer edges; reddish-brown primaries; whitish or reddish-brown edges to secondaries (see Subspp.); wingbars pale brown and inconspicuous. Bill all black; mouth lining yellow to orange. Similar Ash-throated Flycatcher has paler underparts and less yellow on belly; suggestion of a gray collar; underside of tail reddish brown with dark tips. See Voice for helpful ID clues. **Juv:** (Jun.–Nov.) Reddish-brown edges to primaries and secondaries; cinnamon wingbars and uppertail coverts; some reddish-brown edges to tail. Underparts paler overall and less contrasting than in adult. **Flight:** Reddish-brown primaries; lack of reddish brown in ad. tail. May hover while feeding. **Hab:** Large trees along streams. **Voice:** Calls include a sharp upslurred *wheeet,* a plaintive downslurred *peeew* (sometimes with an introductory note like *peeteeew*), a rolling *prrrrrr,* and a more complex *pidipeewpidi.* Dawn-song a continuous and variable combination of these sounds.

Subspp: (2) •*olivascens* (s.e. CA–NM, vag. to w. TX) ad. has little or no reddish brown in tail, secondaries, or wingbars, but does have dull reddish-brown primaries. •*lawrenceii* (vag. to s. TX) ad. has reddish-brown primaries, wingbars, and edges of secondaries and thin reddish-brown edges to outer tail feathers.

Adult TX/06

Adult CA/05

Adult CA/04

Adult NM/06

Ash-throated Flycatcher 1

Myiarchus cinerascens L 8"

Shape: Slim body, big head, bushy crest, and fairly large bill typical of Myiarchus flycatchers. Comparatively, Ash-throated is "average," being medium-sized, moderately large-headed, with medium-sized bill. **Ad:** Pale gray throat and breast blend evenly into very pale whitish-yellow belly; transition between gray and yellow underparts may be whitish. May have faint pale gray collar over back of neck. Tail extensively reddish brown beneath with distinctive dark brown tips (some ad. may lack dark tips); reddish-brown primaries; whitish or pale orange-buff edges to secondaries; wingbars whitish and fairly conspicuous. Bill all black. See Voice for helpful ID clues. **Juv:** (Jun.–Nov.) Like ad. but with more reddish brown in tail (may not have dark tip); reddish-brown edges to secondaries and upperwing coverts; paler overall below. **Flight:** Pale underparts; reddish-brown tail and primaries. **Hab:** Semiarid areas with some large trees. **Voice:** Calls higher-pitched than those of Brown-crested Flycatcher, include a 2-part *pitjeer,* a quick *pit* and *pitbrik,* and a raspy rolling *jrrrt;* dawn-song a continuous and variable combination of these sounds.

Subspp: (1) •*cinerascens.*

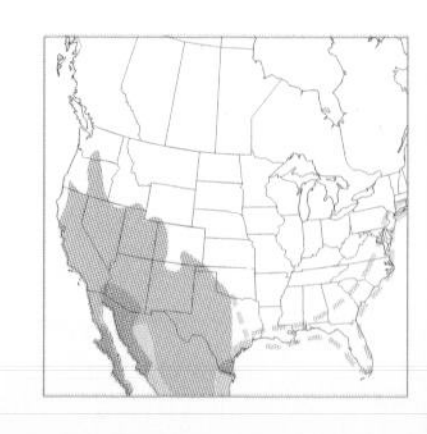

Adult, magister AZ/04

Adult TX/05

Adult, magister AZ/05

Adult TX/06

Brown-crested Flycatcher 1
Myiarchus tyrannulus L 8–9"

Shape: Slim body, big head, bushy crest, and fairly large bill typical of Myiarchus flycatchers. Comparatively, Brown-crested is large or largest (see Subspp.), with long heavy bill. Its short primary extension past tertials equals ⅓ length of longest tertial (in similar Great Crested Flycatcher the primary extension is ⅔ length of longest tertial). **Ad:** Pale gray throat and breast with fairly sharp transition to pale yellow belly. Inner webs of outer tail feathers with extensive reddish brown extending to tail tip and separated from shaft by darker brown line (darker brown line absent on Great Crested); wingbars and tertial edges dusky (not as white or wide as Great Crested's). Bill all black; mouth lining pink or dull yellow. See Voice for helpful ID clues. **Juv:** (Jun.–Nov.) Like ad. but with reddish-brown wingbars. **Flight:** Extensive reddish brown on tail; pale yellow belly. **Hab:** Streamside areas with large trees, arid areas with saguaro cactus. **Voice:** Typical call a sharp *whit;* other calls include a rapid *jweweweweet,* short upslurred *jweet,* downslurred *jwerr* or *wit-jwerr,* sharp *jeert;* dawn-song a continuous and variable combination of these sounds. Sounds lower-pitched than those of Ash-throated Flycatcher.

Subspp: (2) •*magister* (CA–NM) large, with very large bill; larger than Great Crested. •*cooperi* (TX, vag. to FL) smaller, about size of Great Crested.

Adult CA/02

Adult CA/02

Adult CA/01

Nutting's Flycatcher 5

Myiarchus nuttingi L 7¼"

Accidental Mexican vag. to AZ and s. CA. **Shape:** Slim body, big head, bushy crest, and fairly large bill typical of Myiarchus flycatchers. Comparatively, Nutting's is relatively small, compact, short-tailed, and short-billed. Body may appear comparatively more rotund than other "slim" Myiarchus species. **Ad:** Similar to Ash-throated Flycatcher but no darker brown tips to underside of tail; more contrast between darker gray breast and brighter yellow belly (with no white transitional zone); no faint gray collar; orange edges to secondaries. Also see Shape and Voice. **Juv:** Difficult to distinguish from juv. Ash-throated except by calls. **Flight:** Pale underparts; reddish-brown tail and primaries. Darker gray breast and brighter yellow belly than similar Ash-throated Flycatcher. **Hab:** Arid areas with some large trees, often near streams. **Voice:** Calls include a loud upslurred *weeep* and an extended jumble of twitterings and gurglings.

Subspp: (1) •*inquietus.*

Adult TX/06

Adult MN/06

Adult TX/04

Great Crested Flycatcher 1

Myiarchus crinitus L 8"

Shape: Slim body, big head, bushy crest, and fairly large bill typical of Myiarchus flycatchers. Comparatively, Great Crested is medium-sized with fairly long bill. Its long primary extension past tertials equals ⅔ length of longest tertial (in similar Brown-crested Flycatcher the primary extension is only ⅓ length of longest tertial). **Ad:** Dark gray lower face, throat, and upper breast contrast strongly with deep yellow belly; gray more restricted to upper breast than in other Myiarchus species; tail extensively reddish brown beneath with no dark tips; smallest tertial more broadly edged white, forming noticeable white stripe on lower back; wingbars whitish and fairly conspicuous. Bill dark with pale base to lower mandible; mouth lining yellow-orange. See Voice for helpful ID clues. **Juv:** (May–Nov.) Like ad. but with cinnamon to buffy wingbars. **Flight:** Extensive reddish brown in tail; extensive bright yellow belly. **Hab:** Open deciduous woodlands, woods edges. **Voice:** Calls include a loud upslurred *weeep* or *kaweeep,* short *prrrt,* loud *kaweet weeer,* raspy *jjreep;* dawn-song a continuous and variable combination of these sounds. 🎧**61**.

Subspp: Monotypic.

Adult BAH/02

Adult BAH/03

Adult BAH/02

Adult BAH/02

La Sagra's Flycatcher 3

Myiarchus sagrae L 8½"

Rare West Indies vag. to s. FL; accidental to AL. **Shape:** Slim body, big head, bushy crest, and fairly large bill typical of Myiarchus flycatchers. Comparatively, La Sagra's is more big-headed and slim-bodied with long thin bill. Primary projection past tertials short, about ½ bill length; other Myiarchus flycatchers have longer primary projection, about equal to bill length (except Brown-crested's, which is similarly short). Often perches leaning forward. **Ad:** Less colorful than other Myiarchus species (but similar to Ash-throated), with dull reddish brown on primaries; pale gray breast; whitish belly with little or no yellow tones; reddish brown on tail variable, from very little to some. Wingbars and tertial edges whitish and fairly prominent. Bill all dark; mouth lining pale yellow. **Juv:** Like ad. but secondaries may also have dull reddish brown. **Hab:** Dense second growth. **Voice:** Calls include a high-pitched, upslurred, whistled *weeet,* a rapid *widit,* buzzy downslurred *jeew,* and more complex *weeet widiweeer.*

Subspp: (2) •*lucaysiensis* (Bahamian vag. to FL) larger, more reddish brown in tail; •*sagrae* (Cuban vag. to Gulf Coast) smaller, less reddish brown in tail.

Adult, *texanus* TX/11

Adult, *texanus* TX/04

Adult AZ/07

Adult AZ/06

Great Kiskadee 2

Pitangus sulphuratus L 9¾"

Shape: Large heavyset flycatcher with a long thick bill, large head, and short tail. **Ad:** Bold black and white stripes on head; white chin and throat; bright yellow underparts. Wing and tail feathers strongly edged reddish brown; yellow patch on black crown occasionally visible. **Juv:** (May–Sep.) Like ad. but no yellow crown patch and more reddish brown on tail and wings. **Flight:** Strong, sometimes slightly undulating with bursts of flaps; can hover briefly; may dive into water. Bright reddish-brown wings and tail; yellow belly and underwing coverts. **Hab:** Trees and woods near water. **Voice:** Song a loud repeated *jjik kajeeer* or *jjik jjik kajeeer;* calls include a squeaky *jeeeah* and short raspy *jjrt.*

Subspp: (2) •*texanus* (TX) has slightly longer tail, deep yellow underparts; •*derbianus* (vag. to AZ–NM) has slightly shorter tail, paler yellow underparts.

Sulphur-bellied Flycatcher 2

Myiodynastes luteiventris L 8"

Shape: Fairly large flycatcher with a large thick bill, fairly long primary extension past tertials, and moderate-length tail. **Ad:** Heavily streaked brown and white overall with extensively reddish-brown tail and rump; variable pale yellow wash on underparts; yellow patch on black crown occasionally visible. **Juv:** (Jun.–Nov.) Like ad. but with reddish brown on wing and covert feather edges; little or no yellow in crown. **Flight:** Bright reddish-brown tail and rump; dark brown wings and back. **Hab:** Subtropical deciduous forests, wooded streamsides. **Voice:** Calls include short repeated *jt jt jt* and very high-pitched and variable squealing, like *skweeeyah.*

Subspp: Monotypic.

Adult GUA/11

Adult TRI/09

Adult GUA/11

Tropical Kingbird 2

Tyrannus melancholicus L 9¼"

Shape: Nearly identical in shape to Couch's Kingbird. Sturdy, large-headed, large-billed kingbird with a fairly long slightly notched tail; bill longer than or equal to lore + eye. In similar Western and Cassin's Kingbirds, bill is shorter (bill length = length of lore) and tail is proportionately shorter and square-tipped or only slightly notched. **Ad:** Gray head and nape with darker lore and auricular and paler throat and cheek. Upper breast olive, shading to yellow on rest of underparts. Wings and tail brownish black with thin whitish edges to secondaries, tertials, wing coverts, and tail feathers (this can wear off to varying degrees). Back grayish washed olive-green. Distinguished from Couch's Kingbird by calls (see Voice). Reddish-orange crown patch, usually concealed. **Juv:** (May–Oct.) Like ad. but browner above, with buffy margins to feathers of tail and uppertail coverts; no red crown patch. **Flight:** Concolor tail, wings, and back. **Hab:** Open areas with scattered trees, shrubs; woods edges, roadsides; often near water. **Voice:** Swiftlike high-pitched chittering; also common call a repeated very high-pitched *tseepee tseepee tseepee*. Very different from lower-pitched buzzy *kt'jweeer* call of Couch's Kingbird.

Subspp: (3) •*satrapa* (vag. to East Coast) has strongly notched tail, deep yellow belly, chin and throat concolor grayish white; •*occidentalis* (s. AZ, TX, vag. to both coasts) has moderately notched tail, pale yellow belly, chin and throat concolor whitish; •*melancholicus* (poss. vag. to CA) has white chin contrasting with dark gray throat.

IDENTIFICATION TIPS

Kingbird Species ID: Kingbirds are fairly large flycatchers that belong to the genus *Tyrannus.* Most have rather large thick bills and medium-length tails, and they tend to perch conspicuously as they look for insects. Two species, the Fork-tailed and Scissor-tailed Flycatchers, have extremely long tails; four species, Tropical, Couch's, Cassin's, and Western Kingbirds, are similarly colored, with grayish heads and bright yellow bellies; the remaining three species, Thick-billed, Gray, and Eastern Kingbirds, are mostly dark gray above and white below. In this group, look closely at the length and thickness of the bill and the patterns of colors on the breast and belly.

Adult TX/04

Adult TX/05

Adult TX/05

Couch's Kingbird 2

Tyrannus couchii L 9¼"

Shape: Nearly identical in shape to Tropical Kingbird. Couch's bill may average slightly shorter and thicker, but there is much overlap with Tropical. Bill longer than or equal to lore + eye (in Western and Cassin's Kingbirds, bill length = length of lore). **Ad:** Identical to Tropical Kingbird; best distinguished by calls (see Voice). **Juv:** (May–Oct.) Identical to Tropical juv., except for calls. **Flight:** Concolor tail, wings, and back. **Hab:** Open areas with scattered trees, shrubs; woods edges, roadsides. **Voice:** Common call a midpitched buzzy *kt'jweeer* or rising *suet suet suet kt'jweeer.* Very different from high-pitched lisping *tseepee tseepee* call of Tropical Kingbird.

Subspp: Monotypic. **Hybrids:** Possibly Scissor-tailed Flycatcher.

Adult AZ/05

Adult AZ/05

Adult AZ/05

Cassin's Kingbird 1

Tyrannus vociferans L 9"

Shape: Compared to Couch's and Tropical Kingbirds, bill shorter (bill length = length of lore); head smaller; tail shorter and only slightly notched or square-tipped. **Ad:** Dark gray head and breast contrast with white dash from chin to below eye; breast fairly dark gray, contrasting with bright yellow belly and undertail coverts; tail dark brown with pale tip and thin pale outer edges of outermost feathers (Western Kingbird has broader white edges to tail and no white tip); back dark gray. Reddish-orange crown patch, usually concealed. Similar Western Kingbird's pale gray throat and breast contrast weakly with white dash from below bill to eye. Similar Tropical and Couch's have whitish throat, greenish-yellow breast, dark yellow belly and undertail coverts. **Juv:** (May–Aug.) Like ad. but may have brownish wash to upperparts; lacks crown patch. **Flight:** Brownish tail contrasts with grayish back. **Hab:** Streamsides, canyons, open woods. **Voice:** Song a series of hoarse notes getting louder and ending in a sputtering; calls include loud *jigweeer,* also buzzy *jjrreep.*

Subspp: (1) •*vociferans.*

Adult CA/05

Adult CA/05

Adult MN/05

Adult AB/–

Adult TX/05

Adult TX/04

Western Kingbird 1

Tyrannus verticalis L 8³/₄"

Shape: Relatively small-headed short-billed kingbird with a relatively short square-tipped tail. Bill length = length of lore. **Ad:** Pale gray head, breast, and upperparts contrast slightly or blend with whitish throat; lore blackish; auricular often faintly darker than head; pale gray breast blends with bright yellow belly. Tail blackish with broad white outer edges to outermost feathers (can wear thin by late summer) and dark tip. Tertials broadly edged with white. Reddish-orange crown patch, usually concealed. Similar Cassin's Kingbird has brownish tail with faint pale edges to outer feathers and pale tip, and tertials with thin pale brown edges. **Juv:** (Jun.–Oct.) Like ad. but much paler beneath; lacks reddish-orange crown patch. **Flight:** Black tail contrasts with pale gray back. **Hab:** Open areas with trees or shrubs, grasslands, ranch yards, suburbs, streamsides. **Voice:** Calls include a harsh *kit kit kit,* a more extended *kit kit kitdit kadeet kitterdit,* a jumbled chatter, and a more whistled *tuweet tuweet kitterdeew.* 🎧**62**.

Subspp: Monotypic. **Hybrids:** Eastern Kingbird, Scissor-tailed Flycatcher.

Eastern Kingbird 1

Tyrannus tyrannus L 9"

Shape: Large, slender, relatively small-headed kingbird with a fairly short thin bill. Bill length = length of lore. Relatively long primary projection past tertials. **Ad:** Black head and dark gray upperparts contrast sharply with bright white underparts; black tail broadly tipped with white (may wear thin by late summer); wing feathers and coverts finely edged with white (wears away by late summer). Breast has variable pale gray band. Orange to red crown patch, usually concealed. **Juv:** (Jun.–Oct.) Like ad. but upperparts may have reddish-brown tinge; lacks crown patch. **Flight:** Broad white tip to black tail; mostly dark underwings. **Hab:** Open areas with trees. **Voice:** Dawn-song 2 complex phrases roughly alternated, like *t't'tzeer t't'tzeer t'tzeetzeetzee;* calls include a rapidly repeated 2-part call, chittering, like *kitterkitterkitter;* also a clipped 2-part *kt'zee kt'zee,* a short *tzit.* 🎧**63**.

Subspp: Monotypic. **Hybrids:** Western Kingbird.

Adult AZ/07

Adult FL/05

Adult AZ/07

Adult AZ/07

Adult FL/05

Adult FL/05

Thick-billed Kingbird 2

Tyrannus crassirostris L 9½"

Shape: Large-bodied, short-tailed, big-headed kingbird with a long, broad-based, deep-based bill. Bill longer than lore + eye; bill thickness at base twice diameter of eye. Hindcrown often with small sharp peak. **Ad:** Dark grayish-brown upperparts; darker lore and auricular contrast sharply with white throat and breast; belly variably washed with pale yellow; yellow crown patch, usually concealed. **Juv:** (May–Sep.) Like ad. but with reddish-brown edges to wing and uppertail coverts and more yellow below; lacks crown patch. **Flight:** All dark gray above; no white band at tail tip. **Hab:** Streamsides with large trees. **Voice:** Well-spaced harsh *chweebur, chweejjjjr;* also extended harsh chittering, sometimes ending in *chweebur* and a questioning *chaweeet*.

Subspp: Monotypic.

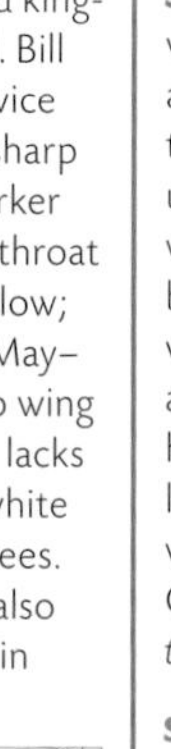

Gray Kingbird 2

Tyrannus dominicensis L 9½"

Shape: Large, big-headed, thick-necked kingbird with a very long, thick, spikelike bill. Bill noticeably longer than lore + eye; thickness at base twice diameter of eye. Notched tail. **Ad:** Dark gray upperparts; darker lore and auricular contrast with whitish underparts (lacks any yellow); sides of breast may have grayish wash; wing coverts usually with fine whitish edges. Orange crown patch, usually concealed. **Juv:** (Jun.–Nov.) Like ad. but may have brownish wash to upperparts. **Flight:** Very long bill; all dark gray upperparts. **Hab:** Open areas with some trees, coastal areas, suburbs. **Voice:** Calls include a raspy *tip't'chee t'cheeerrry;* also harsh *ts'tseeer* and *ts'cheee*.

Subspp: (1) •*fugax*.

Adult m. TX/05

Adult f. FL/02

Juv. FL/02

Scissor-tailed Flycatcher 1

Tyrannus forficatus L 10–15"

Shape: Body and wings similar in shape to Western Kingbird's (bill fairly short and = length of lore), but tail much longer and very strongly forked. On average, f. tail slightly shorter than that of m., but broad overlap; juv. tail much shorter, but still noticeably long. **Ad:** Very pale grayish-white head, breast, and back; dark wings with coverts and tertials edged white; flanks, undertail coverts, and underwing coverts pinkish to orangish; axillaries salmon. Red crown patch, usually concealed. F. may be paler than m. **Juv:** (May–Jan.) Like ad. but with shorter tail; brownish wash to upperparts; buff instead of pink or orange underparts. **Flight:** In flight, forked tail often held closed, sometimes spread out into long V. Bright salmon axillaries on ad. **Hab:** Open areas with scattered trees. **Voice:** Dawn-song a series of *wip wip wip* interspersed with *wip-perrcheee;* calls include series of *wip wip wip wip,* also short *pik.*

Subspp: Monotypic.

Adult, *savana* NH/11

Adult, *savana* NH/11

Fork-tailed Flycatcher 3

Tyrannus savana L 10–16"

Rare Cent. and S. American vag. to possibly anywhere in NAM; most frequently, so far, in e. NAM. **Shape:** Very long forked tail. On average, f. tail slightly shorter than that of m., but broad overlap. Juv. tail much shorter, but still noticeably long. **Ad:** Black cap contrasts with bright white underparts and gray back; dark wings with coverts and tertials edged gray. Yellow crown patch, usually concealed. M. has P10 strongly notched; f. has P10 just tapered or slightly notched. **Juv:** (Jan.–Jun.) Like ad. but tail shorter and wing and uppertail coverts edged cinnamon and washed with brown. **Flight:** Note white underwing coverts; long tail often spread out into long V during flight. **Hab:** Open areas with scattered trees. **Voice:** Calls include a very high short *pik,* a series of harsh grating notes like *jjjee jjjee jjjee,* and a rapid series of percussive sounds like *tr'tr'tr'tr'tr.*

Subspp: (2) •*savana* (main vag. to USA from SAM); black crown contrasts mildly with dark gray back, white collar thin and indistinct, underparts may have yellow wash, flanks grayish, ad. m. with P8 strongly notched at tip. •*monachus* (vag. to TX from Cent. America); black crown contrasts strongly with pale gray back, white collar more complete and distinct, underparts mostly all white with no gray or yellow, ad. m. P8 not notched at tip.

Adult m. MEX/05

Adult f. TX/06

Rose-throated Becard 3

Pachyramphus aglaiae L 7¼"

Shape: Medium-sized, large-headed, short-tailed flycatcher-like bird with a somewhat bushy crest and short stubby bill. **Ad:** M. gray overall with blackish crown; red throat patch. F. brownish overall with blackish crown; back brown; tail reddish brown; underparts and collar tan. **Juv:** (May–Aug.) Reddish-brown back, wings, tail; blackish crown; buffy underparts. **1st Yr:** M. mixed brown and gray on upperparts; belly tan; some rose on throat. **Flight:** M. with light gray underparts and red throat; f. pale below, brown above, dark cap. **Hab:** Streamside woods. **Voice:** High downslurred *tseeer,* sometimes interspersed with sputtering sounds.

Subspp: (2) Geographically separate. •*albiventris* (AZ–NM) has short bill; f. underparts buffy white; m. back pale gray and throat patch pale rose. •*gravis* (s. TX) has long bill; f. underparts rich buff; m. back dark brownish gray and throat patch deep rose.

Shrikes

Vireos

Adult, Continental Grp. FL/03

Adult, Continental Grp. MT/06

Juv. CA/09

Adult CA/12

Adult, stretching, Continental Grp. NM/06

Loggerhead Shrike 1

Lanius ludovicianus L 9"

Shape: Like Northern Shrike except slightly smaller; bill stubby, thick, slightly hooked; bill shorter than lore (longer than lore in Northern). **Ad:** Medium to dark gray upperparts contrast with whitish underparts (see Subspp.); breast clear and unbarred (rarely faintly barred in 1st yr.); black mask, wings, and tail. Bill all black all year; black mask may narrow in lores or stay wide and extends just over top of bill; nasal tufts dark. F. upperparts slightly darker than male's. Similar Northern Shrike has narrower dark mask over lore; pale gray lower mandible all year; faint barring on breast; less contrast between paler gray back and light breast; whitish nasal tufts. **Juv:** (Apr.–Sep.) Much like ad. but with whitish to buff wingbar on greater coverts; upperparts and breast faintly washed brown and with indistinct barring on breast; bill very short and may be slightly paler at base; mask less extensive than on adult. Acquires ad. plumage by 1st winter but may retain some buff tips on coverts through 1st yr. **Flight:** Undulating with rapid wingbeats and short glides; may hover while hunting. White base of primaries creates distinct patch above and below; tail graduated, tips increasingly white toward outer feathers; trailing edge of secondaries whitish. **Hab:** Open areas with trees or shrubs; roadsides, hedgerows. **Voice:** Song includes a variety of fairly musical phrases (often repeated), like *chidlip chidlip tseedeedee julip julip*, mixed with harsh sounds; calls include a variety of harsh sounds like *chhhht, tk-tk tk, jjaah.*

Subspp: (7) Divided into two groups. **California Island Group:** Generally darker upperparts, less white on primary bases. •*anthonyi* (Channel Is. CA), •*mearnsi* (San Clemente Is. CA), •*grinnelli* (San Diego Co. CA). **Continental Group:** Generally paler upperparts, more white on primary bases. •*miamensis* (s. FL), •*ludovicianus* (s. LA–NC south to cent. FL), •*excubitorides* (AB–SK south to AZ–w. TX), •*mexicanus* (rest of continental range).

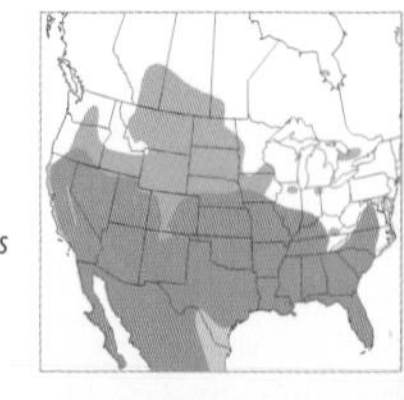

Adult –/12

Adult MN/12

Juv./1st winter CA/01

Juv./1st winter BC/02

Juv. NH/12

Northern Shrike 1

Lanius excubitor L 10"

Shape: Medium-sized, large-headed, short-necked bird with a fairly long tail. Bill rather short, thick, and markedly hooked at tip; bill slightly longer than lore (shorter than lore in Loggerhead Shrike). Similar Loggerhead Shrike smaller in size with slightly shorter less hooked bill; tail length proportionately the same. **Ad:** Pale gray upperparts blend with whitish underparts; breast finely barred with darker gray; black mask, wings, and tail. Bill black with basal third of lower mandible paler in fall and winter, all black rest of year (may remain paler throughout year in f.). Black mask narrows in lores and does not extend over top of bill; nasal tufts usually white (blackish in Loggerhead). M. larger than female. Similar Loggerhead has wider dark mask over lore; dark lower mandible all year; unbarred breast; more contrast between darker back and light breast; blackish nasal tufts. **Juv/1st Winter:** (Jun.–Mar.) Mask brownish, indistinct; upper mandible brownish; back brownish to gray; underparts distinctly barred and buffy to gray; pale patch on primary bases reduced and buffy or absent; bill considerably shorter than adult's at first. **Flight:** Undulating with rapid wingbeats and short glides; often hovers while hunting. White base of primaries creates distinct patch above and below; tail graduated, tips increasingly white toward outer feathers; trailing edge of secondaries whitish. **Hab:** Open areas near groups of trees or shrubs. **Voice:** Song a variety of short whistled phrases like *pidip tulit peetoo,* mixed with harsh notes; calls include a rolling *djreeet,* a harsh *chhhht.*

Subspp: (1) •*borealis.*

Adult, *lucionensis** HKO/04

Juv. IND/02

Brown Shrike 4

Lanius cristatus L 7"

Casual Asian vag. to West Coast; once to Atlantic CAN. **Shape:** Small, large-headed, broad-necked bird with a moderate-length tail and short, thick, hooked bill. **Ad:** Brown cap and back; brighter reddish-brown rump; dark mask through eye; underparts white with buffy sides of breast and flanks. Bill black. M. mask black, flanks unbarred; f. mask less distinct, flanks lightly barred. **Juv:** Like ad. f., but bill with dark tip and pale base, mask smaller, flank and breast feathers tipped darker, wing feathers with pale buff margins. **Flight:** Long tail, blunt head; wings all brown (no white patches as in other shrikes). **Hab:** Open areas near groups of trees or shrubs. **Voice:** Song soft warbles interspersed with harsh sounds; calls include a raspy *jjt jjt jjt jjt*.

Subspp: (1) Most likely •*cristatus*. **L. c. lucionensis* (KOR–e. CHI), gray crown, used for good photograph.

IDENTIFICATION TIPS

Vireos are a group of small birds with short narrow tails, broad necks, rather large heads, and fairly thick blunt-tipped bills. They are often inconspicuous as they move about tree canopies or shrub layers gleaning insects off leaves or eating berries. The sexes are alike.

Species ID: There are roughly two groups of vireos: those with eyelines and pale eyebrows (usually lacking obvious wingbars) and those with a large supraloral dash connected to a large eye-ring, creating "spectacles" (usually have white wingbars). Look closely at the patterns on the face and head and look for the colors of the underparts.

Adult TX/04

Adult, *griseus* OH/05

White-eyed Vireo 1

Vireo griseus L 5"

Shape: Small broad-necked vireo with a relatively long slightly hooked bill; bill about length of lore; tail moderately long. **Ad:** Yellow "spectacles" from bill to around white eye; throat whitish; nape gray; 2 broad (when fresh) white to yellowish wingbars; flanks variably bright yellow to olive (see Subspp.). **Juv:** (Apr.–Aug.) Like ad. but with buffy wash to head, spectacles, and wingbars; eye dark. **1st Winter:** (Sep.–Mar.) Like ad. but with dark grayish-brown eye. **Flight:** Yellowish underwing coverts. **Hab:** Dense shrubs, open areas with shrubs. **Voice:** Song a varied warble with a harsh *chek* before and usually after, like "Quick! Give me the beer check!"; calls include a nasal *jjheeee,* a short *jwut,* and a downslurred *djew.* Mimicked calls of other birds may occur in song.

Subspp: (3) •*micrus* (s. TX) small, with greenish flanks; •*maynardi* (s. FL) large, with large bill, yellowish flanks; •*griseus* (rest of range) large, with small bill, bright yellow flanks.

Adult BAH/04

Adult BAH/04

Thick-billed Vireo 4

Vireo crassirostris L 5"

Casual Caribbean vag. to FL. **Shape:** Small relatively long-billed vireo; bill blunt-tipped and about length of lore. Tail moderately long. **Ad:** Large supraloral yellowish dash separate from white to yellowish partial eye-ring behind and below eye and broken at top (White-eyed Vireo has continuous yellow from supraloral area to around eye); 2 broad whitish wingbars; throat and underparts concolor dull olive; upperparts fairly uniform greenish gray. **Flight:** Pale below, olive-green above; bold yellow supraloral dash. **Hab:** Dense shrubs. **Voice:** Song and calls similar to those of White-eyed Vireo, but song more raspy and calls longer and well spaced.

Subspp: (1) •*crassirostris.*

Adult, *arizonae* AZ/04

Adult, *bellii* OH/04

Adult TX/05

Adult, *bellii* OH/06

Adult CA/04

Bell's Vireo 1

Vireo bellii L 4¾"

Shape: Small, slender, large-headed, relatively long-tailed (see Subspp.) vireo with a long slightly hooked bill. **Ad:** Rather plain; continually pumps or waves tail. Whitish supraloral dash joins with thin whitish eye-ring broken in front and back (not true spectacles); lore subtly darker; 2 inconspicuous wingbars, lower one more evident; underparts whitish with some to no yellow on flanks (see Subspp.); upperparts gray to olive (see Subspp.); lower mandible pinkish to yellowish. Molt occurs in late summer on breeding grounds; birds brightest during fall migration, palest during spring and summer breeding. **Juv:** (May–Aug.) Similar to ad. but brownish above; head markings and wingbars may be more distinct in summer than faded adult's. **Flight:** Slightly undulating; may hover briefly. **Hab:** Dense shrubby areas, streamsides. **Voice:** Song a harsh rushed jumble of notes about 2 sec. long and with slight crescendo to end; calls include scolding *nyeh nyeh nyeh* and rising *zreee.*

Subspp: (4) •*pusillus* (cent.–s. CA) has long tail, often waved side to side; is palest of subspp., with gray upperparts, grayish flanks with yellow wash when fresh. •*arizonae* (s. NV–s.w. UT to s. CA–s.w. NM) has long tail, often bobbed up and down; olive-gray back, white to pale yellow flanks. •*medius* (s.w. TX–s.cent. NM) is similar to *arizonae* but has shorter tail, slightly more yellow to flanks. •*bellii* (rest of range to east) has short tail, greenish-olive back, yellowish-green flanks.

Adult m. TX/04

Adult f. TX/04

Black-capped Vireo 2
Vireo atricapilla L 4½"

Shape: Very small moderately long-tailed vireo with a thin slightly hooked bill. **Ad:** Dark head with prominent, thick, white spectacles; eye-ring slightly broken at the top; eye reddish. Back olive; underparts white with pale olive to yellow flanks; bill black. Two yellowish wingbars. M. with solid black cap; f. with dark gray cap, sometimes with black outline to spectacles. **Juv/1st Winter:** (May–Mar.) Like ad. f. but with dark brown eye (until Jan.); buffy wash to less well defined spectacles; grayish-green upperparts. Some birds with whiter spectacles may be m. **1st Summer:** (Apr.–Sep.) M. with black cap but gray nape; f. like ad. **Flight:** Relatively long tail; yellow underwing coverts and flanks; dark cap. **Hab:** Shrublands. **Voice:** Song a well-spaced series of fairly scratchy short complex phrases like *jheee wididip, dijidip, didip'zeeer;* calls include a short *tsidik* and rising *dzreee.*

Subspp: Monotypic.

Adult NM/05

Adult NM/06

Gray Vireo 2
Vireo vicinior L 5½"

Shape: Medium-sized long-tailed vireo with a deep-based slightly hooked bill. Primary projection past tertials short, about bill length (longer than bill length in similar Plumbeous Vireo). **Ad:** Often flicks or wags tail. Upperparts plain gray; underparts paler gray with whitish belly. Distinct thin white eye-ring; thin pale lore. One thin whitish wingbar on greater coverts. Fresh-plumaged birds (late summer–fall) may have slight greenish wash to flanks and lower back; worn birds (spring/summer) all gray and wingbar may be thin or absent. **Juv:** (May–Aug.) Like ad. but with distinct wingbar and brownish tinge overall. **Flight:** Direct; usually short and low. Long tail; grayish overall. **Hab:** Dry foothills, shrubby areas in open pinyon-juniper or oak woodlands. **Voice:** Song a spaced series of varied, slightly buzzy, mid-pitched, whistled phrases like *jeree jeroo jeray jerah;* calls include a harsh *jjaar,* a harsh descending trill of whistles, a short chatter, and a short whistled *twit.*

Subspp: (2) Mild differences; may not be valid. •*californicus* (CA–n.w. AZ) has darker gray upperparts; •*vicinior* (UT–s. AZ and east) has paler gray upperparts.

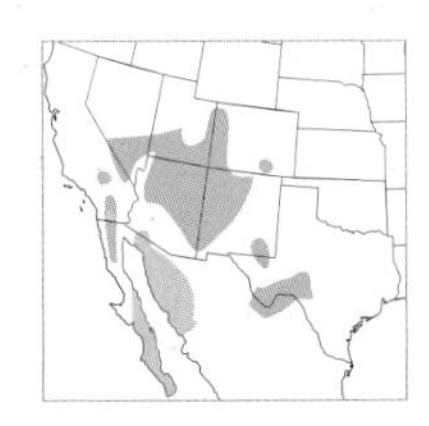

Adult TX/05

Adult OH/04

Yellow-throated Vireo 1

Vireo flavifrons L 5½"

Shape: Medium-sized, big-headed, broad-necked, short-tailed vireo with a rather thick bill. Long primary projection past tertials (2 x bill length). Distinguished from similar Pine Warbler by shorter tail, thicker bill, longer primary projection, heavyset look. **Ad:** Yellow spectacles; bright yellow throat and breast contrast with white belly; 2 white wingbars; dark eye. **Juv:** (May–Aug.) Brownish wash to upperparts; throat pale yellow. **Flight:** Short and direct. Yellow foreparts; olive back contrasts with gray rump. **Hab:** Mature deciduous forests. **Voice:** Song a series of well-spaced, short, burry, whistled phrases like *jeejoo jeejay jeeharee;* calls include series of harsh notes like *cheechecheche,* a soft *whit,* and a drawn-out *nyeeet.*

Subspp: Monotypic. **Hybrids:** Blue-headed Vireo.

Adult NM/05

Adult NM/05

Plumbeous Vireo 1

Vireo plumbeus L 5¾"

Shape: Medium-sized, broad-necked, relatively short-tailed vireo with a slightly hooked bill. Primary extension past tertials long (2 x bill length). Similar Cassin's Vireo has shorter tail and smaller bill. **Ad:** Gray above; whitish below with little or no yellow tones (pale yellow wash on lower flanks may be visible fall/spring on some birds). Distinct white spectacles (broken below lore); edges of secondaries gray (greenish in Cassin's Vireo); 2 distinct white wingbars. Not always possible to distinguish from dull Cassin's, especially when latter is in worn summer plumage. **Juv:** Like ad. but with brownish wash over back. **Flight:** Direct; usually short. Short tail; mostly gray. **Hab:** Open coniferous forests with shrub understory; higher elevations. **Voice:** Song a series of well-spaced, short, buzzy, whistled phrases like *jareer jaret jeerio jareeeb jeeer* (similar to Yellow-throated, Gray, and Cassin's Vireos); calls include a harsh chattered *chechechechee,* a soft *wurp,* and a soft warbled *loodle.* 🎧**64**.

Subspp: (1) •*plumbeus.*
Hybrids: Cassin's Vireo.

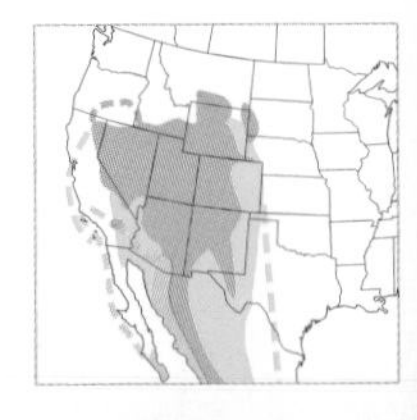

Adult CA/04

Adult OR/08

Cassin's Vireo 1

Vireo cassinii L 5½"

Shape: Medium-sized, relatively short-tailed vireo with a slightly hooked bill. Tail shorter than in similar Plumbeous Vireo. Primary extension past tertials long (2 x bill length). **Ad:** Dull gray above, with olive-green tones to back and rump (grayish when worn); whitish below, with olive-yellow sides of breast and flanks. Whitish spectacles (broken below lore) and throat contrast mildly with dull gray head (stronger contrast in Blue-headed Vireo, weaker contrast in Plumbeous); edges of secondaries greenish (whitish in Plumbeous); 2 whitish to yellowish wingbars. Less contrasting plumage overall than similar Blue-headed. M. more contrasting than f. Not always possible to disting. from Plumbeous and Blue-headed Vireos, especially in worn plumage. **Juv:** (May–Aug.) Like ad. but with brownish wash over back. **Flight:** Direct; usually short. Short tail; yellowish flanks. **Hab:** Open deciduous or mixed coniferous forests. **Voice:** Song a series of well-spaced, short, buzzy, whistled phrases like *jareer jaret jeerio jareeeb jeeer* (similar to Yellow-throated, Gray, and Plumbeous Vireos); calls include a harsh chattered *chechechecheee.* 🎧**65**.

Subspp: (1) •*cassinii.* **Hybrids:** Plumbeous Vireo.

Adult TX/05

Adult, *solitarius* OH/04

Blue-headed Vireo 1

Vireo solitarius L 5½"

Shape: Medium-sized relatively short-tailed vireo; similar shape to Cassin's Vireo, but bill slightly longer, slightly hooked. Primary extension past tertials long (2 x bill length). **Ad:** Bluish-gray to gray head; olive to grayish-olive back and rump (see Subspp.); bright white below with bright yellow sides of breast and flanks. Bright white spectacles and throat contrast strongly with dark gray face; edges of secondaries greenish; 2 bold yellowish-white wingbars. More contrasting plumage overall than similar Cassin's Vireo. Not always possible to distinguish from Cassin's, especially in worn plumage. **Juv:** (May–Aug.) Like ad. but with brownish wash over back. **Flight:** Direct; usually short. Short tail; yellowish flanks. **Hab:** Extensive forests of variable types; adaptable. **Voice:** Song a series of well-spaced, short, whistled phrases like *teeyoo taree tawit teeyoh* (song sweeter than that of Plumbeous and Cassin's Vireos). Calls include a harsh chattered *chechechecheee* and a nasal *neah neah neah.* 🎧**66**.

Subspp: (2) •*solitarius* (s. PA–NJ north to BC–NS) has medium-sized bill, greenish back; •*alticola* (WV–MD south) has longer heavier bill, darker grayish-green back. **Hybrids:** Yellow-throated Vireo.

Adult, Interior Grp. AZ/02

Adult, Interior Grp. AZ/04

Adult, Pacific Grp. CA/11

Hutton's Vireo 1

Vireo huttoni L 5"

Shape: Small, large-headed, short-tailed vireo with a relatively thick, blunt, slightly hooked bill (similar Ruby-crowned Kinglet has thin finely pointed bill). Primary projection past tertials less than 2 x bill length. Crown rounded (peak or crest on similar Empidonax flycatchers). **Ad:** Grayish to olive-green above (see Subspp.); paler grayish to olive below; pale lore; thin pale eye-ring broken at top; bill all dark. Two whitish wingbars, with darkest portion of wing between the 2 bars; legs and feet bluish gray. Similar Ruby-crowned Kinglet has darkest portion of wing at base of secondaries, forming a dark area below greater covert wingbar; also legs and especially feet are orangish. **Juv:** (Apr.–Jul.) Like ad. but with brownish tinge to upperparts; buffy wingbars. **Flight:** Dull olive overall with contrasting white wingbars. **Hab:** Evergreen forests, including conifers, live oaks. **Voice:** Song a series of harshly whistled phrases, each one repeated several times, like *jureeet jureeet jureeet jeeur jeeur;* calls include a raspy ascending *jreee,* a short *tik,* and a whinnylike *rrree'ee'ee'ee.*

Subspp: (8) Two geographic groups that may deserve species status. **Pacific Group:** Small; greenish-olive back. •*obscurus* (BC–n. CA), •*parkesi* (n.w. CA–cent. CA), •*huttoni* (cent. CA–s.w. CA), •*unitti* (Santa Catalina Is. CA), •*sierrae* (n.cent.–s.cent. CA), •*oberholseri* (e.cent.–s.w. CA). **Interior Group:** Large; grayish to grayish-olive back. •*stephensi* (AZ–NM), •*carolinae* (s.w. TX).

Adult TX/04

Adult, Western Grp. CA/09

Adult, *gilvus* OH/05

Adult, Western Grp. CA/05

Adult, *gilvus* OH/05

Warbling Vireo 1

Vireo gilvus L 5½"

Shape: Medium-sized slender vireo with a relatively short slightly hooked bill and moderate-length tail. **Ad:** A plain vireo. Upperparts grayish with greenish tinge; underparts whitish with variable yellowish wash on flanks and undertail coverts; no wingbars. Whitish eyebrow arches up over eye; faint gray eyeline; pale lore. Similar Philadelphia Vireo has shorter tail, brightest yellow on throat and central breast, and a darkish lore. **Juv:** (May–Jul.) Like ad. but with brownish wash to upperparts; white underparts; pale thin brownish wingbar on greater coverts.

Flight: Pale whitish belly and underwing coverts contrast with dark flight feathers. **Hab:** Deciduous woodlands, usually near water. **Voice:** Song a series of rapidly slurred whistles lasting about 2 sec. (more mellow in Eastern Group of subspp.; harsher in Western Group); calls include a nasal *neeyah,* a short *kit,* and a short chatter.

Subspp: (3) Divided into two groups. **Western Group:** •*swainsoni* (AK–CA) small, with short thin bill, dull olive-gray crown and back. •*brewsteri* (s. ID–w. SD south to AZ–s.w. TX) larger, with short thin bill; dull gray crown contrasts with olive-gray back. **Eastern Group:** •*gilvus* (AB–LA and east) large, with longer thicker bill, grayish back and crown.

Adult NH/09

Adult TX/05

Adult TX/04

Adult TX/04

Philadelphia Vireo 1

Vireo philadelphicus L 5¼"

Shape: Medium-sized vireo with a relatively short slightly hooked bill and short tail. **Ad:** Upperparts olive to grayish olive; underparts yellowish, brightest on throat and central breast, paler on flanks. Whitish eyebrow; gray eyeline; dark lore. Plumage brightest after molt (late summer–fall). Often shows a faint wingbar. Similar Warbling Vireo has longer tail, brightest yellow only on sides of breast, and pale lore. **Juv:** (Jun.–Aug.) Like ad. but with brownish wash to upperparts; eyebrow yellowish buff. **Flight:** Pale yellowish underparts and underwing coverts. **Hab:** Young deciduous woods and woodland edges. **Voice:** Song a series of 2–3-syllable clear, whistled, well-spaced phrases like *teeawit teeyoo turaweet* (similar to Red-eyed Vireo song; rate of delivery varies with breeding stage); calls include a nasal drawn-out *myaah* and a short harsh *tchet*.

Subspp: Monotypic. **Hybrids:** Red-eyed Vireo.

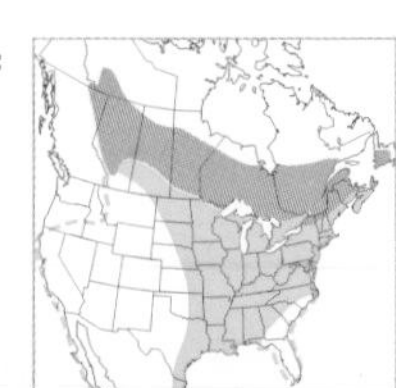

Red-eyed Vireo 1

Vireo olivaceus L 6"

Shape: Large short-tailed vireo with a medium-length slightly hooked bill. Crown shallow, giving sleek-headed look. Fairly long primary extension past tertials (1½ x bill length). **Ad:** Back olive-green contrasting with bluish-gray crown; underparts mostly white with pale yellow on sides of breast, flanks, and undertail coverts. Dark eyeline; straight whitish eyebrow bordered above by thin blackish line on either side of gray crown. Dark red eye. **Juv:** (May–Aug.) Like ad. but with brownish wash to upperparts; indistinct wingbars; eye dark brown (red by late winter). **Flight:** Yellow mostly on rear flanks and undertail coverts; short tail. **Hab:** Older deciduous woods, suburban or rural. **Voice:** Song a series of 2–3-syllable clear, whistled, well-spaced phrases like *teeawit teeyoo turaweet* (similar to Philadelphia Vireo song; rate of delivery varies with breeding stage); calls include nasal drawn-out *myaah*, rasping drawn-out *tjjjjj*, short harsh *tchet*. **67**.

Subspp: (2) •*caniviridis* (WA–n.cent. ID to east OR) upperparts grayish olive; •*olivaceus* (rest of range) upperparts brighter green. **Hybrids:** Philadelphia Vireo.

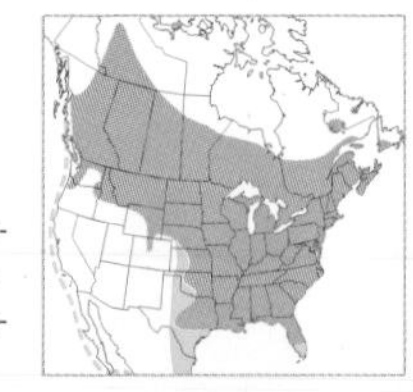

Adult PAN/02

Adult –/09

Yellow-green Vireo 3

Vireo flavoviridis L 6¼"

Rare Mex. species that may sometimes breed in s. TX; vag. farther n. in TX and to NM, AZ, CA. **Shape:** Same shape as Red-eyed Vireo but bill averages slightly larger (variation in bill size overlaps broadly with Red-eyed bill length). **Ad:** Back olive-green contrasting with bluish-gray crown; underparts mostly white with bright to dull yellow on sides of breast, flanks, and undertail coverts (see Subspp.). Yellow may extend from sides of breast up sides of neck to auriculars and may almost meet across breast (yellow more extensive than on similar Red-eyed). Dark eyeline and lore; grayish eyebrow bordered above by thin dark line (can be indistinct or absent, see Subspp.). Dark red eye. **Juv:** (May–Aug.) Like ad. but with brownish wash to upperparts; indistinct wingbars; eye dark brown (changes to red by late winter). **Flight:** Yellow extensive along sides and onto underwing coverts; short tail. **Hab:** Woodlands. **Voice:** Song a series of high-pitched, scratchy, whistled phrases, reminiscent of a House Sparrow, like *cheeup cheeri cheeup;* calls include a buzzy nasal *nyeeah* and a short chatter.

Subspp: (2) •*hypoleucus* (vag. to CA–AZ) has smaller bill, dull yellow on underparts; •*flavoviridis* (s. TX, vag. FL) has larger bill, brighter yellow on underparts.

Adult FL/04

Adult FL/04

Black-whiskered Vireo 2

Vireo altiloquus L 6¼"

Shape: Same shape as Red-eyed Vireo but bill averages longer and is more strongly hooked. **Ad:** Back brownish olive contrasting slightly with brownish to grayish crown; underparts mostly white with pale yellow mostly on rear flanks and undertail coverts. Thin dark whisker through malar area (variably distinct); dark eyeline and lore; whitish to buffy eyebrow; face pattern duller than in Red-eyed. Dark red eye. **Juv:** (Jun.–Aug.) Like ad. but with brownish wash overall; indistinct wingbars; eye dark brown (changes to red late winter). **Flight:** Pale below, dark above; pale yellow mostly on rear flanks and undertail coverts. **Hab:** Coastal mangroves. **Voice:** Song a series of harsh high-pitched whistles, like *chip chilip chilip;* calls include a nasal *nyeeah,* a short *kwit,* and a chatter.

Subspp: (2) Vary in size, bill size, color of eyebrow. •*barbatulus* (FL) small, with small bill (still larger than in Red-eyed), white eyebrow, gray crown; •*altiloquus* (vag. to FL, LA) larger, with larger bill, buffy eyebrow, brownish-gray crown.

Jays

Scrub-Jays

Magpies

Crows

Ravens

Adult, *canadensis* ME/01

Adult, worn, *canadensis* MB/06

Adult, worn, *canadensis* ME/01

Adult OR/07

Juv. OR/07

Gray Jay 1

Perisoreus canadensis L 11½"

Shape: Relatively short-tailed jay with a very short bill. Feathers appear loose and fluffy; slightly larger than Blue Jay. **Ad:** Gray upperparts; paler gray to whitish below. Whitish face and forehead; variably gray from nape and crown to eye (see Subspp.). **Juv:** (Apr.–Aug.) Slate-gray head and upperparts; slightly paler gray below; pale malar streak variable, from bright white to inconspicuous. **Flight:** Alternates flaps with long glides. **Hab:** Boreal and montane coniferous forests. **Voice:** Calls varied and usually fairly quiet; include a musical *quer quer quer,* a percussive *shek shek shek* and *ch'ch'ch,* a downslurred *wheeew,* a level whistled *weeet,* and a harsh *jewk jewk.*

Subspp: (6) Because individuals vary greatly and subspp. are clinal, only broad generalizations can be made on subspecies. •*pacificus* (AK–YT south to n.w. BC) and •*obscurus* (s.w. BC–CA) have darkest and most extensive gray on nape and crown, which extends to above eyes (*obscurus* has unique white shafts to back feathers). •*canadensis* (across N. and throughout E.), •*bicolor* (s.e. BC–n.e. OR to MT), and •*albescens* (s.e. YT–e.cent. BC to n.e. WY–n.w. MN) have less extensive and often paler gray on nape and crown, extending to just behind eyes. •*capitalis* (e. ID–cent. CO south) has palest and least extensive gray on just the nape.

Adult, *frontalis/paralia* OR/10

Adult, *diademata* AZ/06

Adult, *bromia* NH/10

Adult, *bromia* NH/05

Adult, *bromia* NH/09

Steller's Jay 1

Cyanocitta stelleri L 11½"

Shape: Relatively short-tailed long-crested jay. **Ad:** Dark jay with a blackish head, breast, and upper back; rest of body, wings, and tail dark blue, with thin black barring on wings and tail; no white on wings or tail. **Juv:** (May–Aug.) Similar to ad. but gray head, back, and underparts. **Flight:** Blackish head and back; blue wings, rump, and tail. **Hab:** Mature forests of mountains and coasts, feeders. **Voice:** Calls mostly harsh and loud; include a rapid-fire *shushushushook,* drawn-out *chjjj chjjj chjjj,* grating *gr'r'r'r'r,* and some hawk imitations. Voices of coastal birds can be noticeably different from Rockies birds. **68**.

Subspp: (8) Differences weak, subsp. classification needs reworking. •*stelleri* (coastal s. AK–cent. BC), •*carlottae* (Queen Charlotte Is. BC), and •*paralia* (coastal s. BC–cent. OR) have either no or few blue streaks on forehead. •*carbonacea* (w.cent. CA) and •*frontalis* (cent. OR–s. CA) have some pale blue streaking on forehead. •*annectens* (cent. BC–s. SK south to n.e. OR–n.w. OR), •*macrolopha* (n. UT–s. WY south to NV–CO), and •*diademata* (AZ–s.w. TX) have whitish streaking on forehead, latter two with bold white dash above eye. **Hybrids:** Blue Jay.

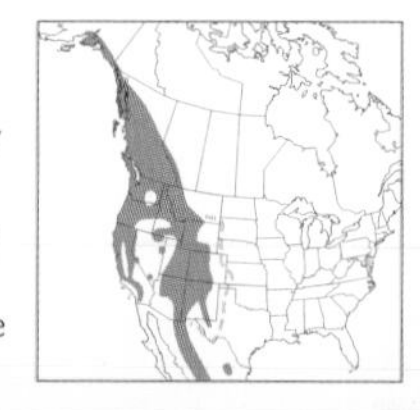

Blue Jay 1

Cyanocitta cristata L 11"

Shape: Relatively short-tailed moderately crested jay. **Ad:** Blue upperparts boldly spotted with white on wings and outer tail; gray breast, whitish belly; vertical black loral bar; complete black collar. **Juv:** (May–Aug.) Like ad. but with grayish crest and back; gray lores; thinner white wingbar. **Flight:** Secondaries and outer tail feathers broadly tipped white. **Hab:** Deciduous woods. **Voice:** Calls varied and loud; include a harsh *jaaay jaaay,* a bell-like *tooloodle tooloodle,* a grating *gr'r'r'r'r,* a soft *kueu kueu,* a squeaky *wheedelee,* and hawk imitations. **69**.

Subspp: (3) Differences weak. •*bromia* (cent. AB–n.e. NE and east) upperparts bright blue, much white on tertials; •*cyanotephra* (s.e. WY–NE and south) upperparts pale blue, much white on tertials; •*cristata* (s. IL–VA and south) upperparts medium blue, least white on tertials. **Hybrids:** Steller's Jay.

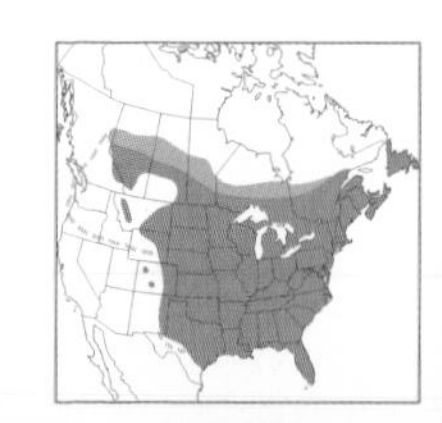

Green Jay 2

Cyanocorax yncas L 10½"

Shape: Long-tailed jay with a fairly short bill. **Ad:** Unmistakably brightly colored. Bright green back, wings, and tail; blue head; black breast. Pale green belly; yellow undertail coverts and outer tail feathers. **Juv:** (May–Jul.) Like ad. but with brownish wash to head and throat; belly pale yellow. **Flight:** Bright yellow outer tail feathers contrast with rest of green upperparts. Flight with deep wingbeats, short glides. **Hab:** Open woodlands, brushy thickets, feeders. **Voice:** Variety of harsh calls, including *cjjj cjjj cjjj, jreet jreet, sheet sheet sheet, jeer jeer,* and *gr'r'r'r'r.*

Subspp: (1) •*glaucescens.*

Brown Jay 3

Psilorhinus morio L 16½"

Limited to vicinity of Rio Grande in s. TX. **Shape:** Very large, long-tailed, magpielike jay with a rather short thick bill. **Ad:** Dark brown upperparts and breast; pale brownish wash over whitish belly; bill black. **Juv:** (May–Jul.) Like ad. but with yellow bill, feet, and legs. 2nd- to 3rd-year birds may have bill mottled with yellow and black patches. **Flight:** Dark brown with long tail. **Hab:** Dense riparian woods. **Voice:** Calls include a squealing downslurred *chweeer chweeer chweeer* given at different pitches and intensities, and a more hornlike *tooer tooer tooer.*

Subspp: (1) •*palliatus.*

Adult FL/02

Adult FL/02

Adult CA/05

Florida Scrub-Jay 2

Aphelocoma coerulescens L 11"

Limited to scrub oak habitats of cent. FL. **Shape:** Long-tailed jay with a large bill. **Ad:** Blue upperparts except for gray back; gray belly with blue undertail coverts; blue necklace surrounding whitish throat; whitish forehead; blue cheek. **Juv:** (Mar.–Aug.) Grayish head, back, and wing coverts; blue primaries, secondaries, and tail; whitish forehead; partial gray necklace. **Flight:** All blue above except for gray upper back. **Hab:** Scrub oaks and brush. **Voice:** Calls harsh; include a clipped *jreet jreet,* a slightly more drawn-out *jreee jreee,* and a percussive *tikatikatika.*

Subspp: Monotypic.

Island Scrub-Jay 2

Aphelocoma insularis L 13"

Limited to Santa Cruz Is., off s. CA. **Shape:** Large relatively long-tailed jay with a large bill. **Ad:** Dark blue upperparts except for brownish back; gray belly with blue undertail coverts; blue necklace surrounds whitish throat; blue forehead; black mask surrounds cheek, eye, and extends over base of upper mandible. **Juv:** (Mar.–Aug.) Grayish head, back, and wing coverts; blue primaries, secondaries, and tail; gray forehead; partial gray necklace. **Flight:** All dark blue above except for gray upper back. **Hab:** Chaparral and oak woodlands. **Voice:** Calls harsh (similar to Western Scrub-Jay's); include scratchy *chee chee chee,* rising *jreeet,* and percussive *gr'r'r'r.*

Subspp: Monotypic.

Adult, Interior Grp. AZ/03

Adult, Pac. Coastal Grp. CA/03

Adult OR/07

Juv. OR/08

Western Scrub-Jay 1

Aphelocoma californica L 11½"

Shape: Long-tailed jay with a large bill. **Ad:** Blue upperparts except for brownish or grayish back (see Subspp.); grayish or whitish belly with whitish or blue-tinged undertail coverts; blue or dusky-streaked necklace surrounds bright white throat; blue forehead; slate-gray mask surrounds cheek and eye but does not extend over base of upper mandible (as in Island Scrub-Jay). White eyebrow conspicuous to very faint. **Juv:** (Apr.–Aug.) Grayish head, back, and wing coverts; blue primaries, secondaries, and tail; gray forehead; hint of gray necklace. **Flight:** All dark blue above except for gray upper back. **Hab:** Dense brushy areas, gardens, woodlands. **Voice:** Calls harsh; include scratchy *chee chee chee,* rising *jreeet,* and percussive *gr'r'r'r.* **70**.

Subspp: (6) Two groups; differences moderate. **Pacific Coastal Group:** More contrastingly colored—deeper blue above with dark gray upper back, blackish mask, conspicuous white eyebrow, distinct blue breastband. •*californica* (WA–w.cent. CA), •*superciliosa* (s. OR–w. NV–s.cent. CA), •*obscura* (s.w. CA). **Interior Group:** Less contrastingly colored—lighter blue above with medium-grayish upper back, grayer underparts, duller gray mask, faint whitish eyebrow, faint gray breastband composed of short streaks. •*woodhouseii* (s.e. OR–s.w. ID and south), •*suttoni* (s.e. ID–s. WY and south), •*texana* (cent. TX).

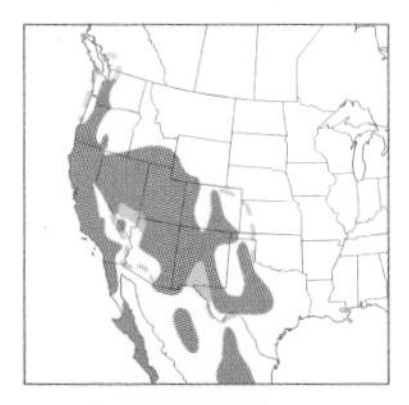

Clark's Nutcracker 1

Nucifraga columbiana L 12"

Shape: Large, stocky, long-winged, crestless jay with a long deep-based bill. Long primary projection past tertials. **Ad:** Body pale gray or tan-gray overall except for white undertail coverts; wings and tail black except for white tips to secondaries and white outer tail feathers. Whitish on face near bill and around eye. **Juv:** (Apr.–Sep.) Like ad., but can have more obvious brownish wash over gray body and lacks white on face. **Flight:** Gray body; black wings and tail; white trailing edge of secondaries; white outer tail feathers. **Hab:** Mountain coniferous forests. **Voice:** Calls harsh; include raspy *chjjjt chjjjt,* a *kwee kwee,* a drawn-out *jeee,* a rattle, and a downslurred *kweeer.*

Subspp: Monotypic.

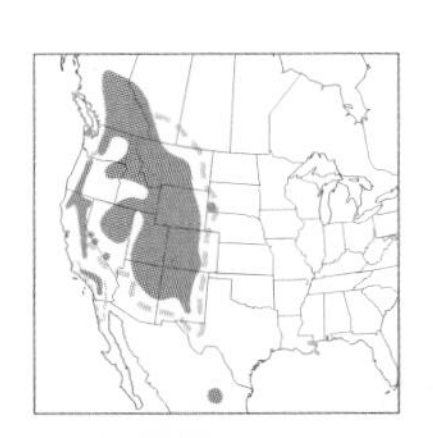

Adult, *arizonae* AZ/11

1st yr., *arizonae* AZ/02

Adult, *arizonae* AZ/03

Juv., *arizonae* AZ/07

Mexican Jay 2

Aphelocoma ultramarina L 11½"

Shape: Fairly large crestless jay with a short blunt bill and relatively short broad tail. Similar Western Scrub-Jay has longer tail; Pinyon Jay has sharper-pointed bill and proportionately shorter tail. **Ad:** Blue above (sometimes with grayish wash over back; see Subspp.); grayish below, sometimes with paler whitish belly; throat may be whitish or gray like breast (see Subspp.); head and face generally concolor, but can have suggestion of grayish wash over side of cheek and ear. Similar Pinyon Jay blue overall with more sharply pointed bill and shorter tail. Lacks white eyebrow of Scrub-Jays. **Juv:** (May–Sep.) Like ad. but grayish blue above. Bill black like ad. or whitish to pinkish, gradually changing to black over 2 yrs. (see Subspp.). **Flight:** Long-winged broad-tailed appearance. **Hab:** Pine-oak-juniper woodlands. **Voice:** Calls include a harsh rising *jreek jreek jreek* and a grating *gr'r'r'r* (*A. u. couchii* only).

Subspp: (2) Differences moderate; distinct ranges. •*arizonae* (AZ–NM) larger; grayish blue above; gray throat and breast; bill blotched with white to pink over first 2 yrs.; cooperative breeder. •*couchii* (s.w. TX) smaller; brighter blue above; white throat that contrasts with gray breast; bill black just after fledging; monogamous breeder.

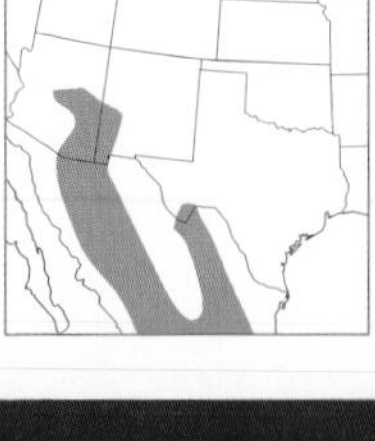

Adult, *cyanocephalus* NM/11

Adult, *cassini* OR/07

Juv., *cassini* OR/07

Juv., *cassini* OR/08

Pinyon Jay 1

Gymnorhinus cyanocephalus L 10½"

Shape: Fairly small jay with a short tail and fairly long, finely pointed, spikelike bill. **Ad:** Blue overall; underparts slightly duller blue; throat paler, sometimes whitish; lore dark; undertail coverts bluish to whitish. Similar Mountain Bluebird smaller with much shorter bill. Similar Mexican Jay blue above and gray below with more blunt-tipped bill and proportionately longer tail. **Juv:** (Mar.–Aug.) Like ad. but grayer blue overall. **Flight:** All blue with short tail. **Hab:** Pinyon-juniper and open dry pine woodlands. **Voice:** Calls include a nasal, nuthatch-like *kwoi kwoi kwoi,* descending *kwakwakwa,* high trill, and rattle.

Subspp: (3) Differences subtle. •*cyanocephalus* (MT–SD south to n. NM) pale blue with shorter slightly decurved bill; •*rostratus* (s.e. CA) and •*cassini* (rest of range) darker blue with longer straight bill.

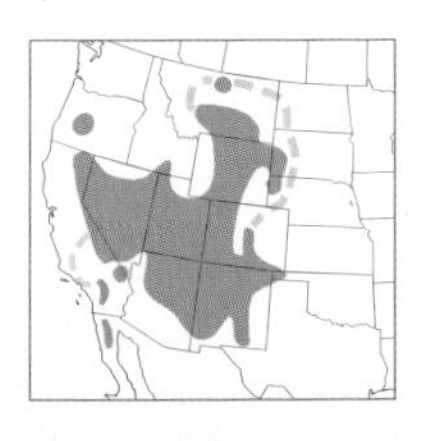

Adult CO/06

1st yr. CO/07

Adult CA/03

Black-billed Magpie 1

Pica hudsonia L 19"

Shape: Large very long-tailed bird with thick crow-like bill. Wings broad and rounded; tail graduated. **Ad:** Head, back, and undertail coverts black with a blue gloss; white scapulars and belly. Wings and tail black with iridescent blue gloss; white may be seen near tips of primaries on perched bird. Bill black. M. may have longer central tail feathers than f. **Juv:** (May–Oct.) Like ad. but wings and tail less iridescent; wing coverts greenish; white scapulars and belly washed with brown, looking "dirty"; no white near tips of primaries. **1st Yr:** No white near tips of primaries; bill may have whitish tip; facial skin may be bare around eye in summer. **Flight:** White on inner webs of primaries. **Hab:** Riparian thickets, parks, farmyards. **Voice:** Calls include a harsh *jeet* given singly or in a series, *jeejeejeejeet;* song soft, varied, babbling. 🎧71.

Subspp: (1) •*hudsonia.*

Adult CA/11

Yellow-billed Magpie 2

Pica nuttalli L 16½"

Limited to cent. CA. **Shape:** Slightly smaller and more compact than Black-billed Magpie, but overall proportions similar, with very long tail and large bill. **Ad:** Black head, back, and undertail coverts; white scapulars and belly; wings and tail black with iridescent blue-green gloss. Bill yellow; yellow bare skin from base of bill to below eye, sometimes surrounding eye. **Juv:** (May–Sep.) Like ad. but black areas have a brownish cast; belly creamy. **Flight:** White along inner webs of primaries flashes from above and below. Central tail feathers longest. **Hab:** Open oak woodlands. **Voice:** Similar to Black-billed Magpie. Calls include a harsh *jeet* given singly or in a series, *jeejeejeejeet;* also has a soft, varied, babbling song.

Subspp: Monotypic.

Adult ENG/07

Adult CYP/09

Eurasian Jackdaw 4

Corvus monedula L 14½"

Casual European vag. to Northeast. **Shape:** Compact broad-necked member of the crow genus with a short, deep-based, conical bill. Long primary projection past tertials. Walks with upright posture on relatively short legs. **Ad:** Glossy black overall with paler gray nape and sides of neck; pale eye. **Flight:** More pointed wings than on other crows; stubbier bill. **Hab:** Parks, suburbs, woods. **Voice:** Call a *kwerk kwerk*.

Subspp: (2) •*monedula*; •*spermologus*.

Adult, *hargravei* NM/12

Adult, *brachyrhynchos* NH/09

American Crow 1

Corvus brachyrhynchos L 17½"

Shape: Bill not noticeably massive (as in ravens); tail evenly rounded at tip rather than being noticeably longer in the center and wedge-shaped (as in ravens); slightly larger on average than Fish and Northwestern Crows. All NAM crows best distinguished by range, calls, and behavior. **Ad:** All black; back and crown look subtly scaled due to glossy violet centers to feathers; underparts iridescent except for belly. **Juv:** (May–Aug.) May appear flat brownish black with no gloss; grayish iris. **1st Yr:** (Sep.–Jul.) Wings and tail wear to brown and contrast subtly with black coverts and body feathers. **Flight:** Tail gently rounded at tip (raven tail wedge-shaped). Flies with rowing wing action. Generally does not soar and circle to rise on thermals—a behavior typical of Fish Crows and ravens. **Hab:** Fields, woods, suburbs, landfills. **Voice:** A wide variety of short and drawn-out *caws*, also a grating rattle; subsp. *hesperis* has slightly lower harsher calls on average. 🎧**72**.

Subspp: (4) Differences weak, mostly in size. •*hargravei* (s. ID–AZ–NM) large, with small bill; •*brachyrhynchos* (NT–s. TX–NJ and north) large, with long bill; •*pascuus* (cent. FL south) large, with medium-length bill; •*hesperis* (3 separate ranges: cent. BC–s. SK south to s. CA; CO–w. KS; LA–VA–n. FL) smallest, with small bill. Since *hesperis* is the smallest American Crow subsp. and its range overlaps with the generally smaller Northwestern and Fish Crows, using size distinctions in the identification of these three spp. is difficult.

Adult BC/02

Adult AK/02

Adult MEX/05

Northwestern Crow 1

Corvus caurinus L 16"

Shape: Tail evenly rounded rather than being noticeably longer in the center and wedge-shaped (as in ravens); bill not noticeably massive (as in ravens). Slightly smaller than American Crow on average, but at edge of range, distinction from American Crow may be impossible, especially since the smallest subsp. of American Crow (*hesperis*) is the one that abuts that range. Majority of Northwestern Crow range seems to be avoided by American Crow. All crows best identified by range and call. **Ad:** All black with glossy blue sheen overall. **Juv:** (May–Aug.) Before 1st prebasic molt in fall, may appear flat brownish black with no gloss and have a grayish iris. **1st Yr:** (Sep.–Jul.) Wings and tail wear to brown, contrast subtly with black coverts and body feathers. **Flight:** Tail gently rounded at tip. **Hab:** Coastal forests and intertidal zones. **Voice:** A wide variety of short and drawn-out *caws*, also a grating rattle; generally slightly lower and harsher than American Crow's (except for subsp. *hesperis;* see American Crow, Voice).

Subspp: Monotypic.

Tamaulipas Crow 3

Corvus imparatus L 14½"

Rare; only species of crow in southernmost TX. **Shape:** Smallest of NAM crows; bill relatively small, thinner at base than in other crows. **Ad:** All black with glossy sheen overall. All crows best identified by range and call. **Juv:** (May–Aug.) Like ad. but with less iridescence; brownish-black wings and tail. **Flight:** Small crow with relatively long tail. **Hab:** Open areas, landfills. **Voice:** Calls include a croaking *grow grow*.

Subspp: Monotypic.

Adult FL/01

Adult FL/03

Adult FL/03

Fish Crow 1

Corvus ossifragus L 15"

A crow of mostly tidal regions from MA to TX. Extending range inland along major rivers. **Shape:** Averages smaller than American Crow; tail gently rounded at tip. **Ad:** All black with glossy sheen overall. No scalelike effect on back or contrasting dull flat plumage as on American Crow, but this is difficult to see in the field. Best ident. by habitat (mostly coastal) and call. **Juv:** (May–Aug.) May appear flat brownish black, with no gloss and grayish iris. **1st Yr:** (Sep.–Jul.) Wings and tail wear to brown, contrast subtly with black coverts and body feathers. **Flight:** Large groups often rise by circular soaring on thermals, then stream off. **Hab:** Coastal, near water, farmlands, towns, landfills. **Voice:** Calls generally short, nasal, and higher-pitched than in American Crow; include short *anh* and double *anh anh,* also a rattle. **73**.

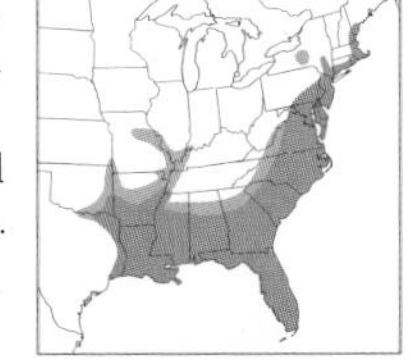

Subspp: Monotypic.

Adult NM/04

Adult NM/04

Adult NM/04

Chihuahuan Raven 1

Corvus cryptoleucus L 19½"

Shape: Midway in size between smaller American Crow and typically much larger Common Raven. Disting. from similar American and Tamaulipas Crows by size, massive blunt-tipped bill, long thin shaggy throat feathers, long nasal bristles covering 2/3 of culmen. Disting. from Common Raven with difficulty based on shape, since a smaller subsp. of Common Raven (*sinuatus*) abuts its range. **Ad:** All black with glossy sheen overall; white bases to otherwise black neck and upper breast feathers seen during preening or ruffling by wind. Nasal bristles extensive and usually cover half or more of upper mandible (usually less than half in Common Raven, but difficult to see). Separated from Common Raven best by color of base of neck feathers (gray in Common Raven), voice, and habitat. **Juv:** (Jun.–Sep.) Brownish black with little glossy sheen. **1st Yr:** (Oct.–Jul.) Wings and tail wear to brownish. **Flight:** Similar to Common Raven, but tail not as prominently wedge-shaped; wings long. Frequently soars (American Crow rarely soars). **Hab:** Arid grasslands, agricultural fields (Common Raven uses wide range of habitats). **Voice:** Calls generally higher-pitched and less varied than Common Raven's; include a slightly rising *craah craah.*

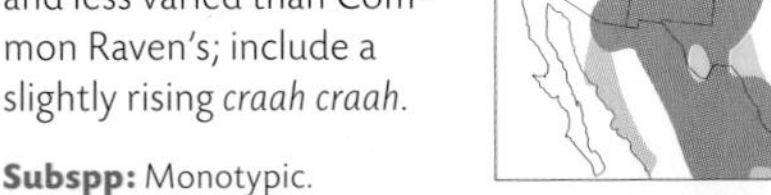

Subspp: Monotypic.

Adult, *clarionensis* CA/11

Adult, *principalis* NH/09

Adult, *principalis* NH/09

Adult, *principalis* NH/09

Common Raven 1

Corvus corax L 24"

Shape: Extremely large; bill deep-based and massive; thin feathers of neck very long and shaggy; nasal bristles usually cover about half length of culmen (half or more in Chihuahuan Raven). Bill size and shape, length of neck feathers, and overall size vary (see Subspp.). **Ad:** All black with a glossy sheen overall. Base of neck feathers pale gray to gray (white in Chihuahuan Raven)—seen during preening or ruffling by wind. Separated from Chihuahuan Raven by color of base of neck feathers, voice, and habitat. **Juv:** (Jun.–Sep.) Like ad. but brownish black. **1st Yr:** (Oct.–Jul.) Wings and tail wear to brownish. **Flight:** Tail graduated and wedge-shaped at tip; long tail and prominent neck, head, and bill project at either end. Wings relatively long, narrow, and often quite pointed at tip. Often seen soaring much like a large buteo, also rising on thermals in a spiral fashion. **Hab:** Wide range of habitats, usually with cliffs. **Voice:** Wide variety of generally low-pitched croaks and other sounds; include a low *growk,* drawn-out *kraaah,* and rattles. 🎧**74**.

Subspp: (4) Differences subtle. •*kamtschaticus* (w. AK) large, with deep-based bill; •*principalis* (s. BC–w. NC and north) large, with long bill; •*sinuatus* (e. WA–MT south to s.e. AZ–TX) medium-sized, with fairly small thin bill; •*clarionensis* (w. WA–CA–s.w. AZ) smallest, with short thinner bill, and long nasal bristles like Chihuahuan Raven.

Larks

Martins

Swallows

Adult m., Western Rufous Grp. CA/04

Adult f., Western Pale Grp. CO/06

Adult m., Eastern Dark Grp. ME/11

Adult f. TX/03

Adult f., Eastern Dark Grp. NH/09

Juv. AB/07

Horned Lark 1

Eremophila alpestris L 7¼"

Shape: Very slender-bodied bird and pointed at both ends; long wings; flattened crown; relatively small, thin, sharp-pointed bill. "Horns" (small projecting feathers) on back of crown, generally larger on males. **Ad:** Distinct head pattern—wide dark bar from lore through cheek; whitish to deep yellow eyebrow and throat; dark crown with black "horns." Dark breastband on upper breast; rest of breast variably white to yellow, streaked to unstreaked (see Subspp.); belly white. M. tends to have more distinct, larger black markings on head and breast, and longer "horns," than f. Plumage differences (among sexes and subspp.) slightly masked in fall and winter by buff tips to feathers; clearer in spring and summer when these wear off. **Juv:** (Apr.–Aug.) Fine white spotting on dark crown and back; facial pattern only suggested and black breastband absent. Whitish wingbar and pale edges to tertials. **Flight:** Wings long, pale underneath; contrasting tail blackish with thin white outer edge; head visible in flight. **Hab:** Open fields and barrens with sparse vegetation. **Voice:** Song is several short notes followed by a rising trill, like *treet treet trititititeet,* given in aerial display or from ground; common flight call *tsee titi.*

Subspp: (21) Divided into three groups. **Western Rufous Group** (coastal BC–coastal CA–s.w. NM): Small, yellow face and throat, reddish-brown to pinkish-tinged nape and collar; widespread subsp. is •*rubea.* **Western Pale Group** (AK through interior of coastal provinces and states east to w. NT and w. TX): Large, white to pale yellow face and throat, grayish-brown nape and collar; most widespread subsp. is •*arcticola.* **Eastern Dark Group** (e. NT–e. TX and east): Small to large, pale to deep yellow face and throat, dark brown to reddish brown on nape and collar; most widespread subsp. is •*alpestris.*

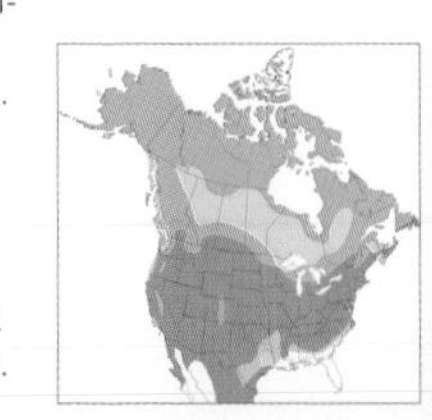

Adult, *arvensis* ENG/06

Adult, *arvensis* ENG/06

Sky Lark 3

Alauda arvensis L 7¼"

Introduced to Vancouver Is. BC; rare to w. AK (where it has bred) and vag. to CA. **Shape:** Somewhat sparrowlike but slightly larger, with smaller head, slender bill, broad-based triangular wings; substantial crest, when raised, gives head triangular peak. **Ad:** Whitish eye-ring and eyebrow; dark streaked crown, brown-streaked auricular with dark outline. Chin white; necklace of fine streaks across buffy breast. **Juv:** (Jun.–Aug.) Like ad. but pale edges to back and wing feathers, creating scaled look; eyebrow wider and paler. **Flight:** Often takes short fluttering flights with tail partly lowered and spread. Tail dark with white outer feathers; secondaries and inner primaries with white trailing edge. **Hab:** Open fields, pastures with short vegetation. **Voice:** Song a long (2–5+ sec.) musical mixture of warbling and slurred notes given during aerial display; calls include a short liquid *chururk*.

Subspp: (2) Differences slight. •*arvensis* (Vancouver BC) introduced from Europe; •*pekinensis* (rare in w. AK is., vag. to CA) migrant and breeder; darker and more heavily streaked above than *arvensis*.

Adult URU/12

Adult MA/10

Brown-chested Martin 5

Progne tapera L 7½"

Accidental austral migrant from SAM to e. NAM. **Shape:** Much larger than a swallow, with long fairly deep-based inner wing and long tapered outer wing. Long undertail coverts wrap up around sides of rump; tail slightly forked; wing tips just reach tail tip on perched bird. **Ad:** Brown cap and upperparts; white throat, belly, and undertail coverts; breast washed brown with a variably distinct darker brown breastband, sometimes with an extended line of brown dashes down center of breast (see Subspp.) Fine whitish edges to coverts and secondaries. **Juv:** Like ad. but with grayish sides to throat. **Flight:** Often stays low as it flies about feeding; flight fast and effortless; much larger than other swallows with which it might feed. **Hab:** Clearings, open areas. **Voice:** Bubbling song.

Subspp: (1) •*fusca* (austral mig. to n. SAM) has defined breastband with line of dashes down center of breast.

Adult m., *subis* FL/03

1st-yr. m., *subis* FL/03

Adult f., *subis* FL/03

1st-yr. f., *subis* FL/03

Adult m., *subis* FL/03

Adult f., *subis* FL/03

Purple Martin 1

Progne subis L 8"

Shape: Largest common member of swallow family. Wings long, bill relatively large, tail forked when closed. **Ad:** M. all dark; black wings and tail; glossy bluish-black body. F. black wings and tail; silvery forehead, collar; sooty throat; underparts pale gray with fine blurry darker spotting; undertail coverts dark-centered with wide silvery margins. **Juv:** (Apr.–Aug.) Both sexes have sooty throat; whitish belly with fine dark streaks; white undertail coverts with fine dark streaks. **1st Yr:** (Sep.–Aug.) F. like juvenile. M. like juv. but with a few to many dark purple feathers on throat and breast. **Flight:** Larger, longer-winged, and with head and neck projecting farther in front of wings than other swallows. Wingbeats slower, and may do more soaring than other swallows. **Hab:** Open areas near water (often near human habitations), open mountain woods with snags, deserts with saguaro cacti. **Voice:** Song is a series of paired notes, creating a gurgling sound 2–6 sec. long (given by both sexes); m. song includes a grating *gr'r'r'r* (at end in *P. s. subis,* interspersed in *P. s. hesperis*). Calls include a hoarse *cher* or *cher cher* and a *zweeet* (upslurred in *subis,* downslurred in *hesperis*). **75**.

Subspp: (3) Distinct ranges and habitats. •*arboricola* (coastal BC–CA–NM) is large, nests mostly in tree holes (some in birdhouses); •*hesperis* (s. AZ deserts) is smaller, nests in holes of large cacti; •*subis* (rest of range) is medium-sized, nests in human-made birdhouses and gourds.

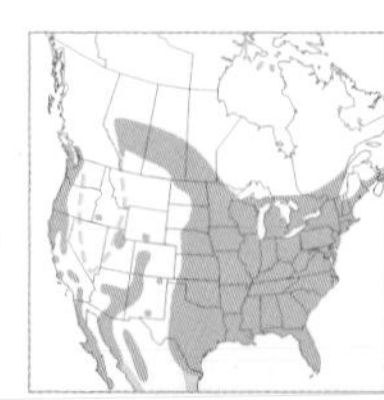

Adult NH/07

1st-yr. f. NH/05

Adult NH/05

Juv. NH/08

Tree Swallow 1

Tachycineta bicolor L 5¾"

Shape: Medium-sized broad-necked swallow. In flight, wings roughly triangular with broad base; tail shallow-forked when closed. On perched bird, wings project just past tail (similar Violet-green Swallow's wings project well beyond tail). **Ad:** Above, iridescent bluish green (m. brightest, f. varies with age; see below). Below, white; dark area on head extends from bill to just below eye and down side of neck; clear demarcation on head between dark and light areas. In flight, white triangular spur extending from rear flanks onto sides of rump separates this species from other swallows at all ages (except Violet-green, which has a narrow white band on rump sides, almost meeting across top of rump). **Juv:** (May–Sep.) Above, grayish brown; below, clear white with brownish wash across breast forming faint band, palest in the center (juv. Bank Swallow with clearly defined dark breastband with downward extension in middle). Clear demarcation of light and dark area on face (more blended on juv. Violet-green). Perched, disting. from similar Violet-green juv. by wing/tail lengths (see Shape). **1st Yr:** (Sep.–Aug.) M. like adult. F. like ad. but grayish brown above with no to over half feathers iridescent blue. **Flight:** Broad-based triangular wings; relatively long tail. Uniformly dark above except for white spurs on

sides of rump. **Hab:** Open areas with snags, usually near water. **Voice:** Song a few introductory notes followed by a short bubbling warble, like *tseetseetsee chirdup chirdup;* calls include *chee,* a double *cheedeep,* and *chee chee chee* (in a long series). 🎧**76**.

Subspp: Monotypic. **Hybrids:** Cliff Swallow.

IDENTIFICATION TIPS

Swallows are mostly small birds that spend the majority of their time flying through the air to catch insects. Their wings are broad-based and sharply pointed at the tips; when they are perched, you can see their long primary extension past the tertials (about half the length of their closed wing).

Species ID: The Purple Martin is the largest of the swallows and the only one that can be all dark beneath (adult m.). For most other swallows, look at the color of the upperparts: some are mostly brown, others are iridescent blue or green, others have reddish-brown rumps. Also look closely at the patterns of light and dark on the underparts and at whether and how light areas extend onto the rump.

Adult m., *thalassina* CA/07

Adult m., *thalassina* CA/06

Adult f., *thalassina* CA/07

Adult f., *thalassina* CA/06

Violet-green Swallow 1

Tachycineta thalassina L 5¼"

Shape: Small, relatively long-winged, short-tailed swallow. On perched bird, wings project well past tail at all ages (project just past tail on Tree Swallow). **Ad:** Back greenish; wings and tail blackish. Narrow white rectangular patch extends from rear flanks well onto rump (making bird look almost white-rumped); this separates it from all our other swallows at all ages. M. has violet-green crown; white on auriculars extending up over back of eye; purple rump. F. has brownish-green crown; dusky auriculars that shade gradually to pale throat (not strongly demarcated as in Tree Swallow); sometimes thin white line over eye; sooty rump. **Juv:** (May–Oct.) Grayish-brown upperparts; whitish underparts; dusky face fading to paler throat. See Ad. and Shape for Tree Swallow comparison. **Flight:** Narrow wings; short tail. Uniformly dark above except for white bands on sides of rump. **Hab:** Open woodlands, deciduous or mixed. **Voice:** Calls include a midpitched *choit,* a repeated *cheecheechee,* and a high-pitched *tseet.* **77**.

Subspp: (2) •*thalassina* (all of USA–CAN range) large; •*brachyptera* (vag. from MEX to s. AZ) small.

Adult, *psammochrous* CA/01

Adult, *psammochrous* CA/06

Adult, *serripennis* NH/05

Adult, *psammochrous* CA/06

Northern Rough-winged Swallow 1

Stelgidopteryx serripennis L 5½"

Shape: Medium-sized, short-tailed, relatively broad-winged swallow. On perched bird, wings reach tail tip; head appears small and well defined by fairly thin neck; tail broad and squared-off. **Ad:** Above, uniformly brown (see Subspp.); below, dusky face and throat gradually shade to pale whitish belly and undertail coverts. **Juv:** (May–Oct.) Like ad. but with faint cinnamon wash over throat and upperparts; cinnamon tips to greater coverts create rusty wingbar (lesser coverts may be paler or cinnamon). **Flight:** Above, concolor brown (similar Bank Swallow with paler back and rump than upper back and wings; also see Subspp.); below, dusky throat shades to pale whitish belly. Rump all dark with no white. Tail square to slightly rounded. **Hab:** Near rocky outcrops, bridges, or embankments for nesting. **Voice:** Song is a harsh chattering, 3–5 sec. long; calls include spaced short buzzes like *djjt, djjt, djjt* and an explosive *bjeeert*.

Subspp: (2) •*psammochrous* (cent. CA–s.e. TX) has pale grayish-brown upperparts; crown and rump may be lighter. •*serripennis* (rest of range) has uniformly dark brown upperparts.

Adult, riparia CA/08

Juv., riparia EUR/10

Adult, riparia CA/08

Juv., riparia EUR/10

Bank Swallow 1

Riparia riparia L 5¼"

Shape: Small, narrow-winged, relatively long-tailed swallow. Wings reach tail tip on perched bird; tail relatively thin, slightly notched. **Ad:** Above, subtly bicolored—pale brown rump and lower back contrast with darker wings; below, white except for wide brown breastband, which often extends as variably thick line down center of breast. **Juv:** (May–Oct.) Like ad., but upperpart feathers finely margined paler and throat may have buffy wash. **Flight:** Most rapid wingbeats of our swallows. Subtly bicolored above, with paler lower back and rump, and darker upper back, wings, and tail (see Northern Rough-winged subsp. *psammochrous*). Rump all dark with no white. Breastband can be hard to see in rapid flight, but white throat and cheek often stand out. **Hab:** Lowland areas near water. **Voice:** Calls include single *jit* or double *jijit* (more clipped than Northern Rough-winged's call); song just a long rapid series of notes like powerline zapping sounds.

Subspp: (2) •*riparia* (all of NAM range); •*diluta* (vag. to AK) has paler upperparts and indistinct breastband.

Adult, *pyrrhonota* UT/06

Adult CA/03

Juv., *pyrrhonota* UT/07

Adult, *pyrrhonota* NH/05

Adult, *pyrrhonota* MT/06

Juv. CA/08

Cliff Swallow 1

Petrochelidon pyrrhonota L 5½"

Shape: Relatively large-headed broad-necked swallow with short slightly notched to squared tail and broad wings that just reach tail tip on perched bird. **Ad:** Above, dark back and tail contrast with pale buffy to orange-buff rump; below, breast pale buff, belly whitish. Bluish-black crown, dark reddish-brown face, and reddish-brown to black throat create dark-hooded look (Cave Swallow looks dark-capped because of its dark crown and pale face and throat); usually has small whitish to pale buff "headlight" on forehead (see Subspp.). **Juv:** (Jun.–Dec.) Like ad. pattern but duskier on head and upperparts; forehead dark brown, face and throat brownish. **Flight:** Dark-hooded look; pale buffy to orange-buff rump; usually white forehead. Short tail; comparatively broad wings. **Hab:** Varied habitats, open areas with cliffs, bridges, buildings where nests built. **Voice:** Song composed of squeaking and grating notes (similar to but less musical than Barn Swallow); calls include a downslurred *tzeer* and a rough *jjup*.

Subspp: (4) •*pyrrhonota* (cent. CA–VA and north) large, with buff to white forehead; •*tachina* (s. CA across n. AZ and n. NM to s.w. TX) medium, with buff to cinnamon forehead; •*swainsoni* (s.e. AZ–s.w. TX) small, with darkest forehead—brownish to reddish brown; •*ganieri* (OK–s. TX and east) medium-sized, with buff to whitish forehead. **Hybrids:** Tree and Barn Swallows.

Adult, *pelodoma* TX/03

Adult, *pelodoma* TX/06

Juv., *pelodoma* TX/09

Juv., *pelodoma* TX/06

Cave Swallow 1

Petrochelidon fulva L 5½"

Shape: Relatively large-headed broad-necked swallow with short slightly notched to square tail and broad wings that just reach tail tip on perched bird. Like Cliff Swallow. **Ad:** Above, dark back and tail contrast with reddish-brown to deep buff rump; below, breast pale buff, belly whitish. Dark bluish-black crown and dark reddish-brown forehead contrast with pale buffy throat and face, creating capped look. **Juv:** (Jun.–Dec.) Like ad. pattern but wings and tail brownish, tertials broadly tipped with white. **Flight:** Dark-capped look; deep buff to reddish-brown rump; dark reddish-brown forehead. Short tail; comparatively broad wings. **Hab:** Open areas; often nests in culverts or under bridges. **Voice:** Calls include an upslurred *vweet* and downslurred *tseeah.*

Subspp: (3) •*pelodoma* (TX and regular fall vag. in East, accidental in AZ) large and palest, with buff to pale reddish-brown throat and rump; •*fulva* (s. FL and vag. up East Coast) small, with moderately dark throat and rump; •*cavicola* (vag. to FL from Cuba) medium-sized with cinnamon throat and reddish-brown rump. **Hybrids:** Barn Swallow.

Adult m., *erythrogaster* TX/04

Adult, *erythrogaster* OH/05

Adult f., *erythrogaster* OH/05

Adult, *erythrogaster* OH/05

Juv., *erythrogaster* TX/07

Juv., *erythrogaster* CA/08

Barn Swallow 1

Hirundo rustica L 6¾"

Shape: Large narrow-winged swallow with a long deeply forked tail. On perched bird, tail projects well past wings. **Ad:** Above, iridescent bluish black; throat and forehead pale to deep brick-orange; rest of underparts whitish to orange (see Subspp.); partial to complete dark breastband (see Subspp.). M. has glossy blue upperparts; dark reddish-brown breast; very long tail streamers. F. has dull blue upperparts; paler breast; shorter tail streamers. **Juv:** (Jun.–Dec.) Like ad. but upperparts brownish, underparts paler. Tail short, less deeply forked; generally reaching to wing tip on perched bird. **Flight:** Long deeply forked tail has white subterminal band across outer feathers. More graceful and streamlined flier than other swallows; seems to slice through the air. **Hab:** Open areas. **Voice:** Song a continuous twitter interspersed with grating sounds; calls include short *chevit* or *chevitvit,* a short clear whistle, and a *cheveet*. 🎧**78**.

Subspp: (4) •*erythrogaster* (all of NAM range except Gulf is.) medium-sized, with partial breastband, buffy to orange underparts; •*insularis* (LA–AL is.) like *erythrogaster* but with whitish to buffy underparts; •*rustica* (Eurasian, vag. to AK) large, with complete wide breastband, white to pale buff underparts; •*gutturalis* (Asian, vag. to w. AK–BC) small, with complete or partial narrow breastband, white to pale buff underparts. **Hybrids:** Cliff and Cave Swallows.

Adult BAH/04

Adult BAH/04

Bahama Swallow 4

Tachycineta cyaneoviridis L 5¾"

Casual vag. from Bahamas to s. FL. **Shape:** Medium-sized, relatively short-winged, long-tailed swallow. Tail deeply forked. On perched bird, wings slightly to considerably shorter than tail. **Ad:** Above, dark bluish green; below, white; sharp demarcation on head where dark cap meets white cheek (sometimes a thin black line). White more extensive on cheek than in Tree Swallow; does not extend above eye as in m. Violet-green Swallow. No white on sides of rump (as in Tree and Violet-green). Sexes similar, but f. may be browner above and slightly duskier below. **Juv:** (May–Aug.) Like ad. but upperparts brownish; underparts uniformly whitish to dusky (no darker band across breast as in juv. Tree Swallow). **Flight:** White underwing coverts at all ages unique among our swallows. Fairly long deeply forked tail. **Hab:** Open woodlands, pine forests. **Voice:** A clipped *chit* or *chichit*.

Subspp: Monotypic.

Adult, *urbicum* EUR/05

Adult, *urbicum* EUR/05

Common House-Martin 4

Delichon urbicum L 5"

Casual vag. to w. AK and e. CAN. **Shape:** Small swallow with shape similar to Tree Swallow but tail more deeply forked. **Ad:** Above, bluish black with pale indistinct collar and bright white rump; below, all white. M. underparts white; f. underparts grayish. **Juv:** Like ad. but duller brownish black above; tail less deeply forked. **Flight:** Bright white rump contrasts with rest of blackish upperparts; whitish underwing coverts. Long deeply forked tail. **Hab:** Open areas. **Voice:** Song a rough scratchy twittering; calls include short *chrit*.

Subspp: (2) •*lagopoda* (East Asian vag. to w. AK) has white uppertail coverts (with white rump), slightly forked tail; •*urbicum* (w. Eur. vag. to e. CAN) has blackish uppertail coverts (with white rump), deeply forked tail.

Chickadees

Titmice

Nuthatches

Wrens

Adult TX/06

Adult, worn TX/05

Adult GA/02

Adult OH/11

Carolina Chickadee 1

Poecile carolinensis L 4¾"

Shape: Smaller, shorter-tailed, and slightly smaller-headed than Black-capped Chickadee. **Ad:** Plumage overall less contrasting and less colorful than Black-capped's (especially when in fresher plumage during winter and spring). White cheek may blend to gray sides of nape; gray to whitish margins (see Subspp.) to greater coverts, tertials, secondaries, and tail feathers; gray to pale buff flanks; gray back. Where range overlaps with similar Black-capped Chickadee, identification challenging; subsp. characteristics merge, and hybrids occur. Similar Black-capped has all-white cheek and sides of nape; white margins to greater coverts, tertials, secondaries, and tail feathers; rich buff flanks; and sometimes olive tinge to back; see also Shape, Voice, and Range. **Flight:** Bouncing flight path; note white cheek. **Hab:** Variety of woodlands, gardens, feeders. **Voice:** Song 3–5 high-pitched whistled notes like *fee bay fee boo* (Black-capped song just 2 notes); calls include *deedeedee, chickadeedee, chebechebechay, tseedleedeet,* and quiet *tseet.* Most calls higher, harsher, and more rapid than those of Black-capped; in region of overlap, one species may learn the other's song, in which case calls (which are innate) may be more helpful. 🎧**79**.

Subspp: (4) •*carolinensis* (n.e. AR–s.e. VA and south) small, flanks grayish buff, secondary edges grayish; •*atricapilloides* (KS–cent. TX) large, flanks grayish buff, secondary edges whitish; •*agilis* (s. LA–s.e. TX–s. AR) medium-sized, flanks pale grayish, secondary edges whitish; •*extimus* (MO–n.e. VA and north) medium-sized, flanks buffy, secondary edges whitish, making distinction from nearby Black-capped difficult. **Hybrids:** Black-capped Chickadee.

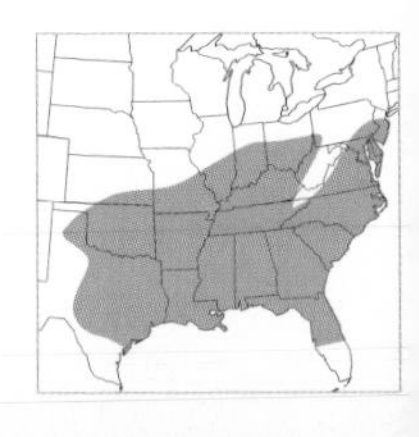

Adult, Eastern Grp. NH/03

Adult, worn, Eastern Grp. ME/06

Adult, Eastern Grp. NH/09

Adult, Northwestern Grp. BC/02

Black-capped Chickadee 1

Poecile atricapillus L 5¼"

Shape: Comparatively long-tailed large-headed chickadee. **Ad:** Black cap and bib; white cheek and side of nape; back gray, or brownish to olive gray; greater coverts, secondaries, tertials, and tail with bold white margins when in fresh plumage in winter and spring; flanks pale to rich buff. Colors brighter and more contrasting in winter and spring, when feathers fresh; more muted in summer, when feathers worn. Where range overlaps with Carolina Chickadee, identification difficult; see that species for comparisons. **Flight:** Bouncing flight path; note white cheek. **Hab:** Variety of woodlands. **Voice:** Song 2 high-pitched whistled notes like *feebee* (Carolina song 2–5 notes); calls include *deedeedee, chickadeedee, tseedleedeet, chebechebechay,* and quiet *tseet.* Most calls lower-pitched and slower than those of Carolina; in region of overlap, one species may learn the other's song, in which case calls (which are innate) may be more helpful. 🎧**80**.

Subspp: (10) Divided into three groups. **Eastern Group** (s.e. MB–e. KS and east): Medium-sized, back moderately dark, fairly distinct white covert edges; main subsp. is •*atricapillus.* **Interior Western Group** (AK–MB south to CO–cent. KS): Largest, palest flanks and back, distinct white covert edges; main subsp. is •*septentrionalis.* **Northwestern Group** (s.w. BC–n.w. MT south to n. CA): Smallest, darkest overall, indistinct covert edging; main subsp. is •*occidentalis.* **Hybrids:** Carolina and Mountain Chickadees.

Adult, *baileyae* CA/05

Adult, *gambeli* NM/11

Adult, worn, *gambeli* CO/07

Mountain Chickadee 1

Poecile gambeli L 5¼"

Shape: Comparatively longer-winged, shorter-tailed, and longer-billed than Black-capped Chickadee. **Ad:** Black cap, white eyebrow (eyebrow may be less distinct in late summer when worn); wings plain gray with no distinct pale edges to greater coverts, secondaries, or tertials; flanks pale gray, sometimes with olive or buff tinge; cheek white. **Flight:** Bouncing flight path; note white cheek and white eyebrow. **Hab:** Mountain coniferous forests, feeders. **Voice:** Song 2–6 whistled notes on either nearly same or different pitches, like *fee fee fee* or *fee bay bay* (with regional variations), higher-pitched and faster than Black-capped's. Calls include *deedeedee, chickadeedee, tseedleedeet, chebechebechay* (all raspier than Black-capped's), and quiet *tseet.*

Subspp: (3) •*gambeli* (s.e. ID–cent. MT south to s. AZ–s.w. TX) large, back gray with olive tinge, flanks grayish; •*baileyae* (YT–AB south through cent. OR to s. CA) small to medium-sized, back dark gray with buff tinge, flanks grayish with wash of olive or buff; •*inyoensis* (rest of range) medium-sized, back pale gray, flanks buffy gray. **Hybrids:** Black-capped and Boreal Chickadees.

Adult AZ/07

Adult AZ/08

Mexican Chickadee 2

Poecile sclateri L 5"

Shape: Similar to Black-capped Chickadee—comparatively long-tailed, large-headed. **Ad:** Large black bib extends down onto upper breast; rest of underparts dark gray and similar in color to back and wings; greater coverts may have pale tips, creating thin wingbar. **Flight:** Bouncing flight path; note white cheek and gray belly. **Hab:** Coniferous forests. **Voice:** Song is short whistled notes; calls include buzzy and hissing sounds like *dzee dzee* and *shhh.*

Subspp: (1) •*eidos.*

Adult, *neglectus* CA/11

Adult, *littoralis* ME/02

Adult, *rufescens* BC/11

Adult AK/03

Chestnut-backed Chickadee 1

Poecile rufescens L 4¾"

Shape: Small relatively short-tailed chickadee. **Ad:** Distinctive chestnut back and rump; flanks chestnut to gray; breast and belly whitish. White cheek patch narrower toward the nape than in other chickadee spp. **1st Yr:** (Sep.–Aug.) Molt limits often visible on greater coverts—new inner coverts fresh and white-edged, old outer ones worn and buff-edged. **Flight:** Bouncing flight path. **Hab:** Coniferous and deciduous forests, shrubby forest edges, parks, gardens. **Voice:** Seems to lack whistled song of other chickadees; calls include *seebeeseebee, chickadeedee, tseedleedeet*, and *tseet*; faster and buzzier than Black-capped's.

Subspp: (3) •*rufescens* (AK–n. CA to MT) has bright reddish-brown back and flanks; •*neglectus* (coastal cent. CA) has pale reddish-brown wash on back and flanks; •*barlowi* (cent.–s. CA) has reddish-brown back, gray to brownish-olive wash on flanks.

Boreal Chickadee 1

Poecile hudsonicus L 5½"

Shape: Large relatively short-tailed chickadee. **Ad:** Brownish cap; white cheek; gray side of nape; flanks strongly pinkish buff when fresh; little contrast between cap and back colors; wing feathers without pale edging. Similar Gray-headed Chickadee (range limited to n. AK) has white margins to wing and tail feathers, grayish-brown cap, white cheek and sides of nape. **Juv:** (Jun.–Sep.) Can be paler-colored than ad., leading to possible confusion with Gray-headed Chickadee, except for cheek color and feather margins, which are like ad. **Flight:** Bouncing flight path. **Hab:** Boreal forests. **Voice:** Lacks whistled song of Black-capped; calls comparatively raspy, include *tsi tsi tsa jweee, tsijwee, tseet, chit,* and a short trill.

Subspp: (5) Two groups. **Grayish-backed Group:** Grayish backs. •*stoneyi* (n. AK–n. NT), •*columbianus* (s. AK–WA–MT). **Brown-backed Group:** Brownish backs, sometimes with tawny or olive tinge. •*farleyi* (s. NT–AB–s.w. MB), •*littoralis* (s. QC–n.e. NY and east), •*hudsonicus* (cent. AK–MN–s. ON and north). **Hybrids:** Mountain Chickadee.

Adult, *lapponicus** FIN/04

Adult, *lapponicus* FIN/04

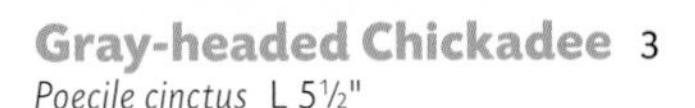

Gray-headed Chickadee 3

Poecile cinctus L 5½"

Shape: Large relatively short-tailed chickadee. **Ad:** Grayish-brown cap; white cheek and side of nape; whitish margins to greater coverts, secondaries, and tail feathers. Flanks cinnamon when fresh, paler when worn; little contrast between cap and back colors. **Juv:** (Jun.–Sep.) Like ad., but can be duller. **Flight:** Bouncing flight path. **Hab:** Willows and spruces. **Voice:** Lacks whistled song of Black-capped Chickadee; calls relatively low-pitched and drawn-out, include *deee deee deee, ch'deee,* and *tseet.*

Subspp: (1) •*lathami.* **P. c. lapponicus* (n. EUR) used for good photographs.

Adult, *vandevenderi* AZ/02

Adult, *phillipsi* AZ/11

Bridled Titmouse 2

Baeolophus wollweberi L 5¼"

Shape: Small relatively long-crested titmouse. **Ad:** Gray overall with distinctive facial pattern: black throat; whitish cheek and gray crest outlined with black. Underparts buffy to whitish (see Subspp.). **Juv:** (May–Aug.) Like ad., but head pattern less distinct. **Flight:** All gray with black and white lines on head. **Hab:** Oak and pine woodlands, sycamores, feeders. **Voice:** Song a series of repeated midpitched whistles, like *peeyer peeyer peeyer;* calls include very high-pitched *tseeyer tseeyer,* chattering *ch'ch'ch'ch,* rapid sibilant *tseetseetsee,* and short *tsit.*

Subspp: (2) •*vandevenderi* (cent. AZ–s.w. NM) back dark gray, underparts whitish; •*phillipsi* (s.e. AZ) back with bright olive tinge, underparts buffy.

Adult CA/01

Adult CA/01

Oak Titmouse 1

Baeolophus inornatus L 5¾"

Shape: Relatively short-tailed small-crested titmouse; slightly shorter-billed than Juniper Titmouse. **Ad:** Plain gray overall with a brownish wash to back and crown; bill gray; crest when down inconspicuous. Similar Juniper Titmouse has longer bill, grayer back and crown, and different song. **Juv:** Like ad. **Flight:** All gray. **Hab:** Oak and oak-pine woodlands, gardens. **Voice:** Song a midpitched varied series of 3–7 two-note whistled phrases, like *peeyer peeyer;* calls include a trill, a sputtered *tsika-deedee,* a sibilant *see see see,* and a short soft *tsit.*

Subspp: (3) •*inornatus* (OR–cent. CA) has short bill, flanks may be tinged pale brown; •*affabilis* (s.w. CA) has medium-length bill, flanks tinged brown; •*mohavensis* (s.e. CA) has medium-length bill, flanks light gray.

Adult, *ridgwayi* NM/11

Adult, *ridgwayi* NM/11

Juniper Titmouse 1

Baeolophus ridgwayi L 5¾"

Shape: Relatively short-tailed small-crested titmouse; slightly longer-billed than Oak Titmouse. **Ad:** Plain gray overall with gray back and crown; bill gray; crest when down inconspicuous. Distinguished from similar Oak Titmouse by back color, range, shape, and voice. **Juv:** Like ad. **Flight:** All gray. **Hab:** Open mixed woods with junipers; also oaks and pinyon-juniper woodlands. **Voice:** Song a rapid trill of midpitched whistles on same pitch (Oak Titmouse has separate phrases that are slightly higher-pitched); calls include *jawee jawee,* a chattering *ch'ch'ch,* and a soft *tsit.*

Subspp: (2) •*zaleptus* (OR–CA–NV) has medium-length bill, may have buff on flanks; •*ridgwayi* (ID–w. OK–TX) has long bill, pale gray flanks.

Adult OK/12

Adult NH/07

Adult NH/09

Juv. NH/08

Tufted Titmouse 1

Baeolophus bicolor L 6½"

Shape: Large, relatively long-tailed, moderately crested titmouse. **Ad:** Forehead black; crest gray; bill blackish; gray above; whitish face and underparts; flanks orange-buff. Where range overlaps with Black-crested Titmouse, hybrids occur, creating a range of combined characteristics, most commonly a dark gray crest and white forehead. **Juv:** (May–Aug.) Like ad. but gray forehead and paler face. **Flight:** Gray above, white below, orange-buff flanks. **Hab:** Deciduous woodlands, suburbs, feeders. **Voice:** Song 2–4 whistled 2-part phrases like *peter peter peter;* calls include *jwee jwee jwee, tsikajwee jwee, seeseesee,* and soft *tsit*. **81**.

Subspp: Monotypic. **Hybrids:** Black-crested Titmouse.

IDENTIFICATION TIPS

Chickadees and Titmice are small woodland birds that are often in small flocks in winter and readily come to feeders. Titmice have a small crest and are mostly gray, sometimes with buffy flanks. Chickadees lack a crest and are mostly gray with a dark cap and throat. The sexes look alike.

Species ID: For chickadees, look carefully at the head and face pattern, the colors of the cap and back, and the margins of the secondaries and their coverts. For titmice, look for the presence and location of black on the face and head.

Adult TX/11

Black-crested Titmouse 2

Baeolophus atricristatus L 6½"

Shape: Large, relatively long-tailed, moderately crested titmouse. **Ad:** Forehead whitish; crest black; bill blackish; gray above; whitish face and underparts; flanks orange. Where range overlaps with Tufted Titmouse, hybrids occur, creating a range of combined characteristics, most commonly a dark gray crest and white forehead. **Juv:** (May–Aug.) Pale forehead and gray crest. **Hab:** Deciduous woods, shrubby areas, gardens, feeders. **Voice:** Song 5–7 whistled slurred notes like *peerpeerpeer-peerpeer;* calls include *jwee jwee, seeseeseesee,* and soft *tsit* (generally higher-pitched than those of Tufted Titmouse).

Subspp: (3) •*palodura* (s.w.–n. TX) has black crest and sometimes nape, forehead usually white; •*castaneafrons* (s.w.–cent. TX) and •*atricristatus* (s. TX) have blackish crest, whitish to pale cinnamon forehead. **Hybrids:** Tufted Titmouse.

Adult m., *ornatus* TX/05

Adult f., *acaciarum* CA/06

Juv., *acaciarum* CA/08

Verdin 1

Auriparus flaviceps L 4½"

Shape: Very small fairly long-tailed bird with short, sharp-pointed, conical bill. In part, distinguished from juv. Lucy's Warbler by more conical bill and proportionately longer tail; from juv. Bushtit and N. Beardless-Tyrannulet by longer bill and straight culmen. **Ad:** Body gray overall; face, throat, fore-crown yellow; lore dark; bill blackish. Lesser coverts reddish brown, but often hidden under scapulars. M. tends to have brighter, more extensive yellow on head and throat and brighter rufous lesser coverts than female. Active gleaner, often flicking tail up. **Juv:** (Apr.–Sep.) Pale gray overall; pale lore; no yellow; pale base to lower mandible. **Flight:** Tiny; gray with yellow head. **Hab:** Desert scrub areas, often along washes or streams. **Voice:** Song a whistled *tsitewtewtew tsitewteetee;* calls include an explosive *steeet.*

Subspp: (2) Differences slight.

•*acaciarum* (w. NM and west);

•*ornatus* (e. NM and east).

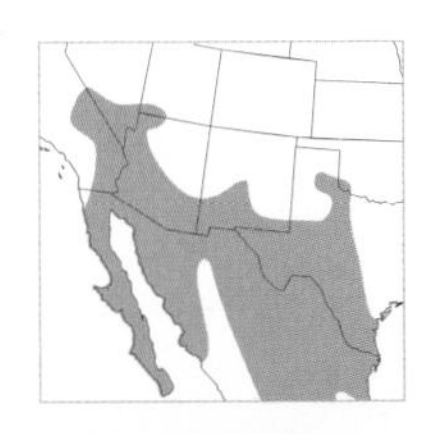

Adult m., Brown-crowned Grp. CA/10

Adult f., Brown-crowned Grp. CA/10

Adult m., Gray-crowned Grp. TX/04

Adult m., Gray-crowned Grp. AZ/04

Adult m., Gray-crowned Grp. NM/11

Bushtit 1

Psaltriparus minimus L 4½"

Shape: Very small, round-bellied, long-tailed bird with a short stubby bill. Culmen strongly curved. **Ad:** Grayish above; paler gray below, sometimes with brownish wash; crown varies from gray to brown (see Subspp.); auriculars can have no to considerable black (see Subspp.). M. dark eye; f. pale gray, white, or yellow eye (starting a month after fledging; before that it is dark). Species is polymorphic in amount of black on auriculars, varying by geography, sex, and age—extensive black more common in south, males, and juveniles. **Juv:** (Apr.–Aug.) Like ad. with fluffier undertail coverts. **Flight:** Gray overall, very small body, very long tail. **Hab:** Oak and riparian woods, scrub, gardens. **Voice:** Calls, given frequently in flocks, include high-pitched loud *pseeet,* a clipped *chichit,* a raspy *tzzee tzzee,* and a short *tsit.*

Subspp: (6) Three groups. **Brown-crowned Group** (along Pacific Coast): Brown crown, tinged brown underparts, no black on auriculars of juv. males. •*saturatus* (BC–s.w. WA), •*minimus* (n.w. OR–s. CA), •*melanurus* (s.w. CA), •*californicus* (s. OR–s.cent. CA). **Gray-crowned Group** (interior portion of range): Gray crown, gray underparts, occasional black on auriculars of juv. males (mostly in south). •*plumbeus* (e. OR–s.e. CA east to OK–cent. TX). **Black-eared Group:** Gray crown, and all juv. males have black on auriculars. •*dimorphicus* (s.w. TX).

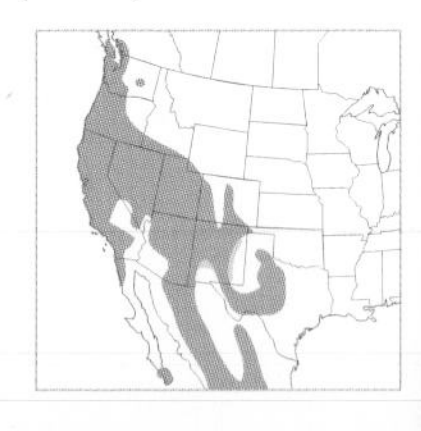

Adult, Interior West Grp. AZ/12

Adult m., *carolinensis* NH/07

Adult, Interior West Grp. OR/08

Adult f., *carolinensis* OH/10

White-breasted Nuthatch 1

Sitta carolinensis L 5¾"

Shape: Small, very short-tailed, large-headed, broad-necked bird with a long to medium-length thin bill (see Subspp.) and long shallow crown. Larger with proportionately longer bill than other nuthatches. **Ad:** Gray back and wings; white face surrounds dark eye and contrasts with dark forehead, crown, and nape; rusty undertail coverts; flanks variably whitish to gray to brownish gray (see Subspp.). Blackish markings on greater coverts and sometimes tertials (see Subspp.). F. has dull gray back, often a brownish tinge to margins of secondaries and greater coverts, and usually gray (sometimes blackish, see Subspp.) crown; m. has bluish-gray back, bluish margins to secondaries and greater coverts, and jet-black crown. **Juv:** (May–Aug.) Like ad. but paler overall with wider wingbar on greater coverts. **Flight:** Short tail with white band across corners; rusty undertail coverts. **Hab:** Mature deciduous and mixed woods. **Voice:** Eastern subspp. give a low-pitched and well-spaced *ank, ank, ank* and a rapid *werwerwerwer* "song"; Interior Western subspp. tend to give a higher-pitched and faster series of nasal notes, like *nye nye nye nye;* Pacific Coastal subspp. give well-spaced calls that are higher-pitched and raspier, like *eeer, eeer.* 🎧**82**.

Subspp: (5) Divided into three groups. **Pacific Coastal Group:** •*aculeata* (Pacific Coast from BC to CA) a medium-dark subspecies; bill medium length, flanks and breast sides tinged brownish gray, back medium gray. **Interior West Group:** •*tenuissima* (west of Rocky Mts. toward coast), •*nelsoni* (east of Rocky Mts. to western edge of Plains), and •*oberholseri* (s.w. TX) are darkest subspecies; bill longish, flanks and breast sides whitish to gray with little or no brown, back dark gray. **Eastern Group:** •*carolinensis* (rest of range to east) is palest subspecies; bill short and thick, flanks and breast sides white to grayish, back light gray, extensive black markings on tertials; f. crown variably gray to dull black (like m.), the latter more common in Southeast.

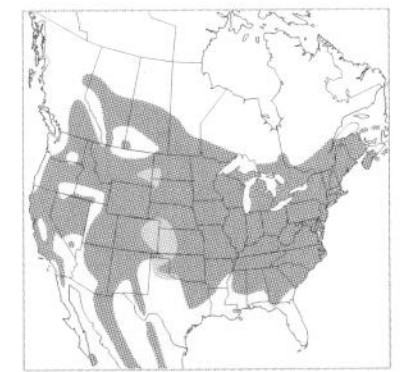

Adult m. QC/02

Adult f. NH/12

Adult f. TX/11

Red-breasted Nuthatch 1

Sitta canadensis L 4½"

Shape: Small, very short-tailed, large-headed, broad-necked bird with a medium-length thin bill and long shallow crown. Smaller than White-breasted Nuthatch; relatively smaller head than Pygmy and Brown-headed Nuthatches. **Ad:** Rusty to reddish orange below; bluish gray above; dark crown; white eyebrow; wide black eyeline from bill to back. M. with black crown, deep orange underparts; f. with bluish-gray crown, pale orange underparts. **Juv:** (Jun.–Aug.) Like ad. but duller, with slightly less contrast in plumage; sexes sometimes distinguishable as in adult. **Flight:** Very small; gray; orange underparts; orange underwing coverts. **Hab:** Evergreen forests of mountains and the North. **Voice:** Song a nasal *meeep meeep meeep;* calls include low-pitched *beer beer beer,* chattering *chr'r'r'r'r,* a soft *ink ink,* and a buzzy *djeee djeee.* 🎧**83**.

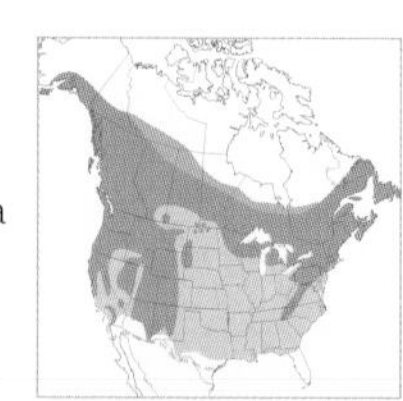

Subspp: Monotypic.

Adult, *melanotis* AZ/05

Adult CA/03

Adult, *melanotis* OR/08

Pygmy Nuthatch 1

Sitta pygmaea L 4¼"

Shape: Very small, relatively large-billed, large-headed nuthatch. **Ad:** Bluish-gray back; grayish to brownish crown; variably darker eyeline; indistinct whitish patch on nape; flanks variably washed buff to grayish buff. Sexes alike. Similar to Brown-headed Nuthatch except for crown color, voice, and range. **Juv:** (May–Aug.) Like ad. but upperparts dull brownish. **Flight:** White band across each corner of short tail, underparts pale buff. **Hab:** Pine forests. **Voice:** Song a series of rapid, 2-part, high-pitched whistles like *tidi tidi tidi tidi;* calls include a short series of sharp whistles, like *peet peet, peet peet peet,* also a soft *imp.*

Subspp: (3) •*pygmaea* (cent. coastal CA) small; dusky eyeline contrasts slightly with crown, extensively buffy flanks. •*leuconucha* (s. CA) large; grayish eyeline and crown, pale grayish-buff flanks. •*melanotis* (rest of range) small; blackish eyeline contrasts strongly with crown, pale grayish to buff flanks.

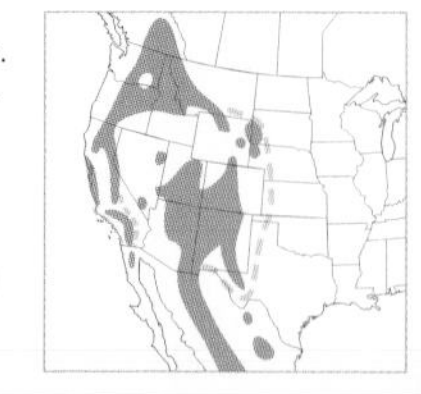

Adult, *pusilla* TX/01

Adult brown morph TX/03

Adult reddish morph, *americana* NH/05

Adult, *pusilla* TX/02

Brown-headed Nuthatch 1

Sitta pusilla L 4½"

Shape: Very small, relatively large-billed, large-headed nuthatch. **Ad:** Bluish-gray back; brown to chestnut crown (paler when worn, see Subspp.); dark eyeline; distinct whitish patch on nape; underparts pale, sometimes washed buff centrally and washed gray on flanks. Sexes alike. Similar to Pygmy Nuthatch except for crown color, voice, and range. **Juv:** (Apr.–Jul.) Like ad. but paler with grayish crown. **Flight:** Small patch of white in each corner of short tail. **Hab:** Pine forests. **Voice:** Calls include a repeated squeaky *chyudit chyudit* interspersed with rapid squeaky bursts of twittering, a sharp *pit,* and a harsh *jjeee.*

Subspp: (2) •*caniceps* (cent. FL) has brown crown; •*pusilla* (rest of range) has chestnut crown.

Brown Creeper 1

Certhia americana L 5¼"

Shape: Small broad-necked bird with long pointed tail feathers and thin, relatively long, downcurved bill. **Ad:** Brown above with small white spots and indistinct streaking; whitish throat and belly; buffy undertail coverts. Whitish eyebrow; broad brown eyeline. Buffy to grayish base of primaries and secondaries creates zigzag pattern on closed wing and wing-stripe in flight. This species is polymorphic, with gray, brown, and reddish morphs, creating different shades to the upperparts. Habitually creeps up tree trunk, then flies down to base of new tree to repeat process. **Juv:** (May–Aug.) Like ad. but underparts may have dusky spots or faint bars. **Flight:** Long tail; buffy wing-stripe; chestnut rump. **Hab:** Mature woodlands. **Voice:** Song regionally variable; can be a series of very high-pitched whistles alternating high-low, like *tseeesuuu tseeesay tseeeseee;* may also have faster notes in center like *tseee tudi wedo tseee.* Calls include a quavering, very high, whistled *tseee* or *tseeeseee,* and a short *tzit.*

Subspp: (9) Differences slight; complicated by polychromaticism. •*americana* (MN–n. WV and north), •*nigrescens* (s. WV–e. TN–w. NC), •*montana* (cent. BC–SK south to AZ–w. TX), •*albescens* (s.e. AZ–s.w. NM), •*phillipsi* (w.cent. CA), •*stewarti* (Queen Charlotte Is.), •*occidentalis* (coastal s. AK–cent. CA), •*alascensis* (s.cent. AK), •*zelotes* (s.cent. OR–w. NV–s. CA).

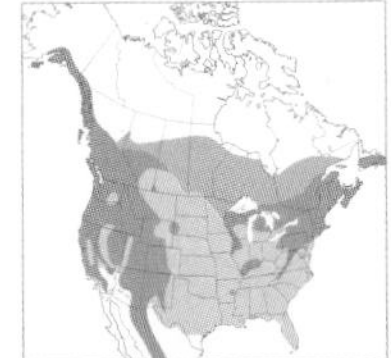

Adult AZ/05

Adult, *obsoletus* CA/02

Adult NM/11

Adult, *obsoletus* NM/05

Cactus Wren 1

Campylorhynchus brunneicapillus L 8½"

Shape: Large, relatively long-tailed, large-headed wren with a sturdy thrasherlike bill. **Ad:** Reddish-brown crown; bold white eyebrow; variably spotted breast (see Subspp.), streaked back; barred wings and tail; dark red to orange eye. **Juv:** (Apr.–Aug.) Less spotting on throat and upper breast; buffer below; pale gray eye. **Flight:** Rounded tail heavily barred and tipped with white; tail held partially spread in flight. **Hab:** Deserts, desert scrub, coastal sage scrub. **Voice:** Song a long series of short harsh *chuts* or *churits*, getting louder toward the end; calls mostly low-pitched harsh sounds and include a harsh rattling chatter, a quacking *krak krak*.

Subspp: (3) •*sandiegense* (coastal s. CA) has unstreaked throat, brown back and crown, no central concentration of spots on breast; •*anthonyi* (s.cent. CA–s.w. UT–cent. AZ) has streaked throat, reddish-brown crown, pale grayish back, central concentration of spots on breast; •*couesi* (e. AZ–TX) like *anthonyi* but with dark brown crown and darker brownish-gray back.

Rock Wren 1

Salpinctes obsoletus L 6"

Shape: Medium-sized fairly short-tailed wren with relatively long, thin, slightly downcurved bill. **Ad:** Grayish-brown upperparts finely and variably speckled with white; pale belly with buffy flanks; whitish throat and chest with fine gray streaking; faint whitish to buffy eyebrow. Plumage characteristics strongest in fresh plumage (winter), faint in worn plumage (summer). Often bobs body up and down while standing; otherwise scoots in and out of rock crevices. **Juv:** (Apr.–Aug.) Like ad. but may be paler; eyebrow indistinct; throat and chest with little or no streaking; bill often shorter. **Flight:** Distinctive buffy terminal band on outer tail feathers; cinnamon rump. **Hab:** Arid rocky areas, rockslides. **Voice:** Song a changing series of slightly buzzy whistled phrases, usually repeated several times, like *zizeee zizeee zizeee, zitchew zitchew, zoey zoey zoey zoey,* a little like song of Northern Mockingbird (possible occasional mimicry of other birds); calls include a clear *tikeer* and a buzzy trilled *dzeeeee*.

Subspp: (2) •*pulverius* (San Clemente, San Nicolas Is. off s. CA) has short thin bill; •*obsoletus* (rest of range) has long thicker bill.

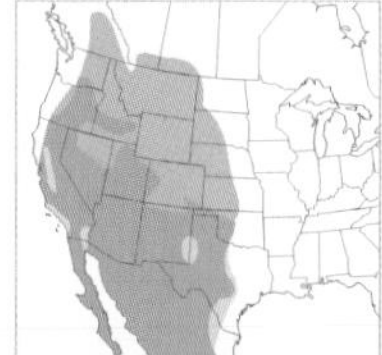

Adult CA/05

Adult CA/05

Canyon Wren 1

Catherpes mexicanus L 5¾"

Shape: Medium-sized wren with a long head, flattened crown, and long, thin, decurved bill. **Ad:** Basically bicolored; rich brown to reddish brown over most of body, with contrasting bright white throat and upper breast. Fine white speckling on upperparts and variable dark barring on belly. Tail bright rufous with fine dark bars. **Juv:** (Apr.–Aug.) Little or no white speckling on back, nor dark barring below. **Flight:** All reddish brown except for white throat; tail brighter reddish brown than rest of body. **Hab:** Arid and semi-arid regions with rocky crevices and outcrops. **Voice:** Song a long series of descending notes, slowing in delivery at end, like *teeteeteetititi tuh tuh too, too, too,* often ended with buzzy calls from male or female; calls include a buzzy *dzeee dzeee* given as bird bobs.

Subspp: (4) Differences slight; much individual variation. •*griseus* (s. OR–n.w. UT north), •*punctulatus* (n.cent.–s.cent. CA), •*conspersus* (NV–e. CO south to s. CA–TX), •*pallidior* (cent. ID–SD south to n.e. UT–n. CO).

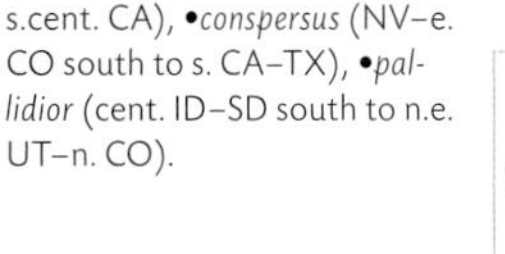

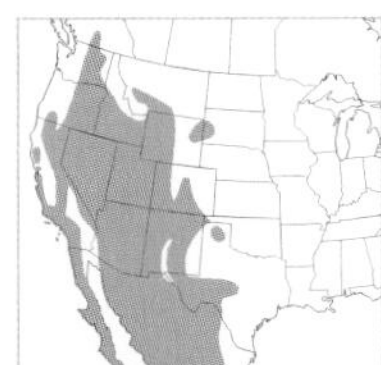

Adult TX/05

Adult, *ludovicianus* OH/11

Carolina Wren 1

Thryothorus ludovicianus L 5½"

Shape: Medium-sized, deep-bellied, large-headed, broad-necked wren with a fairly heavy, slightly downcurved bill. **Ad:** Variably warm brown above; pale to rich buff below; bright white to buffy eyebrow; undertail coverts barred brown and white. **Juv:** (Apr.–Sep.) Like ad. but less richly colored, with no barring on undertail coverts and small buff tips to wing coverts. **Flight:** Evenly reddish brown above. **Hab:** Shrubby areas, gardens. **Voice:** Song a loud burst of melodic notes, like "teakettle teakettle teakettle" (sings all year), song ending often overlapped with *churrr* call of female; calls varied and include a chattering *churrr,* a 2-part *tidik,* and a raspy *djjjj.* 🎧**84**.

Subspp: (6) Differences slight; much individual variation. •*ludovicianus* (IA–s.e. TX and east) fairly large, with short bill; •*miamensis* (s.e. GA through FL) large, with long bill; •*oberholseri* (cent. TX) fairly small; •*lomitensis* (s. TX) small; •*burleighi* (Cat Is. MS) fairly small; •*nesophilus* (Dog Is. FL) medium-sized.

Adult, Pacific Coastal Grp. CA/03

Adult, Interior West Grp. NM/11

Bewick's Wren 1

Thryomanes bewickii L $5\frac{1}{4}$"

Shape: Medium-sized, long-tailed, proportionately small-bodied wren with variable length, slightly downcurved bill. Tail strongly graduated at tip. **Ad:** Conspicuous whitish eyebrow; central tail feathers barred; upperparts unmarked reddish brown to grayish (see Subspp.); underparts unmarked grayish, often with paler throat; undertail coverts strongly barred. Some polymorphism, with gray and reddish-brown morphs. Tail often flicked side to side. **Juv:** (Apr.–Aug.) Like ad. but breast faintly mottled or barred; undertail coverts fluffy, with little or no barring. **Flight:** Plain brownish above except for barred central tail feathers and shorter white-tipped outer tail feathers. **Hab:** Open areas with shrubs, streamside woodlands, gardens. **Voice:** Song a mixture of notes, buzzes, and trills (a bit like Song Sparrow's); calls include a raspy *jjjeee* and short *chik.* 🎧**85**.

Subspp: (10) Divided into three groups. **Eastern Group:** Medium-sized, upperparts reddish brown. •*bewickii* (e. NE–AR and east). **Pacific Coastal Group:** Small, upperparts dark brown. •*calophonus* (coastal BC–OR), •*marinensis* (n.w. CA), •*spilurus* (w. cent. CA), •*drymoecus* (s.cent. OR–w. NV–s.w. CA), •*charienturus* (s. CA). **Interior West Group:** Fairly large, upperparts pale grayish to pale reddish brown. •*eremophilus* (e. CA–cent. CO south to AZ–TX), •*cryptus* (e. CO–w. OK), •*pulichi* (w.cent. OK–KS), •*sadai* (s. TX).

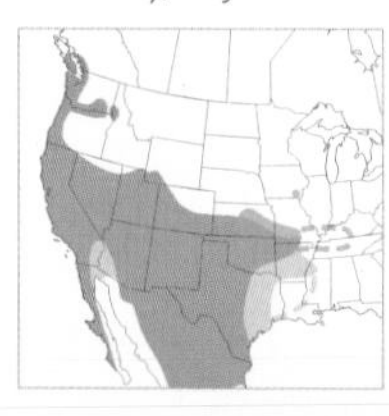

Adult, *aedon* NH/08

Adult, *parkmanii* NM/06

House Wren 1

Troglodytes aedon L $4\frac{3}{4}$"

Shape: Fairly small wren with rather "average" proportions for a wren; bill thin and slightly downcurved; tail and bill both medium length. **Ad:** Plain-faced; sometimes faint eyebrow extending just behind eye; upperparts dull brownish to grayish brown; often dark barring on back (see Subspp.). Underparts paler; sometimes barring on rear flanks; undertail coverts brownish with darker barring. **Juv:** (May–Aug.) Like ad. but with little or no barring on undertail coverts and flanks; may be more reddish brown on wing coverts and belly than adult. **Flight:** Very plain and concolor above, with no white marking on tail. **Hab:** Shrubby areas, gardens. **Voice:** Song a cascading jumble of fairly harsh trills lasting 2–3 sec.; calls include rapid chattering *ch'ch'ch'ch.* 🎧**86**.

Subspp: (3) •*parkmanii* (cent. TN–cent. ON and west) has grayish-brown back and crown, back usually barred; •*aedon* (e. TN–e. ON and east) has slightly reddish-brown back and crown, back usually unbarred; •*cahooni* (Mexican vag. to s.e. AZ) has pale brown throat and breast (but is a separate subsp. from the Mexican "Brown-throated Wren," subsp. *brunneicollis*).

Adult, *hiemalis* ME/04

Adult, *hiemalis* OH/05

Winter Wren 1

Troglodytes hiemalis L 4"

Shape: Medium-sized wren with short straight bill. Broad neck, deep belly, and very short tail almost always cocked over rump create a very round look to body. **Ad:** Reddish-brown to dark brown upperparts; paler buffy to brown belly; contrastingly whitish throat. Pale buffy eyebrow; dark barring on wings, tail, and flanks. Variable buffy or whitish spotting on back and wing coverts. Similar Pacific Wren more concolor overall with only slightly paler throat and little or no spotting on back and wing coverts. See Voice and range for further distinctions. **Juv:** (May–Aug.) Like ad. but barring on body less distinct; eyebrow may be fainter; breast may be mottled with dark. **Flight:** Very small, "no tail," paler throat. **Hab:** Variable; old-growth forests, northern barrens, often near water. **Voice:** Calls of Winter and Pacific Wrens differ markedly. Winter's call is more musical, slightly longer, lower-pitched, like *jeet-jeet, jeet-jeet, je'je'jeet*. Pacific's call is unmusical, harsh, shorter, higher-pitched, like *cht-cht, cht-cht, ch'ch'cht*. Songs of both species a very long, high-pitched, variable twittering lasting 5–10 sec., with the song of Winter Wren only slightly more musical and slightly faster.

Subspp: (2) •*hiemalis* (WA–PA north to e. BC–NF), •*pullus* (WV–GA in Appalachian Mts.).

Adult, Western Grp. BC/06

Adult, Alaska Is. Grp. AK/07

Pacific Wren 1

Troglodytes pacificus L 4"

Shape: Either relatively large wren with fairly long straight bill or our smallest wren with a short straight bill (see Subspp.). Broad neck, deep belly, and very short tail almost always cocked create a very round look to body. **Ad:** Uniformly pale brown to dark brown (see Subspp.) with only slightly paler throat and breast. Pale buffy eyebrow; dark barring on wings, tail, and flanks. See similar Winter Wren and Voice. **Juv:** (May–Aug.) Like ad. but barring on body less distinct; eyebrow may be fainter; breast may be mottled with dark. **Flight:** Very small, "no tail," uniformly colored. **Hab:** Variable; old-growth forests, northern barrens, often near water. **Voice:** Call is unmusical, harsh, high-pitched, like *cht-cht, cht-cht, ch'ch'cht*. Song is a very long, fast, high-pitched, unmusical, variable twittering lasting 5–10 sec. See Winter Wren for comparison.

Subspp: (7) Divided into two groups. **Western Group:** Small, all dark brown. •*pacificus* (s. AK–s.w. and s.cent. CA), •*salebrosus* (cent. BC–w. MT–e. OR), •*helleri* (Kodiak Is. AK). **Alaska Island Group:** Large, all pale brown. •*alascensis* (Pribilof Is.), •*meligerus* (Attu, Buldir Is.), •*kiskensis* (Kiska–Unalaska Is.), •*semidiensis* (Semidi Is.).

Adult ND/06

Adult OH/07

Sedge Wren 1
Cistothorus platensis L 4½"

Shape: Small, fairly short-billed, fairly short-tailed wren; bill fairly thick throughout and length roughly equals lore + eye. Similar Marsh Wren has much longer bill that is clearly greater than lore + eye. **Ad:** Crown and back rich brown variably streaked with white and black; underparts rich buff except for paler chest and whitish throat; wings strongly barred; undertail coverts buffy and unbarred. Buffy eyebrow; dark eyeline. Tail often flicked up. Similar Marsh Wren has longer bill, no whitish streaking on the crown, and less barring on the wings. **Juv:** (Jun.–Aug.) Like ad. but less streaking on back; little or no streaking on crown; underparts with pale buffy wash except on whitish throat. **Flight:** Streaked back, orangish rump. **Hab:** Marshes, wet meadows. **Voice:** Song several short harsh sounds followed by a dry trill, like *cht cht ch't't't't;* calls include a *chut* or *ch'ch'chut* and a short buzzy *dzzrt.*

Subspp: (1) •*stellaris.*

Adult ND/06

Adult CA/12

Marsh Wren 1
Cistothorus palustris L 5"

Shape: Medium-sized long-legged wren with a medium-sized slightly downcurved bill; bill is fairly thin throughout and length is clearly greater than lore + eye. **Ad:** Distinct whitish eyebrow; crown dark brown to blackish and unstreaked; upper back blackish with fine white streaks; wings dark and mostly unbarred; flanks and belly reddish brown; throat whitish. Similar Sedge Wren has smaller bill, white streaking on crown, and prominent barring on wings. **Juv:** (May–Aug.) Like ad. but duller and slightly darker overall; little or no streaking on back; crown dark grayish brown. **Flight:** Reddish-brown rump and upperwing coverts. **Hab:** Marshes. **Voice:** Song a liquid, squeaky, rattling trill; calls include a short *chek* or *ch'ch'ch'chek.*

Subspp: (13) Classification in need of revision. Differences slight, with the exception of •*griseus* (e. SC–e.cent. FL), "Worthington's Wren," which is grayish overall and with barred flanks (other subspp. with no barring). Attempts to reclassify subspp. in larger groups by song analysis may be more fruitful. Eastern subspp. have a simpler song repertoire than western subspp.

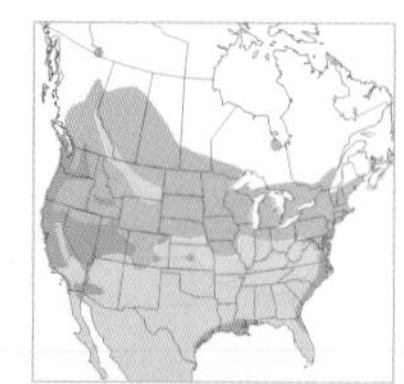

Dipper

Kinglets

Old World Warblers

Gnatcatchers

Old World Flycatchers

Adult, *unicolor* AK/03

Adult, *unicolor* YT/04

American Dipper 1
Cinclus mexicanus L 7½"

Shape: Medium-sized, deep-bellied, short-tailed bird with a relatively small head and fairly short thin bill. Appears rounded and heavy-bodied. **Ad:** Dark gray overall with variable brownish wash on head; bill blackish; legs pale pinkish. Swims and walks underwater; tail often raised. **Juv:** (May–Sep.) Like ad. but bill yellowish to pinkish with dark tip; throat pale; underparts mottled with gray, cinnamon, and some white; wing feathers finely tipped white. **Flight:** Usually low and along streams; rapid bursts of wingbeats alternate with short glides. **Hab:** Rushing mountain streams, high-elevation lakes. **Voice:** Song a long series of high-pitched whistled phrases containing repeated syllables like *taweet taweet, treeoo treeoo, tabit tabit;* can be soft and hard to locate. Calls include a buzzy *bzeeet* either singly or in a series.

Subspp: (2) •*unicolor* (throughout range) has brownish head; •*mexicanus* (vag. to s.e. AZ) has darker brownish head.

Adult, *jocosus** HKO/03

Red-whiskered Bulbul 2
Pycnonotus jocosus L 7"

Asian species; populations in CA and FL from escaped caged birds. **Shape:** Fairly small long-tailed bird with a long thin crest that is always raised and sometimes pointed forward. **Ad:** Blackish-brown head; dark brown upperparts; red-and-white cheek patch; white throat and breast with brown partial breastband; belly buffy to white; undertail coverts red. **Juv:** Like ad. but with no red on cheek; pinkish to orangish undertail coverts. **Flight:** Long tail with large white tips on all but central feathers. **Hab:** Agricultural and suburban habitats. **Voice:** Rich varied whistling like *cheepurdletee;* calls include a rapid *pidijou* and a harsh *jeeer.*

Subspp: (1+) •*emeria* (FL); possibly a different subsp. in CA. **P. j. jocosus* (s. China) used for good photograph.

Adult m., *satrapa* MI/06

Adult f., *satrapa* OH/11

Golden-crowned Kinglet 1

Regulus satrapa L 4"

Shape: Small, broad-necked, short-tailed bird with a short, thin, finely pointed bill. **Ad:** Grayish underparts; olive back and wings. Strongly patterned head: central crown yellow (f.) or orange (m.) flanked by blackish lateral stripes that meet over forehead; broad white eyebrow; dark eyeline. White wingbar on tips of greater coverts; dark bar on secondaries just below it; rest of primaries and secondaries edged greenish. Constantly flicks wings. **Juv:** (Jun.–Aug.) Like ad. but with gray rather than yellow central crown. **Flight:** Fast and fluttering; may hover while feeding. **Hab:** Coniferous woods in summer; deciduous or mixed woods in winter. **Voice:** Song a series of very high-pitched *tseees*, often rising in pitch and sometimes ending in short chatter; calls include a high-pitched *dzeee* (much like Brown Creeper).

Subspp: (3) •*olivaceus* (s. AK–s.w. OR) is small with medium-length bill, greenish-olive back. •*apache* (s. AK–s. YT south to CA–NM) is small with long bill, yellowish-olive back. •*satrapa* (AB–WI and east) is large with small bill, grayish-olive back.

Adult CA/11

Adult m. CA/11

Ruby-crowned Kinglet 1

Regulus calendula L 4¼"

Shape: Small, broad-necked, short-tailed bird with a short, thin, finely pointed bill. Similar Hutton's Vireo has deeper-based bill with more curved culmen. **Ad:** Grayish olive above; pale buffy (sometimes with olive wash) below. Plain-faced look, with broad pale crescents in front of and behind eye. Two white wingbars (usually only lower one visible), with dark area on base of secondaries below lower wingbar. Legs and especially feet orangish. M. has red crown patch that is usually concealed; f. no red patch. Constantly flicks wings. Similar Hutton's Vireo has eye-ring broken at top, no darker bar on wing below wingbar, and bluish legs and feet. **Juv:** (Jun.–Aug.) Like ad. but wingbars buffy; no red patch. **Flight:** Bursts of wingbeats; often hovers while feeding. **Hab:** Mostly coniferous or mixed forests, shrubby edges. **Voice:** A series of high-pitched *tseees* followed by lower, repeated, whistled phrases, like "see see see, look at me, look at me, look at me"; calls include a harsh *chidit* or *ch'ch'ch'ch*.

Subspp: (2) •*grinnelli* (s. AK–BC) small with dark upperparts; •*calendula* (rest of range) large with medium-dark upperparts.

Adult JAP/06

Adult CHI/05

Middendorff's Grasshopper-Warbler 4

Locustella ochotensis L 6"

Casual Asian vag. to w. AK. **Shape:** Fairly small, broad-necked, shallow-crowned bird with a moderately thick-based bill and fairly long strongly graduated tail. **Ad:** Upperparts dull brown with faint dusky streaks; rump slightly warmer; tail tipped pale. Underparts whitish with buffy flanks and sides of breast, the latter also slightly streaked. Whitish eyebrow; dark eyeline; pale cheek. Bill with upper mandible dark, lower orange. Legs pale. Brighter in fresh fall plumage with pale yellow wash to underparts. Generally secretive and on ground, except when singing. **Flight:** Brown above with warmer rump; strongly graduated tail. **Hab:** Open grassy areas, shrubs. **Voice:** Calls include a percussive *trit't't't.*

Subspp: Monotypic.

Dusky Warbler 4

Phylloscopus fuscatus L 5¼"

Casual Asian vag. to w. AK islands, s.w. AK, and CA, mostly in fall. **Shape:** Small, compact, fairly short-tailed, shallow-crowned bird with a thin fine-pointed bill. Primary projection past tertials short (about equal to bill length). **Ad:** Upperparts concolor, unmarked grayish brown (no wingbar); underparts dingy whitish with dull buff on flanks and undertail coverts. Pale whitish eyebrow; dark eyeline and lore; thin white arc below eye. Bill dark above, orangish below. Legs dark brown. Flicks wings and usually stays on or near ground. **Flight:** Bursts of wingbeats; underwing coverts brownish. **Hab:** Shrubby areas, woods. **Voice:** Song a series of high-pitched trills; calls include a sharp *tshik.*

Subspp: Monotypic.

Adult HKO/02

Adult CHI/05

Yellow-browed Warbler 4

Phylloscopus inornatus L 4½"

Casual Asian vag. to w. AK islands. **Shape:** Small broad-necked bird with a fairly short thin bill and moderate-length tail. **Ad:** Dull olive above and whitish below; long, prominent, whitish to yellowish eyebrow above a thin dark eyeline; 2 whitish to yellowish wingbars, the lower longer and wider; tertials sooty with thin white edges; dark bases of secondaries show as a thin dark bar below the lower wingbar as in Ruby-crowned Kinglet. Sexes alike. Often flicks wings. **Flight:** Pale whitish below with prominent whitish to yellowish eyebrow. **Hab:** Trees and scrub. **Voice:** Call a thin rising *tseeyeeet*.

Subspp: Monotypic.

Adult AK/–

Adult CA/09

Arctic Warbler 2

Phylloscopus borealis L 5"

Shape: Small, compact, fairly short-tailed, shallow-crowned bird with a moderately thick-based fine-pointed bill. Primary projection past tertials long (1½–2 x bill length). Similar Dusky Warbler's primary projection is short. **Ad:** Upperparts olive to brownish olive; underparts whitish to yellowish (see Subspp.). Thin whitish eyeline extends well back onto nape and may curve up at end; dark eyeline; thin white arc below eye. Short thin wingbar on greater coverts; may be small wingbar on median coverts but can be worn away or hidden. Bill dark with orangish base to lower mandible. Legs brownish pink. Similar Dusky Warbler is brownish above and has no wingbars. **Juv:** (Jun.–Aug.) Like ad. but more brownish overall. 1st-fall and -winter birds are brighter greenish above with a pale yellow wash below, compared with ad. **Flight:** Bursts of wingbeats; may hover; underwing coverts yellowish. **Hab:** Small trees and shrubs. **Voice:** Song a rapidly repeated harsh sound, like *chirchirchirchir;* calls include a short buzzy *djjt*.

Subspp: (2) •*kennicotti* (w. AK) has short wings and bill; underparts white with slight yellow tinge. •*xanthodryas* (vag. to AK) has long wings and bill; underparts yellow.

Adult m., summer TX/04

Adult winter, *caerulea* QC/10

Adult, *caerulea* FL/02

Adult winter, *caerulea* GA/02

Blue-gray Gnatcatcher 1

Polioptila caerulea L 4½"

Shape: Small, small-bodied, long-tailed bird with a medium-length thin bill. Legs fairly long and very thin. Tail slightly rounded; difference in length between shortest and longest tail feathers about half width of closed tail (about width of closed tail in similar Black-capped Gnatcatcher). **Ad:** Grayish above; whitish below; thin white eye-ring; broad white tertial edges. Undertail appears almost all white (see Subspp.). M. with bluish-gray upperparts; in summer, has thin black eyebrow, mostly in front of eye and extending to forehead. F. similar but has no eyebrow and may be tinged brown on back. In late summer, when outer tail feathers are molting, undertail may appear mostly black. Tail often flicked side to side. **Juv:** (May–Jul.) Like ad. f. but crown brownish gray. **Flight:** Outer tail feathers mostly white. **Hab:** Variable; deciduous woodlands and shrublands, often near water; pinelands. **Voice:** Song a fairly long and varied high-pitched twittering, a little like a high-pitched Catbird song; calls include a thin nasal *tzyeee tzyeee* (eastern birds) and lower-pitched raspy *jeeep jeeep* (western birds).

Subspp: (3) •*caerulea* (KS–cent. TX and east); m. back bluish gray, undertail white through undertail coverts, m. eyebrow thin. •*obscura* (w. TX–WY and west); m. back darker, undertail mostly white but black just before undertail coverts, m. forehead more extensively black leading to eyebrow. •*deppei* (s. TX) like *caerulea* but smaller.

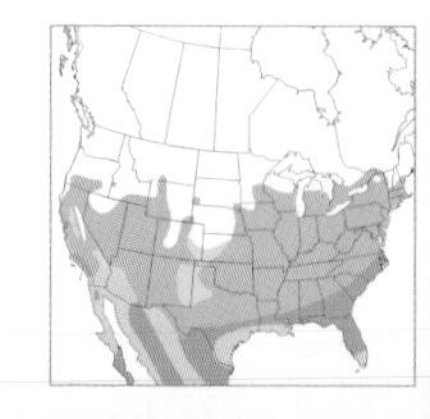

Adult m., summer CA/04

Adult winter CA/11

California Gnatcatcher 2

Polioptila californica L 4½"

Shape: Small, small-bodied, long-tailed bird with a short thin bill. Legs fairly long and very thin. **Ad:** Dark gray above with brownish tinge to back; slightly paler gray below with pinkish-buff flanks (southern populations slightly paler overall). Undertail appears mostly black with thin white tips and edges to outer 2 feathers on each side. M. in summer has black cap and indistinct eye arc just below eye; in winter, thin black eyebrow or black blotches over eye. F. has faint brownish wash overall except for grayish head; no black on head. Tail often flicked side to side. Similar Black-tailed Gnatcatcher is slightly paler overall and has broad white tips to black tail; also note range and Voice. **Juv:** (May–Jul.) Like ad. f. **Flight:** Outer tail feathers mostly black. **Hab:** Dense low shrubs in coastal sage scrub. **Voice:** Calls include a drawn-out raspy *jeyeer jeyeer* that descends at end, a soft *jerr jerr,* and a harsh *tshhhh.*

Subspp: (1) •*californica.*

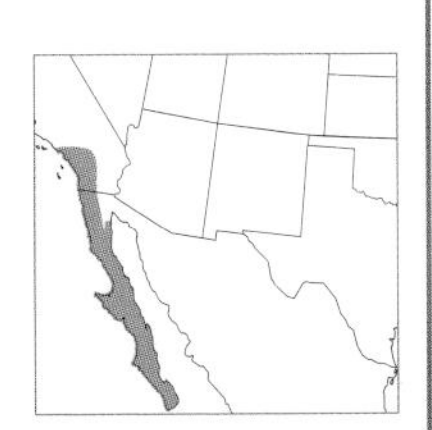

Adult m. AZ/–

Adult f. AZ/08

Black-capped Gnatcatcher 3

Polioptila nigriceps L 4¼"

Rare Mexican species seen in s.e. AZ, where it may breed. **Shape:** Similar to Blue-gray Gnatcatcher but with substantially longer bill and more abruptly rounded tail. Difference in length between shortest and longest tail feathers about equal to width of closed tail (about half width of closed tail in similar Blue-gray Gnatcatcher). **Ad:** Gray above; grayish white below. Undertail appears almost all white with thin black area at base next to undertail coverts. M. in summer has black cap that extends all the way below eye; in winter, thin black eyebrow or black blotches over eye. F. with brownish back and no black on head; indistinct thin whitish eye-ring. **Juv:** (May–Jul.) Like ad. female. **Flight:** Outer tail feathers mostly white. **Hab:** Arid areas with dense shrubs, canyons. **Voice:** Song a slow varied mixture of warbling and harsh sounds; calls include a raspy *djeeer djeeer,* a short *dzit.*

Subspp: (1) •*restricta.*

Adult m., summer, *lucida* AZ/04

Adult f., summer, *lucida* CA/01

Adult m., winter, *lucida* CA/01

Black-tailed Gnatcatcher 1

Polioptila melanura L 4½"

Shape: Small, small-bodied, long-tailed bird with a short thin bill. Legs fairly long and very thin. Like California Gnatcatcher in shape. **Ad:** Gray above; dull whitish to pale gray below. Undertail appears mostly black with thick white tips and thin white edges to outer 2 feathers on each side. M. in summer has black cap and white eye-ring; in winter, thin black eyebrow or black blotches over eye. F. with faint brownish wash overall except for grayish head; no black on head. Tail often flicked side to side. **Juv:** (May–Jul.) Like ad. f. **Flight:** Outer tail feathers mostly black. **Hab:** Arid areas with sparse shrubs. **Voice:** Calls include a repeated, harsh, hissing *pshhh pshhh pshhh,* a harsh short *jjjj jjjj,* a short chattered *ch'ch'ch'ch,* and a high thin note, *pseet.*

Subspp: (2) •*lucida* (s. NV–s. AZ and west) has paler back, less white on tail tips; •*melanura* (s. NM and east) has slightly darker back, more white on tail tips.

Adult JAP/10

Gray-streaked Flycatcher 4

Muscicapa griseisticta L 6"

Casual Asian vag. to w. AK. **Shape:** Fairly small, large-headed, broad-necked bird with a short blunt bill and relatively short tail. Primary projection past tertials very long (3 x bill length). Legs short. **Ad:** Dull brown above with darker primaries and secondaries; whitish edging to tertials. Thin white eye-ring; pale supraloral dash. Underparts white with distinct dark streaking on throat, across breast, and along flanks. Undertail coverts white and unstreaked. Similar Dark-sided Flycatcher has whitish underparts with brownish wash on flanks and sides of breast; variable dark streaking on breast and flanks; undertail coverts with subtle dark spots. **Juv:** Like ad. but darker above with pale edges to wing coverts and tertials. 1st fall with thin white wingbar. **Flight:** Long pointed wings; relatively short tail. **Hab:** Open woods, woods edges. **Voice:** Calls include a high thin *seeet.*

Subspp: Monotypic.

Adult m. UAE/03

Juv./1st fall HKO/09

Adult m. UAE/03

Adult f. UAE/12

Juv./1st fall HKO/09

Taiga Flycatcher 4

Ficedula albicilla L 5¼"

Casual vag. from Asia to w. AK, accidental in CA. **Shape:** Small, deep-bellied, broad-necked, short-tailed bird with a short thin bill and high rounded crown. Legs short. **Ad. Summer:** Upperparts brown; underparts grayish white. Head gray to brown with thin white eye-ring; very thin pale wingbar on greater coverts. Legs black. Distinctive tail black with white bases to outer tail feathers; longest uppertail coverts black. M. has gray head, red throat bordered below by gray wash; f. has brown head, white throat, and gray wash to upper breast. Tail often flicked up. **Ad. Winter:** Like ad. summer female. **Juv:** Like ad. summer female. **Flight:** Brown back; tail black with white side patches. **Hab:** Open woods, woods edges. **Voice:** Calls include dry rattles and a short high *tseee*.

Subspp: Monotypic.

Dark-sided Flycatcher 4

Muscicapa sibirica L 6"

Casual Asian vag. to w. AK. **Shape:** Fairly small, large-headed, broad-necked bird with a short blunt bill and relatively short tail. Primary projection past tertials very long (3 x bill length). Legs short. **Ad:** Dull brown above with darker primaries and secondaries; whitish edging to tertials. Thin white eye-ring; pale supraloral dash. Underparts whitish with brownish wash on flanks and sides of breast; some indistinct streaking within wash on breast, faint in center of breast. Thin, pale, whitish partial collar. See Gray-streaked Flycatcher for comparison. **Juv:** Like ad. but darker above with pale edges to wing coverts and tertials and pale spots on back. 1st fall with thin buffy wingbar. **Flight:** Long pointed wings; relatively short tail. **Hab:** Open woods, woods edges. **Voice:** Song a varied string of whistles and trills; calls include short trills and a high-pitched *tseeyeeer*.

Subspp: (1) •*sibirica*.

Thrushes

Adult m. JAP/06

Adult m. AK/06

Adult f. AK/06

Adult m. AK/06

Siberian Rubythroat 3

Luscinia calliope L 6"

Rare Asian vag. to AK. **Shape:** Small, deep-chested, deep-bellied, short-tailed thrush with a thin fine-pointed bill. **Ad:** Warm brown above (warmest on wings and rump); mostly buffy below (strongest on flanks) except for white vent and undertail coverts; whitish eyebrow contrasts with dark lore; black bill. M. has bright ruby throat; white malar streak; grayish upper breast. F. has white to pinkish throat; indistinct white malar streak; buffy breast. Secretive, stays in dense shrubbery. **1st Winter:** M. like ad. m. but paler red throat; f. like ad. **Flight:** Plain brown above. **Hab:** Open areas with shrubs. **Voice:** Song a rambling melodic mix of high and low whistled notes (may include imitations); calls include a low harsh *chak* and a whistled *peeet*.

Subspp: Monotypic.

Bluethroat 2

Luscinia svecica L 5½"

Shape: Small, deep-chested, deep-bellied, short-tailed thrush with a thin fine-pointed bill. **Ad:** Brownish upperparts; whitish belly and undertail coverts; broad white to buffy eyebrow from bill to nape; dark lore. Distinctive reddish-brown base to black outer tail feathers seen during tail flicking or when wings drooped. M. has bright blue bib bordered on chest with fine black, white, then broad chestnut bands; blue bib has central oval chestnut patch (breast pattern obscured in fall and winter due to pale edges to fresh feathers). F. has necklace of dark streaks on whitish breast; usually some red or blue on throat; pale submoustachial dash bordered by dark malar streak. Secretive; runs along ground under cover; tail often cocked or flicked upward. **Juv:** Like ad. f. but with finely streaked breast. **Flight:** Low. Brownish wings and back; outer tail feathers black with wide chestnut bases. **Hab:** Moist woods, brushy streamsides. **Voice:** Song a long series of repeated notes and sounds, each sound repeated many times and tending to get faster toward the end (often includes imitations); calls include a harsh *chak*, a throaty croaking *prurk*, and a whistled *weeet*.

Subspp: (1) •*svecica*.

Adult m. HKO/01

Adult f. CHI/05

1st-yr. m. CHI/05

Red-flanked Bluetail 4

Tarsiger cyanurus L 5½"

Casual Siberian vag. to w. AK, CA; most vagrants 1st-yr. birds. **Shape:** Small, broad-necked, fairly deep-bellied thrush with a medium-length tail and thin fine-pointed bill. Long primary projection past tertials (about 2½ x bill length). **Ad:** M. with blue upperparts (brightest on sides of crown, lesser coverts, tail base) except for white supraloral dash; underparts mostly white except for orangish flanks and partial blue collar. In fall, m. upperparts duller overall, with olive-brown wash especially over back. F. warm brown above (warmest on wings) except for bluish tail, white eye-ring; white throat contrasts with brown face and brownish breast; flanks orange, rest of underparts whitish. Tail and wings often flicked. **1st Yr:** F. like ad.; m. much like f. but with suggestion of a collar and hints of blue on upperparts. **Flight:** Both sexes have strong blue at tail base. **Hab:** Coniferous woods. **Voice:** Song a repeated whistled *cheedeet chur churdur;* calls include a harsh *chek* or *chechek* and a high-pitched whistled *heeet heeet.*

Subspp: (1) •*cyanurus.*

Adult m. HKO/04

Adult f. HKO/04

Stonechat 4

Saxicola torquatus L 5¼"

Casual Asian vag. to w. AK. **Shape:** Small, broad-necked, fairly deep-bellied thrush with a short tail and short thin bill. Short primary projection past tertials (about 1½ x bill length). Tends to stand erect. **Ad:** Thin white patch on lower scapulars sometimes visible on perched bird (see Flight). M. with black hood; partial white collar and rich buff breast and flanks; black wings and tail. F. brown above with blackish streaking on crown and back; buffy eyebrow; dusky auriculars; underparts mostly buffy, contrasting slightly with whitish throat. **1st Winter:** Like ad. f. but m. may have traces of black in auriculars. Rump buffy (white in ad.). **Flight:** Rump pale; tail black; white patch on scapulars. **Hab:** Open areas with sparse vegetation. **Voice:** Song is a short, high-pitched, harsh twittering; calls include a high-pitched *sweeet* and a low *chht chht.*

Subspp: (1) •*stejnegeri,* "Siberian Stonechat."

Adult m., summer EUR/04

1st-summer m., *oenanthe* AK/07

1st winter ENG/09

Adult f. EUR/04

1st winter EUR/10

Northern Wheatear 2

Oenanthe oenanthe L 5¾"

Shape: Deep-chested short-tailed look, with a defined head and medium-sized sturdy bill typical of thrushes. Comparatively, Northern Wheatear is small, long-legged, short-billed, has long primary projection past tertials. **Ad. Summer:** M. with black auricular and lore (creating a mask); white eyebrow; gray crown and back; black wings; white underparts with variable buffy wash over throat and breast (see Subspp.) that fades to white over summer. F. has buffy auricular with fine dark streaking and grayish lore; whitish to buffy eyebrow; brownish wash to gray crown and back; dark brown wings; buffy underparts. **Ad. Winter:** M. like summer but with brownish wash to gray upperparts; buffy to grayish edges to secondaries and greater coverts; reduced dark mask. F. like summer but with strong brownish or cinnamon wash to upperparts; buffy eyebrow; greater coverts with thin or no buffy edging. **1st Winter:** Both sexes like winter ad. f. but with broad buffy edges to secondaries and greater coverts; pale eyeline with little contrast to rest of head. **1st Summer:** M. with dark brown wings contrasting slightly with blackish greater coverts. F. like ad. female. **Juv:** (Jun.–Aug.) Dull buffy brown overall with pale spotting on breast and back; wings and coverts broadly edged warm brown. **Flight:** White rump; white tail with bold black inverted T formed by black terminal tail band and black central tail feathers (also seen as bird bobs and fans tail). **Hab:** Barren rocky areas. **Voice:** Song a rich jumble of high-pitched repeated whistles and sounds (often includes imitations); calls include *took* or *tooktook,* a whistled *weeet,* and a rattling *tk'tk'tk'tk.*

Subspp: (2) •*oenanthe* (AK to n.w. NT) smaller, with less buff overall; •*leucorhoa* (n.e. NT, vag. to NF) 15–20% larger, with more buff overall.

Adult m., *sialis* NH/03

Adult f. TX/04

Adult f., *sialis* NH/03

Juv., *sialis* NH/09

Eastern Bluebird 1

Sialia sialis L 7"

Shape: Fairly small, broad-necked, fairly deep-bellied, short-legged thrush with a short tail and short bill. Moderate primary projection past tertials (about 2 x bill length). **Ad:** M. upperparts bright blue; underparts rich reddish brown on throat, breast, and flanks, contrasting with bright white belly and undertail coverts; red on breast extends in an arc under auriculars. F. head, back, and wing coverts vary from brownish gray to grayish blue; sometimes back is more brownish and contrasts with gray or bluish head; wings and tail bright blue; white or pale reddish-brown throat contrasts with grayish cheek; pale reddish brown of breast extends back in an arc under auriculars, setting them off; pale reddish-brown breast and flanks contrast strongly with bright white belly and undertail coverts; whitish arc behind eye. **Juv:** (Apr.–Aug.) Brownish gray above with whitish streaks; grayish breast spotted with white; belly pale; wings and tail blue. **Flight:** Direct flight; underwings pale with faint wing-stripe. **Hab:** Farmland, woods edges, open woods. **Voice:** Song a short, rich, variable warble preceded by a few short harsh sounds, sometimes likened to "cheer cheerful charmer"; calls include a *turwee* or *turawee* and a short chattered *ch'ch'ch'ch*. 🎧**87**.

Subspp: (4) •*sialis* (MB–AZ and east) small; short bill; m. upperparts deep purplish blue. •*fulva* (s.e. AZ) large; m. upperparts medium blue; m. and f. underparts pale reddish brown. •*nidificans* (s. TX) large; m. upperparts dark blue; m. and f. underparts dark reddish brown. •*grata* (s. FL) like *sialis* but with slightly longer bill. **Hybrids:** Mountain Bluebird.

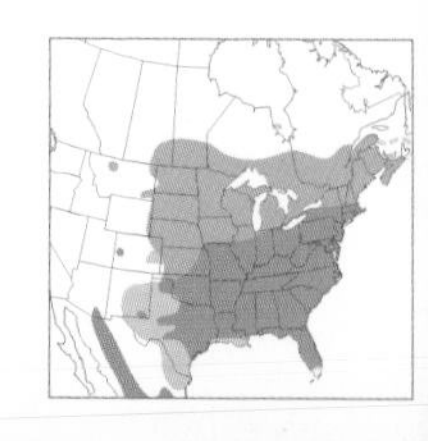

Adult m. CA/11

Adult f. CA/04

Adult m. NM/11

Juv. CA/06

Western Bluebird 1

Sialia mexicana L 7"

Shape: Fairly small, broad-necked, fairly deep-bellied, short-legged thrush with a short tail and short bill. Moderate primary projection past tertials (about 2 x bill length). **Ad:** M. upperparts mostly dark blue; back with no to substantial chestnut; scapulars chestnut; throat and head entirely blue; breast and flanks rich chestnut; belly and undertail coverts grayish to bluish. F. head and throat all grayish with no strong contrasts (no reddish-brown arc under auriculars as in Eastern Bluebird); underparts lack sharp contrasts; breast and flanks dingy reddish brown with gray tinge; belly and undertail dull gray; wings and tail dark blue; whitish partial eye-ring behind eye. **Juv:** (Apr.–Aug.) Like Eastern Bluebird juv.; best distinguished by seeing parent feed them. **Flight:** Direct flight; less contrast between belly and breast than in Eastern Bluebird. Underwing with strong paler wing-stripe. **Hab:** Farmland, woods edges, open woods. **Voice:** Song a mixture of 3 elements in varying order, like *ch'ch'ch churrr chup;* calls include *chweer* and short harsh *ch'ch'ch.* 🎧**88**.

Subspp: (3) Differences weak. •*occidentalis* (BC–MT south to CA–NV) small; m. breast dark reddish brown. •*bairdi* (UT–CO south to AZ–s.w. NM) large; m. breast pale reddish brown. •*jacoti* (s.e. NM–TX) medium-sized. **Hybrids:** Mountain Bluebird.

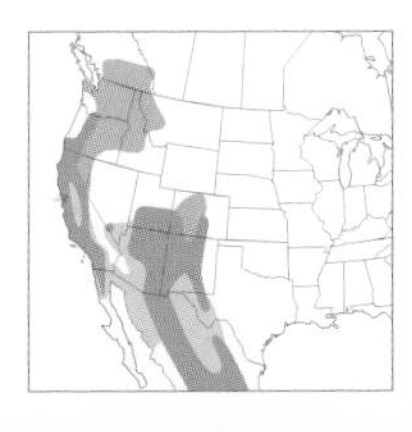

Adult m. MT/06

Adult f. NM/11

Adult f. NM/12

Juv. OR/08

Mountain Bluebird 1

Sialia currucoides L 7¼"

Shape: Fairly small, broad-necked, fairly deep-bellied, short-legged thrush with a short tail and short bill. Long primary projection past tertials (3–4 x bill length). Wings, tail, legs, and bill all proportionately longer than on Eastern and Western Bluebirds. **Ad:** M. striking turquoise blue overall except for variably silvery-gray belly and undertail coverts. F. least colorful of the f. bluebirds; head, back, breast, and flanks generally concolor grayish to grayish brown; sometimes a tinge of reddish brown to breast (but not flanks); throat generally paler than rest of head; vent and undertail coverts white, contrasting with belly and flanks; wings and tail turquoise (not dark blue as in other two bluebird spp.). **Juv:** (May–Aug.) Like Eastern Bluebird juv.; best distinguished by seeing parent feed them. **Flight:** Long wings; generally concolor. Often hovers over ground to hunt. **Hab:** Open areas. **Voice:** Song a low-pitched mixture of rough *churr* sounds, like *churr churr-churr chp;* calls similar to these elements. 🎧**89**.

Subspp: Monotypic. **Hybrids:** Eastern and Mountain Bluebirds.

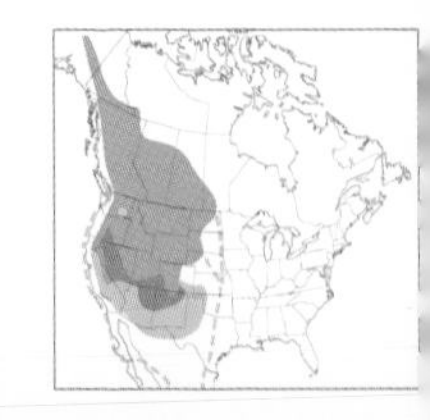

Adult NM/11

Adult NM/11

Townsend's Solitaire 1

Myadestes townsendi L 8½"

Shape: Medium-sized, long-tailed, deep-bellied, fairly small-headed member of the thrush family with a short pointed bill. Legs relatively short. **Ad:** Smooth gray overall with bold white eye-ring; buffy patches on closed wing formed by buffy base of inner primaries and outer secondaries; also sometimes buffy wingbar from buff tips to greater coverts. Tail has white ends to outermost feathers. **Juv:** (Jun.–Sep.) Spotted breast and back; buffy patches on wing; comparatively long tail. **Flight:** Buffy wing-stripe across base of flight feathers seen from above and below; tail may be flicked open during flight, revealing white ends and outer web on outermost feathers. **Hab:** Open coniferous woods, junipers. **Voice:** Song a robinlike mixture of short whistled phrases, but slightly higher-pitched; calls include a high, thin, quavering whistle like the sharpening of a metal saw, also a harsh *tjjik*.

Subspp: Monotypic.

IDENTIFICATION TIPS

Thrushes are a large family of small to medium-sized birds generally with deep chests, long wings, and fairly short thin bills. Those that forage on the ground have fairly long legs. Below are aids to identifying three major groups within the thrushes.

- **Bluebirds** are much loved because of their bright colors and use of nest boxes to raise young. The sexes look different, with the females having paler body colors and usually a whitish eye-ring (strongest behind eye). The males of our three species—Eastern, Western, and Mountain Bluebirds—are easily distinguished by the patterns and hue of blue on their head and throat. Females can be more difficult to distinguish and vary individually in the intensity of their colors. To distinguish among them, look first at primary extension past tertials (much longer in Mountain Bluebird) and then at the pattern of blue, red, and white on the throat, face, and underparts.
- **Catharus Thrushes** are secretive woodland birds with beautiful songs that belong to the genus *Catharus.* They are brownish to reddish brown above and whitish to grayish below, with varied amounts of dark spotting on their breast and flanks. Check to see if the head and tail are concolor or if one is more reddish brown than the other; look at the color, shape, and extent of the spots on the breast; and look for streaking or buffiness on the face. Their songs are easily distinguished. The sexes look alike.
- **Robins** and other birds in the genus *Turdus* all have the distinctive shape of the familiar backyard American Robin: they are deep-chested, broad-necked, fairly short-tailed birds. They spend much time on the ground looking for insects and earthworms to eat; they also eat berries in winter. Looking closely at the pattern of their faces and underparts will enable you to readily distinguish them. This group includes many vagrants from Siberia, Europe, and Mexico.

Veery 1

Catharus fuscescens L 7"

Shape: Deep-chested short-tailed look, with a defined head and medium-sized sturdy bill typical of Catharus thrushes. **Ad:** Above, concolor dull to bright reddish brown. Below, spots limited to upper breast and are reddish to brown, small, and indistinct to distinct; upper breast washed with buff to pinkish buff; flanks pale gray; central belly white. Face relatively plain; can have grayish streaking on cheek; eye-ring grayish and indistinct; malar streak reddish brown and inconspicuous. **1st Yr:** (Sep.–Aug.) Like ad., but some greater coverts may be tipped buff (when fresh) and paler than replaced ones. **Flight:** Even reddish-brown tones to upperparts. **Hab:** Moist deciduous woods, often near streams. **Voice:** Song a lovely descending spiraling whistle like *tooreeyur-reeyur-reeyur-reeyur* (similar Swainson's Thrush song ascends); calls include a downslurred *veeer* or *veerit*. **90**.

Subspp: (5) In general, western subspp. are darker, duller, with distinct breast spots; eastern subspp. are brighter, with less distinct breast spots. •*salicicola* (NM–CO–BC) has dark-brown-tinged-rusty upperparts, dull buff breast, distinct brown spots. •*levyi* (AB–MI) is similar but has slightly more reddish tones. •*fuliginosus* (QC–NF) has deep reddish-brown upperparts, brighter wash on breast, distinct warm brown breast spots. •*fuscescens* (ON–MD–ME) has buff on breast, with indistinct breast spots; •*pulichorum* (WV–GA) is bright on breast, with distinct spots.

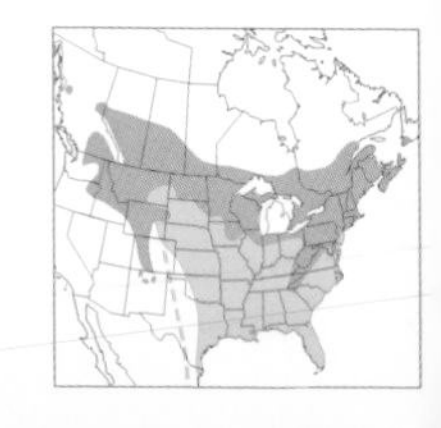

Adult TX/04

Adult, Olive-backed Grp. OH/05

Adult, Olive-backed Grp. OH/05

Adult, Russet-backed Grp. CA/05

Adult, Olive-backed Grp. OH/05

Swainson's Thrush 1

Catharus ustulatus L 7"

Shape: Deep-chested short-tailed look, with a defined head and medium-sized sturdy bill typical of Catharus thrushes. **Ad:** Above, generally concolor grayish brown with either olive or reddish tinge. Buffy eye-ring and supraloral bar create a buffy "spectacled" look. Also buffy are the throat, upper breast, and light areas of face (such as the crescent under the auriculars). Below, spots on breast extend indistinctly onto flanks—dark brown and triangular on upper breast, becoming oval and indistinct along flanks; belly whitish; flanks brownish gray to brownish. **1st Yr:** (Sep.–Aug.) Like ad., but some greater coverts may be tipped buff (when fresh) and paler than replaced ones. **Flight:** Upperparts concolor dull grayish brown. **Hab:** Coniferous and deciduous woods, riparian woodland (in West). **Voice:** Song a lovely ascending spiraling whistle like *yooray yooree yooreeyee* (similar Veery song descends); calls include a flat *kweeet,* a short *kwip,* a 2-part *kw'purr,* and a high whistled *peeet*.

Subspp: (6) Divided into two groups. **Russet-backed Group** (along Pacific Coast): Reddish tinge to upperparts, smaller sparser spots, less prominent "spectacles." •*ustulatus* (coastal s. AK–n. CA); •*oedicus* (cent.–n. CA); •*phillipsi* (Queen Charlotte Is., BC). **Olive-backed Group** (rest of range): Olive to grayish tinge to upperparts, larger darker spots, prominent "spectacles." •*incanus* (w. AK–NT south to n. AB–n. BC); •*swainsoni* (BC to NF and ME); •*appalachiensis* (NH and NY south).

Adult AK/06

Adult, *aliciae* OH/05

1st yr. TX/04

Adult, *aliciae* OH/05

Gray-cheeked Thrush 1

Catharus minimus L 7¼"

Shape: Deep-chested short-tailed look, with a defined head and medium-sized sturdy bill typical of Catharus thrushes. Comparatively slightly larger, with long primary extension past tertials. **Ad:** Above, generally concolor dull grayish brown, sometimes with olive tinge. Below, spots on breast extend onto upper belly and flanks—blackish brown, triangular, and distinct on upper breast, becoming gray, oval, and indistinct on upper belly and flanks; flanks and sides of breast variably washed brownish gray; belly white; may be slight buffy wash on breast. Face relatively plain grayish to brownish; some white streaking on auriculars; lore pale gray; variably gray around eye; usually a pale crescent behind eye, but no distinct complete eye-ring. Bill with upper mandible dark, lower mandible yellowish with variable dark tip. **1st Yr:** (Sep.–Aug.) Like ad., but some greater coverts may be tipped buff (when fresh) and paler than replaced ones. **Flight:** Upperparts concolor dull brown with no reddish-brown contrasts. **Hab:** Dense tall shrubs. **Voice:** Song a fairly high-pitched series of spiraling whistles; middle phrase tends to ascend, final phrase tends to descend (the opposite in Bicknell's Thrush; higher and harsher than Veery's song). Calls include a downslurred *jeeer.*

Subspp: (2) •*minimus* (NF–nearby mainland) warmer brown above, yellow portion of lower mandible more extensive (past nostril) and brighter; •*aliciae* (rest of range) duller gray-brown above, yellow portion of lower mandible shorter and duller.

1st yr. NH/06

Adult NH/06

1st yr. NH/06

Bicknell's Thrush 2

Catharus bicknelli L 6¾"

Slightly smaller close relative of Gray-cheeked Thrush with limited breeding range in the Northeast, south of Gray-cheeked breeding range. Separated from Gray-cheeked best by breeding range and song. Nonsinging migratory birds are difficult or impossible to identify due to subtlety of clues, individual variation, and subspecies variation in Gray-cheeked Thrush and are best labeled "gray-cheeked type." **Shape:** Deep-chested, short-tailed look, with a defined head and medium-sized sturdy bill typical of Catharus thrushes. Comparatively smaller with shorter primary extension past tertials than Gray-cheeked. **Ad:** First decide if you have a "gray-cheeked type" thrush, then try to decide if it is Gray-cheeked or Bicknell's. See Gray-cheeked Thrush for identification of the "gray-cheeked type." Bicknell's best distinguished from Gray-cheeked by song (see Voice). Bicknell's generally has warmer brown tail that subtly contrasts with brown back (not as much as in Hermit Thrush), while Gray-cheeked is nearly concolor. Bicknell's tends to have reddish-brown base to primaries, whereas Gray-cheeked (subsp. *aliciae*) has paler grayish bases to primaries. Yellow portion of Bicknell's lower mandible tends to be half or more length of bill and bright yellow to yellow-orange, while Gray-cheeked yellow portion tends to be half or less and duller yellow. These subtle distinctions best observed when comparing with Gray-cheeked subsp. *aliciae;* less valuable when comparing with Gray-cheeked subsp. *minimus,* which can have warm brown upperparts, a contrasting reddish tinge to tail, reddish-brown bases to primaries, and longer, brighter yellow portion to lower mandible (but see Shape). **1st Yr:** (Sep.–Aug.) Like ad., but some greater coverts may be tipped buff (when fresh) and paler than replaced ones. **Flight:** Upperparts concolor dull brown with no reddish-brown contrasts. **Hab:** Higher-elevation woods with mixed conifers and hardwoods. **Voice:** Song very similar to that of Gray-cheeked Thrush, but middle phrase tends to descend and final phrase tends to rise (the opposite in Gray-cheeked); calls include *tsweer* or *tsweet.*

Subspp: Monotypic.

Adult, Western Lowland Grp. CA/04

Adult, Western Lowland Grp. CA/10

Adult, Northern Grp. OH/05

Adult, Northern Grp. NH/05

Adult, Western Mountain Grp. CA/04

Adult, Northern Grp. OH/05

Hermit Thrush 1

Catharus guttatus L 6¾"

Shape: Deep-chested, short-tailed look, with a defined head and medium-sized sturdy bill typical of Catharus thrushes. Comparatively smaller and stockier than other Catharus thrushes. **Ad:** The only Catharus thrush to regularly winter in NAM; has the distinctive and helpful habit of flicking its tail quickly up and slowly down. Above, reddish-brown tail and uppertail coverts and reddish-brown primaries contrast with rest of duller brown upperparts. Below, fairly dense dark spots on white or pale buffy breast; throat white; flanks buffy or grayish (see Subspp.). Distinct thin white eye-ring. **1st Yr:** (Sep.–Aug.) Like ad., but some greater coverts may be tipped buff (when fresh) and paler than replaced ones. **Flight:** Tail is most reddish portion of upperparts. **Hab:** Wide variety of woods and edge habitats. **Voice:** Lovely flutelike series of rising and falling notes with an extended first note of changing pitch (similar Wood Thrush has no extended first note); calls include a rising *zweee* and a short *chup*. **91**.

Subspp: (13) Divided into three groups (only one representative of each group mentioned here). **Western Lowland Group** (8 subspp. along Pacific Coast): Small, upperparts dark to pale brown, gray flanks, extensive spotting on white breast. •*guttatus* (s. AK–w. BC). **Western Mountain Group** (3 subspp. in Rocky Mts.): Large, upperparts pale grayish brown, extensive spotting with large spots on white breast, gray flanks. •*auduboni* (s.e. WA–s. MT south to NM–TX). **Northern Group** (2 subspp. in North and East): Medium-sized, upperparts reddish brown, flanks and undertail coverts with buffy wash, dark spotting on buffy breast. •*faxoni* (NT–s. AB east to NF–MD).

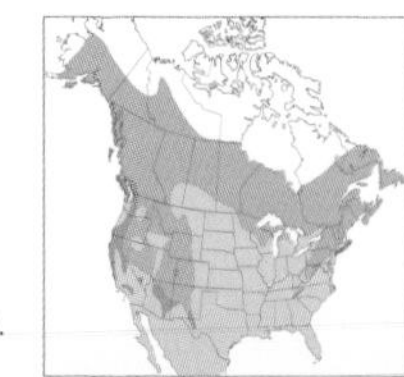

Adult OH/05

Adult TX/05

Adult OH/05

Adult OH/05

Wood Thrush 1

Hylocichla mustelina L 7¾"

Shape: Deep-chested short-tailed look, with a defined head and medium-sized sturdy bill typical of Catharus thrushes. Compared to similar Catharus thrushes, is larger, deeper-bellied, shorter-tailed, with deeper-based bill. **Ad:** Above, bright reddish brown on head and nape shading to duller reddish brown on rump and tail. Below, large black spots on white breast continue onto flanks. Face with bold black-and-white streaking on auriculars; white eye-ring. **1st Yr:** (Sep.–Aug.) Like ad., but some greater coverts may be tipped buff (when fresh) and paler than replaced ones. **Flight:** Upperparts most bright red on head. **Hab:** Mixed deciduous woods with shrub understory. **Voice:** Lovely flutelike series of rising and falling notes with an introductory stutter, *but but but eeyolay, but but but aholee* (similar Hermit Thrush song has extended first note); calls include a low murmuring *bwubwubwub,* a louder *bweebwee-bweeb,* and a high, squeaky, whistled *eeee.* 🎧**92**.

Subspp: Monotypic.

Adult SIN/12

Adult SIN/12

Eyebrowed Thrush 3

Turdus obscurus L 8½"

Rare Asian spring migrant through w. AK; casual in fall; accidental to CA. **Shape:** Deep-chested and short-tailed, with a defined head and medium-sized sturdy bill typical of thrushes. Shaped much like a small American Robin. **Ad:** Distinct white eyebrow (more distinct in m.); dark lore; white patch below lore and eye; whitish unmarked undertail coverts; white on belly extends up onto lower central breast in an inverted V or U; legs yellowish, pale. M. brown above with gray head, cheeks, and throat; chin white; central upper breast grayish; flanks orangish. F. concolor brown on back, nape, and crown; face and throat grayish and finely streaked with white; upper breast and flanks reddish brown (flanks less so than in m.). Similar American Robin can have partial to complete whitish eyebrow (especially 1st yr.); differs in having spotted undertail coverts; breast and forebelly usually orange, hindbelly may have limited white in center; dark legs; streaked white throat. **1st Winter:** White tips to greater coverts create thin wingbar; m. similar to ad. female. **Flight:** Underwings pale gray; bright white central belly and undertail coverts; no white tips to outer tail feathers (most, but not all, American Robins have white tips to outer tail feathers). **Hab:** Coniferous forests. **Voice:** Song a series of whistled notes followed by softer twittering or chattering; calls include a high-pitched *tseee,* a clipped *tsip-tsip,* and a lower *tchup.*

Subspp: Monotypic.

Adult m., *eunomus* JAP/01

Adult m., *eunomus* JAP/03

Dusky Thrush 4

Turdus naumanni L 9½"

Casual Asian spring vag. to w. AK, w. BC, and WA. **Shape:** Deep-chested and short-tailed, with a defined head and medium-sized sturdy bill typical of thrushes. About size and shape of American Robin. **Ad:** Strong subspecies differences in plumage colors (see Subspp.), but overall pattern similar. Dark auricular patch bordered by broad pale eyebrow above and pale face in front and below. Underparts variably spotted by dark-centered feathers with paler margins (centers black or orange; see Subspp.); can be suggestion of darker band across breast. M. generally with more contrasting plumage than female. See Subspp. for more clues. **Juv:** Like ad. female. **Flight:** Bright reddish-brown underwings; whitish belly. **Hab:** Open woods. **Voice:** Song a series of melodic whistles followed by a soft trill; calls include a harsh *tcher tcher.*

Subspp: (2) Intermediates between the two subspecies occur. •*naumanni* (vag. to w. AK): Eyebrow, foreface, and throat orangish; underparts feathers with orange centers and pale margins; rump and outer tail feathers orangish; rest of upperparts grayish brown; legs pale; sexes similar. •*eunomus* (vag. to AK, BC, YT, WA): Eyebrow, foreface, and throat white; underparts feathers with blackish centers and whitish margins; f. with concolor brown upperparts; m. with blackish upperparts and contrasting reddish-brown wings; legs dark.

Adult

Adult ENG/01

Adult FIN/01

Fieldfare 4

Turdus pilaris L 10"

Casual Eurasian vag. to Northeast (occas. farther inland and AK). **Shape:** Deep-chested short-tailed look, with a defined head and medium-sized sturdy bill typical of thrushes. Slightly larger-bodied than American Robin. **Ad:** Above, pale gray head, lower back, and rump contrast with dark reddish-brown back and inner wings. Below, throat and upper breast orange streaked with black; belly grayish with blackish chevrons along flanks. Foreface blackish; auricular gray; thin white eyebrow; thin white malar streak. White marginal underwing coverts often visible at bend of folded wing. **1st Yr:** (Sep.–Aug.) Like ad., but outer greater coverts tipped buff (when fresh) and paler than replaced inner ones. **Flight:** Pale gray rump and head contrast with dark back and tail; underwing coverts bright white. **Hab:** Open woods, shrubby edges. **Voice:** Song a mixture of harsh chatters and warbles; calls include a harsh *chechecheche* and a high-pitched *tseee.*

Subspp: Monotypic.

Adult ENG/01

1st yr. ENG/01

Redwing 4

Turdus iliacus L 8¼"

Casual Eurasian vag. to Northeast, also WA. **Shape:** Deep-chested short-tailed look, with a defined head and medium-sized sturdy bill typical of thrushes. Smaller, slightly larger-bodied, and shorter-tailed than American Robin; bill short and fine-pointed for a thrush. **Ad:** Above, plain brown. Below, whitish with heavy brown streaking on throat, breast, and flanks; conspicuous deep orange on flanks. Broad buffy to white eyebrow; white submoustachial line. **1st Yr:** (Sep.–Aug.) Like ad., but outer greater coverts tipped buff (when fresh) and paler than replaced inner ones. **Flight:** Below, dark orange flanks and underwing coverts. **Hab:** Mixed woods, shrubby streamsides. **Voice:** Song a variable mixture of whistled notes and squeaky twittering; calls include a *chirup* and a high-pitched *tseee.*

Subspp: (2) •*iliacus* (Asian vag. to West Coast) is small, paler-backed; streaking on underparts thin and brownish. •*coburni* (Eur. vag. to East Coast) is slightly larger, darker-backed; streaking on underparts thick and blackish.

Adult CRI/12

Adult PAN/12

Clay-colored Thrush 3

Turdus grayi L 9"

Breeds in southernmost TX; casual n. and w. in TX. **Shape:** Deep-chested short-tailed look, with a defined head and medium-sized sturdy bill typical of thrushes. Slightly smaller with proportionately longer bill and shorter primary projection past tertials than American Robin. **Ad:** Olive-brown above; buffy brown below; throat paler with faint streaking. Bill yellowish; iris dark reddish; legs gray. **1st Yr:** (Jul.–Oct.) Like ad., but some greater coverts may be tipped buff (when fresh) and paler than replaced ones. **Flight:** Orangish underwing coverts brighter than buffy underparts. **Hab:** Moist lowland woods, shrubby edges of fields, gardens. **Voice:** Song a series of fairly low-pitched whistled phrases, some repeated several times; calls include a shrill *scree-yeeah* and a high-pitched *seee.*

Subspp: (1) •*tamaulipensis.*

Adult TX/02

Adult TX/02

Adult TX/03

White-throated Thrush 4

Turdus assimilis L 9½"

Casual Mex. vag. to s. TX. **Shape:** Deep-chested short-tailed look, with a defined head and medium-sized sturdy bill typical of thrushes. **Ad:** Upperparts and face concolor dark brown with a clear, thin, yellowish to orangish eye-ring; throat mildly to heavily streaked black; small white patch on upper breast sometimes visible; rest of underparts dull grayish brown; legs dull pinkish to orangish. **Flight:** Underwing coverts pale orange. **Hab:** Moist subtropical woods. **Voice:** Song a series of fairly low-pitched whistled phrases, some repeated several times; calls include a buzzy *dzrrrk* and a short *kek.*

Subspp: (1) •*assimilis.*

Adult TX/03

Adult AZ/04

Adult CA/03

Rufous-backed Robin 3

Turdus rufopalliatus L 9¼"

Rare Mex. vag. to Southwest. **Shape:** Deep-chested short-tailed look, with a defined head and medium-sized sturdy bill typical of thrushes. **Ad:** Warm orangish-brown back and wing coverts contrast with gray head and rump. Foreface blackish (lacking the white eye arcs of American Robin); thin eye-ring yellow to orange; throat white with dark streaks extending onto upper breast; breast and flanks orangish; central belly and undertail coverts white. **Flight:** Orange upperwing and underwing coverts contrast with gray flight feathers. **Hab:** Variable; woods, gardens, dense shrubs. **Voice:** Song a mellow series of low-pitched whistles and short warbles; calls include a *chur chur,* a harsh *churchurchurk,* and a whistled burry *preeyur.*

Subspp: (1) •*rufopalliatus.*

Adult m. BC/01

Adult f. CA/01

Varied Thrush 1

Ixoreus naevius L 9½"

Shape: Deep-chested short-tailed look, with a defined head and medium-sized sturdy bill typical of thrushes. Comparatively small-headed, deep-bellied, and shorter-tailed. **Ad:** Dark crown and face, with long orange eyebrow extending mostly behind eye; 2 bold orange wingbars on dark wings. Underparts mostly orange with variable dark breastband; undertail coverts whitish with a few dark spots. M. upperparts dark gray to blue-gray; breastband blackish; breast and belly deep orange. F. upperparts brownish gray; wings and tail brown; breastband brown to dark gray and faint to well marked. **Juv:** (May–Aug.) Like ad. f. but with mottled orange breast, no breastband, and white belly. **1st Yr:** Breastbands paler than ad.; outer greater coverts tipped brownish (when fresh) and paler than replaced inner ones. **Flight:** Underwing dark with pale wing-stripe through base of flight feathers; upperwing with orange stripe on secondary greater coverts and base of primaries. **Hab:** Moist coniferous and mixed woods. **Voice:** Song a haunting series of flat, often slightly buzzy, extended whistles, each well spaced and on a different pitch; calls include a series of dry *chk*s, a short trilled *vreee,* and a short harsh *chrrr.* 🎧**93**.

Subspp: (4) Differences in female. •*meruloides* (AK–NT) f. palest above and below; •*naevius* (coastal s. AK–n. CA) and •*carlottae* (Queen Charlotte Is. BC) f. medium dark above and below; •*godfreii* (cent. BC–e. WA–MT) f. medium dark above and below.

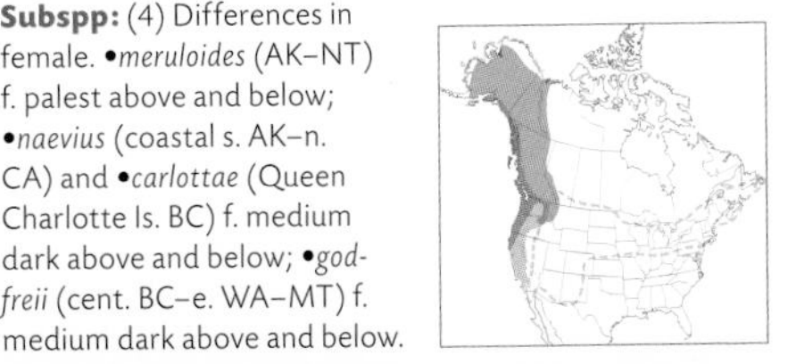

Adult m., summer, *migratorius* NH/07

Juv., late stage, *migratorius* NH/09

Adult f., winter FL/02

Juv., early stage, *migratorius* NH/08

Adult f., winter GA/02

Adult FL/02

American Robin 1

Turdus migratorius L 10"

Shape: Deep-chested short-tailed look, with a defined head and medium-sized sturdy bill typical of thrushes. **Ad:** Above, pale brown to blackish. Below, breast, flanks, and belly pale orange to dark reddish orange; vent and undertail coverts white with variable dark centers on coverts; throat white with dark streaks. Face dark with variable white markings around eye; varies from fragments of white eye crescents to almost complete eye-ring broken in front at the dark lores; from no eyebrow to a distinct whitish eyebrow (mostly f. and imm.); usually has some supraloral spot just in front of eye. Subsp. variation in overall darkness can make sexing difficult. In general, m. has black head contrasting with dark gray to dark brown back and more richly colored reddish-orange breast and belly. Female is usually more muted; brown crown blends to paler brown or grayish back, and breast is pale orange, often with pale feather edges on upper portion. Fresh fall/winter plumage has whitish edges to feathers of breast and belly and sometimes crown and nape; rear belly may look white centrally (white usually not as extensive and usually does not reach upper breast as in Eyebrowed Thrush). Legs dark. **Juv:** (May–Sep.) Like ad. overall but with buff spots on back, large dark spots on reddish breast and undertail coverts. **1st Yr:** Generally paler than ad.; sometimes outer greater coverts tipped buff (when fresh) and paler than replaced inner ones. **Flight:** From below, reddish breast contrasts with white rear belly and undertail coverts. Often white tips to outer tail feathers (see Subspp.). **Hab:** Variable; woods, gardens. **Voice:** A series of short whistled phrases with some high scratchy notes, sometimes likened to "cheer-up cheerily, cheer-up cheerily"; calls include a low *tuk* or *tuktuktuk,* a sharp *teeek,* an extended *teecheecheechoochooch,* a thin *seee.* **94**.

Subspp: (5) In general, white tips to tail feathers vary geographically: most white in East, least white (to none) in West. •*nigrideus* (n. QC–NF) medium-sized, darkest upperparts; •*migratorius* (AK–cent. BC east to QC–NJ) medium-sized, medium-dark upperparts; •*achrusterus* (s.e. OK–MD and south) small, medium-dark upperparts; •*propinquus* (s.cent. BC–s. SK south to s. CA–w. TX) large, pale to medium-dark upperparts, may have no white tips to tail; •*caurinus* (coastal s. AK–n. OR) small, pale upperparts, may have no white tips to tail.

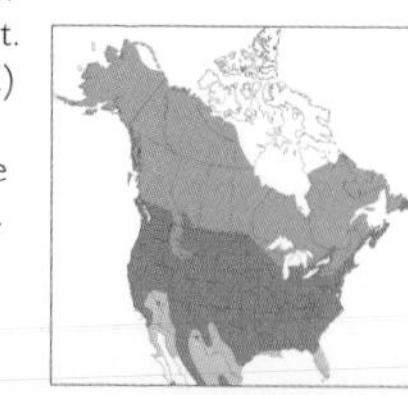

Adult m. AZ/07

Adult m. AZ/07

Adult m. AZ/07

Aztec Thrush 4

Ridgwayia pinicola L 9¼"

Casual Mex. vag. to Southwest. **Shape:** Deep-chested short-tailed look, with a defined head and medium-sized sturdy bill typical of thrushes. Comparatively long-necked and short-tailed. **Ad:** Upperparts, head, and breast mostly brown except for white uppertail coverts; belly whitish. Closed wing dark, strongly patterned with white—greater coverts edged white, base of primaries white, secondaries with broad whitish tips. M. upperparts and breast dark brown; f. upperparts and breast finely streaked lighter brown. **Juv:** (Jul.–Sep.) Wing and tail like ad.; body overall brownish, heavily streaked with white to pale yellow on head, back, and breast; belly pale with indistinct streaking. **Flight:** Uppertail coverts form broad white V; tail tipped white; dark wings have white patch at base of primaries and whitish trailing edge. **Hab:** Moist canyon forests. **Voice:** Calls include a nasal *sweeeyah,* a rising *seeyeep,* and a buzzy *zrip.*

Subspp: Monotypic.

Wrentit

Catbird

Mockingbirds

Thrashers

Starling

Wagtails and Pipits

Waxwings

Phainopepla

Adult CA/10

Adult CA/10

Wrentit 1

Chamaea fasciata L 6½"

Shape: Small, large-headed, broad-necked, round-bodied bird with a short thin bill, long legs, and very long, thin, rounded tail. **Ad:** Upperparts brown to grayish brown; underparts slightly paler with variable tawny to pinkish wash on breast and flanks; pale eye. Tail often cocked. **Juv:** (Apr.–Aug.) Like ad. **Flight:** Note long tail. **Hab:** Chaparral, dense shrubs. **Voice:** Song an accelerating series of short whistles, like *pit pit pitpit pipipipipit* (m.) or even *pit pit pit pit* (f.); calls include a rattling *ch'ch'ch'ch,* a short *chuk,* and a soft *meeah.*

Subspp: (5) Northern subspp. tend to be darker with shorter tails, southern subspp. paler with longer tails. •*phaea* (coastal OR–n. CA) and •*margra* (s.cent. OR) dark with short tail; •*rufula* (n.w. CA) medium-dark with short tail; •*fasciata* (w.cent. CA) medium-dark with long tail; •*henshawi* (n.cent. CA–s.w. CA) pale with long tail.

Adult, *carolinensis* NH/09

1st yr. FL/02

Gray Catbird 1

Dumetella carolinensis L 8½"

Shape: Slim rather flat-crowned bird with a fairly long, broad, slightly rounded tail. **Ad:** Slate gray overall with a thin black cap and inconspicuous reddish-brown undertail coverts; tail blackish. Tail often cocked. **Juv:** (Jun.–Aug.) Like ad., but body grayish brown, lacks black cap, undertail usually paler reddish brown. 1st yr. like adult, but with brownish primaries. **Flight:** Gray upperparts with black tail and cap. **Hab:** Shrubs, woods edges, suburban plantings. **Voice:** Song a series of whistled sounds and noises, each given once (may be some imitations), interspersed with scratchy *meew;* calls include an extended catlike *meeew,* a soft *kwut,* a harsh ratchety *ch'ch'ch'ch.* 🎧**95**.

Subspp: (2) •*ruficrissa* (w. MB–n.w. TX and west) has paler undertail coverts; •*carolinensis* (rest of range to east) has darker undertail coverts.

Adult GA/03

Adult GA/03

Juv. CA/08

Adult GA/03

Northern Mockingbird 1

Mimus polyglottos L 10½"

Shape: Slim, flat-crowned, long-tailed, long-legged bird with a fairly thick relatively short bill. **Ad:** Gray above; whitish below; 2 white wingbars; white base to primaries creates a patch on edge of folded wing. Indistinct gray eyeline; yellowish eye. **Juv:** (Apr.–Mar.) Like ad. but with fine spotting on throat and breast and gray to dark eye; some faint streaking on belly may persist through winter (which can suggest Bahama Mockingbird, which has streaking on flanks). **Flight:** Distinctive large white patches on outer wing; white outer tail feathers (Bahama Mockingbird lacks both these; outer tail feathers are white only at tips). **Hab:** Open areas with shrubs, gardens, parks. **Voice:** Song a long series of whistles and sounds (often imitations of other birds' songs and calls), each repeated 3 or more times; calls include a raspy *cjjjj,* a repeated harsh *chik,* a rapid *ch'ch'chik,* and a deeper *chewk.* 🎧**96**.

Subspp: Monotypic. **Hybrids:** Bahama Mockingbird.

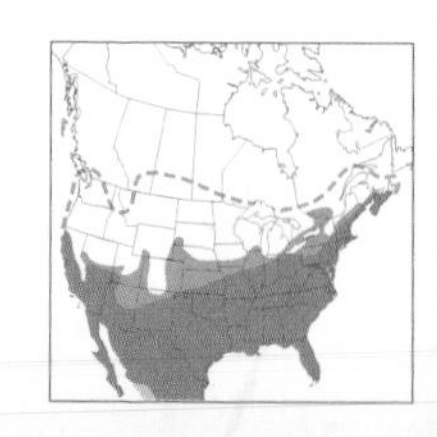

Adult BAH/02

Adult BAH/04

Bahama Mockingbird 4

Mimus gundlachii L 11"

Casual Caribbean visitor to s. FL. **Shape:** Like Northern Mockingbird but a little larger, with a slightly bigger body. **Ad:** Brownish gray above, with indistinct darker streaking on nape and back; whitish below, with distinct dark streaking on lower belly, flanks, and undertail coverts. Strong malar streak; faint whitish eyebrow. Underside of folded tail white only at tip. Similar N. Mockingbird has streaking only on belly (juv.) and white primary patches; underside of folded tail all whitish; less secretive behavior. **1st Yr:** (Sep.–Aug.) Like ad. but no streaking on back or nape and streaking below indistinct and usually limited to flanks and undertail coverts. Similar juv. N. Mockingbird has streaking or spotting mostly on throat and breast, has white wing patches and white outer tail feathers. **Flight:** No white patch on outer wing; tail all dark with white tips to outer feathers. Similar N. Mockingbird has white wing patches and white outer tail feathers. **Hab:** Shrubs in open areas, parks, gardens. **Voice:** Song a varied mixture of whistles and harsh notes (less repetition than N. Mockingbird); calls include a harsh *chewk.*

Subspp: (1) •*gundlachii.* **Hybrids:** Northern Mockingbird.

Adult AZ/01

Adult MEX/01

Blue Mockingbird 5

Melanotis caerulescens L 10"

Accidental Mexican stray to AZ and TX. **Shape:** Medium-sized bird with a fairly long tail, fairly small head, and relatively long straight bill. **Ad:** Steel-blue overall except for black mask covering face and auriculars; paler blue streaking on head and throat. Blue color looks brilliant to dull depending on light conditions. Eye dark reddish brown. **Juv:** (Apr.–Aug.) Dull gray with bluish wings and tail. **1st Yr:** Like ad. but blue less vivid; lacks paler streaking on head and throat. **Flight:** All blue; fairly long tail, short wings. **Hab:** Shrubby woodland understory. **Voice:** Song a varied series of whistles and sounds with much repetition (1–4 times), but less repetition than N. Mockingbird; calls include a *chooleep,* a percussive *pitik,* and a low *chewk.*

Subspp: (1) •*caerulescens.*

Adult TX/09

Adult, *rufum* ME/05

Adult, *rufum* TX/05

Brown Thrasher 1

Toxostoma rufum L 11"

Shape: Medium-sized, long-tailed, long-legged bird with a long shallow crown and sturdy bill typical of thrashers. Comparatively, Brown's tail very long; bill relatively short and bottom edge of lower mandible fairly straight; primary extension past tertials short (about lore length), but still longer than in most other thrashers. Similar Long-billed Thrasher has longer, slightly downcurved bill and very short primary extension. **Ad:** Upperparts rich reddish brown with 2 white wingbars; underparts whitish with thick brownish streaking on breast and flanks (flanks may be pale buff); undertail coverts whitish and largely unmarked. (Similar Long-billed has white underparts with blackish streaking; often has dark-centered undertail coverts.) Brownish-gray cheek; yellow to yellowish-orange eye. Basal half of lower mandible pale (usually only extreme base pale in Long-billed or bill all dark). **Juv:** (Jun.–Sep.) Like ad. but with buffy spotting on upperparts; gray eye. **Flight:** Pale buffy tips to outer tail feathers (may be worn away by summer). Flights short and low between cover. **Hab:** Thickets, shrubs, woods edges. **Voice:** Song a long series of musical phrases and harsh notes often given in pairs, usually with pauses between, like *teeahwee-teeahwee teeoo-teeoo chay-chay* (often contains imitations of other birds); calls include a kisslike smack and a harsh rising *peeyee.* **97**.

Subspp: (2) •*rufum* (ON–s.e. TX and east) smaller with deep reddish-brown upperparts; •*longicauda* (rest of range to west) larger with paler upperparts.

Adult TX/12

Adult TX/11

IDENTIFICATION TIPS

Catbird, Mockingbirds, and Thrashers all belong to the family Mimidae. Many are known for their rambling imitations of other birds' songs. They are fairly large birds with long tails and fairly long legs, and the thrashers (except Sage Thrasher) have long decurved bills. Most eat insects gathered on the ground, and many take advantage of plentiful berries in fall and winter.

Species ID: Identification of Gray Catbird and the mockingbirds is straightforward, but several of the thrashers look much alike. The thrashers can be divided into those with spots or streaks on their breast and those with a clear breast; note that the spots on some species can become worn and indistinct in mid- to late summer. In addition, pay close attention to their shape, including overall length, length and curvature of bill, and length of tail. Many have distinctive calls.

Long-billed Thrasher 2

Toxostoma longirostre L 11½"

Shape: Medium-sized, long-tailed, long-legged bird with a long shallow crown and sturdy bill typical of thrashers. Comparatively, Long-billed's tail very long; bill moderately long and bottom edge of lower mandible slightly downcurved; primary extension very short (shorter than lore). Similar Brown Thrasher has shorter straighter bill and slightly longer primary extension. **Ad:** Upperparts dark brown with 2 white wingbars; underparts white with heavy blackish streaking; undertail coverts whitish, often with dark centers. Dark gray cheek; yellowish-orange to orange eye. Small to no pale area at base of lower mandible. See Brown Thrasher for comparison. **Flight:** Dark brown above; no pale tips to outer tail feathers. **Hab:** Woods and thickets. **Voice:** Song similar to Brown Thrasher's but without pauses, with less repetition, and with more harsh notes; calls include a sharp smack and a rising whistled *tooeee.*

Subspp: (1) •*sennetti.*

Adult NM/04

Adult AZ/02

Adult AZ/02

Bendire's Thrasher 2

Toxostoma bendirei L 10"

Shape: Medium-sized, long-tailed, long-legged bird with a long shallow crown and sturdy bill typical of thrashers. Comparatively, Bendire's is smaller, with shorter bill, shorter tail, and slightly peaked crown. Bill fairly straight along lower mandible and clearly longer than lore + eye. Similar Sage Thrasher has bill about equal to lore + eye. Similar Curve-billed Thrasher has longer slightly downcurved bill; longer neck; longer shallower crown. **Ad:** Upperparts brownish gray; underparts paler brownish gray with brownish spotting on breast and flanks. Spotting on upper breast tends to be chevrons (Curve-billed has indistinct round spots, especially on sides of upper breast); spots most distinct on fresh plumage in fall through spring. Flanks tend toward brown (more grayish in Curve-billed). Bill with pale base to lower mandible (all dark in Curve-billed at all ages). **Juv:** (Apr.–Aug.) Like ad. but with cinnamon margins to greater coverts and tertials; reddish-brown hue to upperparts; no white tips on tail. Similar Curve-billed juv. has short all-dark bill and distinctively different call note. **Flight:** Flights easy and of normal height. Small white tips to outer tail feathers. **Hab:** Open arid areas, especially with grasslands. **Voice:** Song a long varied series of fairly screechy phrases and whistled notes without pauses and with much repetition (2–5 times per phrase); calls include a *chek chek chek* and a rising whistled *tooeee.*

Subspp: Monotypic.

Adult, *palmeri* AZ/03

Adult, *palmeri* AZ/05

Adult, *palmeri* AZ/03

Adult, *oberholseri* TX/11

Curve-billed Thrasher 1

Toxostoma curvirostre L 11"

Shape: Medium-sized, long-tailed, long-legged bird with a long shallow crown and sturdy bill typical of thrashers. Comparatively, Curve-billed is medium-sized, with moderate-length slightly downcurved bill and moderate-length tail. Similar Bendire's Thrasher is smaller with shorter straighter bill and shorter neck. **Ad:** Upperparts brownish gray (sometimes thin wingbars; see Subspp.); underparts paler with grayish-brown spotting on breast and flanks. Spotting on upper breast tends to be indistinct and round, especially on sides of upper breast (Bendire's has distinct chevrons); spots most distinct on fresh plumage in fall through spring. Flanks tend toward grayish (more brownish in Bendire's). Bill usually all dark. See Voice. **Juv:** (Apr.–Aug.) Like ad. but with cinnamon margins to greater coverts and tertials; reddish-brown hue to upperparts. Distinguish from juv. Bendire's by call note. **Flight:** Flight jerky and generally low. Pale tips to outer tail feathers can be large or small (see Subspp.). **Hab:** Desert areas, often with dense shrubs; suburban yards. **Voice:** Song a varied mixture of whistles and sounds given without pauses, with much repetition, and often includes imitations of other birds; calls include a distinctive strong *wit-weeet* or *wit-weeet-wit,* and a chattering *churrr.* 🎧**98**.

Subspp: (2) •*palmeri* (w.cent.–s. AZ) has grayish breast, making spotting less distinct; thin grayish or no wingbars; indistinct or no white tips to outer tail feathers. •*oberholseri* (rest of range) has white breast, making spotting more distinct; distinct whitish wingbars; white tips to outer tail feathers.

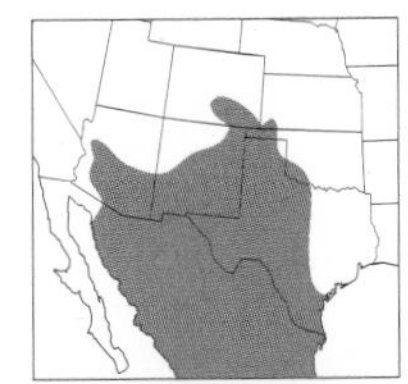

Adult, *redivivum* CA/01

Adult, *sonomae* CA/01

California Thrasher 2

Toxostoma redivivum L 12"

Shape: Medium-sized, long-tailed, long-legged bird with a long shallow crown and sturdy bill typical of thrashers. Comparatively, California is large, with fairly long tail and long strongly downcurved bill. Similar Crissal Thrasher has longer tail, longer bill, and proportionately bigger head. **Ad:** No spots on breast. Mostly grayish brown overall with orangish vent and undertail coverts. Head with white throat, narrow thin malar streaks, pale thin eyebrow; dark eye; fine black streaks across the auriculars. **Juv:** (Feb.–Aug.) Like ad. but with buffy margins to greater coverts and tertials. **Flight:** Flight generally short, can appear labored. Dark tail with pale tips to outer feathers. **Hab:** Chaparral and other dense brush. **Voice:** Song a varied mixture of whistles and sounds given at a leisurely rate (slower than in most other thrashers) with some repetition, often includes imitation of other birds; calls include a rising *chireep,* a soft *tchup,* and a harsh *chak*. 🎧**99**.

Subspp: (2) •*sonomae* (n.-cent. CA) mostly concolor chest and belly; •*redivivum* (cent.–s. CA) gray chest that contrasts slightly with buffy belly.

Adult CA/01

Adult NV/04

Crissal Thrasher 2

Toxostoma crissale L 11½"

Shape: Medium-sized, long-tailed, long-legged bird with a long shallow crown and sturdy bill typical of thrashers. Comparatively, Crissal is large, very long-tailed, with a long, relatively thin, strongly downcurved bill. Similar California Thrasher has slightly shorter tail, slightly shorter bill, and proportionately smaller head. Similar Le Conte's Thrasher is smaller, has shorter tail, shorter bill, proportionately smaller head. **Ad:** No spots on breast. Mostly grayish overall with deep reddish-brown undertail coverts. Head strongly patterned in throat area with broad black malar streak; whitish throat; pale eye. **Juv:** (Mar.–Aug.) Like ad. but with buffy margins to greater coverts and tertials. **Flight:** Flights generally short; can appear somewhat labored. Long tail and relatively short wings. **Hab:** Along dry desert creekbeds, also mesquite and chaparral. **Voice:** Song a varied mixture of musical to harsh whistled phrases, with little pause, some imitation of other birds; calls include a distinctive *ch'jurijurijuri* or *jurijuri.*

Subspp: (2) •*coloradense* (s.e. CA–s.w. AZ) slightly paler than •*crissale* (rest of range).

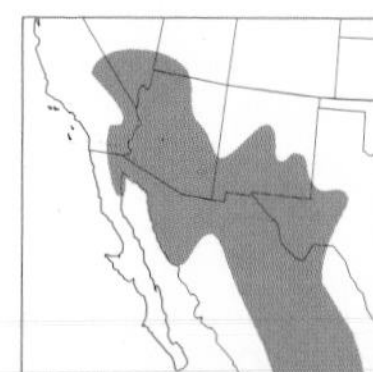

Adult CA/03

Adult CA/03

Adult CA/03

Juv. CA/04

Le Conte's Thrasher 2

Toxostoma lecontei L 11"

Shape: Medium-sized, long-tailed, long-legged bird with a long shallow crown and sturdy bill typical of thrashers. Comparatively, Le Conte's is medium-sized and long-tailed, with a long strongly down-curved bill. Similar Crissal Thrasher has longer tail and bill. **Ad:** No spots on breast. Mostly very pale sandy gray overall, with a whitish throat, darker tail, and pale orangish undertail coverts. Head with little pattern; dark lore; thin dark malar streak; dark eye. Often runs with tail cocked. **Juv:** (Feb.–Aug.) Like ad. but with brownish wash to upperparts; pale lore. **Flight:** Flights usually short and low. Pale back and wings contrast with darker tail. **Hab:** Desert areas with sparse vegetation. **Voice:** Song a varied series of whistles given in fairly unhurried manner, with much repetition and some imitation of other birds; calls include a short, rising, whistled *tsuweep.*

Subspp: (1) •*lecontei.*

Adult NM/11

Adult AZ/02

Adult, worn OR/08

Adult NM/06

Sage Thrasher 1

Oreoscoptes montanus L 8½"

Shape: Medium-sized, long-tailed, long-legged bird. Compared to other thrashers, Sage is small and relatively long-winged, with a smaller head and short fairly straight bill. Bill about length of lore + eye. In similar Bendire's Thrasher, bill length much longer than lore + eye. **Ad:** Grayish brown above; whitish below with thick streaking on breast and belly; variable buffy wash on flanks in fresh plumage; thin dark malar streaks border white chin; yellow to amber eye. By late summer, streaking on underparts and thin white wingbars can become faint and indistinct and bird can appear darker. **Juv:** (Jun.–Sep.) Like ad. but with dark streaking on back; less streaking on belly; buffy margins to tertials; eyes dark brown. **Flight:** Flights short and low between cover. Large white tips to outer tail feathers (seen from below on perched bird). **Hab:** Sage, desert scrub. **Voice:** Song a continuous rambling series of pleasant whistles and sounds, often quite long; calls include a harsh *chek* and a high harsh *churrr.*

Subspp: Monotypic.

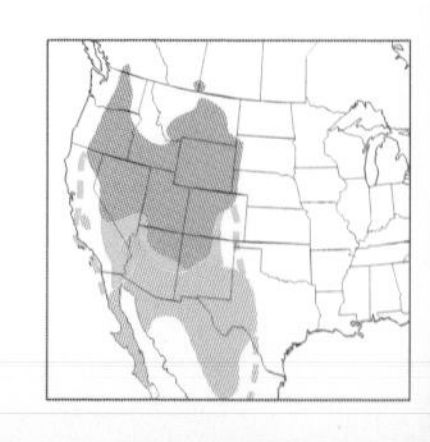

Adult m., summer NJ/01

Juv. molting to ad. winter TX/09

Juv. ON/06

Adult winter CA/10

Adult IL/03

European Starling 1

Sturnus vulgaris L 8½"

Shape: Stocky short-tailed bird with a rather large head and relatively long spikelike bill. Wings fairly long and pointed. **Ad. Summer:** Glossy black iridescent plumage overall, with variable light spotting on flanks and belly. Bill bright yellow; starts turning dark after breeding (Aug.). M. has bluish base to lower mandible (Jan.–Jul.). F. has pinkish base to lower mandible (Jan.–Jun.). **Ad. Winter:** Glossy black overall, with fine white or buffy tips to body feathers and buffy margins to all flight feathers. The dots and margins wear off gradually over winter to produce the summer plumage. Bill black (Aug.–Dec.); starts turning yellow in Dec. **Juv:** (May–Oct.) Smooth mouse-gray overall, with a paler throat; buffy margins to wing feathers; dark bill. Often in large flocks by mid–late summer. **Flight:** Note triangular wings with somewhat translucent flight feathers. **Hab:** Urban and suburban areas. **Voice:** Song a rambling mix of squeals and whistles, with some imitation of other birds; calls include a continuous chortling done with bill closed, a harsh *djjj,* a high-pitched squeal, and a whistled downslurred *wheeeuuu.* 🎧**100**.

Subspp: (1) •*vulgaris.*

Adult JD/02

Siberian Accentor 4

Prunella montanella L 5"

Casual Asian vagrant to w. AK and other states and provinces in the Northwest. **Shape:** Small, broad-necked, deep-chested bird with fairly short, thin, fine-pointed bill and moderate-length notched tail. **Ad:** Facial pattern with bill shape is unique. Dark crown and mask; warm orange-buff eyebrow and throat; pale arc under eye. Underparts mostly buffy, with faint dark spotting on breast sides; brownish streaking along flanks; undertail coverts whitish. Back streaked brown to warm brown; grayer collar; dotted whitish wingbar on greater coverts. Feeds mostly on ground or stays secretive in shrubs. **Flight:** Pale belly, unmarked yellowish throat, and bold facial pattern. **Hab:** Low forests near rivers, bogs, tundra. **Voice:** Song a short high-pitched warble; calls include a high, thin, whistled *tseeseeree.*

Subspp: (1) •*badia.*

Adult m. JAP/06

Adult f. JAP/09

Gray Wagtail 4

Motacilla cinerea L 7¾"

Casual Eurasian migrant to w. AK is.; also a vag. to CA and BC. **Shape:** Medium-sized, very long-tailed, long-legged, small-headed bird with a short, thin, fine-pointed bill. Ankles often seen below body. Deep chest creates slightly front-heavy look. Tail longer than that of other wagtails. **Ad:** Gray head and back; yellow rump, belly, and undertail coverts; black wings and tail (lacks wingbars or wing panels of other wagtails); broad white edges to tertials; thin white eyebrow. M. with yellow breast and black throat; f. with whitish throat and breast. Wags tail. **Juv:** Like ad. f. but with 2 buffy wingbars; buffy wash across breast. **Flight:** Undulating flight. White base to secondaries creates white wing-stripe above and below; outer tail feathers white. **Hab:** Open areas along streams or rivers. **Voice:** Song a rapid series of high short whistles, like *seeseeseesee;* calls include a sharp *dzeezee* and *suwit.*

Subspp: (1) •*cinerea.*

Adult m., summer, *tschutschensis* AK/06

Adult f., summer, *tschutschensis* AK/06

Adult m., summer, *tschutschensis* AK/06

Juv., *tschutschensis* AK/07

Eastern Yellow Wagtail 2

Motacilla tschutschensis L 6½"

Shape: Small, fairly long-tailed, long-legged, small-headed bird with a short, thin, fine-pointed bill. Ankles often seen below body. Deep chest creates slightly front-heavy look. **Ad. Summer:** Lemon-yellow underparts; olive to dark grayish upperparts with thin whitish wingbars. Dark grayish head with thin white eyebrow and small white arc under eye. Dark tail with outer feathers mostly white. F. similar to m. but slightly duller. Wags tail. **Ad. Winter:** Similar to summer, but duller with more white on throat and upper breast. **Juv:** (Jun.–Aug.) Dull brownish above; whitish to buffy yellow below. Buffy eyebrow; buffy arc below eye; dark malar streak may extend across upper breast, outlining whitish throat. **1st Winter:** Like pattern of ad. but washed brown above; below, off-white to washed with pale yellow; auricular brownish with dark line on lower edge. **Flight:** Undulating flight. White outer tail feathers; yellowish belly. **Hab:** Barren areas with some shrubs; stream edges, open slopes, short-grass areas. **Voice:** Song a high-pitched sibilant *tsiusiusiusiu;* calls include a buzzy *dzeeer* and a high-pitched *ts'weee.*

Subspp: (2) •*tschutschensis* (AK–YT) has medium yellow underparts, speckled sides of breast, yellow throat. •*similima* (Aleutian and Pribilof Is.) has brighter yellow underparts, no speckling on breast, white throat.

Adult m., summer, *lugens* JAP/07

Adult m., summer, *ocularis* HKO/04

Adult winter, *lugens* JAP/01

Adult m., summer, *alba* ICE/07

White Wagtail 3

Motacilla alba L 7¼"

Shape: Medium-sized, long-tailed, small-headed, long-legged bird with a short, fairly thin, fine-pointed bill. Ankles often seen below body. Deep chest creates slightly front-heavy look. **Ad. Summer:** White forecrown and face; thin dark eyeline extending from bill to dark hindcrown and nape; large black bib on breast; white belly. Tail black with white outer feathers. Back gray or black (see Subspp.); flight feathers mostly white or blackish (see Subspp.); wing coverts mostly white. Wags tail. **Ad. Winter:** Like ad. summer but black bib replaced by dark loop on breast; may have yellowish tint to face. **1st Yr:** See Subspp. descrip. below. **Flight:** Strongly undulating. Long tail with white outer feathers; ad. has mostly white or mostly dark flight feathers (see Subspp.). **Hab:** Bare ground, disturbed areas, shorelines. **Voice:** Song short well-spaced twittered phrases; calls include a short explosive *tsivit* or *tsivilit*.

Subspp: (3) Subspp. *lugens* and *ocularis* used to be considered 1 species, which was then split into 2 species, (Black-backed and White), which are now 1 species again. Intergrades occur. Identify to subspecies with caution, using multiple clues for confirmation.

M. a. lugens (AK is., vag. farther south). **Ad. Summer:** White chin, white flight feathers and upperwing greater and median coverts, slightly longer and thicker bill than *ocularis*; m. with black back, f. with gray back (may be some black on scapulars). **Ad. Winter:** Similar to ad. summer but black bib replaced by dark loop on breast, back dark gray, sometimes with black flecking in male. **1st Winter:** Generally white forehead contrasting with gray to blackish crown (some imm. f. may have gray forehead); often shows molt limits, with the worn, brownish outer greater coverts contrasting with fresh, grayer inner greater coverts. Median and greater coverts more uniformly whitish, creating effect of a pale panel (imm. *ocularis* has 2 wingbars). May have distinct yellow wash to face. **1st Summer:** F. very similar and perhaps indistinguishable from ad. summer *ocularis* except for white chin. M. like f. but with some black flecking on back. Median and greater coverts more uniformly whitish, creating effect of a pale panel (imm. *ocularis* has 2 wingbars).

M. a. ocularis (w. AK, vag. farther south). **Ad. Summer:** Gray back, black chin, dark gray flight feathers with white upperwing greater and median coverts, slightly shorter and thinner bill than *lugens*; sexes similar. **Ad. Winter:** Similar to ad. summer but black bib replaced by dark loop on breast and chin white. **1st Winter:** Like ad. winter but with gray crown and forehead; often shows molt limits, with the worn, white-tipped outer greater coverts contrasting with fresh, buffy-edged inner greater coverts. Median and greater coverts paler-tipped, creating the effect of 2 wingbars (rather than white panel of ad.). **1st Summer:** Similar to ad. summer but median and greater coverts paler-tipped, creating the effect of 2 wingbars (rather than white panel of ad.).

M. a. alba (European vag. to QB, NC, FL, and possibly NF). Differs from the other 2 subspp. in having an all-white face with no black lore or black eyeline; has 2 thin white wingbars at all ages and seasons, and no black crown in winter.

Adult HKO/02

Adult HKO/01

Adult RUS/05

Adult AK/08

Adult PHI/03

Adult PHI/03

Olive-backed Pipit 3

Anthus hodgsoni L 6"

Rare Eurasian migrant to w. AK is.; accidental to CA and NV. **Shape:** Small short-tailed bird rather pointed at both ends. Short fine-pointed bill; long shallow crown. **Ad:** Distinctive buffy lore; whitish eyebrow; small white-and-black patch at back of auricular; crown with dark streaks. Back olive-brown and faintly streaked darker; flanks and breast buffy and streaked dark; belly white; 2 thin white wing-bars. Secretive, pumps tail as it feeds on ground. **Flight:** Note short tail, streaked breast, and white belly. **Hab:** Open woodlands. **Voice:** Song a varied series of trills mixed with repeated notes; calls include a buzzy *spiz* and a short *stit*.

Subspp: (1) •*yunnanensis*.

Pechora Pipit 4

Anthus gustavi L 5½"

Casual Asian vag. to w. AK is. **Shape:** Fairly small, short-tailed bird with a short, sharp-pointed, thick-based bill and small head. Primary projection past tertials length of bill (other pipits have practically no primary extension). **Ad:** Brown above, whitish below, heavily streaked on chest and flanks; thin white streaks along either side of dark brown back; white chin, buffy breast, white belly; 2 white wing-bars. Head with streaked crown and nape; eye-ring broken in back and front; pale lower mandible; dark loral line and malar patch. **Flight:** Streaked sides, white belly, buffy breast. **Hab:** Open areas with some trees or shrubs. **Voice:** Song is mechanical trills with harsh double calls like *trrrr, chiitsa, trrrr;* calls include a short *tzip;* quieter than other pipits.

Subspp: (1)•*gustavi*.

Adult m. AK/06

Adult EUR/10

1st winter EUR/10

Red-throated Pipit 3

Anthus cervinus L 6¼"

Shape: Fairly small short-tailed bird with a short sharp-pointed bill and relatively small head. **Ad:** Unstreaked reddish-orange throat; strong streaking on back and flanks (American and Sprague's Pipits streaked on back *or* flanks)—back and rump streaked with black, buff, and white; flanks streaked blackish on pale buff background. Bill dark with lower mandible yellowish at base. M. bright reddish-orange on throat, face, and breast, with little or no streaking on breast; f. reddish-orange on throat with streaking across breast. Secretive among grasses; pumps tail as it feeds on ground. **1st Winter:** Like ad. f. but more heavily streaked overall and more contrasting face pattern; heavy dark malar patch; variable buffy wash below (esp. when fresh); no red on throat, face, or breast (Aug.–Mar.), except sometimes a few red feathers (probably m.). **Flight:** Undulating; tail dark with outermost feather on each side all white (2 outer tail feathers all white on Sprague's and American). **Hab:** Short grass, fields. **Voice:** Song a series of slow trills with varied quality and pitch, like *sisisisisisi trerererere tewtew-tewtew;* calls include a high, thin, downslurred *tsseeah* or *speee.*

Subspp: Monotypic.

Adult ND/06

Adult MT/06

Sprague's Pipit 2

Anthus spragueii L 6½"

Shape: Fairly small fairly short-tailed bird with a moderately long sharp-pointed bill and relatively small head. Similar American Pipit has longer tail. **Ad:** Strongly streaked black and whitish on back; relatively unstreaked buffy belly and flanks; fine dark streaking across upper breast. Crown streaked, but face rather plain and buffy with large eye and whitish eye-ring. Legs pale, pinkish. **Juv:** (Jun.–Aug.) Similar to ad. but back feathers blackish with buffy margins, creating scaled look. **Flight:** Undulating flight. **Hab:** Open grasslands with few or no shrubs. **Voice:** Song a series of short, high, thin whistles slightly descending at end, like *tseetseetseetsutsutsu;* calls include a *skweet* given in flight or on ground.

Subspp: Monotypic.

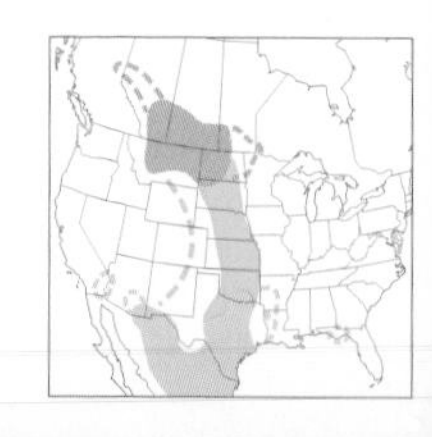

Adult summer, worn, Western Montane Grp. CO/07

Adult summer, Buff-bellied Grp. ME/05

Adult winter, Asian Grp. HKO/10

Adult winter, Buff-bellied Grp. OH/10

American Pipit 1

Anthus rubescens L 6½"

Shape: Fairly small moderately long-tailed bird with a short sharp-pointed bill and relatively small head. Similar Sprague's Pipit has shorter tail. **Ad. Summer:** Back grayish with faint or no streaking (sometimes dotted); breast and flanks buffy, heavily streaked to unstreaked (see Subspp.). Grayish auriculars and crown contrast with whitish to buffy eyebrow (see Subspp.); eyebrow can be indistinct. Outermost tail feathers white; next to outermost broadly tipped white. Legs usually dark (except subsp. *japonicus,* which can be pale). Bobs tail while feeding on ground. **Ad. Winter:** Generally slightly more streaked and less buffy than summer. **Juv:** (Jun.–Aug.) Like ad. but back streaked with black; buffy edges to back and wing feathers. **Flight:** Undulating flight; drops down to ground in stepped fashion; white on outer tail feathers. **Hab:** Short-grass or plowed fields, sandbars or mud at water's edge. **Voice:** Song a long series of steadily repeated sounds like *tseer tseer tseer tseer* given during rise and descent of flight display (delivered faster on descent); calls include a rapid *tseeseet* or *tseeseeseet* given in flight or on ground and a short *tsit.*

Subspp: (5) Divided into three groups. **Buff-bellied Group:** •*geophilus* (AK–NT), •*pacificus* (BC–AB south to cent. OR), •*rubescens* (n. AK–NF) are buffy to whitish below with dark streaking on breast and flanks, brownish-gray upperparts, dark legs. **Western Montane Group:** •*alticola* (western mts. from MT south) is pinkish buff to orangish buff to reddish below with faint or no streaking, gray to pale gray upperparts, dark legs. **Asian Group:** •*japonicus* (visitor to AK, CA) has white underparts with heavy dark streaking on breast and flanks, dark grayish-brown upperparts, heavy dark malar mark, pinkish legs.

Adult, *pallidiceps* NH/12

Adult, *pallidiceps* NH/01

Adult, *pallidiceps* NH/01

1st-yr. f., *pallidiceps* MN/01

Bohemian Waxwing 2

Bombycilla garrulus L 8¼"

Shape: Medium-sized, broad-necked, sleek-feathered bird with short legs, short tail, and a long thin crest. Cedar Waxwing smaller. **Ad:** Grayish back and belly; reddish-brown to grayish face (see Subspp.) and reddish-brown undertail coverts; closed wings with 2 white dashes (tips of primary greater coverts and outer tips of secondaries); yellowish outer tips to primary tips; variable number of small red waxy projections to secondaries. Yellow tail tip. Little or no white on forehead. Waxy projections vary by age and sex; young female has least, older male most. On f., black throat blends into gray breast; on m., black throat sharply defined from breast. **Juv:** (Jun.–Oct.) Like ad. but with grayish head; blurry dusky streaking on breast and flanks. 1st-yr. f. may have 0–5 waxy projections on wing and reduced yellow on tail; 1st-yr. m. and older birds have 4 or more waxy projections. **Flight:** Bursts of wingbeats interspersed with glides. Yellow tail tip; dark undertail; white wingbar, and white trailing edge to secondaries. **Hab:** Open coniferous or deciduous woods, often near water, in summer; fruit trees in winter. **Voice:** No song; calls include a drawn-out, dry, buzzy, high-pitched *dzeee* (slightly lower-pitched and rougher than similar *dzeee* call of Cedar Waxwing) and a clear sibilant *seee* much like Cedar Waxwing.

Subspp: (2) •*pallidiceps* (all NAM range) has reddish-brown head, dark gray flanks; •*centralasiae* (vag. to w. AK) has grayish wash on head, pale gray flanks.

Adult FL/02

Adult FL/02

1st-yr. f. FL/02

Juv. MN/10

Cedar Waxwing 1

Bombycilla cedrorum L 7"

Shape: Medium-sized, broad-necked, sleek-feathered bird with short legs, short tail, and short thin crest. Bohemian Waxwing larger with longer crest. **Ad:** Brownish-gray head, back, and breast; pale yellow belly; white undertail coverts. Wings with no white or yellow markings; variable number of small, red, waxy projections on secondary tips; thin whitish inner edge to tertials. Yellow tail tip. Thin white line across forehead. Waxy projections vary by age and sex; young f. least to older m. most; m. has more extensive black on chin. **Juv:** (Jun.–Nov.) Like ad. but head grayish with white streaks on forehead; blurry dusky streaking on underparts; whitish undertail coverts. 1st-yr. f. has 0–3 waxy projections on wing, reduced yellow on tail; 1st-yr. m. has 1–7 waxy projections. **Flight:** Bursts of wingbeats interspersed with glides. Yellow tail tip, whitish undertail, no wingbar. **Hab:** Open woods with shrubs, edges, often near water. **Voice:** No song; calls include a drawn-out, slightly buzzy, very high-pitched *dzeee* (may be slightly higher than similar *dzeee* call of Bohemian Waxwing) and a clear, sibilant, very high-pitched *seee* much like Bohemian Waxwing. 🎧**101**.

Subspp: (2) •*cedrorum* (ON–MO and east) darker overall, reddish-brown wash to head and breast; •*larifuga* (MB–KS and west) lighter overall, grayish wash to head and breast.

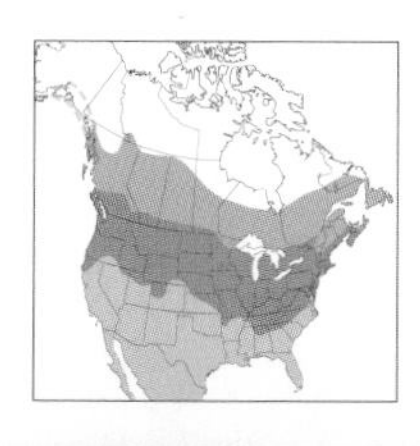

Adult m. AZ/01

Adult f. AZ/03

Adult m. CA/03

Adult f. NM/11

Phainopepla 1

Phainopepla nitens L 7¾"

Shape: Medium-sized, long-tailed, thin-billed bird with a long bushy crest. Legs short. **Ad:** M. glossy black overall with dark red eye. F. dull gray overall with thin whitish edges to all wing feathers; edges on coverts create suggestion of 2 thin wingbars; dark red eye. Often flicks tail. **Juv:** (Apr.–Jul.) Like ad. f. but slightly paler and with buffy to whitish edges to wing feathers; buffy to whitish wingbars; eye dark brown. **1st Yr:** M. may have mix of gray and glossy black feathers; f. like adult. **Flight:** Slow, buoyant, with zigzags. M. has base of primaries white, showing as large patches above and below in flight. **Hab:** Open and arid woods in canyons, on hillsides, along streambeds; near mistletoe. **Voice:** Song a varied series of well-spaced low-pitched phrases, each variably repeated (may include imitations), like *jawee, jawee, churr, churr, churr, tweetoo tweetoo;* calls include a soft rising *hoi* and a harsh *churrr.*

Subspp: (2) •*lepida* (all of NAM range) small; •*nitens* (s.w. TX, vag. NM–CA) large.

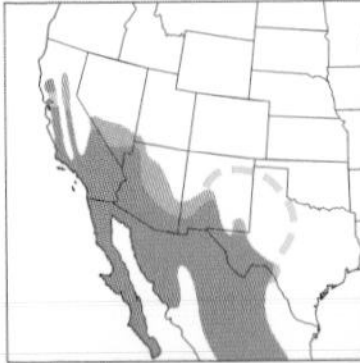

Olive Warbler

Wood-Warblers

Adult m., *arizonae* AZ/05

Adult f., *arizonae* AZ/05

Adult m., *arizonae* AZ/05

1st-summer m., *arizonae* AZ/04

Olive Warbler 2

Peucedramus taeniatus L 5"

Shape: Long thin bill with slightly curved culmen; long wings; long notched tail. **Ad:** M. has pumpkin-orange head and upper breast; black "mask" consisting of black auricular and lore; gray upperparts; 2 white wingbars and white spot at base of primaries on folded wing; yellow-olive edges to secondaries; whitish lower belly and undertail coverts; extensive white in outer tail feathers. F. similarly patterned but with duller yellow head and upper breast; mottled gray ear patch; gray nape and back; thinner wingbars; smaller white patch at base of primaries (often hidden); grayish-buff lower breast. **1st Yr:** M. duller than adult m. and with mottled black cheek; yellower head and breast (1st summer may show traces of orange on head and breast); more brownish olive above. F. similar to adult f. but with duller yellow head and breast; narrower wingbars; smaller, obscured primary patch; more brownish olive above. Similar imm. f. Hermit Warbler has shorter bill, essentially an all-yellow face, no white at base of primaries. **Flight:** Extensive white on outer tail feathers; yellow to orange hood. **Hab:** Coniferous forests above 7,000 ft.; some down-slope movement in winter. **Voice:** Loud series of 2–3-syllable notes, very similar to a titmouselike *peeta peeta peeta peeta*; call a short soft *teeu*.

Subspp: (2) •*arizonae* (s.e. AZ–s.w. NM) has slightly paler gray back and mostly gray belly and flanks; •*jaliscensis* (vag. to s.w. TX) has dark gray back, gray belly, brownish flanks.

Adult m. TX/04

Adult f. TX/04

Adult m. CT/05

Adult f. TX/04

Golden-winged Warbler 2

Vermivora chrysoptera L 4¾"

Shape: Relatively long sharply pointed bill; moderate-length tail. **Ad:** M. bright yellow crown; face pattern of black throat, large black patch through eye and cheek; gray back; yellow wing patch; gray upperparts; whitish underparts; large white spots on outer tail feathers. F. paler; yellow crown variably tinged olive; gray cheek and throat; 2 yellow wingbars (rarely, yellow wing patch); olive-gray upperparts; grayish breast sides. **1st Winter:** M. like ad. m. but with olive tinge on crown and back. F. like ad. f. but duller; olive-yellow crown; olive upperparts. **Flight:** Extensive white on outer tail feathers, mostly gray underparts. **Hab:** Shrubby early-successional habitats, swamps and bogs with forest edge. Blue-winged Warblers, tolerant of a wider variety of successional habitats, eventually replace Golden-wingeds in areas of range overlap. Golden-wingeds are declining and range has moved northward. **Voice:** Primary song several buzzy notes with the first higher than the others, like *bzeee zay zay zay.* Alternate song just like Blue-winged's alternate song, with fast variable short musical notes after and/or before buzzy notes. Call like Blue-winged's, a sharp *tsik.*

Subspp: Monotypic. **Hybrids:** Blue-winged Warbler. See Blue-winged Warbler for details.

Adult m. OH/04

Adult f. TX/04

Adult m. OH/05

1st summer TX/04

1st-winter f. MI/09

Blue-winged Warbler 1

Vermivora cyanoptera L $4^{3}/_{4}$"

Shape: Relatively long sharply pointed bill; moderate-length tail. **Ad:** M. buttery-yellow crown contrasts with short striking black eyeline; wings bluish gray with 2 white or yellow-tinged wingbars; olive-yellow back and rump; yellow underparts contrast with white undertail coverts; large white spots on outer tail feathers. F. with variable olive tinge on crown, duller eyeline, and thinner wingbars; duller yellow below. All ages and sexes have pinkish-brown bill with dusky culmen in winter, all-dark bill in summer. Similar Yellow and Prothonotary Warblers lack dark eyeline and white wingbars. **1st Summer:** Similar to respective adults but slightly duller. **1st Winter:** Similar to adult f. but duller and crown and forehead with more olive; m. has blackish lore, yellowish forehead, olive crown; f. has dark gray lore and olive forehead and crown. **Flight:** Extensive white on outer tail feathers; yellow underparts contrast with white undertail and undertail coverts. **Hab:** Varied early-successional habitats such as second-growth forest and open areas with extensive brushy undergrowth. **Voice:** Primary song 2 buzzy low-pitched notes, the 2nd lower, like *bzeee bzay;* almost sounds like words "blue-winged." Alternate song has buzzy notes preceded and/or followed by variable, fast, short musical notes (similar to Golden-winged's alternate song). Call a sharp *tsik.*

"Brewster's Warbler" CT/06

Blue-winged × Golden-winged hybrid TX/05

"Brewster's Warbler" TX/04

"Lawrence's Warbler" OH/05

Subspp: Monotypic. **Hybrids:** Golden-winged and Kentucky Warblers. Hybridization with Golden-winged Warbler results in offspring with varying traits of both species. These hybrids can then breed with either species, producing many intermediate plumage variations. First-generation hybrids (Golden-winged x Blue-winged) usually show the dominant characteristic of the thin black eyeline (from the Blue-winged) and have whitish underparts variably washed with yellow. These hybrids, called "Brewster's Warblers," are the more common of the two main hybrid types. Second-generation "Brewster's" can have white underparts and yellow wingbars. A certain segment of the population of Blue-winged and Golden-winged Warblers carries some of the mtDNA (mitochondrial DNA) of the other species. Backcrosses of these "impure" individuals with "Brewster's Warbler" can show the recessive characteristics of yellow underparts (from the Blue-winged) and black auricular and throat (from the Golden-winged). These backcrosses, called "Lawrence's Warblers," are rarer than "Brewster's"; they can look a little like a Golden-winged Warbler with a yellow belly. Hybrids almost always mate with adults of either species (not with other visible hybrids), and many variations occur. Sexing hybrids is difficult due to this variation, but females are generally duller in color and have less contrast in their plumage overall. Hybrids use the song of either Blue-winged or Golden-winged Warblers or may also use a combination of elements from the songs of both species.

Adult m., summer TX/04

Adult f., summer TX/04

Adult m., summer TX/05

Adult f., summer TX/05

Adult m., winter NY/09

Adult winter f. or 1st winter MI/09

Tennessee Warbler 1

Oreothlypis peregrina L 4¾"

Shape: Relatively short-tailed large-headed warbler. **Ad. Summer:** M. with pale gray crown; dull gray face with white eyebrow; dark eyeline; bright olive-green upperparts; whitish underparts with distinctive white or whitish undertail coverts. A very small portion of males have a few orange crown feathers, not visible in field. F. similar but olive wash on crown; dull eyeline; light yellow on eyebrow, throat, and underparts (can be extensive or limited); white undertail coverts. **Ad. Winter:** M. like summer but crown and sides of neck with olive mottling; variable light yellow wash on underparts; flanks grayish or pale olive. F. crown greenish; eyebrow washed yellow; dull eyeline; upperparts olive-green; strong yellowish wash on underparts; undertail coverts white, rarely tinged yellow. Birds in fall can have faint, thin, yellowish wingbars and pale tips to primaries. **1st Winter:** Like adult winter f. but often with heavier yellow wash on underparts. Some winter adults and 1st-winter birds cannot be reliably aged in the field. Similar winter Orange-crowned Warbler has split eye-ring, yellow (not white) undertail coverts as brightest part of underparts, indistinct breast and flank streaking (sometimes hard to see), and longer tail; pale patch at bend of wing, when visible, a good clue. Similar Philadelphia and Warbling Vireos larger, slower, heavier-billed; Philadelphia Vireo has yellowish undertail coverts. **Flight:** Short-tailed; long-winged; whitish undertail coverts contrast with pale or yellowish underparts. **Hab:** Boreal woodlands with areas of shrubs and small deciduous trees, spruce bogs. **Voice:** 2–3-part trill, the last part fastest, reminiscent of a car trying to start—*tsit, tsit, tsit, tsit, tsut tsut tsut teeteeteeteetee.* Call a sweet *chip.*

Subspp: Monotypic. **Hybrids:** Nashville Warbler.

Adult m., summer, *ruficapilla* NH/05

Adult summer, *ruficapilla* OH/05

Adult m., summer, *ridgwayi* CA/04

Winter m., *ruficapilla* NY/09

Summer f. TX/05

Winter f., *ruficapilla* OH/10

Nashville Warbler 1

Oreothlypis ruficapilla L 4½"

Shape: Small short-tailed warbler. **Ad. Summer:** M. with gray head; conspicuous white eye-ring; rufous crown patch (usually hidden); olive-green back, wings; bright yellow throat, breast; whitish on vent area separates yellow of underparts from yellow undertail coverts; greenish edges to flight feathers. F. with muted grayish head, duller yellow below; small, mostly concealed rufous crown patch small and indistinct. Similar Orange-crowned Warbler duller, with split eye-ring, eyeline; Virginia's Warbler has less yellow below and never yellow on throat; Connecticut Warbler is larger, with different shape and behavior, duller underparts; all lack Nashville's greenish-edged flight feathers. **Ad. Winter:** Both sexes duller overall than in summer. M. with crown patch more concealed, grayer-tinged back, and more olive wash to underparts. F. with grayer back, brown wash to crown, and underparts only mildly olive. **1st Winter:** Similar to but duller overall than winter adults. M. with grayish-olive head; whitish or buffy eye-ring; rufous crown patch (can be absent); paler yellow underparts; belly may be white; yellow undertail coverts; wing and tail feathers browner. F. dullest of all with no crown patch; grayish-olive head; buffy eye-ring; variable pale yellow below; some with much white in lower belly and vent. **Flight:** Grayish back with greenish rump and wings. **Hab:** Dry or wet open woodlands, coniferous forests, edges of swamps or spruce bogs, successional habitats. **Voice:** Series of 2-syllable notes, followed by shorter lower trill, like *see-it see-it see-it, titititi.* Subsp. *ridgwayi* has less distinct trill. Call a metallic *pink.*

Subspp: (2) •*ridgwayi* (n.w. MT–s.cent. CA and west) has brighter yellow breast; brighter olive-yellow rump contrasts with more grayish back, more white on belly; pumps tail much more than *ruficapilla.* •*ruficapilla* (MB–NF and south) has duller breast and rump, less white on belly. **Hybrids:** Tennessee and Black-throated Blue Warblers.

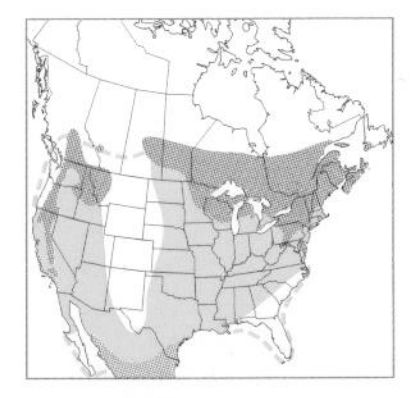

Adult summer, *celata* MB/06

Adult summer, *lutescens* CA/04

Adult summer, *celata* MB/06

Adult summer, *lutescens* CA/04

Orange-crowned Warbler 1

Oreothlypis celata L 4¾"

Shape: Relatively long tail projects well past undertail coverts; bill thin, sharply pointed, slightly decurved. **Ad. Summer:** Eastern subsp. is one of our dullest unmarked warblers (western subspp. are brighter yellow). Yellow undertail coverts are most distinctive part of bird on all ages and sexes, can be very pale in palest birds. M. has split whitish or yellowish eye-ring; faint eyebrow; short dark eyeline; plain dusky olive upperparts; small whitish mark sometimes visible at bend in wing; dull yellow to olive underparts; blurry streaking on breast sides, sometimes additionally on underparts; orange crown patch usually concealed by olive feather tips (may be visible in threat or alarm displays or after bathing). F. similar, but duller above with less yellow below; split eye-ring pale yellow or whitish. Similar Yellow and Wilson's Warblers are distinguished from subsp. *lutescens* by plainer faces, noticeable dark eye; Philadelphia Vireo larger, slower moving, has thicker bill. **Ad. Winter:** Gray or grayish-brown tinge on upperparts (especially on subsp. *celata*); duller below. Similar winter Tennessee Warbler has brighter green back, no underpart streaking,

Winter, *celata* MI/09

Winter CA/09

Winter, *celata* MI/09

Winter, *orestra* TX/01

Winter TX/01

no split eye-ring, distinctive white (not yellow) undertail coverts, shorter tail, straighter bill. **1st Winter:** F. dullest of all plumages; underparts pale grayish with faint yellow wash. **Flight:** Yellow undertail coverts contrast with dull underparts in duller subspp. **Hab:** Woodland groves with dense undergrowth, thickets, forest edges, brushy areas, chaparral. **Voice:** High-pitched trill that trails off and usually drops in pitch; call a metallic sharp *chet.* **102**.

Subspp: (4) •*celata* (AK–cent. AB and east) dullest subsp.; grayish to olive upperparts; underparts grayish to dull olive-yellow. •*lutescens* (s.e. AK–s. CA) brightest subsp.; upperparts and underparts bright yellowish olive. •*orestra* (s. YT–e. CA–s.w. TX) largest subsp.; intermediate in brightness between *celata* and *lutescens*. •*sordida* (Channel Is., CA) somewhat duller than *lutescens*.

Adult m. NM/06

Adult m. CA/05

Female CO/08

Adult TX/05

Adult MEX/12

Virginia's Warbler 1

Oreothlypis virginiae L 4½"

Shape: Fairly long-billed and long-tailed warbler. **Ad:** M. has gray upperparts; bold white eye-ring; rufous crown patch (can be concealed); yellow breast patch; whitish or pale grayish underparts; yellow undertail coverts and olive-yellow rump. F. similar, with paler yellow breast patch, smaller crown patch. Bobs tail. Similar Nashville Warbler more yellow below; always some yellow on throat and olive on back and wings. Similar Colima Warbler larger with brownish flanks and tawny undertail coverts. **1st Winter:** M. less yellow on breast; brownish wash above. F. reduced or no yellow on breast, no crown patch; brownish tinge on grayish upperparts. **Flight:** Grayish above with olive-yellow rump. **Hab:** Dry woodlands, canyons, brushlands, usually at 4,000–9,000 ft. **Voice:** Song a series of 1–2-part sounds, rising or on same pitch, like *sooweet sooweet sooweet, sweet, sweet;* call a metallic *pink*.

Subspp: Monotypic.

Colima Warbler 2

Oreothlypis crissalis L 5¼"

Shape: The largest and longest-tailed warbler in the genus *Vermivora*. **Ad:** Gray head with white eye-ring; rufous crown patch (often concealed); brown wash on back and flanks; orangish to golden undertail coverts; dull olive rump. Sexes similar. Similar Virginia's Warbler is smaller with shorter bill and tail, has no brown wash on flanks, undertail coverts yellow, bobs tail. **Flight:** Large warbler; grayish overall with orangish to golden undertail coverts. **Hab:** Breeds locally in canyons in Chisos Mts. TX, from about 6,000 to 7,700 ft. **Voice:** Musical trill, sometimes with last 2 notes downslurred.

Subspp: Monotypic.

Adult m. AZ/04

Adult f. AZ/05

1st winter AZ/09

Adult, *sodalis* AZ/02

Adult, *sodalis* AZ/02

Lucy's Warbler 1

Oreothlypis luciae L 4¼"

Shape: Small warbler; moderate-length bill and tail. **Ad:** M. has pale gray upperparts with paler face; dark eyes; white lores; eye-ring indistinct; dark chestnut crown patch and uppertail coverts (not always visible); whitish underparts (washed with buff in winter). F. similar with paler uppertail coverts; crown patch smaller and mostly obscured. Bobs tail. Similar Bell's Vireo has wingbars, heavier bill, longer tail. Similar juv. Verdin has pale base to lower mandible, longer tail. **1st Winter:** Similar to juv. but no buffy wingbars. **Juv:** (May–Jul.) Similar to adult f. but brownish-gray back, buff-tinged underparts, buffy wingbars, and pale reddish-brown rump. Similar Colima and Virginia's Warblers have no chestnut rump; Virginia's has yellow undertail coverts; imm. f. Yellow Warblers of dullest subspp. have olive-yellow edges to wings and pale yellow tail spots. **Flight:** Gray overall, reddish-brown rump. **Hab:** Woodland and streamside mesquite and cottonwood areas of arid Southwest. **Voice:** Two or 3 rapid series of sweet notes on different pitches; call a sharp *pink*.

Subspp: Monotypic.

Crescent-chested Warbler 4

Oreothlypis superciliosa L 4¼"

Casual Mexican vag. to s. TX and s. AZ. **Shape:** Small warbler with a broad neck and fairly long tail; appears somewhat large-headed. **Ad:** M. has blue-gray head; white eyebrow; small white crescent below eye; greenish back; yellow throat; yellow breast with crescent-shaped chestnut mark; gray wings and tail; whitish belly and undertail coverts. F. slightly paler overall with smaller chestnut mark on chest. Similar Tropical and Northern Parulas have white wingbars, lack white eyebrow; similar Rufous-capped Warbler has rufous crown, no chest mark. **1st Yr:** Like adult f. but may have reduced or no chestnut mark on chest. **Flight:** Strong white eyeline, mostly yellow below. **Hab:** Highland forests. **Voice:** Song is a buzzy trill; call is *sik*.

Subspp: (2) •*sodalis* (vag. to AZ and TX) paler overall with less contrast; •*mexicana* (possible vag. to s. TX) more brightly colored overall with more contrast.

Adult m. TX/05

Adult m., *americana* CT/05

Adult f. OH/05

1st-summer f. TX/04

1st-winter m. OH/09

1st-winter f. OH/09

Northern Parula 1

Parula americana L 4¼"

Shape: Small short-tailed warbler with a fairly long bill; appears somewhat large-headed. **Ad:** M. throat and breast yellow, with variable black and chestnut breastbands (slightly veiled in winter); head and upperparts bluish gray except for yellowish-green upper back; short white eye crescents, black lores; black upper mandible, yellow lower; 2 prominent white wingbars; white lower belly and undertail coverts. F. similar with chestbands paler or absent or just a chestnut wash on breast; lore gray. Similar Tropical Parula lacks white eye crescents, defined dark chestbands, is more extensive yellow below. Similar imm. f. Magnolia Warbler has complete eye-ring, distinct tail pattern where large white spots form partial band, more extensive yellow below. **1st Winter:** Similar to adult f. but paler and with greenish-tinged upperparts; 1st-winter m. not distinguishable from adult winter f.; f. lacks breastbands and is strongly washed with green on upperparts. **Flight:** Very small; mostly gray with yellow throat and breast. **Hab:** Forests with Spanish moss (in South) and *Usnea* lichens (in North and West); these are used in their nests. **Voice:** Primary song of subsp. *americana* is a buzzy ascending trill that ends with an abruptly lower note, *zeeeeeeeee-yup;* subsp. *ludoviciana* primary song ends with less intense, buzzier, slurred note. Alternate song for both a series of complex buzzy notes; call a musical, sweet, sharp *djip.* 🎧**103**.

Subspp: (2) Mainly based on differences in primary song (see Voice). •*ludoviciana* (s.w. ON–MI south to e. TX–s.w. AL) is small with short bill; m. underparts with more extensive breastbands and brighter yellow. •*americana* (s.e. ON–PE south to s.cent. AL–FL) has larger bill; m. underparts with less extensive breastbands and duller yellow. **Hybrids:** Yellow-rumped and Yellow-throated Warblers; the latter results in so-called "Sutton's Warbler," which resembles Yellow-throated but with Northern Parula's greenish back and less streaking on flanks.

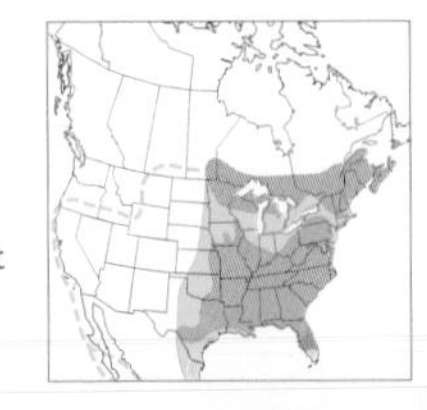

Adult m., *nigralora* TX/04

Adult m., *pulchra* MEX/01

Tropical Parula 3

Parula pitiayumi L 4¼"

Rare local breeder in s. TX; vag. to AZ, LA, MS, and CO. **Shape:** Small short-tailed warbler with a fairly long bill; appears somewhat large-headed. **Ad:** M. has bright yellow throat and breast, with variable orange wash across breast; black lores and variable black on foreface; 2 bold white wingbars; blue-gray upperparts with yellowish-green patch on back; white undertail coverts. F. has yellow throat and breast, with light or no wash of orange; gray lores; no black on face. Similar Northern Parula has bold white eye crescents; yellow on underparts extends only to upper belly; blue-gray of face comes more onto sides of throat; males have chestnut and black breastbands. **1st Yr:** (Aug.–Jul.) Like ad. f. but duller; upperparts washed with olive-green; little or no orangish wash on breast. **Flight:** Yellow throat, breast, and upper belly; dark upperparts. **Hab:** Open stands of live oak with *Tillandsia* bromeliads, which are used in nest construction. **Voice:** Song a buzzy rising trill ending with a separate buzzy note, similar to the ending note of the song of Northern Parula subsp. *ludoviciana*. Also has complex variable alternate song. Call note a weak *chik*.

Subspp: (2) •*pulchra* (vag. to s. AZ) is larger, has wider white wingbars on greater coverts, deeper color on underparts. •*nigralora* (s. TX) is smaller, has thinner wingbars, paler colors below.

IDENTIFICATION TIPS

Warblers are small and generally colorful birds of woodland habitats. Most have a fine thin bill used to glean insects off leaves and branches; a few species catch aerial insects; some feed mainly on the ground. Many are neotropical migrants and seen by bird-watchers most often in spring and fall, when their arrival at various warbler hot spots is highly anticipated.

Species ID: Warblers are often considered difficult to identify, but several approaches can help. First, over half our warbler species have roughly one plumage all year; male, female, and immatures look roughly the same and do not change with the seasons. In some species, females and immatures may have slightly less color and contrast than adult males; the drabbest are 1st-winter females. The difficulty with some of these species is that they look very similar to each other. This is certainly true of Northern and Louisiana Waterthrushes.

In about a third of our warblers, adult males and females look substantially different from each other but neither sex changes with the seasons. In some cases, the adult female is a muted version of the adult male (with less color and contrast), and immatures of both sexes may look much like a duller version of the adult female. Again, the 1st-winter females are the drabbest.

The final group is the most complex, and comprises six main species: Bay-breasted, Blackpoll, Cape May, Chestnut-sided, Magnolia, and Yellow-rumped Warblers. In these, adult male and female look substantially different and each has a different plumage in summer and winter. Thus there are four adult plumages for each species, plus immature plumages as well. Immatures in these species look much like the adult female, and 1st-winter males may be difficult (or impossible) to distinguish in the field from adult winter females. Here too, 1st-winter females are the drabbest, and they sometimes lack most of the distinguishing characteristics of the adults.

Warblers' plumages have traditionally been called "fall" and "spring" as opposed to the terms "winter" (or nonbreeding) and "summer" (or breeding). In fact, their "fall" plumage is kept all winter and their "spring" plumage all summer, just as with other birds. Because of this, we have used the terms "winter" and "summer" for warbler plumages, as we have for the plumages of all other birds.

Adult m., Northern Grp. TX/04

1st winter, Northern Grp. NH/09

Adult f., Northern Grp. TX/04

1st winter, Northern Grp. MI/08

Adult f., Northern Grp. MI/05

Adult m., Mangrove Grp. TX/05

Yellow Warbler 1

Dendroica petechia L 4¾"

Shape: Medium-sized compact warbler with long undertail coverts and fairly short tail. **Ad:** All ages and sexes have pale yellow edges to wing feathers and distinctive yellow tail spots (only *Dendroica* to have yellow tail spots). Often bobs tail. M. all yellow; pale eye-ring; variable chestnut streaks on underparts (typically less prominent in winter) and sometimes chestnut on the head (see Subspp.); black eye stands out against yellow face. F. similar but duller with thinner or no chestnut streaking below. Similar Orange-crowned (subsp. *lutescens*), imm. Hooded, and imm. Wilson's Warblers all lack yellow tail spots. **1st Winter:** Similar to ad. f. but paler with more greenish-yellow crown and face; underparts dull yellowish to almost all grayish with limited chestnut streaking or none; base of lower mandible pinkish. Broad overlap between the sexes and subspecies variation make sexing difficult. **Flight:** Yellow underwing and yellow tail spots. **Hab:** Shrubby areas, wet thickets, especially with willows. **Voice:** A musical series of notes with accent on the last several notes, sounding like *sweet, sweet, sweet, sweeter-than-sweet*. Alternate song is unaccented, longer, and variable and can sound very similar to the alternate songs of Chestnut-sided and Magnolia Warblers and American Redstart. Call a variable sweet *tchip*. **104**.

Subspp: (10) Divided into three groups. **Northern Group:** Male with no reddish brown on crown. •*rubiginosa* (s. AK–s.w. BC), •*brewsteri* (w. WA–w. CA), •*banksi* (cent. AK–YT), •*morcomi* (s. AK–s.cent. CA–n. TX), •*sonorana* (s. NV–s.e. CA–s.w. TX), •*parkesi* (n. AK–n. ON), •*amnicola* (e. YT–n.e. BC–NF), •*aestiva* (s.e. AB–s.cent. OK to PE–SC). **Mangrove Group:** Male with reddish-brown wash on head. •*oraria* (n. MEX–s. TX). **Golden Group:** Male with broader reddish streaking on underparts and sometimes with reddish-brown wash on crown. •*gundlachi* (s.w. FL is.).

Adult m., summer MI/05

1st-winter m. OH/09

Adult f., summer NY/05

1st-winter f. OH/09

Chestnut-sided Warbler 1

Dendroica pensylvanica L 5"

Shape: Relatively stocky warbler with a medium-length tail (usually held up at an angle) and relatively short bill. **Ad. Summer:** M. has bright lemon-yellow crown contrasting with blackish nape; black lores and malar bar; white cheeks; underparts white with considerable chestnut on sides of breast and flanks; 2 yellowish wingbars; streaked back. F. has greenish-yellow crown and nape; less chestnut on sides; lores and malar bar are slaty, less extensive than male's; greenish upperparts with some black streaking. Often moves about with tail cocked. **Ad. Winter:** Lime-green crown and upperparts; gray face; bold white eye-ring; yellowish wingbars; whitish underparts; less chestnut on sides of underparts than in spring, but amount of chestnut ranges from the most on ad. m. to some or none on ad. female. **1st Winter:** M. can be indistinguishable in field from ad. winter female; can have some to no chestnut on sides of underparts; has less extensive streaking on upperparts than ad. male. F. with indistinct streaking on back and no chestnut on flanks. **Flight:** Chestnut flanks and extensive white in outer tail feathers. **Hab:** Shrubby undergrowth in second-growth deciduous woodlands, shrublands. **Voice:** Primary song is several pairs of notes followed by 2 emphatic end notes, "pleased, pleased, pleased to meet you." Alternate song type lacks emphatic ending and can be similar to Yellow Warbler's unaccented song. Call a musical sweet *chip*. **105**.

Subspp: Monotypic.

Adult m., summer TX/04

1st-summer m. ME/05

1st-summer m. OH/05

1st-summer f. TX/05

Adult f., summer TX/05

Adult winter OH/09

Magnolia Warbler 1

Dendroica magnolia L $4^{3}/_{4}$"

Shape: Relatively short-tailed and short-billed warbler. **Ad. Summer:** M. with gray crown, black face mask, and bold white eyebrow behind eye; underparts bright yellow with heavy black streaking, sometimes forming a necklacelike band on upper breast; broad white wingbars joined as patch; black back; yellow rump; white undertail coverts; unique undertail pattern of white tail with dark tip; uppertail has 2 white side patches almost forming a band. Often rapidly fans tail. F. has dark gray mask; olive back with black streaks; less black streaking below; 2 white wingbars. **Ad. Winter:** M. head gray with thin, complete, white eye-ring; hint of eyebrow; no black on head; can have hint of pale gray band across upper breast; narrow white wingbars; black spots on back; less streaking below. Ad. f. similar to winter m., with olive on crown; less streaking below. **1st Summer:** M. similar to adult m., less extensive black on back. F. has white eye-ring; gray face may have hint of mask; almost no eyebrow; reduced streaking on underparts. **1st Winter:** Similar to winter ad. f. but duller and with pale gray band across upper chest; sexes nearly identical; f. may have very faint streaking on breast and flanks and no black spots on back. Similar imm. Prairie Warbler lacks eye-ring, grayish breastband, and yellow rump; has white tail spots; wags tail more. 1st-summer f. Magnolia Warbler may resemble f. Kirtland's Warbler, but Kirtland's lacks eye-ring, yellow rump, and

Adult m., winter MI/09

1st winter OH/09

Undertail OH/09

Magnolia's tail pattern; Kirtland's is also larger and bobs tail. **Flight:** White band across midtail; yellow rump. **Hab:** Dense stands of young spruce, fir, hemlock, and pine in coniferous forests or woodlands and forest edges. **Voice:** Song is variable short series of musical notes, *weety, weety, wee-chee;* also has short unaccented version of song. Call a long nasal *clenk.* 🎧**106**.

Subspp: Monotypic.

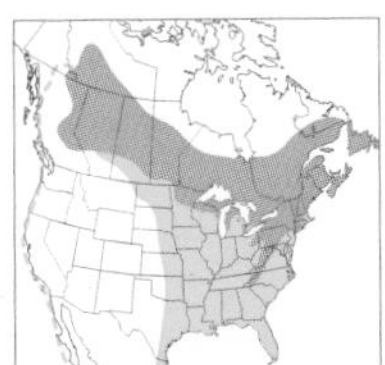

Adult m. TX/04

Adult f. TX/04

Golden-cheeked Warbler 2

Dendroica chrysoparia L 4¾"

Shape: Fairly large-headed warbler with moderate-length bill and tail. **Ad:** M. has golden-yellow face with black eyeline extending from bill through eye to nape; black crown, throat, upperparts, and upper breast; heavy blackish streaks on flanks; white belly and undertail coverts. F. similar but paler than m., with streaked olive-green crown and upperparts; yellow of upper chin mixes with black on upper throat; black lower throat and upper chest. Similar Black-throated Green Warbler has yellow face with olive cheek, yellow patch on vent. **1st Winter:** M. similar to adult female. F. dullest of all, but eyeline, which connects to nape, still distinct on yellow face; very pale yellow to whitish throat; smudge of mottling on sides of breast; little streaking on sides. **Flight:** Extensive white in outer tail feathers; golden cheek. **Hab:** Mixed woodlands with the presence of Ashe Juniper. **Voice:** Song is a variable, buzzy, whistled *zee dee sidee-zee,* harsher than Black-throated Green Warbler's. Call a medium loud *chup.*

Subspp: Monotypic.

Adult m., summer MB/06

Adult f., summer FL/05

Adult m., summer MI/05

Adult m., winter MI/09

Cape May Warbler 1

Dendroica tigrina L 4¾"

Shape: A short-tailed warbler with a thin, finely pointed, slightly downcurved bill. **Ad. Summer:** M. has bright chestnut-orange ear patch surrounded behind and below by bright yellow; black eyeline; yellow to chestnut eyebrow; yellow underparts with heavy black streaking on breast and flanks; extensive white on median and greater coverts, creating a white wing patch; yellow rump; white undertail coverts; white tail spots; all plumages have at least a hint or more of yellow on side of neck behind ear patch and greenish edges to flight feathers. F. duller than m., with grayish ear patch surrounded by paler yellow; yellow below with thin black streaking; thin white wingbars; olive-yellow rump. **Ad. Winter:** Similar to summer ad. but duller. M. with reduced chestnut on ear patch, less solid white on median coverts; f. with pale yellow on neck and underparts, broken white on median coverts. **1st Summer:** M. like summer m. but crown more olive with black streaks; gray eyeline; some olive in the yellow on sides of neck; smaller white wing patch. F. duller than summer f. with more white on throat and breast. **1st Winter:** M. similar to adult winter f. but usually brighter; ear patch usually lacks chestnut in cheek. F. drabbest of all; grayish above with pale gray streaking below; little to almost no yellow on sides of neck, behind cheek, and on breast; olive-yellow rump. Similar dull winter

1st-summer m. OH/05

1st-winter f. OH/09

1st-winter f. MI/08

1st-winter f. OH/09

"Myrtle" Yellow-rumped Warbler has bright yellow rump; winter Palm Warbler has less streaking below, yellow undertail coverts, pumps tail; imm f. Pine Warbler is larger, more brown above, has no olive-yellow rump. **Flight:** Ad. m. has yellow rump and white wing covert patch. **Hab:** Boreal coniferous forests with spruce, where its populations rise or fall in accordance with rise and fall of spruce budworm outbreaks. **Voice:** Song a high-pitched *seet seet seet seet*. Alternate song is somewhat lower-pitched slurred notes and sounds similar to Bay-breasted Warbler's song. Call a high thin *seet*.

Subspp: Monotypic. **Hybrids:** Blackpoll Warbler.

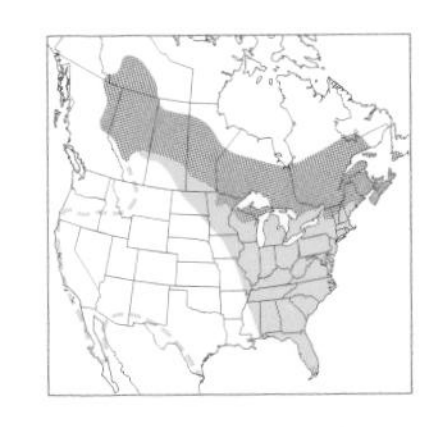

Adult m., summer MI/06

Adult m., summer OH/05

Adult f., summer OH/05

1st-summer m. NH/06

Winter f. NH/08

1st-summer f. MI/05

Black-throated Blue Warbler 1

Dendroica caerulescens L 5"

Shape: Somewhat large-headed broad-necked warbler. **Ad. Summer:** Almost all plumages (except sometimes 1st-summer and 1st-winter f.) have a white patch at base of primaries, creating the look of a "pocket handkerchief." M. with black face, throat, and flanks; upperparts dark blue to blackish blue; underparts white; white spots on outer tail feathers. F. grayish or brownish olive above; thin white eyebrow and lower eye arc stand out on grayish-olive face; buffy to grayish-buff underparts. **Ad. Winter:** M. similar to summer with more olive on back. F. more buff below. **1st Summer:** M. very similar to summer m. but duller flight feathers, slight brownish in upperparts. F. like adult summer f. with small or no white wing patch. **1st Winter:** Similar to respective adults, but m. feathers in black areas tipped whitish and some individuals have olive wash to blue upperparts. F. more brownish above; faint yellowish eyebrow; yellowish buff below; pocket handkerchief smaller or absent. **Flight:** Ad. has white patch at base of primaries; on m. white underparts and underwings contrast with black throat. **Hab:** Upland, mixed woodlands with dense understory. **Voice:** Song is 3–5 buzzy notes, with the last usually upslurred, like *zoo zoo zoo zeeee*. Call a hard *chik* or *kik*. 🎧**107**.

Subspp: Monotypic. **Hybrids:** Nashville Warbler.

Adult m., *halseii* NM/06

Adult f. CA/04

1st-summer f. CA/04

1st-summer m. CA/04

Black-throated Gray Warbler 1

Dendroica nigrescens L 4¾"

Shape: Medium-sized warbler with a moderate-length tail, broad neck, and fairly thick bill. **Ad:** All plumages have 2 white wingbars; extensive white on outer tail feathers; white undertail coverts; and variable yellow supraloral spot. M. has dramatic black-and-white head pattern with small yellow supraloral dot in front of eye; black throat; gray upperparts with dark streaks; 2 white wingbars; white underparts with dark streaking on sides. F. variable; gray auriculars; short black streak through gray central crown; sides of crown blackish; white chin and throat with black band across lower throat, may have black mottling on throat. Similar Black-and-white Warbler has black crown with central white stripe, black-spotted undertail coverts; similar summer m. Blackpoll Warbler has white cheek. **1st Winter:** M. similar to adult m. but finer streaks on back, mottled black on throat, wings brownish, some gray streaking on central crown. F. has white throat and breast; upperparts more brownish gray; buffy flanks with faint streaks. **Flight:** Mostly gray above; extensive white on outer tail feathers; striking head pattern. **Hab:** Coniferous, mixed, oak, and semiarid woodlands. **Voice:** Song a series of variable buzzy notes with last note emphatic, higher, like *weezy, weezy, weezy, wheeet.* Call a nonmusical sharp *tup.*

Subspp: (2) •*nigrescens* (s.w. BC–w.cent. CA) averages smaller, has slight brown wash on upperparts in fall, often less white on tail; •*halseii* (s. BC–WY south to s.cent. CA–s.w. TX) averages larger, has gray upperparts, often more white on tail. **Hybrids:** Townsend's Warbler.

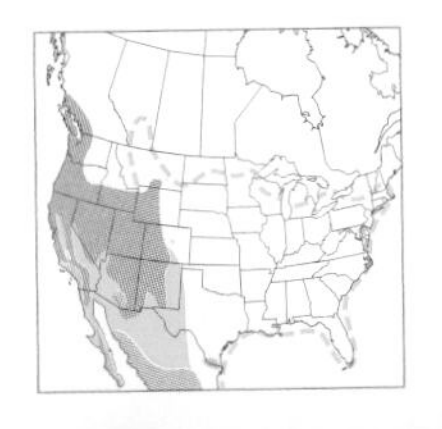

Adult m., summer, *coronata* OH/05

Winter, *coronata* OH/10

Adult f., summer, *coronata* OH/05

Winter, *coronata* FL/03

Adult f., summer, *coronata* OH/05

1st-winter f., *coronata* FL/03

Yellow-rumped Warbler 1

Dendroica coronata L 5¼"

One of the most abundant warblers seen on migration. This species was formerly considered 2 full species—"Myrtle Warbler" and "Audubon's Warbler." These are now considered subspecies of the Yellow-rumped Warbler. The 2 subspp. are distinct (although intergrades occur) and are described separately here.

"Myrtle Warbler"

Dendroica coronata coronata

Shape: Fairly large warbler with a long tail, long wings, and moderate-length bill. **Ad. Summer:** Both sexes have yellow rump and yellow patches on either side of the breast; they also have a whitish throat that extends below lower edge of auricular and white outer tail spots. M. has black mask; white crescent below eye; thin white eyebrow; yellow crown patch on gray crown; white wingbars; heavy black streaks on breast, sometimes forming a patch. F. similar to male, but with brown back; reduced crown patch; dark gray or brownish mask; less extensive streaking on underparts. **Ad. Winter:** Similar to summer adults, but browner and duller overall with fine brownish streaking on breast and flanks; mask brownish; yellow breast patches reduced. **1st Winter:** Similar to ad. winter. 1st-winter m. and winter f. practically indistinguishable. 1st-winter female can be very drab, with yellow on breast sides faint or absent. **Flight:** Note bright yellow rump, relatively large size. **Hab:** Coniferous or deciduous forests. **Voice:** Song a weak, variable, musical trill; call a sharp loud *check*. 🎧**108**.

"Audubon's Warbler"

Dendroica coronata auduboni

Shape: Fairly large warbler with a long tail, long wings, and moderate-length bill. **Ad. Summer:** Both sexes have yellow rump and yellow patches on either side of the breast; white outer tail spots. M. has yellow throat; white eye crescents above and below eye; plain face with little or no white eyebrow; yellow crown patch; black on breast and black streaking on flanks; 2 white wingbars and white edges of

Adult m., summer, *auduboni* AZ/05

Adult m., winter, *auduboni* CA/10

Adult f., summer, *auduboni* CA/04

Adult winter f. or 1st winter, *auduboni* CA/10

Adult f., summer, *auduboni* CA/04

greater coverts create white wing patch. F. similar to m., but with gray face; reduced crown patch; bright to pale yellow throat; brownish gray above. **Ad. Winter:** Similar to summer adults, but paler and browner overall with dusky streaking on breast and flanks; yellow patches on breast sides often reduced. **1st Winter:** M. and adult winter f. indistinguishable; 1st-winter f. has pale yellow to buffy-whitish throat; yellow side patches faint or nearly absent; buffy-brown breast with blurry streaking. **Flight:** Note bright yellow rump, relatively large size. **Hab:** Coniferous or deciduous forests. **Voice:** Song a slow, weak, variable trill. Call is like *chent*.

Similar Species: Similar winter Palm Warbler has yellowish rump but bright yellow undertail coverts and pumps its tail; similar winter Magnolia Warbler has yellow rump but is yellow below; similar winter Cape May Warbler has greenish-yellow rump and is smaller; similar imm. f. Pine Warbler lacks yellow rump.

Subspp: (2) •*coronata* (w. AK–n. BC to NF–MA, south to WV in Appalachians), "Myrtle Warbler"; •*auduboni* (s.e. AK–s.w. SK to s. CA–w. TX), "Audubon's Warbler." See descriptions above. **Hybrids:** Northern Parula and Grace's, Townsend's, Pine, and Bay-breasted Warblers.

IDENTIFICATION TIPS

Distinguishing Myrtle and Audubon's Warblers in Winter Plumage: Myrtles show at least a hint of an auricular patch and have a thin white eyebrow and whitish throat that extends up behind the auricular. Audubon's throat is straight-sided and variably yellowish, can be whitish. Call notes differ. Intergrades occur.

Adult m., summer, *virens* CT/05

1st-summer f., *virens* OH/05

Adult m. TX/04

Adult f., summer TX/04

1st winter, *virens* OH/09

Black-throated Green Warbler 1

Dendroica virens L 4¾"

Shape: Fairly large-headed warbler with moderate-length bill and tail. **Ad. Summer:** All ages and sexes have small yellow wash on vent area. M. solid black chin, throat, upper breast; black streaks on sides and flanks; yellow face; olive eyeline; olive-yellow ear patch; olive-green crown, back, and rump; back with dark streaks; 2 white wingbars; white or sometimes pale yellowish belly; white spots on outer tail feathers. F. has whitish to yellowish chin; black of throat and chest veiled with white feather tips; thinner streaking on sides and flanks. Similar Golden-cheeked Warbler's eyeline connects to its dark nape, and its vent lacks the yellow wash of the Black-throated Green. Some hybrid Hermit x Townsend's Warbler females are similar to female Black-throated Greens but lack yellow on vent. **Ad. Winter:** M. very similar to summer, but black throat more veiled with white, more pale yellowish on lower breast. F. like summer, but more white on throat, blurry side streaks. **1st Summer:** Male's throat mostly black with some yellow or whitish mottling. F. similar to adult summer female; has whitish to yellowish chin and throat with a little black mottling on sides of chest. **1st Winter:** M. very similar to adult winter female; in general, may have more blackish on throat. F. has yellowish chin, whitish throat; can have faint dusky marks on sides of throat and chest; no visible back streaks; hint of pale yellowish on lower chest and vent; thin blurry streaks on sides and flanks. **Flight:** Extensive white on tail, yellowish-green cheek. **Hab:** Coniferous and mixed deciduous forests. **Voice:** Distinctive primary song *zee zee zee zoo zeee,* with last note highest and emphasized. Alternate song variable with less emphatic ending, like *zoo, zee, zoozoo, zee*. Call a sharp *tsik.* **109**.

Subspp: (2) •*virens* (most of range) slightly larger and with longer bill than •*waynei* (coastal VA–SC). **Hybrids:** Townsend's Warbler.

Adult m., summer BC/06

Winter f. CA/01

Adult f., summer TX/04

1st-winter m. CA/11

Winter m. CA/01

1st-winter f. CA/08

Townsend's Warbler 1

Dendroica townsendi L 4¾"

Shape: Fairly large-headed warbler with moderate-length bill and tail. **Ad. Summer:** M. has black cheek patch surrounded by yellow face; black chin and throat; black streaks on sides and flanks; bold black streaks on greenish back; 2 white wingbars; yellow breast and white belly; extensive white on outer tail feathers. F. similar but with black areas duller slate; less black on throat; finer streaking on back, breast, sides, and flanks; median secondary coverts with thin black shafts at tip. **Ad. Winter:** M. like summer but black on crown, throat, breast, and sides variably veiled with yellow feather tips. F. similar to ad. summer female, except black on throat and back streaks obscured by yellowish feather tips. **1st Summer:** M. like ad. summer male but with reduced spotting on back; some with green on crown and less black on throat. F. like ad. summer female but usually with less extensive or more-veiled black on throat. **1st Winter:** M. similar to ad. winter female but with some small black spots on back. F. much paler, with little or no black on lower throat, no streaks on back, blurred streaking on sides and flanks. Similar Black-throated Green Warbler lacks black cheek patch, yellow breast; similar imm. f. Blackburnian Warbler has pale streaks on sides of olive back, pale yellow to orangish tones on throat and breast. **Flight:** Extensive white on tail, yellow breast, and yellow-and-black face. **Hab:** Coniferous forests. **Voice:** Primary song a variable series of whistled and buzzy notes, with several higher-pitched buzzy notes at end, like *weazy weazy weazy tweea*. Alternate song a series of short buzzy notes with rising or changing pitch, ending like *zwee zwee zwee zee*. Call a high sharp *tip*.

Subspp: Monotypic. **Hybrids:** Yellow-rumped ("Myrtle"), Black-throated Gray, Hermit, and Black-throated Green Warblers. Hybridizes regularly with Hermit Warbler in WA–OR where ranges overlap. Hybrids vary but usually have all-yellow face like Hermit Warbler and underparts like Townsend's. Less commonly, hybrids have head pattern and back color of Townsend's, with whitish breast with limited streaking. In general, hybrids in WA sing like Townsend's, in OR sing like Hermit.

Adult m., summer CA/05

Winter f. CA/08

1st-summer f. CA/05

1st-winter f. CA/09

Hermit Warbler 1

Dendroica occidentalis L 4¾"

Shape: Fairly large-headed warbler with moderate-length bill and tail. **Ad. Summer:** M. has yellow face and forehead; black throat and upper breast; bold black streaks on gray back; 2 white wingbars; extensive white on outer tail feathers; white underparts. F. similar but paler; faint gray wash on auriculars; yellowish chin; black on throat more limited or even absent in some 1st-summer females; olive-tinged black feathers on forehead and nape; olive-gray back with faint streaks; faint blurry side streaks. Similar f. Olive Warbler flicks wings, has longer bill and white patch at base of primaries, lacks throat patch; similar Black-throated Green Warbler has yellow vent; similar f. and imm. Golden-cheeked Warblers have black eye-stripe. **Ad. Winter:** M. like summer but throat feathers with whitish tips, more olive on crown and auriculars. F. like summer; amount of black in throat varies, more olive on back. **1st Winter:** M. very similar to ad. female; olive on crown and auriculars; mottled throat; black spots on olive back; some buff on underparts. F. palest of all; extensive olive on crown and auriculars; no black on throat, plain olive-brown back; buffy wash on flanks. **Flight:** Extensive white on outer tail feathers, relatively unmarked yellow face. **Hab:** Mountainous coniferous forests. **Voice:** Song is variable. Primary song is series of high-pitched notes followed by descending phrase, ending in buzzy notes; alternate song is several buzzy notes with rising and then falling ending phrase. Call is a flat *chip*.

Subspp: Monotypic. **Hybrids:** Black-throated Gray and Black-throated Green Warblers. Hybrids with Townsend's Warbler occur regularly (see Townsend's Warbler for description).

Adult m., summer MB/06

Adult f., summer MI/06

Adult m., summer NY/05

1st-winter m. MI/09

Blackburnian Warbler 1

Dendroica fusca L $4^{3}/_{4}$"

Shape: Fairly small, large-headed, broad-necked warbler with a short tail. **Ad. Summer:** All ages and sexes have triangular ear patch; eyebrow that connects around ear patch to color on sides of neck; white wingbars; pale stripes ("braces") on sides of back (sometimes hard to see); and white undertail coverts and outer tail feathers. M. has fiery orange face, throat, and small patch in black crown; black auricular; large white wing patch; streaks on sides and flanks. F. similar but with paler orange face, throat, and crown patch; grayer auricular and crown; streaked back; pale short streaking on sides and flanks; 2 broad white wingbars. **Ad. Winter:** M. with paler orange face and throat; 2 white wingbars. F. similar to summer, but paler orange head markings; more olive upperparts. **1st Winter:** M. similar to winter ad. f., but black eyeline extends across top edge of cheek patch; yellower throat; heavier back and flank streaking. F. palest of all, with yellow to pale buff or whitish face and throat; gray cheek patch; faint streaks on sides and flanks; buffier "braces." Similar imm. f. Cerulean Warbler is pale lemon below; eyebrow does not connect around ear patch to color on sides of neck. **Flight:** M. has distinct orange breast and throat. **Hab:** Tall coniferous or mixed coniferous-deciduous forests. **Voice:** Primary song is a very high-pitched series of notes followed by an even higher (going almost above human hearing range) upslurred ending. Unaccented song is higher in middle, lower at end. Call is a *chip*.

Subspp: Monotypic. **Hybrids:** Black-and-white Warbler.

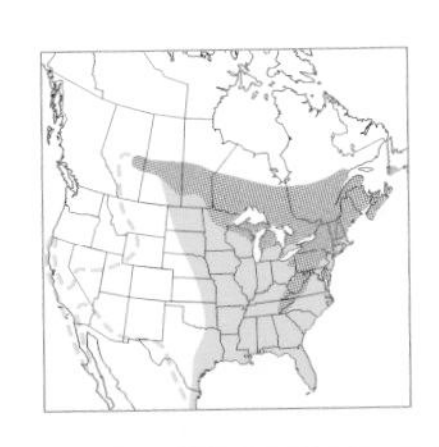

Adult, *dominica* FL/01

1st yr., *albilora* TX/04

Adult, *albilora* TX/04

Adult, *dominica* FL/11

Yellow-throated Warbler 1

Dendroica dominica L 5¼"

Shape: Medium-sized warbler with a long thin bill (see Subspp.). **Ad:** Sexes very similar. M. has bright yellow throat and upper breast; bold, black, triangular face patch extending down to black streaks on breast sides; yellow (or white in some subspp.) supraloral stripe, white eyebrow; white neck patch; black forehead; gray crown; 2 white wingbars. F. very similar to male, but less black on forehead, finer streaking on flanks. Frequently creeps along branches. Similar Grace's Warbler has mostly yellow eyebrow, no large black face patch, no white patch on side of neck. **1st Yr:** Similar to adults but duller; slightly browner back; less black on forehead; paler face patch; buff wash on flanks. Imm. f. dullest, with least black on forehead. **Flight:** Bright yellow throat contrasts with white belly. **Hab:** Floodplain forests, sycamore bottomlands, pine forests. **Voice:** Song is a variable descending series of whistles that usually ends with a higher note. Call is a sweet, rich *chip.*

Subspp: (2) •*dominica* (s.e. NJ–s. FL) has long bill; variable amount of yellow in supraloral area; usually all-yellow chin, sometimes with white on upper chin; inhabits pine forests. •*albilora* (rest of range to west) has moderate-length bill; supraloral area, usually white, can sometimes have yellow tinge; small white area on upper chin can sometimes extend a bit down side of throat; may have more black on crown; inhabits floodplains. **Hybrids:** Northern Parula; this rare hybrid, called "Sutton's Warbler," has a greenish back and no streaks on flanks.

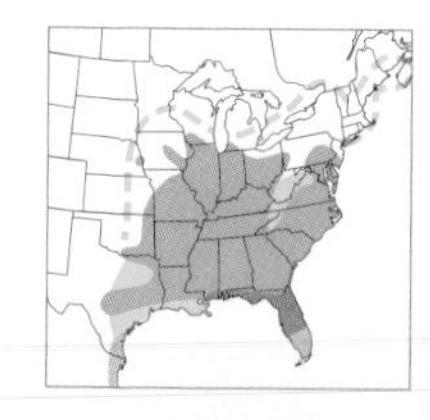

Adult m. NM/05

Adult AZ/04

Adult m. MI/05

Female MI/06

Grace's Warbler 1
Dendroica graciae L 4¾"

Shape: Small warbler with short thin bill and medium-length tail. **Ad:** M. has bold yellow throat and upper breast; yellow eyebrow that is white behind eye; yellowish crescent below eye; black lores and thin moustachial line; gray above with black streaks on back and crown; 2 white wingbars; black streaks on flanks; white belly and undertail coverts; extensive white tail spots. F. similar to male, but lores and moustachial line gray; less black on crown; back browner with finer streaks; thinner streaking on flanks. **1st Winter:** M. similar to ad. f., but upperparts washed brown and with finer streaks; less black on crown and face; paler yellow throat and chest; buffy flanks. F. dull grayish brown above with no streaks; no black on crown or face; buffy flanks. Similar Yellow-throated Warbler has stronger black face pattern, eyebrow all white or with a little yellow in front of eye. **Flight:** Yellow throat and white belly; extensive white in outer tail feathers. **Hab:** Montane pine and mixed forests. **Voice:** Variable rising series of notes, usually 2-parted *chew chew chew chew chew chew chee chee chee chee.* Call is a soft sweet *chip.*

Subspp: (1) •*graciae.*
Hybrids: Yellow-rumped Warbler (subsp. *auduboni*).

Kirtland's Warbler 2
Dendroica kirtlandii L 5½"

This rarest N. American warbler is an endangered species and only nests in jack pines. **Shape:** Large warbler with a relatively long tail and thick bill. **Ad:** M. is yellow below, with black streaks on flanks and white undertail coverts; above, blue-gray with black back streaks and faint wingbars; face with black lores and foreface and white broken eye-ring. F. similar but with no black on face; upperparts tinged brown; less distinct streaks on flanks. Pumps tail up and down. Similar f. Magnolia Warbler does not pump tail, has yellowish rump, distinct white pattern on tail. Similar Prairie Warbler smaller and has yellow face. Similar Palm Warbler has yellow undertail coverts. **1st Winter:** Similar to ad. f.; paler yellow below with less distinct streaks; upperparts and face more brown; imm. f. is brownest of all. **Flight:** Large; mostly yellow below, with white undertail coverts and white tail spots. **Hab:** Requires highly specialized nesting habitat of large stands of dense young jack pines that are 7–15 yrs. old. **Voice:** Loud series of low-pitched notes, rising in pitch and volume at end; resembles song of Northern Waterthrush. Call a descending *chip.*

Subspp: Monotypic.

Adult m., *pinus* OH/04

Adult f. or 1st-yr. m., *pinus* NH/09

1st-yr. m., *pinus* CT/06

Adult f., *pinus* TX/02

Adult f., *pinus* NY/04

1st-yr. f., *pinus* TX/03

Pine Warbler 1

Dendroica pinus L 5¾"

Shape: Fairly large heavily built warbler with a fairly thick bill and long tail. **Ad:** M. has yellow throat, breast, and upper belly, with white lower belly and undertail coverts; yellow supraloral dash connecting to yellow broken eye-ring; neck color extending around olive auricular; variable streaking on breast sides; upperparts olive-green and unstreaked; 2 bold white wingbars; large white spots on outer tail feathers; dark legs. F. similar but paler with duller yellow throat and breast; pale yellow supraloral streak and yellow to whitish split eye-ring; duller wingbars; flanks sometimes with faint streaking; olive-brown tinge on upperparts. Similar winter Blackpoll Warbler has streaked back; neck color not extending around dark ear patch; white tips to primaries; brighter white wingbars; pale yellow or pinkish legs with bright yellow soles of feet; tail extending less beyond undertail coverts. Similar winter Bay-breasted Warbler often has hint of pale chestnut on flanks; some back streaking (but can be faint); buffy undertail coverts; and dark legs. Similar Yellow-throated Vireo is larger, slower, with hooked bill and bright yellow "spectacles" (supraloral stripe and complete eye-ring). **1st Yr:** M. brownish above; dull wingbars; buffy wash on flanks; yellow on throat and breast. F. similar but can be very dull grayish brown overall with limited pale yellow on breast; buff to whitish belly; buff flanks. Aging and sexing can be difficult. **Flight:** Relatively large; dull yellow throat and breast, whitish belly and undertail coverts. **Hab:** Pine or pine-hardwood forests. **Voice:** Song is a fast trill, similar to Chipping Sparrow's but more musical. Call a slurred rich *chip*. 🎧**110**.

Subspp: (2) •*florida* (resident cent.–s. FL) has long bill, less streaking on flanks; •*pinus* (rest of range) has shorter bill, more streaking on flanks. **Hybrids:** Yellow-rumped Warbler (subsp. *coronata*).

Adult m., *discolor* TX/04

1st-winter f., *discolor* NH/08

Female, *discolor* OH/05

1st-winter m., *discolor* CA/11

Prairie Warbler 1

Dendroica discolor L $4^3/_4$"

Shape: Small round-headed warbler with small bill and long tail. **Ad:** M. has bright yellow face and underparts; black eyeline; yellow crescent under eye bordered below by black edge to cheek; dark spot on side of neck (true for all ages); bold black flank streaks; yellowish wingbars; olive upperparts with chestnut streaks (not always visible) on back. F. similar but face markings olive, occasionally with some black at edge of cheek; less flank streaking; less chestnut on back. Bobs tail. Similar Pine Warbler is larger, lacks yellow crescent under eye, no black on face or chestnut on back. Similar imm. winter Magnolia Warbler has complete white eye-ring, pale grayish chestband, yellowish rump, and white tail band across most of tail. **1st Winter:** Both sexes have narrow whitish eye arcs, lacking in adults. Some males almost as bright as ad. m., others paler with little black on face; grayish wash to auriculars. F. duller; less discernible face pattern, has no black on face; grayish crescent under eye; neck spot is olive-gray; brownish-olive upperparts; yellowish underparts, with indistinct streaking on flanks and pale yellow chin and undertail coverts. **Flight:** Mostly yellow underneath with extensive white in outer tail. **Hab:** Dry brushy areas, old fields, successional habitats, young pine plantations, mangrove swamps. **Voice:** A rising series of rapid buzzy notes. Call is a musical *tsip*.

Subspp: (2) •*paludicola* (resident coastal s. FL) has less streaking on back and underparts, is paler yellow below. •*discolor* (rest of range) has more reddish-brown streaking on back, more black streaking on underparts, brighter yellow underparts.

Adult summer, *palmarum*, "Western" MI/05

Adult summer, *hypochrysea*, "Yellow" CT/04

Winter, *palmarum*, "Western" FL/03

Adult summer, *hypochrysea*, "Yellow" ME/06

Winter, *palmarum*, "Western" FL/01

Winter, *hypochrysea*, "Yellow" NH/10

Palm Warbler 1

Dendroica palmarum L 5½"

There are 2 distinctive subspecies, treated below separately. Note that intergrades between the 2 occur in a zone south of James Bay; they have varying amounts of yellow below. Similar Kirtland's and Prairie Warblers pump tail but have no contrasting yellow undertail coverts.

"Western Palm Warbler"

Dendroica palmarum palmarum

Breeds Hudson Bay and to the west; winters TX–NJ–FL, also on West Coast. **Shape:** Fairly large and relatively long-tailed warbler. **Ad. Summer:** All ages have whitish belly that contrasts with bright yellow undertail coverts. Sexes similar. M. has chestnut crown; dark eyeline; bright yellow eyebrow; yellow throat; grayish-streaked underparts (sometimes tinged with faint yellow); olive-yellow rump. F. not always distinguishable from m., tends to have less chestnut on crown, paler yellow areas. Pumps tail. **Ad. Winter:** Similar but duller and browner overall; whitish eyebrow; crown with little or no chestnut; whitish throat; dusky streaking below, sometimes with tinge of yellow on belly; contrastingly bright yellow undertail coverts. For Flight, Habitat, and Voice, see "Yellow Palm Warbler."

"Yellow Palm Warbler"

Dendroica palmarum hypochrysea

Breeds QC and east; winters primarily on Gulf Coast LA–n. FL. **Shape:** Fairly large and relatively long-tailed warbler; slightly larger than *palmarum*. **Ad. Summer:** All ages have bright yellow underparts that do not contrast with yellow undertail coverts. Sexes similar, with chestnut crown; yellow eyebrow; chestnut streaking on breast and flanks; yellowish rump. Pumps tail. **Ad. Winter:** Duller overall; little or no chestnut in crown; little or no contrast between yellow underparts and yellow undertail coverts; keeps yellow eyebrow. **Flight:** Yellow undertail and yellowish rump; white on outer tail feathers. **Hab:** Bogs ringed with spruce, tamarack, cut-over jack pine woods. Despite its name only spends time in palms on parts of winter range. **Voice:** Song is a weak buzzy trill that may change in pitch. Call is a sharp *tsup*.

Adult m., summer ME/05

1st-summer f. OH/05

Adult f., summer OH/05

Winter OH/09

1st-summer f. TX/04

Winter MI/09

Bay-breasted Warbler 1

Dendroica castanea L 5¼"

Shape: Fairly large heavyset warbler with long primary projection; tail appears short and does not extend much beyond undertail coverts. **Ad. Summer:** All ages and sexes have dark legs and 2 bold white wingbars. M. has rich chestnut ("bay") on throat, crown, upper breast, and flanks; black face; buff patch on side of neck. F. varies considerably; less chestnut on head and sides; mottled gray cheek; some are bright, almost like 1st-summer male, with lots of chestnut on crown, but differentiated from m. by buffy split eye-ring. **Ad. Winter:** M. is olive-yellow above with some blackish streaking on back; little or no chestnut in crown; pale eyebrow and eye; grayish eyeline; buffy throat and breast; chestnut on flanks; white-tipped tertials, primaries, and secondaries; grayish rump and uppertail coverts; buff undertail coverts. F. is duller but often not separable from duller adult m. or 1st-winter m.; lacks any chestnut; flanks warm buff. **1st Summer:** M. has chestnut crown and, like ad. m., lacks split eye-ring; cheeks less black, more mottled; paler chestnut underparts. F. variable; some quite similar to ad. f., others are very dull with no chestnut on crown; pale buffy throat and sides. **1st Winter:** M. similar to adult m., but flanks with less chestnut or just warm buff; finer streaking on back. F. olive-yellow above with faint or no back streaks; whitish or yellowish buff below; flanks yellowish or pale buff; pale buffy undertail coverts. Distinguishing winter imm. Bay-breasted and Blackpoll Warblers is challenging. Blackpoll has straw-colored legs (which can have dark sides) and yellow soles of feet; yellow-olive (not pale chestnut or warm buff) sides and flanks which are streaked (Bay-breasted's are unstreaked); back is brighter yellow-olive and back streaks are more noticeable; white or pale yellowish undertail coverts are longer than Bay-breasted's and contrast with yellowish underparts. Similar winter m. Pine Warbler has unstreaked back; neck color extending around dark ear patch; less bold wingbars; long tail extending well beyond undertail coverts; shorter primary projection. **Flight:** Fairly large and heavyset; white tail spots. **Hab:** Boreal coniferous and mixed forests, especially where outbreaks of spruce budworm, a prime food, occur. **Voice:** Series of high-pitched, sometimes doubled notes, like *seetzy seetzy seetzy*, similar to Black-and-white Warbler's. Call a loud *chip* similar to Blackpoll's.

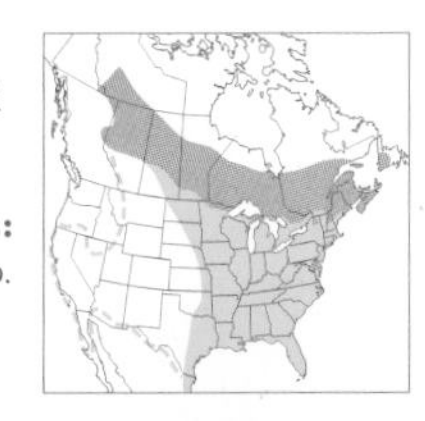

Subspp: Monotypic. **Hybrids:** Yellow-rumped Warbler (subsp. *coronata*), Blackpoll Warbler.

Adult m. TX/05

1st winter OH/09

Summer f. OH/05

Adult f., winter MI/09

Adult f. TX/04

Blackpoll Warbler 1

Dendroica striata L 5½"

Shape: Fairly large sleek-bodied warbler with a long primary projection, long undertail coverts, and short tail. Slightly thinner bill than Bay-breasted. **Ad. Summer:** M. has distinct chickadee-like head pattern of black cap and white cheeks; black malar streak; black streaking on back and flanks; bold white wingbars; yellow or orangish-yellow legs with yellow soles of feet. F. quite variable—some are heavily streaked on crown, suggesting a cap, and some are more extensively streaked across breast and flanks; underparts whitish to yellowish; upperparts olive-gray to olive-yellowish with streaked back; dark eyeline; pale eyebrow; dusky cheek; a few streaks in malar areas; legs and feet yellow. Similar m. Black-and-white Warbler, with striped crown, has different head pattern from male Blackpoll. **Ad. Winter:** Majority of individuals cannot be reliably aged or sexed in fall and winter; upperparts olive-green with black streaks; faint yellowish eyebrow and dark eyeline; pale yellow or whitish underparts with thin or indistinct olive streaking on flanks; white undertail coverts; pale or dark-sided legs and always yellow soles of feet. Similar winter Bay-breasted Warbler has dark legs and feet; chestnut or warm buff sides and flanks which are unstreaked; buff vent and undertail coverts. Similar winter Pine Warbler has unstreaked back; yellowish neck color extending around dark ear patch; less bold wingbars; long tail extending well beyond undertail coverts; shorter primary projection. **1st Winter:** Usually indistinguishable in the field from adult winter. **Flight:** Fairly large warbler with long white undertail coverts. **Hab:** Conifer forests, especially spruce-fir forests. **Voice:** Very high-pitched repeated *seet* notes, loudest in middle; sounds like someone shaking tiny maracas. Call a loud *chip*.

Subspp: Monotypic. **Hybrids:** Cape May and Bay-breasted Warblers.

Adult m., summer OH/04

Winter f. CA/09

Adult f., summer OH/05

Trans. to 1st summer FL/02

Black-and-white Warbler 1

Mniotilta varia L 5¼"

Shape: Fairly large warbler with a short tail and fairly long, thin, slightly downcurved bill. Elongated hind claw helps grip tree trunks. **Ad. Summer:** M. with distinct heavy black-and-white streaking throughout; white central stripe on black crown; black cheek and throat (some may have limited white on chin); black marks on undertail coverts. F. like ad. m. but cheek is grayish buff with a black eyeline above it; throat all whitish; flanks with some buff and narrow streaking. Creeps around tree trunks and along branches like a nuthatch. Similar Black-throated Gray and summer m. Blackpoll Warblers have all-black cap with no central white stripe. **Ad. Winter:** M. like summer ad. m. but has variable amount of white on chin and upper throat, from extensive to none; also has variable amount of black on auriculars. F. like spring ad. f. but some with more buffy flanks. **1st Summer:** M. similar to ad. m. but less to almost no black on chin and throat. F. similar to ad. female. **1st Winter:** M. differs from winter ad. m. in having gray, not black, in cheek; throat mostly white. F. like winter ad. f. with fainter streaking on underparts, buffier flanks. **Flight:** Short tail; white belly and black-spotted undertail coverts. **Hab:** Deciduous and mixed woodlands with some mature trees. **Voice:** Song is a long series of high, repeated, 2-note phrases, like *wheesy wheesy wheesy*. Call a sharp *pit*. **111**.

Subspp: Monotypic. **Hybrids:** Blackburnian and Cerulean Warblers.

Adult m. OH/04

Summer f. OH/05

Adult m. OH/05

1st-summer m. OH/05

Cerulean Warbler 2

Dendroica cerulea L 4½"

Shape: Small, long-winged, short-tailed warbler with the shortest tail extension past undertail coverts of all the warblers. **Ad:** M. has beautiful cerulean blue upperparts with black streaks (some males brighter blue than others); 2 bold white wingbars; white underparts with a black breastband and streaked flanks. F. has turquoise-blue or pale bluish upperparts with no streaking; light bluish crown; whitish or yellowish eyebrow; whitish lower eye arc; whitish underparts washed yellow on throat sides and breast; blurred grayish streaks on flanks; no breastband; 2 white wingbars. **1st Summer:** M. with less bright blue crown; more grayish upperparts; thinner or broken breastband; white eyebrow behind eye. F. similar to ad. f.; some have more greenish upperparts. **1st Winter:** M. has bluish-green upperparts with streaks sometimes limited to sides of back; whitish underparts tinged yellow, with moderate streaks on sides of breast; breastband just on sides of breast or lacking; whitish eyebrow broader behind eye. F. with little if any blue; greenish or grayish-olive unstreaked upperparts; yellow-tinged eyebrow broadens behind eye; extensive yellow wash on throat sides, breast, and flanks; blurry streaking on flanks. Similar drab imm. f. Blackburnian Warbler is larger and longer-tailed, has different head pattern of dark triangular cheek patch surrounded by face color, streaked back with buffy-white "braces." **Flight:** Very short tail; thin breastband on ad. male. **Hab:** Large old-growth deciduous forests, especially near water. **Voice:** Series of rising, rapid, buzzy notes, ending with higher note, like *zray zray zray zree*. Call a sharp *chip*.

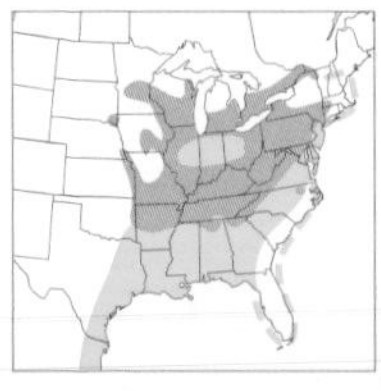

Subspp: Monotypic. **Hybrids:** Black-and-white Warbler.

Adult m. TX/04

1st-summer m. CT/06

1st-summer f. MI/05

Adult f. OH/05

1st winter OH/09

American Redstart 1

Setophaga ruticilla L 5"

Shape: Medium-sized warbler with a very long tail and rather flat broad bill; has obvious rictal bristles on either side at base of bill. **Ad:** M. all black above with striking orange patches on sides of breast, at base of primaries and secondaries, and along the upper two-thirds of outer tail feathers; some with more black on lower breast and belly; thin buffy edges to body feathers in winter. F. has olive upperparts (slightly more brownish in winter); yellow or orangish-yellow breast, wing, and tail patches; white lores; white underparts. Very actively moves about with frequent tail-fanning and drooped wings; often darts out to catch aerial insects. Vaguely similar Painted and Slate-throated Redstarts lack yellow or orange and have white in tail. **1st Summer:** M. very similar to ad. f., but most have some black splotches on head and breast; has black in lores and darker uppertail coverts. F. like ad. f. with smaller yellow wing patch. **1st Winter:** M. similar to adult f., some nearly identical; usually more orangish-yellow side patches; no black in lores. F. like ad. f. but more olive upperparts and palest yellow side patches of all plumages. **Flight:** Long tail with distinctive yellow or orange patches on the sides. **Hab:** Deciduous and mixed woodlands, second-growth habitats. **Voice:** Song varies considerably both among and within individuals; series of high variable notes, usually with an accented downslurred ending note. Can sound like Magnolia, Yellow, and Chestnut-sided Warblers. Call a high, slurred, sweet *chip*. **112**.

Subspp: Monotypic. **Hybrids:** Northern Parula.

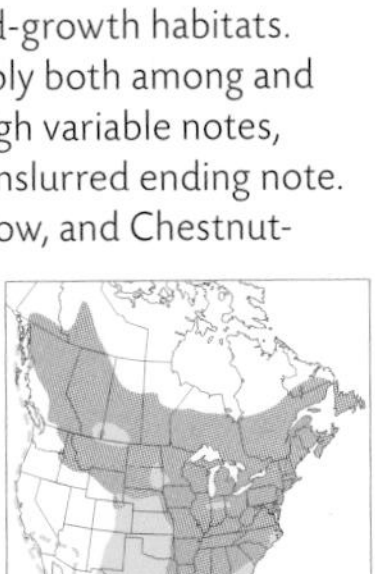

Adult m. TX/04

Adult m. OH/05

Adult f. OH/06

Prothonotary Warbler 1
Protonotaria citrea L 5¼"

Shape: Rather large and large-headed broad-necked warbler with a long bill and short tail; dark eyes appear large on face. **Ad:** M. has butter-yellow to orange head and underparts; greenish back; unmarked blue-gray wings and tail; white undertail coverts and large white tail spots. F. similar but with duller yellow underparts; olive wash to crown and nape; smaller tail spots. Similar Blue-winged Warbler has dark eyeline and whitish wingbars. **1st Winter:** 1st-winter and winter adults have pale bills (bill dark in summer). M. similar to ad. female. F. dullest of all plumages, with olive wash on crown, face, and neck. **Flight:** Extensive white in tail; underparts all yellow except for white undertail coverts. **Hab:** Wooded swamps, floodplain forests, where it is the only eastern warbler that nests in cavities. **Voice:** Song is a series of clear loud *zweet* notes primarily on one pitch. Call a loud *tink*.

Subspp: Monotypic.

Adult OH/04

Adult TX/04

Worm-eating Warbler 1
Helmitheros vermivorum L 5¼"

Shape: Medium-sized broad-necked warbler with a long thick-based bill. **Ad:** Sexes similar, having warm buff head with black stripes along sides of crown and through eyes; olive-brown upperparts; warm buff below, brightest on throat and breast. Similar Swainson's Warbler lacks the black head stripes and rich buff underparts. **1st Yr:** Very similar to adult but with warm brownish fringes to tertials. **Flight:** Underparts all warm buff; no white spots on tail. **Hab:** Wooded hillside slopes and ravines, where it often forages in curled dead leaf clusters in trees. **Voice:** Dry trill on one pitch with a sewing-machine-like quality, similar to Chipping Sparrow's song but drier. Call a dry *chip*.

Subspp: Monotypic.

Adult TX/04

Adult SC/04

Swainson's Warbler 2

Limnothlypis swainsonii L 5½"

Shape: Large, elongated, broad-necked warbler; noticeably long thick-based bill has straight culmen and mild upturn on end of lower mandible; flat forehead; fairly short tail; large feet. **Ad:** Sexes similar; warm brown crown and nape; long whitish to yellowish-buff eyebrow and dark brown eyeline; muted brown back; brown plain wings; unstreaked whitish to pale yellow underparts; grayish flanks; fleshy pink legs and feet. Forages on ground by shuffling through leaf litter while doing a peculiar vibrating motion of its rear end. Similar Worm-eating Warbler has striped forehead; similar Louisiana and Northern Waterthrushes are streaked below. **Flight:** Pale below; plain and unmarked. **Hab:** Wooded lowland swamps, canebrake, also highland rhododendron thickets in southern Appalachians. **Voice:** Heard more than seen; ringing loud song that starts with 2–4 (usually 3) slurred whistles, which can sound like beginning of Louisiana Waterthrush's song but ends with several clear descending notes. Louisiana's song ends with twittering sputtering notes. Call a loud *tisp*.

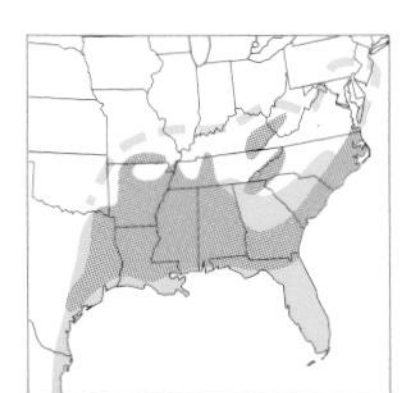

Subspp: Monotypic.

Adult CT/05

Adult, *aurocapilla* OH/05

Ovenbird 1

Seiurus aurocapilla L 5¾"

Shape: Large, broad-necked, deep-chested warbler with fairly long legs and long toes. **Ad:** Sexes similar; 2 black crown-stripes border a central orange streak (which is sometimes hard to see); white eye-ring surrounds big dark eye; olive-brown upperparts; creamy underparts with bold black splotches organized in streaks on breast and sides; olive wash on flanks; pinkish legs. Often walks with wings drooped and tail cocked. Similar Northern and Louisiana Waterthrushes lack eye-ring, lack black crown-stripes and orange crown patch, and both bob tail. **1st Winter:** Similar to adult but with warm brown edges to tips of tertials. **Flight:** Creamy underparts with heavy streaking on breast and flanks. **Hab:** Mature deciduous or mixed forests with dead leaf litter; builds domed ground nest, hence "oven" in its name. **Voice:** Ringing long series of usually 2-part notes that increase in loudness, like *teacher teacher teacher teacher.* Call is sharp loud *chit*. **113**.

Subspp: (3) •*cinereus* (s. AB–CO–NE) has grayish wash on upperparts; •*aurocapilla* (e. BC–NS–GA) has greenish wash on upperparts; •*furvior* (NF) has brownish wash on upperparts.

Adult NH/06

Adult CT/05

Adult OH/08

Adult MI/05

Northern Waterthrush 1

Parkesia noveboracensis L 5¾"

Shape: Rather large elongated warbler that is chunky in front with a more rounded head and shorter neck than similar Louisiana Waterthrush; fairly long heavy bill is usually shorter than bill of Louisiana Waterthrush; long tail usually extends farther past the undertail coverts than Louisiana's. (Northern's tail extension averages 0.55 in., with a range of 0.47 to 0.91 in.; Louisiana's tail extension averages 0.43 in., with a range of 0.35 to 0.55.) **Ad:** Sexes similar. Brown or olive-tinged brown above; evenly colored eyebrow is yellowish buff or whitish and usually of even width throughout, but tapering at end; mostly uniformly colored underparts are pale yellowish buff or whitish; narrow malar streaks; throat usually has thin streaks; dark, heavy, elongated dashes are arranged into streaks on underparts; flanks can have pale buffy-brown wash; legs pink to pinkish brown. Walks with up-and-down tail bobbing. **Flight:** Dark and unmarked above; pale below, with heavy streaking on breast and flanks. **Hab:** Wooded ponds, swamps, willow thickets, lakeshores, streams, near still or slow-moving water. **Voice:** Loud series of ringing notes with downslurred ending. Call a metallic, hollow, very loud *chink,* similar to Louisiana Waterthrush's. **114**.

Subspp: Monotypic.

Adult NH/05

Adult NY/04

Adult OH/03

Adult NH/05

Louisiana Waterthrush 1

Parkesia motacilla L 6"

Shape: Larger more elongated warbler with flatter head and longer neck than similar Northern Waterthrush; longer bill and thicker elongation at rear end (because tail is shorter past undertail coverts) than Northern Waterthrush; tail extends less, an average of 0.43 in., past undertail coverts (Northern's extends an average of 0.55 in. but may extend up to 0.91 in.). **Ad:** Sexes similar. Upperparts olive-brown to gray-brown; two-toned eyebrow is grayish or buff in front of eye, becomes white above eye; eyebrow usually appears broader behind eye, where it can extend well back on neck; narrow malar streaks; usually no streaks on throat, but a small percentage of birds (about 10%) have small throat marks; underparts whitish or creamy white with medium-brown dashes forming blurred, loose streaks on underparts and flanks; flanks washed, sometimes strongly, with salmon buff or yellowish buff; bubble-gum-pink legs brighter than Northern's; walks with tail bobbing (often in a circular motion), involving more of whole rear end than Northern. **Flight:** Dark and unmarked above; pale below, with heavy streaking on breast and flanks. **Hab:** Wooded ravines near fast-flowing mountain brooks and streams. **Voice:** Song starts with 3–4 loud downslurred notes, like *Louise, Louise, Louise,* ending with warbling twitter. Call a hollow loud *chink* or *chunk,* similar to Northern Waterthrush's. 🎧**115**.

Subspp: Monotypic.

IDENTIFICATION TIPS

Comparison of Northern and Louisiana Waterthrushes: Start with shape and behavior. Louisiana is bigger, more elongated, with larger bill; tail extends less past undertail coverts than Northern's. Louisiana bobs tail (often in a circular manner) and uses more of rear end than Northern. Northern's eyebrow solid yellow or solid white; Louisiana's is bicolored, with buff in front, white above and behind eye. Length and thickness of eyebrows not always a reliable clue, since individuals vary greatly. Northern's throat usually with fine streaks; Louisiana's usually clear white. Northern is white or yellowish buff below with heavy dark streaking, sometimes pale buff-brown on flanks; Louisiana is creamy white below with somewhat paler streaking and yellowish or salmon wash on flanks. Louisiana's legs are brighter pink than Northern's. The two species' songs are distinctive; their call notes are quite similar.

Adult m., summer MB/06

Adult winter f. or 1st winter NY/09

Adult f., summer IL/05

1st winter NY/09

Adult f., summer IL/05

1st winter NY/09

Connecticut Warbler 2

Oporornis agilis L 5½"

Shape: Large, deep-chested, long-billed warbler with a large eye, long legs, and long toes. Has long primary extension past tertials, very long undertail coverts, and short tail projection past coverts. **Ad. Summer:** Complete, prominent, white or buff eye-ring present in all ages and sexes (can occasionally be thinly broken at rear). M. has gray hood extending down to upper breast and lightest on throat; bold white eye-ring; yellow underparts; olive tinge to flanks. F. similar, but hood is duller, more brownish-olive; throat whitish or buffy gray; eye-ring whitish or buffy. Skulky warbler that walks, not hops, on ground. Similar Mourning and MacGillivray's Warblers are smaller, with shorter undertail coverts, longer primary projection, and hop, not walk; Mourning can have a thin eye-ring, but it is broken at both front and rear (Connecticut's eye-ring is sometimes broken, but only at rear); female and imm. Mournings usually have yellow of breast breaking up through hood onto the throat (Connecticut's hood is complete and solid across breast, sometimes appearing as a band); MacGillivray's all have distinct white eye-rings broadly broken in front and rear of eye, forming eye crescents, and have brighter yellow underparts than Connecticut. Similar f. Common Yellowthroat is smaller and its olive flanks contrast with yellow undertail coverts and white belly. Similar Nashville Warbler is smaller, has yellow throat and breast and no hood. **Ad. Winter:** Similar to summer ad. but olive-brown wash to hood, upper breast, and upperparts. **1st Summer:** M. similar to ad. m. but some with duller hood and olive on crown. F. similar to ad. f., may be duller with paler throat. **1st Winter:** Similar to winter ad. f. but with buffy eye-ring; m. with variable gray on forehead and breast, whitish or grayish throat; some imm. f. have very brownish hoods and buffy or yellow-buff throats. **Flight:** Dark above, yellow below with defined dark hood. **Hab:** Spruces, tamarack bogs, mixed woods, open poplar woods, mixed forests, jack pine barrens. **Voice:** Halting, loud, repeated phrases, like *Connect-i-cut Connect-i-cut Connect-i-cut*. Call a metallic *chink*.

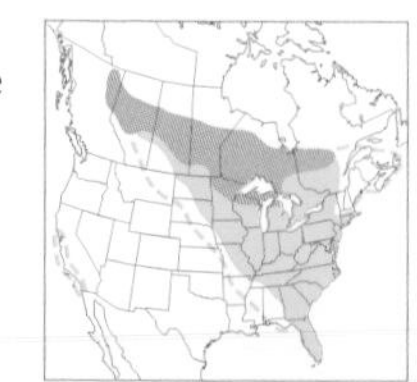

Subspp: Monotypic.

Adult m., summer ON/07

Adult f., summer TX/05

1st winter NY/10

1st-winter m. NY/10

Mourning Warbler 1

Oporornis philadelphia L 5¼"

Shape: Usually has shorter tail extension past undertail coverts than similar MacGillivray's Warbler. **Ad. Summer:** Plumage is variable. M. has dark gray hood; variable black-mottled patch on upper breast and sometimes on throat; usually dark gray or, rarely, black lores; no complete eye-ring (a small percentage of birds have a thin, white, broken eye-ring); olive upperparts and bright yellow underparts, including yellow undertail coverts. F. similar to m., but paler gray hood; no black mottling on upper breast; thin, broken, whitish eye-ring on some; pale gray or whitish-gray throat sometimes shows a hint of yellowish; faint whitish-gray supraloral streak; gray chest contrasts with yellow underparts. Distinguishing Mourning from similar MacGillivray's Warbler is one of the most difficult warbler ID challenges. MacGillivray's all have thick, short, blunt, white eye crescents (never a complete ring) above and below the eye. Mourning Warblers can have thin dull white or yellowish eye-rings that are thinly broken in front of and behind eye or that form an almost complete eye-ring. M. MacGillivray's Warbler has black lores, m. Mourning's lores usually gray, rarely black. MacGillivray's f. and imm. have complete, not broken, bibs and usually have no yellow in throat. Connecticut Warblers have prominent white eye-rings that are complete; only rarely is the male's thinly broken at the rear (never the front). Connecticuts have thicker shape, larger bill, shorter tail projection; are browner above, dingier yellow below; and walk, not hop. F. and imm. Common Yellowthroats are smaller and longer-tailed and have whitish or pale brown (not yellow) on belly. **Ad. Winter:** Similar to summer, but crown with olive-brownish wash. F. often with thin, almost complete, whitish eye-ring; whitish, pale gray, or buffy throat, sometimes with hint of yellow. **1st Summer:** F. can show thin, nearly complete, white eye-ring; is duller and more brownish than ad. female. **1st Winter:** Similar to ad. female; many 1st-winter m. and f. not distinguishable from one another. Thin yellowish-buff supraloral streak; thin variable eye-rings are whitish or yellowish white, and can be broken or almost complete; brownish-olive hood; throat color varies from bright yellow to grayish buff with almost no yellow; throat color usually breaks through hood onto yellow of underparts, giving broken-bib effect. Some 1st-winter m. may show some black feathers on throat and chest, appearing like a complete breastband; bright yellow lower sides of breast and flanks. **Flight:** Dark above, yellow below, with defined dark hood. **Hab:** Dense undergrowth of moist woods, brushy second-growth areas, montane bogs. **Voice:** Variable loud series of 2-syllable phrases with ending phrases lower-pitched, *churry churry churry chorey chorey.* Call a scratchy *chiup*, often with a slight buzzy quality.

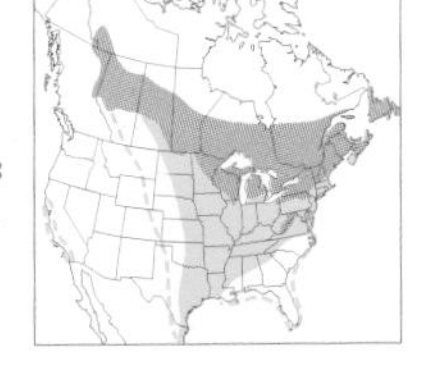

Subspp: Monotypic. **Hybrids:** Common Yellowthroat, Canada and MacGillivray's Warblers.

Adult m., summer, *monticola* AZ/05

Adult f., summer, *tolmiei* CA/05

Adult m., summer, *monticola* AZ/05

Adult f., summer, *tolmiei* CA/05

MacGillivray's Warbler 1

Oporornis tolmiei L 5¼"

Shape: Like Mourning Warbler, but has longer tail that extends farther past undertail coverts. **Ad. Summer:** All ages and sexes have thick, blunt, white eye crescents above and below eye. M. has blue-gray hood; black lores that connect across top of bill; white upper tip of chin; slate-black mottling (not a patch) on throat and upper breast; olive upperparts; bright yellow underparts, including undertail coverts; olive wash on sides and flanks. F. similar but with paler gray hood; no black lores; faint pale gray supraloral streak; pale gray throat with no black; yellow underparts. When on ground, hops. Distinguishing MacGillivray's from similar Mourning Warbler is one of the most difficult warbler ID challenges. Mourning has shorter tail, longer undertail coverts. Adult Mourning lacks bold white eye crescents of MacGillivray's; instead can have thin dull white or yellowish eye-rings that are thinly broken in front of and behind eye or that form an almost complete eye-ring; m. has black breast patch. Imm. Mournings usually have a yellowish throat that breaks through lower edge of hood onto underparts. Similar Connecticut Warbler has thicker shape, larger bill, shorter tail projection; it is browner above and dingier yellow below; and it walks (MacGillivray's hops). Similar f. and imm. Common Yellowthroats are smaller and longer-tailed and have whitish or pale brown (not yellow) on belly. **Ad. Winter:** Similar to spring ad. but more olive-brown on crown; f. throat more buffy. **1st Winter:** Similar to ad. female; pale gray-buff supraloral streak; thick white eye crescents may, in rare instances, form a nearly complete eye-ring; more olive-brown hood becomes grayish and extends solidly across upper breast with no extension of throat color onto breast; throat usually pale gray

Winter, *tolmiei* CA/09

Adult m. OH/05

Winter, *tolmiei* CA/09

Adult f. TX/04

or gray-buff or, rarely, in imm. female, yellowish; yellow below with olive wash on sides of breast and flanks. M. may have mottled darker gray lower edge of hood. **Flight:** Dark above, yellow below with defined dark hood. **Hab:** Dense deciduous thickets, often along streams. **Voice:** Song is short, 2-parted: several short or trilled notes, followed by buzzy variable notes, sometimes similar to Mourning Warbler's. Call a loud *tsik,* sharper or harder than the slightly buzzier call note of Mourning Warbler.

Subspp: (2) •*tolmiei* (CA–SD and north) fairly bright yellow below; •*monticola* (NV–s. WY and south) slightly duller below, with more olive on flanks. **Hybrids:** Mourning Warbler.

Kentucky Warbler 1

Oporornis formosus L 5¼"

Shape: Broad-necked, deep-chested, short-tailed warbler with a fairly short thick bill. **Ad:** Bright yellow "spectacles" formed by yellow eyebrow that wraps partway behind eye stand out on all ages and sexes. M. has dark triangular cheek patch that extends down side of neck; dark crown that becomes gray toward rear; olive-green upperparts; bright yellow underparts. F. similar with less black on crown and sides of neck. Similar Common Yellowthroat lacks yellow spectacles; similar Hooded Warbler has yellow face with no dark cheek patch, also has extensive white tail spots. **1st Winter:** Similar to ad. f.; imm. f. dullest of all plumages, with olive crown; dusky reduced cheek patch; olive tinge on flanks. **Flight:** All dark olive above and yellow below; no white on tail. **Hab:** Deciduous woodlands with dense moist understory. **Voice:** Song a rolling series of 2-syllable notes, like *churry churry churry* (like Carolina Wren's, but not as melodic). Call is a rich low *chunk.*

Subspp: Monotypic. **Hybrids:** Blue-winged Warbler.

Adult m. TX/04

1st-winter m., Eastern Grp., *trichas* CT/10

Female TX/04

Winter f., Eastern Grp., *trichas* NH/09

Common Yellowthroat 1

Geothlypis trichas L 5"

Shape: Fairly deep-chested broad-necked warbler with short wings and long rounded tail, which is often cocked. **Ad:** M. varies in width of band bordering the black mask and amount and brightness of yellow on underparts (see Subspp.). M. has "masked bandit" look; black mask has grayish-white border above and behind; very rarely, yellow supraloral streak through mask; olive-brown upperparts; bright yellow throat and breast; paler belly; brownish wash on flanks; bright yellow undertail coverts. F. similar but lacks mask (a few can show hint of mask); suggestion of pale eyebrow; faint eye-ring; olive-brown cheeks that contrast sharply with yellow throat and upper breast; pale lower belly; brownish flanks; pale yellow undertail coverts. Commonly cocks and flicks open tail. Similar f. and imm. Oporornis warblers have different shape (larger, heavier-bodied, shorter-tailed), solid yellow underparts; Connecticut has bold white eye-ring, walks on ground; Kentucky has yellow spectacles. **1st Yr:** M. similar to adult f. but with variable amount of black feathers in mask area (mask can be obscured by pale gray feather tips, which wear off by midwinter, revealing more black); pale eye-ring; dull yellow throat and breast; creamy belly; flanks with heavy brownish wash; yellow undertail coverts. F. dullest of all plumages, with very little yellow; pale buff eye-ring; throat and breast yellowish buff; upperparts brownish; extensive brownish wash on flanks; some whitish on lower belly; yellowish-buff undertail coverts. **Flight:** Bright yellow breast and undertail coverts; no white in tail. **Hab:** Dense brushy habitats near wet areas, drier habitats with dense understory. **Voice:** Song is repetitive phrases that sound like *whichity whichity whichity* or, going with bandit theme, "your money, your money, your money." Call a loud, sharp, distinctive *tchat*. **116**.

Subspp: (11) Divided into three groups. **Eastern Group:** Moderate size, mask border pale gray, yellow on front includes throat and upper breast. •*insperata* (s. TX), •*trichas* (e. ON–e. TX east to NF–e.cent. VA), •*typhicola* (AL–s.e. VA–s.cent. GA), •*ignota* (s. LA–FL–e. SC). **Western Group:** Large, mask border whitish, yellow on front limited to mostly throat (except see *chryseola*). •*yukonicola* (s. YT–n. BC); •*campicola* (BC–s.e. WA east to w. ON–MN), •*occidentalis* (e. OR–e. CA–CO–n.e. TX), •*scirpicola* (s. CA–s.w. UT–s.w. AZ); •*chryseola* (s.e. AZ–s.w. TX) has mask border tinged yellow and extensive yellow on underparts. **Pacific Coastal Group:** Mask border whitish, and yellow on front from throat to midbelly. •*arizela* (s.e. AK–w.cent. CA), •*sinuosa* (San Francisco Bay region). **Hybrids:** Mourning Warbler, Gray-crowned Yellowthroat.

Adult m. MEX/01

Adult m. TX/03

Adult m. TX/04

1st-yr. f. NY/05

Gray-crowned Yellowthroat 4

Geothlypis poliocephala L 5½"

Casual in s. TX. **Shape:** Relatively large, large-headed, long-tailed warbler with a long thick bill; culmen noticeably curved. **Ad:** M. has thick black lores that extend to just below eye; split white eye-ring; grayish crown; bicolored bill with dark upper mandible, pinkish-yellow lower mandible; grayish or brownish-olive upperparts; yellow throat and breast; creamy belly; brownish wash on sides of breast and flanks; pale yellow undertail coverts. F. similar but lores gray and less extensive; brownish-gray head. Similar Common Yellowthroat has smaller, thinner, straighter, all-dark bill; similar Yellow-breasted Chat is larger with thicker bill, white spectacles. **1st Winter:** Dark lores less extensive and may be absent on f.; brownish wash on upperparts and underparts. **Flight:** Long tail, dark above, all yellow below. **Hab:** Casual in s. TX in dense tall grasses with scattered trees, shrubs. **Voice:** Melodic variable warble similar to song of Indigo Bunting, Blue Grosbeak. Various calls, including a grating *chee-dee dee-deet* and descending *jeeu jeeu jeeu.*

Subspp: (1) •*ralphi* (s. TX). **Hybrids:** Common Yellowthroat.

Hooded Warbler 1

Wilsonia citrina L 5¼"

Shape: Medium-sized warbler with fairly long and broad tail that is often flicked open. **Ad:** M. has dramatic black hood and throat; yellow face and forehead; dark lores; large eye; olive-green back; bright yellow underparts. F. similar to male but hood pattern varies: on some it is almost as dark and extensive as male's but with olive on rear crown; most f. have just a dark border along crown front, down neck sides, and sometimes onto throat. Frequently flicks open tail, showing extensive white on outer feathers. Similar f. Wilson's Warbler smaller, smaller-billed, lacks white in tail. **1st Yr:** (Aug.–Jul.) M. like ad. m. but hood feathers tipped yellow; f. has no black on crown or throat, face bordered with olive. **Flight:** Yellow underparts, extensive white in tail, m. with black hood. **Hab:** Forests with well-developed shaded understory, hardwoods in north, also swamps in south. **Voice:** Loud, musical, whistled notes with emphasized end, like *wee-ta wee-ta wee-tee-yo.* Call a metallic *chink.*

Subspp: Monotypic.

Adult m., *chryseola* CA/05

1st-yr. m. CA/04

Adult f., *pusilla* NY/10

1st-yr. f., *pusilla* OH/05

Wilson's Warbler 1

Wilsonia pusilla L 4³/₄"

Shape: Relatively small warbler with a long, rounded, slender tail; relatively short bill. **Ad:** Amount of black in cap varies by age, sex, and subspecies, and there is some overlap, making it difficult to age and sex some birds. M. has black, shiny, beanielike cap; yellow forehead, eyebrow, and eye-ring; olive-green cheek; olive-green upperparts; entirely yellow underparts. F. similar, but olive crown has little to some black, which, when present, is usually restricted to front half of crown; olive to yellow forehead and lores. **1st Yr:** Similar to adult. M. has some olive-green mottling in cap, especially on the hindcrown; f. has no black in cap; crown and forehead are olive to yellow; thin yellowish eye-ring; duller yellow underparts. Similar imm. f. Hooded Warbler is larger, with bigger eye, dusky lores, more yellow cheek, and has white in tail. Similar Orange-crowned Warbler has dark eyeline, more pointed bill, blurry streaks below. Similar imm. Yellow Warbler is larger, has yellow tail spots, pale yellow edges to wing feathers, shorter tail. **Flight:** All-yellow body, darker wings and tail, no white in tail. **Hab:** Willow and alder thickets near water, moist woodlands with dense ground cover. **Voice:** Chattering trill that may drop in pitch; varies regionally. Call a low flat *timp*.

Subspp: (3) •*chryseola* (s.w. BC–s.w. CA) is bright yellow overall; cap more extensively black by age and sex. •*pileolata* (AK–n. AB south to e. CA–cent. NM) is bright yellow overall with greenish wash on upperparts; cap moderately black by age and sex. •*pusilla* (e. AB–NF south to MN–NS) is dull olive-yellow overall; cap has least black by age and sex.

Adult m. CT/05

Adult m. MI/05

Adult f. TX/05

Winter f. NY/09

Canada Warbler 1

Wilsonia canadensis L 5¼"

Shape: Somewhat long-tailed and long-winged warbler. **Ad:** Variable plumage can make aging and sexing difficult. M. has striking necklace of black streaks that stands out against yellow underparts; yellow supraloral dash connects to bold whitish and/or yellowish eye-ring, simulating "spectacles"; black on forehead, crown, lores, and line extending down throat edge (black areas reduced on 1st-summer m.); blue-gray upperparts; white undertail coverts. F. similar but black necklace less distinct; little or no black on grayish crown or face. Wing tips often drooped, tail often cocked and flicked. **1st Winter:** M. similar to ad. f.; no black on grayish crown; auriculars may have some black; olive back; more diffuse necklace. F. paler with virtually no black in plumage; olive wash on grayish head and face; thin yellowish supraloral streak and whitish eye-ring; olive-brown-washed upperparts; faint to no grayish necklace streaks on breast. Similar dull Nashville Warbler has bright yellow undertail coverts; similar Magnolia Warbler has white wingbars, large white tail patches; Kentucky Warbler is entirely yellow below, no breast streaks, has eye-ring; Kirtland's Warbler is streaked on back and sides, not breast. **Flight:** Gray above, yellow below, with white undertail coverts. **Hab:** Dense moist understory of mature deciduous or mixed woodlands, shrubby areas near streams and swamps. **Voice:** Variable musical warbling that starts with a distinct, separate, but sometimes quiet *chip*. Call a loud *chik*.

Subspp: Monotypic. **Hybrids:** Mourning Warbler.

Adult m. NM/06

Adult m. NM/05

Red-faced Warbler 2

Cardellina rubrifrons L 5"

Shape: Relatively small warbler with a short thick bill, long pointed wings, and long tail. **Ad:** M. has unique and dramatic facial pattern; red face, forehead, sides of neck, and upper breast; black cheeks and back of crown; white nape, rump, and underparts; gray upperparts; some birds have pinkish lower breast from fall through early spring. F. similar with paler, more orangish face. Acrobatically moves about and often flicks tail. **1st Yr:** Similar to ad. f. but more brownish upperparts; m. face may be orange or as bright red as ad. male's; f. plumage dullest, with paler orange face; gray breast. **Flight:** Gray above, with white rump; white below, with reddish throat. **Hab:** Montane canyons and ravines with conifer-oak or deciduous woods, usually above 5,900 ft. **Voice:** Variable series of loud notes with accented ending. Call a sharp *chup.*

Subspp: Monotypic.

Adult m. AZ/04

Juv. AZ/06

Painted Redstart 2

Myioborus pictus L 5½"

Shape: Fairly large warbler with a long tail and short broad bill. **Ad:** M. has distinct red, white, and black plumage; black body; white crescent below eye; red lower breast and belly; large white wing patch; white outer tail feathers; mottled black-and-white undertail coverts. F. very similar to male; may sometimes have paler orangish-red belly. Frequently fans tail, wings partly open, as it forages along branches. Similar Slate-throated Redstart has slaty upperparts, no white wing patch or under-eye crescent. **Juv:** (Jun.–Aug.) Looks like adult but lacks the red belly. **Flight:** Extensive white on tail and upperwings; black throat and red belly. **Hab:** Mountainous pine-oak forests. **Voice:** Low-pitched, musical, variable series of 2-syllable phrases ending with one or several inflected notes. Call a low *chidiyew.*

Subspp: (1) •*pictus.*

Adult MEX/12

Slate-throated Redstart 4

Myioborus miniatus L 5¼"

Casual to s. AZ, s.e. NM, and s. and w. TX. **Shape:** Medium-sized warbler with a short broad bill and long graduated tail. **Ad:** M. has slate-black forehead, lores, chin, throat, and flanks; dark chestnut crown patch; slaty-gray upperparts; bright red breast and belly; dark mottling on lower breast and undertail coverts; large white spots on outer tail. F. sometimes not distinguishable from male; usually has grayer face and throat, paler red underparts; smaller chestnut crown patch. Frequently fans tail. Similar ad. and juv. Painted Redstart have large white wing patches; white arc under eye; more white in tail. **Juv:** (Jun.–Aug.) Similar to adult, but no red below; paler slaty upper- and underparts; warm brown mottling on undertail coverts. **Flight:** Extensive white on tail; no white on upperwings; black throat and red belly. **Hab:** Montane coniferous and mixed forests. **Voice:** Variable series of thin high notes speeding up at end. Call a high *chip*.

Subspp: (1) •*miniatus*.

Adult MEX/04

Fan-tailed Warbler 4

Euthlypis lachrymosa L 5¾"

Casual to s.e. AZ; accidental to w. TX. **Shape:** Large warbler with a long, broad, graduated tail. **Ad:** Black forehead and face with white supraloral spot and white eye arcs; golden-yellow crown patch; gray upperparts; yellow underparts with orange-yellow breast; olive-tawny flanks; white tip to outer tail feathers. Tail is frequently fanned and swung; often walks on ground. Similar Yellow-breasted Chat lacks yellow crown patch; has less yellow below. **1st Yr:** Much like ad.; some may have duller black on face. **Flight:** Large long tail with white outer tips; dark above, orangish yellow below. **Hab:** Damp, shaded, rocky, steep canyons and ravines. **Voice:** Series of sweet notes that go up or down at end. Song similar to Swainson's Warbler's. Call a high *seep*.

Subspp: Monotypic.

1st winter TX/02

Golden-crowned Warbler 4

Basileuterus culicivorus L 5"

Casual to s. TX, accidental to s.e. TX, e.cent. NM. **Shape:** Fairly deep-chested warbler with long tail and short wings. **Ad:** Distinct head pattern of broad yellow to orangish central crown-stripe between 2 black lateral crown-stripes; whitish eyebrow; whitish, split, narrow eye-ring; grayish-olive cheek; olive-gray upperparts; yellow underparts. Similar Wilson's and Orange-crowned Warblers lack crown-stripes. **1st Winter:** Very similar to adults, some with duller crown-stripes. **Flight:** Long tail and short wings; yellow below and olive-grayish above. **Hab:** Humid understory of evergreen or deciduous forests, second-growth habitats. **Voice:** Clear whistled notes with upward ending. Call a buzzy sometimes repetitive *chuk*.

Subspp: (1) •*brasherii* (e. MEX).

Adult AZ/03

Adult TX/04

Rufous-capped Warbler 3

Basileuterus rufifrons L 5"

Rare to w. and cent. TX and s.e. AZ. **Shape:** Fairly small, long-tailed, short-billed warbler; bill is thick and has curved culmen. **Ad:** Sexes alike. Unique head pattern with rufous cap and cheek; bold white eyebrow and moustachial area; dark lores. Olive-gray above; yellow throat and breast; white belly; grayish-buff flanks. Often cocks and waves long tail. **Juv:** (May–Sep.) Similar to ad. but with brownish throat and buffy wingbars. **Flight:** Long tail; dark above; yellow throat and upper breast, whitish belly. **Hab:** Brushy or second-growth habitats in foothills. **Voice:** Fast variable series of *chip* notes. Call a *chit,* sometimes in a series.

Subspp: (2) •*caudatus* (vag. to s.e. AZ) has pale reddish-brown auricular; yellow on underparts extends to midbelly. •*jouyi* (vag. to s.w. and cent. TX) has rich reddish-brown auricular; yellow on underparts less extensive.

Adult m., *virens* OH/05

Adult f., *auricollis* CA/05

Adult f., *virens* CT/01

Adult m., *auricollis* AZ/05

Yellow-breasted Chat 1

Icteria virens L 7"

Shape: The largest wood-warbler; broad-necked; thick blunt-tipped bill with strongly curved culmen; long tail. **Ad:** M. has white "spectacles" and moustachial stripe; black lores; blackish bill with gray base to lower mandible; olive-gray head; golden-yellow throat and breast; olive upperparts; olive-gray flanks with whitish belly and undertail coverts. F. similar to m. but duller; gray lores; brownish bill with yellowish-pink base to lower mandible; paler buff flanks. Similar Common Yellowthroat is much smaller, has black mask (m.); Yellow-throated Vireo is smaller, has yellow spectacles, white wingbars. **1st Yr:** Similar to ad. f. but duller and browner olive above. **Flight:** Large with long tail; dark above; below, yellow throat and breast, white belly and undertail coverts. **Hab:** Dense thickets, tangles, and brushy edges in dry or moist areas. **Voice:** Skulky; heard more than seen. Song is more like a mockingbird's than a warbler's, but slower, with a wide variety of whistles, rattles, squeaks, scolds, mews. Call is a grating *chack*.

Subspp: (2) •*auricollis* (s. SK–w. TX and west; vag. to GA): white moustachial stripe is large and wide, extends to behind eye; breast is fairly bright yellow. •*virens* (e. SD–e. TX and east): white moustachial stripe is small and narrow, generally stops below eye; breast is fairly dull yellow.

Bananaquit

Tanagers

Towhees

Sparrows

Juncos

Buntings

Cardinals

Grosbeaks

Adult, *sharpei**

CAY/03

Bananaquit 4

Coereba flaveola L 4½"

Casual Caribbean vag. to the Keys and s. FL mainland. **Shape:** Small, stocky, short-tailed, relatively long-legged bird with short, deep-based, sharp-pointed, downcurved bill. **Ad:** Dark crown and mask; broad white eyebrow and throat; yellow belly; white undertail coverts. Pinkish to red gape. Small white patch along edge of folded wing (pale bases of outer primaries). **Juv:** (Apr.–Aug.) Like ad. but paler gray above; paler yellow below; eyebrow may be pale yellow, rump patch duller yellow. **Flight:** Yellow rump; white tips to outer tail feathers; small white patch at base of primaries. **Hab:** Variable but always near flowers; woods, gardens. **Voice:** Song a series of high-pitched hissing sounds, like *sister sister sister;* calls include a short *seent.*

Subspp: (1) •*bahamensis.* **C. f. sharpei* (Cayman Is.) used for good photograph.

IDENTIFICATION TIPS

Tanagers are some of our most beautifully colored birds. They are medium-sized compact birds with fairly short tails and moderate-length thick bills with somewhat blunt tips. They glean insects off leaves and add fruit to their diet in fall through spring.

Species ID: Tanagers are strongly sexually dimorphic in summer, with most males being reddish and females greenish yellow. In winter, adult m. Scarlet and Western Tanagers change color and look like adult females; 1st-winter birds look much like the adult female; and 1st-year males will look blotched yellow and red as they gradually acquire their adult plumage. Wing color can be helpful for several reasons. For the Summer and Hepatic Tanagers the wings are strongly edged with the color of the body and blend with it; for other species the wings are brownish to black and contrast with the body color. Also, in winter, when m. Scarlet and Western Tanagers have a similar body color to the female, the color of the wings can identify the sex—males have black wings, females brownish to sooty wings. Male Summer and Hepatic Tanagers, once in ad. plumage, remain red all year.

Adult m., *hepatica* AZ/04

Adult f., *hepatica* AZ/06

Adult m., *hepatica* AZ/06

Adult f., *dextra* TX/06

Hepatic Tanager 2

Piranga flava L 8"

Shape: Medium-sized relatively large-headed bird with a fairly thick blunt-tipped bill; moderate-length tail. Similar Summer Tanager has proportionately larger bill on average (see Subspp.). Small pointed "tooth" on cutting edge of upper mandible (lacking in Summer Tanager; less pronounced in Scarlet Tanager). **Ad:** Upper mandible blackish, lower mandible gray; lore dark gray to sooty; wing mostly concolor with body. M. bright red to dull brick-orange overall (brightest on crown and throat) with variable grayish wash to auriculars, back, flanks, and wings (see Subspp.); may have a few green or yellow feathers. F. bright to dull yellowish olive overall with variable grayish wash to auriculars, back, flanks, and wings (see Subspp.); throat and undertail coverts may have orangish wash; older f. may have a few red feathers. Both sexes may be distinguished from Summer Tanager by their dark bill, dark lores, variable wash of gray on auriculars, back, and flanks, and brightest red on crown and throat. **1st Yr:** (Sep.–Aug.) Generally similar to ad. f., but m. usually has some red feathers or red patches, especially on head (similar to some older females and indistinguishable without knowing exact age of bird). **Flight:** Underwing coverts match body color in each sex. **Hab:** Open pine and pine-oak woods, forest edges, scattered trees. **Voice:** Song a series of 2-part slightly burry phrases, like *jerit jeeroo jeweet jereer;* calls include a single *chup* and a rising nasal *weent* in flight.

Subspp: (2) •*hepatica* (n.w. NM–s. AZ and west) generally large with larger bill proportionately; dull overall colors, lots of gray wash. •*dextra* (CO–w. TX) generally small with smaller bill proportionately; bright overall colors, minimal gray wash.

Adult m., *rubra* OH/04

Adult f. TX/05

Adult m. TX/04

Adult f. TX/04

1st-summer m. TX/04

Adult f. TX/04

Summer Tanager 1

Piranga rubra L 7¾"

Shape: Medium-sized relatively large-headed bird with a deep-based, fairly long, blunt-tipped bill. Similar Hepatic Tanager has proportionately smaller bill on average (see Subspp.) and small "tooth" on cutting edge of upper mandible (just a subtle shallow notch). **Ad:** Bill uniformly dull yellowish to horn color; lore concolor with head; wing mostly concolor with body. M. fairly evenly bright to dark rosy red overall (see Subspp.). F. bright yellow to greenish or brownish yellow overall, some with an orangish wash to throat and undertail (see Subspp.). Both sexes may be distinguished from Hepatic Tanager by their bill and lore color and by lack of gray wash on auriculars, back, and flanks; distinguished from Scarlet Tanager by bill color and lack of contrasting dark wing. **1st Yr:** (Sep.–Aug.) F. much like ad. f. Male similar to ad. f. but with a few red feathers (Sep.–Feb.), variably mixed red and greenish yellow (Mar.–Aug.). **Flight:** Underwing coverts same color as body. **Hab:** Deciduous or pine-oak forests, woods edges, open woods. **Voice:** Song a series of 2-part, slightly burry, whistled phrases delivered at variable tempos, like *turit jeroo taree jeray;* calls include a distinctive clipped call like *kituk* or *kitituk* and a whining *cheew.*

Subspp: (3) •*cooperi* (UT–s.w. TX and west) is large; m. bright rosy red, f. bright yellow. •*ochracea* (e.cent. AZ) is large; m. darker red, f. more greenish yellow. •*rubra* (s.e. NE–cent. TX and east) is small with proportionately smaller bill and shorter tail; m. dark rosy red, f. dull yellow with orangish wash. **Hybrids:** Scarlet Tanager.

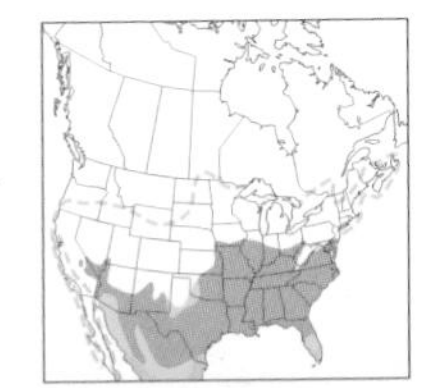

Adult m., summer TX/04

1st-summer m. TX/05

Adult m., winter NJ/09

1st-summer f. TX/04

Scarlet Tanager 1

Piranga olivacea L 7"

Shape: Slightly smaller than similar Hepatic and Summer Tanagers, with a fairly short blunt-tipped bill; bill shorter than or about equal to lore (generally longer than lore in Summer Tanager). Subtle "tooth" on cutting edge of upper mandible. **Ad. Summer:** Bill fairly evenly colored horn to grayish horn; lore concolor with head; dark wings contrast with body. M. brilliant red to orangish red overall except for black wings and tail (may be some red in median coverts). F. yellow on face and underparts (brightest on undertail coverts); olive above; wing coverts edged olive when fresh; primaries and secondaries dull sooty brown with fine olive edging when fresh; sometimes orangish wash on parts of body; sometimes thin yellow wingbar on greater coverts. **Ad. Winter:** M. like ad. f. but with wings and tail all black with little or no olive edging on coverts or flight feathers (body mixed yellow and red during molts). F. like summer f. but with more olive edging on wing feathers. **1st Yr:** F. similar to ad. female but brownish flight feathers contrast with sooty tertials and wing coverts. M. like ad. m. in respective seasons, but flight feathers are dark brown and contrast with black wing coverts. **Flight:** Underwing coverts whitish, contrasting with body color (concolor with body in other tanagers). **Hab:** Older forests with heavy canopy, wooded suburbs. **Voice:** Song a series of slightly burry, whistled phrases delivered fairly quickly, like *jurit jeroo jaree jerilay;* calls include a distinctive *chip-burr,* a single harsh *chip,* and a drawn-out rising *sweeyee.* 🎧**117**.

Subspp: Monotypic. **Hybrids:** Summer and Western Tanagers.

Adult m., summer CA/05

1st-summer m. TX/04

Adult m., winter OR/08

Adult f., summer (pale) OR/08

Western Tanager 1

Piranga ludoviciana L 7¼"

Shape: Fairly small proportionately short-billed tanager. **Ad. Summer:** Bill horn-colored to dusky yellow; sometimes upper mandible darker than lower. Two distinct wingbars, the upper yellow (extensively on m.), lower whitish. M. head orangish red; nape, rump, and underparts bright yellow; back, wings, and tail black. F. head and rump dusky olive overall to olive-yellow (head sometimes with orange wash on face), contrasting with gray back; underparts yellow to pale grayish yellow, with paler belly; wings and tail dusky. **Ad. Winter:** M. like summer but body duller overall; red on head limited to a pale wash on face or absent; back feathers tipped greenish yellow when fresh. F. like summer f.; no orange wash on face. **1st Winter:** M. like winter ad., but flight and tail feathers brownish, contrasting with black coverts and tertials; upper wingbar whitish to pale yellow. F. like winter ad., but dull yellowish olive to more grayish overall; brightest yellow on undertail coverts; little contrast between back and head or back and rump; upper wingbar whitish. **1st Summer:** Female like adult summer but with little to no red on face. M. like adult summer m., but red on head mostly on face, flight feathers and sometimes tail brownish, contrasting with blackish wing coverts, and belly often whitish. **Flight:** Underwing coverts concolor with body (white in Scarlet Tanager); tail all dark, contrasting with yellowish rump (tail corners white and rump darkish in Flame-colored Tanager). **Hab:** Open deciduous or mixed deciduous-coniferous woods. **Voice:** Song a series of slightly burry, whistled phrases delivered fairly quickly, like *jurit jeroo jaree jerilay* (much like Scarlet Tanager's); calls include a fast slightly rising *chideet* or *chidereet* and a rising *wee-yeet*. 🎧**118**.

Subspp: Monotypic. **Hybrids:** Scarlet and Flame-colored Tanagers.

Adult m., *townsendi* FL/02

Adult f. BAH/03

Adult m., *townsendi* FL/02

Adult f. BAH/05

Adult m., *zena* BAH/05

Western Spindalis 3

Spindalis zena L 6¾"

Rare Caribbean visitor to s. FL mainland and the Keys. **Shape:** Fairly small, deep-chested, large-headed bird with a short deep-based bill. **Ad:** Sexes strikingly different. M. with black crown, mask, and malar streak separated by white eyebrow and lower cheek; throat and breast deep orangish yellow; belly, flanks, and undertail white; greater coverts broadly white; edges of secondaries and tertials broadly white; outer tail feathers mostly white; back greenish or black (see Subspp.). F. very plain dark gray above; paler below with olive wash on breast; greater coverts and tertials edged whitish; small white patch at edge of folded wing (formed by white bases of primaries). **Flight:** M. with conspicuous white on central upperwing and white sides to tail; f. with small white patch at base of primaries. **Hab:** Forest edges, fruiting trees. **Voice:** Calls include a high *seee* or *seeeseeeseee*.

Subspp: (3) •*zena* (Bahamian vag. to FL) m. with black back; •*townsendi* (Bahamian vag. to FL) m. with green back; •*pretrei* (Cuban accidental to Key West) m. with bright green back.

Adult m. AZ/04

Adult f. AZ/06

Flame-colored Tanager 3
Piranga bidentata L 7¾"

Rare Mexican visitor to s.e. AZ, where it occasionally breeds; also has occurred in w. TX. **Shape:** Fairly large tanager with slightly larger bill and longer tail than Western Tanager. **Ad:** Dark grayish bill (upper mandible often darker); 2 broad wingbars, white to yellowish; broad white tips to tertials; variable blackish edge to auriculars; back grayish, streaked blackish. M. bright orange to reddish-orange head and underparts; f. usually deep yellow head and underparts, sometimes with orange wash. **1st Yr:** Head and underparts dull yellow to olive-yellow; no orange wash. **Flight:** Outer tail feathers tipped white; rump not strongly contrasting with back; underwing coverts orangish yellow. **Hab:** Wooded mountain canyons. **Voice:** Song a series of burry, whistled phrases delivered fairly quickly, like *jujit jejoo jajee jerilay* (burrier than Scarlet Tanager's); calls include an explosive *ch'dit*.

Subspp: (1) •*bidentata*. **Hybrids:** Western Tanager.

Male, *sharpei* TX/03

Adult f., *sharpei* TX/03

White-collared Seedeater 3
Sporophila torqueola L 4¼"

Shape: Very small, large-headed, broad-necked bird with a short deep-based bill and short rounded tail. Culmen strongly rounded. **Ad. Summer:** M. has black wings and tail, with 2 white wingbars; white patch on wing formed by whitish base to primaries; head and back grayish olive with some black mottling; partial collar and underparts mostly pale to rich buff (see Subspp.), except for whitish throat and some variable mottling on breast; whitish lower eye crescent; bill black. F. has brown wings and tail; pale olive-brown to olive-yellow on head and back; underparts pale buff; 2 whitish wingbars; no white wing patch. **Ad. Winter:** M. similar to summer but with brownish wash over head and upperparts and rich buff on underparts; bill pinkish to dusky. F. like summer f. **1st Yr:** (Aug.–Jul.) M. looks similar to ad. f. but has white wing patch at base of primaries. F. like ad. **Flight:** White patch at base of primaries on m. seen from above and below. Mildly undulating flight path. **Hab:** Dense grasses and reeds near shrubs or small trees. **Voice:** Song several whistled notes followed by a sweet trill, like *seeoo seeoo seeoo tr'r'r'r'r'*; calls include a short *tik* and repeated *chip*.

Subspp: (2) •*sharpei* (MEX north to s. TX) m. pale below with mottled breastband, f. with wingbars. •*torqueola* (from captivity in CA and AZ) m. rich buff below with broad dark breastband, f. without wingbars.

Adult m., *pusillus* COL/–

Adult f., *pusillus* CRI/02

Yellow-faced Grassquit 4

Tiaris olivaceus L 4¼"

Casual Caribbean and Central American vag. to s. FL and s. TX. **Shape:** Very small large-headed bird with a short, deep-based, conical bill and short rounded tail. Line of straight culmen continuous with shallow crown. **Ad:** Both sexes have similar distinctive head pattern, with yellow eyebrow, throat, and lower eye crescent; dark gray bill. M. blackish on rest of face and on breast to a variable degree (see Subspp.); upperparts dark olive-green; underparts grayish. F. concolor pale olive-green overall with traces of m. face pattern. **1st Yr:** M. with slightly less black on face; f. similar to adult. **Flight:** Dark overall with bright yellow chin. **Hab:** Grassy and weedy areas. **Voice:** Song a very high-pitched soft trill; calls include a short *tsik*.

Subspp: (2) •*olivaceus* (Caribbean vag. to FL) m. has limited black patch on upper breast, greenish auriculars. •*pusillus* (e. MEX. to COL and w. VEN, vag. to TX) m. has extensive black on breast and upper belly, blackish auriculars.

Adult TX/02

Adult TX/03

Olive Sparrow 2

Arremonops rufivirgatus L 6¼"

Shape: Broad-necked relatively long-billed sparrow with a medium-length rounded tail. **Ad:** Very plain. Olive wings and tail; grayish face and underparts; undertail coverts gray to olive. Brown lateral crown-stripes; paler grayish central crown-stripe; brownish eyeline; thin broken eye-ring; gray upper mandible, paler lower mandible with white spot at base. **Juv:** (Apr.–Aug.) Dusky streaking on breast, neck, and back; indistinct buffy wingbars. **Flight:** Generally short low flights within cover. **Hab:** Dense forest understory. **Voice:** Song an accelerating repetition of *chip* or downslurred *tseew,* like *tseew, tseew, tseew tseewtseewtseew;* calls include a soft *tsip* or *tsitsitsip* and a longer *tseeer.*

Subspp: (1) •*rufivirgatus*.

Adult CA/07

Adult AZ/04

Adult AZ/04

Juv CA/07

Green-tailed Towhee 1

Pipilo chlorurus L 7¼"

Shape: Medium size, deep belly, long rounded tail, and short conical bill typical of towhees. Comparatively small and proportionately short-tailed. **Ad:** Gray face; bright reddish-brown crown; white chin and submoustachial region separated by dark malar streak; white supraloral dot. Back, wings, and tail olive-green; breast gray; belly whitish. **1st Winter:** Like ad. but reddish-brown crown may be limited to forecrown and feathers tipped buff or gray. **Juv:** (Jun.–Aug.) Heavily streaked on brownish upperparts and whitish underparts. **Flight:** Olive-yellow underwing coverts. Flights usually short, with tail pumping. **Hab:** Dry shrubby areas. **Voice:** Song several well-spaced whistled notes followed by 1–2 varied short trills, like *teer tweet tr'r'r'r ch'ch'ch'ch;* calls include an ascending *meeyee,* a buzzy *dzeee,* and a short *tik.*

Subspp: Monotypic. **Hybrids:** Spotted Towhee.

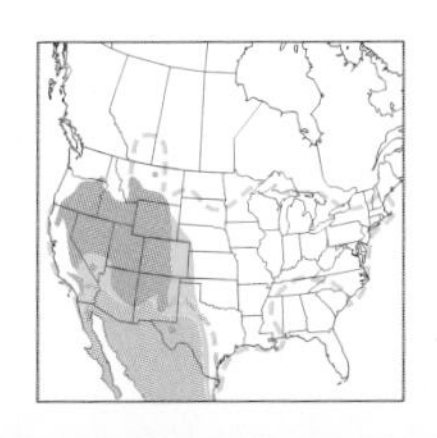

Adult m. CA/05

1st yr. BC/07

Adult f., Coastal Grp. BC/02

Juv. OR/08

Spotted Towhee 1

Pipilo maculatus L 8½"

Shape: Medium size, deep belly, long rounded tail, and short conical bill typical of towhees. Comparatively medium-sized with proportionately long tail. **Ad:** Dark hood and upperparts contrast with reddish-brown flanks and white belly; variable white spots on greater and median coverts and scapulars (see Subspp.); no white patch on base of primaries. M. and f. very similar; m. has black hood and upperparts, f. grayish-brown to blackish hood and upperparts. Similar Eastern Towhee has white patch at base of primaries and lacks white spots on tips of wing coverts and scapulars. **Juv:** Brownish overall on body; heavy streaking; no white wing patch on primaries. M. has black wings and tail, f. dark brown wings and tail. **Flight:** White tips to outer tail feathers; dark hood contrasts with white belly. **Hab:** Dense shrubby areas. **Voice:** Song 1 or more introductory notes followed by a harsh trill (mostly Interior Group) or just a harsh trill (mostly Coastal Group). Calls include a short *chk* or *ch'ch'ch'chk,* a harsh downslurred *jheeah* (mostly Interior Group) or harsh upslurred *jawee* (mostly Coastal Group), and a high-pitched *tseeawee.* **119**.

Subspp: (9) Two groups. **Coastal Group:** Sparse white spotting on back and wings, small white tips on outer tail feathers, darker flanks; less sexual dimorphism. •*oregonus* (BC–s.w. OR), •*falcifer* (n.w.–w.cent. CA), •*megalonyx* (w.cent.–s.w. CA), •*clementae* (s. CA is.). **Interior Group:** Heavy white spotting on back and wings, large white tips to outer tail feathers, paler flanks. •*falcinellus* (s. OR–s.cent. CA), •*arcticus* (AB–ND south to n.e. CO–NE), •*curtatus* (s. BC–e.cent. CA east to n. ID–cent. NV), •*montanus* (s.e. CA–AZ–cent. CO), •*gaigei* (s.e. NM–TX). **Hybrids:** Eastern and Green-tailed Towhees.

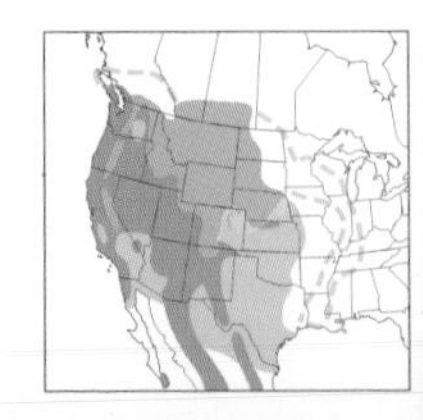

Adult m., *erythrophthalmus* NH/04

Adult m., *alleni* FL/0.

Adult m., *erythrophthalmus* NH/04

Adult f., *erythrophthalmus* NH/10

Eastern Towhee 1

Pipilo erythrophthalmus L 8½"

Shape: Medium size, deep belly, long rounded tail, and short conical bill typical of towhees. Comparatively small with medium-length tail. **Ad:** Dark hood and upperparts contrast with reddish-brown flanks and white belly; white patch on wing created by white bases of primaries. Eyes dark or pale (see Subspp.). M. has black hood and upperparts, f. brown hood and upperparts. **Juv:** (Jun.–Aug.) Brownish overall with heavy streaking on breast and flanks, white wing patch evident. M. has black wings and tail, f. brown wings and tail. **Flight:** White patch at base of primaries; white tips to outer tail feathers; dark hood contrasts with white belly. **Hab:** Open shrub understory of deciduous forests or pines, field edges, gardens, coastal scrub. **Voice:** Song 1 or more introductory notes followed by a trill, like *drink your teeeee* or *your teeee*. Calls include a high buzzy *dzeee* and a rising *chewink* or more whistled *sweek* (mostly FL birds). 🎧**120**.

Subspp: (4) •*erythrophthalmus* (OK–cent. VA and north) has red eye, extensive white on tail corners. •*canaster* (LA–s.w. TN to cent. SC–n.w. FL) and •*rileyi* (AR–LA east to s. VA–n. FL) are similar, with reduced white on tail, orangish to yellowish eye. •*alleni* (cent.–s. FL) has least white on tail, yellowish to white eye. **Hybrids:** Spotted Towhee.

Adult TX/05

Adult TX/04

Canyon Towhee 1

Melozone fusca L 8½"

Shape: Medium size, deep belly, long rounded tail, and short conical bill typical of towhees. Comparatively medium-sized with proportionately long tail. **Ad:** Dull grayish brown overall with cinnamon undertail coverts. Dull reddish-brown crown; whitish eye-ring; small pale supraloral spot; faint malar streak. Buffy throat; dark breast mark; scattered streaking across breast. Distinguished from similar California Towhee by range; similar Abert's Towhee has black face. **Juv:** (Apr.–Aug.) Like ad. but with dusky blurry streaking on breast and flanks; belly whitish. **Flight:** Small white tips to outer tail feathers. **Hab:** Variable; arid shrubby areas in canyons or uplands, open woodlands. **Voice:** Song an introductory call followed by a repeated phrase at a relaxed pace, like *chup, chee-chee-chee-chee* or *chirp, tooee-tooee-tooee.* Calls include a short *chedep,* a harsh percussive chatter, and a high *tseee.*

Subspp: (3) •*mesoleucus* (AZ–w. TX) is small with large deep-based bill, reddish-brown crown. •*mesatus* (n.e. NM–w. OK north) is large with smaller bill, reddish-brown-tinged crown. •*texanus* (cent.–s.w. TX) is small with long thin bill, mostly brown crown.

Adult CA/10

Adult CA/01

California Towhee 1

Melozone crissalis L 9"

Shape: Medium size, deep belly, long rounded tail, and short conical bill typical of towhees. Comparatively large with a very long tail. **Ad:** Dull grayish brown overall with orangish undertail coverts. Hint of reddish brown on throat, lore, and under eye. Faint to strong malar streak. Similar Canyon Towhee has separate range, often a breast dot, more contrasting head patterns (including pale supraloral spot). **Juv:** (Apr.–Aug.) Like ad. but with dusky streaking on breast and flanks; belly brownish; faint reddish-brown wingbars. **Flight:** No white on outer tail feathers. **Hab:** Dense shrubs, streamsides, gardens. **Voice:** Song a halting but accelerating series of metallic chips like *tink tinktink ti'ti'ti'tiink;* calls include a short *tssp* or *ts'ts'ts'tsp,* a metalic *chink,* and a buzzy *dzeee.*

Subspp: (6) •*bullatus* (OR–n. CA); •*carolae* (cent. CA) large; •*petulans* (n.w.–w.cent. CA); •*crissalis* (w.cent.–s.w. CA); •*eremophilus* (s.e. CA) medium-sized; •*senicula* (s.w. CA) small.

Adult, *dumeticolus* CA/05

Adult AZ/05

Abert's Towhee 1

Melozone aberti L 9½"

Shape: Medium size, deep belly, long rounded tail, and short conical bill typical of towhees. Comparatively large. **Ad:** Dull brown overall with varying reddish to pinkish tinge (see Subspp.); black face contrasts with pale bill; undertail coverts pale cinnamon. **Juv:** (Apr.–Aug.) Like ad. but with indistinct streaking on underparts. **Flight:** No white on outer tail feathers; black face; pale bill. **Hab:** Shrubby streamsides. **Voice:** Song a halting but accelerating series of harsh notes, like *chink chinkchink ch'ch'ch'chink;* calls include a *peek* and a high thin *seeep.*

Subspp: (2) •*dumeticolus* (UT–s.w. AZ and west) upperparts faintly reddish, underparts faintly cinnamon; •*aberti* (cent. AZ and east) upperparts grayish, underparts faintly pinkish.

IDENTIFICATION TIPS

Towhees are medium-sized, deep-chested, long-tailed birds with fairly long legs and short, deep-based, conical bills. They eat both seeds and insects, mostly from the ground, where they frequently hunt by jumping forward and then back to scrape away the leaf litter.

Species ID: Two towhees, Eastern and Spotted, look similar, have bright reddish-brown flanks, and are relatively colorful; they can be distinguished by the amount of spotting on their coverts and back and presence of white at base of primaries. The Green-tailed Towhee, with its rufous crown and greenish upperparts, is also distinctive. The other 3 towhees are relatively drab and can be distinguished through a careful look at their head and breast patterns.

IDENTIFICATION TIPS

Sparrows are small birds with short conical bills and varied-length tails. They are birds of primarily grasslands, fields, and open edges, where they feed mostly on seeds and some insects. Most are brownish with streaked backs, and they can look quite similar. Fortunately there are several large genera that have subtle but distinctive shapes. Becoming familiar with these shapes can help you place an individual sparrow into one of these groups, or genera; then you can look for plumage clues to complete your identification.

Species ID: There are 12 genera of sparrows in North America. Only 5 have 3 or more species, and these are the ones that are most useful to know to use this generic approach.

- ***Peucaea:*** Fairly large sparrows with shallow crown and long rounded tail. Except for singing males during breeding (Cassin's sings in flight), they are very secretive and when flushed take short low flights before diving back into vegetation. They rarely give calls during flight. *Rufous-winged, Cassin's, Bachman's, and Botteri's Sparrows.*
- ***Spizella:*** Small to medium-sized sparrows with high rounded crown, short conical bill, and fairly long notched tail. These are fairly conspicuous sparrows that often feed in flocks on the ground; when disturbed they tend to fly up to higher vegetation and look around. They often give calls during flight. They include several species that come to bird feeders. *American Tree, Chipping, Clay-colored, Brewer's, Field, and Black-chinned Sparrows.*
- ***Ammodramus:*** Small, broad-necked, large-headed, flat-crowned sparrows with a fairly large deep-based bill and short tail. These are very secretive sparrows of grasslands and marshes that tend to be alone or in pairs and generally stay low in vegetation; when flushed (a last resort for this genus) they tend to take short low flights and then dive down or stay perched for a few seconds before diving down in cover. They rarely give calls during flight. *Grasshopper, Baird's, Henslow's, Le Conte's, Nelson's, Saltmarsh, and Seaside Sparrows.*
- ***Melospiza:*** Medium-sized to large sparrows with rather average proportions; they are slightly deep-bellied and have a medium-sized bill, rounded crown, and fairly long rounded tail. These sparrows are easily seen in brushy areas and marshes; when flushed or curious they tend to fly up to higher perches for long periods and give short alarm calls. Some come regularly to bird feeders. *Song, Lincoln's, and Swamp Sparrows.*
- ***Zonotrichia:*** Large, deep-bellied, deep-chested, broad-necked sparrows with a fairly small conical bill, rounded crown, and fairly long, slightly notched tail. These sparrows are easily seen and often feed in small flocks out on open ground; when flushed they fly up to nearby higher vegetation and may give short alarm calls. Many come regularly to bird feeders. *Harris's, White-crowned, and Golden-crowned Sparrows.*

Adult AZ/04

Adult AZ/02

Adult AZ/12

Adult AZ/08

Rufous-winged Sparrow 2

Peucaea carpalis L 5¾"

Shape: Fairly large sparrow with a shallow crown and long rounded tail typical of the genus *Peucaea*. Bill comparatively deep-based and short (similar Rufous-crowned Sparrow has longer thinner bill). **Ad:** Gray face with reddish-brown crown and eyeline; 2 short whisker marks off base of bill (moustachial and malar streaks) with the submoustachial region between them gray like rest of face; thin white eye-ring. Bill pinkish, slightly darker on culmen. Distinctive reddish-brown lesser coverts (may be hidden by scapulars). Clear gray breast; 2 thin white wingbars. Similar Rufous-crowned Sparrow has one "whisker," no wingbars, bolder eye-ring, all-dark bill. **Juv:** Like ad. but finely streaked on head, breast, and flanks. **Flight:** Slightly undulating. Pale corners to tail. **Hab:** Desert scrub, arid grasslands. **Voice:** Song an accelerating series of short downslurred whistles, like *cheer cheer cheercheer ch'ch'ch'ch,* or 2–4 sharp calls followed by a trill. Calls include a short *tsip* and a repeated *tzeet tzeet tzeet.*

Subspp: (1) •*carpalis.*

Adult NM/05

Adult, worn AZ/08

Adult AZ/08

Adult TX/05

Adult AZ/11

Cassin's Sparrow 1

Peucaea cassinii L 6"

Shape: Large sparrow with shallow crown and long rounded tail typical of *Peucaea* genus. Bill comparatively short (similar Botteri's Sparrow's bill long, deep-based). **Ad:** Mottled grayish-brown face with few distinct markings (finely streaked brown crown; indistinct pale eyebrow; thin buffy eye-ring). Underparts grayish with dark streaking on rear flanks; back grayish with dark brown spots; tertials black with fine bright white margins; uppertail coverts long (visible beyond wings) with black tips; central tail feathers pale gray with faint or distinct crossbars. Similar Botteri's Sparrow has unstreaked flanks; brownish tertials with buffy edges; no black tips on uppertail coverts; unmarked dark central tail feathers. Feather patterns clearest from early winter through May; later often obscured by feather wear. Rare reddish morph has more reddish brown on upperparts, wings, and tail. **Juv:** (May–Sep.) Like ad. in feather patterns, but buffier overall and with fine streaks on breast and flanks. **Flight:** During breeding often sings while flying up then fluttering down. Pale whitish corners to tail. **Hab:** Grasslands with shrubs. **Voice:** Song 1–5 short whistles followed by a trill followed by 2 whistled notes, each preceded by a high-pitched *tsee,* like *tseet tseet tr'r'r'r'r'r tsee-say tseesooo.* Calls include a high *chip* and a chittering.

Subspp: Monotypic.

Adult, *arizonae* AZ/07

Adult, *arizonae* AZ/08

Adult, *arizonae* AZ/07

Adult, *arizonae* AZ/08

Botteri's Sparrow 2

Peucaea botterii L 6"

Shape: Large sparrow with shallow crown and long rounded tail typical of *Peucaea* genus. Comparatively, bill long and deep-based (similar Cassin's Sparrow has short bill). **Ad:** Mottled grayish-brown face with few distinct markings (finely streaked brown crown; indistinct pale eyebrow; thin buffy eye-ring; thin brown eyeline). Underparts grayish brown and unstreaked; back grayish brown with dark brown streaks; tertials dark brown with buffy margins; uppertail coverts long (visible beyond wings) with black shaft streaks; central tail feathers brown and unbarred. Feather patterns (those described) clearest from early winter through May; after that may be obscured due to feather wear. See Cassin's Sparrow for comparison. **Juv:** (May–Sep.) Similar to ad. in feather patterns, but buffier overall and with fine streaks on breast and flanks. **Flight:** Outer tail feathers with buffy tips. Does not do rising song flight (as Cassin's Sparrow does). **Hab:** Tallgrass areas with scattered shrubs. **Voice:** Song a variable number of halting calls followed by an accelerating trill and sometimes ending with a single note, like *tik, tadeet tadeet, chilip tr'r'r'r'r'r'r tseer;* calls include a *tsip tsip* and a chatter.

Subspp: (2) •*arizonae* (AZ) has more reddish wash to brown upperparts, slightly darker grayish breast; •*texana* (TX) has more grayish-brown upperparts, paler grayish to whitish breast.

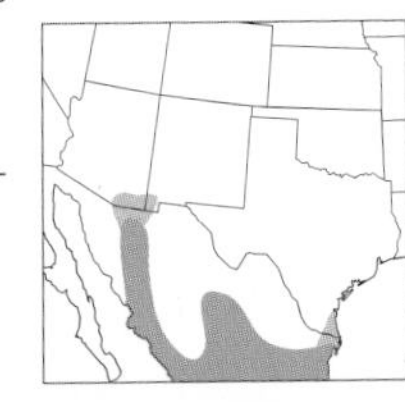

Adult TX/03

Adult FL/03

Adult FL/03

Adult FL/03

Bachman's Sparrow 2

Peucaea aestivalis L 6"

Shape: Large sparrow with shallow crown and long rounded tail typical of *Peucaea* genus. Bill comparatively long and deep-based (similar Cassin's Sparrow's bill short). **Ad:** Warm buffy-gray face with subtle reddish-brown markings (reddish-brown crown and eyeline; grayish eyebrow); clear buffy breast contrasts with whitish belly. Broad reddish-brown streaks on gray back and nape; no wingbar on reddish-brown greater coverts. Distinguished from similar Cassin's and Botteri's also by range, habitat. **Juv:** (May–Sep.) Like ad. but with black streaking on breast and crown. **Flight:** Brown tail has paler brown outer tips. Short fluttering flights with tail pumped. **Hab:** Open pine woods with palmettos and scrub. **Voice:** Song an alternation of beautiful drawn-out whistles and languid trills, like *tsoooo tretretretre tseeee cherchercher cher;* calls include a high *tsip*.

Subspp: (3) •*illinoensis* (TX–IN–n.w. FL) back feathers not dark-centered; •*aestivalis* (s.e. SC–cent. FL) and •*bachmani* (rest of range) back feathers dark-centered.

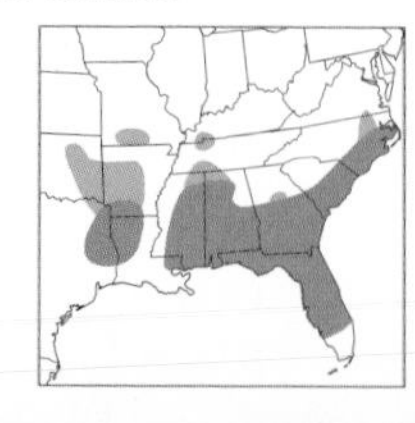

Adult, Pacific Coastal Grp. CA/03

Adult, Pacific Coastal Grp. CA/04

Adult, Pacific Coastal Grp. CA/05

Adult, Pacific Coastal Grp. CA/01

Adult, Southwest Desert Grp. AZ/07

Rufous-crowned Sparrow 1

Aimophila ruficeps L 6"

Shape: Large sparrow with shallow crown and long rounded tail. **Ad:** Gray face with reddish-brown crown and eyeline; one dark "whisker" mark (malar streak) below contrastingly paler submoustachial region; bold white eye-ring. Bill usually dark. Clear gray breast; no white wingbars. Similar Rufous-winged Sparrow has 2 "whiskers," wingbars, thin eye-ring, pinkish bill. **Juv:** (May–Oct.) Like ad. but buffier overall, with streaking on breast and flanks, dark streaking in crown, and indistinct malar streak. **Flight:** Short, direct, labored flights. Tail all dark.

Hab: Arid rocky hillsides with sparse grass. **Voice:** Song a short variable tumble of notes, like a short House Wren song; calls include a distinctive *deer deer deer.*

Subspp: (5) Two groups. **Pacific Coastal Group:** Small; upperparts reddish brown with buff or gray streaking. •*ruficeps* (cent.–s. CA), •*obscura* (s. CA is.), •canescens (s.w. CA). **Southwest Desert Group:** Large; upperparts grayish with brown to reddish-brown streaking. •*scottii* (AZ–w. TX), •*eremoeca* (s.e. CO–cent. TX).

Adult AZ/08

Adult AZ/08

Adult AZ/08

Adult AZ/07

Five-striped Sparrow 3

Amphispiza quinquestriata L 6"

Shape: Large sparrow with shallow crown and long rounded tail. Bill comparatively long and deep-based. **Ad:** Dark gray head boldly marked with bright white eyebrows, moustachial streaks, and throat (making up the "five stripes"); white crescent under eye. Dark gray underparts with black central breast dot and whitish central belly; dark brown upperparts. **Juv:** (Jun.–Sep.) Markings similar to ad. but body paler gray; lacks black on throat; no dark breast dot. **Flight:** Dark overall. **Hab:** Shrubby hillsides. **Voice:** Song a single note followed by a repeated note or short phrase, like *chip sleeslee, tsip chewchew;* calls include a low *churp* and high *tseet.*

Subspp: (1) •*septentrionalis.*

Adult, *arborea* NH/01

Adult, *arborea* OH/02

Adult, *arborea* OH/12

Juv., *ochracea* AK/07

American Tree Sparrow 1

Spizella arborea L 6¼"

Shape: Medium-sized sparrow with high rounded crown, short conical bill, and notched tail typical of *Spizella* genus. Comparatively, our largest *Spizella*. **Ad:** Bright reddish-brown crown; face gray with bright reddish-brown eyeline; broken white eye-ring; two-toned bill (dark above, yellow below). Pale gray breast and belly; warm buffy flanks; dark blurry central breast spot. Wings and back reddish brown; 2 white wingbars. Similar winter Chipping Sparrow smaller, duller, with lightly streaked crown, black eyeline, grayish-buff flanks, and no breast spot. **Juv:** (Jul.–Oct.) Similar to ad., with two-toned bill but heavily streaked breast. **Flight:** Undulating. **Hab:** Open shrubby areas in summer; weedy edges in winter. **Voice:** Song a 2-sec. phrase of sweet whistles and warbling; calls include a soft *teeahleet* and a high-pitched *tink*. 🎧**121**.

Subspp: (2) •*ochracea* (AK–cent. NT–BC) is paler with more gray on back, less reddish brown on flanks; tertials edged white. •*arborea* (rest of range) has more reddish brown on back and flanks; tertials edged buff.

Adult summer IA/05

Adult winter, *arizonae* NM/11

Adult summer, *passerina* NH/04

1st winter, *arizonae* NM/11

Adult winter, *passerina* OK/12

Juv., *passerina* NH/07

Chipping Sparrow 1

Spizella passerina L 5½"

Shape: Small deep-bellied sparrow with high rounded crown, short bill, and long notched tail typical of *Spizella* genus. **Ad. Summer:** Bright reddish-brown crown (with variable whitish streak on forehead); white eyebrow; black eyeline and lore; gray cheek and whitish throat; bill dark gray. Breast and belly fairly uniform silvery gray; 2 thin white wingbars. **Ad. Winter:** Crown reddish brown with fine black streaks and variably distinct gray central crown-stripe; buffy eyebrow; dark eyeline and dark lore; thin whitish eye-ring broken in front and back; brownish cheek; faint or no moustachial streak, but thin indistinct malar streak; bill orangish. Breast and belly fairly uniform silvery gray; gray rump. Similar Clay-colored Sparrow has brown crown with complete gray central stripe, whitish eyebrow, pale lore, thin complete eye-ring (stronger below), dark moustachial line, white moustachial region, brown rump. **1st Winter:** Like ad. winter, but crown dark beige with black streaks (no reddish brown); underparts washed with beige (rather than silvery); may have retained juv. streaking on breast (especially w. subspp.). **Juv:** (May–Nov.) No reddish brown on crown; streaked underparts. **Flight:** Undulating. Gray rump contrasts with brown back. **Hab:** Open woods, parks, gardens. **Voice:** Song an extended fairly harsh trill all on one pitch; calls include a short *seet* and rapid series of *chips*. **122**.

Subspp: (3) •*stridula* (s.w. BC–s.w. CA) and •*passerina* (s. TX–cent. ON and east) ad. winter generally dark with contrasting plumage; •*arizonae* (s. AK–w. ON–s. CA–s.w. TX) ad. winter paler (more like ad. winter Clay-colored and Brewer's Sparrows). **Hybrids:** Clay-colored and Brewer's Sparrows.

Adult summer TX/04

Adult summer MN/05

1st winter CA/10

Adult winter TX/01

Clay-colored Sparrow 1

Spizella pallida L 5½"

Shape: Small deep-bellied sparrow with high rounded crown, short stout bill, and long notched tail typical of *Spizella* genus. **Ad. Summer:** Generally buffy with fairly strong plumage contrasts. Crown brown with black streaks; distinct whitish central crown-stripe; whitish eyebrow; buffy lore; auricular buffy, outlined above with thin dark eyeline, below by dark moustachial streak; thin eye-ring most distinct below eye; thin malar and sub-moustachial streaks; bill pinkish. Wide gray collar contrasts slightly with buffy sides of breast (may be some streaking on central nape); brown rump. **Ad. Winter:** Like ad. summer, but overall buffier, duller, and with more blended features. Similar ad. winter Chipping Sparrow has broken eye-ring, dark lore, reddish-brown crown, variable central crown-stripe, gray rump. See Brewer's Sparrow for comparison. **1st Winter:** Like ad. winter, but may have buffier breast almost forming a band. **Flight:** Brown rump. **Hab:** Shrublands, shrubby and weedy borders, burns. **Voice:** Song a series of 2–5 fairly drawn-out buzzes; calls include a high *tsip*.

Subspp: Monotypic. **Hybrids:** Chipping, Field, and Brewer's Sparrows.

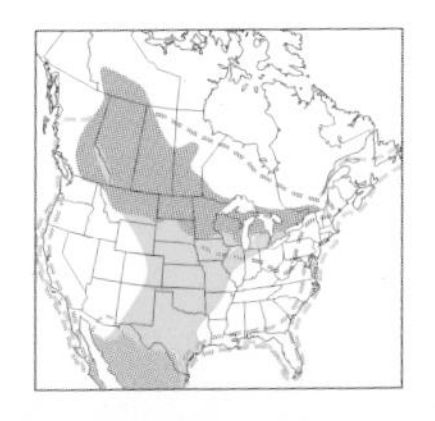

Adult summer, *breweri* MT/06

Adult winter, *breweri* AZ/02

Adult summer, *breweri* WY/07

Adult, *taverneri* YT/–

Brewer's Sparrow 1

Spizella breweri L 5½"

Shape: Small deep-bellied sparrow with high rounded crown, short stout bill, and long notched tail typical of *Spizella* genus. Bill comparatively shorter than in Clay-colored and Chipping Sparrows (see Subspp.). **Ad. Summer:** Generally grayish brown without strong plumage contrasts. Crown brown with black streaks (rarely indistinct gray central crown-stripe); grayish eyebrow not strongly contrasting with brown auricular; thin dark eyeline; pale lore; distinct complete white eye-ring; thin and fairly indistinct moustachial and malar streaks. Gray collar with extensive fine dark streaking; finely streaked brown rump. Similar Clay-colored Sparrow has more contrasting face pattern, faint eye-ring (stronger below), whitish eyebrow, gray central crown-stripe, bolder malar and moustachial streaks, little or no streaking on gray collar. **Ad. Winter:** Like ad. summer but buffier below. **Flight:** Undulating. Brown rump. **Hab:** Variable; open areas with scattered shrubs, coniferous woodlands with brushy understory; subsp. *taverneri* in stunted willows at tree line. **Voice:** Song a long series of trills, buzzes, and whistled notes; calls include a high rising *tseeyp*.

Subspp: (2) •*taverneri* (s.w. AK–s. AB–n.w. MT), "Timberline Sparrow," is darker overall, with heavy black streaking on nape and back, more contrasting facial marks (similar to Clay-colored Sparrow), whitish eye-ring, short bill. •*breweri* (rest of range) is paler overall, with less contrast in plumage, thin black streaking on nape and back, distinct bright white eye-ring, large bill. **Hybrids:** Chipping, Black-chinned, and Clay-colored Sparrows.

Adult, *pusilla* NH/04

Adult TX/04

Adult TX/03

Adult OK/12

Field Sparrow 1

Spizella pusilla L $5^3/_4$"

Shape: Small deep-bellied sparrow with high rounded crown, short stout bill, and long notched tail typical of *Spizella* genus. **Ad:** A gray and reddish-brown sparrow with a strong white eye-ring and pinkish to orangish bill. Reddish-brown cap with variable gray central stripe (see Subspp.); gray face; complete white eye-ring; variable reddish-brown eyeline (behind eye); gray collar. Reddish-brown mantle; variable rusty wash over gray breast and flanks (see Subspp.); 2 thin white wingbars. **Juv:** (May–Oct.) Like ad. but underparts streaked, wingbars buffy. **Flight:** Undulating. Gray rump, reddish-brown back. **Hab:** Old fields with scattered small trees. **Voice:** Song a series of sweet downslurred whistles, starting slow and accelerating into a trill, like *tseet tseer tseertseer tseer'r'r'r'r.* Calls include a short harsh *tchip,* a high-pitched trill, and a drawn-out *tseee.*

Subspp: (2) •*arenacea* (MT–ND south to OK) is mostly gray on face and underparts (may lack rusty eyeline and auricular patch); crown-stripe indistinct. •*pusilla* (rest of range) has strongly brownish facial markings, grayish-brown chest; crown-stripe distinct. **Hybrids:** Vesper and Clay-colored Sparrows.

Adult m., summer CA/04

Adult winter TX/03

Adult f., summer CA/05

Adult winter CA/09

Black-chinned Sparrow 1

Spizella atrogularis L 5¾"

Shape: Small deep-bellied sparrow with high rounded crown, short stout bill, and long notched tail typical of *Spizella* genus. Tail comparatively longer than in other Spizella sparrows. **Ad. Summer:** Plain dark gray head and body with a pinkish bill; back and wings reddish brown with black streaks. M. has extensive black on chin, throat, and foreface; f. with black less extensive to just darker gray on chin and throat. **Ad. Winter:** M. and f. both have from a few spots to no black on throat and chin; older males may have the most spots. **Juv:** Like ad. but with no black on face, indistinct streaking on underparts, faint buffy wingbars. **Flight:** Undulating. Grayish overall with reddish-brown wings and back. **Hab:** Arid shrublands, sage. **Voice:** Song a series of high-pitched whistles starting slow and accelerating to a very rapid trill that usually ascends in pitch; calls include a high *tsip*.

Subspp: (2) •*caurina* (w.cent. CA) has dark gray head and underparts; •*cana* (rest of range) has paler gray head and underparts. **Hybrids:** Brewer's Sparrow.

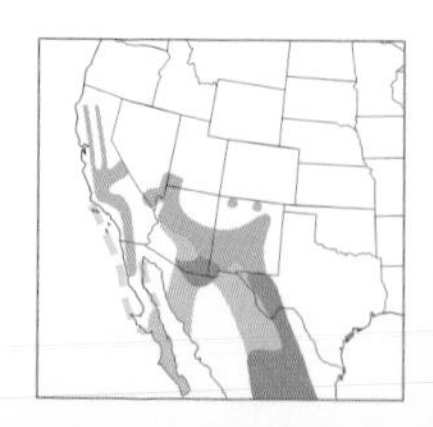

Adult, *confinus* UT/06

Adult, *gramineus* ME/06

Adult, *gramineus* ME/06

Adult OR/08

Vesper Sparrow 1

Pooecetes gramineus L 6¼"

Shape: Medium-sized, relatively small-headed, slender-billed sparrow with a moderately long slightly notched tail. **Ad:** Heavily streaked brown sparrow with distinctive markings not immediately apparent. Bold whitish eye-ring; brown auricular paler centrally; whitish moustachial region wraps around auricular patch; thin malar streak. Distinctive reddish-brown lesser coverts only seen in flight or on perched bird when wrist is held outside the scapulars. White underparts with fine black streaking on breast and flanks. Outer tail feathers white. **Juv:** (Jun.–Sep.) Broad buffy wingbars and no reddish brown on the lesser coverts; streaking on underparts. **Flight:** White outer tail feathers. Often flies from ground to perch when disturbed. **Hab:** Pastures, agricultural fields, sagebrush. **Voice:** Song starts with several slow downslurred whistles, which speed up into a varied warble; calls include a sharp *tsip* and a rising *zeeyeet*.

Subspp: (4) •*affinis* (coastal WA–OR) small, comparatively short-tailed; •*confinis* (BC–e. CA to ON–NE) and •*altus* (s.w. part of range) large, long-tailed; •*gramineus* (MN–MO and east) medium-sized, long-tailed. **Hybrids:** Field Sparrow.

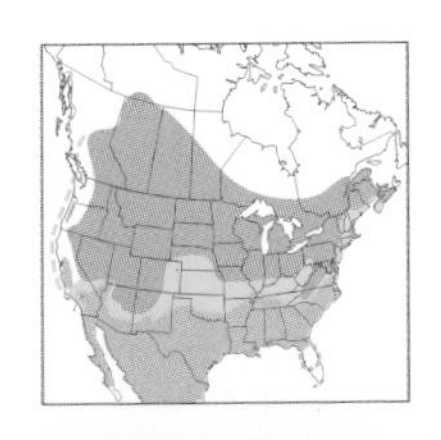

Adult, *strigatus* NM/06

Adult TX/04

1st winter, *strigatus* CA/09

1st winter, *grammacus* ME/09

Lark Sparrow 1

Chondestes grammacus L 6½"

Shape: Medium-sized sparrow with a large deep-based bill and long rounded tail. Primary projection past tertials longer than bill. **Ad:** Distinctive boldly white-and-dark streaked head pattern. Reddish-brown crown with broad white central stripe; reddish-brown auricular enclosing a broad white eye crescent and small white patch at the rear; bold black malar streaks flanked by white throat and white moustachial region; dark central spot on white breast. **1st Winter:** Like ad. but less rich reddish brown on face; may have some dark streaks on breast. **Juv:** (Jun.–Sep.) Similar markings to ad. (note white eye crescent), but streaked underparts and dotted whitish wingbars. **Flight:** Outer tail feathers white; others (except for central ones) tipped white. **Hab:** Variety of woodlands, brushy and weedy areas. **Voice:** Song a fairly short blend of single notes, trills, and warbles, often starting with repeated clear whistle; calls include a high *tsink*. **123**.

Subspp: (2) •*strigatus* (MB–w. TX and west) has paler head markings, narrow dusky streaks on back. •*grammacus* (MN–e. TX and east) has dark head markings, wide black streaks on back.

Adult, *deserticola* CA/01

Adult, *deserticola* CA/05

Adult NM/05

Juv., *deserticola* AZ/07

Black-throated Sparrow 1

Amphispiza bilineata L 5½"

Shape: Medium-sized, large-headed, broad-necked sparrow with a relatively long squared-off tail typical of *Amphispiza* genus. Compared with Sage Sparrow, Black-throated has larger bill, shorter tail. **Ad:** Striking dark-and-white head pattern. Dark gray head with bold white eyebrow; white lower eye arc; white moustachial region; black throat and pointed bib. Grayish-brown back and pale grayish underparts unstreaked. Tail black with white tips to outer feathers. **Juv:** (May–Sep.) Like ad. (with complete eyebrow) but head pattern paler and throat white. Back and breast with thin dark streaks; indistinct buffy wingbars. **Flight:** Streaked head, brown body and wings, black tail with white outer tips. **Hab:** Arid hillsides with scattered shrubs. **Voice:** Song a high-pitched varied mix of whistled notes and short trills, usually starting with a few harsh, soft, introductory *chips*. Calls include a low *chup* and a high *tink*.

Subspp: (3) •*deserticola* (s.w. WY–s.w. NM and west) and •*opuntia* (s.e. CO–s.e. NM east to OK–w. TX) are large, dark-backed, with small white tail spots. •*bilineata* (cent.–s. TX) is small, paler-backed, with large white tail spots.

Adult, Coastal Grp., *belli* CA/02

Adult, Coastal Grp., *belli* CA/01

Adult, Interior Grp. NM/11

Adult, Coastal Grp., *belli* CA/01

Adult CA/03

Sage Sparrow 1

Amphispiza belli L 6"

Shape: Medium-sized, large-headed, broad-necked sparrow with a long squared-off tail typical of *Amphispiza* genus. Compared with Black-throated Sparrow, Sage has shorter bill, longer tail. **Ad:** Medium-gray head with contrasting white supraloral dash; white eye-ring; white submoustachial region; variable malar streak (see Subspp.) borders white throat. Central underparts white and unstreaked with black breast dot; flanks and sides of breast buffy with fine streaking. Fine white edges to outer tail feathers when fresh. Runs along ground with tail held high; often flicks tail up. **Juv:** (May–Aug.) Head pattern like ad. (with white supraloral dot), but less distinct; underparts more heavily streaked. **Flight:** Dark above with blackish tail. **Hab:** Arid areas with sage, saltbrush, or similar shrubs. **Voice:** Song a short varied mixture of notes or a slower series of burry whistles; calls include a high *tsit*.

Subspp: (5) May be two separate species. Divided into two groups. **Coastal Group** ("Bell's Sparrow"): Small; dark gray head and back with no or some indistinct streaks; broad black malar streak, small supraloral spot. •*belli* (w.cent.–s.w. CA), •*clementeae* (San Clemente Is. CA). **Interior Group:** Large; pale gray head, pale brown back with distinct streaks; less distinct malar streak, larger supraloral spot. •*canescens* (cent. CA–s.e. CA–w. NV), •*nevadensis* (e. CA–WY–NM), •*campicola* (WA–OR–ID).

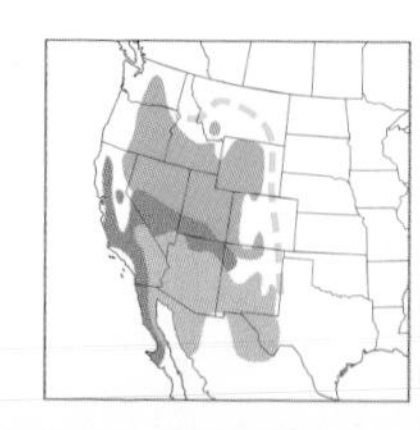

Adult m., summer MT/06

Adult f., summer CA/08

Adult m., winter TX/12

Male, winter TX/12

Lark Bunting 1

Calamospiza melanocorys L 7"

Shape: Large, fairly big-headed, broad-necked sparrow with a large, deep-based, conical bill and fairly short slightly notched tail. Long tertials cover short primaries, leaving almost no primary projection. **Ad. Summer:** M. black with white wing patch (formed by white greater and median coverts); bluish-gray bill. F. upperparts brown with dark streaks; underparts whitish with brown streaks; bluish-gray bill; white to creamy patch along edge of folded wing (formed by broad white edges to greater coverts); whitish submoustachial streak extends below auricular; variably bold malar streak; whitish eyebrow. **Ad. Winter:** M. streaked like summer f. but darker; blackish mottling on face and throat; large white wing panel; heavier streaking on breast; blackish centers to tertials. Molting m. a mixture of brown and black. F. much like ad. summer f. but a little paler and with a buffy to creamy wing patch. **Juv:** Much like ad. f. but with buffy margins to wing and back feathers, creating a scaled look. **1st Winter:** (Jun.–Sep.) Both sexes much like ad. f. **Flight:** White patch on wing coverts (m.); white tips to outer tail feathers. **Hab:** Short-grass areas, agricultural fields, desert brush. **Voice:** Song a slow series of varied phrases, each repeated several times, intermixed with high metallic trills, like *witeo witeo witeo toowee toowee t't't't't't't*. Calls include a high *tsip* given singly or in a series and a whistled *heew*.

Subspp: Monotypic.

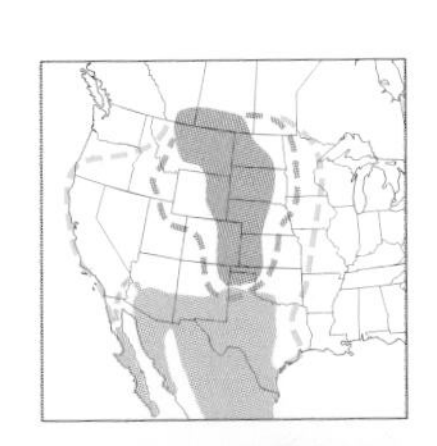

Adult, Pacific Coastal Grp., *beldingi* CA/05

Adult, Western Grp. AK/06

Adult, Large-billed Grp., *rostratus* CA/01

Adult, Western Grp. CA/05

Savannah Sparrow 1

Passerculus sandwichensis L 5½"

Shape: Medium-sized fairly small-headed sparrow, mostly with a fine-pointed bill (see Subspp.), rounded crown, and short notched tail. **Ad:** Heavily streaked with brown except for white throat, central belly, and undertail coverts; eyebrow variably colored, from all yellow to just yellow supraloral area to whitish and no yellow. Streaking on breast thin and distinct; often coalesces into a central breast dot. Crown streaked brown with thin whitish distinct to indistinct central stripe. **Flight:** White belly and, in some cases, pale outer tail feathers. **Hab:** Grassy meadows and agricultural fields. **Voice:** Song a few harsh introductory calls followed by 2 buzzy high-pitched trills on different pitches, like *ts ts ts tseeeee chaaay* (last note lower); calls include a high *tsip*. In Large-billed Group (see Subspp.) song is 3 drawn-out buzzes, like *dzeeee dzoooo dzaaaay.*

Subspp: (14) May be dichromatic in all subspp.—some birds paler, others dark grayish brown. Five groups. **Pacific Coastal Group:** Mostly small, small-billed, dark pat-

Adult, Eastern Grp. NH/06

Adult, Eastern Grp. ME/04

Adult, Eastern Grp. MB/06

Adult, Eastern Grp. GA/02

Adult, Ipswich Grp., *princeps* NH/10

terned; darker farther south. •*crassus* (s. AK); •*brooksi* (s.w. BC–n.w. CA); •*alaudinus* (w.cent. CA); •*beldingi* (s.w. CA), "Belding's Savannah Sparrow," is small, dark-colored with blackish streaking, has deep yellow eyebrow. **Large-billed Group:** •*rostratus* (breeds MEX, winters also s.cent. CA, Salton Sea), "Large-billed Savannah Sparrow," is medium-sized, has very large bill (= lore + eye) with slightly curved culmen, pale grayish-brown back with indistinct streaking, little or no yellow in eyebrow. **Ipswich Group:** •*princeps* (breeds Cape Sable Is. NS, winters on East Coast, s. ME–FL), "Ipswich Sparrow," is large, has relatively large bill, very pale grayish-brown back, faint brownish streaking on breast and flanks, little or no yellow in eyebrow. **Western Group:** Mostly large, fairly pale, bill medium to large. •*sandwichensis* (e. Aleutian Is.–w. AK), •*anthinus* (n.w. AK–cent. BC east to w. NT–n.w. MB), •*nevadensis* (cent. BC–e. CA east to cent. MB–n. AZ), •*rufofuscus* (cent. AZ–cent. NM). **Eastern Group:** Medium-sized, dark, small-billed. •*oblitus* (e. MB–MN–MI), •*mediogriseus* (n. ON–IL east to w. QC–NJ), •*labradorius* (e. QC–NF), •*savanna* (NS–PE). **Hybrids:** Grasshopper Sparrow.

Adult, *pratensis* ME/05

Adult, *perpallidus* ID/06

Adult, *pratensis* ME/06

Adult, *pratensis* ME/06

Juv., *pratensis* OH/07

Grasshopper Sparrow 1

Ammodramus savannarum L 5"

Shape: Small, broad-necked, large-headed, flat-crowned sparrow with a fairly large deep-based bill and short tail typical of *Ammodramus* genus. **Ad:** Blackish-streaked lateral crown-stripes with prominent whitish central stripe; dull buffy to grayish-brown face relatively unmarked except for thin brown line behind eye; eyebrow orange to buffy in front of eye, grayish white behind eye; complete whitish eye-ring. Breast largely unstreaked bright buff to pale buffy brown (may be slight dusky streaking on sides of breast and flanks); belly whitish. Median and greater coverts dark-centered near tips, creating suggestion of dark wingbars. Lemon-yellow edge to wing near wrist. **Juv:** (May–Aug.) Similar to ad. but with fine dark streaking across breast. **Flight:** Flights often short and fluttering as it drops into grasses. **Hab:** Grasslands interspersed with shrubs. **Voice:** Song high and thin with a few introductory ticks followed by a thin insectlike buzzing, like *tst tska dzssssss.* Calls include a *tsit* and rapid *tsidit*.

Subspp: (4) Ranges fairly distinct. •*perpallidus* (w. ON–cent. TX and west) has small slender bill, grayish back, and pale buff-brown breast. •*ammolegus* (s.e. AZ) is largest overall, with long slender bill, gray back with bright reddish-brown streaks, bright buffy breast. •*pratensis* (eastern half of range except FL) is small with a stout bill, gray back, buffy breast. •*floridanus* (cent. FL) is small with large bill, brown back, buffy breast. **Hybrids:** Savannah Sparrow.

Adult ND/06

Adult MT/06

Adult MT/06

Adult MT/06

Baird's Sparrow 2

Ammodramus bairdii L 5½"

Shape: Small relatively large-billed sparrow with a large head, broad neck, and flat crown typical of the genus *Ammodramus.* **Ad:** Dark-streaked lateral crown-stripes with wide orange to buff central stripe; face with ochraceous cast, especially in eyebrow; auriculars with 2 dark blotches along lower rear edge; dark malar and moustachial streaks. Upper breast pale buff with necklace of fine dark streaks (rarely making a central spot); fine streaks along flanks; rest of breast and belly whitish. Back blackish with buff or white streaks; pale buff collar with fine blackish streaks. **Juv:** (Jul.–Oct.) Like ad. but more scaled appearance on back, more streaking on breast. **Flight:** Outer tail feathers with pale edges. **Hab:** Grasslands interspersed with shrubs. **Voice:** Song varied but usually starts with repeated whistles followed by a variable trill; calls include a high *tsit.*

Subspp: Monotypic.

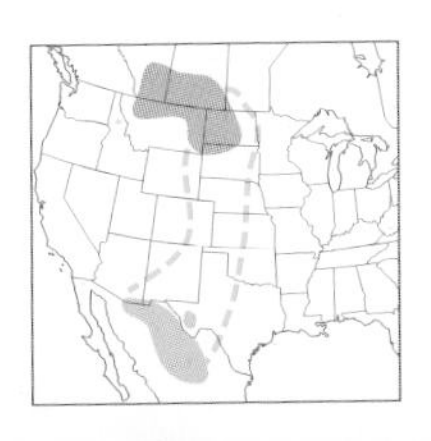

Adult, *henslowii* OK/07

Adult, *henslowii* OH/07

Adult, *henslowii* OH/05

Adult, *henslowii* OH/05

Henslow's Sparrow 2

Ammodramus henslowii L 5"

Shape: Small relatively large-billed sparrow with a large head, broad neck, and flat crown typical of the genus *Ammodramus.* Comparatively, Henslow's has very deep-based bill. **Ad:** Dark lateral crown-stripes with buffy central stripe; olive cast to face and neck with little contrast between eyebrow, auricular, and collar; thin white eye-ring; thin moustachial and malar streaks. Fine dark streaking across buffy upper breast and buffy flanks; broad dark streaks on back appear scaled when fresh; tertials and coverts edged reddish brown. Similar Le Conte's Sparrow has finely streaked grayish collar, tertials edged white, orangish face. **Juv:** (Jun.–Oct.) Like ad. but may lack malar streak and streaking on somewhat olive breast; scapulars strongly edged buff. **Flight:** Note reddish brown on wings. Flights usually short and zigzag. **Hab:** Prairies, grasslands interspersed with shrubs. Winters also in pine savanna. **Voice:** Song a short *tsilik* or *tseeleet* given at widely spaced intervals; calls include *tsip* and drawn-out *seee.*

Subspp: (2) •*henslowii* (e. ON–cent. TN and west) has small thin bill, fairly pale upperparts; •*susurranus* (NY–e. WV and east) has long thick-based bill, fairly dark upperparts.

Adult MN/06

Adult MB/06

Adult MN/06

Adult OH/10

Le Conte's Sparrow 1

Ammodramus leconteii L 5"

Shape: Small slender-billed sparrow with a large head, broad neck, and flat crown typical of the genus *Ammodramus*. **Ad:** Colorful sparrow. Thin, dark lateral crown-stripes with wide white central stripe; orange eyebrow and submoustachial region surround grayish auricular and lore; thin or no malar streak. Breast and flanks slightly duller orange-buff with fine but distinct black streaks on sides and flanks and few or none across breast; belly white. Reddish-brown streaking on gray collar; tertials edged white. See Nelson's, Saltmarsh, and Henslow's Sparrows for comparison. **Juv:** (Jul.–Nov.) Like ad. but not as bright orange on face and underparts; fine streaking on breast. **Flight:** Pale overall with dark streaked back. Flights short, weak. **Hab:** Grassy fields, wet meadows. **Voice:** Song a high raspy buzz with short introductory sounds, like *crkt dzeeeeezit;* calls include *tseeet.*

Subspp: Monotypic. **Hybrids:** Nelson's Sparrow.

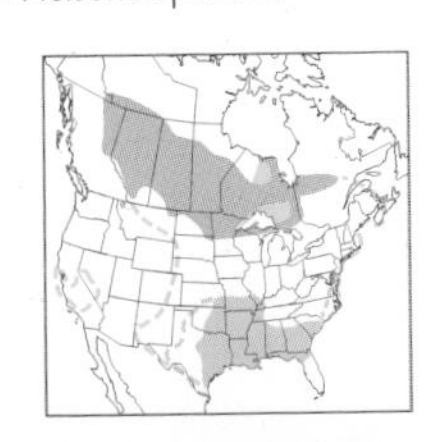

Adult TX/10

Adult ME/07

Adult, *subvirgatus* ME/06

Adult CT/09

Nelson's Sparrow 1

Ammodramus nelsoni L 5"

Shape: Small slender-billed sparrow with a large head, broad neck, and flat crown typical of the genus *Ammodramus.* Comparatively sharp-pointed tail feathers. Bill length varies by subspecies. **Ad:** Both Nelson's and Saltmarsh Sparrows have a dark gray crown with dark brown lateral stripes; bright to dull orange face surrounding a gray auricular; orangish lore; and gray unstreaked collar. Nelson's throat has thin, dark, indistinct malar streaks bordering a whitish to buffy throat that does not contrast strongly with its orange face; its broad orangish eyebrow has no distinct streaking. Substantial variation in underparts by subspecies. *A. n. nelsoni* has bright orange breast and flanks sharply demarcated from and strongly contrasting with its white belly; *subvirgatus* and *alterus* have whitish to pale buff breast that contrasts somewhat with whitish belly and buffy to orangish flanks. In all subspp., brownish streaking on breast and flanks is fine and pale on breast and broad and blurry on flanks. Similar Saltmarsh Sparrow has distinct but thin malar streaks bordering a white throat that strongly contrasts with the orange face; fine black streaking in its orange eyebrow behind the eye; pale buffy breast and flanks with distinct blackish streaking. Similar Le Conte's Sparrow has little or no streaking on breast, gray lore, pale central crown-stripe, and streaked gray collar. **Juv:** (Jul.–Sep.) Similar to ad., but buffy orange below with little or no streaking. **Flight:** Short and low. **Hab:** Coastal salt marshes, wet prairies, and inland marshes. **Voice:** Song a harsh unmusical *crt tshhhhhhhjut;* calls include a short *tik* and sibilant *tssst.*

Subspp: (3) •*nelsoni* (n.w. MN–n.e. SD and west) has short bill, dark brown back with bright white streaks, bright orange face and buff breast, distinct dark streaking on breast and flanks. •*subvirgatus* (coastal s.e. QC and NS south to e.cent. ME) has long bill, grayish back with faint gray streaks, dull buffy face and breast, indistinct grayish streaking on breast and flanks. •*alterus* (s. Hudson Bay) has short bill, medium brown back with dull whitish streaks, dull orange face and buff breast, indistinct grayish streaking on breast and flanks. **Hybrids:** Saltmarsh and Le Conte's Sparrows.

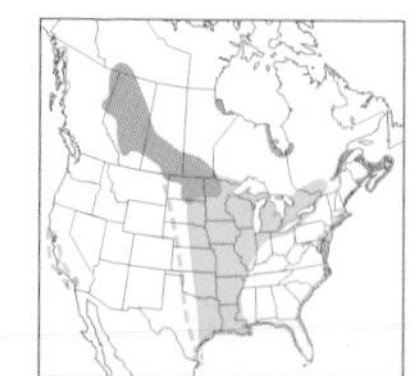

Adult, *caudacutus* ME/05

Adult, *caudacutus* ME/06

Adult NJ/05

Juv., *caudacutus* ME/09

Saltmarsh Sparrow 1

Ammodramus caudacutus L 5¼"

Shape: Small long-billed sparrow with a large head, broad neck, and flat crown typical of the genus *Ammodramus*. Comparatively sharp-pointed tail feathers. **Ad:** Both Saltmarsh and Nelson's Sparrows have a dark gray crown with dark brown lateral stripes; bright to dull orange face surrounding a gray auricular; orangish lore; and gray unstreaked collar. Saltmarsh has thin, dark malar streaks bordering a white throat that contrasts strongly with its orange face; its broad orangish eyebrow has distinct, fine, dark streaking behind the eye. Breast and flanks are light buff to pale orangish and distinctly streaked blackish. Similar Nelson's Sparrow has indistinct malar streaks bordering a buffy throat that does not strongly contrast with the orange face; no black streaking in its orange eyebrow; and brownish streaking on breast and flanks that is fine and pale on breast and broad and blurry on flanks. Similar Le Conte's Sparrow has little or no streaking on breast, gray lore, pale central crown-stripe, streaked gray collar. **Juv:** (Jul.–Sep.) Like ad., but buffy orange below with streaking. **Flight:** Short and low. **Hab:** Coastal salt marshes. **Voice:** Song more varied and slightly more musical than that of Nelson's with some whistled notes preceding harsh trills that occur at various pitches. Calls include a short *tik* and sibilant *tssst*.

Subspp: (2) •*caudacutus* (n. NJ north) has large bill, bright orange face; •*diversus* (s. NJ south) has smaller bill, dull orange face. **Hybrids:** Nelson's and Seaside Sparrows.

Adult, Atlantic Grp., *maritimus* NH/07

Adult, Cape Sable Grp. FL/04

Adult, Atlantic Grp., *maritimus* NJ/05

Adult, Cape Sable Grp. FL/04

Adult, Gulf of Mexico Grp. TX/08

Seaside Sparrow 1

Ammodramus maritimus L 6"

Shape: Stocky, long-billed, short-tailed sparrow with a large head, broad neck, and flat crown typical of the genus *Ammodramus.* Comparatively large with long spikelike bill and rounded tail. **Ad:** Dark sparrow with a bright white throat and white to yellow submoustachial region separated by dark malar streak; yellow supraloral dash; dark gray bill. Breast white to dark gray and heavily streaked (see Subspp.). **Juv:** (May–Aug.) Head like ad., may lack strong malar streak, has buffy breast and flanks with fine streaking, whitish belly. **Flight:** Short-tailed; dark overall with brownish wings. **Hab:** Coastal salt marshes. **Voice:** Song includes harsh and musical introductory notes to a long harsh trill; calls include a *chek chek* and a downslurred *tew tew tew.*

Subspp: (7) Divided into four groups. **Atlantic Group:** Dark grayish-olive back and belly with indistinct darker streaking. •*maritimus* (ME–NC), •*macgillivraii* (NC–FL). **Gulf of Mexico Group:** Gray back and buffy breast, both with distinct dark streaks. •*sennetti* (s. TX), •*fisheri* (e. TX–w. FL), •*peninsulae* (w.cent. FL). **Cape Sable Group:** Dark olive back and white breast, both with distinct dark streaking. •*mirabilis* (s. FL), "Cape Sable Seaside Sparrow." **Dusky Group:** Blackish upperparts. •*nigrescens* (extinct e.cent. FL), "Dusky Seaside Sparrow." **Hybrids:** Saltmarsh Sparrow.

Adult, *georgiana* MI/05

Adult, *nigrescens* DE/07

Adult, *georgiana* OH/05

Adult, *georgiana* MI/05

Swamp Sparrow 1

Melospiza georgiana L 5¾"

Shape: Medium-sized sparrow with rather average proportions. Medium-sized bill, rounded crown, slightly deep-bellied, with a fairly long rounded tail typical of *Melospiza* genus. Comparatively, Swamp is relatively large-headed, broad-necked, short-tailed. **Ad. Summer:** Generally dark sparrow with reddish-brown wings and rich buffy-brown flanks. Crown reddish brown with little or no dark streaking; may have white central streak on forehead; face gray with dark line behind eye; throat white and set off by thin malar streaks. Dark gray collar. Breast gray with variable indistinct dusky streaking; can be hint of central breast dot. M. crown brighter reddish brown than f. crown. **Ad. Winter:** Like ad. summer but duller overall. Crown more streaked with dark brown or black; gray central crown-stripe; face and underparts more washed with brown; breast with blurry dusky streaking. **Juv:** (Jun.–Aug.) Like ad. but more heavily streaked overall, no central crown-stripe. **Flight:** Upperparts dark with reddish-brown wings and rump. **Hab:** Marshes, bogs, marshy edges of lakes, moist thickets. **Voice:** Song a variable trill of modulated notes, like *tweetweetweetweetwee* or *syusyusyusyu;* calls include a short *tsit,* a harsher *chit,* a thin *zeeet*, and a soft *chip*.

Subspp: (3) •*ericrypta* (BC–QC–NF and north) has small bill, pale brown flanks; •*georgiana* (SD–MO to n. NJ–NB) has small bill, dark brown flanks; •*nigrescens* (s. NJ–MD) has large bill, grayish-olive flanks.

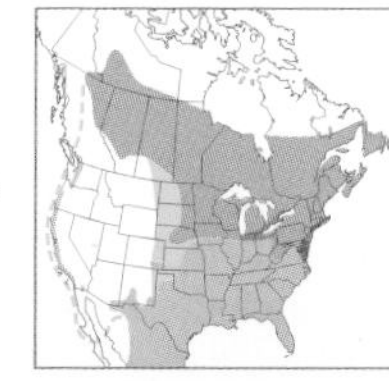

Adult, Sooty Grp. CA/12

Adult, Slate-colored Grp. CA/10

Adult, Sooty Grp. BC/02

Fox Sparrow 1

Passerella iliaca L 7"

The Fox Sparrow has four distinctly marked subspecies groups. These are treated separately below, and some evidence suggests they should be separate species.

Coastal Northwest Sooty Group ("Sooty Fox Sparrow"). **Shape:** Very large, deep-bellied, deep-chested sparrow with rounded crown, relatively short bill, and moderate-length slightly notched tail typical of *Passerella* genus. **Ad:** Darkest of the Fox Sparrows. Concolor dark sooty to warm brown overall with extensively dark flanks and sides of breast; heavy streaking covers most of breast and flanks; back relatively unstreaked. Whitish eye crescents; yellow lower mandible; no wingbars. Usually no light gray or bright reddish brown in plumage. **Juv:** Like ad. but duller with buffy underparts. **Flight:** Uniformly dark above. **Hab:** Coastal and montane shrubby areas, gardens. **Voice:** Song a series of clipped, often buzzy, whistles, well spaced at beginning; calls include a sharp *chit,* rising *seeet,* short *tsack,* and a *tchup.*

Subspp: (7) •*unalaschcensis* (e. Aleutian Is.–w. AK), •*ridgwayi* (Kodiak Is.), •*sinuosa* (s.cent. AK), •*annectens* (s.cent.–s.e. AK), •*townsendi* (s.e. AK–Q. Charlotte Is.), •*chilcatensis* (s. AK–w.cent. BC), •*fuliginosa* (w.cent. BC–n.w. WA).

Western Slate-colored Group ("Slate-colored Fox Sparrow"). **Shape:** Very large, deep-bellied, deep-chested sparrow with rounded crown, relatively short bill, and moderate-length slightly notched tail typical of *Passerella* genus. Comparatively long-tailed. **Ad:** Concolor gray head and back contrast with brown wings and reddish-brown tail. Flanks and breast heavily streaked with brown, back relatively unstreaked. Bill comparatively small and slender. **Juv:** Like ad. but duller with buffy underparts. **Flight:** Grayish back contrasts with reddish-brown wings and tail. **Hab:** Montane shrubby areas. **Voice:** Song a series of whistled notes on different

Adult, Thick-billed Grp. CA/05

Red Grp. NH/12

Adult, Thick-billed Grp. CA/10

Adult, Red Grp. CT/11

pitches, with accent on alternate notes, some notes with burry quality; calls include a short *tsack*.

Subspp: (4) •*olivacea* (s.w. BC–e. WA), •*schistacea* (s.e. BC–s.cent. AB to e. OR–cent. CO), •*swarthi* (s.e. ID–s.e. UT), •*canescens* (e. CA–e.cent. NV).

Western Thick-billed Group ("Thick-billed Fox Sparrow"). **Shape:** Very large, deep-bellied, deep-chested sparrow with rounded crown, relatively short bill, and moderate-length slightly notched tail typical of *Passerella* genus. Comparatively very large-billed, relatively long-tailed. **Ad:** Similar in plumage to Slate-colored Fox Sparrow but with a much longer, deeper-based bill. Bill usually all gray (most often yellowish in other Fox Sparrows). **Juv:** Like ad. but duller with buffy underparts. **Flight:** Grayish back contrasts with reddish-brown wings and tail. **Hab:** Shrubby areas. **Voice:** Song a varied series of whistles and warbles, suggesting the song of a House Finch; calls include a metallic *tsink* and a high thin *seet*.

Subspp: (3) •*megarhyncha* (s.cent. OR–cent. CA), •*brevicauda* (n.w. CA), •*stephensi* (s.cent. CA).

Northern Red Group ("Red Fox Sparrow"). **Shape:** Very large, deep-bellied, deep-chested sparrow with rounded crown. Comparatively short-tailed. **Ad:** Generally bicolored reddish brown and gray on head and back; reddish-brown auriculars, crown, and streaking on back. Wings and tail reddish brown. **Juv:** Like ad. but duller with buffy underparts. **Flight:** Reddish-brown tail and wings contrast with gray rump. **Hab:** Shrubby areas, often along streams. **Voice:** Song a series of rich whistles given at a leisurely pace; calls include a rising *seeyeet* and a short *tsack*.

Subspp: (3) •*iliaca* (n.e. MB and east) breast spotting heavy and bright reddish brown; •*zaboria* (cent. AK–cent. MB) breast spotting fairly sparse and dull reddish brown; •*altivagans* (BC–AB) breast spotting fairly heavy and dull reddish brown.

Adult, Eastern Grp. NH/05

Adult BC/02

Adult, Eastern Grp. OH/03

Adult CA/12

Song Sparrow 1

Melospiza melodia L $4^3/_4$–$6^3/_4$"

Shape: Medium-sized to large sparrow with rather average proportions: medium-sized bill, rounded crown, slightly deep-bellied, with a fairly long rounded tail typical of *Melospiza* genus. **Ad:** Strongly streaked sparrow over most of its body. Much variation (see Subspp.). In general, dull brown crown with dull gray central stripe; broad gray eyebrow; dark brown eyeline; whitish eye-ring; gray to clay-colored moustachial region; broad, dark, strongly flared malar streak. Heavy streaking on breast and flanks; usually an irregular breast dot. **Juv:** (Apr.–Sep.) Heavily streaked overall with only hint of ad. facial pattern. **Flight:** Long tail pumps up and down during short-distance flights. **Hab:** Shrubby areas, often near water. **Voice:** Song starts with several well-spaced phrases or notes followed by some trills and other notes, like "maids maids put on your tea kettle kettle." Calls include a soft *tsip,* a high *seeet,* and a louder deeper *tchup.* **124**.

Subspp: (29) Divided into six groups. Strongly clinal, with much individual variation. One of the most polytypic species in North America. Subspecies and their groups best identified by breeding range; at other times of year sometimes may not be separable. The descriptions of groups and their representative subspp. do not represent all of the variations and are just general guidelines for distinctions. **Eastern Group** (n.e. BC to MO and east): (4) Medium-sized; head and upperparts gray and brown without strong contrasts; wings and tail warm brown, underparts whitish with coarse black and/or brown streaking; bill stout. Main subsp. •*melodia.* **Interior Western Group** (Rocky Mts. from s.e. BC to AZ): (5) Small to medium size; head and upperparts overall pale gray and reddish with minimal streaking; underparts whitish with sparse reddish-brown streaking; bill slender. Main subsp. •*montana.* **California Mainland Group** (coastal s.w. OR–s.w. CA): (8) Small; overall more sharp contrasts on head and underparts due to brighter whites and distinct black streaking; generally bright white chin and supraloral and submoustachial regions; chest and belly white with distinct black streaking; wings and tail dull brown to warm reddish brown; bill slender. Main subsp. •*gouldii.* **California Island Group** (Channel Is. CA): (3) Small; generally pale gray and brown

Adult CA/03

Adult, Alaska Island Grp. AK/03

above; below, white with sparse distinct black streaking; slender bill. Main subsp. •*clementae*. **Northwest Coastal Pacific Group** (s.cent. AK–w.cent. OR): (5) Large; overall very dark, with belly the only bright area on the bird; head and back all very dark brown, with little pattern showing; throat usually dark; flanks strongly washed with dark gray and heavily streaked with dark brown; bill long and slender. Main subsp. •*rufina*. **Alaska Island Group** (AK Is.): (4) Largest among Song Sparrows; overall mostly dark; head and upperparts dull gray with indistinct reddish-brown streaking; greatest contrasts on head due to contrastingly white chin and supraloral and submoustachial regions; underparts mostly dark gray with indistinct dark brown streaking; bill long and slender. Main subsp. •*insignis*. **Hybrids:** White-crowned Sparrow.

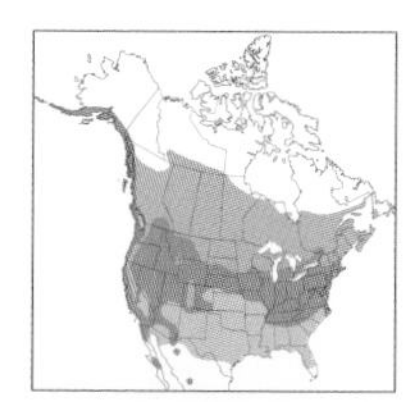

Adult, *lincolnii* NH/10

Adult TX/01

Lincoln's Sparrow 1

Melospiza lincolnii L 5¾"

Shape: Medium-sized sparrow with rather average proportions. Medium-sized bill, rounded crown, slightly deep-bellied, with a fairly long rounded tail typical of *Melospiza* genus. Comparatively, Lincoln's has thin fine-pointed bill, relatively short tail, relatively small head. Alert posture includes slightly peaked crown. **Ad:** Buffy breast and flanks with distinct, fine, black streaking contrast sharply with clear white belly and whitish throat. Often has a breast spot. Lateral crown-stripes dark reddish brown streaked with black; central crown-stripe gray. Broad gray eyebrow; thin buffy eye-ring; buffy submoustachial region; relatively thin blackish malar streak. **Juv:** (Jun.–Aug.) Like ad. but more heavily streaked overall. **Flight:** Grayish brown overall with demarked buffy breast and flanks. **Hab:** Boglike areas, shrubby streamsides. **Voice:** Song a bubbly series of warbles, some a little buzzy, often rising in middle and falling at end. Calls include a short *tsit* and harsher *chit*.

Subspp: (3) •*gracilis* (s.e. AK–s.w. BC) small, underparts with broad streaking; •*alticola* (OR–MT and south) large, underparts with fine streaking; •*lincolnii* (w. AK–e. WA and east) medium-sized, underparts with fine streaking.

Adult white morph MI/05

Adult tan morph OH/10

Adult tan morph NH/10

Adult white morph NH/10

1st-yr. white morph NH/10

White-throated Sparrow 1

Zonotrichia albicollis L 6¾"

Shape: Large, deep-bellied, deep-chested, broad-necked sparrow with fairly small conical bill, rounded crown, and fairly long slightly notched tail typical of the *Zonotrichia* genus. Comparatively, White-throated is slightly smaller, larger-headed, flatter-crowned. **Ad. Summer:** Lateral crown-stripes black to brown; central crown-stripe white, gray, or buff; broad white, gray, or tan eyebrow; variably bright yellow supraloral dot or dash. White throat, finely outlined in black, contrasts with gray breast and cheeks. Brown flanks and gray breast range from unstreaked to fairly heavily streaked; occasionally a hint of a breast dot. Median and greater coverts tipped with white or buff, creating 2 thin dotted wingbars. **Morphs:** Two color morphs, also substantial individual variation. In general, older male white morphs have the brightest, most contrasting plumage and 1st-yr. female tan morphs have the least contrasting plumage. White morphs have white eyebrow and central crown-stripe; more black in lateral crown-stripe; bright white throat patch with faint or no malar streak divisions. Tan morphs have buffy eyebrow and central crown-stripe; more brown in lateral crown-stripes; dull white throat patch with noticeable malar streak divisions. **Ad. Winter:** White morphs become slightly duller and buffier, making distinctions between morphs sometimes subtle; brightest birds are most likely white morphs. **1st Yr:** Like ad. but with browner, duller, less contrasting head pattern, duller throat, and heavy streaking on breast and flanks. **Flight:** Colorful head of light morphs. **Hab:** Brushy understory of coniferous woods, shrubby edges. **Voice:** Song a lovely series of 3–7 long whistles, the first usually lower-pitched, the later ones often quavering; calls include a drawn-out *tseeet,* a high *pink,* and a short *spik.* 🎧**125**.

Subspp: Monotypic. **Hybrids:** Golden-crowned and Harris's Sparrows, Dark-eyed Junco.

Adult summer IL/05

Adult winter IL/03

Adult summer MB/06

1st winter CA/10

Harris's Sparrow 1

Zonotrichia querula L 7½"

Shape: Large, deep-bellied, deep-chested, broad-necked sparrow with fairly small conical bill, rounded crown, and fairly long slightly notched tail typical of *Zonotrichia* genus. Harris's is our largest sparrow. **Ad. Summer:** Black crown, face, and chin surround pinkish-orange bill; rest of face pale gray. Breast and belly white; flanks brownish and streaked; pendant necklace of dark dots across upper breast. **Ad. Winter:** Like ad. summer but reduced black on head and wash of warm brown on face. **1st Winter:** (Sep.–Apr.) Like ad. winter but no large areas of black on head; throat white. **Flight:** Large size; white belly. **Hab:** Open spruce woods, forest edges, burns. **Voice:** Song 2–3 clear whistles all on one pitch; calls include a harsh *tseeek,* a chatter, and a thin *tseeet*.

Subspp: Monotypic. **Hybrids:** White-throated and White-crowned Sparrows.

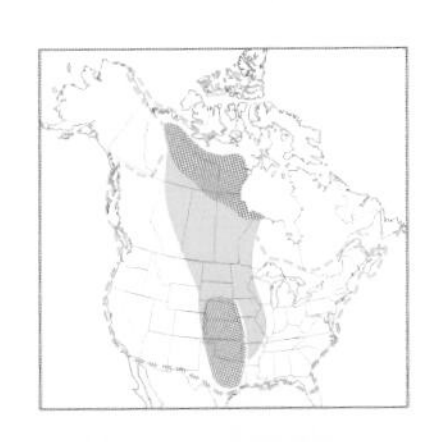

Adult (dark lore), *leucophrys* ME/05

Adult (pale lore), *gambelii* MB/06

1st winter NH/10

Adult, *pugetensis* BC/02

1st winter NH/10

White-crowned Sparrow 1

Zonotrichia leucophrys L 7"

Shape: Large, deep-bellied, deep-chested, broad-necked sparrow with fairly small conical bill, rounded crown, and fairly long slightly notched tail typical of *Zonotrichia* genus. Comparatively, White-crowned is small-headed, large-bodied. Raised crown creates peak at back of head. **Ad:** Striking black and white stripes on head created by white central crown-stripe and eyebrows, black lateral crown-stripes and eyelines; supraloral area black or gray (see Subspp.). Rest of face, collar, and breast pale gray and unmarked. Bill pinkish orange to yellowish; flanks gray to brown (see Subspp.). Two dotted white wingbars. **1st Winter:** (Sep.–Apr.) Like ad. but black and white head stripes replaced by grayish-brown central crown-stripe and eyebrow and rich brown lateral crown-stripes; bill pale pinkish to yellowish. Similar 1st-winter Golden-crowned Sparrow has indistinct lateral crown-stripes, indistinct eyeline, trace of yellow on forehead, gray bill. **Flight:** Streaked back; grayish-brown rump. **Hab:** Shrubby areas, open areas with scattered short trees, gardens. **Voice:** Song variable by subsp., often starts with a long whistle, then short warbles or whistles of varying quality, like *zoooo zeeee jeje* or *zoooo jipel jipel je* or *zeee zoozoo jee*. Calls include a sharp *pink* and a rising *seeyeet*. 🎧**126**.

Subspp: (5) •*leucophrys* (n. ON–NF), "Eastern," and •*oriantha* (s.w. AB–e. CA–n.w. NM), "Mountain," have black supraloral area, brownish-pink bill. •*gambelii* (AK–NT south to BC–n. MB), "Gambel's," has pale supraloral area, pinkish bill, pale gray breast and flanks, no malar streak. •*pugetensis* (s.w. BC–n.w. CA), "Puget Sound," and •*nuttalli* (w.cent. CA), "Nuttall's," have pale supraloral region, yellowish bill, brownish breast and flanks, often a thin malar streak. **Hybrids:** Song, Golden-crowned, and Harris's Sparrows.

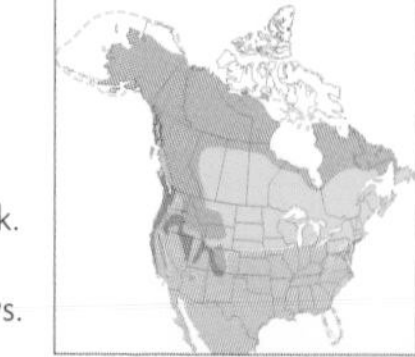

Adult summer CA/05

Adult summer AK/06

1st winter BC/02

Adult winter BC/02

1st winter CA/02

Golden-crowned Sparrow 1

Zonotrichia atricapilla L 7¼"

Shape: Large, deep-bellied, deep-chested, broad-necked sparrow with fairly small conical bill, rounded crown, and fairly long slightly notched tail typical of *Zonotrichia* genus. Comparatively very large. **Ad. Summer:** Dark sparrow. Black forehead and lateral crown-stripes surround front and sides of golden forecrown; otherwise, plain dull gray on face and underparts, with a wash of brown along flanks. Upperparts dull brown, streaked darker on back, plain on rump. Two dotted wingbars. Bill gray above, yellow below. **Ad. Winter:** Like ad. summer, but black areas on crown mixed with brown; yellow forecrown duller. **1st Winter:** Like ad. except face mostly plain brown and unmarked; bill gray; lateral crown-stripes brown and slightly darker than rest of face; usually a hint of yellow on forehead. See 1st-winter White-crowned Sparrow for comparison. **Flight:** Fairly dark overall with light belly and golden crown. **Hab:** Shrubby areas, often on hillsides, gardens. **Voice:** Song a series of 3 clear whistled notes, usually descending in pitch, sometimes written "oh dear me"; calls include a high *tseep,* a sharp *pink,* and a chatter. 🎧**127**.

Subspp: Monotypic. **Hybrids:** White-throated and White-crowned Sparrows.

Adult m., Slate-colored Grp. MN/01

Juv., Oregon Grp. OR/08

Adult f., Slate-colored Grp. NH/10

Dark-eyed Junco 1

Junco hyemalis L 6¼"

Shape: Large, deep-chested, broad-necked sparrow with a short conical bill, high rounded crown, and medium-length slightly notched tail. **Ad:** Unstreaked; pale bill; dark lore; generally strong contrast between darker grayish to brownish upperparts (including breast and flanks) and white belly and undertail coverts. Outer tail feathers variably white, varying by subsp. and by age and sex (young f. having least white and older m. having most). **Juv:** (May–Sep.) Heavily streaked overall except on white belly and undertail coverts. Streaking stops abruptly at belly; bill pinkish. **Flight:** Dark bird with bright white outer tail feathers. **Hab:** Open woods, brushy edges, gardens. **Voice:** Song an extended musical trill on one note; calls include a single or repeated *kew,* a short buzzy *zeeet,* a high-pitched *tsip* or *tsitsitsip*, and a hard sharp *tik.* **128**.

Subspecies are divided into five groups. Each group described separately below. **Hybrids:** White-throated Sparrow.

Slate-colored Group ("Slate-colored Junco"; *Hyemalis* Group). **Ad:** Generally concolor (except for *cismontanus*) dark slate gray to brownish gray overall except for contrasting white belly and undertail coverts. Bill pinkish to bluish white (see Subspp.). M. darkest gray without brown wash; f. with variable brownish wash on head and central back. About 3% of Slate-colored Juncos may have some white tips to wing coverts (see White-winged Group). **1st Winter:** Gray areas variably paler than ad.; light

Adult m., Oregon Grp. NM/11

Adult m., Oregon Grp. NM/12

Adult m., Oregon Grp. BC/02

Adult f., Oregon Grp. NM/12

to strong brownish wash over all dark areas. Molt limits sometimes occur; this leaves older outer coverts with pale tips, suggesting a short faint wingbar.

Subspp: (3) •*hyemalis* (n.w. AK–cent. AB to NF–MA) has pale pink to white bill. •*carolinensis* (Appalachian Mts.) is similar to *hyemalis* but larger, paler, with grayish bill. •*cismontanus* (s. YT–cent. BC–w.cent. AB; winter occurrences in East), "Cassiar Junco": M. with blackish-gray hood that contrasts with gray back, wings, and flanks; f. with gray head that blends to or slightly contrasts with brownish back, wings, and flanks; appears intermediate between other Slate-colored subspp. and Oregon Juncos; f. Oregon usually has sharper contrast between gray head and brownish back, giving it a more distinctly hooded look, and it has pinkish-brown, rather than gray, flanks.

Oregon Group ("Oregon Junco"). **Ad:** Smaller than most other subspp. Variably gray (f. and 1st winter) to black (m. and some 1st-winter m.) hood contrasts sharply with reddish-brown upperparts and white belly, creating a distinctly hooded look. Pinkish brown on flanks not as extensive and, in f. and 1st winter, lores not as contrastingly blacker as in Pink-sided group.

Subspp: (5) Strongly clinal. In general, paler and duller farther inland from the Pacific Coast and south in the group's breeding range. Four subspp. breed mainly down the West Coast: •*oreganus* (s.e. AK–w.cent. BC), •*simillimus* (s.w. BC–w.cent. OR), •*thurberi* (s.w. OR–n.w. CA), •*pinosus* (w.cent. CA). The fifth, •*shufeldti* (cent. BC–AB to n.e. OR–MT), may interbreed with Pink-sided Junco.

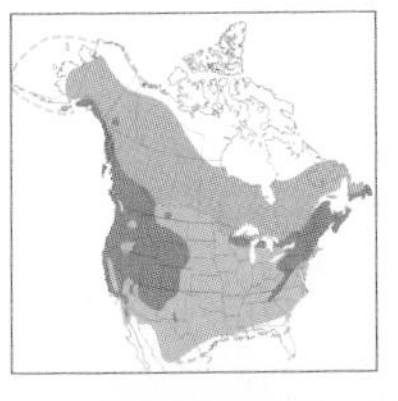

[*continued* →]

Adult, Pink-sided Grp. NM/11

Adult, Gray-headed Grp. NM/11

Adult, Pink-sided Grp. NM/12

Adult, White-winged Grp. CO/03

[continued]

Pink-sided Group ("Pink-sided Junco"). **Ad:** Pale gray hood with slightly paler throat; lores contrastingly blackish. Pinkish cinnamon on flanks extensive, sometimes meeting across breast and covering most of belly; back brownish; substantial white on tail. Similar Oregon Junco has more contrastingly hooded look, less contrast in lores; color on flanks is less extensive and reddish brown. Similar Gray-headed Junco has less extensively grayish flanks, reddish-brown (rather than dull brown) back.

Subspp: (1) •*mearnsi* (s.e. AB–s.w. SK to e. ID–n.w. WY).

Gray-headed Group ("Gray-headed Junco"; *Caniceps* Group). **Ad:** Generally medium gray overall with bright reddish-brown back; little contrast between head, breast, flanks, and paler whitish belly, which can be tinged gray; strongly blackish lores. Outer 3 tail feathers white.

Subspp: (2) •*caniceps* (s. ID–s. WY to s.e. CA–n. NM) has all-pinkish bill, concolor gray throat and head. •*dorsalis* (n.cent. AZ–s.w. TX), "Red-backed Junco," has two-toned bill with upper mandible darker, throat paler than rest of head.

White-winged Group ("White-winged Junco"). **Ad:** Larger than other subspp. Similar to Slate-colored Junco but paler slate gray and with more white in tail; thin white wingbars on median and greater coverts. Outer 3 tail feathers mostly to entirely white (Slate-colored has only outer 2 tail feathers white). F. and some males may lack wingbars; some individuals may have contrastingly darker lores.

Subspp: (1) •*aikeni* (s.e. MT–n.e. WY to w. SD and n.w. NE).

Yellow-eyed Junco 2

Junco phaeonotus L 6¼"

Shape: Large, deep-chested, broad-necked sparrow with a short conical bill, high rounded crown, and medium-length slightly notched tail. **Ad:** Gray head and rump; back, greater coverts, and tertials reddish brown; eye yellow, partially surrounded by black lores (creates "fierce" look); bill dark above, pale yellowish to horn below; white outer feathers to otherwise gray tail. **1st Yr:** Like ad. but tertials edged with brown; iris may still be dark olive-gray to grayish yellow at first (Oct.–Mar.). **Juv:** (May–Sep.) Similar to ad. but gray areas finely streaked darker; iris dark; bill can be all dark. **Flight:** White outer tail feathers; reddish-brown back and inner wing. **Hab:** Higher-elevation pine and oak forests. **Voice:** Song usually has a central drawn-out buzzy note flanked by introductory and closing phrases, like *sibesibesibe dzeeee tsibet;* calls include a *tsit* and a high *tseeep.*

Subspp: (1) •*palliatus.*

Adult m., summer CO/07

Adult f., summer MT/06

Adult m., summer CO/06

Juv. CO/07

McCown's Longspur 2

Rhynchophanes mccownii L 6"

Shape: Medium-sized, short-tailed, short-legged bird with a short conical bill, fairly small flat-crowned head, and defined neck typical of *Calcarius* genus. Comparatively, McCown's has longest, thickest-based bill, shortest tail, moderate primary extension past tertials (about 2 x bill length). **Ad. Summer:** M. with distinctive gray face, black crown and moustachial streak, black bill; median coverts reddish brown, tipped cinnamon or whitish; underparts grayish with black breast patch (when worn, belly can be mostly blackish). F. plain-faced with few distinctive marks; light brown crown finely streaked with black; buffy to whitish eyebrow; brown auricular with little or no dark border at back; whitish throat; bill dusky to pinkish with darker tip; secondary coverts dark, tipped with cinnamon, no reddish brown. **Ad. Winter:** Like ad. summer f.; in general, palest, most plain-faced winter longspur; dull brown overall; indistinctly marked; much white on tail (see Flight). M. has reddish-brown median coverts, tipped cinnamon; may have hint of black patch on breast. F. has buffy breast with streaking mostly on sides (some on upper flanks); median coverts dark, tipped cinnamon, with no reddish brown. Similar Chestnut-collared Longspur has smaller duskier bill, shorter primary projection, dark border to back of auriculars, no reddish brown on median coverts, no dark T on spread tail (see Flight), brownish bill. **1st Winter:** Like ad. winter but buffier overall. **Flight:** Undulating. Dark on central tail feathers and tips of outer feathers forms an inverted blackish T on spread tail (seen best upon landing). **Hab:** Sparsely vegetated short-grass plains in summer; bare dirt areas in winter. **Voice:** Song composed of several musical twittered phrases; calls include a short dry rattle and a *pink*, which can be mixed with the rattle.

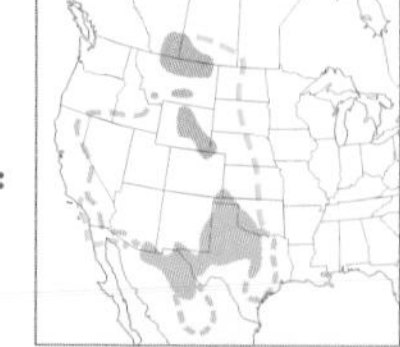

Subspp: Monotypic. **Hybrids:** Chestnut-collared Longspur.

Adult m., summer, *alascensis* AK/06

Adult winter OH/02

Adult f., summer, *alascensis* AK/06

1st winter NH/10

Adult winter OH/01

Juv., *alascensis* AK/07

Lapland Longspur 1

Calcarius lapponicus L 6¼"

Shape: Medium-sized, short-tailed, short-legged bird with a short conical bill, fairly small flat-crowned head, and defined neck typical of *Calcarius* genus. Comparatively, short bill, long primary projection past tertials (3 x bill length). **Ad. Summer:** Reddish-brown greater coverts (creating a reddish-brown wing panel); white belly with heavy streaking on flanks; limited white on dark tail (outer feather on each side all white, next one partially white). M. has black crown, face, throat, and bib; whitish eyebrow; orangish-yellow bill; reddish-brown nape. F. has warm brownish auricular strongly bordered with dark; crown and central nape finely streaked with black; variable reddish-brown nape patch; variable dark spotting on breast (can suggest a breastband). **Ad. Winter:** In general, has the most contrast between dark back and white belly of our winter longspurs. Reddish-brown panel on greater coverts; white belly with heavy streaking on flanks; strong dark border to auricular; tail mostly dark with some white on outer 2 tail feathers. M. breast usually more heavily mottled black; otherwise sexes quite similar. Similar Smith's Longspur is strongly buffier below, has pale spot on neck sides below auricular, lacks reddish-brown wing panel, lacks bold streaking on flanks, has more white on outermost tail feathers, and has shorter primary projection. **1st Winter:** Like ad. winter but usually underparts more buffy; breast markings fainter. **Flight:** Only outermost tail feather on each side all white, next is tipped with white, rest dark. **Hab:** Moist tundra in summer; bare dirt and short-grass areas in winter. **Voice:** Song a somewhat harsh jumble of twittering; calls include a percussive rattle, *tew* or *jit* calls sometimes mixed with rattle, and a *kiteeoo*.

Subspp: (3) •*alascensis* (AK–w. NT) has paler back, less dark streaking; •*subcalcaratus* (cent. NT and east) has darker back, dense streaking; •*coloratus* (w. AK is.) has dark back, heavy streaking.

Adult m., summer MB/06

1st winter CA/09

Adult winter CA/10

1st winter NJ/10

Smith's Longspur 2

Calcarius pictus L 6¼"

Shape: Medium-sized, short-tailed, short-legged bird with a short conical bill, fairly small flat-crowned head, and defined neck typical of *Calcarius* genus. Comparatively, Smith's is longer-tailed and more slender-billed, and has moderately long primary projection past tertials (2 x bill length). **Ad. Summer:** Strongly buffy underparts; both outer tail feathers all white. M. has black-and-white "helmet" and buffy-orange underparts and nape; back strongly streaked with black; white patch on lesser coverts. F. strongly warm buff overall, especially face and underparts; auricular with central paler spot and moderately bordered with dark; breast buffy with distinct to indistinct fine streaking. **Ad. Winter:** Like ad. summer f.; in general, buffier overall, especially on belly, than any other winter longspur. Similar Lapland Longspur has whitish belly, reddish-brown wing panel, bold streaking on flanks, less white in outermost tail feathers. **Flight:** Warm buffy below; white outer tail feathers in mostly dark tail; white lesser coverts on m. **Hab:** Short grasses in tundra in summer; short grassy areas in winter. **Voice:** Song a rapid series of warblerlike whistles (like Chestnut-sided Warbler) with a strong ending like *wichew;* calls include a short percussive rattle.

Subspp: Monotypic.

Adult m., summer CO/06

Adult winter CA/11

Adult winter NV/10

Adult f., summer CO/06

Juv. CO/07

Chestnut-collared Longspur 1

Calcarius ornatus L 6"

Shape: Medium-sized, short-tailed, short-legged bird with a short conical bill, fairly small flat-crowned head, and defined neck typical of *Calcarius* genus. Comparatively, Chestnut-collared has smaller bill, shortest primary projection past tertials (less than 1½ x bill length). **Ad. Summer:** M. has black crown; broad white eyebrow; black eyeline that wraps around back of pale auricular; bright reddish-brown collar; grayish bill; black breast and belly. F. streaked brown above; grayish white below with indistinct streaking; pale creamy eyebrow; indistinct dark eye-stripe; pale auricular with dark rear border; may be some reddish brown on collar. **Ad. Winter:** Both sexes dull grayish brown; streaked above, indistinctly streaked below; facial pattern indistinct; dark rear border to auriculars; grayish-brown bill; tail extensively white with dark triangle on central trailing edge. M. with hint of black belly showing through buffy feather tips. Similar McCown's Longspur has larger bill, longer primary projection, little or no dark border to auriculars, blackish T pattern on tail tip, reddish-brown tips to median coverts (m.). **Flight:** Tail mostly white with brown to black triangle at tip. Makes short song flights during breeding. **Hab:** Short-grass prairies in summer; areas with sparse vegetation, agricultural fields in winter. **Voice:** Song a lovely series of slurred descending whistles; calls include a short *kittle* or *kidit* and a short rattle.

Subspp: Monotypic. **Hybrids:** McCown's Longspur.

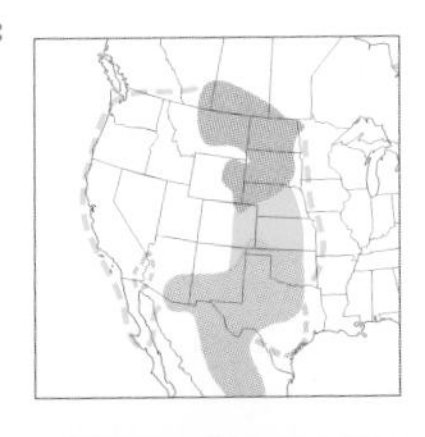

Adult m., summer, *townsendi* AK/07

Winter, *nivalis* NH/11

Adult f., summer AK/06

Juv., *townsendi* AK/07

Winter, *nivalis* OH/01

Adult, *nivalis* NH/11

Snow Bunting 1

Plectrophenax nivalis L 6¾"

Shape: Medium-sized, short-legged, broad-necked, front-heavy bunting with a short conical bill and short notched tail. Long primary projection past tertials. **Ad. Summer:** Striking white-and-black plumage; bill dark. M. all white except for black back, extensive black on primaries, tertials, tail, and at bend of wing. F. similar but with dusky streaking on crown; back streaked with black to blackish brown; more black on coverts and secondaries. **Ad. Winter:** Mostly white with variable bright buffy wash on head, nape, sides of breast, and flanks; back pale gray to light brown with darker streaks; scapulars warm brown; bill orange. Sexes similar in winter, but m. tends to have little or no dusky streaking on tan crown and mostly white primary coverts; f. has some blackish streaking on tan crown and white base to black primary coverts. **1st Winter:** Like ad. winter but streaking on crown heavier and more black on primary coverts for each sex. **Flight:** Long wings black at tip, white at base; mostly black tail; dark rump. **Hab:** Tundra and barren rocky areas in summer; sparse fields and shores in winter. **Voice:** Song a fairly long series of varied, fairly harsh, whistled notes, many repeated; calls include a rapid *chididit* or *chidididit*, a harsh *djeeet*, and a whistled *tew*.

Subspp: (2) •*townsendi* (w. AK is.) is large; •*nivalis* (rest of NAM range) is small with buff-tinged body. **Hybrids:** McKay's Bunting.

Adult m., summer AK/07

Adult m., summer AK/07

Adult m., summer AK/07

Adult winter BC/01

McKay's Bunting 2

Plectrophenax hyperboreus L 6¾"

Shape: Identical to Snow Bunting. **Ad. Summer:** Very similar to Snow Bunting but more extensively white and less black in all plumages. Bill dark. M. all white (including back) except for black tips of primaries, tips of central tail feathers, and tertials. Similar m. Snow Bunting has black back. F. similar but with dusky specks on forecrown; back white streaked with black; white rump; more black on coverts and secondaries. Similar f. Snow Bunting has dusky head, more black on wings and tail, dark rump. **Ad. Winter:** Mostly white with bright buffy wash on head and nape and on sides of breast and flanks. Back much paler than in Snow Bunting; rump white (dark in Snow Bunting). Bill orange. Sexes similar in winter. **Flight:** Long wings black at tip, white at base; mostly white tail with dark subterminal spot; white rump. **Hab:** Rocky tundra in summer; coasts in winter. **Voice:** Like Snow Bunting's. Song a fairly long series of varied, fairly harsh, whistled notes, many repeated; calls include a rapid *chididit* or *chidididit,* a harsh *djeeet,* and a whistled *tew.*

Subspp: Monotypic. **Hybrids:** Snow Bunting.

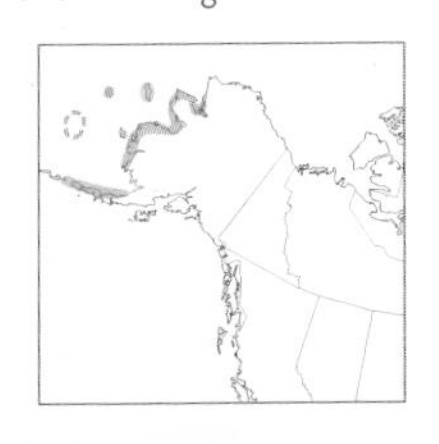

Adult summer JAP/05

Adult winter FRA/12

Little Bunting 4

Emberiza pusilla L 5"

Casual Eurasian vag. to w. AK islands; accidental to CA. **Shape:** Small, short-tailed, short-legged, broad-necked bird with sharp-pointed conical bill and rounded crown typical of the *Emberiza* genus. Often has short crest raised, creating higher slightly peaked crown. **Ad. Summer:** Dark brown crown with paler reddish-brown central stripe; reddish-brown auricular bordered above and behind by a thin dark line; whitish eye-ring; dark bill. Upperparts brown/black streaked with brown; underparts whitish with dark brown streaks on breast and flanks; 2 whitish wingbars. Pale reddish-brown throat may indicate m.; whitish throat may indicate f. **Ad. Winter:** Like summer, but crown brown streaked with black. **Flight:** Distinct thin white edges to outer tail feathers visible in flight. **Hab:** Tundra with low shrubs. **Voice:** Song a fairly high-pitched mixture of short warbles, repeated notes, and short trills; calls include a short *tsik*.

Subspp: Monotypic.

Adult m., summer, *rustica** FIN/06

Adult f., summer, *rustica** FIN/06

Adult winter JAP/01

Rustic Bunting 3

Emberiza rustica L 6"

Casual Eurasian vag. to coastal AK–CA. **Shape:** Small, short-tailed, short-legged, deep-bellied, broad-necked bird with sharp-pointed conical bill and rounded crown typical of the *Emberiza* genus. Often has short crest raised, creating higher slightly-peaked crown. **Ad. Summer:** Underparts whitish with bright reddish-brown streaking on upper breast and flanks. M. has bright reddish-brown upperparts; blackish crown; white eyebrow; black auricular with central paler spot. F. similarly patterned, but crown and auricular are brown; upperparts duller reddish brown. **Ad. Winter:** Like ad. summer f. but f. may have buffy eyebrow. **Juv:** Like ad. f. but duller, buffier. **Hab:** Wet coniferous woods. **Voice:** Song a varied series of rich, low, whistled notes; calls include a fairly low *chit*.

Subspp: (1) •*latifascia*. **E. r. rustica* (n. Eurasia to Japan) used for good photograph.

Adult m., summer –/06

Adult f., summer, *incognita** KAZ/07

Adult m. TX/01

Adult f. TX/03

Reed Bunting 4

Emberiza schoeniclus L 6¼"

Casual Eurasian vag. to w. AK islands. **Shape:** Fairly small, short-tailed, short-legged, deep-bellied, broad-necked bird with a conical bill and rounded crown typical of the *Emberiza* genus; bill with well-curved culmen. Often has short crest raised, creating higher slightly peaked crown. Comparatively, Reed has stubby thick-based bill. **Ad. Summer:** M. has striking black hood with bold white moustachial region; broad white collar; upperparts warm brown and streaked; underparts white with little or no streaking. F. has rusty brown wing coverts; soft reddish-brown streaked upperparts; warm buffy underparts with fine streaking on breast and flanks; dark brown crown with grayish central stripe; buffy eyebrow; pale brown auricular; buffy to whitish moustachial region bordered below by black malar stripe. **Ad. Winter:** Like summer f. **1st Winter:** Both sexes similar to ad. female; f. has brownish rump; m. may have more grayish rump. **Hab:** Shrubby areas. **Voice:** Song short with several harsh notes, faster in middle; calls include a downslurred *seeyoot* and a short *dzoo*.

Subspp: (1) •*pyrrhulina*. **E. s. incognita* (RUS to KAZ) used for good photograph.

Crimson-collared Grosbeak 4

Rhodothraupis celaeno L 8¾"

Casual Mexican visitor to s. TX; often attracted to berry trees or, sometimes, fruit at feeders. **Shape:** Medium-sized, big-headed, broad-necked bird with a long, broad, rounded tail and short deep-based bill with culmen. **Ad:** Black hood extends onto breast; bill gray. M. has crimson nape, breast, and belly, with variable black spotting on belly; undertail coverts reddish with black subterminal bands; back, wings, and tail black; 2 fine pinkish to red wingbars (may be inconspicuous or possibly worn off). F. olive-yellow overall (except for black hood), slightly brighter on nape, breast, and belly. **1st Yr:** F. like ad. f. but black hood reduced to blackish face and throat. M. (Oct.–Mar.) like 1st-yr. female; m. (Apr.–Aug.) has full black hood, olive-yellow with variable patches of black on back, variable patches of red on underparts. **Flight:** Underwing coverts olive or red, like underparts of respective sexes. **Hab:** Shrubby woods, edges. **Voice:** Song a fairly low-pitched accelerating warble ending with an upslurred whistle; calls include a high-pitched rising and falling *seeyoo*.

Subspp: Monotypic.

Adult m., canicaudus TX/02

Adult f., canicaudus TX/–

Adult m., cardinalis FL/01

Juv., superbus AZ/07

Northern Cardinal 1

Cardinalis cardinalis L 8¾"

Shape: Medium-sized, fairly long-tailed, broad-necked bird with a variable length (see Subspp.), pointed crest, and deep-based conical bill. **Ad:** M. pale to deep red overall, with variable grayish wash to back and wings (see Subspp.); black rectangular patch around bill base and chin; bill orangish red. F. brown above with reddish edges to wing and tail feathers; buffy brown on chest and flanks; crest tipped reddish; indistinct blackish area around bill base; bill orangish red. **Juv:** (Apr.–Sep.) Like ad. f., but duller brown overall with a blackish bill, no red on crest, and little or no black feathers around bill base. **Flight:** Slightly undulating; wings and tail reddish, brightest below. **Hab:** Woods edges, shrubs, hedgerows, gardens, feeders. **Voice:** Song a varied series of slurred whistles at varying speeds, like *woit woit woit woit cheer cheer toowee toowee toowee,* sometimes ending with a soft *churrrr.* Calls include a metallic *chip* and a harsh *kwut.* 🎧**129**.

Subspp: (4) •*cardinalis* (ND–e. CO–n. TX–AL and east) is small, with relatively short bill and crest; m. bright red. •*magnirostris* (e.cent. TX–s. LA) is medium-sized, with long crest and bill; m. deep red. •*superbus* (AZ–s.w. NM) is large, with long crest and large bill, back extensively tinged grayish; m. pale red. •*canicaudus* (OK–s.e. TX west to s.e. NM) is medium-sized, with short crest, large bill, back tinged grayish; m. medium red.

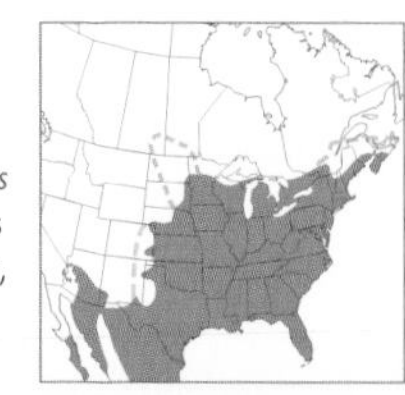

Adult m., summer, *sinuatus* TX/05

Adult f., winter, *sinuatus* TX/12

Adult m., winter, *sinuatus* TX/03

Adult f., winter, *fulvescens* AZ/02

Pyrrhuloxia 1

Cardinalis sinuatus L 8¾"

Shape: Medium-sized, long-tailed, broad-necked bird with a long, thin, pointed crest and short deep-based bill. Culmen strongly and abruptly downcurved. **Ad. Summer:** M. pale gray overall with deep red on face and central underparts; wings and tail reddish underneath; crest red; bill yellow. F. similar but more brownish gray overall; lacks red on face and underparts; bill dull grayish yellow. **Ad. Winter:** Like summer but bill dull horn. In fresh plumage, gray feather tips may obscure or completely hide red on breast and belly of male; these wear off over winter, revealing red underneath. **Juv:** (May–Sep.) Like ad. f. but has thin pale whitish wingbars, underparts may be buffier, bill is blackish gray (color may be kept into Feb.), and crest has little to no red. **Flight:** Wings and tail brightest red from below. **Hab:** Brushy areas, hedgerows, desert scrub, mesquite. **Voice:** Song a varied series of slurred whistles at varying speeds, like Northern Cardinal's but shorter and higher-pitched. Calls include a metallic *chip*.

Subspp: (2) •*fulvescens* (AZ) has grayish-brown upperparts; m. underparts pale brownish and pinkish red. •*sinuatus* (NM–TX) is darker overall; m. underparts dark gray and red.

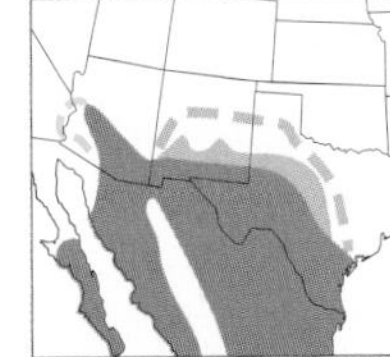

Adult m., summer TX/04

1st-summer m. NH/06

Adult f., summer TX/04

1st-winter m. NH/09

Rose-breasted Grosbeak 1

Pheucticus ludovicianus L 8¼"

Shape: Fairly large, large-headed, broad-necked bird with a short deep-based bill and relatively short tail typical of *Pheucticus* genus. **Ad. Summer:** Bill whitish to pale horn. M. has black hood and back; crimson red triangular bib; white belly and undertail coverts; large white patch at base of black primaries. F. has broad white eyebrow bordered by dark brown crown and auriculars; underparts whitish (buffy on breast when fresh) with variable brown streaking across breast and along flanks; upperparts streaked brown; wings brown with white wingbars and small white patch at base of primaries. Similar Black-headed Grosbeak ad. f. and 1st yr. have dark bill (especially upper mandible); buffy eyebrow on ad. f. when fresh (white in f. Rose-breasted); deep warm buff underparts with little or no streaking in central breast (streaking even across breast in ad. f. and 1st-yr. Rose-breasted). **Ad. Winter:** M. like summer but variable buffy fringes to feathers of head, back, and rump; often has whitish line behind eye. F. like summer but underparts warm buff with more extensive, coarser streaking. **1st Winter:** M. like winter ad. f. but with variable reddish wash over breast. F. like winter ad. f. but has little or no white patch at base of primaries. **1st Summer:** M. like ad. summer m. but primaries and secondaries brown, contrasting with black coverts; may have thin pale edges to head and back feathers. F. like ad. summer female but with little or no white patch at base of primaries. Distinguished from Black-headed f. as mentioned above. **Flight:** M. has large white patch at base of primaries, pink underwing coverts; f. has small white patch at base of primaries, yellow to salmon underwing coverts. **Hab:** Variable; woods, shrubby edges of woods or water, parks, gardens. **Voice:** Song an extended fairly rapid series of mellow whistled notes, like *toweet toweer taweeyah tawoo;* calls include a metallic squeaky *skeet* and a high drawn-out *eeee.* **130**.

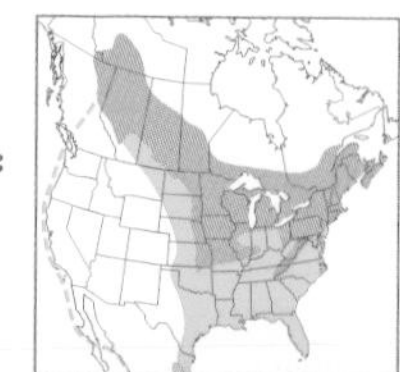

Subspp: Monotypic. **Hybrids:** Black-headed Grosbeak.

Adult m., summer AZ/05

1st-summer m. AZ/05

Adult f. AZ/05

1st-winter m. AZ/08

Black-headed Grosbeak 1

Pheucticus melanocephalus L 8¼"

Shape: Fairly large, large-headed, broad-necked bird with a short deep-based bill and relatively short tail typical of *Pheucticus* genus. **Ad. Summer:** Bill with dark gray upper mandible, pale gray lower mandible. M. has black head; body mostly orange with white undertail coverts; wings and tail black with 2 white wingbars, white patch at base of primaries, and broad white tips to outer tail feathers. F. has broad buffy (when fresh) to white eyebrow bordered by dark brown crown and auriculars; underparts deep warm buff with variable fine dark streaking on sides of breast and along flanks (little or no streaking in central breast); upperparts streaked brown; wings brown with white wingbars and small white patch at base of primaries. Similar Rose-breasted Grosbeak ad. f. and 1st yr. have pale bill; white eyebrow; whitish underparts with variable thicker brown streaking across breast and along flanks. **Ad. Winter:** M. similar to summer ad. m. but crown, head, and back feathers edged with pale brown. F. like summer ad. f. **1st Winter:** M. similar to ad. f. but head pattern more contrasting, with black crown-stripes; mixture of blackish (ad.) and brown (imm.) feathers on wings and/or coverts. F. like ad. f. but little or no white at base of primaries. **1st Summer:** M. like ad. m. but has brown primaries and secondaries that contrast with black wing coverts; may retain white eyebrow of 1st winter. F. like ad. f. but with little or no white patch at base of primaries. **Flight:** Yellow underwing coverts on all ages and sexes; adult m. has large white patch at base of primaries, adult f. has small white patch at base of primaries. **Hab:** Variable; open woods, river edges, mountain canyons, parks, gardens. **Voice:** Song an extended fairly rapid series of whistled notes similar to Rose-breasted's song, but each whistle more quickly delivered and more spaced; calls include a sharp *chik*. 🎧**131**.

Subspp: (2) •*maculatus* (s.w. BC–s. CA) has small bill; ad. m. with faint to distinct orangish eyebrow. •*melanocephalus* (rest of range) has large bill; ad. m. with little or no orangish eyebrow. **Hybrids:** Rose-breasted Grosbeak.

Adult m., worn CA/08

Adult m. TX/02

Adult f. AZ/03

Adult f. TX/02

Yellow Grosbeak 4

Pheucticus chrysopeplus L 9¼"

Casual Mexican visitor to Southwest, mostly s.e. AZ. **Shape:** Fairly large, large-headed, broad-necked bird with a short deep-based bill and relatively short tail typical of *Pheucticus* genus. Comparatively, Yellow Grosbeak is large; bill proportionately much deeper-based; line of culmen continuous with forehead. **Ad:** Yellow head and body; black to sooty wings and tail with 2 white wingbars; dark gray bill; tertials, outer tail feathers, and uppertail coverts tipped white. M. has yellow head and yellow auriculars, black back with yellowish mottling; f. has yellow head with streaked crown and nape, dusky auriculars, olive back with dusky spotting, sooty (rather than black) wings and tail. **1st Yr:** F. like adult female. M. like ad. f. but with larger white patch along edge of folded wing (formed by base of primaries). **Flight:** Yellow underwing coverts and variable white patch at base of primaries (larger in m. than f.). **Hab:** Woods, shrubby edges. **Voice:** Song a short varied series of strong slurred whistles; calls include a sharp *tink* and a softly whistled *hooee*.

Subspp: (1) •*chrysopeplus.*

Blue Bunting 4

Cyanocompsa parellina L 5½"

Casual Mexican vag. to s. TX; generally secretive but may come to feeders; has been seen in LA. **Shape:** Small, stocky, fairly short-tailed bird with a short conical bill; shape similar to that of Indigo Bunting but with more curved culmen. **Ad:** M. dark blue overall with slightly lighter blue highlights on crown and nape, malar area, lesser and median coverts; bill blackish; lores appear blackish. F. is rich deep brown overall, slightly paler below, with no other markings; bill blackish. **1st Yr:** M. like ad. m. but with brownish secondaries and primaries; f. like ad. female. **Hab:** Brushy edges, woods understory. **Voice:** Song a few short notes followed by a rapid, high-pitched, varied warbling; calls include a metallic *tsink*.

Subspp: (1) •*lucida.*

Adult m. TX/04

1st-yr. m. TX/04

1st-yr. m. TX/04

Adult f. TX/04

Blue Grosbeak 1

Passerina caerulea L 6¾"

Shape: Small, broad-necked, deep-chested bird with a short, deep-based, conical bill and moderately long tail. **Ad:** M. bright blue overall with black foreface and 2 wide reddish-brown wingbars; bill with dark gray upper mandible, silver lower mandible (f. bill similar). M. in fresh plumage (fall and winter) has buff tips to head and body feathers; these wear off by spring. F. warm brown above and paler below; 2 reddish-brown wingbars; faint blue on lesser coverts, rump, and tail. **1st Yr:** M. like ad. f. through winter, then with variable amounts of ad. blue feathers, especially on head. F. like ad. f. **Flight:** Underwing coverts concolor with body. **Hab:** Open areas with a few shrubs and small trees, old fields, brushy edges. **Voice:** Song a long series of short burry whistles with little or no repetition; calls include a sharp *chink*.

Subspp: (4) •*salicaria* (cent. CA–w.cent. NV south to s. CA–s.w. AZ) and •*interfusa* (NV–cent. SD south to s.e. AZ–w. TX) are both medium-sized with medium-length to short bill; f. pale brown, m. medium-pale blue. •*caerulea* (rest of range) is small with medium-length bill; f. dark brown, m. darker blue. •*eurhyncha* (vag. s.w. TX) is large with large bill; f. dark brown, m. dark blue.

Adult m., summer CA/05

1st-yr. m. CO/06

1st-yr. m. CA/05

Adult f., summer CA/05

Lazuli Bunting 1

Passerina amoena L 5½"

Shape: Small, broad-necked, deep-chested bird with a short conical bill and moderately long tail typical of the *Passerina* genus. **Ad. Summer:** M. has brilliant turquoise-blue head, back, and wings with 2 white wingbars; breast and flanks variably orangish; rest of underparts white; bill dark gray; lore blackish. F. pale brown above with 2 thin whitish to buffy wingbars; pale buffy or light brown across unstreaked breast; faint blue tint to rump, lesser coverts; faint contrast between grayish throat and buffy breast. Similar f. Indigo Bunting has dull brown breast (more cinnamon when fresh) with faint blurry streaks, more contrasting whitish throat, buffy indistinct wingbars. Similar f. Varied Bunting has concolor dull brown breast and throat (more cinnamon when fresh), no streaking on breast, no distinct wingbars, more strongly curved culmen. **Ad. Winter:** M. similar to summer but with brown edges to feathers of upperparts and throat. F. like summer, but breast may be more brown, wingbars may be buffy. **1st Yr:** M. like ad. m. but with mix of brown and blue feathers on upperparts and throat and/or some worn brownish feathers on wings. F. like ad. but with less blue on wing coverts and rump. **Juv:** (May–Sep.) Like ad. f. but with distinct fine streaking to underparts (some of this may be partially retained on 1st-winter f.). **Flight:** Bold white wingbars and belly of m. **Hab:** Brushy areas, streamside shrubs. **Voice:** Song a long series of high-pitched slightly buzzy whistles, each repeated 2–5 times; calls include a sharp *pik* and a buzzy *dzeeet*. 🎧**132**.

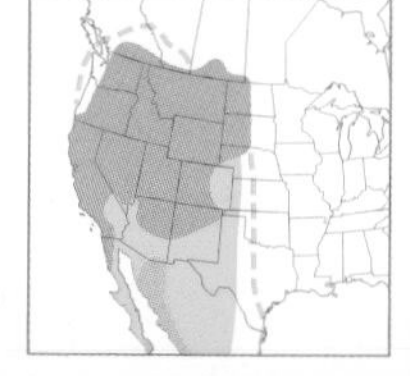

Subspp: Monotypic. **Hybrids:** Indigo Bunting.

Adult m., summer TX/04

1st-yr. m. FL/02

1st-yr. m. FL/03

Adult f. FL/02

Indigo Bunting 1

Passerina cyanea L 5½"

Shape: Small, broad-necked, deep-chested bird with a short conical bill and moderately long tail typical of the *Passerina* genus. **Ad. Summer:** M. all bright deep blue, darkest on head; bill dark; lore blackish. F. brown above with 2 pale, indistinct, buffy wingbars; pale brown below with indistinct streaking across breast and on flanks; whitish throat contrasts with rest of pale brown underparts; faint blue tint to rump, wing coverts, and edges of wing and tail feathers. See Lazuli Bunting for comparison. **Ad. Winter:** M. like summer ad. m. but with variable brownish edges to feathers obscuring blue beneath. F. like summer ad. **1st Yr:** M. with variable mix of brown and blue feathers on body, head, and wings. F. like ad. **Juv:** (May–Aug.) Like ad. f. but with more distinct streaking below; pale bill. **Flight:** M. concolor throughout. **Hab:** Early-successional areas with shrubs and small trees, edges, old fields, roadsides. **Voice:** Song a long series of high-pitched whistles often in pairs; calls include a sharp *tsik* and a buzzing *dzeeet*. **133**.

Subspp: Monotypic. **Hybrids:** Lazuli and Painted Buntings.

Adult m., summer, *pulchra* AZ/07

Adult f., *pulchra* AZ/07

Adult m., summer, *versicolor* TX/05

Adult f., *pulchra* AZ/06

Varied Bunting 2

Passerina versicolor L 5½"

Shape: Small, broad-necked, deep-chested bird with a short conical bill and moderately long tail typical of the *Passerina* genus. Base of lower mandible extends farther back toward front edge of eye than in other *Passerina* buntings, making bill look larger; culmen more sharply curved than in other buntings (except vag. Blue Bunting); comparatively longer-legged than other buntings. **Ad. Summer:** M. with blue head, black lore and chin, reddish nape; rest of body bluish black with variable purplish tones to back and breast; partial red eye-ring. F. grayish brown to reddish brown (see Subspp.) overall, with no wingbars, no streaking below, and throat concolor with rest of underparts (often distinguishable from other f. buntings by this lack of markings and contrast). **Ad. Winter:** M. like summer but feathers with variable pale edging. F. like summer. **1st Yr:** (Nov.–Jun.) M. like ad. f. or with some purple or red feathers. F. like ad. female. **Juv:** (May–Aug.) Like ad. f. but with distinct buffy wingbars. **Flight:** M. very dark. **Hab:** Arid areas with dense vegetation, dry wash edges, canyons. **Voice:** Song a long series of fairly moderate-pitched whistles, delivered more slowly and with fewer paired notes than song of Indigo Bunting; calls include a *chip* and a buzzy *dzzzzt*.

Subspp: (2) •*pulchra* (s.w. NM and west) 1st-yr. birds and ad. f. have reddish-brown wash; ad. m. nape bright red. •*versicolor* (s.e. NM and east) 1st-yr. birds and ad. f. are grayish brown; ad. m. nape dull red. **Hybrids:** Painted Bunting.

Adult m. FL/02

Adult f. FL/02

1st-yr. m. TX/04

Painted Bunting 1

Passerina ciris L 5½"

Shape: Small, broad-necked, deep-chested bird with a short conical bill and moderately long tail typical of the *Passerina* genus. **Ad:** M. has unmistakable blue head, red underparts, bright yellowish-green back. F. a unique bright green above, paler yellowish green below, no markings. **1st Yr:** (Nov.–Jun.) Both sexes like ad. f., but some males may have a few red or blue feathers by spring (only some birds can be sexed). When molting into ad. m. plumage (Jul.–Oct.), all 1st-yr. males will have some red and blue feathers. **Juv:** (May–Sep.) Brownish overall with no other markings (sometimes a green tinge on rump). **Flight:** Red belly of m. obvious; f. all greenish. **Hab:** Shrubs with scattered trees, shrubby edges, woods edges bordering fields. **Voice:** Song a rapid series of short varied whistles with little or no repetition of notes and abrupt changes in pitch; calls include a short sharp *chip* and thin *tseeet*.

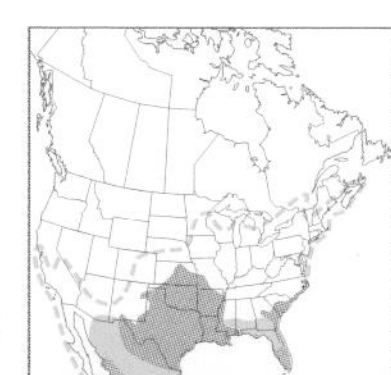

Subspp: Monotypic. **Hybrids:** Indigo and Varied Buntings.

Adult m. IL/06

Adult m. IL/06

Adult f. TX/06

Dickcissel 1

Spiza americana L 6¼"

Shape: Fairly small, deep-chested, large-headed bird with a large deep-based bill and short tail. **Ad:** M. with yellow underparts, black bib, and reddish-brown wing coverts; eyebrow and malar region variably yellow. Winter m. similar but pale edges to bib feathers obscure some black. F. with yellow wash to breast, eyebrow, and moustachial region; thin malar streak and whitish throat; reddish-brown lesser coverts and tips of median coverts; back pale brown with dark streaking. **1st Yr:** M. like ad. female, may have black dots on throat (Mar.–Aug.). F. duller with little or no yellow wash and little or no reddish brown on wing coverts. **Juv:** (May–Aug.) Similar to 1st-yr. f. but with 2 buffy wingbars and fine streaking on breast and flanks. **Flight:** Rapid and direct; often migrates in huge flocks. **Hab:** Grasslands with small shrubs. **Voice:** Song an insectlike, clipped, variable *dick dick dickcissel*; calls include a low buzzy *dzzzrt* and a harsh *chek*.

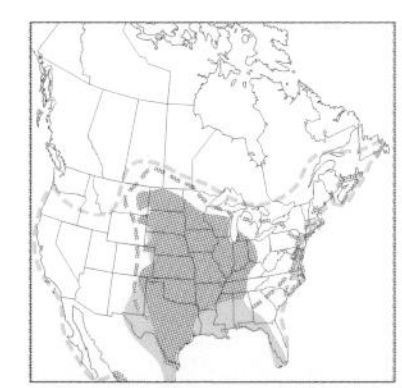

Subspp: Monotypic.

Blackbirds

Meadowlarks

Grackles

Cowbirds

Orioles

Adult m., summer NH/06

Adult winter NH/09

Adult f., summer MN/06

Adult m., summer NH/05

Bobolink 1

Dolichonyx orizyvorus L 7"

Shape: A medium-sized bird with deep-based conical bill, short sharply pointed tail feathers, and long primary projection. F. and winter m. can be distinguished from sparrows by their larger size, slimness, larger bill, and longer primary projection. **Ad. Summer:** M. has black head, back, and underparts contrasting sharply with yellowish-buff nape and white rump and uppertail coverts; white scapulars show as broad swath above folded black wing. Wing feathers edged pale buff after complete spring molt, but this wears off over summer. F. has warm buffy head, broad buffy central crown-stripe bordered by blackish-brown lateral crown-stripes, and dark thin line behind eye; dark back and wing feathers with bold buffy margins, creating scaled look; underparts pale buffy yellow with thin streaking along flanks. **Ad. Winter:** M. and f. alike and much like summer f., but strongly buffy yellow on face and underparts; male bill changes from gray to pinkish like female's. **Juv:** (Jul.–Aug.) Like ad. winter but lacks streaks on flanks. **Flight:** Circling flights over fields, m. flight stalling when he sings. Pointed tail feathers; m. with white rump. **Hab:** Hayfields, grasslands, agricultural areas. **Voice:** Song a jumble of bubbling warbles with some harsh notes intermixed; calls include a repeated *zeep,* a descending *seeyew,* a harsh *chek,* and a *pink* given after breeding season. 🎧**134**.

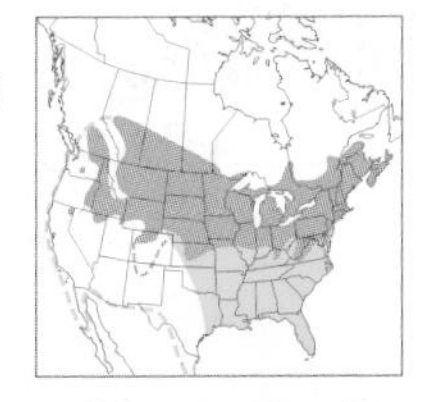

Subspp: Monotypic.

Adult m., summer, displaying, Eastern Grp. NH/04

Adult m., summer TX/03

Adult m., winter NM/01

1st-yr. m., Eastern Grp. NH/04

Red-winged Blackbird 1

Agelaius phoeniceus L 8¾"

Shape: Relatively small blackbird with a fairly long spikelike bill, flat crown, and fairly short tail. Tail length, bill, and body size vary geographically (see Subspp.). **Ad. Summer:** M. all glossy black with bright red marginal coverts ("epaulets") bordered by pale yellow (except in subsp. *californicus,* "Bicolored Blackbird"). F. dark brown overall with heavily streaked back and underparts; pale eyebrow over broad dark eyeline; chin and face washed orangish to pinkish; epaulets orangish to rusty. See Tricolored Blackbird for comparison. **Ad. Winter:** M. like ad. summer but feathers of head, back, and secondary coverts edged with brown to buff; this wears off by spring or summer. F. like ad. summer.

1st Yr: (Aug.–Jul.) M. mainly dull blackish overall; primaries brownish; back and wing feathers edged with buff to pale reddish brown; crown and nape streaked buff; epaulets variably orange and yellow with black mottling; suggestion of a pale eyebrow. Paler edging to feathers reduced through wear by spring and summer. F. like ad. f. but with mostly whitish chin and face (little or no pink or orange) and epaulets paler buffy orange. **Juv:** (May–Sep.) Like 1st-yr. f. but paler and buffier overall. **Flight:** Strong and slightly undulating. Note red epaulets of male. **Hab:** Marshes, meadows, agricultural lands, feeders. **Voice:** Song a few short introductory sounds followed by a drawn-out raspy trill, like *okaleeee;* calls include *ch'ch'ch'chee chee chee chee* by f., a downslurred whistled *tseeyert,* a short *chek,* and a rapid harsh *ch'ch'ch'ch.* 🎧**135**.

Adult f., summer, Eastern Grp. NH/04

Adult m., *californicus*, "Bicolored" CA/04

1st-yr. f. AZ/04

Adult f., *californicus*, "Bicolored" CA/05

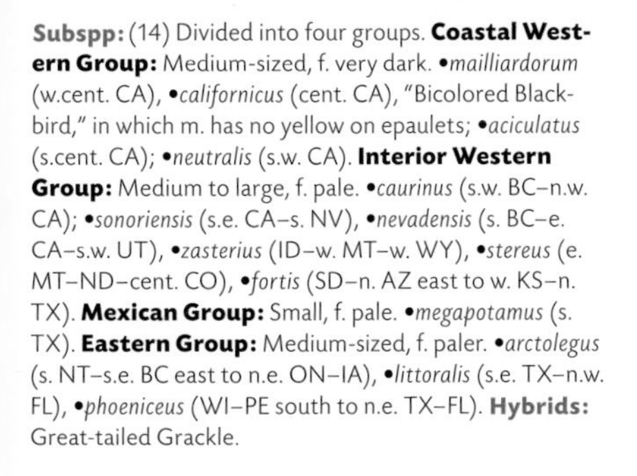

Subspp: (14) Divided into four groups. **Coastal Western Group:** Medium-sized, f. very dark. •*mailliardorum* (w.cent. CA), •*californicus* (cent. CA), "Bicolored Blackbird," in which m. has no yellow on epaulets; •*aciculatus* (s.cent. CA); •*neutralis* (s.w. CA). **Interior Western Group:** Medium to large, f. pale. •*caurinus* (s.w. BC–n.w. CA); •*sonoriensis* (s.e. CA–s. NV), •*nevadensis* (s. BC–e. CA–s.w. UT), •*zasterius* (ID–w. MT–w. WY), •*stereus* (e. MT–ND–cent. CO), •*fortis* (SD–n. AZ east to w. KS–n. TX). **Mexican Group:** Small, f. pale. •*megapotamus* (s. TX). **Eastern Group:** Medium-sized, f. paler. •*arctolegus* (s. NT–s.e. BC east to n.e. ON–IA), •*littoralis* (s.e. TX–n.w. FL), •*phoeniceus* (WI–PE south to n.e. TX–FL). **Hybrids:** Great-tailed Grackle.

Adult m., Eastern Grp. NY/06

Adult m. CA/05

Adult f. CA/05

Adult m. CA/05

Adult f. CA/05

Tricolored Blackbird 2

Agelaius tricolor L 8¾"

Shape: Similar to Red-winged Blackbird but with slightly heavier body, longer thinner-based bill, more pointed wings, slightly longer primary projection. **Ad. Summer:** M. all glossy blue-black with red marginal coverts ("epaulets") bordered by whitish feathers (yellow on m. Red-winged). F. sooty to black with grayish streaking on breast, speckled grayish chin and throat, faint eyeline; any streaking on back grayish or buff (never has reddish-brown tones of f. Red-winged seen in fall and winter); epaulets reddish brown; fairly prominent white wingbar on median coverts. Similar f. Red-winged has orangish chin and throat (ad.), reddish-brown tones on back in fall and winter, thin and less prominent whitish wingbar on median coverts. **Ad. Winter:** M. like ad. summer m., but body feathers tipped gray or buff; epaulet border may be creamy (sometimes hidden). F. like ad. summer female. **1st Yr:** (Sep.–Aug.) M. like ad. winter m. but epaulets brownish red or orangish, not pure red; body may have variable gray streaking. F. like ad. but with no reddish color on epaulets. **Flight:** Direct. White bar on epaulets of m. **Hab:** Marshes, agricultural fields. **Voice:** Song a raspy rising *sho'shhhrrreee;* calls include a short *chuk*.

Subspp: Monotypic.

Adult m. MB/06

Adult f. MT/06

Adult m. displaying IL/05

1st-yr. m. MT/06

Yellow-headed Blackbird 1

Xanthocephalus xanthocephalus L 9½"

Shape: M. almost twice as heavy as f. and about 20% longer. M. large-bodied and relatively short-tailed with a sharp-pointed deep-based bill; line of culmen almost continuous with fairly flat forehead. F. smaller; tail appears average length, bill is thinner-based, and culmen juts from more rounded forehead. **Ad:** M. with bright yellow head and breast and dark lore; rest of body black except for white primary coverts seen on edge of folded wing and in flight. F. mostly dark brown with dull yellow breast, whitish throat, and yellow eyebrow. **1st Yr:** (Aug.–Jul.) M. much like ad. f. but shape is different (see Shape) and generally has more yellow mottling on face and head (at least Aug.–Mar.); primary coverts narrowly tipped with white, lore blackish. F. like ad. f. but can have fine white tips to primary coverts (slightly less than 1st-yr. m.); lore brownish and paler than male's. **Juv:** (Jun.–Aug.) Much like 1st-yr. f., but pale buffy yellow over all of head and breast; whitish throat; and whitish wingbars. **Flight:** Mildly undulating. M. has white patches at base of primaries. **Hab:** Prairie wetlands, cattail marshes, farm fields, occasionally woods edges. **Voice:** Song a few gargled notes followed by a drawn-out, raspy, low-pitched screech, like *gugleko grrraaaah*, or a few introductory notes followed by a short trill. Calls include *cheecheechee* (by f.), a 2-part *chukuk*, and a *check*. 🎧**136**.

Subspp: Monotypic.

Adult summer TX/02

Adult winter TX/04

Adult TX/04

Adult, Southwestern Grp. TX/09

Eastern Meadowlark 1

Sturnella magna L 9½"

Shape: Medium-sized, deep-chested, short-tailed, broad-necked bird with a large head, flat crown, and long, finely pointed, fairly deep-based bill. Tail feathers pointed. **Ad. Summer:** Bright yellow throat and belly with a black V across breast; white eyebrow, yellow supraloral dash, dark eyeline. Best distinguished from similar Western Meadowlark by calls and song (with some exceptions; see Voice). Key plumage clues: Eastern Meadowlark has malar area mostly white; blacker lateral crown-stripes; dark warm brown back with reddish-brown tones (see Subspp.); broad dark barring on wing coverts; dark-centered uppertail coverts; 3–4 outer tail feathers mostly white. Western Meadowlark has mostly yellow malar area; brownish lateral crown-stripes; pale brown back; thin barring on wing coverts; 2 outer tail feathers mostly white. **Ad. Winter:** Yellow-and-black pattern of underparts may be obscured by buffy tips to fresh feathers, which wear off by spring. **Juv:** (May–Sep.) Whitish throat and faint streaking across breast where black V occurs on adult. **Flight:** Alternates glides with stiff rapid wingbeats below the horizontal. White outer tail feathers. **Hab:** Meadows, grasslands, farm fields. **Voice:** Song 3–5 clear descending whistles, like *seeooh seeyeer;* calls include a short buzzy *dzeert,* a more drawn-out *bjeeert,* and a chattering *ch'ch'ch'ch.* **137**.

Subspp: (4) Two groups. **Southwestern Group:** •*lilianae* (cent. AZ–s.w. TX), "Lilian's Meadowlark," is large, with whitish auriculars contrasting more sharply with dark eyeline and more extensive white in tail; similar to Western Meadowlark, being duller and paler brown. •*hoopesi* (s. TX) is medium-sized, with brownish auriculars, extensive white in tail. **Eastern Group:** •*argutula* (s.e. KS–NC and south) is small, with dark upperparts, dark auricular, moderate white in tail. •*magna* (n. TX–VA and north) is large, with dark upperparts, dark auricular, less white in tail. **Hybrids:** Western Meadowlark.

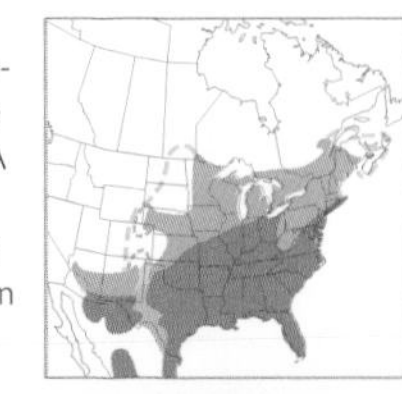

Adult summer CA/05

Adult summer, *neglecta* MT/06

Adult winter CA/09

Western Meadowlark 1

Sturnella neglecta L 9½"

Shape: Identical shape to Eastern Meadowlark. **Ad. Summer:** Bright yellow throat and belly with a black V across breast; white eyebrow, yellow supraloral dash, dark eyeline. Best distinguished from similar Eastern Meadowlark by calls and song. Key plumage clues: Western Meadowlark has mostly yellow malar area; brown lateral crown-stripes; pale brown back; thin barring on wing coverts; 2 outer tail feathers mostly white. Eastern Meadowlark has malar area mostly white; dark warm brown back with reddish-brown tones (see Subspp.); broad dark barring on wing coverts; dark-centered uppertail coverts; 3–4 outer tail feathers mostly white. **Ad. Winter:** Yellow-and-black pattern of underparts may be obscured by buffy tips to fresh feathers, which wear off by spring. **Juv:** (May–Sep.) Whitish throat, streaking across breast where black V occurs on adult. **Flight:** Alternates glides with stiff rapid wingbeats below the horizontal. White outer tail feathers. **Hab:** Meadows, grasslands, farm fields. **Voice:** Song begins with a few clear whistles and ends in rich warble, very different from clear downslurred whistles of Eastern Meadowlark; calls include a low *chuk,* a rattle, and a high-pitched *weeet.* 🎧**138**.

Subspp: (2) •*confluenta* (s.w. BC–s.w. CA) has central tail feathers with black shaft streaks, dark yellow breast; •*neglecta* (rest of range) has central tail feathers with no black shaft streaks, pale yellow breast. **Hybrids:** Eastern Meadowlark.

Adult m., summer ME/06

Adult m., winter MN/10

Adult f., summer NY/05

Adult f., winter IL/09

Rusty Blackbird 1

Euphagus carolinus L 9"

Shape: Medium-sized, large-headed, broad-necked, and deep-bellied blackbird with a relatively short tail and moderate-length thin bill that can appear slightly downcurved but has a fairly straight culmen. Similar Brewer's Blackbird has a smaller head, defined neck, and longer tail appearing as discrete elements; also slightly shorter thicker-based bill with a curved culmen. Distinguished from grackles by its smaller size, thinner bill, and shorter tail. **Ad. Summer:** M. all dull black with little or no greenish gloss; eye yellow. F. slate gray overall; eye yellow. **Ad. Winter:** M. with variable rusty-brown edging to feathers; head mostly rusty except for dark auriculars; wing coverts and tertials rusty-edged; rusty edges wear off over winter to reveal black summer plumage. F. similar but with even more buffy and reddish brown over head and body; pale buffy eyebrow contrasts with black feathers that surround yellow eye; pale throat. M. has black rump; f. gray rump. **1st Yr:** (Sep.–Jul.) Like ad. winter but with even more extensive pale edging to feathers (usually wears off by midsummer). **Juv:** (Jun.–Aug.) Prominent buffy wingbars. **Flight:** Mildly undulating. Short tail. **Hab:** Breeds in spruce bogs, wet woods; winters in bottomland woods and in fields near water. **Voice:** Song a short jumbled phrase followed by a long *eee* sound, like *jibilo tseeee;* calls include a low *chuk*.

Subspp: Monotypic.

Adult m., summer CA/04

Adult f., summer (pale eye) CA/03

Adult f., summer CA/09

Adult f., winter BC/11

Brewer's Blackbird 1

Euphagus cyanocephalus L 9"

Shape: Medium-sized, fairly small-headed, deep-bellied blackbird with a relatively short tail and short thick bill with a curved culmen. See Rusty Blackbird for comparison. Distinguished from grackles by its smaller size, shorter bill, and shorter tail. **Ad. Summer:** M. all black with much glossy violet iridescence on head contrasting with greenish-blue iridescence on body; yellow eye. F. pale to dark gray overall (see Subspp.), darker on wings and tail, with a dark eye (10% of pop. has yellowish eye). **Ad. Winter:** Like summer, but may have slightly paler eyebrow. F. has flat blackish wings; usually dark eye. M. has iridescent black wings; white eye. Similar f. Rusty Blackbird with brown edging on wing feathers. **1st Winter:** Like ad. but body feathers more broadly tipped with brown. **Flight:** Slightly undulating. Slightly longer tail, slimmer body than Rusty Blackbird. **Hab:** Varied; wet meadows, shrubby water edges, roadsides, fields, urban areas. **Voice:** Song a short scratchy *krt jee;* calls include a *chek,* a downslurred *tseeyur,* and a chattering *ch'ch'ch'ch.* **139**.

Subspp: (3) •*minusculus* (s.w. OR–CA) small; f. gray with broad reddish-brown tips to fresh feathers. •*brewsteri* (s.w. NT–n. OR east to WI) large; f. dark gray with narrow reddish-brown tips to fresh feathers. •*cyanocephalus* (rest of range) large; f. pale gray with brown tips to fresh feathers. **Hybrids:** Great-tailed Grackle.

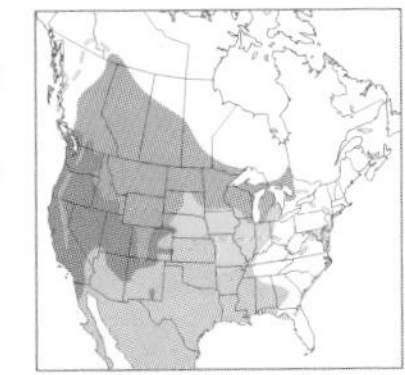

Adult m., *versicolor* TX/04

Adult f., *versicolor* MO/06

Adult m., *stonei* NJ/–

Juv., *versicolor* MO/06

Adult m. FL/02

Adult m., *quiscula* GA/02

Common Grackle 1

Quiscalus quiscula L 12½"

Shape: Our smallest grackle. Large head, slightly rounded crown, broad neck, and proportionately shorter tail create more compact look than in other grackles. Bill large, but less than length of head. Tail long, only slightly graduated (compared to other grackles); m. tail held in shallower V than other grackles during display flights. F. smaller than m. but proportions similar. **Ad:** Both sexes have bluish iridescent hood. M. body and wings are either iridescent bronze contrasting with bluish hood or darker multicolored iridescence blending with hood (see Subspp.); eye yellow. F. body and wings blackish brown with little iridescence. **1st Yr:** (Sep.–Jul.) Like ad. but with slightly shorter wings and tail; less iridescence; eye brownish into Feb. **Juv:** (Jun.–Sep.) Dark brown overall; lacks iridescence; dark eye. **Flight:** Direct or slightly undulating; note tail length and amount of graduation. **Hab:** Open areas with trees, parks, agricultural fields. **Voice:** Song a screechy *reedleeeeek* or *ch'gaskweeek;* calls include a *chack,* a drawn-out *chaaah,* a high-pitched whistled *tseee.* 🎧**140**.

Subspp: (3) •*quiscula* (e. LA–s.e. VA and south), "Florida Grackle"; small with a thin bill, iridescence on m. body bluish to greenish, blending with hood. •*stonei* (cent. LA–CT–w. NC), "Purple Grackle"; large with a moderate-sized bill, iridescence on m. body dark and multicolored, blending with hood. •*versicolor* (rest of range), "Bronzed Grackle"; large with large bill, iridescence on m. body bronze, contrasting with darker hood.

Adult m., *torreyi* GA/03

Adult f., *torreyi* GA/03

Adult f., *westoni* FL/02

1st-winter m. FL/01

1st-winter f. NJ/02

Boat-tailed Grackle 1

Quiscalus major L m. 16½", f. 14½"

Shape: Medium-sized grackle, m. slightly shorter length than American Crow. Large head, slightly rounded crown, and broad neck create more "proportionate" look than in Great-tailed Grackle. Bill large, but not as long as head. M. tail very long, graduated for less than half its length, held in deep V during display flights. F. similar to m. but smaller, with shorter bill and tail. Similar Great-tailed larger; proportionately longer bill, smaller flatter head, and thinner neck; tail graduated for more than half its length. **Ad:** Eye color yellow to brown (see Subspp.). M. glossy black overall with blue-green iridescence on body and wings, violet iridescence on head. F. has dark brown back and wings with some iridescence; pale brown head and underparts; head with contrastingly darker crown, lore, and eyeline, and thin malar streak. **1st Yr:** (Aug.–Jul.) M. like ad. but with less iridescence, especially on wings; f. like ad. but buffier overall. Both have slightly shorter wings and tail than ad.; eye brownish into Feb. and then turns yellow or stays brown (see Subspp.). **Juv:** (Jun.–Sep.) Brownish overall with paler buffy underparts, lacking iridescence; dark eye; substantially shorter tail and wings than adult. **Flight:** Direct; may alternate flapping with glides. Note long tail and proportion that is graduated. **Hab:** Fresh- or saltwater marshes and nearby trees and vegetation, parks, roadside areas. **Voice:** Song a scratchy *b'jeee b'jeee b'jeee* mixed with short low chortles; calls include a whistled *tseet,* a repeated *pup pup pup,* and a *jeer jeer.*

Subspp: (4) •*torreyi* (e. NY–n. FL) large, with relatively short tail, yellow eye; •*westoni* (n.cent. FL south) large, with relatively short tail, mostly brown eye; •*alabamensis* (coastal n.w. FL–MS) slightly smaller than above, with relatively long tail, mostly yellowish eye; •*major* (LA–e. TX) smallest subsp., with proportionately long tail, mostly brownish eye, sometimes yellow eye. **Hybrids:** Great-tailed Grackle.

Adult m. TX/–

Adult f., *nelsoni* CA/01

Adult m., *nelsoni* CA/02

Adult f., *prosopidicola* TX/01

Great-tailed Grackle 1

Quiscalus mexicanus L m. 18", f. 15"

Shape: Our largest grackle. M. appears disproportionately small-headed; massive bill almost as long as head; neck thin in relation to large body; tail very long and graduated for over half its length. F. similar to m. in all ways but less pronounced—does not appear quite so small-headed, long-billed, or long-tailed. M. tail held in deep V during display flights. See Boat-tailed Grackle for comparison. **Ad:** M. black with violet-blue iridescence overall; eye bright yellow. F. dark brown above with some iridescence on back and wings; head and underparts from pale to warm brown (see Subspp.); head shows slightly more contrast than on similar f. Boat-tailed, with paler eyebrow and more prominent malar streaks; eye yellow. **1st Yr:** (Aug.–Jul.) M. like ad. but with less iridescence; f. like ad. but buffier overall. Both have slightly shorter wings and tail than ad.; eye brownish into Feb. **Juv:** (May–Aug.) Like ad. f. but with some streaking on belly; dark eye. **Flight:** Direct; long tail can get blown in wind, angling bird across flight path. Note long tail and proportion that is graduated. **Hab:** Open areas with trees and water, fields, scrub, urban areas. **Voice:** Song of eastern subspp. in 4 parts, alternating mechanical sounds with repeated *chewechewechewe* sounds; song of western subsp. *nelsoni* a series of harsh sounds with a final accented *chweee*. Calls include a sharply rising whistled *toooweeet* (m.), a chatter, and a low *chuk*.

Subspp: (4) •*prosopidicola* (s. NE–s. TX and east) fairly large with relatively short tail, fairly slender bill; f. underparts pale brown. •*mexicanus* (vag. to s. TX) large with thick bill; f. underparts dark brown. •*monsoni* (s.w. TX–s. WY west to s. UT–s.cent. AZ) fairly large with fairly long tail, fairly thin bill; f. underparts grayish brown. •*nelsoni* (w. AZ–CA) small with thin bill; f. underparts pale brown. **Hybrids:** Red-winged Blackbird, Brewer's Blackbird, Boat-tailed Grackle.

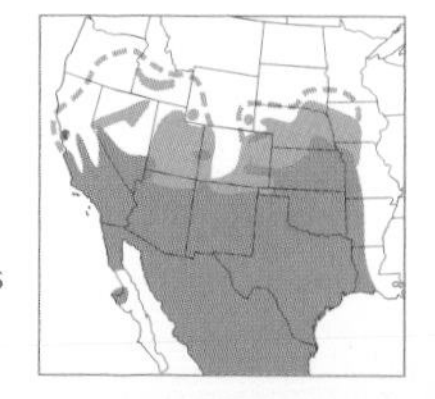

Adult m., *aeneus* TX/05

1st-yr. m., *loyei* AZ/06

Adult m., *loyei* AZ/08

Adult f., *aeneus* TX/04

Juv., *loyei* AZ/09

Bronzed Cowbird 1

Molothrus aeneus L 8½"

Shape: Medium-sized, short-tailed, fairly deep-bellied bird with a fairly short, deep-based, conical bill and flat shallow crown typical of *Molothrus* genus. Comparatively, Bronzed Cowbird has a deeper-based bill, shorter tail. M. is flat-crowned with a neck ruff which, when raised, changes his shape to very broad-necked, large-bodied, and small-headed. F. with more "normal" cowbird proportions but with shorter tail; lacks neck ruff. **Ad:** M. glossy black overall; bronze iridescence on body, bluish iridescence on wings and tail; red eye. F. grayish or dull black to brownish black (see Subspp.) with some iridescence; red eye. **1st Yr:** (Aug.–Jul.) M. may have gray or flat black feathers mixed with glossy black; primaries tipped brownish or washed brownish; eye red. F. grayish brown or flat black to brownish black, no iridescence. **Juv:** (May–Sep.) Blackish overall or uniformly grayish brown with streaking on underparts (see Subspp.). **Hab:** Urban lawns, parks, fields, open woods. **Voice:** Song a soft mixture of high whistles and what sounds like water drops falling into a pool; calls include a harsh *tchuk,* a long series of whistles and harsh trills, and a chattering *ch'ch'ch'ch* (f.).

Subspp: (2) •*loyei* (s.w. NM and west) f. and juv. grayish; •*aeneus* (s. TX and east) f. and juv. brownish black to black.

Adult m., *obscurus* AZ/07

Trans. to adult, *obscurus* AZ/08

Adult f., *obscurus* AZ/05

Juv., *artemisiae* MT/06

Adult f., *ater* NH/05

Brown-headed Cowbird 1

Molothrus ater L 7½"

Shape: Medium-sized, short-tailed, fairly deep-bellied bird with a fairly short, deep-based, conical bill and flat shallow crown typical of *Molothrus* genus. Comparatively shorter-billed, shorter-tailed. Fairly long primary projection past tertials (1½–2 x bill length). Similar f. Shiny Cowbird has longer thinner-based bill; slightly longer tail; shorter primary projection past tertials (equal to bill length). **Ad:** M. with dull brown head and neck; rest of body glossy black; dark eye. F. grayish brown with whitish chin; indistinct streaking on breast; pale edges to secondaries; dark eye. Similar f. Shiny Cowbird warm brown overall; faint paler eyebrow; faint darker eyeline; secondaries with no obvious pale edges. **Juv:** (May–Sep.) Like ad. f. but with more distinct dark streaking on underparts, pale edging to upperpart feathers, and buffy wingbars. **Flight:** Mildly undulating. **Hab:** Pastures, woods edges, urban lawns. **Voice:** Song a short gurgling sound followed by a squeaky whistle, like *bublocomtseee;* calls include a very high double whistle like *pseeeseeee,* a *chuk,* and a chattering *ch'ch'ch'ch* (f.). **141**.

Subspp: (3) •*obscurus* (s. CA–s.w. LA and south) small; f. grayish brown. •*artemisiae* (s.e. AK–w. ON south to n. CA–n. NM) large; f. grayish brown. •*ater* (rest of range to east) medium-sized; f. dark brown.

Adult m. FL/04

Adult f. FL/-

Juv. trans. to adult f. BON/11

Shiny Cowbird 3

Molothrus bonariensis L 7½"

Shape: Medium-sized, short-tailed, fairly deep-bellied bird with fairly short, deep-based, pointy bill and flat shallow crown typical of *Molothrus* genus. Comparatively long-billed, long-tailed. Short primary projection past tertials (equal to bill length). M. appears large-headed, broad-necked; f. appears slightly smaller-headed, less broad-necked. Similar f. Brown-headed Cowbird has shorter deeper-based bill; slightly shorter tail; longer primary projection past tertials (1½–2 x bill length). **Ad:** M. all black with bluish-purple iridescence; dark eye. F. warm brown overall; faint paler eyebrow; faint darker eyeline; throat brownish; secondaries with no obvious pale edges; dark bill. **Juv:** (May–Sep.) Both sexes blackish with faint streaking below; upper mandible dark, lower yellowish. **Flight:** Slightly undulating; short rounded wings. **Hab:** Open areas with trees or shrubs, agricultural areas. **Voice:** Song a soft, repeated, low, bubbling noise followed by a rising scratchy whistle, like *grglk grglk p'ts'tseee;* calls include a *chuk,* a high, thin, whistled *tseee* (often repeated), a chattering *ch'ch'ch'ch* (mostly f.).

Subspp: (1) •*minimus.*

Adult m., *spurius* TX/04

Adult f., *spurius* TX/04

Adult m., *spurius* TX/04

1st-yr. m., *spurius* TX/04

Orchard Oriole 1

Icterus spurius L 7¼"

Shape: Medium-sized, flat-crowned, and broad-necked bird with a long tail and deep-based but fine-pointed bill typical of *Icterus* genus. Comparatively, Orchard is smaller with a shorter more squared-off tail and a short, thin-based, slightly downcurved bill. **Ad:** M. easily identified by combination of chestnut belly and rump and black hood, back, wings, and tail. F. concolor yellow to greenish-yellow below; slightly darker above with a gray or grayish-green back; in rare instances older f. has some chestnut or black feathers on wing coverts or elsewhere. Similar f. Hooded Oriole is duller overall with slightly paler belly; also larger with longer thicker-based bill and longer more graduated tail. **1st Yr:** (Sep.–Aug.) M. has uniformly greenish-yellow underparts, black foreface and throat, and usually some telltale chestnut or black feathers on body where they will occur on adult. Similar 1st-yr. m. Hooded Oriole has no black on foreface (but black on throat) and no black or chestnut body feathers; similar 1st-yr. m. and ad. f. Bullock's Oriole have gray belly and black on head restricted to central throat. 1st-yr. f. much like ad. f. (but no chestnut or black as rarely occurs in older f.). **Flight:** Often low, with rapid wingbeats. **Hab:** Orchards, open woods, parks, streamside groves. **Voice:** Song is a lovely series of varied whistles with a few buzzy sounds intermixed; calls include a chattering *ch'ch'ch'ch,* a downslurred *teew,* a *chuk,* and a *ch'peeet.* 🎧**142**.

Subspp: (2) •*spurius* (all NAM range) body black and chestnut in m., deep yellow in f. •*fuertesi* (Mex. vag. to s. TX) body black and yellowish orange in m., pale yellow in f.

Adult m., *nelsoni* AZ/07

Adult f., *nelsoni* TX/05

Adult m., *nelsoni* CA/04

1st-yr. m., *nelsoni* CA/04

Hooded Oriole 1

Icterus cucullatus L 8"

Shape: Medium-sized, flat-crowned, and broad-necked bird with a long tail and deep-based but fine-pointed bill typical of *Icterus* genus. Comparatively, Hooded has a thin slightly downcurved bill and long strongly graduated tail. **Ad:** M. is distinctive with yellow to orange body (see Subspp.), black foreface and throat. F. upperparts dull grayish green; underparts washed pale yellow, slightly paler on the belly. Similar f. Orchard Oriole has concolor underparts and is smaller with a shorter thinner-based bill and shorter more squared-off tail; similar f. Bullock's has all-gray belly. **1st Yr:** (Sep.–Aug.) M. has black throat, pale lore, yellow head, paler yellow belly. In addition to shape differences, similar 1st-yr. m. Bullock's has gray belly, black lore, black throat-stripe, suggestion of eyeline; similar 1st-yr. m. Orchard has black foreface and throat, rarely some chestnut. F. like ad. female but with shorter bill, making her even more like f. Orchard Oriole. **Flight:** Flight direct with full, moderately fast wingbeats. **Hab:** Suburban or rural areas, often with palms. **Voice:** Song a very rapid series of short whistles mixed with harsh sounds; calls include a whistled *weet* (unlike Orchard Oriole's *chuk*), a short *chew,* and a rapid brief chatter.

Subspp: (3) •*nelsoni* (CA–s.w. TX) large; ad. f. pale yellow, ad. m. orangish yellow. •*cucullatus* (s.cent. TX) medium-sized; ad. f. orangish yellow with gray flanks, ad. m. orange. •*sennetti* (s. TX) small; ad. f. pale yellow, ad. m. orangish yellow.

Adult m. CA/10

Adult m. MEX/03

Streak-backed Oriole 4

Icterus pustulatus L 8¼"

Casual Mexican visitor to s.e. AZ, where it has bred; also seen in CA, AZ, OR, NM, TX, CO, and WI. **Shape:** Medium-sized, flat-crowned, and broad-necked bird with a long tail and deep-based but fine-pointed bill typical of *Icterus* genus. Comparatively, Streak-backed has thick-based straight bill and short rounded tail. **Ad:** M. has reddish-orange body (deepest on head); black foreface and throat; orange back with black streaks; bold white margins to greater coverts and flight feathers, creating whitish wing. F. similar, but orangish yellow with darker orange on sides of throat and cheek. **1st Yr:** (Sep.–Aug.) M. dull orange with indistinct streaking on back. F. greenish yellow with deeper orange on sides of throat; faint or no streaking on back. **Flight:** Direct with full, moderately fast wingbeats. **Hab:** Wooded streamsides, brushy thickets. **Voice:** Song a varied but slightly halting warble; calls include a rising *weet* or *ch'weet,* a *chek,* a chattering *ch'ch'ch'ch,* and a long series of clear whistles.

Subspp: (1) •*microstictus.*

Adult FL/04

1st yr. FL/01

Spot-breasted Oriole 2

Icterus pectoralis L 9½"

Introduced from Central America. **Shape:** Medium-sized, flat-crowned, and broad-necked bird with a long tail and deep-based but fine-pointed bill typical of *Icterus* genus. Comparatively large, with thick-based slightly downcurved bill and moderate-length strongly graduated tail. **Ad:** Orange body, black throat and foreface, black spotting on sides of breast; greater coverts all black with no wingbar (unique among our orioles); small white patch at base of primaries on folded wing; broad white edges to tertials and secondaries. Sexes alike. **1st Yr:** (Sep.–Aug.) Like ad. but body paler orange and may lack spots on sides of breast; greater coverts dark with faint or no buffy tips; lacks white primary patch; tertials still strongly edged white. **Flight:** Direct. **Hab:** Gardens, parklike settings. **Voice:** Song a series of rich whistles, many repeated; calls include a short *ch't'chk,* a nasal *nyeahh,* and a *wip.*

Subspp: (1) •*pectoralis.*

Adult TX/01

Adult TX/02

1st yr. TX/02

1st yr. TX/05

Altamira Oriole 2

Icterus gularis L 10"

Shape: Medium-sized, flat-crowned, and broad-necked bird with a long tail and deep-based but fine-pointed bill typical of *Icterus* genus. Comparatively large, with short very thick-based bill. **Ad:** Orange head and body; black triangle through lore, black central throat. Lesser coverts orange, forming bar on upperwing; greater coverts black with broad white tips; small white patch at base of primaries on folded wing; broad white edges to tertials. Sexes alike. As well as shape differences, similar ad. m. Hooded Oriole has black lesser coverts and wider black throat patch, lacks white primary patch. **1st Yr:** Like ad. but less brightly colored body; back greenish yellow instead of black; wings and tail sooty; no distinct white patch on primaries. **Flight:** Direct, sometimes with jerky wing motion. **Hab:** Open woodlands, brushy areas, wooded streamsides. **Voice:** Song a slow series of rich slurred whistles, some repeated; calls include a single whistled *teeu,* a chatter, and a 2-part *teekew.*

Subspp: (1) •*tamaulipensis.*

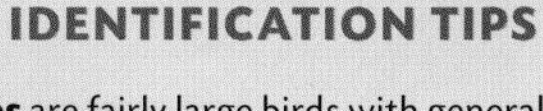

IDENTIFICATION TIPS

Orioles are fairly large birds with generally long, thick-based, but sharply pointed bills and long tails. Most are combinations of black with orange, russet, or yellow. They eat caterpillars, fruit, and insects and use plant fibers to make their characteristic woven suspended nests.

Species ID: Most of our orioles are sexually dimorphic, with adult males being orangish and females more yellowish; several southern resident species have sexes that look alike. The patterns of the head, breast, and back are the best way to distinguish adult males; females demand a closer look and the use of shape, especially length of tail and shape of bill. Confusing matters slightly, 1st-yr. males tend to look much like the adult female but with some suggestion of the adult male's head pattern; also, some females can take on some of the colors and patterns of adult males in later years.

Adult m. CA/05

1st-yr. m. CA/05

Adult f. CA/05

1st-yr. f. CA/05

Bullock's Oriole 1

Icterus bullockii L 9"

Shape: Medium-sized bird with a long tail and deep-based but fine-pointed bill typical of *Icterus* genus. Shape like that of Baltimore Oriole. **Ad:** Varies much less than similar Baltimore Oriole. M. distinctive with black crown and nape; orange eyebrow; black lores and eyeline; white median and greater coverts create large white patch on closed wing; tail orange with black tip and black central feathers. F. has dull orangish-yellow head and breast; belly extensively gray; suggestion of dark eyeline; gray lores; chin white or sometimes centrally black. Disting. from 1st-yr. m. by gray rather than black lore, no black on crown. Similar ad. f. Baltimore has mostly concolor underparts (except for some paleness on vent area); usually some to extensive dark mottling on head; pale lores; no eyeline. Similar f. Orchard and f. Hooded Orioles have yellow to yellowish-green underparts, downcurved bill; Orchard has concolor underparts, Hooded has slightly paler belly. **1st Yr:** (Sep.–Aug.) M. has yellow to orangish head and breast and pale belly; black lores; suggestion of dark eyeline; some black on chin; and occasionally some black feathers on crown. F. is similar to ad. f. but with less yellow on head and breast; may be almost entirely grayish with just a yellow wash on head. **Flight:** Strong, direct. Ad. m. disting. from Baltimore by white upperwing coverts, dark tip on orange tail. **Hab:** Deciduous trees near openings, parks, gardens, streamsides. **Voice:** Song a short rapid series of fairly harsh notes mixed with short chattering; calls include a *chet,* a *weet,* and a rough chattering *ch'ch'ch'ch.* **143**.

Subspp: (2) •*parvus* (s. CA–s. NV–s.w. AZ) small; •*bullockii* (rest of range) large. **Hybrids:** Baltimore Oriole.

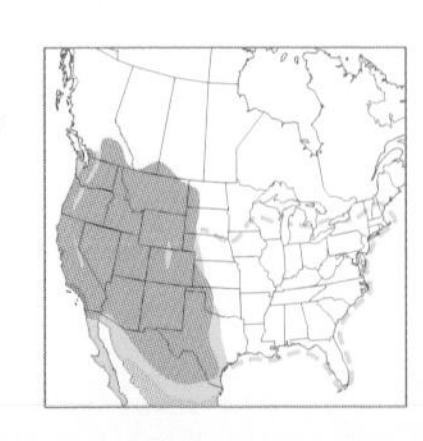

Adult m. OH/05

1st-yr. m. TX/04

Adult f. (heavily marked) OH/05

1st-yr. m. NY/–

Adult f. (mildly marked) OH/05

1st-yr. f. OH/05

Baltimore Oriole 1

Icterus galbula L $8\frac{3}{4}$"

Shape: Medium-sized broad-necked bird with a long tail and deep-based but fine-pointed bill typical of *Icterus* genus. Comparatively medium-sized, with medium-length straight bill and short fairly squared-off tail. Bullock's Oriole has similar shape. **Ad:** Plumage more variable than on other orioles. To recognize adults, check greater coverts—they should be uniformly tipped white, creating complete wingbar. M. has orange body with solid black hood and back; outer tail feathers broadly orange at tips; central tail feathers black; lesser coverts orange, forming a bar. F. has orange underparts, slightly paler on the vent area; head and back vary from dusky orange to slightly mottled with black to almost all black with traces of orange (never solid black like m.); central tail feathers dusky orange, outer tail feathers orange; lesser coverts blackish. See Bullock's Oriole for comparison. **1st Yr:** To recognize 1st yr., check greater coverts for molt limits—newer coverts are larger, fresh, darker, and tipped white; older ones are paler, worn, with no white tips, resulting in a shortened wingbar. M. head and back from dusky orange to mottled black to pure black; central tail feathers sooty to black; lesser coverts mottled orange. F. head and back range from smooth dusky orange to mottled black; central tail feathers orangish (not sooty to black as in 1st-yr. m.); new wing coverts and/or tertials sooty (not black as in 1st-yr. m.). Dull 1st-yr. f. similar to f. Bullock's. **Flight:** Strong, direct. Ad. m. distinguished from Bullock's by orange lesser coverts and orange corners to tail; Bullock's has black lesser coverts, white greater and median coverts, black bar across tail tip. **Hab:** Deciduous trees near openings, parks, gardens, streamsides. **Voice:** Song a varied series of clear whistles, often with repetition; calls include a single upslurred whistled *weeet,* a chattering *ch'ch'ch'ch,* a short harsh *cher,* and 2-note whistled alarm call, like *teeetooo.* **144**.

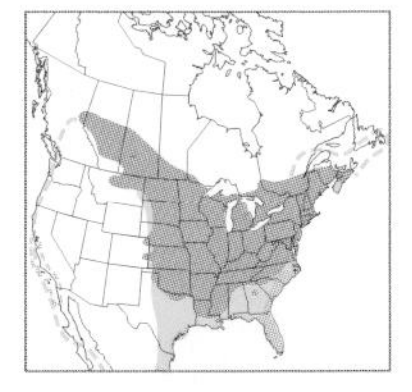

Subspp: Monotypic. **Hybrids:** Bullock's Oriole.

Adult m. AZ/04

Adult f. AZ/05

Adult f. AZ/04

1st-yr. m. NM/05

Scott's Oriole 1

Icterus parisorum L 9"

Shape: Medium-sized broad-necked bird with a long tail and deep-based but fine-pointed bill typical of *Icterus* genus. Comparatively large, with long straight bill and short tail. **Ad:** M. has black hood, back, and breast; deep yellow belly and bar on upperwing; outer tail feathers yellow at base for half their length. F. olive-green head and back; yellow-olive to deep yellow belly; variable black or sooty gray on face, throat, upper breast, and back. See Audubon's Oriole for comparison. **1st Yr:** (Sep.–Aug.) M. like ad. m. but black back feathers with pale grayish tips; primaries brown (black in ad.); body can be dusky yellow. F. like adult. **Flight:** Slightly undulating, generally low. Note male's yellow base to outer tail feathers. **Hab:** Arid and semiarid woods, often with yucca. **Voice:** Song a rich rapid series of low-pitched whistles; calls include a rapid *ch'cht* and a low *chuk*.

Subspp: Monotypic.

Adult TX/03

Adult TX/12

1st yr. TX/02

Audubon's Oriole 2

Icterus graduacauda L 9½"

Shape: Medium-sized flat-crowned, and broad-necked bird with a long tail and deep-based but fine-pointed bill typical of *Icterus* genus. Comparatively, Audubon's is large, with very deep-based straight bill, long graduated tail; flat crown blends with line of culmen. **Ad:** Deep yellow overall (including back) except for black hood; tail all black. Sexes alike (m. may have more extensive hood, brighter-colored back). Similar ad. m. Scott's Oriole has black back, more extensive black hood, yellow base to outer tail feathers (ad. Audubon's tail all black). **1st Yr:** Like ad. but crown washed with green, back greenish, outer tail feathers edged dusky green, wings brownish. **Flight:** Strong and direct. All-black tail, yellowish back. **Hab:** Open woodlands, brushy areas, wooded streamsides. **Voice:** Song a noticeably slow and low-pitched series of well-spaced whistles with little repetition; calls include a rising, 2-part, whistled *too-ee* and a nasal *nyeehnk nyeehnk*.

Subspp: (1) •*audubonii*.

Finches

Grosbeaks

Crossbills

Redpolls

Siskins

Goldfinches

Old World Sparrows

Adult m., summer DEN/06

Adult f., winter ITA/12

Common Chaffinch 4

Fringilla coelebs L 6"

Casual Eurasian vag. to Northeast; most sightings here and elsewhere in NAM may be of birds escaped from captivity. **Shape:** Fairly small, deep-bellied, broad-necked bird with a slender conical bill and moderate-length notched tail. Crown rather flat when feathers are relaxed, peaked and triangular when raised. **Ad:** Closed wing patterned with 2 bold whitish wingbars (the upper one sometimes hidden) and small white patch at base of primaries. M. (brightest in summer) has gray crown and collar around rosy brown face with small black forehead; back, breast, and belly rosy brown; outer tail feathers white. F. pale brown below, warm brown on back and crown, bold wing and tail pattern mentioned above. **Juv:** (Apr.–Sep.) Like ad. f. **Flight:** Undulating; white outer tail feathers and bold white wingbars. **Hab:** Open woods, parks, gardens. **Voice:** Song a rapid, energetic, accelerating series of short whistles ending with a flourish (and sometimes a downward *jeeer*), like *jee jee jee sisisichitteryjeeer.* Calls include a *fink* and a *yup.*

Subspp: (1) •*coelebs.*

Adult m., *erythrinus** TUR/08

Adult f., *erythrinus** FIN/06

Common Rosefinch 4

Carpodacus erythrinus L 5¾"

Casual Asian vag. to w. AK islands. **Shape:** Small, deep-bellied, broad-necked, fairly large-headed finch with a short notched tail and short, deep-based, conical bill with a curved culmen. Primary projection past tertials about 2 x bill length. Fairly large dark eye. **Ad:** M. extensively deep rose-red on head, breast, back, and rump; little or no facial pattern; 2 whitish to reddish wingbars. F. streaked grayish brown overall with little contrast; little facial pattern; 2 thin whitish wingbars. Both sexes lack pale eye-ring. **Juv:** Like ad. f. but with some dull olive wash on face and upperparts. **Flight:** Fairly long pointed wings and stubby bill. **Hab:** Wet woods, shrubby streamsides, parks. **Voice:** Song a short series of 2–3 slurred whistles, like *vidyoo veeri voiyoo;* calls include a short upslurred *tooee.*

Subspp: (1) •*grebnitskii.* **C. e. erythrinus* (e. EUR to w. Asia) used for good photograph.

Adult m., summer FIN/06

Adult m., winter ENG/11

Adult f., summer FIN/05

Adult f., winter ENG/12

Brambling 3

Fringilla montifringilla L 6¼"

Rare visitor from Asia to w. AK, casual farther south. **Shape:** Fairly small, deep-bellied, broad-necked bird with a slender conical bill and moderate-length notched tail. Crown rather flat when feathers are relaxed, peaked and triangular when raised. Long primary projection past tertials. **Ad. Summer:** Closed wing boldly patterned with 2 white to orangish wingbars, small white patch at base of primaries, orangish lesser coverts; white rump. M. with black head and back, orangish-brown throat and breast, large dark spots along whitish rear flanks; bill black. F. face plain grayish brown with dark brown lateral crown-stripes continuing down nape; breast orangish; faint spots along flanks; bill black. **Ad. Winter:** M. similar to summer, but pale edges to head and back feathers create mottled grayish appearance; bill dusky to pale with dark tip. F. like summer f., but bill orangish with dark tip. **Flight:** Undulating; white rump, orangish lesser wing coverts. **Hab:** Open mixed coniferous and deciduous woods, fields. **Voice:** Song a single, harsh, drawn-out whistle, like *dzrrreee*, often repeated; calls include a harsh *jeeak*.

Subspp: Monotypic.

Adult m., summer, Brown-cheeked Grp. CA/05

Adult m., winter, Gray-cheeked Grp. AK/03

Adult f., summer, Brown-cheeked Grp. CA/05

Adult f., winter, Gray-cheeked Grp. NM/12

Gray-crowned Rosy-Finch 1

Leucosticte tephrocotis L 6¼"

Shape: Fairly small, deep-bellied, small-headed bird with a short conical bill and moderate-length notched tail. Long primary projection past tertials; crown peaked when crest raised. Bill longest among rosy-finches. **Ad:** Breast feathers warm brown to cinnamon-brown without blackish centers and with variable thin pale fringes; blackish forecrown; pale silvery-gray hindcrown, which extends to eye and sometimes through cheek (see Subspp.), contrasts strongly with surrounding warm brown back; limited rose on brownish rear belly. M. with bright pink on wings and more extensive bright pink on belly. F. with duller less-extensive pink on wings and belly; forecrown less extensively black. Both sexes have bill blackish in summer, yellowish in winter (f. bill color changes to yellow about 1 mo. earlier than m.). Similar Black Rosy-Finch has shorter bill, black-centered breast feathers, little or no rose on underparts. **1st Winter:** Like ad. but greater secondary coverts and primary edges mostly to all whitish (ad. pink in these areas); little or no pink on underparts. **Juv:** (Jun.–Aug.) Grayish brown overall with pinkish to buff wingbars. **Flight:** Undulating, can be zigzag; pinkish rump. **Hab:** Alpine areas, tundra in summer; open areas, mountain meadows, fields in winter. **Voice:** Song a descending series of harsh notes, like *cheew cheew cheew cheew;* calls include a repeated downslurred *jeew.*

Subspp: (6) Divided into two groups. **Gray-cheeked Group:** Large to medium-small; gray cheek. •*littoralis* (w.cent. AK–s.w. YT south to n.cent. CA), •*griseonucha* (Aleutian and Kodiak Is. AK), •*umbrina* (Pribilof Is. AK). **Brown-cheeked Group:** Medium-small to small; brown cheek (more like Black Rosy-Finch). •*tephrocotis* (cent. YT–s.cent. BC–n.w. MT), •*wallowa* (n.e. OR), •*dawsoni* (e. CA). **Hybrids:** Black Rosy-Finch.

Adult m., summer WY/06

Adult f., winter NM/12

Adult m., winter NM/12

Black Rosy-Finch 2

Leucosticte atrata L 6¼"

Shape: Fairly small, deep-bellied, small-headed bird with a short conical bill and fairly short notched tail. Long primary projection past tertials. Bill medium-length among the rosy-finches. **Ad:** Breast feathers with blackish centers and gray to brown fringes; blackish forecrown; pale silvery-gray hindcrown extends to eye and contrasts strongly with surrounding dark gray to black collar and cheek; little or no rose on belly. M. forecrown and body very black with bright rose on wings. F. forecrown and body feathers broadly edged with gray, creating a cool gray wash overall; paler pink on wings. Both sexes have bill blackish in summer, yellowish in winter (m. bill turns black sooner than f.). Similar Gray-crowned Rosy-Finch has longer bill, warm brown–centered breast feathers, more rosy underparts. **1st Winter:** Like ad. but greater secondary coverts and primary edges mostly to all whitish (adults pink in these areas); little or no pink on underparts. **Juv:** (Jun.–Aug.) Grayish brown overall with pinkish to buff wingbars. **Flight:** Undulating, can be zigzag; pinkish rump. **Hab:** Tundra or alpine areas with cliffs, rock slides in summer; open areas, feeders in winter. **Voice:** Song like that of other rosy-finches (see Gray-crowned); calls include a repeated downslurred *jeew.*

Subspp: Monotypic. **Hybrids:** Gray-crowned Rosy-Finch.

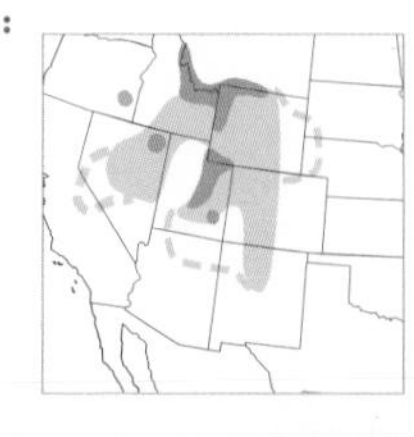

Adult m., summer CO/06

Adult f., winter NM/12

Adult m., winter NM/12

Brown-capped Rosy-Finch 2

Leucosticte australis L 6¼"

Shape: Fairly small, deep-bellied, small-headed bird with a short conical bill and fairly short notched tail. Long primary projection past tertials. Bill shortest among the rosy-finches. **Ad:** Breast feathers cold brown without blackish centers and with variable gray to buffy fringes; extensive rose on lower breast and belly. M. with blackish forehead blending to dark gray of hindcrown, which extends to eye; this gray area on face does not contrast strongly with brown cheek. Extensive pink on belly; deep pink on wings. F. with brownish forehead that blends to grayish-brown hindcrown and rest of face; little or no contrast on face; dull pink on belly, less extensive than on m.; pale pink on wings. Both sexes have bill blackish in summer, yellowish in winter. Similar Black and Gray-crowned Rosy-Finches have pale gray napes that contrast strongly with rest of body. **1st Winter:** Like ad. but greater secondary coverts and primary edges mostly to all whitish (ad. pink in these areas); little or no pink on underparts. **Juv:** (Jun.–Aug.) Grayish brown overall with pinkish to buff wingbars. **Flight:** Undulating, can be zigzag; pinkish rump. **Hab:** Tundra or alpine areas with cliffs, rock slides in summer; open areas, feeders in winter. **Voice:** Song like that of other rosy-finches (see Gray-crowned); calls include a repeated downslurred *jeew.*

Subspp: Monotypic.

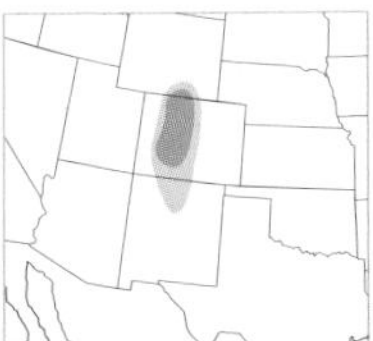

Adult m. MN/01

Adult m. MI/01

Adult f. NH/12

Adult f. NH/12

Pine Grosbeak 1

Pinicola enucleator L 9"

Shape: Medium-sized, deep-chested, deep-bellied, broad-necked bird with a fairly long notched tail and short deep-based bill with a curved culmen. **Ad:** Two broad white wingbars on blackish wings (upper wingbar may be hidden); broad white edges to tertials and secondaries; blackish bill. M. with variable shades and extent of red to orangish red on head, back, breast, and sometimes belly (see Subspp.); rest of body gray. F. body mostly gray with olive, bronze, or reddish brown on head and rump (sometimes a hint on breast). **1st Yr:** Like ad. f. (Sep.–Mar.); in 1st summer (Apr.–Aug.), m. can have a few red body feathers. **Juv:** (Jun.–Sep.) Mostly brown with buffy wingbars. **Flight:** Strong, slightly undulating. **Hab:** Open spruce and fir woods in summer; also any areas with tree fruits and seeds, suburbs in winter. **Voice:** Song a rapid squeaky warble; calls vary geographically and include a high *titweet* and a *peew peew.*

Subspp: (6) •*leucurus* (cent. AK–n.e. BC east to NF–CT) variably sized, m. breast and flanks pale red to red; •*montanus* (cent. BC–AZ and NM) large, m. breast dark red, flanks mostly gray; •*californicus* (e. CA) medium-sized, m. breast dull red, flanks mostly gray; •*flammula* (s.cent. AK–n.w. BC) medium-sized, m. breast and flanks bright red; •*carlottae* (w.cent.–s.w. BC is.) small, m. breast and flanks dark red; •*kamtschatkensis* (vag. to w. AK) medium to small, m. breast and flanks bright red. **Hybrids:** Purple Finch.

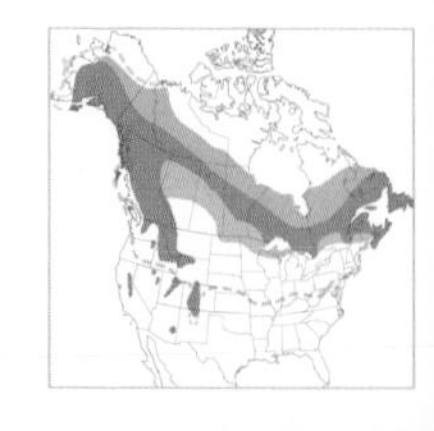

Adult m. CO/06

Adult f. CA/03

Adult m. NM/10

Adult f. CO/07

Cassin's Finch 1

Carpodacus cassinii L 6¼"

Shape: Fairly small moderately slim bird with relatively short notched tail and a fairly short conical bill with a straight culmen, making it look very sharp-pointed (see Subspp.). Long primary projection past tertials 2–3 x bill length (less in House and Purple Finches); exposed primary tips unevenly spaced (more even in Purple Finch). Bushy crest often raised. Similar Purple Finch has deeper belly, shorter bill with a curved culmen, shorter primary projection; similar House Finch has slimmer belly, shorter bill, shorter primary projection. **Ad:** Both sexes have thin pale eye-ring. M. with pale rosy wash over head, breast, and rump, brightest and darkest on crown; fairly contrasting dark brownish streaking on pinkish-brown back; underparts white with variable fine dark streaking on flanks and undertail coverts. F. with thin pale eye-ring; streaked brown overall; indistinct whitish eyebrow and moustachial region on brown face; fine blackish streaking on white underparts; undertail coverts finely streaked. See Purple Finch for comparison. **1st Yr:** Both sexes like ad. f., but some 1st-yr. m. may have pinkish tinge. **Flight:** Undulating; m. with grayish underwing coverts. **Hab:** Open coniferous forests. **Voice:** Song a rapid, short, musical warble, slightly higher-pitched and longer than song of Purple Finch; calls include a simple *keeup* and rapid *tidilip*.

Subspp: (2) •*vinifer* (n. CA–w. ID and north) has longer bill, darker plumage; •*cassinii* (s.e. CA–e. ID and south) has shorter bill, paler plumage.

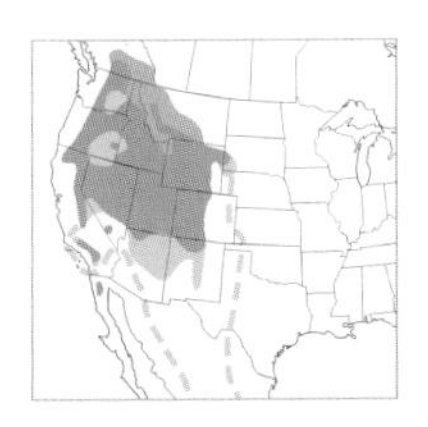

Adult m., *purpureus* MI/05

Adult f., *purpureus* NH/–

Adult m., *purpureus* NH/–

Adult f., *purpureus* NH/11

Purple Finch 1

Carpodacus purpureus L 6"

Shape: Fairly small, deep-bellied, broad-necked, fairly large-headed bird with a short conical bill with slightly curved culmen and a relatively short notched tail. Primary projection past tertials 1½–2 x bill length (2–3 x in Cassin's Finch); distance between exposed primary tips fairly even (uneven in Cassin's). See Cassin's Finch for comparison. **Ad:** Both sexes lack pale eye-ring. M. with bright raspberry wash evenly over head, breast, and rump; weakly contrasting brownish streaking on maroon back; may be a few blurry brown streaks on flanks, otherwise belly, flanks, and undertail coverts white to pinkish and unstreaked. F. with distinct to indistinct whitish eyebrow and moustachial streak (see Subspp.) on brown face; thick brown streaking on white to creamy underparts; slightly darker brown streaking on brown back; undertail coverts streaked or unstreaked (see Subspp.). Similar Cassin's Finch has thin pale eye-ring; f. with indistinct whitish eyebrow and moustachial region, blackish streaking on underparts (including undertail coverts), contrastingly black streaking on brown back; m. with rosy wash brightest and darkest on crown. **1st Yr:** Both sexes like ad. f., but some 1st-yr. m. may have pinkish or yellowish tinge. **Flight:** Undulating; m. with pinkish underwing coverts. **Hab:** Mostly coniferous woods, but also mixed forests, edges, orchards, gardens, feeders. **Voice:** Song a rapid, short, musical warbling; calls include a sharp flat *pik* and whistled *tweeoo.* **145**.

Subspp: (2) •*californicus* (n.w. WA–s.w. CA) small; culmen generally straight; f. with less contrasting whitish eyebrow and moustachial region than f. *purpureus* (thus, more like f. Cassin's Finch or even House Finch), faint olive tinge overall, blackish streaks on undertail coverts; m. head dull red. •*purpureus* (rest of range to north and east) large; culmen generally curved; f. with bold and contrasting white eyebrow and moustachial region, grayish tinge overall, few or no streaks on undertail coverts; m. head bright red. **Hybrids:** Pine Grosbeak.

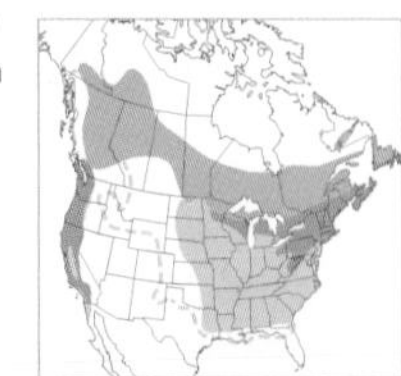

Adult m. AZ/07

Adult f. NM/12

Adult m. NH/11

Adult f. TX/01

House Finch 1

Carpodacus mexicanus L 6"

Shape: Fairly small, slender, small-headed bird with a generally short, stubby, conical bill (see Subspp.) and relatively long slightly notched tail. Culmen noticeably curved (mostly straight in Cassin's and Purple Finches). Similar Purple and Cassin's Finches have deeper belly, longer bill, less curved culmen, and are larger-headed and more broad-necked, with deeply notched tail. **Ad:** Both sexes lack pale eye-ring. M. with variable orangish-red (occasionally yellow) on head, breast, and rump and generally brownish auriculars; whitish belly and flanks strongly streaked with brown; back brownish with indistinct streaks. F. head finely streaked with brown and with weak facial pattern; dusky streaking on underparts blurry and not strongly contrasting with dull whitish background; no orangish red on plumage. **1st Yr:** M. may be like ad. f. or ad. m. (see Subspp.), or may have just limited orangish red on rump; f. like ad. f. **Flight:** Slightly undulating; m. with grayish underwing coverts. **Hab:** Extremely variable; usually edges, in wild or suburban areas, feeders. **Voice:** Song a long rapid warble, usually ending with a harsh slurred *jeeyee* or *jeeer;* calls include a *ch'weet*. **146**.

Subspp: (4) •*frontalis* (s.w. BC–s.w. CA and MT–AZ east) has small bill, sparse streaking on underparts; 1st-yr. m. and ad. m. with orangish red or yellow on head and breast. •*solitudinus* (e. WA–ID south to n.w. AZ) has small bill, sparse streaking on underparts; 1st-yr. m. with little or no red, ad. m. with reduced red on breast. •*clementis* (Channel Is. CA) and •*potosinus* (s.cent. TX) have large bill, moderate to heavy streaking on underparts; 1st-yr. m. and ad. m. with extensive red on head and breast.

Adult m. OR/08

Juv., trans. to adult m. OR/08

Adult f. OR/08

Juv. OR/08

Red Crossbill 1

Loxia curvirostra L 6¼"

Shape: Fairly small, short-tailed, short-legged, big-headed, broad-necked bird with a distinctive heavy bill with tips that cross each other. Tail deeply notched; primary projection very long (2½ x bill length). **Ad:** Lacks bold white wingbars. M. body mostly all red or reddish orange, sometimes a mix of red and yellowish feathers; wings and tail blackish; throat and breast usually concolor. F. brownish or yellowish green, often brightest on rump, crown, and breast; throat usually grayish; may have thin, indistinct, buffy or grayish wingbar on greater coverts (but not like broad, well-defined, white wingbar of White-winged Crossbills, which also have white tips to tertials). In both sexes, undertail coverts whitish with dark chevrons. **1st Yr:** M. may be reddish or orangish mixed with some brown; f. usually brownish or greenish with little yellow. **Juv:** (Jan.–Sep., can breed in winter) Brownish overall, with heavy brown streaking on whitish underparts; thin buffy wingbars. May breed in this plumage. **Flight:** Undulating. Long pointed wings and fairly short notched tail. **Hab:** Coniferous woods. **Voice:** Song a series of varied, clipped, high-pitched phrases, like *jit jit jit tokity tokity chewy chewy;* calls include a *kip kip kip.*

Subspp: Classification still in flux, complicated by nomadism; may be several species. Using overall size, bill size, and 10 flight-call "types," populations may be divided into five groups. **Small Group:** Type 3 (all of range) is small; bill small. **Medium-sized Group:** Types 1, 2, 4, and 10 (all of range), plus types 5 and 7 (only in West), are medium-sized; bill medium-small to medium-large. **Large Group:** Type 6 (only extreme Southwest) is large; bill large. **Newfoundland Group:** Type 8 (NF) is medium-sized; bill medium; darker-plumaged. **South Hills Group:** Type 9 (s.cent. ID) is sedentary. **Hybrids:** Pine Siskin.

Adult m. NH/03

Adult f. QC/01

Adult m. QC/01

Juv. AK/06

White-winged Crossbill 2

Loxia leucoptera L 6½"

Shape: Fairly small, short-tailed, short-legged, big-headed, broad-necked bird with a distinctive fairly thin bill with tips that cross each other. Tail deeply notched; primary projection very long (2½ x bill length). Distinguished from Red Crossbill by slightly thinner-based bill, slightly longer tail. **Ad:** Two broad well-defined white wingbars; white tips to tertials (lacking on Red Crossbill). M. head, rump, and most of underparts pinkish red; back all black to mostly red; rear flanks paler and strongly to indistinctly streaked darker. F. variably faintly greenish (sometimes yellowish) on crown, back, rump, and breast; otherwise, grayish-brown upperparts and pale underparts darkly streaked with brown. **1st Yr:** M. mixed brown with pinkish red or orangish yellow; f. brownish or faintly greenish. **Juv:** (Jan.–Sep., can breed in winter) Brownish overall with heavy brown streaking on whitish underparts; distinct broad white wingbars. May breed in this plumage. **Flight:** Undulating. Long pointed wings and short notched tail. **Hab:** Coniferous woods. **Voice:** Song a very long series of dry trills on varying pitches, like *tre'e'e'e'e, jr'r'r'r'r'r, cha'a'a'a'a;* calls include a repeated *chudut,* a rising *veeyeet,* and repeated *chet* notes.

Subspp: (1) •*leucoptera.*

Adult m., fresh MN/12

Adult f., fresh MN/12

Adult m., spring MI/03

Adult f., spring MI/03

Adult m., worn AK/07

Adult f., worn AK/07

Common Redpoll 1

Carduelis flammea L 5¼"

Shape: Small, deep-bellied, small-headed, broad-necked bird with a relatively thin-based, sharp-pointed, conical bill and moderate-length strongly notched tail. Distinguished from very similar Hoary Redpoll (with caution and adding plumage clues) by slightly longer and proportionately narrower-based bill, creating a more acute point; slightly more sloping forehead; and rounded crown (when crest is relaxed). **Ad:** Both Common and Hoary Redpolls have red forecrown, black chin and lore, 2 white wingbars. Key clues to distinguishing the species (besides shape) are undertail coverts, rump, flank streaking, wingbar width, and color of the scapular edges. Common Redpoll has white undertail coverts with 2 to many broad dark streaks; a dusky or whitish rump with extensive dark streaking (except older m., which can have bright pinkish-red rump); extensive broad dark streaking on the flanks; moderately wide white tips to greater secondary coverts; and brownish edges to scapulars, creating a dark brown back. M. has pinkish red to bright red on breast and flanks, the feathers tipped with white when fresh and wearing away over the year, revealing more deep red. F. generally with little or no pink on breast and flanks; generally darker than male on back. Similar Hoary Redpoll paler overall;

1st winter QC/02

Adult f. (rump streaks) MI/03

Juv. AK/07

1st winter IL/01

whitish undertail coverts with 0–3 thin dark streaks; whitish rump with little or no streaking; a few thin dark streaks on flanks; broad white tips to greater secondary coverts; grayish edges to scapulars, creating a more frosty look to back; ad. m. with wash of pink on breast and flanks; ad. f. with no pink. **1st Yr:** Often shows molt limits, with inner greater coverts fresher, darker, and more broadly tipped with white than older outer coverts. M. with faint wash of pink on breast and flanks (may not be distinguishable in field from ad. f.); f. with no red on breast and flanks. **Juv:** (Jul.–Aug.) Generally streaked brown overall without red or black on head; buffy wingbars. **Flight:** Undulating; streaked or pinkish rump. **Hab:** Low coniferous or mixed scrub, tundra in summer; open areas, edges, feeders in winter. **Voice:** Song a long series of repeated short calls and trills; calls include a rising scratchy *jeeyeet* and a husky *chew chew chew*.

Subspp: (2) •*rostrata* (n. NT) is large; bill deep-based with slightly curved culmen; streaking on flanks heavy and dark brown. •*flammea* (rest of range) is smaller; bill slender; streaking on underparts thinner and medium brown. **Hybrids:** Hoary Redpoll, Pine Siskin.

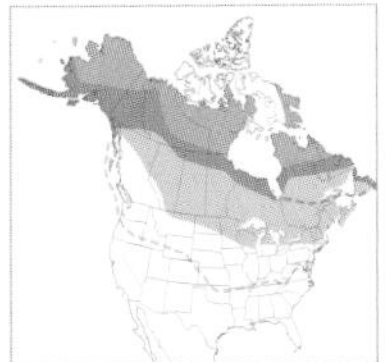

Adult m. AB/02

Adult m. AK/04

Adult m. AB/02

Adult f. NH/01

Hoary Redpoll 2

Carduelis hornemanni L 5½"

Shape: Small, deep-bellied, small-headed bird with a relatively thick-based, sharp-pointed, conical bill and moderate-length strongly notched tail. Distinguished from Common Redpoll (with caution and adding plumage clues) by slightly shorter, proportionately deeper-based bill, creating more broadly angled tip and stubbier look; slightly steeper forehead and flattened crown (when crest is relaxed); head and broad neck blended into body. Shorter bill and steeper forehead create an impression of a "pushed-in" bill. **Ad:** Both Common and Hoary Redpolls have red forecrown, black chin and lore, 2 white wingbars. Key clues to distinguishing the species are undertail coverts and rump, both difficult to see except at close range. In Hoary Redpoll, both sexes have white undertail coverts with 0–3 thin indistinct streaks and whitish rumps with moderate to no dark streaking. Additionally, upperparts whitish to grayish with grayish-brown streaking; white edges of greater coverts thicker; and flanks variably with a few thin dusky streaks. M.

Adult f. AK/04

Adult f. AK/04

Adult f. AK/04

Adult f. ME/02

with pinkish limited to upper breast, the feathers tipped with white when fresh (fall and winter) and wearing away over the year, revealing more color. F. generally without pink on breast and flanks; generally slightly darker than male. For comparison with very similar Common Redpoll, see that account. **1st Yr:** Often shows molt limits, with inner greater coverts fresher, darker, and more broadly tipped with white than older outer coverts. **Juv:** (Jul.–Aug.) Generally streaked brown overall without red or black on head; bill less stubby than on adult. **Flight:** Undulating; usually unstreaked whitish rump. **Hab:** Low coniferous or mixed scrub, tundra in summer; open areas, edges, feeders in winter. **Voice:** Sounds similar to Common Redpoll. Song a long series of repeated short calls and trills; calls include a rising scratchy *jeeyeet* and a husky *chew chew chew.*

Subspp: (2) •*hornemanni* (n. NT) large, large-billed, very broad-necked, whitish overall with little streaking. •*exilipes* (rest of range) smaller, smaller-billed, browner with heavier dusky streaking. **Hybrids:** Common Redpoll.

Adult, *pinus* NH/05

Adult, *pinus*, "green morph" NH/01

Adult, *pinus*, "green morph" NH/01

Adult, *vagans* CA/09

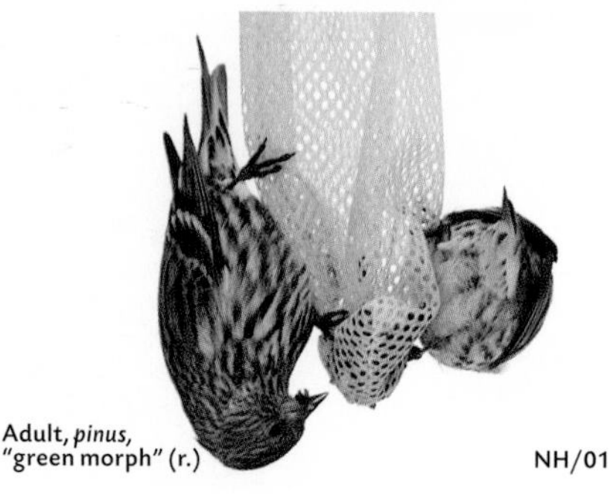

Adult, *pinus*, "green morph" (r.) NH/01

Pine Siskin 1

Carduelis pinus L 5"

Shape: Small, slim, fairly small-headed bird with a short well-notched tail and slender, fine-pointed, conical bill with straight culmen (rarely, slightly decurved). **Ad:** Head and body finely streaked with brown; underparts dull whitish with streaking across breast and along flanks; yellow edges to primaries and secondaries. Two thin buffy to white wingbars on median and greater coverts (greater coverts have yellow edges, and base of secondaries underneath is also yellow, so that when white to buffy wingbars wear off the greater coverts, the bird can appear to have a wide lower "wingbar" that is yellowish). Great variation in amount of yellow on individuals; males and adults tend to have more yellow in plumage, but this varies (see Subspp.); birds cannot be accurately sexed in the field (see Subspp.). One percent of birds are more extensively colored with yellow shadings (the so-called green morph); they have more yellow on wing, side of nape, flank, and often undertail coverts. **Juv:** (Apr.–Sep.) Heavily streaked brown overall; wide buffy wingbars; very little yellow. **Flight:** Undulating; variably wide yellow stripe through wing, dark rump. **Hab:** Coniferous woods, seed-bearing trees, feeders. **Voice:** Song a long mixture of fairly harsh chirps on varied pitches with some repetition; calls include a *checheche,* a buzzy rising *dzzzzeeee,* and a flat *tseeyeet.* 🎧 **147**.

Subspp: (3) •*vagans* (AB–w. TX and west) small, with light to moderate brown streaking, moderate yellow in wings; •*pinus* (SK and east) small, with moderate to heavy blackish streaking, moderate yellow in wings; •*macroptera* (Mex. vag. to Southwest) large, with light dusky streaking, extensive yellow in wings. **Hybrids:** Red Crossbill, Common Redpoll.

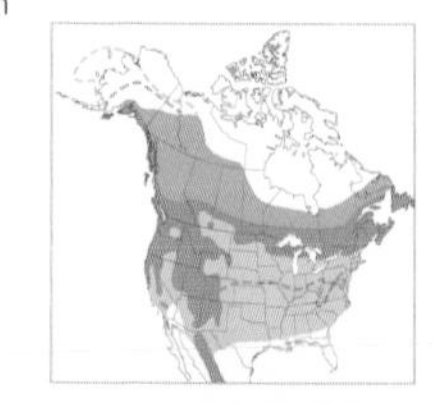

Adult m. FIN/05

Adult f. JAP/03

Adult f. ENG/10

Eurasian Siskin 5

Carduelis spinus L 5"

Accidental Eurasian vag. to w. AK islands; birds seen in ON, ME, MA, and NJ may be escapes from captivity. **Shape:** Small, slim, fairly small-headed bird with a short well-notched tail and slender, fine-pointed, conical bill, sometimes with slightly decurved culmen. **Ad:** M. with black cap and chin (partially veiled by gray in fresh plumage); unstreaked yellow face and breast; 2 broad yellow wingbars; slight dark streaking and yellow wash on flanks; gray bill. F. with 2 broad yellowish wingbars; yellowish eyebrow extends around dusky auricular and onto sides of breast; white underparts with fine dark streaking limited to sides of breast and flanks, leaving central underparts clear; bill gray; yellow wash to streaked rump; white undertail coverts. Similar f. Pine Siskin has thin whitish wingbars; little or no yellow on eyebrow and face; streaked whitish underparts; white undertail coverts (can be yellowish in green-morph Pine Siskin). **Flight:** Undulating; extensive yellow wash to upperparts, broad yellow band across wings. **Hab:** Coniferous woods in summer; woods edges, fields, feeders in winter. **Voice:** Song a rambling series of chirps and short whistles; calls include a downslurred *tiyoo* and a rising *toowee.*

Subspp: Monotypic.

Adult m. CA/05

Adult f. CA/05

Lawrence's Goldfinch 2

Carduelis lawrencei L 4¾"

Shape: Small, slim, small-headed bird with a short notched tail and short, fairly deep-based, conical bill. **Ad:** Gray body; golden patch on breast; black wings with golden wingbars (or whole coverts); golden edges to primaries and secondaries; pinkish bill. M. with black foreface and forecrown; bright and extensive golden on breast and wings. F. with all-gray head; reduced and duller golden on breast and wings. Similar winter f. American Goldfinch lacks golden edges to primaries. **1st Winter:** M. similar to ad. m. but with partial black on foreface and forecrown. F. like ad. f. **Juv:** (May–Aug.) Like ad. f. but with brownish wash overall; indistinct dusky streaking on breast and belly. **Flight:** Undulating; whitish underwing coverts, golden patches on wings, yellowish rump. **Hab:** Open live oak woodlands, brushy fields, chaparral; often near water. **Voice:** Song a clipped and halting series of short high whistles and chirps (many imitations of other birds' calls); calls include a bell-like *teetoo* and a soft *tsip.*

Subspp: Monotypic.

Adult m., *psaltria* TX/04

Adult m., *hesperophilus* CA/03

Adult f., *hesperophilus* CA/03

Adult f., *psaltria* TX/06

1st-winter m., *hesperophilus* CA/03

Lesser Goldfinch 1

Carduelis psaltria L $4\frac{1}{2}$"

Shape: Small, slender, small-headed bird with a short notched tail and short, fairly deep-based, conical bill. Proportionately shortest tail of our three goldfinches. **Ad:** M. underparts, including undertail coverts, plain yellow; black cap and dark back (either all black or greenish with dark streaks; see Subspp.); large white patch at base of primaries; 1 white wingbar. F. with bright to dull yellow underparts, including undertail coverts; olive upperparts; very small white patch at base of primaries; 1 whitish wingbar on greater coverts (a second indistinct bar on median coverts sometimes hidden); bill dark horn to grayish; little or no white on tail. Similar f. American Goldfinch lacks large white patch on primaries, has pinkish bill in summer, white undertail coverts, 2 well-defined wingbars. **1st Winter:** M. like ad. f. but with some black feathers in crown, contrast between black tertials and brownish wings, and large white patch at base of primaries. F. like ad. f. **Juv:** (May–Aug.) Like ad. f. but washed with buff overall, wingbars buffy, white primary patch small or absent. **Flight:** Undulating; dark underwing coverts, white patch at base of primaries larger on m. **Hab:** Variable; woods, edges, gardens, feeders. **Voice:** Song a long continuous series of chips and whistles, often with many repetitions and imitating other birds' calls; calls include a plaintive *bearbee* and a harsh *chit chit chit.*

Subspp: (3) •*psaltria* (CO–s.w. NM and east) m. upperparts mostly black to all black; •*hesperophilus* (n.e. UT–s.e. AZ and west) m. upperparts vary geographically from west to east, with increasing black streaking on green back; •*mexicanus* (s. TX) back all black. **Hybrids:** American Goldfinch.

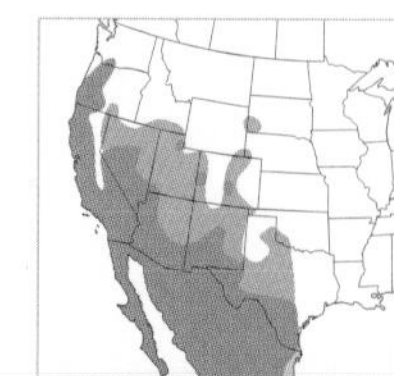

Adult m., summer, *tristis* ME/06

Adult m., winter, *tristis* NH/12

Adult f., summer, *tristis* MN/07

Adult f., winter TX/01

1st yr., *tristis* MN/06

American Goldfinch 1

Carduelis tristis L 5"

Shape: Fairly small, slim, somewhat small-headed bird with a fairly long notched tail and short conical bill. **Ad. Summer:** Pale pinkish-orange bill. M. with black cap; bright yellow body with white undertail coverts; 2 white wingbars on black wings. F. with greenish-yellow crown; bright yellow underparts with white undertail coverts; dusky yellow upperparts; 2 white wingbars on dark brown wings. Similar f. Lesser Goldfinch has yellow undertail coverts; white patch on base of primaries; dark gray bill; shorter tail. **Ad. Winter:** Both sexes variably pale grayish to grayish or brownish yellow (brightest yellow on throat); white undertail coverts; 2 whitish wingbars; dark gray bill. M. with black wings, yellow lesser coverts, and sometimes small white patch at base of primaries. F. with dark brown wings, dull brownish lesser coverts, no white patch at base of primaries. **1st Yr:** Much like ad. in respective seasons, but inner greater wing coverts darker and more broadly tipped with white to whitish. **Juv:** (Jul.–Oct.) Brownish to brownish yellow overall except for whitish undertail coverts; wings with wide buffy wingbars and wide buffy edges to secondaries and tertials. **Flight:** Undulating. Both sexes with whitish underwing coverts; summer m. with white rump. **Hab:** Weedy fields, open wetland shrubs, hedgerows, gardens, feeders. **Voice:** Song a long continuous warbling (30 sec. or more) and a short sputtering warble lasting only 2–3 sec.; calls include a *titewtewtew,* a *bearbee* or *bearbee bee,* and a rising *sweeyeet.* 🎧**148**.

Subspp: (4) Divided into two groups. **Eastern Group:** Medium-sized; summer m. with extensive black cap and bright yellow body. •*tristis* (cent. ON–e. CO and east). **Western Group:** Small to large; summer m. with smaller black cap, paler yellow body. •*jewetti* (s.w. BC–s.w. OR) undertail coverts may be tinged brown; •*salicamans* (w. CA); •*pallida* (rest of range in West). **Hybrids:** Lesser Goldfinch.

Adult m. JAP/02

Oriental Greenfinch 4

Carduelis sinica L 6"

Casual Asian vag. to w. AK. **Shape:** Fairly small, deep-chested, broad-necked finch with a short notched tail and deep-based conical bill with curved culmen. **Ad:** Dark brown or olive-brown overall with large yellow primary patch and largely white tertials on black wings; rump and base of outer tail feathers yellowish; bill pinkish. M. has gray crown and nape; f. brown crown and nape. **Juv:** Like f. but paler below and with dark streaks. **Flight:** Large yellow patch at base of primaries. **Hab:** Open woods, parks. **Voice:** Song a series of musical trills on varying pitches; calls include high-pitched twittering and a buzzy *jweeee.*

Subspp: (1) •*kawarahiba.*

Adult m., *europea** DEN/06

Adult f., *europea** FIN/04

Eurasian Bullfinch 4

Pyrrhula pyrrhula L 6½"

Casual Eurasian visitor to w. AK islands and occasionally the AK mainland. **Shape:** Fairly small, deep-bellied, broad-necked finch with a very short deep-based bill. Culmen abruptly downcurved. **Ad:** Black cap and chin; gray back and nape; white rump and undertail coverts; black wings with broad whitish wingbar on greater coverts; black bill. M. with rosy red chest and belly; f. with grayish-buff chest and belly. **Juv:** (Jun.–Oct.) Like ad. f. but with all-grayish-buff head (lacks black cap and chin). **Flight:** White rump contrasts with black tail and gray back. **Hab:** Variable; mixed woods, gardens, feeders. **Voice:** Calls include a repeated whistled *tyooo.*

Subspp: (1) •*cassinii.* **P. p. europoea* (n. EUR) used for good photograph.

Adult m., *vespertinus* NH/06

Adult f., *vespertinus* NH/06

Adult f. ME/04

Juv., *vespertinus* MN/09

Evening Grosbeak 1

Coccothraustes vespertinus L 8"

Shape: Medium-sized, short-legged, large-headed bird with a very large, deep-based, conical bill and relatively short notched tail. Line of culmen fairly continuous with flattened forehead and shallow crown. **Ad:** In both sexes, bill pale greenish in spring and summer, paler in winter. M. with bright yellow eyebrow and forehead, contrasting with rest of dark head; underparts (including undertail coverts) deep yellow; wings and tail black except for white tertials, inner secondaries, and inner greater coverts. F. with grayish head and back separated by dull yellowish collar; pale gray to buffy-gray chest and belly; white undertail coverts; blackish wings with grayish tertials, inner secondaries, and inner greater coverts; white patch at base of primaries. **Juv:** (Jun.–Oct.) Like ad. f. but browner overall and with yellow wash to inner greater coverts; as in ad., juv. f. has white patch at base of primaries, m. has none. **Flight:** Undulating; m. with broad white patch to inner wing; f. with white patch at base of primaries. **Hab:** Coniferous and mixed woods. **Voice:** Call a ringing *keep keep,* sometimes with a burry quality; when heard from a flock, sounds a little like sleigh bells. 🎧**149**.

Subspp: (3) •*brooksi* (BC–cent. NM and west) has long bill; f. back medium brown, m. eyebrow thin. •*montanus* (s.e. AZ) has long bill; f. back pale brown, m. eyebrow thin. •*vespertinus* (AB and east) has short bill; f. back dark brown, m. eyebrow broad.

Adult m., summer JAP/05

Adult m., winter JAP/01

Hawfinch 4

Coccothraustes coccothraustes L 7"

Casual Eurasian visitor to w. AK islands. **Shape:** Fairly large short-tailed finch with a large head, broad neck, and thick body. Bill is moderately long and very deep-based. Inner primaries have unusual extensions at tip, perpendicular to rest of feather. **Ad. Summer:** Brown head with distinctive black pattern enclosing eye, lore, base of bill, and chin. Underparts grayish brown; wings mostly black with a bluish sheen; primary greater coverts whitish at tips; secondary greater coverts warm brown with whitish bases. Bill grayish black. Sexes alike except that f. has grayish panel on secondaries. **Ad. Winter:** Like summer but bill pale yellow to silvery. **Juv:** Somewhat like ad. but belly with dark spots. **Flight:** Big-billed, white or grayish wing patches, white tip to very short tail. **Hab:** Woods, fruit trees. **Voice:** Call a loud, sharp *piks*.

Subspp: (1) •*japonicus*.

Adult –/06

Juv. MO/08

Eurasian Tree Sparrow 2

Passer montanus L 6"

Shape: Fairly small, large-headed, broad-necked bird with a short stout bill. **Ad:** Black bill, lore, and throat; brown crown; white auricular and cheek with large black central spot; underparts grayish with buffy flanks; small white tips to median and greater coverts create 2 dotted wingbars. Bill black in summer, paler in winter. Sexes alike. **Juv:** (May–Sep.) Like ad. but buffier overall. **Flight:** Direct; short wings and tail. **Hab:** Closely associated with human activity, buildings, parks, cities, farms, feeders. **Voice:** Song a series of 2-part and single chirps; calls include a sharp *pik* and a *chet*.

Subspp: (1) •*montanus*.
Hybrids: House Sparrow.

Adult m., summer GA/03

Adult f. GA/03

Adult m., summer GA/03

Adult m., trans. to summer GA/03

House Sparrow 1

Passer domesticus L 6¼"

Shape: Fairly small, large-headed, broad-necked bird with a medium-length stout bill. **Ad. Summer:** M. with black bill, lore, chin, and bib contrasting with grayish cheek and underparts; dark gray crown; rich brown on auriculars extends to nape; back and wings streaked with rich browns and buff; median coverts with broad white tips, creating single upper wingbar; bill black. F. with brown crown; pale buffy eyebrow and lore; thin dark brown eyeline; buffy-gray underparts; wings and back similar to m. but with less rich brown; bill pale. **Ad. Winter:** M. with pale edges to chin and breast feathers, partially concealing black bib (these wear off over winter, revealing black bib); yellowish base to bill. F. like summer female. **Juv:** (May–Oct.) Like ad. f. but buffier overall. **Flight:** Direct; short wings and tail. **Hab:** Closely associated with human activity, buildings, parks, cities, farms, feeders. **Voice:** Song a series of 2-part varied chirps, like *chirup chireep chirup;* calls include a *chirup,* a churring *ch'ch'ch'ch,* and a nasal 2-part *kwerkwer.* **150**.

Subspp: (1) •*domesticus*. **Hybrids:** Eurasian Tree Sparrow.

Acknowledgments

No field guide of this magnitude can be produced alone, and we have many people to thank. First and foremost we are grateful to all of the superb photographers who have agreed to let us use their beautiful images of birds. There has never been a greater collection of photographs of the species and plumages of North American birds, and we hope the photographers are proud of their work as it appears in the guide. Their names and the photographs they took are all listed in the photo credits.

We also want to give a special thanks to Doug Wechsler at VIREO, Visual Resources for Ornithology, and his assistants Dan Thomas, Matt Sharp, and Jason Fritts. VIREO is a fabulous online resource for photographs of birds and was extremely helpful to us throughout this project.

Next we would like to thank the expert readers who took the time to carefully review the manuscript. Their corrections and additions helped improve this guide immensely. Readers of the complete manuscript included Paul Lehman, Marshall Iliff, and Tony Leukering. In addition, we had experts read sections on particular groups of birds: Brian Wheeler read hawks, Kevin Karlson read shorebirds, Brian Patteson read seabirds, Cameron Cox read waterfowl, and Alvaro Jaramillo read gulls and terns.

We also had several people go over the photographs and captions to make sure that all was accurate, and we thank them for their careful work and excellent suggestions. Marshall Iliff reviewed all photos and captions in the first, second, and fourth quarters of the book (pages 3–357 and 568–761). Alvaro Jaramillo also checked photos and captions of the third and fourth quarters (pages 358–761), and additional experts reviewed photos and captions in specific sections: Brian Wheeler checked hawks, Kevin Karlson reviewed shorebirds, Brian Patteson checked seabirds, and Alvaro Jaramillo reviewed gulls, terns, and flycatchers.

Of course, we take full responsibility for any errors that may have inadvertently remained in the text or in the photos and their captions.

We are grateful to Peter Pyle for helping us use the information on subspecies from his books *Identification Guide to North American Birds, Part I* and *Part II.* As we say elsewhere, these books constitute one of the great ornithological works of our time.

Finally, we want to thank the people at Hachette Book Group for encouraging us to take on this project and supporting us as we completed it. Terry Adams, vice president and publisher of Little, Brown paperbacks, executive editor Tracy Behar, and senior managing editor Mary Tondorf-Dick were particularly supportive over a long period of time, and we want to give them our deep thanks. Also, thanks to members of the production staff, especially Pamela Schechter, whose skill and experience were essential in pulling together the final product.

After us, the two people who did the most work on this book were Laura Lindgren of Laura Lindgren Design and Peggy Leith Anderson, executive copy editor at Little, Brown and Company. Laura created a beautiful, purposeful design, then spent well over a year arranging (and rearranging) the more than 3,400 photographs and the 800 pages of manuscript, transforming them into the layout of the final guide. She had the patience of Job and the quick intelligence and discipline to handle the immense amount of detail with accuracy, all the while being a pleasure to work with. No small feat!

We owe our greatest thanks to Peg Anderson, who oversaw the project from the time we first handed it in to final publication. Her uncanny ability to copyedit, remember detail, catch inconsistencies, organize photos, deal with what must have seemed like an endless stream of corrections coming from us, make suggestions on design, content, and layout, and keep the project on track was just short of superhuman. Not to mention keeping a wonderful and warm sense of humor throughout. We could not have done this field guide without her, and our affection for her has deepened as we together moved the project toward completion.

And because we, Lillian and Don, are a married couple and a team in every way, we want to thank each other for the tremendous hard work, patience, humor, and love that sustained us through this immense undertaking. We were able to create an atmosphere and work environment where each of us could contribute the best of our individual skills and abilities, resulting in the best guide we could make.

Yours in birding,
Don and Lillian Stokes

PEGGY HOWARD

Glossary

Accidental – Descriptive term associated with ABA birding code 5, signifying five or fewer North American records of a bird's occurrence, with fewer than three in the last 30 years.

Accipiter – Member of the genus *Accipiter*, a small genus of hawks.

Alcid – Member of the family Alcidae, which includes puffins, murres, and guillemots.

Alula – A small group of feathers near the base of the primaries at the bend in the wing.

Ankle – The first major joint in a bird's leg seen below the body.

Auricular – Feathers just behind the eye and around the ear.

Axillaries – Feathers between the base of the underwing and the top of the flank; the "armpit."

Belly – Area of the body between the chest and undertail coverts.

Bicolored – Containing two distinct colors.

Breast – Chest.

Breastband – Band of feathers across the chest, usually darker.

Buteo – Member of the genus *Buteo,* a large genus of hawks.

Call – A short instinctive vocalization of birds; see also **song.**

Casual – Descriptive term associated with ABA birding code 4, signifying six or more North American records of a bird's occurrence, with three or more in the last 30 years.

Cere – Raised fleshy area at base of upper mandible on hawks, eagles, falcons, and their relatives.

Chevrons – Small V-shaped marks on the bodies of some birds.

Chin – The area just below the base of the bill.

Clinal – A gradual transition in plumage between two subspecies over a geographic area.

Concolor – Of uniform color.

Coverts – Smaller feathers that cover the base of major wing and tail feathers.

Culmen – The lengthwise ridge along the midline of the upper mandible.

Crest – Long feathers of the crown that may be raised.

Decurved – Curved downward.

Dihedral – Upward angle from horizontal.

Ear patch – Feathers around the ear; see also **auricular.**

Eclipse plumage – Nonbreeding plumage of male ducks.

Empid – Member of the genus *Empidonax,* a group of small flycatchers.

Eye crescent – Pale crescents above and/or below eye.

Eye-ring – A fine ring of feathers around the eye.

Eye-stripe – A stripe before and/or after the eye.

Facial disk – A roundish area of feathers around the face, such as on owls and harriers.

Flank – Side of a bird's body just below the wing.

Flight feathers – The main feathers of the tail and wings, called **primaries** and **secondaries** on the wing.

Gonydeal angle – The expansion of the lower mandible near the tip, prominent in gulls.

Gonys – The prominent lengthwise ridge along the midline of the lower mandible of some birds, notably gulls.

Gorget – Iridescent throat of a hummingbird.

Greater coverts – The longest coverts over the base of secondaries or primaries.

Gular pouch – An expandable sac on a bird's throat.

Hood – Dark feathering over the head and neck of a bird.

Hybrid – The result of successful interbreeding between two different species or hybrids.

Immature – Any plumage between juvenal and adult; also, a young bird in that plumage.

Inner wing – The part from the bend in the wing to the body.

Introduced – A species that has been released by humans into an area where it did not previously live.

Juvenal plumage – The first main plumage of a bird that has left the nest.

Juvenile – A bird in its juvenal plumage.

Lesser coverts – The smallest of the coverts over the base of the secondaries or primaries.

Lore – The area between the base of the bill and the front of the eye.

Malar streak – A streak on the side of the throat (below the submoustachial streak).

Mandible – Either the upper or lower part of the bill.

Mantle – The upper back.

Marginal coverts – Small feathers along the leading edge of the inner wing.

Median coverts – The medium-sized coverts over the base of the secondaries or primaries, between the greater and lesser coverts.

Mirror – A white patch near the tip of dark flight feathers; mostly refers to gulls.

Molt – A period of normal loss and growth of feathers.

Molt limits – The presence of both retained and replaced feathers as a result of incomplete molting of a group of feathers; often observed in the greater coverts of smaller birds and indicates a 1st-winter or 1st-year bird.

Morph – One of two or more consistent variations in plumage color showing up in a species, such as light and dark morphs or red and gray morphs, especially in seabirds and hawks.

Moustachial streak – A streak along the lower edge of the auricular.

Nape – Back of the neck.

Orbital ring – A thin fleshy circle directly around the eye, sometimes brightly colored, such as in gulls.

Patagial bar – In some birds, a thick dark bar on the leading edge of the underwing extending from the body to the wrist.

Primaries – The long flight feathers of the outer wing.

Primary projection – The amount that the tips of the primaries extend past the tertials or past the tail in a perched or standing bird.

Rare – Descriptive term associated with ABA birding code 5, referring to species that occur in North America in low numbers but annually.

Resident (res.) – A bird that remains in the same geographic area all year.

Rump – The area between the lower back and uppertail coverts.

Secondaries – The long flight feathers of the inner wing.

Sideburn – A dark line on some falcons extending from below the eye through the cheek.

Song – A partially learned and complex vocalization typical of birds in the latter half of the phylogenetic order (about shrikes to finches).

Spectacles – Generally refers to pale eye-rings that connect to pale lores, looking like eyeglasses.

Speculum – Iridescent or whitish group of secondaries; term usually used in reference to ducks.

Supraloral – The area just over the lore, such as a paler supraloral dot or dash.

Tarsus – The lower, longer part of the visible leg on birds.

Tertials – The innermost feathers of the wing, covering the secondaries on a bird at rest.

Tibia – The shorter, upper part of the visible leg on birds.

Ulnar bar – A diagonal bar angling across the inner wing from the wrist to roughly the innermost trailing edge of the wing.

Undertail coverts – The small feathers that cover the underside of the base of the tail feathers.

Uppertail coverts – The small feathers that cover the upperside of the base of the tail feathers.

Vagrant (vag.) – A bird that has wandered outside its normal migration route.

Vent – Area between the legs.

Vermiculation – A detailed wavy pattern on feathers.

Wattle – Pendant fleshy skin that hangs from the base of the bill.

Wing window – A pale translucent area of (usually) the outer wing.

Wing-stripe – A streak traversing the length of the wing, often paler than the rest of the wing.

Wrist – The visible bend in the wing.

Key to Codes for States, Provinces, and International Locations

U.S. States and Territories

AK – Alaska
AL – Alabama
AR – Arkansas
AZ – Arizona
CA – California
CO – Colorado
CT – Connecticut
DC – District of Columbia
DE – Delaware
FL – Florida
GA – Georgia
HI – Hawaii
IA – Iowa
ID – Idaho
IL – Illinois
IN – Indiana
KS – Kansas
KY – Kentucky
LA – Louisiana
MA – Massachusetts
MD – Maryland
ME – Maine
MI – Michigan
MN – Minnesota
MO – Missouri
MS – Mississippi
MT – Montana
NC – North Carolina
ND – North Dakota
NE – Nebraska
NH – New Hampshire
NJ – New Jersey
NM – New Mexico
NV – Nevada
NY – New York
OH – Ohio
OK – Oklahoma
OR – Oregon
PA – Pennsylvania
PR – Puerto Rico
RI – Rhode Island
SC – South Carolina
SD – South Dakota
TN – Tennessee
TX – Texas
UT – Utah
VA – Virginia
VT – Vermont
WA – Washington
WI – Wisconsin
WV – West Virginia
WY – Wyoming

Canadian Provinces and Territories

AB – Alberta
BC – British Columbia
LB – Labrador*
MB – Manitoba
NB – New Brunswick
NF – Newfoundland*
NS – Nova Scotia
NT – Northwest Territories
NU – Nunavut
ON – Ontario
PE – Prince Edward Island
QC – Quebec
SK – Saskatchewan
YT – Yukon

Countries and International Locations

AAR – Antarctica
ALB – Albania
ALD – Aldabra
ANT – Antigua
ARG – Argentina
ARU – Aruba
ATL – Atlantic Ocean
ATP – Antipodes
AUK – Aukland Is.
AUS – Australia
BAH – Bahamas
BAR – Barbados
BEL – Belize
BOL – Bolivia
BON – Bonaire
BOR – Borneo
BRA – Brazil
CAN – Canada
CAY – Cayman Is.
CHI – Chile
CLI – Clipperton Is. (e. Pacific)
COL – Colombia
CRI – Costa Rica
CUB – Cuba
CYP – Cyprus

**For the purposes of this guide, the province of Newfoundland and Labrador was treated as two separate entities.*

ENG – England
EST – Estonia
EUR – Europe
FIN – Finland
GAL – Galápagos
GBR – Great Britain
GRE – Greece
GRN – Greenland
GUA – Guatemala
HKO – Hong Kong
HON – Honduras
HUN – Hungary
ICE – Iceland
IND – India
INO – Indian Ocean
ITA – Italy
JAP – Japan
KAZ – Kazakhstan
KOR – Korea
LAP – Lapland
LHI – Lord Howe Is. (s.w. Pacific)
LIT – Lithuania
MEX – Mexico
MOR – Morocco
NAM – North America
NET – Netherlands
NIC – Nicaragua
NOR – Norway
NZE – New Zealand
OMA – Oman
PAC – Pacific Ocean
PAN – Panama
PER – Peru
PHI – Philippines
POR – Portugal
RUS – Russia
SAF – South Africa
SAM – South America
SAT – South Atlantic
SCO – Scotland
SIB – Siberia
SIN – Singapore
SKO – South Korea
SLO – Slovakia
SOR – South Orkney Is.
SPA – Spain
SWI – Switzerland
TDC – Tristan da Cunha
TDF – Tierra del Fuego
THA – Thailand
TRI – Trinidad and Tobago
UAE – United Arab Emirates
UKM – United Kingdom
URU – Uruguay
USA – United States of America
VEN – Venezuela
VIS – Virgin Islands
WAL – Wales
YEM – Yemen

Photo Credits

Numbers refer to pages; letters refer to position on the page. Going down: T=top, M=middle, B=bottom; UM=upper middle, LoM=lower middle. Going across: L=left, C=center, R=right; LC=left center, RC=right center. Page numbers in *italics* identify photos that also appear on the inside covers of the book.

A. Allen/Cornell Lab of Ornithology
449TL, 449BL

Lynn Barber
703TR

P. Basterfield/Windrush
352BL

Cliff Beittel
15ML, 46TL, 128BL, 133BL, 147TL, 206R, 318TL

Vic Berardi
166BL, 166TR

R. Brooks/Windrush
348TL

Mark Chua
391TC, 391TR

Richard Crossley
18BL, 30BR, 36TL, 42BR, 43BL, 116BR, 159BR, 175TRC, 209BL, 209BR, 212TR, 223B, 231BR, 232TR, 237MR, 241TL, 242BL, 243MR, 243BR, 247MR, 249T, 265BR, 266BR, 272ML, 272BR, 284BR, 294ML, 294MR, 294BL, 299BR, 320TR, 321TL, 321MR, 349ML, 724ML

Mike Danzenbaker
4BR, 7BR, 10TL, 19TR, 28TR, 41BR, 42MR, 43BR, 45TR, 51MR, 51BR, 58BR, 73BR, 75BR, 77ML, 91TRC, 91TR, 91BR, 94TL, 98TR, 100TL, 101TL, 101ML, 101BL, 102TC, 105TR, 105TL, 105BL, 106BR, 124BR, 135BR, 183TL, 187BC, 210BR, 214R, 219BL, 220TR, 224TR, 225BR, 227TR, 234TL, 234BR, 239BR, 240BR, 241TR, 241BR, 242BR, 243BL, 248TR, 249B, 254BR, 255B, 259BR, 260TL, 260TR, 261TL, 261ML, 261BR, 262TR, 262MR, 263MR, 263BL, 263BR, 265BL, 265BC, 267BR, 272TL, 272MR, 277BR, 279BR, 281BR, 283TR, 283BR, 284BL, 285TR, 286BR, 287MR, 287BR, 307TR, 307ML, 307BR, 307BL, 317MR, 323MR, 323BR, 328TL, 329BL, 345BR, 356BL, 364BR, 364BL, 365BL, 366TL, 367TL, 368BL, 369BR, 370BL, 378BR, 390TL, 390BL, 393TC, 394BR, 400BR, 401TL, 402TLC, 405BR, 406TL, 407MR, 407BR, 407BL, 410BR, 411TR, 413L, 414TR, 414BR, 415TR, 415BL, 433TC, 442TL, 442TLC, 455TR, 457TL, 457TR, 459BC, 460BL, 461TL, 461TR, 465TL, 466BL, 466BR, 467BR, 471B, 473BR, 476BR, 477TL, 477R, 477BL, 478TL, 487TR, 487BR, 505TL, 505BL, 514BR, 514TR, 515TR, 515BR, 516TR, 516BR, 520TL, 520TR, 520BL, 520BR, 542L, 549L, 551TL, 551BL, 551BR, 562BR, 564BL, 565BR, 567TL, 567BL, 567R, 571BL, 579TL, 580TR, 580BR, 582TL, 584BL, 584BC, 598TR, 640BL, 648BL, 648BR, 671BR, 679TL, 679BL, 698TR, 701BR, 702TL, 702BR, 755ML, 760TL

J. Davies/Windrush
344TL

Don Delaney
752TL, 752BL

Don DesJardin
298TL, 309BL, 315TL, 317BL, 323BL, 347BL

Henry Detwiler
640TR

Roger Eriksson
592BR, 594BR, 597TL, 597BL, 602BL, 602MR, 605TL, 606BR, 607BL, 608BR, 615BR, 617TR, 621BR, 622BR, 625MR

Heather Forcier
8TL, 8ML, 43MR

Thomas Grey
431BL

Martin Hale
241ML, 260BR, 263ML, 268TL

Kwang Kit Hui
342BR

Marshall Iliff
270MR, 370BR

Alvaro Jaramillo
310BR, 314BR, 315LoML, 317TL, 322BL, 328BR, 330BL, 330BR, 330TR, 335LoML, 337MR, 337BR, 339BL, 339BR, 339TR, 413R, 511TR

Kevin Karlson
9BL, 15MR, 15TL, 16ML, 24BL, 26BL, 39TL, 39BL, 43TR, 45ML, 45BR, 46UML, 46LoMR, 53BR, 56MR, 73ML, 73BL, 73MR, 75TL, 182TL, 186TL, 186BL, 212ML, 215BR, 225TR, 238TR, 242BC, 247TL, 252TL, 252MR, 252BR, 255TR, 256TL, 256BR, 257BR, 258BL, 264UML, 264TR, 266ML, 269ML, 271MR, 273BR, 273TR, 276TL, 276TR, 278BR, 282TL, 282BL, 285MR, 286BL, 286MR, 287TL, 287TR, 295TL, 305B, 310TL, 310TR, 312TL, 329TL, 332TL, 333TL, 334BR, 339TL, 341ML, 343TR, 343ML, 344ML, 344BL, 344BR, 344MR, 346BL, 347T, 347BR, 347MR, 351TL, 351BR, 352LoML, 352BR, 353ML, 353BR, 355TR, 363UMR, 366BL, 475TL, 475TR, 475BL, 475BR, 495TR, 495BR, 571TL, 648TR, 696BR, 697BR, 698BR

Phil Kelly
368BR

Nick Kontonicolas
364TC

Peter Latourette
613BL, 613TR

Jerry Ligouri
152BL, 157ML, 157TR, 157BC, *158TL,* 158TR, 159TR, 164BL, 164TR, 166BL, 166TR, 170TR, 170BL, 170BR, 171TR, 176TL, 178BR, 180TL, 180TR, 180BL, 181TC, 184TL, 184BR, 184BL, 185TL, 185BL, 186TR, 188BR, 190TL, 190BL, 190BR, 192TR, 192BR, 193T, 193BR

Jeff Lynch
290TL, 293TR, 316TR, 336TL, 336R

Bruce Mactavish
290TR, 293BL, 298TR, 299TL, 299TR, 320TL, 327BL, 335TR

Bud Marschner
752TR, 753BL, 753TR, 753TL

Garth McElroy
48BL, 61TR, 296TL, 301TL, 324BL, 393BL, 403TRC, 403TR, 462TR, 462BR, 464TL, 494TL, 497TL, 497BL, 510ML, 515BL, 524TL, 533TR, 535TL, 535BL, 537TL, 559R, 572BL, 585TR, 586TL, 602TR, 604TR, 629BR, 629TL, 652TL, 653TL, 653BL, 655T, 661BL, 663TL, 668BL, 670BR, 675TR, 681TL, 682TL, 690TL, 692B, 697MR, 722TL, 726BR, 741TL, 741BL, 748BR, 749TL, 757TL, 757TR

Steve Mlodinow
12BR, 13TR

Alan Murphy
392TL, 392BL, 392TLC, 403TLC, 404TR, 407TL, 410TR, 416TR, 425TL, 425TC, 428TL, 437BR, 453TR, 453BR, 455TL, 458BL, 458BR, 468BR, 469TR, 470BR, 470BL, 472TL, 476BL, 487TL, 489TL, 494BL, 494TR, 494BR, 497BC, 497BR, 502BR, 503TR, 503BL, 510MR, 519BL, 522TL, 530BL, 532BL, 533TL, 534TR, 554BR, 556BR, 556TR, 579TC, 591TL, 591TR, 592TR, 594T, 594TR, 595ML, 596BR, 597MR, 597BR, 600TL, 600MR, 602TL, 604TL, 604BL, 609BL, 609TR, 609BR, 611ML, 612ML, 612BL, 616BL, 618BR, 619TL, 621BL, 622BL, 625TL, 626TL, 626BR, 627TL, 634TL, 634BL, 635TR, 636TR, 637BL, 640BR, 645ML, 645MR, 647BL, 647TR, 650TR, 650BR, 652BR, 658ML, 665BR, 667BL, 673TR, 673BR, 682BL, 687BR, 695B, 712TR, 713BL, 716TR, 728TL, 728TR, 731TL, 731TR, 737TL, 747BR, 757BC

Judd Patterson
146BL

Brian Patteson
88BC, 90TRC, 90TR, 90BR, 92L, 92LC, 94BL, 98BR, 105BR, 106BL

Jeff Poklen
297R, 309MR, 361TR, 370ML

Clair Postmus
480TR, 480BR, 480BRC, 638BR, 641BR

Wayne Richardson
211TR, 211BR, 216BR, 219BR, 250BR, 269BR

J. Roberts/Windrush
348BL

Robert Royse
iB, 297BL, 300BL, 300BR, 319TR, 334TL, 334TR, 345BL, 398TL, 398TR, 430TL, 459TR, 459BL, 460BR, 461BL, 461BC, 463TR, 465TR, 469TL, 472BR, 487BL, 488TR, 488BR, 489TR, 489BR, 490TR, 490BL, 491BR, 492TL, 492BL, 492R, 493BL, 493ML, 497TR, 501TL, 502BL, 522BR, 524ML, 525TL, 527TL, 527TR, 527BR, 530TR, 531BR, 534BL, 534BR, 535BR, 536BL, 536BR, 538BL, 538TR, 541TR, 541BL, 541BR, 553BL, 554TR, 555T, 555B, 558TL, 574B, 575TL, 578TL, 578BR, 584BR, 585BR, 588TR, 588BR, 590BR, 592TL, 592BL, 593BR, 596TL, 596BL, 598TL, 603TL, 606TL, 606BL, 609TL, 611TL, 615TL, 617TL, 617BL, 618TL, 620TL, 623TL, 624TL, 624BR, 626ML, 626TR, 628BR, 629BL, 630TL, 632TL, 637TR, 638TL, 638BL, 638TR, 645TL, 651BL, 651TR, 657TR, 659TR, 659BR, 660BL, 660TR, 660BR, 661TL, 662TL, 662TR, 663BL, 663TR, 664TR, 664MR, 666TL, 670TL, 672MR, 672TR, 672BL, 673TL,

675BL, 676BR, 677BL, 677TR, 677BR, 678BL, 678TR, 679BR, 679TR, 683TL, 683BL, 683BR, 686BL, 687TL, 688TL, 690TR, 693TL, 694TL, 695TL, 695TR, 697BL, 697TR, 698TL, 700BL, 721TL, 723BL, 744TR, 747BL, 748TL, 750TL, 750TR, 750ML, 750MR, 751TR

Larry Sansone
379BL

Brian Small
iT, 4TL, 4ML, 4BL, 4MC, 7TL, 7ML, 9TR, 10BL, 10ML, 10MR, 11BR, 14TL, 14MR, 14BR, 14TR, 18TL, 18ML, 19TL, 19ML, 20TL, 21TL, 26TL, 27TL, 27ML, 27TR, 28TL, 28ML, 32TL, 32ML, 33TL, 33ML, 34BL, 34TL, 35TL, 35ML, 37TL, 37TC, 42TL, 43ML, 44TL, 47TL, 49ML, 51TL, 53TL, 53ML, 53TR, 54TL, 54TR, 56TL, 60TL, 60BL, 60ML, 60BC, 62TL, 64TL, 64ML, 65TL, 65TR, 65B, 66TL, 66TR, 66MC, 66MR, 66BR, 70TL, 70BL, 70TR, 70BR, 71ML, 73TR, 75BL, 75TR, 76TL, 76TR, 77TR, 78TL, 78BL, 79TL, 80BL, 80TR, 81TL, 81BL, 81TR, 81BR, 82TR, 89RC, 89R, 99TL, 109TL, 110ML, 114TL, 114BL, 114BLC, 114TR, 117TL, 117TRC, 120TR, 120BR, 121BL, 121TR, 124TL, 124TC, 132BR, 134BR, 137TL, 137BL, 147BL, 147BR, 148TC, 148TR, 148BC, 148BR, 149MLC, 149TR, 149BRC, 151TL, 151ML, 152BL, 152TC, 152TR, 152BR, 156TR, 160TL, 160TR, 161TL, 161TR, 161BR, 163TL, 163BL, 163TR, 170TL, 171TL, 171BL, 172TL, 172TR, 172BR, 173TL, 176ML, 176MR, 177ML, 178TL, 178TR, 178BL, 179TL, 182TR, 182LR, 187BL, 188TL, 189TR, 196TL, 196TR, 196BR, 198TR, 200T, 205BR, 210TL, 212TL, 213TL, 213R, 215TL, 216BL, 218TL, 221TL, 221BL, 221TR, 222ML, 222TL, 229TR, 233TL, 233TR, 237BL, 238TL, 242TR, 244ML, 246TL, 246TR, 246BR, 247ML, 247BL, 250TL, 250ML, 250TR, 251TL, 251TR, 251BR, 252BL, 253TL, 253BL, 253TR, 254TL, 254BL, 256TR, 256ML, 258TL, 258ML, 259TL, 259TR, 264TL, 265TL, 265ML, 266TL, 267TL, 267ML, 267TR, 267MR, 268TR, 270TL, 270TR, 270MR, 271TL, 271ML, 271TR, 272BL, 273ML, 276BL, 277TR, 277BL, 278TL, 278BL, 279T, 279BL, 281TL, 282TR, 283L, 285TL, 285BL, 286TL, 286ML, 286TR, 291BR, 291TL, 294TL, 304BR, 304TL, 304TR, 306TL, 306BL, 307TL, 308TL, 309UML, 309TL, 309BR, 310BL, 314TL, 314BL, 315BL, 316TL, 316BL, 322TL, 329UML, 329LoML, 331BL, 332BL, 336BL, 343TL, 343BL, 344TR, 346TL, 346TR, 356TL, 356TR, 356ML, 357TL, 357BL, 369MR, 373TRC, 378TR, 379TC, 380TR, 382ML, 384TC, 384BR, 385TL, 385BL, 395TL, 397TR, 399TL, 399TR, 400TRC, 401TRC, 402TL, 402TRC, 404TL, 404TC, 408BL, 408BR, 417TL, 420BR, 420TL, 422BL, 422TR, 423BR, 423TR, 423TL, 423BL, 426TL, 426TR, 426BL, 426BR, 427TL, 427BL, 428TR, 428BR, 429TL, 429TR, 429BL, 429BR, 431TR, 432BR, 434BR, 437TL, 437TRC, 437TLC, 438L, 438RC, 438R, 438LC, 439LC, 440TL, 440TLC, 440TR, 440BR, 441TLC, 441TRC, 441TR, 442TRC, 442BR, 442TR, 443TL, 443TRC, 443TR, 444TL, 444TC, 444BR, 444BL, 444BC, 445TL, 445TC, 445BL, 445TR, 446TLC, 447TL, 447TLC, 447TRC, 450BC, 450BR, 450TL, 451B, 453TL, 453BL, 453BRC, 458TL, 458TR, 459TL, 459BR, 460TL, 460TR, 461BR, 462TL, 462BL, 464BL, 467BL, 468TL, 468BL, 469BL, 470TL, 472TR, 474BR, 474TR, 474BLC, 476TL, 476TR, 478BL, 478TR, 478BLC, 478BR, 479TL, 479TLC, 479BC, 480TL, 480BL, 480BLC, 481TL, 484MR, 484TR, 488TL, 488ML, 489BL, 491TR, 493TL, 499L, 499TR, 499B, 500R, 501TR, 501BR, 502TL, 503BR, 504TL, 504ML, 507TR, 507MR, 507BR, 514TL, 514BL, 515TL, 517ML, 519TL, 522TR, 523TR, 527BL, 529L, 529MR, 530ML, 531TL, 532TR, 532MR, 533BL, 534TL, 535TR, 536TL, 538TL, 541TL, 544TL, 545TL, 545BL, 546TL, 552TR, 553TL, 557TL, 558BL, 560TL, 560BL, 560TR, 561TR, 565BL, 572TL, 573B, 574TL, 575TR, 575BR, 576TL, 577TR, 578BL, 580TL, 584TL, 584TR, 586BR, 590TR, 591BR, 592MR, 593TR, 594ML, 594MR, 595BL, 596TR, 601T, 602ML, 604MR, 606TR, 608TL, 611BR, 613ML, 614BL, 618MR, 624BL, 632BR, 633BR, 636TL, 641TR, 644TL, 645TR, 645BR, 646TR, 649TL, 649BL, 651TL, 651BR, 653TR, 654BL, 654TR, 654BR, 657BL, 658BL, 658TR, 659TL, 659BL, 661BR, 662BR, 664BR, 667TR, 668TL, 669TR, 669BL, 671TL, 671BL, 671TR, 672BR, 674BL, 674TR, 674BR, 676BL, 677TL, 678BR, 680BL, 681TR, 682TR, 682BR, 684TL, 684TR, 685BL, 687BL, 689BL, 692TL, 693TR, 693BR, 694BL, 696TR, 697ML, 698BL, 699TL, 699BL, 700ML, 704TR, 705TR, 706TL, 706BL, 708BL, 709BR, 709BL, 709TR, 710BL, 711TL, 712TL, 713BR, 718BR, 719TR, 719BR, 720BL, 720TL, 726BL, 727TL, 727BL, 728BR, 729TL, 730TL, 730BL, 730TR, 730BR, 734TL, 734BL, 734TR, 734BR, 735TR, 736TL, 736BL, 736TR, 741BR, 742BL, 742R, 743BL, 743R, 744TL, 745TR, 747TL, 747TR, 748BL, 748TR, 756TR, 756MR, 756BR

Hans Spiecker
599TR, 599BR

Lloyd Spitalnik
594BL, 595MR, 603BL, 618BL, 629TR, 630TR, 630MR, 630BR, 631BR, 631BL, 635BR, 636BL

Bob Steele
11TL, 11TR, 117TR, 117BR, 281TR, 292TL, 309TR, 349TL, 360TR, 361TL, 362MR, 364TL, 365TR, 366TR, 371BR, 372BL, 373TR, 375TL, 375BR, 376L, 376BR, 385ML, 400TLC, 404BR, 415BR, 419TR, 421TL, 421ML, 424BR, 430TR, 430BR, 432TR,

440BL, 453BLC, 454B, 455B, 463BL, 463BR, 464BR, 466TL, 466TR, 467TR, 470TR, 472BL, 484BL, 485ML, 485BL, 488BL, 493TR, 493BR, 501BL, 504BL, 506TL, 508TL, 516BL, 516TL, 517BL, 517TR, 517BR, 519BR, 523BR, 529BR, 537BR, 538BR, 543BR, 551TR, 553TR, 553BR, 569BL, 577TL, 577BR, 579BL, 581BR, 597TR, 611BL, 611TR, 613BR, 613MR, 614TR, 614BR, 619BR, 623R, 633TL, 633BL, 657BR, 661TR, 663BR, 665BL, 668BR, 670BL, 673BL, 676TR, 676MR, 681BR, 684BL, 685TL, 689BR, 691TL, 691BR, 699TR, 700TL, 700MR, 701TR, 708TL, 718BL, 721TR, 721BR, 731BR, 731BL, 732TL, 750BL, 750BR, 751BL, 754BL

Lillian Stokes
v, x, xii, xvii, xviii, *xix,* xx, xxi, xxiiT, *xxiiB, xxiii,* xxiv, 4MR, 9TL, 9ML, 12BL, 12ML, 13TL, 13BR, 14ML, 16TR, 16TL and inset, 17BR, 22TL, 22ML, 23BL, 23TR, 23MR, 23BR, 25TL, 25ML, 25BL, 25TR, 25BR, 29ML, 29TL, 41LoML, 44TC, 44TR, 50TL, 50ML, 50MR, 51TR, 51UML, 51BL, 52TL, 52ML, 52BL, 59TL, 59BL, 67TL, 67BL, 67TR, 67BR, 68T, 71TL, 76ML, 76BL, 82TL, 82ML, 93TR, 93BR, 96TL, 96BL, 96TR, 96BR, 97TRC, 97TR, 97BR, 112TL, 112BL, 112BR, 113TL, 113ML, 113BL, 113TR, 113MR, 113BR, 113MC, 115TL, 115BL, 115MR, 115TC, 115TR, 115BR, 119TL, 119BL, 119TR, 119BR, 119BC, 120TL, 124BL, 125TL, 125ML, 125BL, 125BR, 125TC, 125TR, 126ML, 126TL, 126BL, 126TR, 126BR, 127TC, 127TR, 127BR, 128TL, 128TR, 128BR, 129TL, 129BL, 129TR, 129BR, 130TL, 130MR, 130TR, 130BL, 131TL, 131ML, 131TR, 131BL, 131BR, 131MR, 132TL, 132TR, 134TL, 134BL, 134TR, 135TL, 135BL, 135TR, 135BC, 136BL, 138TL, 138ML, 138BL, 138TR, 138BR, 138MR, 139TL, 139ML, 139BL, 139BR, 139TR, 140TR, 140MR, 140BR, 142TR, 142MR, 142BR, 143TL, 143BL, 143TR, 143BR, 144TC, 144TR, *144TL,* 144BL, 146TL, 146BL, 150TR, 152BR, 152TL, 152ML, 161BC, 164TL, 164BR, 165TL, 165BRC, 165BR, 167BL, 169T, 174TR, 184TR, 185TR, 185BR, 197TR, 197BC, 199T, 202TL, 202BL, 202BR, 203TL, 203BL, 203TR, 204TL, 204BL, 204MR, 205L, 205TR, 206TL, 206BL, 210ML, 210BL, 210MR, 215BL, 215TR, 216ML, 216TR, 217TL, 217ML, 218ML, 218BL, 218TR, 220TL, 220BL, 223TL, 224TL, 224BL, 225TL, 225BL, 229BL, 229BR, 231TL, 231BL, 236TR, 237T, 237BR, 238BL, 238LR, 239TL, 239BL, 243TL, 246BL, 251BL, 252TR, 256BL, 257ML, 257BL, 257TR, 258TR, 258BR, 259UML, 259LoML, 259BL, 259MR, 264LoML, 264BL, 264MR, 264BR, 265TR, 269BL, 271BL, 271BR, 276BR, 277TL, 284TR, 296TR, 296BL, 296BR, 297TL, 302TL, 302TR, 302BL, 302BR, 303UML, 303MR, 303TL, 303BL, 303TR, 305TL, 306TR, 306BR, 312R, 312BL, 313TL, 313TR, 313ML, 313MR, 313BL, 313BR, 318TR, 318BL, 319TL, 319UML, 319LoML, 319BL, 319BR, 319UMR, 326TL, 326R, 326BL, 327TL, 327TR 327UML, 327LoML, 327LoMR, 335TL, 335UML, 335MR, 335BR, 345ML, 350TL, 350BL, 350TR, 350BR, 352TL, 352TR, 352MR, 353TL, 353MR, 353BL, 354TL, *354UML,* 354MR, 354LoML, 354BR, 355UML, 355MR, 355LoML, 357TR, 357MR, 357UML, 357BR, 357LoML, 363TR, 363BR, 364MR, 366BR, 374TL, 374TR, 374MR, 374BR, 378TL, 378BL, 381TL, 381TR, 381BL, 381BR, 383TL, 383ML, 383BL, 383TR, 383BR, 385TR, 385MR, 385BR, 388TL, 388BL, 392BR, 392TR, 392TRC, 398BL, 398BC, 399BR, 402BR, 402TR, 403BR, 405BL, 405TR, 425BR, 435B, 439RC, 439R, 440BR, 444TR, 445BR, 445BC, 446TL, 448TL, 448TC, 448BL, 448BR, 450TC, 450TR, 450MR, 468BC, 481TR, 481ML, 481BL, 481BR, 484TL, 485BR, 498TR, 498BC, 498BR, 500TL, 500BL, 505BR, 507TL, 507ML, 507BL, 508BL, 508TR, 508BR, 510BL, 511BR, 512TL, 512ML, 512TR, 512MR, 512BL, 512BR, 513TL, 513TR, 513BL, 513BR, 517MR, 519ML, 519TR, 519MR, 522BL, 523TL, 523BL, 528TR, 528BR, 528BL, 531TR, 532ML, 536TR, 537BL, 544BL, 544BR, 552TL, 552BR, 552BL, 556TL, 556BL, 557TR, 557MR, 557BR, 558TR, 558BR, 560ML, 560BR, 561TL, 561BL, 561BR, 566TR, 566BR, 566ML, 566BL, 566TR, 566TL, 569BR, 570TL, 570TR, 570BR, 587TL, 587BL, 587TR, 595TR, 604ML, 607TL, 608BL, 610TL, 610ML, 610MR, 610BR, 612TR, 616TL, 618TR, 619TR, 620ML, 620BL, 620BR, 621ML, 623BR, 624TR, 627BR, 634BR, 648TL, 648ML, 653BR, 664ML, 675TL, 675BR, 675MR, 685TR, 686TL, 688BL, 688ML, 688BL, 690MR, 690BR, 700BR, 700TR, 704TL, 706TR, 711BR, 711BL, 711TR, 713ML, 713TL, 715TL, 715TR, 715BR, 716TL, 716BR, 717TL, 724BL, 724BR, 725TL, 725BL, 725TR, 725MR, 728BL, 735TL, 735ML, 735BR, 735BL, 744BR, 744BL, 748TR, 748BL, 754TL, 754TR, 754MR, 754BR, 759TL, 759TR, 761TL, 761BL, 761TR, 761BR, 762, 792

Brian Sullivan
42TR, 47TR, 86RC, 86R, 91TL, 91TLC, 91BL, 93TL, 93TLC, 93BL, 94TR, 94BR, 95ML, 97TL, 97TLC, 98TL, 98BL, 98TRC, 103BR, 217BL, 222BL, 237TR, 254TR, 269TR, 285LoML

Mark Szantyr
12MR

J. Tanner/Cornell Lab of Ornithology
449TR, 449BR

Stan Tekiela
32BR, 56TR, 130BR, 166TL, 364TR, 367ML, 368UML, 370TL, 373ML, 379TL, 400TL, 425BL, 425BR, 485TL, 564BR

Glen Tepke
367BL, 369TL, 369BL, 369TR, 371TL, 371BL, 371MR, 372BR, 373BL, 375TR, 375BL, 404BL

D. Tipling/Windrush
92ML, 311BL

David Tipling
209TR, 222BL, 242MR, 262ML, 280TL, 280BL, 280TR, 301BL, 301BR, 355BL, 373BL, 375BL, 375TR, 404BL

Brian Wheeler
145TL, 145BL, 145TR, 145MR, 145BR, 148TL, 148ML, 150TL, 154TL, 154BL, 154TR, 154BR, 155TL, 155BL, 155TR, 155UMC, 155LoMR, 155UMR, 155BR, 162TL, 171BC, 173BL, 173TR, 173BR, 175TLC, 177TR, 177MR, 179TR, 179BL, 179BR, 181BL, 181TR, 181MC, 181BR, 189BL, 193BL

Jim Zipp
152TL, 152TR, 152BM, 157BR, 158BL, 158BLC, 159TL, 159BL, 159ML, 163BR, 165BL, 174TL, 174BR, 174BL, 175BL, 175BLC, 176TR, 188BL, 191TL, 191TC, 192TL, 192BL, 201BL, 201TR, 343BR, 393TR, 393BR, 397BL, 398BR, 406BL, 448TR, 484BR, 502TR, 524TR, 525TR, 525BR, 543TR, 546ML, 546BL, 575BL, 576TR, 578TR, 586BL, 586TR, 588BL, 588TL, 591BL, 593TL, 595TL, 600TC, 612TL, 618ML, 620TR, 620MR, 621TL, 625TR, 627TR, 628TR, 634TR, 637TL, 639R, 641BL, 666BL, 666TR, 680BR, 685B, 742TL, 752BR

Brian Zwiebel
600TR, 600BL, 600BR, 603TR, 603BR, 604BR, 605ML, 605BL, 607TR, 607BR, 608TR, 608ML, 610BL, 610TR

Visual Resources in Ornithology (VIREO)

George Armistead/VIREO
27BR, 45BL, 83TC, 83TR, 84TL, 84BL, 84TR, 84BR, 85BL, 86LC, 89L, 89LC, 90TL, 90TLC, 99BR, 100TR, 101TR, 104BL, 106TR, 108TL, 108BR, 108ML, 110TL, 140BL, 187TR, 194TC, 231ML, 233BR, 248BR, 303LoML, 320BL, 321ML, 321BL, 321BR, 325MR, 327BR, 335BL, 337TR, 337ML, 340BR, 341BR, 372TL, 683TR

Yuri Artukhin/VIREO
5TR, 5MR, 87ML, 87BL, 243TR, 244MR, 244BR, 292TR, 292MR, 292BR, 292BL, 295MR, 328BL, 329TR, 330TL, 331TR, 331UML, 331MR, 332BR, 333TR, 333MR, 340TR, 360BL, 362BL, 372TR, 583BL

Christian Artuso/VIREO
380BL, 391BL, 545TR, 739BL

Aurélien Audevard/VIREO
702BL

Ron Austing/VIREO
400BL, 406TR, 435BL

Gerard Bailey/VIREO
124ML, 199BL, 753BR

G. H. Balazs/VIREO
151MR, 151BR

E. Bartels/VIREO
110BL

Glenn Bartley/VIREO
26ML, 26TR, 37TR, 43TL, 44ML, 47ML, 47BR, 48MR, 118TR, 137TR, 162TC, 400TR, 439L, 446TRC, 446TR, 525BL, 531BL, 532BR, 537TR, 565MR, 579TR, 613TL, 631TL, 632BL, 647BR, 650BL, 652BL, 690BL, 691BL, 691TR, 693BL, 704BR, 723BR, 726TL, 733BL, 748TL, 748BR

Tony Beck/VIREO
226BL, 655B

Lance Beeny/VIREO
60TR, 60BR

Adrian Binns/VIREO
8BR, 45TL, 61BL, 77TL, 82BR, 99BL, 106TL, 106ML, 107BL, 109TR, 111BL, 124MR, 136TL, 159BC, 209ML, 226TL, 298BL, 324BR, 331BR, 341TR, 342T, 358TR, 384MR, 389L, 394BL, 403TL, 407TR, 418BR, 719TL, 732TR

James Bland/VIREO
57TL

Rick & Nora Bowers/VIREO
8TR, 8MR, 10BR, 24TR, 26BR, 32TR, 37ML, 38TR, 49TL, 50BL, 50TR, 54BL, 54BR, 58TL, 63TL, 68BR, 68BL, 69TR, 71BL, 80BR, 88BR, 108BL, 112TR, 118BL, 198TL, 200BL, 201TL, 210TR, 226R, 266TR, 270ML, 273BL, 304BL, 328TR, 346BR, 379TR, 396TR, 396BR, 396BL, 396TL, 406BR, 410TL, 411TL, 411BL, 411BR, 417BL, 419BL, 421BR, 422BL, 424TL, 424TR, 427TR, 428BL, 432TL, 433TR, 435ML, 437TR, 451TL, 456T, 457BL, 457BR, 465B, 467TL, 471TL, 471TR, 474BL, 478BRC, 479TR, 490TL, 500TL, 506BL, 518TL, 528TL, 540TL, 569TL, 574TR, 585TL, 587BR, 590TL, 593BL, 598BR, 598BL, 599TL, 599ML, 599BL, 601B, 631TR, 639L, 644BL, 654TL, 658TL, 664BL, 664TL, 665TR, 667BR, 672TL, 674TL, 678TL, 705BR, 707TL,

707TR, 707BL, 707BR, 715BL, 717BL, 722TR, 723TR, 724TR, 724MR, 728ML, 732BL, 736BR, 741TR, 743TL, 745BR, 757BL, 757BR, 759BR

Erica Brendel/VIREO
701TL, 701BL

Edward S. Brinkley/VIREO
111BR

John Cancalosi/VIREO
16TR, 58BL, 62TR, 120ML

Paula Cannon/VIREO
169M

Alan & Sandy Carey/VIREO
57TR

Vijay Cavale/VIREO
486R

R. J. Chandler/VIREO
219ML, 272TR

Robin Chittenden/VIREO
262TL

Brian Chudleigh/VIREO
214BL, 227BL, 232BL, 260BL, 275BL, 275TR

Herb Clarke/VIREO
32BL, 38BL, 49TR, 151TR, 287LoML, 418BL, 418TR, 729R

Peter G. Connors/VIREO
188TR, 341TL

Cindy Creighton/VIREO
293BR

Richard Crossley/VIREO
18BL, 30BR, *36TL*, 42BR, 43BL, 116BR, 159BR, 175TRC, 209BL, 209BR, 212TR, 223B, 231BR, 232TR, 237MR, 241TL, 242BL, 243MR, 243BR, 247MR, 249T, 265BR, 266BR, 272ML, 272BR, 284BR, 294ML, 294MR, 294BL, 299BR, 320TR, 321TL, 321MR, 349ML, 724MR

Jim Culbertson/VIREO
53MR, 167BR, 181TL, 278TR, 389R, 395TR, 456B, 518BL, 533TC, 564TR, 564MR, 635BL, 649TR, 668TR, 708TR, 708BR, 732BR

Rob Curtis/VIREO
10TR, 18BR, 21TR, 64TR, 74B, 117TLC, 142BRC, 207MR, 217BR, 225ML, 298BR, 354TR, 408TL, 414TL, 414BL, 422TL, 440TRC, 498BL, 572R, 579BR, 630ML, 630BL, 646TL, 646BR, 689TL, 689TR, 696TL, 709TL, 712BR, 713TC, 713TR, 719BL, 720BR, 722BR, 724TL, 726TR, 733BR, 749BR, 751BR

Richard Day/VIREO
189TL

Rob Drummond/VIREO
268BL

John S. Dunning/VIREO
387TL, 387BL, 434TR, 650TL

Scott Elowitz/VIREO
282ML, 485TR, 616BR

Stuart Elsom/VIREO
74T, 261MR, 269TL, 363LoML, 363BL, 368TR, 511TL

Hanne & Jens Eriksen/VIREO
183TR, 183BR, 187BR, 228BL, 230TR, 234TR, 240BL, 255TL, 261BL, 266BL, 274TR, 348BR, 362TR, 363TL, 547TL, 547BL, 547BC, 703BL, 739TL, 739TR, 758TR

Roger Eriksson/VIREO
220BR

Alec Forbes-Watson/VIREO
107TR, 122TL

Sam Fried/VIREO
30TR, 30MR, 31TL, 39BR, 62BL, 212BL, 267BL, 270BL, 388R, 666BR

Joe Fuhrman/VIREO
31BR, 82BL, 115ML, 156BR, 213L, 287UML, 295BL, 431TL, 450BL, 474TL, 474BL, 557BL, 582BL

Laura C. Gooch/VIREO
18TR

Manuel Grosselet/VIREO
133BR, 432BL, 635TL

Don Hadden/VIREO
95BL, 337BL, 352UML

Martin Hale/VIREO
31TR, 86L, 227TL, 232TL, 240TL, 241ML, 260BR, 263TL, 263ML, 268TL, 284TL, 305TR, 329BR, 348TR, 348ML, 380TL, 380BR, 486L, 540R, 547TR, 547BR, 550TL, 550TR, 550BR, 582TR, 583ML, 585BL

Peter Harrison/VIREO
87TL

Victor Hasselblad/VIREO
563TL, 760TR

John Hoffman/VIREO
71TR

Steven Holt/VIREO
59BR, 117BL, 117ML, 498TL

Julian R. Hough/VIREO
52BR, 212BR, 218MR, 260ML

Steve Howell/VIREO
419TL

Bill Hubbick/VIREO
24TL

John Hunter/VIREO
82MR

Dustin Huntington/VIREO
23TL, 23ML, 505TR, 530BR, 716BL, 745BL

Ian Hutton/VIREO
107TL, 244TR

Marvin Hyett/VIREO
40TR, 110BR, 110MR, 110TR, 121TL, 121ML, 198B, 200BR, 233BL, 321TR, 327UMR, 354BL, 495BL

Santiago Imberti/VIREO
359BL

Jukka Jantunen/VIREO
12TL, 13ML, 15BR, 19BR, 21BR, 24BR, 26MR, 29BR, 34BC, 34BR, 37MR, 44MR, 44BR, 47BL, 48TR, 49MR, 49BL, 49BR, 63BL, 63TR, 76MR, 79BL, 79BR, 197BR, 212MR, 231TR, 238MR, 239TR, 253BR, 282BR, 311BR, 311LoML, 316BR, 322BR, 353TR, 368LoML, 373MR, 379BR, 397BR, 510BR, 540BL, 544TR, 563BL, 686TR

Alvaro Jaramillo/VIREO
95BR

Edgar T. Jones/VIREO
479BL

M. P. Kahl/VIREO
434TC

Kevin Karlson/VIREO
146BR, 349R, 394TL, 649BR, 699BR

Serge Lafrance/VIREO
323TR

Stephen J. Lang/VIREO
112ML, 421BR

Greg W. Lasley/VIREO
22TR, 22BR, 69BR, 77BL, 77BR, 80TL, 85TL, 85TC, 85TR, 114BR, 127TL, 137BR, 144BR, 150BL, 150BR, 160BR, 162TR, 194BL, 194TR, 207TR, 207BR, 216MR, 223TR, 242TL, 355BR, 359BR, 384TL, 386TL, 386BL, 397BC, 401TR, 402BL, 405TL, 409TL, 409TR, 409BL, 409BR, 416BR, 416ML, 434TL, 434BL, 454TL, 463TL, *468TR,* 473TL, 473TR, 473BR, 482L, 482R, 506R, 510TR, 517TL, 524BL, 529TR, 554TL, 573T, 590BL, 602BR, 605TR, 605BR, 616TR, 622TL, 644TR, 644BR, 645BL, 657TL, 658BR, 660TL, 665TL, 669TL, 680TL, 694TR, 696BL, 705TL, 705BL, 710TR, 712BL, 720T, 737TR, 737B, 745TL, 756BL

Peter Latourette/VIREO
7TR, 29TR, 122BR, 122TR, 360TL, 367BR, 376TR, 421TR, 613BL, 613TR

Geoff Lebaron/VIREO
340TL

Jerry Ligouri/VIREO
50BR

Mark Lockwood/VIREO
13MR, 71BR, 703BR

Geoff Malosh/VIREO
287BL

Garth McElroy/VIREO
41UML, 41TL, 48ML, 116TC, 236BL, 258MR, 324TR, 333BL, 351TR, 447TR, 447BL, 454TR, 559TL, 559BL, 560MR, 569TR, 628TL, 667TL, 676TL, 680TR, 687TR, 706BR, 718TL, 718TR, 759BL

Martin Meyers/VIREO
8BL, 37BR, 46UMR, 52TR, 64BR, 98TLC, 165TR, 187TL, 204BR, 217TR, 441BR, 464TR, 565ML, 576BR, 576BL, 699MR, 760BR

Liz Mitchell/VIREO
87TR

Steve Mlodinow/VIREO
8MC, 9BR, 20BL, 108TR, 338ML, 338BL, 338BR, 382BR, 491TL, 729BL

Claude Nadeau/VIREO
42ML, 273MR, 394TR, 530TL, 749BL, 749TR, 751TL

Laure Neish/VIREO
441TL, 652TR

Kayleen A. Niyo/VIREO
694BR

Rolf Nussbaumer/VIREO
29BL, 133TL, 219TL, 395BR, 395BL, 408TR, 417BL, 417TR, 427BR, 433TL, 433BL, 435TL, 564TL, 670TR, 704BL, 756TL

Tony Palliser/VIREO
92RC, 95TL, 95TC, 95TR

Brian Patteson/VIREO
88TL, 93TRC, 99TR, 99MR, 100BR, 111TL, 294BR, 342BL, 358BL, 359TL, 360BR, 363LoMR, 374BL

Dennis Paulson/VIREO
270BR

Jari Peltomäki/VIREO
5BR, 6TL, 6BL, 16BR, 28BL, 33BR, 36BR, 36MR, 38TL, 38ML, 38MR, 38BR, 40ML, 40TL, 40MR, 40BR, 41BL, 41UMR, 46LoML, 46BL, 46BR, 51LoML, 57BR, 62BR, 63BR, 73TL, 156TL, 156MR, 156BR, 209TL, 211TL, 222TL, 227BR, 228BR, 230TL, 234BL, 235TL, 240TR, 241MR, 245BR, 257TL, 262BR, 275BR, 300TR, 390TC, 390TR, 390BR, 442BL, 702TR, 702MR, 739BR, 740TL, 740BL, 755TL, 758BR

Robert L. Pitman/VIREO
83BL, 88BL, 101BR, 102BL, 104TR, 104BR, 104TL, 104ML, 107BR, 109B, 340ML, 368TL

Jeff Poklen/VIREO
294TR, 309LoML, 325TR, 333UML, 334BL, 361BL

Eric Preston/VIREO
92R, 103BL

John Puschock/VIREO
41LoMR, 250MR

Peter Pyle/VIREO
244BL

Patricio Robles Gil/VIREO
248TL, 338TR, 419BR

Robert Royse/VIREO
41TR, 688TL

Sid & Shirley Rucker/VIREO
416TL, 418TL, 424BL, 430BL

Ronald M. Saldino/VIREO
244TL

Kevin Schafer/VIREO
87BR, 102TL, 403BL

Bill Schmoker/VIREO
48BR

Barth Schorre/VIREO
13BL, 384BL

Johann Schumacher/VIREO
52MR, 136BR, 218BR, 229TL, 285BR, 399BL, 435TR, 615BL, 717BR

Robert Shantz/VIREO
82BL, 727BR

Tadeo Shimba/VIREO
5TL, 5BL, 15TR, 31BL, 58TR, 232BR, 235BR, 247TR, 260MR, 268BR, 274B, 387R, 546R, 562TR, 758TL, 760BL

Jorge Sierra/VIREO
121BR, 207TL, 207BL

Rob & Ann Simpson/VIREO
199BR

Brian E. Small/VIREO
66MR, 71BC, 87MR, 340BL, 359TR, 362TL, 397TL, 410BL, 433BR, 440BLC, 479BR, 491BL, 510TL, 554BL, 632TR, 647TL, 710BR, 733TL

Hugh P. Smith Jr./VIREO
204TR

Kevin Smith/VIREO
491BL, 692TR, 733TR

Art Sowls/VIREO
365TL

Bob Steele/VIREO
48TL, 64TC, 78BR, 118TL, 221BR, 222ML, 224BR, 433BC, 524BR, 545BR, 577BL, 598ML, 661ML, 662BL, 686BR, 727ML, 755TR, 755BR

Harold Stiver/VIREO
39TR, 44BL, 111TR, 413RC

Morten Strange/VIREO
263TR, 415L, 543TL, 562TL, 562BL, 583TL

Michael Stubblefield/VIREO
11BL, 16BL, 35BL, 103TR, 299BL, 722BL, 725BR

Brian Sullivan/VIREO
76BR, 118BR, 365BR, 367TR, 373TLC, 518TR, 583TR, 614TL

Warwick Tarboton/VIREO
127BL, 280BR, 348MR

Alan Tate/VIREO
85BR, 337TL, 341BL

Mathew Tekulsky/VIREO
21BL, 431BR, 570BL

Glen Tepke/VIREO
100BL, 247BR, 362BR, 363UML, 373BR, 413LC

David Tipling/VIREO
7BL, 63ML, 74M, 197L, 542R, 543BL, 550ML, 550BL, 563BR, 563ML, 740TR

Ray Tipper/VIREO
219TR, 262BL

Pavel S. Tomkovich/VIREO
274TL

Manabu Totsuka/VIREO
703TL

Casey Tucker/VIREO
42BL

Joseph Turner/VIREO
124TR, 167TR, 216TL, 420BL, 420TR

Tom J. Ulrich/VIREO
59TR, 61TL, 64BL, 66BL

Christie Van Cleve/VIREO
142TL, 142BL, 142BLC

Tom Vezo/VIREO
273TL, 285UML, 291BL, 338TL, 349BL, 371TR, 373TL, 382TL, 451TR, 549MR, 735MR

Gerrit Vyn/VIREO
61BR

Alan Walther/VIREO
723TL

Ross Wanless/VIREO
83TL

Magill Weber/VIREO
34TR

Doug Wechsler/VIREO
28BR, 140TL, 236TL, 236BR, 284ML, 303BR, 345MR, 382TR, 384TR, 386R, 391TL, 495TL, 549TR, 549BR, 581TL, 581BL, 581TR, 583MR, 583BR, 643T, 681BL, 682ML, 691TC, 697TL

James M. Wedge/VIREO
132BL, 355TL, 646BL

Brian K. Wheeler/VIREO
120BL, 146TR, 147TR, 148BL, 149ML, 149BL, 149BR, 149TL, 151BL, 151BC, 157TL, 157BL, 158BRC, 158BR, 160BL, 161BL, 165BLC, 166BR, 167TL, 168TL, 168TR, 168BL, 168BR, 169B, 171BR, 172BL, 174TL, 175TR, 175BR, 176BL, 176BR, 177TL, 177BL, 177BR, 180BR, 181MR, 186BR, 190TR, 191TR, 194TL

Matt White/VIREO
196BL

James Woodward/VIREO
133TR

Steve Young/VIREO
6TR, 9MR, 16MR, 19BL, 20TR, 20BR, 30ML, 30BL, 33TR, 33UMR, 33LoMR, 35TR, 35BL, 36BL, 36TC, 36TR, 36ML, 37BL, 40BL, 46TR, 53BL, 78TR, 79TR, 88TR, 102TR, 116TL, 116BL, 116TR, 116MR, 136TR, 211ML, 211BL, 228TL, 228TR, 230B, 235BL, 235TR, 241BL, 245TL, 245BL, 245TR, 248BL, 250BL, 261TR, 275TL, 281BL, 291TR, 293TL, 295TR, 295ML, 295BR, 300TL, 301TR, 351BL, 358BR, 358TL, 361BR, 551MR, 563TR, 740BR, 755BL

Dale & Marian Zimmerman/VIREO
191BR, 503TL, 571TR, 727TR

Tim Zurowski/VIREO
369ML

Index

Created specifically to accompany *The Stokes Field Guide to the Birds of North America,* the bonus CD contains more than 600 sounds and songs of 150 common birds from the *Stokes Field Guide to Bird Songs* CDs. The MP3 files are easy to download to a wide variety of portable listening devices. The CD also features a booklet in PDF format with song and call descriptions and 150 photographs.

Audio Tracks

Here, listed by audio track number and page number in this guide, are the 150 birds featured on the bonus CD. Track numbers are indicated in the bird accounts at the end of the **Voice** section next to this symbol: 🎧.

1. Canada Goose, 13
2. Wood Duck, 18
3. Mallard, 22
4. Ruffed Grouse, 59
5. Wild Turkey, 68
6. California Quail, 70
7. Gambel's Quail, 70
8. Northern Bobwhite, 71
9. Common Loon, 76
10. Western Grebe, 81
11. Pied-billed Grebe, 82
12. American Bittern, 124
13. Great Blue Heron, 125
14. Black-crowned Night-Heron, 134
15. Yellow-crowned Night-Heron, 135
16. Osprey, 144
17. Bald Eagle, 152
18. Sharp-shinned Hawk, 158
19. Cooper's Hawk, 159
20. Red-shouldered Hawk, 164
21. Broad-winged Hawk, 166
22. Swainson's Hawk, 170
23. Red-tailed Hawk, 174
24. American Kestrel, 184
25. Sora, 197
26. Clapper Rail, 198
27. King Rail, 200
28. Virginia Rail, 201
29. Common Moorhen, 203
30. American Coot, 204
31. Sandhill Crane, 206
32. Black-bellied Plover, 210
33. Killdeer, 217
34. Eastern Willet, 236
35. Lesser Yellowlegs, 238
36. Greater Yellowlegs, 239
37. American Woodcock, 284
38. Laughing Gull, 302
39. Ring-billed Gull, 312
40. Western Gull, 314
41. Herring Gull, 318
42. Common Tern, 350
43. Forster's Tern, 353
44. Eurasian Collared-Dove, 381
45. White-winged Dove, 382
46. Mourning Dove, 383
47. Barn Owl, 396
48. Eastern Screech-Owl, 398
49. Great Horned Owl, 399
50. Barred Owl, 403
51. Common Nighthawk, 409
52. Chimney Swift, 414
53. Belted Kingfisher, 435
54. Red-bellied Woodpecker, 439
55. Downy Woodpecker, 444
56. Hairy Woodpecker, 445
57. Pileated Woodpecker, 448
58. Northern Flicker, 450
59. Black Phoebe, 468
60. Eastern Phoebe, 468
61. Great Crested Flycatcher, 474
62. Western Kingbird, 479
63. Eastern Kingbird, 479
64. Plumbeous Vireo, 490
65. Cassin's Vireo, 491
66. Blue-headed Vireo, 491
67. Red-eyed Vireo, 494
68. Steller's Jay, 498
69. Blue Jay, 498
70. Western Scrub-Jay, 501
71. Black-billed Magpie, 504
72. American Crow, 505
73. Fish Crow, 507
74. Common Raven, 508
75. Purple Martin, 512
76. Tree Swallow, 513
77. Violet-green Swallow, 514
78. Barn Swallow, 519
79. Carolina Chickadee, 522
80. Black-capped Chickadee, 523
81. Tufted Titmouse, 528
82. White-breasted Nuthatch, 531
83. Red-breasted Nuthatch, 532
84. Carolina Wren, 535
85. Bewick's Wren, 536
86. House Wren, 536
87. Eastern Bluebird, 552
88. Western Bluebird, 553
89. Mountain Bluebird, 554
90. Veery, 556
91. Hermit Thrush, 560
92. Wood Thrush, 561
93. Varied Thrush, 565
94. American Robin, 566
95. Gray Catbird, 569
96. Northern Mockingbird, 570
97. Brown Thrasher, 572
98. Curve-billed Thrasher, 575
99. California Thrasher, 576
100. European Starling, 579
101. Cedar Waxwing, 587
102. Orange-crowned Warbler, 596
103. Northern Parula, 600
104. Yellow Warbler, 602
105. Chestnut-sided Warbler, 603
106. Magnolia Warbler, 604
107. Black-throated Blue Warbler, 608
108. Yellow-rumped Warbler, 610
109. Black-throated Green Warbler, 612
110. Pine Warbler, 618
111. Black-and-white Warbler, 623
112. American Redstart, 625
113. Ovenbird, 627
114. Northern Waterthrush, 628
115. Louisiana Waterthrush, 629
116. Common Yellowthroat, 634
117. Scarlet Tanager, 646
118. Western Tanager, 647
119. Spotted Towhee, 652
120. Eastern Towhee, 653
121. American Tree Sparrow, 663
122. Chipping Sparrow, 664
123. Lark Sparrow, 670
124. Song Sparrow, 686
125. White-throated Sparrow, 688
126. White-crowned Sparrow, 690
127. Golden-crowned Sparrow, 691
128. Dark-eyed Junco, 692
129. Northern Cardinal, 704
130. Rose-breasted Grosbeak, 706
131. Black-headed Grosbeak, 707
132. Lazuli Bunting, 710
133. Indigo Bunting, 711
134. Bobolink, 715
135. Red-winged Blackbird, 716
136. Yellow-headed Blackbird, 719
137. Eastern Meadowlark, 720
138. Western Meadowlark, 721
139. Brewer's Blackbird, 723
140. Common Grackle, 724
141. Brown-headed Cowbird, 728
142. Orchard Oriole, 730
143. Bullock's Oriole, 734
144. Baltimore Oriole, 735
145. Purple Finch, 746
146. House Finch, 747
147. Pine Siskin, 754
148. American Goldfinch, 757
149. Evening Grosbeak, 759
150. House Sparrow, 761